PERSPECTIVES ON NEW CROPS AND NEW USES

Proceedings of the Fourth National Symposium

NEW CROPS AND NEW USES
BIODIVERSITY AND AGRICULTURAL SUSTAINABILITY
Phoenix, Arizona, November 8–11, 1998

Under the Auspices of
Association for the Advancement of Industrial Crops (AAIC)
New Uses Council, Inc.
Purdue University Center for New Crops and Plant Products

Co-sponsored by

U.S. Department of Agriculture (USDA), Agricultural Research Service (ARS)
Cooperative State Research Education and Extension Service (CSREES)
Sustainable Agriculture Research and Education (SARE)
Alternative Agricultural Research and Commercialization (AARC)
ARS, U.S. Water Conservation Laboratory

American Oil Chemists' Society
American Society for Horticultural Science
Crop Science Society of America
The Society for Economic Botany

Department of Energy, Oak Ridge National Laboratory
The Thomas Jefferson Agriultural Institute
Procter & Gamble
Food and Agriculture Organization of the United Nations
Desert Botanical Gardens

Perspectives on New Crops and New Uses

Edited by

Jules Janick

Purdue University

Production Manager
Anna Whipkey
Purdue University

Jacket photograph: kiwifruit courtesy of Martin Heffer.

ISBN 0-9615027-0-3
Printed in the United States of America

ASHS Press
600 Cameron Street
Alexandria, VA 22314-2562
telephone 703.836.4606 • fax 703.836.2024 • e-mail ashspress@ashs.org

CONTENTS

FLORAL & LANDSCAPE CROPS

MEDICINAL, AROMATIC, SPICE, & BIOACTIVE CROPS

List of Contributors

A. Agueguia
CNRCIP, IRAD
Dschang
Cameroon

Teklu Andebrhan
Agricultural Research Station
Virginia State University
Petersburg, Virginia 23806

Gregory Anderson
University of Connecticut
Ecology & Evolutionary Biology U43
Storrs, Connecticut 06269-3043

Robert E. Armstrong
AARCC/USDA
0156 South Building
14th and Independence Ave, SW
Washington, DC 20250-0401

T.J. Arnason
Department of Biology
University of Ottawa
30-Marie Curie
Ottawa, Ontario K1N 6N5
Canada

Dennis V.C. Awang
MediPlant Consulting Services
Ottawa, Ontario K1G 3J1
Canada

Phillip C. Badger
DOE Southeastern Regional Biomass
  Energy Program
Tennessee Valley Authority
Muscle Shoals, Alabama 35662-1010

Carmela A. Bailey
Agricultural Materials Program
USDA-CSREES
Aerospace Building Room 873
Washington, DC 20250-2220

D.D. Baltensperger
Panhandle Research and Extension
  Center
University of Nebraska
4502 Ave. I
Scottsbluff, Nebraska 69361-4939

I.V. Bartish
SLU-Balsgård
Department of Horticultural Plant
  Breeding
291 94 Kristianstad
Sweden

C. Bergeron
Department of Biology
University of Ottawa
30-Marie Curie
Ontario Canada K1N 6N5
Canada

Abdelfettah Berrada
Colorado State University
Southwestern Colorado Research
  Center
P.O. Box 233
Yellow Jacket, Colorado 81335-0233

Harbans L. Bhardwaj
Agricultural Research Station
Box 9061
Virginia State University
Petersburg, Virginia 23806

S.F. Blade
Crop Diversification Centre North
Alberta Agriculture
Food and Rural Development
R.R. #6
Edmonton, Alberta T5B 4K3
Canada

Melvin G. Blase
Department of Agricultural Economics
University of Missouri
Columbia, Missouri 65211

Robert Bolksberg-Fireovid
NIST, US Department of Commerce
Administration Building, Room A225
Gaithersburg, Maryland 20899

Michal W. Borys
Universidad Popular Autónoma del
  Estado de Puebla
Escuela de Agronomía
21 Sur 1103
Col. Santiago
72160 Puebla, Pue
Mexico

J. Bouton
Department of Crop & Soil Sciences
University of Georgia
Athens, Georgia 30602-7272

Vicki L. Bradley
USDA, ARS Western Regional Plant
  Introduction Station
PO Box 646402, 59 Johnson Hall
Washington State University
Pullman, Washington 99164-6402

D. Bransby
Department of Agronomy and Soils
Auburn University
Auburn, Alabama 36849

Mark A. Brick
Department of Soil & Crop Sciences
Colorado State University
Fort Collins, Colorado 80523

Dwayne R. Buxton
USDA-ARS-NPS
5601 Sunnyside Avenue
Beltsville, Maryland 20705

Elizabeth J. Callan
Louisiana State University
Agricultural Center
Department of Agronomy
104 Sturgis Hall
Baton Rouge, Louisiana 70803-2110

Paul J. Caswell
Archer Daniels Midland Corporation
4666 Faries Parkway
Decatur, Illinois 62526

Alejandro Ching
Northwest Missouri State University
Alternative Crops Research and
  Development Center
800 University Drive
Maryville, Missouri 64468

Beth M. Christensen
University of Wyoming
Department of Renewable Resources
Laramie, Wyoming 82071-3354

L. Davis Clements
Department of Biological Systems
  Engineering
University of Nebraska
Lincoln, Nebraska 68583

Cobus Coetzee
Agricultural Research Council
Fynbos Unit, P/Bag X1
Elsenburg, 7607
South Africa

B. Conger
Department of Plant & Soil Sciences
University of Tennessee
Knoxville, Tennessee 37996

Jimmie M. Crane
Oregon State University
Department of Crop Science
Crop Science Building 451C
Corvallis, Oregon 97331-3002

Richard A. Criley
Department of Horticulture
University of Hawaii
3190 Maile Way, No. 102
Honolulu, Hawaii 96822

Stafford M.A. Crossman
University of the Virgin Islands
Agricultural Experiment Station
RR 2 10000 Kingshill
St. Croix, US Virgin Islands 00850

N. D'Croz-Mason
Agronomy Department
University of Nebraska
Lincoln, Nebraska 68583-0915

Fred Diem
Virginia Cooperative Extension
Virginia Polytechnic Institute & State
 University
Exmore, Virginia 23350

P. Diemer
FAO
Via delle Terme di Caracalla
00100 Rome
Italy

David A. Dierig
U.S. Water Conservation Laboratory
4331 E. Broadway Road
Phoenix, Arizona 85040

Angela R. Duncan
Axxon Biopharm Inc.
11303 Amherst Ave #2
Silver Spring, Maryland 20902

Hamid Elhag
Department of Pharmacognosy
College of Pharmacy
King Saud University
P.O. Box 2457
Riyadh 11451
Saudi Arabia

Mahmoud M. El-Olemy
Department of Pharmacognosy
College of Pharmacy
King Saud University
P.O. Box 2457
Riyadh 11451
Saudi Arabia

Pablo Eyzaquirre
IPGRI
Via delle Sette Chiese
142, 00145 Rome
Italy

A.R. Ferguson
HortResearch
Private Bag 92169
Auckland
New Zealand

Chad Finn
Northwest Center for Small Fruit
 Research
USDA ARS, Horticultural Crops
 Laboratoy
3420 N.W. Orchard Ave.
Corvallis, Oregon 97330

James L. Fowler
New Mexico State University
PO Box 30003, MSC 3LEY
Las Cruces, New Mexico 88003

Edwin C. French
University of Florida
Agronomy Department
PO Box 110500
Gainesville, Florida 32611-0500

J.L. Galván S.
Universidad Popular Autónoma del
 Estado de Puebla
Escuela de Agronomía
21 Sur 1103
Col. Santiago
72160 Puebla, Pue
Mexico

R.G. Gaudiel
RPG Agri Limited
Coalhurst, Alberta T0L 0V0
Canada

David Glassner
National Renewable Energy Laboratory
1617 Cole Boulevard
Golden, Colorado 80401

Jeffrey Goettemoeller
Northwest Missouri State University
Alternative Crops Research and
 Development Center
800 University Drive
Maryville, Missouri 64468

Catherine M. Grieve
U.S. Salinity Laboratory
450 West Big Springs Road
Riverside, California 92507-4617

P. Griffee
FAO
Via delle Terme di Caracalla
00100 Rome
Italy

Robert L. Guenthner
USDA, ARS Western Regional Plant
 Introduction Station
PO Box 646402
59 Johnson Hall
Washington State University
Pullman, Washington 99164-6402

Fernando R. Guillen-Portal
Agronomy Department
University of Nebraska
Lincoln, Nebraska 68583-0915

Abraham H. Halevy
The Hebrew University of Jerusalem
Department of Horticulture
P.O. Box 12
Rehovot 76100
Israel

Stephan R.P. Halloy
New Zealand Institute for Crop &
 Food Research
Invermay, PB 50034
Mosgiel
New Zealand

Anwar A. Hamama
Agricultural Research Station
Box 9061
Virginia State University
Petersburg, Virginia 23806

Richard M. Hannan
USDA, ARS Western Regional Plant
  Introduction Station
PO Box 646402
59 Johnson Hall
Washington State University
Pullman, Washington 99164-6402

Zhigang Hao
Quality Botanical Ingredients, Inc.
500 Metuchen Road
South Plainfield, New Jersey 07080

Michael J. Havey
Agricultural Research Service–USDA
Department of Horticulture
1575 Linden Drive
University of Wisconsin
Madison, Wisconsin 53706

Charles Heiser
Indiana University
Department of Biology
Bloomington, Indiana 47405

Margo E. Herdendorf
University of Wyoming
Spatial Data and Visualization Center
Laramie, Wyoming 82071-3067

James Hettenhaus
Chief Executive Assistance
3211 Trefoil Drive
Charlotte, North Carolina 28226

Vernon Heywood
Centre for Plant Diversity & Systematics
School of Plant Sciences
The University of Reading
PO Box 221, Whiteknights
Reading RG6 6AS
United Kingdom

Toby Hodgkin
IPGRI
Via delle Sette Chiese
142, 00145
Rome
Italy

Paul Hoepner
Department of Agriculture and
  Applied Economics
Virginia Polytechnic Institute & State
  University
Blacksburg, Virginia 24061

T. Hofmann
Technical University of Munich
Institute of Food Chemistry
Lichtenbergstr. 4
85748 Garching
Germany

Maurice M. Iwu
Bioresources Development and
  Conservation Programme
Axxon Biopharm Inc.
11303 Amherst Ave #2
Silver Spring, Maryland 20902

Jules Janick
Center for New Crops & Plant Products
1165 Horticulture Building
Purdue University
West Lafayette, Indiana 47907-1165

Elton Jefthas
Agricultural Research Council
Fynbos Unit, P/Bag X1
Elsenburg, 7607
South Africa

N. Jeppsson
SLU-Balsgård
Department of Horticultural Plant
  Breeding
291 94 Kristianstad
Sweden

Duane L. Johnson
Department of Soil & Crop Sciences
Colorado State University
Fort Collins, Colorado 80523

Richard C. Johnson
USDA, ARS Western Regional Plant
  Introduction Station
PO Box 646402
59 Johnson Hall
Washington State University
Pullman, Washington 99164-6402

Gary D. Jolliff
Oregon State University
Department of Crop and Soil Science
Corvallis, Oregon 97331-3002

Altin Kalo
Department of Agriculture and
  Applied Economics
Virginia Polytechnic Institute & State
  University
Blacksburg, Virginia 24061

Charles W. Kennedy
Louisiana State University
Agricultural Center
Department of Agronomy
104 Sturgis Hall
Baton Rouge, Louisiana 70803-2110

N. Kerr
Department of Human Ecology
University of Alberta
Edmonton, Alberta T6G 2M8
Canada

Roy N. Keys
University of Arizona
Department of Plant Sciences
Tucson, Arizona 85721

T.B. Kireeva
Department of Botany and Plant
  Ecology
Faculty of Biology and Chemistry
Udmurtian State University
Russia

Steven J. Knapp
Oregon State University
Department of Crop Science
Crop Science Bldg 451C
Corvallis, Oregon 97331-3002

V.S. Krutilina
Moscow Timiriazev Agricultural
  Academy
Department of Soil Science & Plant
  Chemistry
Timiriazevskaya Str. 49
Moscow U-550
Russia

Daniel E. Kugler
CSREES/USDA
901 D St, SW
8th floor Aerospace Building
Washington, DC 20250-2220

Kathryn Lardizabal
Calgene LLC
1920 Fifth Street
Davis, California 95616

Michael W. Lassner
Calgene LLC
1920 Fifth Street
Davis, California 95616

Desmond R. Layne
Department of Horticulture
Clemson University
Clemson, South Carolina 29634-0375

Polly S. Leonhard
56 Oriole Drive
Ephrata, Pennsylvania 17522

Kenneth W. Leonhardt
Department of Horticulture
University of Hawaii
3190 Maile Way, No. 102
Honolulu, Hawaii 96822

Helena Leszczyñska-Borys
Universidad Popular Autónoma del
 Estado de Puebla
Escuela de Agronomía
21 Sur 1103
Col. Santiago
72160 Puebla, Pue
Mexico

W. Letchamo
Herba Medica
P.O.Box 241
Trout Lake, Washington 98650

Thomas S.C. Li
Pacific Agri-Food Research Centre
Agriculture and Agri-Food Canada
Summerland, British Columbia V0H 1Z0
Canada

J. Livesey
Department of Biology
University of Ottawa
30-Marie Curie
Ottawa, Ontario K1N 6N5
Canada

M. Marsella
FAO
Via delle Terme di Caracalla
00100 Rome
Italy

Kathleen A. McKeown
Department of Botany
Box 7612
North Carolina State University
Raleigh, North Carolina 27695-7612

S. McLaughlin
Environmental Sciences Division
Oak Ridge National Laboratory
Oak Ridge, Tennessee 37831

M.M. Meboka
CNRCIP, IRAD
Ekona, Buea
Cameroon

Tadesse Mebrahtu
Agricultural Research Station
Virginia State University
Petersburg, Virginia 23806

James G. Metz
Calgene LLC
1920 Fifth Street
Davis, California 95616

Lin Chau Ming
Department of Horticulture
Agronomical Sciences College
Sao Paulo Statal University
Botucatu - Sao Paulo
CEP: 18.603-970
Brazil

Yosef Mizrahi
Ben Gurion University of the Negev
P.O. Box 653
Beer-Sheva 84105
Israel

Jack Moes
The Great AgVenture Consulting
20 Marquis Crescent
Brandon, Manitoba R7B 3R8
Canada

Ali I. Mohamed
Agricultural Research Station
Virginia State University
Petersburg, Virginia 23806

Candelario Mondragon Jacobo
Instituto Nacional de Investigaciones
Forestales y Agropecuarias
Nogal 259 Fracc. Arboledas
Queretaro, QRO. 76140
Mexico

Monsanto Company
800 Lindbergh Boulevard
St. Louis, Missouri 63167

Mario R. Morales
Center for New Crops & Plant Products
1165 Horticulture Building
Purdue University
West Lafayette, Indiana 47907-1165

J. Bradley Morris
USDA/ARS/PGRCU
University of Georgia
1109 Experiment St.
Griffin, Georgia 30223-1797

Jaber S. Mossa
Department of Pharmacognosy
College of Pharmacy
King Saud University
P.O. Box 2457
Riyadh 11451
Saudi Arabia

B.E. Mote
SC Johnson Polymer
SC Johnson & Son, Inc.
8310 16th Street
Sturtevant, Wisconsin 53177-0902

Larry C. Munn
University of Wyoming
Department of Renewable Resources
Laramie, Wyoming 82071-3354

Denis J. Murphy
John Innes Centre
Norwich, NR4 7UH
United Kingdom

Robert L. Myers
Thomas Jefferson Institute for Crop
 Diversification
601 Nifong Bldg, Suite 5A
Columbia, Missouri 65203

X. Ndzana
CNRCIP, IRAD
Ekona, Buea
Cameroon

L.A. Nelson
Agronomy Department
University of Nebraska
Lincoln, Nebraska 68583-0915

Avinoam Nerd
Ben Gurion University of the Negev
P.O. Box 653
Beer-Sheva 84105
Israel

L. Nyochembeng
CNRCIP, IRAD
Ekona, Buea
Cameroon

S. Nzietchueng
CNRCIP, IRAD
Ekona, Buea
Cameroon

W. Ocumpaugh
Department of Soil & Crop Sciences
Texas A&M University
College Station, Texas 77843

Chris O. Okunji
Division of Experimental Therapeutics
Walter Reed Army Inst. of Research
Washington, DC 20307-5100

O.U. Onokpise
Florida A&M University
Rm 303 South Perry-Paige Building
Martin Luther King Boulevard
Tallahassee, Florida 32307

Stefano Padulosi
IPGRI, CWANA
c/o ICARDA
P.O. Box 5466
Aleppo
Syria

Manuel C. Palada
University of the Virgin Islands
Agricultural Experiment Station
RR 2 10000 Kingshill
St. Croix, US Virgin Islands 00850

D. Parrish
Department of Crop & Environmental
  Soil Science
Virginia Polytechnic Institute & State
  University
Blacksburg, Virginia 24061

Brett E. Patrick
Oregon State University
Department of Crop and Soil Science
Corvallis, Oregon 97331-3002

H.A. Persson
SLU-Balsgård
Department of Horticultural Plant
  Breeding
291 94 Kristianstad
Sweden

R. Neal Peterson
The PawPaw Foundation (PPF)
P.O. Box 1277
Franklin, West Virginia 26807

H. Pfeiffer
FAO
Via delle Terme di Caracalla
00100 Rome
Italy

Winthrop B. Phippen
Center for New Crops & Plant Products
1165 Horticulture Building
Purdue University
West Lafayette, Indiana 47907-1165

Linnette Poindexter
New Mexico State University
PO Box 30003, MSC 3LEY
Las Cruces, New Mexico 88003

E. Barclay Poling
North Carolina State University
Campus Box 7553
NCSU Centennial Campus
Raleigh, North Carolina 27695-7553

Kirk W. Pomper
Land-Grant Program
Atwood Research Facility
Kentucky State University
Frankfort, Kentucky 40601-2355

Gordon M. Prine
University of Florida
Agronomy Department
PO Box 110500
Gainesville, Florida 32611-0500

Roman Przybylski
Department Foods and Nutrition
University of Manitoba
Winnipeg, Manitoba R3T 2N2
Canada

Naveen Puppala
Agricultural Science Center at Clovis
New Mexico State University
Clovis, New Mexico 88101

T. Putter
FAO
Via delle Terme di Caracalla
00100 Rome
Italy

Robert Quinn
HC 77
Boxes 808
Big Sandy, Montana 59520

Janet H. Rademacher
Mountain States Wholesale Nursery
P.O. Box 2500
Litchfield Park, Arizona 85340

Christopher Ramcharan
University of the Virgin Islands
Agricultural Experiment Station
P.O. Box 10,000 Kingshill
St. Croix, US Virgin Islands 00850

Muddappa Rangappa
Agricultural Research Station
Box 9061
Virginia State University
Petersburg, Virginia 23806

Dennis T. Ray
University of Arizona
Department of Plant Sciences
Tucson, Arizona 85721

Emmy Reinten
Agricultural Research Council
Fynbos Unit, P/Bag X1
Elsenburg, 7607
South Africa

Bruce Riddell
High Country Elevators, Inc.
07026 US. Hwy 666
Dove Creek, Colorado 81324

Sérgio F.R. Rocha
Department of Horticulture
Agronomical Sciences College
Sao Paulo Statal University
Botucatu - Sao Paulo
CEP: 18.603-970
Brazil

J.C. Rudelich
SC Johnson Polymer
SC Johnson & Son, Inc.
8310 16th Street
Sturtevant, Wisconsin 53177-0902

A.E. Sama
CNRCI, IRAD
Ekona, Buea
Cameroon

Vassilios Sarafis
CMM
University of Queensland
St. Lucia Queensland 4072
Australia

Ali O. Sari
Aegean Agricultural Research
 Institute
PO Box 9
Menemen, Izmir
Turkey

Melinda S. Schaad
University of Wyoming
Department of Renewable Resources
Laramie, Wyoming 82071-3354

Dan Schafferman
The Volcani Center
Department Genetic Resources and
Seed Research, A.R.O.
Bet Dagan
Israel

Tom Schechinger
Iron Horse Custom Farming
816 Iron Horse Road
Harlan, Iowa 51537

P. Schieberle
Technical University of Munich
Institute of Food Chemistry
Lichtenbergstr. 4
85748 Garching
Germany

Ella Shabelsky
The Volcani Center
Department of Genetic Resources and
 Seed Research, A.R.O.
Bet Dagan, Israel

Hope Shand
Rural Advancement Foundation
 International
PO Box 655
Pittsboro, North Carolina 27312

Michael C. Shannon
U.S. Salinity Laboratory
450 West Big Springs Road
Riverside, California 92507-4617

U.L. Sharanov
Department of Botany and Plant
 Ecology
Faculty of Biology and Chemistry
Udmurtian State University
Russia

James E. Simon
Center for New Crops & Plant Products
1165 Horticulture Building
Purdue University
West Lafayette, Indiana 47907-1165

Bharat P. Singh
Fort Valley State University
Agricultural Research Station
Fort Valley, Georgia 31030-3298

Melpo Skoula
MAICh
Department of Natural Products
PO Box 85
73100 Chania
Greece

Ernest Small
Eastern Cereal and Oilseed Research
 Center
Research Branch, Agriculture and
 Agri-Food Canada
Central Experimental Farm
Ottawa, Ontario K1A 0C6
Canada

Mark W. Stack
Colorado State University
Southwestern Colorado Research
 Center
P.O. Box 233
Yellow Jacket, Colorado 81335-0233

David E. Starner
Virginia Polytechnic Institute & State
 University
Northern Piedmont Agricultural
 Research and Extension Center
PO Box 448
Orange, Virginia 22960

Susan B. Sterrett
Department of Horticulture
Eastern Shore Agricultural Research
 and Extention Center
Virginia Polytechnic Institute & State
 University
Painter, Virginia 23420

Allen Sturko
Manitoba Industry, Trade, & Tourism
500–155 Carlton Street
Winnipeg, Manitoba R3C 3H8
Canada

Akio Suzuki
Seedex, Inc.
1350 Kansas Avenue
Longmont, Colorado 80501

A.O. Tairu
Department of Chemical Sciences
University of Agriculture, Abeokuta
Ogun State
Nigeria

C. Taliaferro
Department Agronomy
Oklahoma State University
Stillwater, Oklahoma 74078

J.T. Tambong
CNRCIP, IRAD
Ekona, Buea
Cameroon

Daniel B. Taylor
Department of Agriculture and
 Applied Economics
Virginia Polytechnic Institute & State
 University
Blacksburg, Virginia 24061

D.L. Trumbo
SC Johnson Polymer
SC Johnson & Son, Inc.
8310 16th Street
Sturtevant, Wisconsin 53177-0902

Varro E. Tyler
Department of Pharmacognosy
Purdue University
P.O. Box 2566
West Lafayette, Indiana 47996-2566

Donald L. Van Dyne
Department of Agricultural
 Economics
University of Missouri
Columbia, Missouri 65211

George F. Vance
University of Wyoming
Department of Renewable Resources
Laramie, Wyoming 82071-3354

Roberto F. Vieira
EMBRAPA/CENARGEN
Caixa Postal 02372
Brasilia, DF 70770-900
Brazil

K. Vogel
USDA Agricultural Research Service
University Nebraska
Lincoln, Nebraska 68583

Shaoke Wang
Seedex, Inc.
1350 Kansas Avenue
Longmont, Colorado 80501

Charles L. Webber III
USDA, ARS, SCARL
P.O. Box 159
Lane, Oklahoma 74555

Anna Whipkey
Center for New Crops & Plant Products
1165 Horticulture Building
Purdue University
West Lafayette, Indiana 47907-1165

Wayne F. Whitehead
Fort Valley State University
Agricultural Research Station
Fort Valley, Georgia 31030-3298

S. Wullschleger
Environmental Sciences Division
Oak Ridge National Laboratory
Oak Ridge, Tennessee 37831

J.G. Wutoh
University of Maryland Eastern
  Shores
Princess Anne, Maryland 21853

Anand K. Yadav
Fruit Biotechnology
Agricultural Experiment Station
Fort Valley State University
Fort Valley, Georgia 31030-4313

Zohara Yaniv
The Volcani Center
Department of Genetic Resources and
  Seed Research, A.R.O.
Bet Dagan
Israel

Jennifer A. Young
USDA-NRCS
820 West Lincoln Avenue
Riverton, Wyoming 82501

PREFACE

David Dierig and Jules Janick

Perspectives on New Crops and New Uses is the fourth volume in a series of symposia proceedings that include *Progress in New Crops*, 1990, Timber Press, Portland Oregon; *New Crops* 1993, Wiley Press, New York; and *Advances in New Crops*, 1996, ASHS Press, Alexandria, Virginia. The present volume is based on the symposium entitled *New Crops and New Uses: Biodiversity and Agricultural Sustainability*, held November 8 to 11, 1998, in Phoenix, Arizona and jointly organized by the Association for the Advancement of Industrial Crops (AAIC), the Purdue University Center for New Crops and Plant Products, and the New Uses Council, Inc. The symposium objectives were to broaden discussions on the relationship of new crops to the topics of global biodiversity, sustainable agriculture, and new uses of alternative and conventional crops. We were also interested in drawing attention to the issues of new crops and new uses as well as to stimulate dialog among farmers, public and private researchers, industry, and policy makers. While the program was national in emphasis, there was a significant international contingent; half of the posters were from overseas contributors. Consequently significant interest was expressed in the organization of an overseas, international symposium, perhaps in 2001.

The conference included about 250 participants representing a broad spectrum of interests and backgrounds. The 4-day program consisted of invited presentations, posters, panel discussions, as well as a tour, dinner and panel discussion at the Desert Botanical Garden. Audience participation throughout was vigorous and contentious with strong views opined around such topics as intellectual property rights, the rights of farmers to save seed, neutraceutical and medicinal crops, and research policy. We were pleased that our reception speaker, Dr. I. Miley Gonzalez, Under Secretary for Research, Education, and Economics, U.S. Department of Agriculture, indicated that he plans to support an initiative included in the 2001 federal budget on new crops and new uses. Most significantly, we discerned a general concensus that the concept of new crops and new uses, was beginning to have an impact beyond the wishes of its champions, with discernable positive effects on US and world agriculture. There was evidence of proactive action taken by industry, the research community, farmers, and marketers for new crops and new uses, and clearly the issue has reached the consciousness of the consuming public. The ten years since the first meeting in Indianapolis in 1988 has witnessed a number of concrete advances such as the establishment of the Association for the Advancement of Industrial Crops (AAIC), incorporation of new crops language in the Farm Bill of 1998, the creation of the Thomas Jefferson Agricultural Institute, and general acceptance of a number of new crops and uses.

We acknowledge the assistance of many of our colleagues for this symposium and this volume of the proceedings. These include Dennis Ray of AAIC and Kennith Foster of the New Crops Council who served as co-conference organizer, Steven McLaughlin, Robert Myers, and James Simon who served on the program committee, Francis Nakayama for editing the Abstracts and Program booklet, Gail Dahlquist for registration, John Nelson for tours, Pernell Tomasi for poster management, and Anna Whipkey for her computer expertise in the preparation of this volume. The many sponsors are listed next to the title page but we would like to express our appreciation for the major supporters especially the United States Department of Agriculture (USDA) and associated organizations including the Agricultural Research Service (ARS), Alternative Agricultural Research and Commercialization (AARC), and the Sustainable Agriculture Research and Education program (SARE), as well as contributions from the Fund for Rural America's Crop Diversification Center and the United Nations Food and Agricultural Organization (FAO) who financed the participation of our keynote speaker, Vernon Heywood.

This volume includes papers from invited participants, some of the panel discussions, as well as those who submitted posters. Contributions include a diverse assortment of crops, topics, and views and, as previous volumes, the volume is divided in three parts:

Part I: PROGRAMS & POLICY which includes three subsections: Biodiversity, New Crops, New Uses; Policy; and Sources of Public Funding.

Part II: CONSERVATION & INFORMATION which includes two subsections: Germplasm Conservation and Crop Information Systems.

Part III: STATUS OF NEW CROPS & NEW USES which includes eight subsections: Cereals & Pseudocereals; Legumes; Oilseed & Industrial Crops; Fiber & Energy Crops; Fruits; Vegetables; Floral & Landscape Crops; and Medicinal, Aromatic, Spice, & Bioactive Crops.

The addresses of all contributors are presented on p. ix to xv. A detailed index to species, crops, and products is found on p. 513 and the index of authors is found on p. 527.

It is our pleasure to dedicate this volume to the late Anson Ellis Thompson in recognition of his impact on the development of new industrial crops. Dr. Thompson, known to his many friends and colleagues as "Tommy," passed away on June 14, 1996 at the age of 72. We miss him. Tommy was internationally recognized and respected for his leadership on new industrial crop development; he initiated and led research in developing cuphea, lesquerella, guayule, and vernonia. He was respected for his enthusiasm, integrity, ideas, energy, and achievements as a researcher and teacher, and for his loyalty and openness as a human being. We hope that this volume as well as our continued efforts in developing new crops and new uses will be a lasting tribute to him.

Anson E. Thompson

PART I
PROGRAMS & POLICY

BIODIVERSITY, NEW CROPS, NEW USES

Trends in Agricultural Biodiversity

Vernon Heywood

"Biodiversity is a non-detachable part of the concept of sustainability. ... biodiversity is essential for agricultural production, as agriculture should be for biodiversity conservation."
Brazilian Government proposal to SBSTTA of the CBD, Second Meeting
Montreal 2–6 September 1996

Although the context of this paper is a "World Biodiversity Update," any attempt at such would be presumptuous if not impossible. What I will attempt, however, is to highlight some of the major developments in biodiversity action and policy that have emerged during the past year or two. In fact, several of these are directly related to agricultural biodiversity, and indeed to the issue of new crops. Then I shall explore the main trends in the appreciation, conservation and sustainable use of what is termed agricultural biodiversity.

The coming into effect of the Convention on Biological Diversity has led to a wide range of activities and initiatives as governments attempt to get to grips with the problems of implementing what is no more than an outline convention. The deliberations of the Conference of the Parties (COP) of the Convention and its Subsidiary Body for Scientific Technical and Technological Advice (SBSTTA) have been criticized by some as spending too much time on issues such as biotechnology, biosafety, and intellectual property rights (Raven 1998) while issues such as the preservation of species have been ignored. Others regard the Convention as providing incentives for countries to conserve and sustainably use their own biodiversity and ensure that the benefits derived from it by third parties are equitably shared (Seyani 1998). Certainly, a great deal of controversy has been generated by matters such as access to genetic resources, farmers' rights, and technology transfer, reflecting different attitudes and perceptions between developing and developed countries. These are serious issues that demand attention and while those of us who are alarmed at the growing rate of biodiversity loss and habitat destruction are naturally impatient and would like to see more urgent steps being taken to stem the losses, progress in reconciling the different positions is being made and at the same time many countries are taking positive steps to put in place mechanisms for the conservation and sustainable use of their biodiversity.

Another key development was the endorsement by governments of the Global Plan of Action (GPA) for the "Conservation and Sustainable Use of Plant Genetic Resources for Food and Agriculture" at the International Technical Conference on Plant Genetic Resources held in Leipzig, 17–23 June 1996 (FAO 1996).

The GPA sets out a global strategy for the conservation and sustainable use of plant genetic resources for food and agriculture with an emphasis on productivity, sustainability and equity (Cooper et al. 1998) and complements the CBD. In fact, a significant development that followed from these two was the convergence of interest between bodies such as FAO and IPGRI and conservation and development organizations and agencies such UNESCO-MAB, IUCN, WWF, and ITDG. On the one hand, the CBD recognized agricultural biodiversity as a focal area in view of its social and economic relevance and the prospects offered by sustainable agriculture for reducing the negative impacts of biological diversity, enhancing the value of biological diversity and linking conservation efforts with social and economic benefits (Decision III/11 Conservation and Sustainable Use of Agricultural Biological Diversity of the Conference of the Parties to the Convention on Biological Diversity). On the other hand, it is recognized that the Global Plan of Action covers a number of multidisciplinary areas such as in situ conservation of wild plants and crop relatives in natural ecosystems that extend the traditional activities of sustainable agriculture and plant genetic resource conservation and that successful implementation will require the development of new partnerships with a range of intergovernmental and non-governmental organizations as well as with indigenous and local communities.

Another area of developing concern is the role of indigenous communities in the conservation and sustainable use of biodiversity. Although the all-pervading influence of human action in modifying biodiversity is widely recognized, and the Global Biodiversity Assessment (Heywood 1985) included a whole section on this topic, the complexity of the social, cultural, ethical, religious, and other human interactions with biodiversity and agroecological systems. As the STAP Expert Group on Sustainable Use of Biodiversity (UNEP 1998) notes, *"Development of sustainable use projects requires a paradigm shift from a focus on protection and the development of protected areas to considering also such skills as dealing with the interaction of socio-economic and ecological systems."*

TRENDS IN BIODIVERSITY

It is exceedingly difficult to assess overall trends in biodiversity since there is little agreement as to which parameters, such as rate of deforestation, species loss, extent and number of protected areas etc., to measure as a baseline. An appropriate core set of indicators is still being developed for the CBD and for use in national reports on biodiversity. Various sets of indicators of biodiversity conservation have been proposed (Reid et al. 1993; Hammond et al. 1995; UNEP 1993; WCMC 1992, 1994) and the reviews, analyses, and tables given in World Resources. A guide to the global environment series produced by WRI are also relevant. It has been agreed by the COP that a Global Biodiversity Outlook, partly based on national reports on measures taken for the implementation of the provisions of the Convention, will be produced but the first issue has not yet been prepared.

A recent overview of the state of the environment is given in the "Global Environment Outlook" published by UNEP (1997). This notes that while significant progress has been made on several fronts in confronting environmental challenges in both developing and industrial regions, from a global perspective the environment has continued to degrade during the past decade (to 1997), and significant environmental problems remain embedded in the socio-economic fabric of nations in all regions. Progress towards a global sustainable future is just too slow. A sense of urgency is lacking.

Perhaps the most alarming recent series of events has been the profound impact of the 1997/8 El Niño current on world climate, leading to disastrous forest fires, storm surges, and floods, in many parts of the world that have resulted in incalculable losses of biodiversity as well as famine, epidemics, death and injury. This is all the more disheartening since at a blow, much careful conservation action and planning has been undone. The ENSO (El Niño Southern Oscillation) phenomenon links climatic anomalies in different parts of the world and their simultaneous appearance in many localities has led Richard Grove (Grove 1998) to de-

Table 1. Summary of emerging global environmental trends (UNEP 1997).

Unsustainable use of renewable resources
> The use of renewable resources—land, forest, fresh water, coastal areas, fisheries, and urban air—is beyond their natural regeneration capacity and is therefore unsustainable.

Increasing greenhouse gas emissions
> Greenhouse gases are still being emitted at levels higher than the stabilization target agreed upon internationally under the UN Framework Convention on Climate Change.

Reduction in natural areas and biodiversity
> Natural areas and the biodiversity they contain are diminishing due to the expansion of agricultural land and human settlements.

Increasing use of chemicals
> The increasing, pervasive use and spread of chemicals to fuel economic development is causing major health risks, environmental contamination, and disposal problems.

Escalating use of energy
> Global developments in the energy sector are unsustainable.

Unplanned urbanization
> Rapid, unplanned urbanization, particularly in coastal areas, is putting major stress on adjacent ecosystems.

Disruption of global biogeochemical cycles
> The complex and little understood interactions among global biogeochemical cycles are leading to widespread acidification, climatic variability, changes in the hydrological cycles, and the loss of biodiversity, biomass, and bioproductivity.

---

scribe them as being teleconnected. Such extreme climatic events and episodes of climatic instability have had the unexpected effect of convincing many of the public and not a few scientists that climate change is a real phenomenon whose effects will seriously impact upon them.

THE COMPONENTS OF AGROBIODIVERSITY

Agricultural biodiversity, also known as agrobiodiversity, is no easier to define than biodiversity itself. It can be generally regarded as biodiversity in an agricultural context and can be described as the variety and variability amongst living organisms (of animals, plants, and microorganisms) that are important to food and agriculture in the broad sense and associated with cultivating crops and rearing animals and the ecological complexes of which they form a part. It is not just a subset of biodiversity but an extension of it in that it embraces units (such as cultivars, pure lines, and strains) and habitats (agroecosystems such as farmers' fields) that are not normally considered or even accepted as properly part of biological diversity. Some authors would in fact exclude artificial diversity such as introduced species in an area (including by implication agricultural crops) from the concept of biodiversity "because it cannot fulfil the full range of societal values that native biodiversity does" (Angermeier 1994).

Agrobiodiversity includes all those species and the crop varieties, animal breeds and races, and microorganism strains derived from them, that are used directly or indirectly for food and agriculture, both as human nutrition and as feed (including grazing) for domesticated and semi-domesticated animals, and the range of environments in which agriculture is practiced. It also includes habitats and species outside of farming systems that benefit agriculture and enhance ecosystem functions.

As with biodiversity proper, it can be considered at three main levels—those of ecological diversity, organismal diversity, and genetic diversity (Heywood 1995), each forming a hierarchy of elements (Table 2).

Agrobiodiversity is by definition the result of the deliberate interaction between humans and natural ecosystems and the species that they contain, often leading to major modifications or transformations: agroecosystems are the product, therefore, of not just the physical elements of the environment and biological resources but vary according to the cultural and management systems to which they are subjected. Agrobiodiversity thus includes a series of social, cultural, ethical, and spiritual variables that are determined by local farmers (in the broad sense) at the local community level. These factors must be taken into account as part of the process of selection and introduction of new or underdeveloped crops, although they are often overlooked.

The components of agrobiodiversity may be summarized as in Table 2. Thus it covers not only the whole gamut of genetic resources (from advanced cultivars to primitive land races, domesticates, semidomesticates, wild relatives) but the diversity of ecosystems and agroecosystems within landscapes that are exploited in some way for agriculture and forestry, and the complex set of human interactions.

The recognition of agrobiodiversity—i.e. diversity in crops, agroecosystems, and approaches—as a concept and as an issue, is a major conceptual breakthrough, reinforced by the CBD and the Global Plan of Action. It signifies, I believe, the emergence of a new paradigm for agriculture that embraces not just the most technologically advanced and efficient farming and production systems, dependant on highly bred or engineered crops and animal breeds, with an emphasis on uniformity and standardization, and based on a very restricted set of species, but recognizes that the great diversity of traditional farming systems and practices in many cultures in different parts of the world and the thousands of species that are locally cultivated or semidomesticated in home gardens or other polycultures, or harvested from the wild in nearby habitats make a major and essential contribution to food security for hundreds of millions of people across the globe.

It has been estimated that more than three million hectares survive under traditional agriculture as raised fields, terraces, swidden fallows, polycultures, home gardens, and other agroforestry systems (Altieri 1998) and while these seldom have the potential to produce marketable surpluses, they do make a major contribution to food security (Heywood 1999) and traditional cropping systems are said to provide as much as 20% of the world's food supply.

It is important, however, not just to pay lip service to the legitimate needs and aspirations of local, small scale farmers and agroforesters whose dependence on diversity has been threatened by the wide-scale conver-

Table 2. The composition and levels of agrobiodiversity.

Agroecological diversity	Organismal diversity	Genetic diversity
biomes	kingdoms	gene pools
agroecological zones	phyla	populations
agroecosystems	families	individuals
polycultures	genera	genotypes
monocultures	species	genes
mixed systems	subspecies	nucleotides
rangelands	varieties	
pastures	cultivar groups	
fallows	cultivars	
agroforestry systems	land races	
agrosylvicultural	breeds	
sylvopastoral	strains	
agrosylvopastoral	pure lines	
home gardens		
forest ecosystems		
managed forests		
plantation forests		
seed forests		
fisheries		
fresh water systems		
marine systems		
habitats		
field		
plot		
crop		

Socio-cultural diversity: human interactions with the above at all levels

sion of the agricultural sector from traditional to modern systems. What we need is to assess, inventory, monitor, and try and understand this diversity of species and approaches in traditional farming systems and in various forms of wild harvesting and extractivism and also capture the traditional knowledge on agriculture, cultural practices, uses and so on which is rapidly disappearing as the older generations die out.

TRENDS IN AGROBIODIVERSITY

Several trends in agrobiodiversity may be discerned (Table 3), most of them outlined or implicit in the FAO Global Plan of Action and CBD/COP decisions. Although a great deal of attention has been focused on the possible impacts of biotechnology and genetically modified (GM) crops on agriculture and society, a topic that I shall simply refer to, it is notable that other trends are concerned with the conservation and sustainable use of biodiversity and agrobiodiversity, especially with the identification and conservation of the genetic variation in plants that are of actual or potential use to agriculture.

Sustainable Agriculture

The challenge facing agriculture over the coming decades in an expanding global economy is to achieve stable production on a sustainable basis, by introducing technologies and management practices that would ensure a healthy environment, stable production, economic efficiency, and equitable sharing of social benefits. With the global population likely to rise by 2.5 billion in the next three decades, FAO estimates that food production must rise by some 75% during that period. As Carter (1998) has observed recently *"Overpopulation, degradation of the environment, and exhaustion of crop land present significant challenges in the*

Table 3. Trends in agrobiodiversity.

Moves towards sustainable agriculture
  Attempts to develop sustainable agricultural production in such a way that its negative impacts on natural biodiversity are minimized.

Bioregional perspective
  Adoption of a broader perspective of agriculture as an element within a broader panorama of bioregional and landscape development.

Inventory needs
  Recognition of the need to survey and inventory those plant and animal resources that may be used in agricultural development.

Genetic resource conservation
  Efforts to conserve genetic resources both in situ and ex situ will be intensified

Importance of on-farm management
  Recognition of the importance of on-farm management of crop genetic diversity in the form of land races and the need to manage and enhance these.

Importance of biodiversity in natural ecosystems
  Recognition of the fact that natural and semi-natural ecosystems contain wild plant species, races, and populations that are of importance for food and agriculture, such as wild relatives of crops, are important sources of material for agroforestry, habitat restoration, and reforestation, and species that are wild harvested and contribute to farm household incomes.

Need to broaden the genetic basis of crops
  Recognition of the very narrow diversity maintained in some crops and the need therefore to widen the genetic base of crops.

Need to cultivate a diversity of crops
  Recognition of the desirability of cultivating a greater diversity of crops and the introduction of new crops as a way of promoting agricultural sustainability.

Contribution of diversity to farm households
  Recognition of the role that a diversity of wild and semi-domesticated species may play in food and livelihood security of farm households and their potential for further development and wider use.

Importance of traditional knowledge
  Recognition of the importance of traditional knowledge about agricultural practices and individual species and the need to record and conserve this knowledge.

The impact of biotechnology and GM crops
  Recognition of the need to assess the effects of biotechnology on agriculture, the need to review mechanisms by which GM crops could be monitored, and impacts on Intellectual property rights.

new millennium. But unlike population booms of the past, this time crop acreage will not rise accordingly with population growth," he said. "For the human species, these simple facts define the major mission of the next century—bringing our relationship with the earth into balance. For those of us working in agriculture, our mission is no less critical—increase productivity simply to buy time."

However, while agricultural productivity has increased in the second half of this century at a greater rate than world population increase, such increases in productivity in the past have been at the expense of wide environmental damage, seldom recognized at the time. Sustainable agriculture, while it cannot be rigidly defined, is widely interpreted as consisting of practices that are ecologically sound, socially responsible, and economically viable (Thrupp 1996). Daily et al. (1998), for example, have drawn attention to the need to take into account the environmental and social costs of agricultural production as well as the direct farming costs. Achievement of high levels of GNP may be at the cost of a depletion of the natural resource base such as mining of the soil, lowering of water table, and impairment of other ecosystem services.

Bioregional Management

Instead of viewing agricultural ecosystems as self-contained units, there is an increasing tendency to adopt a broad landscape or bioregional approach in which all aspects of the landscape are taken into consideration—natural, semi-natural, heavily modified including agricultural, industrial, and urban, protected and unprotected.

As Miller (1996) in a valuable review of the bioregional approach comments, *"Since the landscape is fragmented and much wildland has been converted to other use, the boundaries and coverage of some protected areas may not conform to the size and shape of the ecosystems that are to be maintained and managed. ... Moreover, in landscapes where protected areas have not been established, key genetic, taxonomic, and ecological elements of diversity that once may have been found in wildlands, or extensive farm or forest operations, are now relegated to isolated patches in intensively managed farms, pastures, timber-harvesting sites, and suburban, urban, and industrial areas."* Natural vegetation fragments are now becoming an almost universal component of our landscapes—*"Storm-battered islands in a sea of human settlement"* (Lash 1996)—and their management and that of agroecosystems has to be planned and implemented in the context of large biotically viable regions.

It is at the landscape scale, as Halffter (1998) points out, where the consequences of human action such as deterioration, ecosystem modification and fragmentation as well as pollution are most dramatic. It is also from the landscape perspective that we can analyze the diversity of species not just as a function of the heterogeneity of the biological and physical environment but also as a function of human activity.

Survey and Inventory

The conservation and sustainable use of traditional agrobiodiversity is of course undertaken locally but before action to support or enhance it can be organized, we need to assess and survey what different farming systems exist and which species are currently being cultivated or used by them. As is the case with biodiversity conservation, inventory is an essential to provide a base line for monitoring and subsequent action. An increasing amount of information is being gathered on the diverse types of management systems of agrobiodiversity (e.g. Alcorn 1990; Gadgil and Berkes 1991; Redford and Padoch 1992; Hladik et al. 1993; Altieri 1995, 1998; Gómez–Pompa 1996) but we need a more systematic gathering of this information, perhaps through national biodiversity strategies/action plans.

Well over 6000 species of plants are known to have been cultivated at some time or another and many thousand that are grown locally are scarcely or only partially domesticated, while as many if not more are gathered from the wild. Not surprisingly, most of the partially domesticated or wild-collected species are found in the tropics. For example, the Plant Resources Project of South-East Asia (PROSEA) records nearly 6000 species in its Basic List of species (some of them exotic) used by humankind in that area (Jansen et al. 1993) and assuming similar levels in other tropical regions, we can extrapolate to a figure of 18–25 thousand species for the tropics as a whole. In addition several thousand plant species are used in human activities in Mediterranean and temperate regions of the world.

These figures exclude most of the 25 thousand species that are estimated to have been used or are still in use as herbal medicines in various parts of the world, especially China, tropical Asia, the Indian subcontinent, Africa, and Central and South America, and the many thousands of species that are grown as ornamentals in parks and in public and private gardens and in the horticultural trade.

Gradually efforts are being made to improve this inventory but again action is needed at a national level in the first instance and it is remarkable how little attention has been given in national conservation or biodiversity strategies or action plans to the importance of identifying and listing which species are used by humans.

Conservation of Genetic Resources and Questions of Access

As noted in the Recommendations made by a DIVERSITAS Working Group of Experts on various Articles of the CBD (DIVERSITAS 1998), the wide interpretation of the concept of genetic resources implied in the Convention will require a policy adjustment for both ex-situ and in-situ management. The national and international efforts of the last 30 or so years to sample and maintain genetic diversity of crop and pasture species for future use in plant breeding and crop production, mainly in seed banks or in a smaller number of cases in field gene banks or clonal collections, will continue although with changes in focus.

One of the key questions that has to be addressed when considering the use of wild species as potential new crops is access to their genetic material. The term "access" today normally has the connotation of national sovereignty over natural resources as covered by Article 15 of the CBD. However, in the case of new, underexploited crops and semi/pre-domesticated species, access in the sense of actual availability, irrespective of "ownership" tends to be overlooked. Most gene banks have tended to give little priority to acquiring accessions of "minor" species or crop relatives. In fact, attention has often been drawn to the narrow spectrum of species—mainly the staple crops—that make up the bulk of the main national and international germplasm collections (Heywood 1993). While 15–20 thousand species are represented in gene banks, the numbers and size of the accessions of the vast majority of these is small in comparison with the accessions of land races and cultivars of the main crop and pasture species. The formal genetic resources sector has not been able to devote substantial resources to these other species although in recent years there has been a substantial broadening of approach and this is likely to continue and develop further.

While this work on genetic resource conservation will still concentrate on ex situ approaches, increasing attention will be paid to the in situ conservation of genetic diversity of wild relatives of crops and other wild species that are of importance to agriculture, forestry, and other human needs.

At the grass roots level, genetic conservation is aimed at needs that are normally not covered by large international germplasm centers or even by national systems. Much of the work that has been done is at local level by NGOs with strong farmer participation and by grass roots movements, such as community seed banking (Berg 1996). Most of the wild plant species that are used in local farm households have not been the subject of attention by the formal genetic resources sector although in recent years more attention is being paid to so-called minor or underutilized crops and to the wild relatives of major crops. It is likely that in the future there will be strong demands for these centers, and for FAO and IPGRI, to focus attention on these grass roots needs and for assistance to be given to farm households and local communities to help them obtain, maintain, and conserve genetic material of their locally used crops and semidomesticates. Conventional approaches to germplasm conservation have tended to concentrate on "conserving much and using little" but the interests of the small farmer are more likely to be served by making germplasm available than by storing it in genebanks for potential future use. For these small traditional farmers in the developing world that are cultivating plants under conditions that may be considered marginal, satisfying their needs through the traditional means of breeding for client farmers may not be appropriate or possible and an alternative that has been suggested is breeding with partner farmers—this would involve deploying advanced breeding materials, recombination with local materials and exposing it to evaluation and selection by local farmers (Berg 1996).

Broadening the Genetic Basis of Crops

At the genetic level of agrobiodiversity, the danger of depending on uniform genetic material and the need to broaden the genetic basis of many of our crops has frequently been pointed out. The dangers of depending on uniform genetic material as the basis of production in major crops are now well known. For example, the need for greater genetic diversity in many tropical crops is frequently observed. In yams, for example, the white Guinea yam (*Dioscorea rotundata*) that is West African in origin, a great amount of diversity is being lost throughout Africa and the future of the crop is at risk. Germplasm of wild relatives is proving an important source of new traits that will help restore diversity to the crop. As Hodgkin (1996) notes, the conservation of genetic diversity found within crops is important both for the continuing demand for new improved cultivars and for ensuring that there is sufficient variation in current agricultural systems to prevent disease, epidemics, and other disasters.

The Role of Diversity in Crop Production Systems

The GPA suggest that *"In the future agricultural systems will need to incorporate a broader range of crops including inter alia crops which produce raw material or are sources of energy."* In recent years there has been a greater willingness on the part of some farmers in some parts of the world to increase the range of crops grown. This is a particularly important issue in arid and semi-arid marginal lands and links in with the introduction of new crops and the wider exploitation of underutilized species.

There is growing recognition of the role that diversity plays in traditional farming systems. It is estimated that about 60% of the world's agriculture consists of traditional subsistence farming systems in which there is both a high diversity of crops and species grown and in the ways in which they are grown, such as polycropping and intercropping, that leads to the maintenance of greater or lesser amounts of variation within the crops. As just noted, greater attention is now being paid to the desirability of maintaining genetic variation within crops as a strategy for avoiding losses or failure in the face of disease or other factors and often farmers will retain traditional varieties alongside modern highly bred cultivars (Brush 1995).

New Crops

While development of genetically modified soybean and other major species may play a key role in achieving enhanced productivity that is essential for human survival, we must also look to new crops for part of the solution. New species are being added continuously to our list of economically important cultivated plants but as is abundantly well known the scope for extensive introduction of new crops is strictly limited because of a whole series of technical and socio-economic problems. The contribution of the Center for New Crops and Plant Products and the work of the New Crops Symposia and other similar meetings are indicative of the progress that is being made in this area.

Underutilized Crops

Important initiatives have been started in various parts of the world to explore the potential use of hundreds of local crops that are currently underexploited. An outstanding example is the series "Especies Vegetales Promisorias" (Promising Plant Species) of the countries of the Andrés Bello Convention (Bolivia, Colombia, Chile, Ecuador, Spain, Panama, Peru, and Venezuela). Promising species, in this context, are defined as those that are essentially native, have not been extensively domesticated, are underutilized or little known but with economic potential in the short, medium, and long term and about which basic scientific knowledge is available to validate their status as promising species. Over 1000 species have been identified by this project and over ten volumes published (SECAB 1989), containing a mass of valuable information on taxonomy, geography, ecology, properties, uses, phytochemistry, economic importance, agronomy, and industrialization. A review of neglected crops (*cultivos marginados*) of the New World has been published by FAO (Hernández Bermejo and León 1992).

Several groups have focused on the development of new or underutilized crops, such as the International Centre for Underutilized Crops (ICUC) whose goal is food security, nutrition, and economic welfare of human beings through assessing, developing, and utilizing the biological diversity of underutilized crops and species for sustainable and economic production of food and industrial raw materials. It has published a number of books on genetic resources of underutilized crops (e.g. Anthony et al. 1995, de Groot and Haq 1995; Smartt and Haq 1997).

Various networks have been created such as the Network on Underutilized Fruits for Asia (UTFANET), a regional network of southern and eastern African underutilized crops (SEANUC), and those established under IPGRI's Underutilized Mediterranean Species (UMS) project, e.g. the Rocket (*Eruca, Diplotaxis*) Genetic Resources Network (Padulosi 1995). IPGRI also runs a program for Promoting the Conservation and Use of Underutilized Crops that has published a series of treatments that include information on cultivation, agronomy, production, prospects, and related topics. The MEDUSA network (Heywood and Skoula 1997) for the identification, conservation, and sustainable use of wild plants of the Mediterranean region is concerned with native species of actual or potential importance to agriculture, especially in the semi-arid marginal lands of the area.

Particular attention is being focused on energy and industrial crops. For example, a recent FAO report on potential energy crops for Europe and the Mediterranean (El Bassam 1996) gives a catalog of species, some of them wild, many already cultivated, together with information on their characteristics, cultivation methods and utilization for energy production. The report considers that it is vital to increase the number of plant species that might be introduced into cultivation for this purpose. In addition, many so-called minor species that have been traditionally used as herbs, condiments, and medicinals are now being looked at for possible industrial applications. A report, prepared for the European Commission (Smith et al. 1997), highlights the wide array of species and products that could be developed for industry and energy in Europe.

The Role of Biodiversity in Natural Ecosystems

An important trend is the recognition of the need to consider natural and semi-natural ecosystems in the context of agrobiodiversity. These ecosystems contain wild plant species, races, and populations that are of importance for food and agriculture, such as wild relatives of crops and species, are important sources of material for agroforestry, habitat restoration, and reforestation, and species that are wild harvested and contribute to farm household incomes. They also form part of the landscapes within which agricultural systems are found and provide a whole series of environmental services and functions, such as soil stabilization, water and air quality, on which healthy sustainable agroecosystems depend. It is also now increasingly recognized that alterations in natural ecosystems through accelerated deforestation, logging, or conversion to other uses, may also affect the viability of neighboring agriculture. As a recent report on the effects of climate change and land use on Amazonian forests notes (Laurance 1998), there are alarming synergisms between human land-uses and natural climatic variability. Thus logged or fragmented forests are increasingly susceptible to fire and climatic vicissitudes and fragmentation of forests can lead to a juxtaposition of forest fragments with fire-prone pastures and farmlands. The risks to farming systems under such circumstances are obvious.

Wild and Wild-harvested Species

While large-scale agriculture will continue to focus on monocultures of the major crops, increasing attention will be paid to the role of wild and wild-harvested species in farm systems. Another major source of agrobiodiversity is the tens of thousands of species that are grown in a pre- or semi-domesticated state on home gardens or similar polycultures. As noted above, many thousands more are wild harvested to supplement farm household incomes.

The biodiversity of most of the wild species used in traditional farming systems is usually poorly studied. Even their identification and classification is often unsatisfactory, leading to considerable confusion when the plants or their products are traded. Even less is known about their detailed distribution, the extent, size and diversity of their populations, their breeding behavior, pollination mechanisms and so on. Since most of the species we are concerned with have never been cultivated or are at most semi-domesticated on a local scale, our knowledge of their most basic biology and agronomy is virtually non-existent and we must depend on knowledge developed over long periods by local farming societies. Such indigenous knowledge, as Altieri (1995, 1998), points out often includes very detailed understanding of the physical environment, including weather and soil types. It is essential that this local knowledge, that is itself an important resource, should be recorded and made available for future generations.

On-farm Management

Much of the diversity in agroecosystems is found in the hundreds of thousands of land races that have developed over the centuries in farmers fields, through a process of unconscious selection and saving of seed or vegetative propagules for future planting seasons. These land races are agroecotypes that are adapted to the local ecological, agronomic, social, and cultural traditions. They are increasingly at risk through replacement by modern cultivars as farmers attempt to increase yields.

The term on-farm conservation is applied to the dynamic conservation of genetic diversity in such land races and weedy crop relatives in traditional, usually low-input farming systems and is an area of increasing interest and concern. As stated in the Global Plan of Action (Priority Activity 2), there is a need for better understanding and improvement of the effectiveness of on-farm conservation so that action can be taken to increase its contribution to food production and security. Amongst the research needed is work on the promotion of little known crops, their seed production, marketing, and distribution. On-farm conservation is a form of in situ conservation but is significantly different from in situ conservation of wild species in natural or semi-natural ecosystems and should not be confused with it as has happened frequently in the past.

Agricultural Biotechnology

An inevitable and highly controversial trend in the handling of agrobiodiversity is the growing and developing use of the techniques of agricultural biotechnology. Apart from the well-publicized risks that may be attendant on the cultivation and consumption of genetically modified crops, appropriate technologies are likely to play an increasing role in assessing genetic variability and in the breeding and enhancement of crops.

THE NEGLECT OF AGROBIODIVERSITY AND AGROECOLOGY BY CONSERVATION BIOLOGISTS

Although the World Conservation Strategy (IUCN, UNEP, WWF 1980) published in 1980 recognized the importance of agriculture and included a program for the conservation of zones rich in genetic resources, subsequently mainstream conservation gave them little priority, preferring to concentrate on charismatic or flagship species, or any kind of endangered species, and on protected area systems and tropical forests. Certainly, lip service was paid to the conservation of wild species used by humans in campaigns such as WWF's "Saving the Plants that Save Us," although agriculture was generally viewed as detrimental to the conservation of biodiversity; and occasionally attention was drawn to importance of wild species in tropical forests in terms of their value to extractivists, as a justification for their conservation rather than conversion to other uses. But with the exception of the pioneering work of Altieri (1995, 1998) and others on agroecology and the importance of traditional farming systems not only in terms of cultivated fields but the natural ecosystems that surround them, and Gary Nabhan on local seed savers, conservationists and conservation biologists alike have contrived to avoid these issues.

Only recently (Vandermeer and Perfecto 1997) in an editorial in *Conservation Biology* addressed the issue of the agroecosystem and the need for viewing it under the conservation biologists lens. The failure, they say, of conservation biologists to show much interest in agroecosystems, simply because they are already tainted, is matched by the lack of interest from agroecologists in the biodiversity found in traditional agroecosystems since it has no obvious connection to production. Since we are increasingly concerned with introducing forms of sustainable agriculture that will reduce the adverse effects of agriculture on biodiversity, it is vital that the conservation biology community changes its attitude and starts considering agroecosystems as legitimate areas for study and begin asking the same questions that they do about so-called "natural ecosystems."

The development agencies have not focused much attention on the effects of agricultural development projects on biodiversity and fewer than 2% of 377 agricultural projects financed since 1988 by the World Bank have dealt explicitly with biodiversity (Srivastava et al. 1996)

Vandermeer and Perfecto (1997) state the situation as follows:

The fact is that most of the terrestrial world is, in one sense or another, an agroecosystem. If we are to ignore this ecosystem simply because it does not fall within our romantic notion of "pristineness," we leave the vast majority of the Earth's surface to the husbandry of those who care little about biodiversity preservation. On the other hand, if we are to ignore the preservation of biodiversity per se simply because it does not fit obviously into classical production categories, we leave the preservation of the world's biodiversity to those who refuse to work outside of national parks and nature preserves.

This reluctance to consider agroecosystems is another manifestation of the disdain shown by some for artificial, i.e. human generated biodiversity and who argue that this should not be our concern and that we should not advocate the preservation of diversity so much as protection of ecological integrity.

CONCLUSIONS

To summarize, many of the trends in agrobiodiversity center around the recognition of the benefits that accrue from diversity as such, whether it be in terms of genes in crops, the range of species cultivated, or the range of cultivated systems used. It could be argued that maintenance of biodiversity and achieving agricultural sustainability—the theme of this conference—is as likely to be found in developing strategies to maintain multiple species agroecosystems as in large-scale modern monocultures, despite the imperative facing us of raising staple food production to match the demands of the world's growing population. But there is no contradiction if we remember that 20% or more of human nutrition probably derives from traditional farming systems, not to mention non-food crops such as non-wood forest products and medicinal plants, supplemented by highly diverse systems of wild harvesting. We need to ensure that these traditional systems are not eroded or swept away by political and economic forces that lead to soil erosion, decreased biodiversity on farms, genetic erosion, and loss of traditional knowledge (Altieri 1998). On the contrary considerable efforts must be put into enhancing productivity and intensification of production in these systems so that they too can make a substantial contribution to the overall food demand in a way that does not put the peasant farmers at risk or lead to further environmental degradation.

"Biodiversity is a non-detachable part of the concept of sustainability. ... biodiversity is essential for agricultural production, as agriculture should be for biodiversity conservation." The agricultural and biodiversity conservation sectors must work in partnership and the fact that this is now beginning to happen is perhaps the most important trend of all in the development of agricultural biodiversity.

REFERENCES

Altieri, M.A. 1995. Agroecology: the science of sustainable agriculture. Westview Press, Boulder, CO.

Altieri, M.A. 1998. The agroecological dimensions of biodiversity in traditional farming systems. In: D.A. Posey (ed.), Cultural and spiritual values of biodiversity. UNEP (in press).

Alcorn, J. 1990. Indigenous agroforestry strategies meeting farmers' needs. In: A. Anderson (ed.), Alternatives to deforestation: Steps towards sustainable use of the Amazon Rain Forest. Columbia University Press, New York.

Angermeier, P.I. 1994. Does biodiversity include artificial diversity? Conservation Biol. 8:600–602.

Anthony, K., N. Haq and B. Cilliers, (eds). 1995. Genetic resources and utilization of underutilized crops in southern and eastern Africa. Proceedings of a Symposium held at the Institute for Subtropical Crops, Nelspruit, South Africa. FAO, ICUC, CSC, Nelspruit.

Berg, T. 1996. Dynamic management of plant genetic resources: potentials of emerging grass-roots movements. Studies in Plant Genetic Resources. Study 1. FAO, Rome.

Brush, S.B. 1995. In situ conservation of landraces in centers of crop diversity. Crop Sci. 35:346–354.

Carter, T. 1998. 10th annual soybean conference, CSIRO, Brisbane, Sept. 15, 1998 quoted in Plant Breeding News, Sept. 17, 1998.

Cooper, H.D., C. Spillane, I. Kermali, and N.M. Anishetty. 1998. Harnessing plant genetic resources for sustainable agriculture. Plant Genetic Resources Newslett. 114:1–8.

Daily, G., P. Dasgupta, B. Bolin, P. Crosson, J. du Guerny, P. Ehrlich, C. Folke, A.M. Jansson, B.-O. Jansson, N. Kautsky, A. Kinzig, S. Levin, K.-G. Mäler, P. Pinstrup-Andersen, D. Siniscalo, and B. Walker. 1998. Global food supply: food production, population growth and the environment. Science 281:1291–1292.

de Groot, P. and N. Haq (eds.). 1995. Promotion of traditional and underutilised crops. Report of a Workshop held in Valletta, Malta, June 1992. International Centre for Underutilized Crops. Commonwealth Science Council, London.

DIVERSITAS. 1998. Recommendations on scientific research from a DIVERSITAS Working Group of Experts that should be undertaken for the effective implementation of articles 7, 8, 9, 10 and 14 of the Convention on Biological Diversity, Mexico City, Mexico, 24–25 March 1998. UNEP/CBD/COP/4/Inf. 18.

El Bassam, N. 1996. Renewable energy. Potential energy crops for Europe and the Mediterranean region. Federal Agricultural Research Centre (FAL), Braunschweig, Germany and FAO Regional Office for Europe (REU). REU Technical Series 46. FAO, Rome.

FAO. 1996. Global plan of action for the conservation and sustainable use of plant genetic resources for food and agriculture. FAO, Rome.

Gadgil, M. and F. Berkes. 1991. Traditional resource management systems. Resource Manage. Optimization 18:127–141.

Gómez–Pompa, A. 1996. Three levels of conservation by local people. p. 347–356. In: F. di Castri and T. Younès (eds.), Biodiversity, science and development: Towards a new partnership. CAB Int., Wallingford UK.

Grove, R.H. 1998. Ecology, climate and empire: Colonial and global environmental history, 1490–1940. White Horse Press, Cambridge.

Halffter, G. 1998. A strategy for measuring landscape biodiversity. Biol. Int. 36:3–17.

Hammond, A., A Adriaanse, E. Rodenburg, D. Bryant, and R. Woodward. 1995. Environmental indicators: A systematic approach to measuring and reporting on environmental policy performance in the context of sustainable development. World Resources Inst., Washington, DC.

Hernández Bermejo, E. and J. León. 1992. Cultivos marginados. Otra perspectiva de 1492. FAO, Rome.

Heywood, V.H. 1993. Broadening the basis of plant resource conservation. p. 1–13. In: J.P. Gustafson (ed.), Gene conservation and exploitation. Plenum Press, New York.

Heywood, V.H. (ed.). 1995. Global biodiversity assessment. Cambridge Univ. Press, Cambridge.

Heywood, V.H. 1999. Use and potential of wild plants in farm households. FAO, Rome.

Heywood, V.H. and M. Skoula (eds.), Identification of wild food and non-food plants of the Mediterranean Region. Cahiers Options Méditerranéennes 23, 1997.

Hladik, C.M., A. Hladik, O.F. Linares, H. Oagezy, A. Semple, and M. Hadley (eds.). 1993. Tropical forests, people and food. Biocultural interactions sand applications to development. Man and the Biosphere Series 13. UNESCO, Paris and The Parthenon Publishing Group, Carnforth.

Hodgkin, T. 1996. Some current issues in conserving the biodiversity of agriculturally important species. p. 357–368. In: F. di Castri and T. Younès (eds.), Biodiversity, science and development: Towards a new partnership. CAB Int., Wallingford UK.

IUCN, UNEP, WWF. 1980. The world conservation strategy. Living resource conservation for sustainable development. IUCN, Gland.

Janssen, P.C.M., R.H.M.J. Lemmens, L.P.A. Oyen, J.S. Siemonsma, F.M. Stabast, and J.L.C.H. van Valkenburg (eds.). 1993. Plant resources of South-East Asia. Basic list of species and commodity grouping. Final version. PROSEA, Bogor.

Lash, J. 1996. Preface. In: K.R. Miller, Balancing the Scales: Guidelines for increasing biodiversity's chances through bioregional management. World Resources Inst., Washington, DC.

Laurance, W.F. 1998. A crisis in the making: responses of Amazonian forests to land use and climate change. Trends Ecology Evolution 13:411–415.

Miller, K.R. 1996. Balancing the scales: Guidelines for increasing biodiversity's chances through bioregional management. World Resources Inst., Washington DC.

Padulosi, S. (compiler). 1995. Rocket genetic resources network. Report of the First Meeting, 13–15 Nov. 1994, Lisbon, Portugal. Int. Plant Genetic Resources Inst., Rome.

Raven, P. 1998. Planet of the plants. World Conservation 2/98:22–23.

Redford, K.H. and C. Padoch (eds.). 1992. Conservation of neotropical forests. Working from traditional resource use. Columbia Univ. Press, New York.

Reid, W.V., J.A. McNeely, D.B. Tunstall, D.A. Bryant, and M. Winograd. 1993. Biodiversity indicators for policy-makers. WRI/IUCN. World Resources Inst. Washington, DC.

SECAB. 1989. Especies vegetales promisorias de los paises del Convenio Andrés Bello. Secetaría Ejecutiva del Convenio Andrés Bello (SECAB), Bogotá.

Seyani, J. 1998. Planet of the plants. World Conservation 2/98:22–23.

Smartt, J. and N. Haq (eds.). 1997. Domestication, production and utilization of new crops. Int. Centre for Underutilized Crops, Southampton, UK.

Smith, N.O., I. Maclean, F.A. Miller, and S.P. Carruthers. 1997. Crops for industry and energy in Europe. University of Reading. European Commission Directorate General XII E-2, Agro-Industrial Unit. Office for Official Publications of the European Communities, Luxembourg.

Srivastava, J., N.J.H. Smith, and D. Forno. 1996. Biodiversity and agriculture: Implications for conservation and development. Technical paper 321. World Bank, Washington DC.

Thrupp, L.A. (ed.). 1996. New partnerships for sustainable agriculture. World Resources Inst., Washington, DC.

UNEP. 1993. Guidelines for country studies on biological diversity. UNEP, Nairobi.

UNEP. 1997. Global Environment Outlook. UNEP. Oxford Univ. Press, New York & Oxford.

UNEP. 1998. STAP Expert Group Workshop on Sustainable Use of Biodiversity. Malaysia, 24–28 Nov. 1997 (Reported to 11th STAP meeting, 21-23 Jan. 1998, Agenda Item 6).

Vandermeer, J. and I. Perfecto. 1997. The agroecosystem: A need for the biologist's lens. Conservation Biol. 11:591–592.

WCMC. 1992. Global biodiversity. Status of the Earth's living resources. Chapman & Hall, London.

WCMC. 1994. Bioddiversity data sourcebook. B. Groombridge (ed.), World Conservation Press, Cambridge, UK.

New Crops for Canadian Agriculture

Ernest Small

The designation "new crop" may be applied to virtually any useful plant that in some respect is new. The following categories of newness (which are not all mutually exclusive) are economically important in discussing new crops: (1) gathering new wild crops from nature; (2) cultivating an undomesticated plant not previously grown; (3) domesticating (changing genetically) an undomesticated plant; (4) breeding improved cultivars of domesticated plants; (5) growing crops in new areas; (6) growing crops for new uses; (7) growing crops with new management techniques; (8) selling crops in new markets.

Category 2 and (especially) category 3 represent "new crops" in the narrow sense perhaps most widely understood. Although all domesticated crops originally came from the wild, in recent times the domestication of wild plants relatively infrequently produces crops of notable economic significance. Based on 160 crops grown and/or imported in the US that had a value of at least $1 million, Prescott–Allen and Prescott–Allen (1986) found that only six crops with this value were domesticated since 1900, a rate of success of less than 7 per century. Almost all potentially important new crops for a political or agronomic region are cultivated elsewhere, and indeed the leading domesticated crops of Canada all originated in foreign lands. Crop diversification, especially involving new crops, is considered to be a fundamental area deserving support in Canada (Small, in 1999; note Fig. 1).

The following reviews what is significantly new (for any of the eight kinds of newness pointed out above) with respect to the major classes of crops in Canada, such as cereals, oilseeds, forages, vegetables, and fruits. These crop classes represent arenas of competition, and new crops generally can only compete within one of these arenas. As will be noted, some of these arenas thrive on the introduction of new entities, while others are like clubs that are very hostile to the entry of new members. Given the breadth of the topic, only a limited amount of detail and selected examples will be presented for the crop categories, and minor crops that either have had very limited success in Canada or show little potential will not be discussed or will be given only incidental mention. For those seeking additional information on new crop development in Canada, it should be noted that the following present a wealth of information on the World Wide Web: (1) the federal department of agriculture, i.e. Agriculture and Agri-Food Canada (AAFC), Canada's primary plant breeding institution, which has research stations in all provinces; (2) the provincial agriculture or resource departments; and (3) a wide variety of farm-oriented organizations.

CANADIAN AGRICULTURE

Agriculture is the most important of the industries dealing with the biological resources of Canada, exceeding the value of forestry, fishing, and trapping. Agriculture and the allied agri-food industry are respectively responsible for about 2 and 6% of the Gross National Product (GDP), and about 14% of Canada's employment. In 1996 (the year of the last comprehensive census) the agri-food industry of Canada was worth over $70 billion* (about 8.8% of Canada's

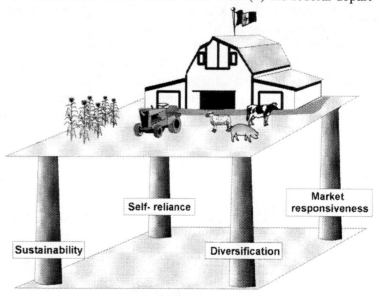

Fig. 1. A vision of agriculture in Canada, conceptualizing crop diversification as one of four essential supporting pillars (Agriculture Canada 1989).

---

*In this paper Canadian dollar figures are reported in Canadian currency, and American dollar figures in American currency.

GDP), of which 24% represented on-farm production, the remainder accounted for by allied food industries, commercial sales, and the food service industry. The equivalent of 70% of Canada's agricultural production, with a value of about $20 billion, was exported in 1996, while the value of imported agricultural and food products was about two-thirds of this. This overall trade surplus shows that Canadian agriculture is dependent on world markets, and suggests that new crops represent an important measure to address market fluctuations and declines. The most important commodities exported by Canada are grains and grain products (35% of total agri-food exports), red meat and live animals (20%), and oilseeds and oilseed products (13%). The US is Canada's most important trading partner, taking about half of Canada's agri-food exports. In return, about 60% of Canada's agri-food imports are from the US. Imports include fruits and nuts (19% of total agri-food imports), vegetables (10%), and red meats (8%). The relatively short growing season of Canada necessitates a wide variety of agricultural imports, most notably hot-region plantation crops such as coffee, tea and spices (10% of all imports).

Fig. 2 shows the distribution of farmland in Canada. More than 40% of Canada is forested; about half of this area is capable of producing timber, and about a quarter is currently managed for timber production. Adverse climate, soil, and other circumstances prevent profitable agriculture in most of the country. Although Canada has close to 10 million square kilometers, making it the world's second largest country, only 67.7 million ha (0.677 million square kilometers) are arable (Canadian Federation of Agriculture 1995; Reid 1995). By comparison, total land in farms in the US in 1997 was 392 million ha, about six times the area used for farming in Canada. (The value of US crops in this year was about 109 billion dollars, about 10 times the value of Canadian production.) The arable area of Canada, about 7% of the country, is equal to about three times the land area of Great Britain. However, it has been estimated that less than 5% of Canada's land is actually capable of producing crops, and most of this is already in production (Environment Bureau 1997a). Moreover, only half of this land capable of producing crops is prime agricultural land, and much of this has succumbed to urban development (Science Council 1991). Canada may already be approaching its upper limit of farmland development (Acton 1995). An additional 6% of Canada's land can be used for grazing (Environment Bureau 1997a).

Limiting Factors for Agriculture in Canada

All crops grown in Canada (with the exception of greenhouse and cultured mushroom crops) and all potentially new crops are strongly constrained by climate and soil factors (Fig. 3). Of course, length of season, distribution of temperature and precipitation, soil fertility, and physical aspects of land are universal determinants of what crops can be grown. Land use inventories that assess the suitability of land for agriculture, forestry, recreation, and wildlife have been in use for several decades (Statistics Canada 1986). For example, the soil regions of Western Canada differ in the capacity to grow crops. The brown soil in the semi-arid region of the Prairies varies considerably from year to year in crop yield depending on degree of drought,

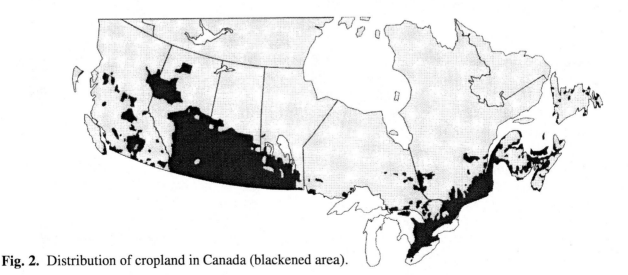

Fig. 2. Distribution of cropland in Canada (blackened area).

while dark brown soil is not as vulnerable to drought. The black soil retains moisture better than the brown soil, is rarely subject to drought, and produces higher yields. The gray soil zone has higher moisture levels, cooler temperatures, and a shorter growing season. Management practices in the different zones are necessary, since climatic conditions influence the susceptibility of crops to disease and pest infestation.

Ideally, knowledge of three factors can be used to produce an excellent identification of what crop should be grown where and when. First, an agricultural knowledge of the growth requirements of crops is necessary. Second, measures of the comparative extent to which local soil, climate, and pests and diseases match the needs of the crops. And third, predictions of markets for the crops, both domestic and foreign. In fact, such detailed budgets are issued by Canada's federal and provincial agriculture departments for various regions (Spak 1998b), and are helpful in making decisions as to what major crops should be planted, especially in the prairie regions described above, for which there are only a few major crops and detailed information is available. In theory, such elegant prediction could be done for every possible crop, so that everyone, everywhere would know exactly what crop to grow. Of course, this is a Utopian scenario, since such detailed knowledge is not available and is only acquired after fairly expensive studies. Nevertheless, it is well to keep in mind that knowledge of both old and new crops is a key to resolving the universal question of farmers, "What should I plant this year?" and indeed the NewCROP website (www.hort.purdue.edu/newcrop) of which this article is a contribution is perhaps the premier source of needed information.

History of Dominant Crops in Canada

Because cereals and oilseeds dominate Canadian agriculture, it is instructive to examine the historical importance of these during this century in Canada. As can be seen from Fig. 4, wheat, oats, barley, maize and flax have been major crops in Canada for at least the last 75 years, and while their relative importance has varied, all have remained prominent for many years. No other cereal has become important in Canada, but canola (rapeseed) and soybeans, discussed below, have been the leading Canadian oilseeds for just the last several decades.

Fig. 5 illustrates the progressive cumulative value of the 68 Canadian crops for which national statistics are compiled (these range from over 4 billion dollars for wheat to less than 2 million dollars for apricots). This shows that the increased value becomes progressively less as one adds crops to the economy, and superficially it suggests that new crops are not needed. This is an incorrect conclusion for the following reasons:

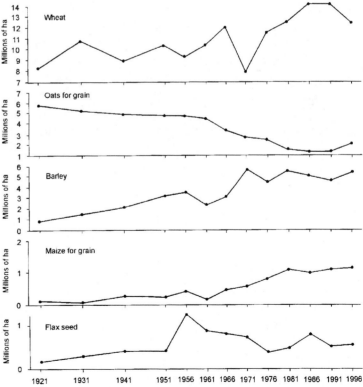

Fig. 3. Principal soil zones of the Western Canadian prairies, Canada's most important agricultural region. These zones heavily influence which crops are advisedly cultivated.

Fig. 4. Areas of the principal cereals and oilseeds grown in Canada for which records are available for all of the comprehensive 10-yearly (1921–1951) and 5-yearly (1951–present) agricultural censuses (data from Statistics Canada 1997a).

(1) one cannot predict the future value of new crops, which are needed for many reasons; (2) the crops with the highest national incomes are not necessarily profitable; (3) some of the most important crops are not suitable for some regions; (4) farm gate values do not measure the very high value of on-farm use of many crops; (5) farm gate values do not measure the very high value-added aspect of many crops.

Crop Specialization by Region

For simplicity, five regions are discussed, as follows (west to east): British Columbia specializes on fruits (particularly apples) and vegetables, and also has strong livestock and dairy production. The prairie provinces are specialized in grain and oilseeed farms, particularly wheat, oats, barley, canola, rye, and flax. Most of Canada's grazing lands occur in the prairies, and there is a very large red meat industry, with Alberta alone producing half of Canada's beef. Ontario and Quebec are the best areas for maize, and there are strong livestock, dairy, and horticultural sectors. Ontario is the center of soybean cultivation, and there is substantial greenhouse and fruit cultivation. The Maritime (Atlantic) area is particularly suitable for forage crops and an associated livestock industry, as well as potato and fruits such as blueberry. Tables 1 and 2 and Fig. 6 provide summary data for Canada's most important crops, for the five regions, based on crop area and farm receipts.

Relative Importance of Types of Crops

The relative percentage crop area categorized by type of crop (based on Table 1) is shown in Fig. 7. Grains occupy over 60% of Canada's farmland. The oilseeds have about 15% of Canada's farmland, but are a higher-value crop. Forages and fodders occupy about 18% and, along with the coarse grains (barley, oats, maize) and presscake from the oilseeds, contribute to the feeding of the very large livestock population. Vegetables and dry legumes (peas and beans) have only about 3% of the land, but represent very high value crops. In the following discussion of what is new in the crop groups, the length of the treatment is approximately proportional to the importance of the groups in Canada.

GRAINS

Canada produces about 5% of the world's wheat, 9.9% of the world's barley, 14% of the world's oats, and 1.4% of the world's maize. Canadian grains are used for domestic food consumption, animal feeds and industrial uses, with about half of cereal production exported. The dominance of Canadian crops by the major cereals has been overwhelming throughout this century. As shown in Fig. 4, the total area seeded to wheat increased from 8.2 million ha to 30.7 million ha between 1921 and 1996. Of the 276,548 farms surveyed in the 1996 census, 94,000 (29.4%) grew wheat. The total area of barley increased almost six times to 5.2 million ha between 1921 and 1996; the total area of maize for grain increased more than 12 times to 1.1 million ha. By contrast, the area occupied by oats decreased by over 70% to 2.0 million ha (with an associated drop in the number of horses and ponies from 3.5 million to 444,000). These four cereals represent over 60% of the area currently devoted to crops in Canada, and over 44% of current total crop farm receipts. Most of Canada's maize is grown in Ontario and Quebec, while most wheat, barley and oats are grown in theprairie provinces (Tables 1, 2). Wheat, barley, and oats account for 65.5% of the crop area and 61% of the farm receipts of the prairie region, and canola accounts for another 12.2% of the crop area and 22.7% of the farm receipts. Such dependence on only four crops in the prairies is of particular concern, and not

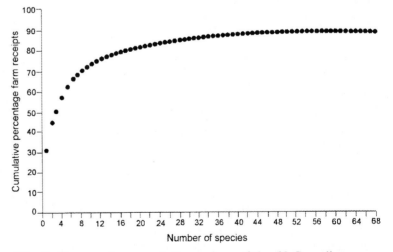

Fig. 5. Progressive cumulative values of the 68 Canadian crops for which national farm gate value or farm receipt statistics are available (compiled from Statistics Canada 1996, 1998a,b).

Table 1. Area of crops grown in Canada in 1996 (based on comprehensive 5-yearly census; excludes Territories; data from Statistics Canada 1997b; in some cases sums for Canada are not additive because of method of data collection).

Crop	British Columbia	Prairie provinces	Ontario	Quebec	Maritime provinces	Canada
Grains						
Wheat	40,146	12,013,427	315,231	34,661	15,801	12,419,264
Barley	45,116	4,877,886	134,688	125,225	58,263	5,241,179
Oats	34,083	1,867,116	39,804	85,106,	18,642	2,044,748
Maize (for grain)	642	29,134	767,142	331,775	3,465	1,132,157
Maize (for silage)	9,636	17,615	119,799	40,149	4,161	191,359
Rye	2,849	156,150	26,500	4,010	2,390	191,899
Triticale	49	25,347	281	180	--	25,857
Mixed grains	2,232	134,503,	113,216	32,019	11,626	293,596
Canary seed	56	248,635	6	55	--	248,752
Buckwheat	27	13,777	2,755	2,875	292	19,726
Oilseeds						
Canola (rapeseed)	25,821	3,480,691	21,571	3,211	141	3,531,435
Soybean	--	666	776,209	96,693	3,323	876,901
Flaxseed	189	591,183	640	90	--	592,104
Mustard seed	--	238,833	13	175	--	239,021
Sunflower seed	109	36,230	433	208	121	37,099
Safflower	--	1,496	--	--	--	1,611
Forages and Fodders						
Alfalfa (including mixtures)	161,485	2,602,126	598,711	210,949	25,112	3,598,383
All other tame hay and fodder	186,487	1,160,016	419,416	670,730	175,839	2,612,488
Forage seed for seed	18,987	158,661	4,820	967	400	183,833
Vegetables						
Potato	3,642	43,883	16,149	18,722	67,912	150,309
Other vegetables (excluding greenhouse)	7,117	7,987	64,131	40,313	8,151	127,697
Dry legumes						
Dry field peas	3,606	531,872	386	388	65	536,319
Lentil	284	303,107	6	--	--	303,401
Dry field bean	27	44,543	43,927	5,133	250	93,949
Fruits and Nuts						
Tree fruits and nuts	10,453	106	19,046	7,958	4,104	41,668
Berries and grape	6,887	1,032	8,480	10,706	12,707	39,812
Miscellaneous						
Tobacco	--	--	27,597	1,831	--	29,428
Sugar beet	--	23,866	85	--	--	23,953
Other field crops	702	25,724	2,638	296	135	29,494
Nursery products	3,213	3,784	10,610	3,500,	415	21,521
Sod	936	4,251	9,525	5,689	1,562	21,964
Christmas trees	9,453	1,852	11,285	12,342	16,138	51,070
Total crop area	574,234	28,645,499	3,555,100	1,745,956	431,015	34,951,977

surprisingly it is this important region, which accounts for 62.6% of crop farm receipts of Canada, that one finds the greatest support in Canada for crop diversification. Indeed, in the last decade there has been a concerted effort among grain producers to diversify production to overcome market fluctuations, drought, early frost, and trade wars.

Table 2. Gross value of farm receipts in Canada for Canadian crops for 1997 (in thousands of Canadian dollars; based on Statistics Canada 1998a). (Note that gross farm receipts include transfer payments from governments for a few crops, and for such crops represent a slight over evaluation of farm gate value).

Crop	Value (Canadian $)					
	British Columbia	Prairie Provinces	Ontario	Quebec	Maritime Provinces	Canada
Grains						
Wheat	11,885	4,127,359	71,325	13,163	3,237	4,226,969
Barley	5,908	919,576	11,612	17,855	6,858	961,809
Maize	--	16,666	425,200	246,010	54	687,930
Oats	1,659	257,705	5,201	8,998	573	274,136
Rye	19	28,967	5,410	--	--	34,396
Canary seed	--	49,650	--	--	--	49,650
Oilseeds						
Canola	6,840	1,974,553	17,196	-	-	1,998,589
Soybean	--	--	726,158	87,403	417	813,978
Flaxseed	--	333,207	--	--	--	333,207
Mustard seed	--	91,214	--	--	--	91,214
Sunflower seed	--	16,950	--	--	--	16,950
Forages and Fodders (n.b.: mostly used on farm, and so farm receipts do not reflect large quantities grown)						
Hay and clover	13,823	63,291	4,996	3,535	141	85,786
Forage, grass seed	4,375	30,395	1,664	76	--	36,510
Vegetables						
Potato	15,839	165,096	53,292	88,018	196,770	519,015
Other vegetables	173,168	76,633	462,139	251,949	37,363	1,001,252
Dry legumes						
Dry peas	--	196,495	--	-	--	196,495
Lentil	--	99,712	--	--	--	99,712
Dry bean	--	11,874	28,640	--	--	40,514
Fruits						
Apple	32,467	--	81,270	25,104	15,160	154,001
Other tree fruits	13,852	--	42,822	-	520	57,194
Strawberry	4,481	3,355	17,351	16,376	7,811	49,374
Other berries & grape	79,517	1,743	45,032	27,716	23,061	177,069
Miscellaneous						
Floriculture & nursery	265,254	109,660	540,893	140,899	65,189	1,121,895
Tobacco	--	--	328,727	22,437	--	351,164
Forest products	32,300	6,565	16,988	58,707	19,990	134,550
Maple products	--	--	10,108	86,457	5,043	101,608
Sugar beet	--	34,483	--	--	--	34,483
Other crops	7,019	84,172	32,354	30,651	14,473	168,669
Total crop value	709,016	8,699,321	2,973,578	1,125,354	396,660	13,903,929
Total livestock	945,324	5,963,192	3,659,289	3,367,328	603,388	14,538,521

Wheat (*Triticum aestivum* L.)

Wheat and rye are the only grains with the potential to make raised (leavened) breads because their gluten content gives strength and elasticity to bread dough; wheat by far is the most important crop for this purpose. Canada is the world's largest exporter of hard red spring wheat, well known for its excellent milling and baking qualities and for its suitability in blending with lower protein wheats. The typical Western Canadian growing season of short cool nights and long, sunny dry days is ideal for the production of consistent, high-protein wheat, and normally this type of wheat is priced at a premium to softer and lower-protein wheats. In 1903, William Saunders developed the 'Marquis' cultivar, which set a standard not surpassed until the 1980s by 'Neepawa' of 1987–88. Canadian breeding of superior hard red spring wheat has since been constant. 'AC Barrie', a new, hard red spring wheat was recently released and has proven to be very popular. The three Canadian prairie provinces are the chief wheat-producing provinces of Canada. Canada produces about 5% of the world's wheat, but because of its relatively small population, exports over 75% of its annual production, and accounts for about 20% of the world's wheat exports.

Although hard red spring wheat dominates Western Canada, there have been concerted attempts to breed other types of wheat cultivars, to meet the changing needs of world markets (Dietz et al. 1998). Only a small amount of the class known as "Western Canadian Soft White Spring" wheat is produced in Canada, in part because it requires irrigation in Western Canada. However, there is a very good market for soft wheats, which go into cookies, cakes, crackers, specialty breads, and noodles. The cultivar 'AC Reed' was released from AAFC at Lethbridge in 1994 in an attempt to expand into the soft spring wheat market. AAFC at Swift Current, Saskatchewan, is responsible for the breeding of several recent wheat cultivars of the "Canadian Prairie Spring class:" 'AC Crystal', a high-yielding semi-dwarf red wheat with intermediate protein, and stronger gluten than its predecessors, improving its milling qualities for bread-making (due for release in 1999); and 'AC Karma' and 'AC Vista', white wheats for the oriental-noodle market.

For 1998–99, Canadian non-durum wheat area, mostly spring wheat, declined to 7.7 million ha, the smallest area since 1972 (Fig. 8). In response to wheat surpluses and associated trade problems, many western Canadian farmers have turned to alternative crops (popularly called "specialty crops"). Indeed, this has been the single most important stimulus to crop diversification in Canada, next to the trend of tobacco replacement (Loughton et al. 1991). Figure 8 shows that while wheat area was decreasing, strong compensating increases occurred in the cultivation of the chief specialty crops of the prairie provinces (common bean, sunflower, canary seed, mustard, lentil, and pea).

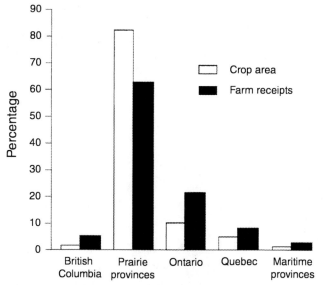

Fig. 6. Relative percentage for crop area and farm receipts for the five regions of Canada discussed in text.

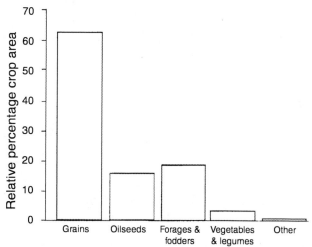

Fig. 7. Relative percentage crop area occupied by the major types of crop in Canada.

21

Durum Wheat (*Triticum turgidum* L. = *T. durum*)

Most types of common wheat can be used to produce bread and Asian-style noodles. For example, bread is produced in most countries from a blend of hard and soft wheats, and where high-protein hard wheat is unavailable, wheat gluten can be added. Durum wheat has an amber yellow endosperm (from which semolina is produced), unlike the white endosperm of common wheat, so that pasta from durum semolina is amber colored. The flavor and cooking qualities of durum pasta are superior, and durum wheat is preferred for the production of pasta products, such as spaghetti and macaroni, and for couscous, the staple food in North Africa. Durum is suited to a dry climate, with hot days and cool nights, and does well under dry conditions. About 8% of the world's wheat production is durum wheat. The leading producers of durum wheat are the European Union, Canada, and the US. Canada is the leading exporter (Lennox 1998). For 1998–99, Canadian durum wheat area rose to a record high 2.9 million ha. In North America, Western North Dakota and southern Saskatchewan are particularly suited to durum wheat, and it is also grown under irrigation in Arizona and the California deserts. Durum wheat, as a crop, compares to common wheat much as alternative and new crops do. It is a relatively high-value commodity with a more stable future in Canada than common wheat. Recently, new technology and consumer taste changes have altered the pasta market toward a stronger, less elastic gluten, particularly in Italy, the main manufacturer of pasta-making equipment, and the country with the most prodigious appetite for pasta. To meet this altered market, new Canadian durum cultivars with stronger gluten content are in the process of being registered for market testing (Anon. 1998b).

Winter Wheat

Winter wheat is the fourth largest crop in Ontario (behind soybean, maize, and tobacco). Ontario is the main producer of winter wheat in Canada, producing about 1 million tonnes annually, using cultivars (e.g. 'Augusta', 'Harus') that can survive the relatively mild winters. Soft white winter wheat is most commonly grown, and this is used for producing soft gluten flour for confectionery products such as cakes, cookies, breakfast cereals, and crackers. Ontario red spring wheat is used primarily for domestic feed, and to a lesser

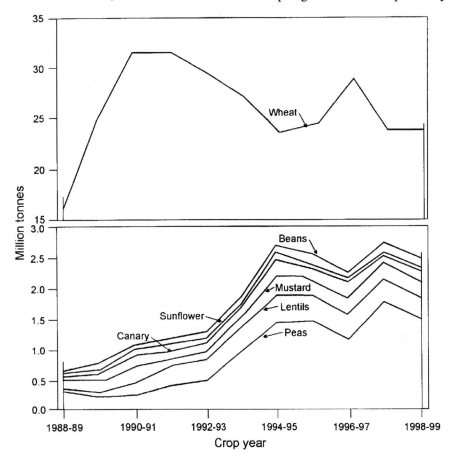

Fig. 8. Recent increasing cultivation of "specialty crops" (common bean, sunflower, canary seed, mustard, lentil, and pea) in response to decreasing wheat cultivation (above) in Canada.

extent for domestic human consumption. Spring wheat is grown somewhat in eastern Ontario, using Western Canadian cultivars, but the quality has often been below the standards demanded by North American millers. In recent years, Ontario wheat growers have been shifting away from traditional soft white winter wheat into both soft red winter and hard red winter wheat cultivars. The shift into hard red wheat is due to the recent availability of new cultivars able to achieve the high protein levels required by the North American milling industry. Improved soft red winter wheat cultivars have also become recently available, and although soft white winter wheat is preferred for the production of breakfast cereals, cakes, and pastry flour, the soft white winter wheat that has traditionally been grown in Ontario suffered a severe fusarium outbreak in 1996, decreasing its popularity. For a review of the changing wheat situation in Ontario, see Lennox (1996) and McKinnon (1997a).

Spelt Wheat (*Triticum spelta* L.)

Spelt is an ancient wheat that has been a staple grain in Ethiopia for centuries. It has become a top-selling organic and health food, grown as a specialist crop, often for people with allergies, and for pasta. Although minor, spelt is increasingly cultivated in Canada, with 825 ha reported in the 1996 Census of Agriculture, mostly in Ontario. Several thousand ha are cultivated in the US.

Triticale (×*Triticosecale* Widdmark)

Triticale is the stabilized hybrid of wheat (*Triticum*) and rye (*Secale*). Poland, Germany, China, and France account for nearly 90% of world triticale production. Globally, triticale is used primarily for livestock feed. In Mexico, which grows the crop, triticale is used mostly for whole-grain tricale breads and tortillas. In the US, triticale is harvested mostly for forage, but there is a small market for pancake mixes and crackers due to a savory, nutty flavor. Ethanol plants will pay a premium for triticale over barley since it has more starch and no hull, making alcohol production more efficient. Although wheat-rye hybrids date back to 1875, it was only in 1953 that the first North American triticale breeding program was initiated, at the University of Manitoba. Although improved cultivars have been bred, triticale has remained unimportant in Canada. However, triticale does well in regions where wheat performs poorly, notably on cold and infertile soils, extremely sandy soils, soils with high levels of boron, salty soils, acidic soils, manganese-deficient soils, and dry soils. Canada does not have large area of such soil types, but there are about 2 million ha of marginal, light mineral, low productivity land in Western Canada where triticale has the potential to displace or supplement traditional feed grain production. Winter triticale is a higher-yielding, earlier-maturing alternative to spring triticale for short season areas of the prairies. 'Pika' and 'Wintri' are the only cultivars found to be suitable for use in Western Canada. Canadian triticale is mostly used for feed and forage. Triticale production in Eastern Canada is growing, with most production in Saskatchewan, followed by Alberta and Manitoba. Forage triticale equals or outperforms barley, oats, rye, and mixed grain in areas of Western Canada, so that there is a reasonable probability that it will become more important. For a review of the triticale situation in Canada, see McKinnon (1996).

Barley (*Hordeum vulgare* L.)

Barley is basically a livestock feed, and is the major feed grain in Canada. Barley is well suited to the Canadian prairies, where most of this crop is grown, while other feed grains such as maize and sorghum are not. Barley is also a good rotation crop with wheat, tends to be higher yielding, matures earlier, and is more resistant to drought and salinity problems. It is also used in brewing beer (see below). Hulless barleys have hulls that are easily removed by threshing, as with wheat. Some hulless cultivars are produced in Canada (such as the two-rowed 'Condor', 'Phoenix', and 'CDC Dawn', and the six-rowed 'Tupper', 'Buck', and 'Falcon'), resulting in more digestible, higher-protein feed, especially for swine and poultry feeding. Pigs and chickens are monogastric (non-ruminant) animals, which are unable to digest the fibrous hull. Barley is useful for most classes of livestock, although poultry lack the enzyme to digest beta-glucans, a water-soluble fiber (this viscous polysaccharide is denatured by adding the enzyme beta-glucanase to the rations; beta-glucans from barley have been shown to reduce cholesterol in people). Although not as palatable as other cereals, some barley is consumed by humans. Barley kernels are polished to remove the inedible part of the grain.

"Pearled" barley is highly polished barley; by contrast, "pot barley" is less polished, hence slightly larger than pearled barley. Pearled barley is used in breakfast cereals and infant foods. Hulless barley grains of course already lack much of the inedible portion. Various health advantages have been claimed for human consumption of barley, including benefits for regulating blood sugar levels in diabetics, and for lowering cholesterol and heart disease. A market is emerging for the fractionated components of barley kernels as pharmaceuticals (see Nutraceuticals). Because it is such a major Canadian crop (third-ranking in terms of farm receipts), breeding of Canadian cultivars is given high priority, and new cultivars are constantly emerging. For example, in Eastern Canada where there are frequent strong wind and rain storms and barley often lodges, recent cultivars (the six-rowed 'AC Alma' and the two-rowed 'AB162-9') have strong stems to keep the plants upright.

Malting Barley

About 10% of world barley production is malted, and the other 90% used as animal feed. Malting barley is simply high-quality barley with appropriate characteristics to produce good malt for making beer. Most barley cultivars can be used to make barley malt, but some are specifically bred for the purpose. Two-row malting barley represents the international industry standard outside North America and also in some South American markets. However, six-row barleys are also used. Most malting cultivars are in fact used for feed. About half of Canadian malting barley is two-row. In Canada, three-quarters of the area seeded to barley consists of malting cultivars, the other quarter being feed cultivars, which tend to have higher yields. Growing conditions in western Canada are usually cooler and drier than the US for producing superior six-row malting barley, and so the US has been Canada's major export market for this type of malting barley in recent years. The variable climate on the Prairies sometimes results in only a proportion of the malting barley crop meeting specific malting barley requirements, but as most of the crop is fed to livestock this generally does not compromise the supply for export. China is an increasing importer of Canadian 2-rowed malting barley. The creation of new two-row barley cultivars with improved malting and agronomic performance desired by the export market has been an important development in Canada in the last two decades. Hulless barley cultivars with higher energy content, discussed above, are not suitable for the malting barley process, so that hulled cultivars continue to be grown in Canada. A recent review of the malting barley situation in Canada is McKinnon (1997b).

Oats (*Avena sativa* L.)

Before 1910, the area seeded to oats often exceeded the area for wheat in Canada, in order to feed horses. Up until 1920, the area for oats was similar to wheat area in Canada, but from the early 1920s to the late 1970s, with the introduction of tractors and the replacement of horse power by machine power, the area seeded to oats decreased steadily until the 1970s. Since the 1980s, Canada has consistently captured a significant share of the world export market. Oats are still used primarily as animal feed, but human consumption is increasing, especially in North America, where oats are considered healthy, especially oat bran (see Nutraceuticals). In the US (and to a lesser extent in Canada), oats is used somewhat for pasture, silage, and haylage, and especially as a cover crop to protect soil, notably on marginal land subject to erosion, and as a nurse crop to protect newly planted forages. The world's leading oat producers are Russia, the European Union, Canada, the US, and Australia. Canada is a leading exporter, in 1997–98 accounting for almost half of all world oat exports, excluding products, 95% of this to the US. In 1997–98, about 1.9 million ha were seed to oats, an area exceeded only by wheat, barley, and canola. Over 90% of production is in Western Canada, and this represents a shift from Eastern Canada where oat cultivation has become less economical as a feedstuff. A further shift has been from the Western to the Eastern Prairies, closer to the major oat market in Minneapolis. The cool growing season on the northern Prairies where oat production is concentrated is a problem, being addressed by continuing plant breeding. Alberta is Canada's major oat producing province, and most of the crop is grown in the north in the grey woodland soil, which is quite acidic, freeing up aluminum, which reduces yield by up to 40%. There is currently research (at AAFC at Lacombe) to breed aluminum-tolerant oats to meet this problem. Oat breeding in Eastern Canada, incorporating new molecular techniques, is a specialty of AAFC at Ottawa. Oats are less likely to be traded than other grains because their bulky nature increases transport costs. Oats are about 25% hull, and offer less energy than barley and maize, limiting their use primarily to starting feedlot rations and for feeding horses. Hulless or naked oats, which lose their hull during harvest, have promise as a feed ingredient. 'AC Lotta', 'AC Percy', and very recently 'Cavena' from AAFC at Ottawa are Canadian cultivars. Hulless oats have higher protein and fat than conventional oats, as much energy as maize, and a better balance of amino acids, and have good prospects as a new Canadian crop. Another new trend is the industrial use of oats in Canada, with the establishment of a processing plant in Saskatoon that refines oat extracts used in products such as animal coat washes and diabetes screening tests. In the past 15 years, oat products have increased from virtually nothing to an estimated 200,000 tonnes for 1997–98. A recent review of trends in the Canadian oat crop is McKinnon (1998).

Maize (*Zea mays* L.)

Maize has a very long history of cultivation in Central Canada, dating back over a thousand years. With the introduction of US higher-yielding maize hybrids in the mid 1950s, commercial production expanded in the southernmost regions of Ontario. New cultivars of maize have been instrumental in Canada in lowering susceptibility to early frosts and avoiding harvest problems. With the continual development of Canadian hybrids for cooler and shorter growing seasons, commercial maize production spread beyond southern Ontario, and today maize is widely cultivated in Ontario and Quebec, with limited production in Nova Scotia, Manitoba, and Alberta. About three-quarters of Canada's maize is produced in Ontario and Quebec. Maize is primarily a feed ingredient in Canada, with Central Canada accounting for the bulk of consumption (barley is the major feed grain in Western Canada). Fodder maize, used mainly for silage, requires less heat units and has a wider growing range than grain maize. Fodder maize is generally grown for on-farm use. There is a trend for increasing food and industrial use of maize in Canada (see discussion of ethanol production, below). Canada is normally a net importer of maize, with western Canada acquiring it from the US. For a review of the maize industry in Canada, see Kurbis (1996b).

Canary Seed (*Phalaris canariensis* L.)

Canada is the world leader in the production and export of canary seed from annual canary grass, used in caged and wild bird food mixtures. Commercial production of canary seed started in the US after World War II, concentrated in Minnesota and North Dakota, and production moved to Manitoba and Saskatchewan to become commercially viable in the early 1980s. In 1996, Canada produced 90% of the world supply of canary seed, about 90% in Saskatchewan, the remainder in Manitoba and Alberta. Canary grass is extremely well adapted to the hard red spring wheat of the Prairies, although more sensitive to heat and drought. Ten to 30% of production is used domestically, the remainder, about 125,000 tonnes, is annually exported, largely to Europe, South America, and the US. Canary seed is suited to and mainly produced in the brown soil zone of Western Canada. Substitute bird seeds with the quality of canary seed are generally unavailable (there is occasional substitution with proso millet from the northern US) and, with increasing interest in birds as pets, long term growth of the industry seems assured. A new cultivar known as 'CDC Maria' was recently developed by the University of Saskatchewan's Crop Development Centre. This is expected to revolutionize the industry, eliminating problems such as itchiness and dust associated with the hairy seed coat of older cultivars. 'CDC Maria' and other "hairless" cultivars are expected to replace the traditional canary seed cultivars, which can not be used by humans for food because of the hairs. Dehulled canary seed can be processed into flour and bran, and in addition to this food potential for humans there is some potential for cosmetic purposes. A recent review of the canary seed situation in Canada is Gray (1997b).

Rye (*Secale cereale* L.)

Rye is a relatively minor cereal in Canada, which is perhaps surprising since it has the ability to withstand unfavorable growing conditions and often thrives where other cereals fail. Although rye can have a higher feed value than barley, the high soluble fiber content (pentosans) reduce feed value for poultry and swine. Although Canada is one of the world's major rye exporters and produces high-quality rye, the world market is small, and Europe grows the crop well, mostly for making bread. In Canada, rye is grown mainly for grain, but also for pasture and hay. Fall rye also provides soil cover from fall through spring. There is limited domestic use of rye in Canada for distilling and for food use, compared to the other cereals discussed above. Several new breeding lines of winter rye were recently bred at AAFC at Lethbridge.

Proso millet (*Panicum miliaceum* L.)

Proso millet has long been a staple grain in Africa, and has been grown as a forage crop. This sorghum relative is used mostly in the pet food and birdseed industries in North America, but may have some potential as a Canadian grain for human consumption. 'AC Prairie Gold', a millet line adapted to prairie growing conditions, was recently released by AAFC at Morden. While only 1000–2000 ha are currently grown in western Canada, the potential for proso millet has been estimated to be 10,000–15,0000 ha or more (Kiehn and Reimer 1992).

False Melic Grass [*Schizachne purpurascens* (Torr.) Swallen]

This widespread native grass of Canada produces large grains. It has been suggested that it could be developed into a special cereal like wild rice (Dore and McNeill 1980), although this would require considerable development.

Wild Rice (*Zizania palustris* L.)

Wild rice (not to be confused with wild forms of *Oryza sativa* L.), Canada's only native cereal, is collected from natural or planted stands, particularly by indigenous people (Aiken et al. 1988; Crop Development Centre 1991). It requires considerable development, but is well suited to Canada. Wild rice is an economically attractive crop in that the supply is limited while market demand is increasing, a premium price can be obtained, and the climate and natural aquatic habitats of portions of Canada provide competitive advantages. Semi-domesticated paddy wild rice is in commercial production in California and Minnesota, and provides competition for Canadian producers. However, a natural advantage for Canada is the availability of extensive shallow lake and river systems, which usually do not require much if any drainage control. For wild rice to expand as a crop in Canada, development of non-shattering, disease-resistant cultivars is needed. Fast-maturing strains would be an added advantage for northern regions so that seeds would mature before frost.

Buckwheat (*Fagopyrum esculentum* Moench)

Common buckwheat is grown in many major grain producing countries, especially Russia and China. Major exporters are China, Brazil, France, the US, and Canada. Japan accounts for almost all of the world's buckwheat imports. Buckwheat has been grown in Canada for many years as a special crop, and is an important cash crop in Manitoba, but production is currently low. However, this crop presents opportunities for diversification and value-added activities because the Japanese market is growing. In Japan, buckwheat flour is employed in combination with wheat flour to prepare buckwheat noodles (*soba*), a traditional dish. In some cases, Japanese noodle manufacturers add ground leaves to the buckwheat flour, producing a green noodle. Only about 10% of Canada's buckwheat production is used domestically for human consumption, but this could increase if processors develop new buckwheat products such as snack foods and flour for crêpes. Buckwheat can be grown as a green manure crop, companion crop, cover crop, and as a source of dark buckwheat honey. The grain and straw can be used for livestock feed, but the nutritive value is lower than that of cereals. The protein in buckwheat flour is of exceptional quality, containing a high amount of lysine, which is deficient in cereals. Foods are prepared from the groats (dehulled seed) or from the flour. The low gluten content of the flour makes it ideal for crêpes, and in mixtures with wheat flour for bread, pancakes, noodles, and breakfast cereals. Groats and grits (groat granules) can be used for porridge and other breakfast cereals. Dehulled groats can be baked or steamed and eaten as a vegetable like rice, or used in appetizers, soups, salads, breads, and desserts. Development of new Canadian cultivars is occurring to counter climatic disadvantages associated with buckwheat production in Canada. The main buckwheat cultivars grown in Canada are 'Mancan' and 'Manor', developed by AAFC at Morden. 'Mancan' is employed as a quality standard by Japanese millers because of its soft white starch. 'Manisoba' is a new, higher-yielding cultivar with a larger seed that is easier to dehull, leaving behind high whole groat content. 'Manisoba' also facilitates popping the seed, like popcorn, to prepare some specialty products. Western Canadian cultivars are not well adapted to eastern Canada, and a high portion of growers in Ontario and Quebec cultivate buckwheat simply as a green manure crop or a cover crop to crowd out weeds. Research is underway to improve buckwheat cultivars in Quebec and Ontario, where lodging is a frequent problem due to excess moisture. Unlike most cereal crops, buckwheat cannot recover from lodging. There has been considerable recent governmental and grower association encouragement to increase production and marketing of buckwheat in Canada. New proposed value-added activities include dehulling, flour-making, noodle making, and roasting for snacks. Buckwheat produces rutin, which increases the elasticity of arteries and prevents their hardening, and is in demand by the pharmaceutical industry. Scientific research on buckwheat is centered at AAFC Morden. Recent achievements include the development of a self pollinating buckwheat with extremely low seed abortion and better frost tolerance. This is expected to result in a new cultivar in the next several years. A high-yielding self-pollinating experimental strain was recently bred in Manitoba (Henckes and Dietz 1997). A good review of buckwheat in Canada is Vincent and Longmuir (1996).

Quinoa (*Chenopodium quinoa* Willd.)

Quinoa originated in the highlands of Peru and Bolivia, where it became a staple crop of the Inca empire. By comparison with most cereals it is rather primitive, requiring dehulling to remove bitter seed coat saponins. Considerable quinoa is sold in Canada as a gourmet item in health food stores, in the form of whole grain, pasta, or flour. Quinoa is considered to have some promise for Canada, and there is a Canadian Quinoa Association (Anon. 1992). Currently available forms are late-maturing, therefore vulnerable to frost, and are also susceptible to insect damage. Quinoa would appear to have some possibility for development through germplasm selection in Canada, but is likely to find a more receptive area of cultivation in other countries. In the US, quinoa seems to represent one of the relatively few apparently successful introductions of a new food plant (Johnson 1990).

Amaranth (*Amaranthus hypochondriacus* L. = *A. leucocarpus* S. Watson)

Grain amaranth, a pseudo-cereal, is another ancient grain used similarly to quinoa. This dietary staple of Aztec and Mayan civilization is still grown in South and Central America, where it originated, and is used as a vegetable in India and China. The seed can be popped like popcorn and flaked like oatmeal, and is notably high in protein. Amaranth is enjoying a renaissance in popularity in North America. It is sold in health food stores, particularly when organically grown, but has achieved little market status. Amaranth has been experimentally cultivated as an annual grain at AAFC at Morden, and has been thought to have fair long-term potential in the southern Canadian prairies (Kiehn and Reimer 1992). Its future in the US has been considered debatable by some (Lehmann 1991), promising by others (see articles in previous proceedings, particularly *Advances in New Crops*, 1990).

OILSEEDS

Oilseeds tend to be higher-value crops than cereals, and are useful as alternatives in crop production and market diversification. Canola, flaxseed, and sunflower seed are particularly considered to be major cash crops for Western Canadian producers, especially when grain markets are poor. In 1996 the farm value of oilseed production in Canada was estimated at $2.883 billion. In addition, oilseed processing contributed $0.5 billion in direct value-added and over $1 billion in spinoff benefits to the Canadian economy. The meal left after oil extraction is also of considerable importance as livestock feed and, as noted below, the grain itself may be a useful animal or human food. In 1997, total vegetable oil production in Canada reached a record level of 1.6 million tonnes, with canola oil accounting for 77%, soybean oil for 17%, linseed oil for 3%, and sunflower oil for 1%. Several of the world's major edible oils, including palm oil, cottonseed oil, peanut oil, coconut oil, olive oil, and palm kernel oil, simply cannot be produced in the climate of Canada, which imports these commodities. Nevertheless, Canada has been a net exporter of vegetable oil since 1992, but this is mostly due to canola exports, especially to the US. In 1997, Canada was responsible for 43% of total world rapeseed/canola oil exports, 10% of linseed oil exports, and less than 1% of soybean and sunflower oil exports. Generally, there is strong regionalization of cultivation of Canada's oilseed species, as noted below. Attempts are underway in Canada to breed new oilseeds (most notably edible oil mustard and edible oil flax, as discussed below). There is also considerable research in Canada to develop new and improved cultivars, which may result in the major oilseeds being cultivated in regions where they are now absent or little grown.

Most of the world protein meal supply is derived from oilseed production. Oilseeds tend to produce meal that differ in their ability to meet the nutritional requirements of the different livestock categories. For example, non-ruminant livestock such as hogs and poultry need high protein feed without the high fiber content suitable for ruminant animals. Small but growing markets for protein meal include the fish feed market (aquaculture) and direct human consumption. Good dietary practice for livestock involves a complementary balance of the base grains of feed rations (such as wheat, barley, and maize), the various high-protein meal supplements, and grass and legume forage/silage crops. Canola meal, Canada's major meal, can be used up to maximum levels in feed rations of 20% for poultry, 15% for grower and finisher hogs, and 25% for dairy cattle.

Canola (*Brassica napus* L., *B. rapa* L.)

Oilseeds in Canada are currently dominated by canola (rapeseed), a high value crop cultivated by some 80,000 farms, that has become Canada's second most important crop after wheat. Canola (a trade-marked name) is primarily used in salad and cooking oils, margarine and shortening, and the mealy residue after the oil is extracted is used in livestock feeds. The 1996 census data show that canola accounts for 75% of all vegetable oils produced in Canada, 87% of salad oils, 49% of margarine oils, and 64% of shortening. The development of the canola industry is the premier example of a successful new crop for Canada. The breeding of new edible oil cultivars occurred as a focussed investment strategy that involved over 200 scientist years, costing $40,000,000, spread over 30 years (Jolliff and Snapp 1988). Canola is a relatively new Canadian crop, having begun with rapeseed cultivation in 1942 in Western Canada as a source of lubricants for the allied war effort. Today, a small area of high erucic acid rapeseed in still produced in Canada to satisfy the industrial market. Limitations of nutritional composition of available wartime cultivars restrained human consumption in Western countries. In the mid-1970s, AAFC and the University of Manitoba produced new cultivars, now known as canola, with less than 2% erucic acid and less than 30 micromoles/g of aliphatic glucosinolates in the meal (current levels have been further reduced, respectively to less than 1% and less than 20 micromoles/g). From 1976 to 1996, the total area of canola increased over five times to 3.5 million ha, representing 10% of Canada's total land in crops. In 1985, The US Food and Drug Administration granted canola GRAS (Generally Recognized as Safe) status, and in the light of its superior nutritional characteristics, canola oil sales to the US increased from virtually nil to over 400,000 tonnes annually. About two-thirds of Canada's exports of canola are to the US. Canola seed and meal sales to the US also increased along with canola oil sales. Canola oil's nutritional properties are responsible for its domination of the salad oil market: of the commercially available edible oils, canola contains the lowest levels of saturated fats (6%), the second highest level of monounsaturated fats (58%), and the highest level of the essential fatty acid, linoleic acid (10%). Globally, Canada produces about 17% of the world's rapeseed.

Given the controversy over the public acceptance of genetically-engineered foods, it is perhaps surprising to learn that close to 50% of the cultivated canola area in Canada consists of transgenic cultivars carrying selective resistance to specific herbicides. There are several new types of canola that are currently being bred and may well have a place in Canadian agriculture. "Super-high erucic acid rapeseed" is a type of genetically modified oilseed. A derivative of erucic acid, eruacmide, is used as a slip agent and plasticizer in the manufacture of plastic films. Other types of products that may be produced include cosmetics, lubricants, pharmaceuticals, plasticizers, and surfactants. "Odyssey 500" high stability oil is another product that may become useful. This is over 20 times more stable than conventional vegetable oils. It remains liquid at lower temperatures than other highly stable oils, and has no flavor or color. It may be useful as a moisture barrier, viscosity modifier, gloss enhancer, anti-duster, and band releaser. Still another new type of canola oil that offers stability, long shelf life, and fresh flavor is a high oleic oil, "Clear Valley 75." This has been praised for its desirable combination of taste and nutrition. It has the lowest level of saturated fats, no trans fatty acids from hydrogenation, and bland neutral taste that makes it ideal for cereals, popcorn, dried fruit, and crackers (Beckman 1998).

The most exciting new prospect in Canada for canola is the breeding of mustard (*Brassica juncea* Coss.) into a new canola species. To date, canola has been represented by the two *Brassica* species *B. napus* (Swede rape) and *B. rapa* (*B. campestris* L., turnip rape). For the past decade, research has been in progress in Canada toward the breeding of a drought-resistant canola-grade mustard. Canola cultivars presently available are not well suited for many of the relatively dry regions of Western Canada. By contrast, mustard cultivars have several advantages: higher-yielding in all but the short season regions of Western Canada, early maturing, more resistant to late spring frosts, more heat and drought tolerant, more resistant to seed shattering, and more resistant to blackleg disease. *Brassica juncea* is in fact used as an edible oil crop in China, India, Russia, and Eastern Europe, where an oil with higher levels of erucic acid is permitted, but this is not accepted in most Western countries. Moreover, the high glucosinolate meal has limited use for animal feed. Researchers at AAFC, Saskatoon have developed a mustard plant whose seeds contain meal and oil indistinguishable from canola (Anon. 1998a). A new cultivar may be available by 1999. Given the spectacular success of canola, this could represent an important new crop that would extend the region of the Prairies where canola-class plants can be grown. For marketing purposes, this new crop could be represented as canola, since the products are about identical. A hurdle that remains is the obtainment from the US of a GRAS designation, a necessity to remove trade restrictions. The fortunes of canola rose dramatically after it obtained GRAS status in 1985.

Soybean [*Glycine max* (L.) Merr.]

Soybean was first cultivated in Canada in 1893, but not in significant amounts until the late 1920s. Most soybeans are currently grown in Ontario (90%) and Quebec (9%). In the mid-1970s, it was almost impossible to find soybean growing in Eastern Ontario because of the inhospitable climate. 'Maple Arrow', a cultivar bred at AAFC Ottawa, provided the key adaptation for soybean to be transformed into the biggest cash crop in Ontario, where it is known as the "miracle crop." Most soybeans are used domestically in Canada, and the increase in the domestic supply has meant that imports from the US are usually equaled by exports. Soybean oil is used in a huge number of products, for example in the manufacture of edible oils, and in industrial products such as paint, varnish, resins, and plastics. Soybean meal is an important livestock feed, although half of Canada's supply is imported (unlike canola which is crushed mainly for its oil, soybeans are processed primarily for the meal). Most of Canada's soybean feed goes to the hog and chicken industries. Due to the presence of enzymes, soybeans must be roasted before being fed to livestock. Canadian research is attempting to eliminate the need for roasting, and has resulted in a reduction in the levels of the deleterious enzymes so that whole unroasted soybeans have become a significant constituent of livestock rations in Eastern Canada. Canadian cultivars have been bred with qualities required by specific soyfood markets of the Asia Pacific region. The large-seeded, white, high-protein types are prized by southeast Asian markets. Cultivars such as 'Special Quality White Hilum Beans' are exported for processing into tofu, natto, misto, and tempe in Asian markets. 'AC Onrei' is a very large-seeded high-protein cultivar suitable for top-quality nigari tofu. 'OX756' is another line produced by AAFC at Harrow designed to expand exports into the premium Asian soy food market. This is low in enzymes that cause a grassy-beany flavor that some consumers dislike. Genetically modified soybean cultivars are prevalent in the US, and are likely also to be established in Canada. As with other genetically modified crops, there is some public resistance to acceptance of human foods produced from transformed plants, especially in Europe, and this may affect the future development of export markets.

Flaxseed (*Linum usitatissimum* L.)

Flaxseed is generally known as linseed outside of North America, where the name flax refers to the fiber form of the crop used for the linen textile industry. Flaxseed was the first oilseed widely grown in Western Canada, and today the fiber form is cultivated only in very small amounts. Canada is the world's largest producer and exporter of flaxseed. Only a small proportion of Canadian flaxseed is crushed domestically. Canadian flaxseed is produced entirely in Western Canada, mostly in Saskatchewan. Flaxseed represents only 1% of the world supply of oilseeds, but as noted in the following is considered to have high potential for increased industrial use, as well as for human food and feed markets. Flaxseed (linseed) oil is a non-edible drying oil used in manufacturing paints, varnishes, linoleum, printing ink, oilcloth, putty, and plastics. The introduction of petroleum-based floor coverings and latex-based paints resulted in a worldwide decrease of the industrial use of linseed oil for paint and floor covering over the last several decades. Nevertheless, industrial use is expected to increase because of the development of new products. The biodegradability and non-allergenic characteristics of linoleum, coupled with quality improvements, have resulted in a resurgence of demand for linoleum in some parts of Europe. There has also been interest in using a linseed oil based concrete sealant. More significantly, there has been recent research into the development of edible oil-type flaxseed or "Linola" as a vegetable oil, and this market is likely to increase in Canada. Linola lines lack the high amounts of omega-3 fatty acids of conventional flaxseed lines, which makes them less nutritional, but they are more stable at high temperatures and less likely to go rancid, and so more competitive in the vegetable oil market. There has been much interest in Canada in the pharmaceutical value of edible linseed. It is well known that hardening of the arteries, heart disease, and strokes have been dietarily linked to overconsumption of saturated fats. It is much less well known that an unbalanced ratio of polyunsaturated fats has the same effects. The ratio of two polyunsaturated fats is considered particularly important—omega-6 and omega-3, recommended in an intake ratio of 3:1. Average dietary ratios in North America range between 12:1 and 20:1. Flaxseed is generally high in alpha linolenic acid, an omega-3 fatty acid, and has an omega-6/omega-3 ratio of 0.3/1, and so is extremely helpful in balancing the ratio to a healthy level. Omega-3 fatty acids lower levels of triglycerides in the blood, thereby reducing heart disease, and also show promise in the battle against inflammatory diseases such as rheumatoid arthritis. Poultry eating feed rations enriched with flaxseed produce eggs that are notably lower in saturated fat in the yolk. Full-fat (whole) flaxseed is in demand by the laying hen market. About 5% of Canadian laying hens are in fact consuming 10–20% flax in their rations, and so producing eggs that are relatively desirable in their balance of polyunsaturated fatty acids. About a dozen Canadian companies are now selling omega-3 eggs, and several US companies are following suit (Henckes 1998a). Dairy cows fed with flaxseed can produce omega-3 enriched milk and butter, and beef and chicken can be similarly enriched, although how practical this is remains to be determined. Crushing flaxseed for linseed oil produces meal/cake that serves as protein supplements in livestock rations, mainly in Western Europe. Flaxseed has been used extensively in baking in Germany and other central European countries, and there is a growing and highly profitable niche for flaxseed bakery products in North America, especially for specialty breads. Since the early 1990s, there has been some cultivation of 'Solin', a light-colored low-linolenic acid type of flaxseed that has a fatty acid profile similar to sunflower oil. New uses for flaxseed fiber are currently being developed. About $20 million of flaxseed fiber and tow were exported from western Canada to the US in 1995, but only 15–20% of available Canadian flaxseed straw is so used because of high transportation costs in moving the flaxseed straw to the processing plant, and the majority of straw is usually burned on the field. There is increasing interest in Canada in using high-quality fiber for fiberboard and similar application (see discussion below), so that a larger market for flax fiber may develop. For a review of flaxseed in Canada, see Beckman (1997).

Mustard [*Brassica juncea* (L.) Czern. & Coss., *Sinapis alba* L.]

Mustard is both a condiment and an oilseed crop, and has been grown in Canada since 1936. There are two species grown, *Brassica juncea* (brown and oriental mustard), and *Sinapis alba* (yellow or white mustard). Mustard has been an exceptional success in Canada, with an average of about 200,000 ha producing an average of 250,000 tonnes of seeds, most of which is exported. How much growth there remains for this crop remains to be seen. A small percentage of Canadian mustard is crushed locally, and some is ground to produce mustard flour, mostly for export. Canada is the world's largest supplier of mustard seed, exporting the seed to Japan, the US, Europe, and Bangladesh for use as a condiment. Nearly 40% of Canada's exports of mustard seed goes to the US, but America is increasing its seeded area. Bangladesh, Canada's second-ranking export destination, crushes mustard seed to produce a hot edible oil that is popular in the Indian sub-continent. A small shift from yellow mustards to brown and oriental mustards has been predicted in Canada (Gray 1998).

Sunflower (*Helianthus annuus* L.)

The first official government breeding program of sunflower in Canada was initiated in 1930. However, as for rapeseed, commercial cultivation began during World War II as a response to the vegetable oil shortage.

Sunflower is grown in relatively small amounts, mainly in southern Manitoba and southeastern Saskatchewan, and it has become a minor "specialty" crop in the cereal areas, serving as an excellent rotation crop for wheat that reduces diseases of the latter. About half of current sunflower production is destined for the confectionery market, 40% is crushed for oil, and 10% is used for bird feed. The residual oil-cake or high-protein meal produced after oil extraction is used for animal feed. About 30% of Canadian production is exported, the US accounting for about 70% of exports, the remainder largely to Germany, Belgium, the Netherlands, and Turkey. Confectionery type seeds have striped hulls, and the largest forms are used for human food. Sunflower seeds can be roasted and salted or baked into bread products for human consumption. Oil-type Canadian sunflower seed cultivars (which can also be used for birdseed) are characterized by black hulls. In the early 1990s, sunola, a short-stemmed drought-resistant type of oilseed sunflower that can be grown as a field rather than a row crop, was introduced into the Canadian prairies, and production of this has since been expanding (Anon. 1994). However, the area of sunflower seed cultivation has been fluctuating, generally declining in Canada for the last decade. Minnesota, North Dakota, South Dakota, and Texas are presently superior sites for growing sunflower. Sunflower has not become a major source of vegetable oil in Canada because it is susceptible to diseases, has a longer growing period than desirable (120–130 days), needs specialized equipment, and is relatively expensive to produce. The latter two problems are due to the need to row crop sunflower, because of its tall height, and this requires specialized seeding and harvesting equipment, which represents additional capital costs to the producer. New types of sunflower are needed to overcome these problems. To some extent recent sunflower hybrids with earlier maturity, increased yields, and shorter stalks have generated some expansion, but Canadian production remains limited due to the high heat and moisture requirements of the plant. The sunflower situation in Canada is reviewed by Christie (1995a).

Safflower (*Carthamus tinctorius* L.)

Safflower is a crop that is deserving of attention because of its versatility. It can be grown for edible oil, meal, or whole seed for dairy cattle, birdseed, and oil for industrial uses. Safflower oil is a wholesome oil, high in polyunsaturated fatty acids, that because of its high linoleic acid content commands a premium price among edible oils, and is competitive from a health viewpoint with canola and olive oil. The cool climate of the Canadian Prairies tends to increase the level of oleic acid (e.g. to over 80%, compared to about 73% in California). Industrial uses are limited, but the drying oil produced by safflower, which is intermediate between soybean and linseed oils, can be used in non-yellowing drying paints, alkyd resins in enamels, and caulks and putties. Because it is a long season crop, safflower extracts water from the soil for a longer period than cereal crops, and the long taproot can draw moisture from deep in the subsoil. These properties can help prevent the spread of dryland salinity, using up surplus water from recharge areas that otherwise would contribute to the development or expansion of saline seeps.

Commercial cultivation of safflower began in Alberta in 1943 when wartime new crop adaptation research was in progress, and currently is concentrated in Alberta and southern Saskatchewan. Production on the Prairies was sporadic from the 1950s through to the 1970s, but in the early 1980s contracts were obtained in southern Manitoba and southern Saskatchewan to produce safflower for processing facilities in Culbertson, Montana. The US cultivars used up to that point were too late in maturing and had severe disease susceptibility, and these problems led to a drastic reduction in cultivation in Manitoba. Most current Canadian cultivars are low in oleic acid and high in linoleic acid, and therefore more appropriate for the birdseed market, and in fact the Canadian safflower crop is currently used as birdfeed, mostly in the US. Oilseed safflower expansion depends on the development of adapted, high-oleic cultivars with high content of oil, improved seedling establishment, and active export efforts and/or a local oil processor becoming established. 'Saffire', the first Canadian safflower cultivar, is a good birdseed cultivar released in 1985, that has a total oil content of only 32%, generally too low for the oilseed market. A more recent (1991) cultivar, 'AC Stirling', is a dual-purpose birdseed/oilseed cultivar averaging 35% oil, and is considered to have the capability of expanding the oilseed market.

Other Oilseeds

Crambe (*Crambe abyssinica* Hochst. ex R.E. Fries = *C. hispanica* L.), a cool-season annual originating from Ethiopia, has been raised in large areas in North Dakota. Meadowfoam (*Limnanthes alba* Hartw.), a winter annual, originates from and is adapted to the Pacific Northwest of the US, where it has been grown. It has also been grown on Vancouver Island. Both crops seem suitable for Canada. It is too early to judge the potential of these experimental oilseed crops, although the relatively large investments in relation to limited commercial success to date in the US and other countries are discouraging.

PULSE CROPS: DRY BEANS AND PEAS

Pulses grown in Canada as dry beans include common bean, lentil, field pea, chickpea, and faba bean. Pulses are low in fat, rich in fiber and complex carbohydrates, and good sources of vitamins, and consumption of these healthy foods has been increasing. Additionally, these legumes fix nitrogen, reducing the amount of nitrogen fertilizer required, and generally improving the yield of crops that follow in a rotation. Some pulses, most notably peas, have become an important livestock feed. Pulses have been the chief new successful crops that have served to diversify Western Canadian agriculture since the 1980s.

Common Bean (*Phaseolus vulgaris* L.)

Dry edible beans have been a commercial crop in Canada since the mid-19th century, and while not particularly new, recent bean cultivars have allowed the area seeded to expand somewhat from traditional growing areas. Demand for dry common bean has been increasing with world population. In Canada, beans are grown mostly in Ontario (which accounts for about 70%), Alberta, Manitoba, and to a lesser extent in Saskatchewan and Quebec, all areas which provide the necessary warm growing season. White and colored beans are produced in about equal amounts, but this represents a noticeable switch from white to colored bean production, primarily due to increased demand for colored cultivars for export. Indeed, colored bean production continues to increase in Canada, because of the creation of new disease-resistant cultivars, strong promotion and market development, and increasing processing capacity in Western Canada. Ontario produces most of Canada's white beans (also known as white pea beans, navy beans, and alubia chicas). Colored beans are grown mostly in Quebec, Ontario, Manitoba, and Saskatchewan, and in Alberta under irrigation. Most (75–85%) of Canada's beans are exported, about half to Europe, and about 40% to the US. Bean yields are quite variable because of the requirement for warmth and sensitivity to adverse weather, so that new cultivars better adapted to Canadian conditions are desirable. For a more detailed analysis of the bean crop in Canada, see Vincent (1995) and Gray (1997a).

Adzuki Bean [*Vigna angularis* (Willd.) Ohwi & Ohashi]

A new adzuki (azuki) bean cultivar, 'AC Gemco', was recently produced by AAFC at Harrow (Ontario), and has genetic consistency that growers have sought, as well as large seed size and high yields. This annual pulse is a major crop in Asia (second only to soybean in Japan), and a limited crop has been produced to date in North America.

Mung Bean [*Vigna radiata* (L.) R. Wilczek]

Another area of growth for beans is the sprout market, especially mung bean. Canada currently imports almost 2 million kg annually of mung beans for sprouting. The cultivar 'AC Harrowsprout' was recently produced by AAFC at Harrow to meet the demand for a domestic supply of mung beans.

Pea (*Pisum sativum* L.)

In 1997–98, Canada produced about 14% of the world's dry peas (about 13 million tonnes), and most of this was exported to Europe. Saskatchewan, Alberta, and Manitoba respectively account for about 70, 20, and 10% of the dry pea seeded area. Field pea has become Canada's sixth most important crop. The area cultivated has been rising for the past 20 years, especially recently, due to an expanding export market, particularly in the European Economic Union, where peas are a traditional feed ingredient for hogs. In eastern Canada, where only 1,000 ha were seeded to peas in 1996, the crop is used mainly for on-farm livestock feed, and the prospects for peas as a new crop in areas other than the prairies seem limited. A detailed analysis of the pea crop in Canada is Skrypetz (1998).

Lentil (*Lens culinaris* Medic.)

Canada produces about 13% of the world's lentils (about 2.9 million tonnes for 1997–98), and is the third largest producing country after India and Turkey. About 85% are grown in Saskatchewan, the remainder in Manitoba and Alberta. Lentil is a relatively new crop for Canada, produced on the prairies in significant quantities only since the late 1960s. Canadian cultivars ('Laird', 'Eston', 'Richlea', and others) are all green with yellow cotyledons, unlike the red-cotyledon lentils that comprise the bulk of the world's lentil production. 'CDC Redwing', a new red cultivar, is considered promising in part because of its disease resistance. Lentils have some prospect for being used as livestock feed, like peas. For additional information on the lentil crop in Canada, see Gray (1998).

Chickpea (*Cicer arietinum* L.)

Chickpeas (garbanzo beans) have a wide variety of food uses, and the lower grades can be used as livestock feed. In 1997–98 Canada produced 14,500 tonnes on 10,500 ha. This crop requires a fairly long growing season and prefers dry conditions because of susceptibility to ascochyta blight (caused by *Ascochyta rabiei*, a devastating seed-borne fungal disease) and a need for heat to set seed. Chickpea is well adapted to the brown soils of Western Canada, and its deep, extensive root system provides good drought tolerance. The Crop Development Centre of the University of Saskatchewan has been concerned with breeding shorter season and ascochyta-resistant cultivars suited to the southern Prairies. Although there is appreciable international competition, there does seem to be a good prospect that chickpea could become a more significant crop in Canada.

Faba Bean (*Vicia faba* L.)

The faba bean is a small-seeded form of broadbean, an ancient vegetable bean of Europe. A smaller-seeded type of broadbean (known as "la gourgane") is grown in commercial amounts in the St. Jean region of Quebec, mainly for soup (Munro and Small 1997). Faba bean cultivation started in western Canada in 1972, and the area under production has fluctuated widely since then. Protein content of 24–30% makes faba bean an attractive on-farm protein supplement for livestock feeding, and there is also good potential for use as silage. While faba bean is a very minor crop in Canada, it could well become more important.

Grass Pea (*Lathyrus sativus* L.)

Grass pea or chickling vetch is a creeping vine. It is the leading pulse crop in Bangladesh, and is also commonly grown in India, and to a lesser extent in the Middle East, southern Europe and some parts of South America. It is usually grown for grain outside of North America, but can be used for fodder. As a pulse, grass pea is very high in protein, but a neurotoxic amino acid is present in wild and most cultivated forms that if consumed in sufficient amounts can cause the irreversible crippling disease known as lathyrism. This toxin to a considerable extent has been bred out of some cultivars (although lathyrism in Asia from consuming grass pea is common). Because of its drought tolerance, grass pea has been judged to have good potential as a future new pulse crop for low rainfall areas of the Canadian prairies, occupying as much as 100,000 ha (Kiehn and Reimer 1992). In the prairies, drought strongly restricts the yield of most current pulse crops. In Canada, grass pea has been used only as a drought-tolerant green manure. 'AC Greenfix', produced by AAFC at Swift Current, was recently released. This annual legume acts as a ground cover alternative to summerfallow, helping to prevent wind and water erosion, as well as adding nitrogen to the soil. For a brief review of grass pea in Canada, see Henckes (1995).

GRASS FORAGE AND FODDER

Forage and fodder plants represent one of the best, although relatively neglected, possibilities for crop diversification in Canada. Most commercial forage production is fed on-farm to cattle, and does not capture the market spotlight. However, forages and processed hays are responsible for much of the 14 billion dollar livestock industry in Canada. World meat consumption is expanding at about twice the rate of population growth, and this has led to a two-thirds increase in world meat trade, three-quarters of this in chicken, most of the remainder in pork (Bateman 1997). Most grains and forage and fodder legumes and grasses used to feed Canadian livestock are grown domestically. There is considerable potential for increasing production of feed grains for export, but it is also sensible to concentrate on improvement of forages for the domestic livestock industry, since forages are much cheaper than grain. Forage seed production in particular recently become a growth industry in Canada, and much of this is due to associated improvements in pollination technology. In addition to benefits for livestock, forages are highly useful for soil conservation and there are some value-added opportunities.

Recent developments in grass and legume forages are documented below. There are some prospects for other kinds of forage. There has been some interest in Canada in developing native halophytes for forage, especially *Atriplex* species. Salt-tolerant forage could be used to exploit the widespread salinized soils in the prairie provinces. Sedges (*Carex*) are the dominant forage in some northern areas of Canada, and have potential for development. It was recently found that *C. praegracilis* Boott has some excellent nutritional properties, as well as adaptation to saline soils (Catling et al. 1994). The least-domesticated of brassicaceous crops, fodder kale and rape (*Brassica* spp.), have received some attention in Canada since they can provide forage well after frost, a distinct advantage in this country (Henckes 1992). This high-protein forage crop deserves more development.

Most forage grasses in Canada are improved European imports of such species as fescues (*Festuca* spp.), orchardgrass (*Dactylis glomerata* L.), timothy (*Phleum pratense* L.), Kentucky bluegrass (*Poa pratensis* L.), and smooth bromegrass (*Bromus inermis* Leyss.) (the latter two are circumboreal). New cultivars are regularly bred in Canada. For example, a winter-hardy orchard grass cultivar for Northern Ontario and parts of Quebec was recently released by AAFC at Ottawa. Native grasses have been referred to as a "sleeping giant," which with proper development can be enormously productive (Henckes 1992). These include bluestems (*Andropogon*), grama grasses (*Bouteloua*), bromegrass (*Bromus*), manna grasses (*Glyceria*), wheatgrasses (*Elymus*, and other genera depending on taxonomic concept), and other species. Slender wheatgrass [*Elymus trachycaulus* (Link) Gould ex Shinners] is the only native Canadian grass to have a cultivar developed for forage purposes. Cultivars of several wheatgrasses were developed at prairie stations of AAFC (Elliott and Bolton 1979). Admittedly native grasses are not generally as productive in pastures as the imported grasses, but they will outperform the latter in habitats that are marginal to agriculture. Native grasses that can be adapted to Canadian agriculture are under study at AAFC at Saskatoon and Ottawa.

Sorghum [*Sorghum bicolor* (L.) Moench]

Sorghum, an African grain that is a major tropical cereal, has been experimentally grown in Canada as a grain and silage crop, often as a hybrid with sudan grass [*S. arundinaceum* (Desv.) Stapf] for forage. Sorghum has not performed well and its future as a Canadian new crop is limited (Chubey 1983; Anon. 1985).

Use of Native Grasses for Grassland Restoration and Maintenance

Native species are used extensively in the US for revegetation, reclamation, wildlife plantings, roadside management, urban landscaping, and permanent cover. In Canada, there is growing interest in using native vegetation for waterway management, roadside maintenance and beautification, the florist and nursery trades, xeriscaping of drought-prone lands, and haylands and grazing lands. Ducks Unlimited Canada, a private, non-profit conservation organization known for its dedication to restoring and managing wetland habitats, has been a key supporter of these efforts, especially for the provision of tall, dense nesting cover for waterfowl. Efforts to date have focussed largely on the use of native grasses for the grasslands in the Canadian prairie provinces, where 70% of North America's waterfowl are produced. Traditional mixtures of introduced grasses and legumes often have not been satisfactory on a long-term basis, and the use of native grasses is increasing. Although many European and Asian grasses thrive under prairie conditions, there is a growing sentiment that native plants provide a more natural and desirable component of the prairie ecosystems. A very strong effort is underway to find and use local sources of native grasses (see Joyce 1993 for a list of species employed). Native Canadian forage grasses currently account for less than 1% of all pedigreed forage grasses grown and inspected in Canada. Although expensive and difficult to acquire in comparison with imported grasses, they are preferred by reclamation agencies to approximate original vegetation in disturbed wild lands. "Ecovar" is a recent term for ecological (not taxonomic) varieties of wild plants that have been chosen because they are particularly useful for conservation or restoration of wild landscapes. Ecovars are not necessarily artificially selected like cultivars, and if artificially selected are not greatly changed from the original populations, retaining much more genetic variability than cultivars. In theory this makes for adaptability. In Canada, there is considerable support for the use of grass ecovars to restore prairie and parkland wildlife habitats. Ecovars that stabilize duck-hunting wetlands, which are very extensive on the prairies, are particularly in demand, and although American ecovars are still predominantly used, Canadian material is becoming more important. A review of the use of wild plants to maintain and improve farm and bordering lands is Environment Bureau (1997b). Joyce (1993) reviews the use of native grasses for conservation and reclamation purposes in Canada.

LEGUME FORAGE AND FODDER

Alfalfa and the true clovers (*Trifolium* species) are very unlikely to be displaced as the dominant legume forages of Canada, but there are several recent additions, such as birdsfoot trefoil (*Lotus corniculatus* L.), and sainfoin (*Onobrychis viciifolia* Scop.) that are well adapted to Canadian conditions and offer good possibilities of improvement by breeding. Sweetclovers (*Melilotus* species) have long been used in western Canada for forage and as a honey crop, but they grow so well as weeds that they may have potential as a low-till crop. There have been continuing efforts to adapt several semi-domesticated legume forages in Canada (Crop Development Centre 1991).

Alfalfa (*Medicago sativa* L.)

Alfalfa, the queen of forages, is the subject of breeding research in Canada, especially in the West. Recent cultivars with high levels of resistance to verticillium wilt and bacterial wilt, main disease threats for the irrigated areas of Western Canada, are 'Barrier' and 'Blue J', produced by AAFC at Saskatoon. 'AC Grazeland' is a recent low-bloat cultivar from AAFC at Saskatoon. In Quebec, annual losses from lack of cold tolerance have been estimated to exceed $10 million, and attempts are underway by AAFC at Sainte-Foy to obtain more cold-tolerant cultivars. The use of genetically-modified alfalfa as a pharmaceutical and industrial products producer is also under study at Sainte-Foy. The alfalfa leafcutter bee, *Megachile rotundata* (F.), has become the dominant pollinator of alfalfa in Canada in recent decades. This has been critical to developing the alfalfa seed industry, since the honey bee is a quite inadequate pollinator of alfalfa in Canada. Not only is Canada now self-sufficient in alfalfa seed production, but it has become the world's chief exporter of alfalfa leafcutter bees (Small et al. 1997). All Canadian alfalfa cultivars contain high amounts of hemolytic saponins, which have been demonstrated to be very detrimental to poultry, and somewhat detrimental to pigs, both classes of livestock whose exports are expected to increase dramatically in Canada. The recent discovery of Turkish strains that are essentially devoid of hemolytic saponins (Small 1996) has the potential of leading to greater usage of alfalfa in the livestock industry.

Cicer Milkvetch (*Astragalus cicer* L.)

Cicer milkvetch has been considered to be a legume with good grazing potential for the last two decades in Canada. A new cultivar, said to be high-yielding, bloat-free, long-lived, easy to grow, appealing to cattle, not appealing to pocket gophers, and lacking serious diseases or insect pests, is scheduled for release from AAFC at Lethbridge in 2000.

Crownvetch (*Coronilla varia* L.)

This winter-hardy, salt-tolerant forage is being developed in Canada. It is used principally in this country to protect steep roadsides against erosion, but has good forage possibilities.

Fenugreek (*Trigonella foenum-graecum* L.)

Fenugreek is best known as a popular Asian seed spice, but it has also been grown as a livestock feed in the Old World. Fenugreek for forage is a quite recent novelty in Canada, where it has proven to be an excellent legume silage, yielding as much dry matter as two cuts of alfalfa, and 16–18% protein (compared with 18–20% for alfalfa) (Small 1997a). although cattle show initial reluctance to consume the strong-smelling fenugreek, they adjust to it in 7–10 days. The advantage over alfalfa possessed by fenugreek is that is an annual, therefore useful in a rapid crop rotation, for example with barley grain for silage. Fenugreek contains compounds with oxytocin activity (hormones that induce milk letdown in humans and animals), and this property may be desirable for dairy cattle. Fenugreek is also a natural source of diosgenin, a precursor of steroids used in the manufacture of birth control pills and other drugs, and it may one day replace present commercial sources [South American yam (*Dioscorea*) species].

Lupins (*Lupinus* species, especially white lupin, *L. albus* L.)

Lupins have attracted some interest in eastern Canada as a frost-tolerant, high-protein, nitrogen-fixing livestock feed. In Western Canada, success has been much more limited. For a brief review of Canadian experience, see Henckes and Leake (1994).

VEGETABLES

Except for potato (discussed below), most of Canada's fresh vegetables are grown in Ontario and Quebec. The vegetable growing area in Canada has been expanding for many years (Fig. 9), and has increased 76% to 128,000 ha between 1951 and 1996. The crops with the most area devoted to cultivation were sweet corn, peas, tomato, carrot, and bean. However, the most lucrative crops on a per hectare basis were shallots ($16,709), lettuce ($16,071), and celery ($15,167)—about twice the per hectare value of tomatoes ($8,644) and much higher than sweet corn ($1,609) and peas ($1,228) (Statistics Canada 1998b).

The relative value of the 23 most important Canadian vegetable crops is shown in Fig. 10. This diversity makes it difficult for any new vegetable to capture much of the market. Nevertheless, in recent years Canadian farmers have been growing more non-traditional vegetables, such as daikon radishes, bok choy, and escarole, a reflection of Canada's changing ethnic mix and more cosmopolitan tastes in food. An extensive analysis of all Canadian vegetables, with information on new cultivars and new trends, is Munro and Small (1997).

The only Canadian native vegetable that has elicited significant interest for domestication is ostrich fern [*Matteuccia struthiopteris* (L.) Todaro]. Young fronds are harvested as "fiddleheads" in Nova Scotia and New Brunswick for the gourmet food market. This crop is gathered from natural stands, but there is some interest in commercial cultivation. Although fern breeding is complicated, abundant genetic variability appears to be present in Canada and could serve to improve an already very attractive and novel crop.

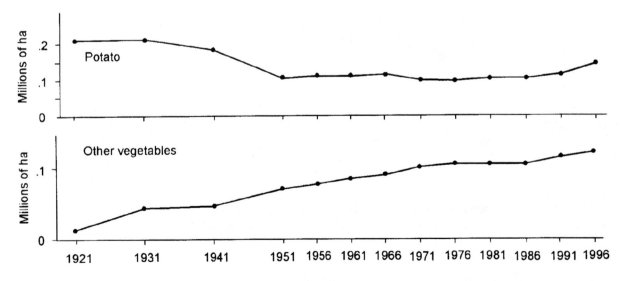

Fig. 9. Area of potato and other vegetables in Canada, based on the comprehensive 10-yearly (1921–1951) and 5-yearly (1951–present) agricultural censuses.

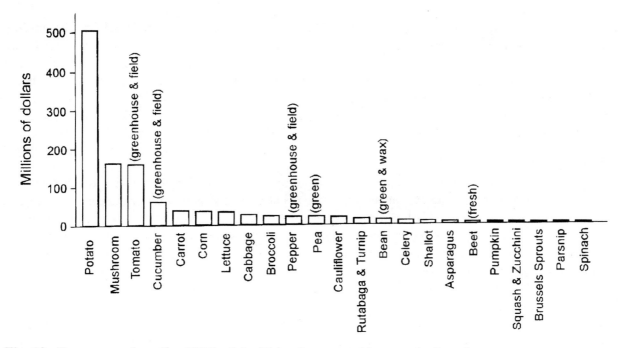

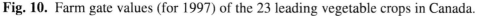

Fig. 10. Farm gate values (for 1997) of the 23 leading vegetable crops in Canada.

Potato (*Solanum tuberosum* L.)

The potato is unusual in that, aside from forages, it is the only crop grown commercially in every province of Canada. Potatoes are cultivated in large amounts in Canada most notably on Prince Edward Island, where most of the arable area of the province is devoted to this single crop. Since the market for potatoes is quite good, there has not been much interest in diversifying the agriculture base on PEI, where farmers are content with this most perfect of food plants. A larger area of Canada was devoted to potato in the 1920s than today, but the crop has been expanding in recent years in all of Canada (Fig. 9), especially in Manitoba and Prince Edward Island. The increases in potato area have been due to increased processing capacity, primarily for export markets. Because of the large volume of the crop, the potato has long been a candidate for new uses. The potato is especially susceptible to disease and insect problems, and is the most chemically intensive crop grown in Canada, so that dependence on it can be somewhat unsettling. Because of the narrow genetic base of the major cultivars, there is an international effort underway to utilize the over 200 tuber-bearing species of *Solanum* in order to make modern potato more resistant to diseases and insects. For more information on potato in Canada, see Munro and Small (1997).

Mushrooms

The white button mushroom [*Agaricus bisporus* (Lange) Imbach] commonly sold in supermarkets is the predominant commercial mushroom. There are only a few hundred mushroom farms in Canada, but their gross sales (close to a million dollars/farm annually) are 10 times that of the average farm in Canada. However, a large investment in buildings is required, as well as a large number of workers. Ontario is the leading production area of fresh and processed mushrooms (over 50%), followed by British Columbia (about 30%) and the prairie provinces (about 15%). There are widespread attempts underway in Canada to establish mushroom farms in other areas. Mushrooms can be cultivated almost anywhere (although proximity to markets is important), and provide a good means of crop diversification. The mushroom industry is useful in using up Canada's large supply of manure.

Specialty mushrooms include enoki, maitake, nameko, oyster, pompom, shiitake, shimeji, wine cap, and others. These unusual mushroom species are valued by ethnic markets in North America, and by trendy restaurants, and therefore provide promise of a growing market in Canada.

Wild edible mushrooms, primarily pine mushrooms, chanterelles, morels, and false morels, have become an important and growing multimillion dollar industry in Canada (Redhead 1992). Although mushrooms are wildcrafted throughout Canada, most of the commercial wild supply is harvested in British Columbia. The pine mushroom [*Tricholoma magnivelare* (Peck) Redhead] is Canada's most important wild mushroom (Redhead 1997). It resembles the Japanese matsutake, which is esteemed in Japan, and over the last decade Canada has exported almost all of its harvest of pine mushrooms in fresh form to Japan. In Japan, the retail price of fresh pine mushrooms can exceed $200/pound, while Canadian pickers can receive up to $100/pound. In 1993 Canadian exports of pine mushrooms to Japan were worth 1,840,000,000 yen (Anonymous 1995; at the time, 1 American dollar was worth about 117 yen). There is also a large chanterelle harvest in Canada, in the West and in the Maritimes. Chanterelles and morels are generally exported to Europe.

CULINARY HERBS

Culinary herbs and spices have been gaining in favor in Canada, and there are now thousands of hectares grown annually. The species are too numerous to document here, and only a few examples will be given. Coriander (*Coriandrum sativum* L.) has proven to grown very well in Saskatchewan, where 3,765 ha were grown in 1996. Over 400 ha of garlic (*Allium sativum* L.) were grown in Canada in 1996, by about 480 farmers, a challenging endeavor because of the climate. Stevia [*Stevia rebaudiana* (Bertoni) Bertoni] is a new crop in Canada, grown mostly experimentally to date, in British Columbia, Alberta, and especially in Ontario (Borie 1998). This perennial herb of Paraguay and Brazil provides a non-nutritive sweetener that is widely used in Asia, although not yet judged acceptable for consumption in North America. It has been developed into an annual crop in Canada, but whether it will be able to compete with the perennial crop grown in California remains to be seen. Detailed information on the more than 100 culinary herbs that are grown in Canada is in Small (1997a).

GREENHOUSE CROPS

The relatively short growing season of Canada, by comparison with more southern countries from which crops can be imported, is a handicap for agriculture. Not surprisingly, glasshouse culture provides an important partial solution. Greenhouse cultivation (including glass, plastic or other protection) is expanding rapidly in Canada. All provinces are experiencing an increase in greenhouse area, with the largest increase in Ontario, especially in the Leamington area. The total area of greenhouse culture in Canada doubled from 1981 to 1996 to 12.7 million square meters (about 1,300 ha). Between 1991 and 1996, the increase was 51%. In 1996, greenhouse growers achieved sales of about a billion dollars. The sale of cut flowers and potted plants represented three-quarters of this (the values of the major species are shown in Fig. 11), and there was also about $200 million in vegetable sales. Tomatoes represent half of Canadian greenhouse vegetables, and generate a gross income of over $500,000 per hectare. Half of the greenhouse tomato crop is grown in Ontario. The very high income, coupled with a large domestic market and a potentially large export market to the US may explain why tomato is such a popular greenhouse vegetable crop. The other major Canadian greenhouse crops—cucumber, lettuce, and pepper, are also very high value crops during the Canadian winter. Among greenhouse trends in Canada are: use of longer lasting covering materials as well as material that allow more light transmission; change to hydroponic/soilless culture (especially rockwool); computerization of controls for the greenhouse environment; and carbon

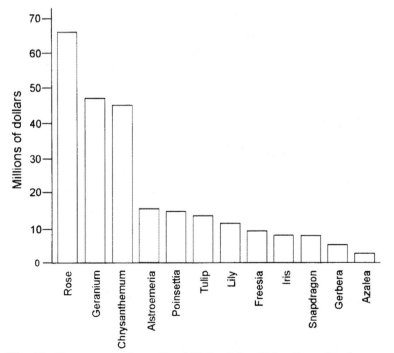

Fig. 11. Farm gate values (for 1996) of the 12 leading glasshouse ornamental crops in Canada.

dioxide supplementation. There has also been a significant switch from growing pink-colored tomato cultivars to red-colored. The latter are more acceptable in the US market, and the recent large greenhouse area increase in Ontario has been for exports to the US.

FRUITS

The values of the 11 leading Canadian fruit crops are shown in Fig. 12 (see Small and Catling 1996 for a review that also includes minor fruits grown in Canada). The most commonly cultivated fruits in Canada are apple, blueberry, grape, strawberry, and raspberry. About a third of the fruit area of Canada is devoted to low bush blueberry production, largely from Quebec eastward. The remaining cultivated fruit area is widely distributed in Canada. As with vegetables (above), the crops with the most area devoted to cultivation do not produce the maximum gross return per hectare. The most lucrative per hectare crops are cranberry ($24,861), kiwi ($23,101), sweet cherry ($10,998), strawberry ($9,859), and peach ($9,138)—all higher than the leading fruits, apple ($5,127), blueberry ($3,284), and grape ($6,706) (Statistics Canada 1998b).

Breeding of new fruit cultivars has been fairly continuous in recent years in Canada. Breeding programs for tree fruits in Canada are discussed by Quamme (1996), and for the small fruits by Jamieson (1996). Numerous raspberry and strawberry cultivars have been released from AAFC at Vancouver, and kiwi (*Actinidia deliciosa* C.S. Liang & A.R. Fergusson) was recently developed as a Canadian crop, although high world production of this fruit has depressed its profitability. AAFC at Vancouver recently introduced into Canada the grape kiwi (also known as winter kiwi and arguta) [*A. arguta* (Siebold & Zucc.) Planch ex Miq.], which is hardy to –25°C and produces abundant clusters of smooth-skinned berries that are more flavorful than full-sized kiwis. There are several cultivars of this relatively new fruit, which is acquiring popularity in British Columbia and the northwestern US.

Cranberry (*Vaccinium macrocarpon* Ait.) is a minor fruit crop in Canada. The center of cranberry cultivation and production is Massachusetts, but large quantities are also raised in the peatlands of British Columbia, as well as New Jersey, Washington, and Oregon. Considerable cranberry culture also occurs in Wisconsin, and in limited degree in Ontario, Quebec, and the maritime provinces (especially New Brunswick and Nova Scotia). Cranberry is receiving increasing attention in Quebec. Cranberry is rather unique among crops in that it has been grown for many years but still demand exceeds supply. About 10% of Canada is covered by peatlands, and there is no shortage of peat with which to construct engineered (i.e. artificially constructed) bogs to grow this crop. There has been concern that environmentally sensitive wetlands might be eliminated or damaged by development of new cranberry bogs, but technology is available to control damage to wetland sites, and even to develop bogs on dryland sites. Canada has the resources to expand its cranberry industry, and this could become the most important new fruit crop in certain areas.

There is a small grape wine industry in British Columbia and southern Ontario, and a very small industry in Quebec and Nova Scotia. Most people are surprised to learn that despite the cold climate of Canada, prize-winning wines are occasionally produced that rival the best wines of the world. Less surprising is the fact that a Canadian specialty is ice-wine, made

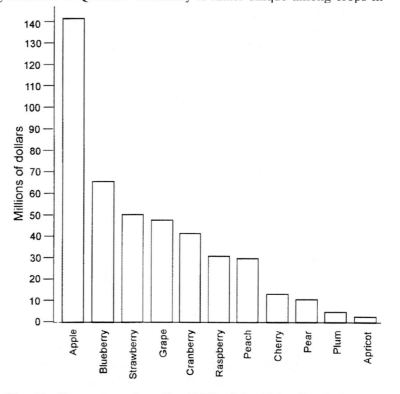

Fig. 12. Farm gate values (for 1997) of the 11 leading fruit crops in Canada.

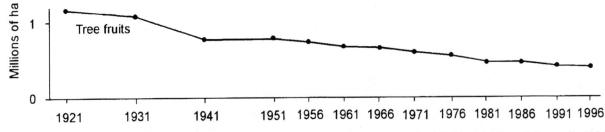

Fig. 13. Area of tree fruits in Canada, based on the comprehensive 10-yearly (1921–1951) and 5-yearly (1951–present) agricultural censuses.

from grapes allowed to freeze on the vine. *Labrusca*-type grape plantations (with germplasm of the North American fox grape, *Vitis labrusca* L.) are being replaced in Canada with *vinifera* grapes (imported cultivars of the European *V. vinifera* L.) for wine-making. Until the 1980s, Canadian wines were made mainly from native *labrusca* cultivars, such as 'Concord' and 'Niagara'; these grapes are grown today mainly for juices, jams, and the fresh markets. *Vinifera* grape cultivars, such as 'Cabernet Sauvignon', are cultivated for wine-making. Hybrids of the two types, for example 'Vidal', are grown in Ontario to make ice wine. For a review of the grape wine industry in Canada, see Read (1994).

There are over 100 wild fruit species gathered in Canada (Turner and Szczawinski 1979). These are well adapted to the climate and soils, and would seem to offer excellent prospects for domestication. In fact, only two wild fruits have received much breeding attention in Canada as new fruits (sandcherries, *Prunus* spp., have been bred as ornamentals). Saskatoon (*Amelanchier alnifolia* Nutt.) is perhaps the best example of domestication of a Canadian indigenous species (see Mazza and Davidson 1993 for a good review). Considerably improved forms were bred by AAFC at Beaverlodge, Alberta, and the annual farm gate value exceeds a million dollars. Although this crop is still in its infancy, it has been predicted that over 4000 ha of saskatoons could eventually be planted on the prairies (Henckes and Dietz 1992). Although still basically a wild crop, lowbush blueberry (*Vaccinium angustifolium* Ait.) has undergone considerable selection at AAFC, Kentville, Nova Scotia. About 9,000 kg of wild lowbush blueberries are harvested annually in Canada, largely from stands that are managed by burning, fertilizing, or pruning. Other possibilities of wild Canadian fruits that deserve consideration for domestication are: lingonberry (*Vaccinium vitis-idaea* L.), with harvests in Newfoundland of the order of 150,000 kg annually (domesticated cultivars from Europe are cultivated to a small extent in Canada); highbush cranberry (*Viburnum trilobum* Marsh.); cherries (*Prunus* spp.); cloudberry (*Rubus chamaemorus* L.); elderberry (*Sambucus canadensis* L.) and currants and gooseberries (*Ribes* spp.). White blister rust, a pathogen of *Ribes* that infects 5-needle pines, has limited currant and gooseberry development in North America, but some cultivars are resistant and can be safely grown.

Tree fruit area has been decreasing very steadily and noticeably for the past 75 years in Canada (Fig. 13), and an explanation is in order. Trees and to a lesser extent bush fruits are a long-term investment that cannot respond flexibly to market fluctuations, i.e. one can't put in another crop easily if market or weather conditions suggest this would be advantageous. However, these are universal problems with woody perennials; below, some special problems in growing such plants in Canada are discussed.

There are ecological reason why most fruits (and also nuts) are competitively disadvantaged in Canada. Most commercial fruits (strawberry is an exception) and nuts (peanuts is an exception) are woody perennials. Woody perennials put a substantial portion of their energy into building up wood and, in deciduous plants, regrowing their foliage annually. In a long-season climate, both the woody tissues and leaf tissues are functionally useful for a long period. By contrast, in short-season climates as found in Canada, growing wood and new leaves annually, and indeed consolidating energy reserves to survive the winter, represent a proportionately greater deflection of energy from fruit production. Still another reason has to do with the greater intensity of insolation in areas near the equator. Moreover, trees near the equator can intercept as much as 95% of the annual incident radiation, while annuals may intercept as little as 20% (Corley 1989). Still another factor favoring woody fruit trees in tropical regions is that the soil is often too infertile to support unfertilized annual crops, so that growing woody plants is highly advantageous. For these reasons, perennial crops are far less common in northern climates such as Canada's in comparison with tropical and subtropical areas.

NUTS

Filberts (*Corylus* spp., also known as hazel nuts), grown in British Columbia, are the only significant nut crop in Canada. A small crop of peanut (*Arachis hypogaea* L.) has been grown for some time in southern Ontario, for the peanut butter crop. Attempts to grown peanut in southern Alberta were unsuccessful because the growing season proved too short. The walnut (*Juglans nigra* L.) and butternut (*J. cinerea* L.) can be grown in Canada, but not competitively (a very limited amount of walnuts have been grown in British Columbia). Cold-hardy forms of the English walnut (*J. regia* L.), particularly the Carpathian walnut, can also be grown in the warmest regions of Canada. At least one grower in Ontario is marketing the Japanese heartnut [*J. ailantifolia* Carrière var. *cordiformis* (Maxim.) Rehd., known also under other names]. Numerous popular nuts can be grown efficiently in warm-temperate and hot regions, and because of their very good keeping qualities and the economy of transportation of these high-value, relatively light commodities, a strong nut industry has not developed in Canada, and indeed does not have good prospects.

AROMATIC CROPS

Interest in Canada in essential oil crops is centered in Western Canada. Crops that can be grown there include: dill (*Anethum graveolens* L.), caraway (*Carum carvi* L.), coriander, spearmint (*Mentha spicata* L.; also Scotch spearmint, *M.* ×*gracilis* Sole), peppermint (*Mentha* ×*piperita* L.), hyssop (*Hyssopus officinalis* L.), horseradish (*Armoracia rusticana* P. Gaertn., B. Mey. & Scherb.), garlic, onion (*Allium cepa* L.), monarda (*Monarda* species and hybrids), fennel (*Foeniculum vulgare* Mill.), fenugreek, summer savory (*Satureja hortensis* L.), sage (*Salvia officinalis* L.), tarragon (*Artemisia dracunculus* L.), chives (*Allium schoenoprasum* L.), anise-hyssop [*Agastache foeniculum* (Pursh) O. Kuntze], parsley [*Petroselinum crispum* (Mill.) Nym ex A.W. Hill], and basil (*Ocimum basilicum* L.). The essential oil industry is in its infancy in Canada (several thousand hectares and several million dollars in total), so the long-term success of most of these species as aromatic sources in Canada still remains to be determined. The majority of these can also be grown as fresh or dried flavoring herbs/spices, and coriander grown as a spice has had reasonable success. However, the market for aromatic and spice crops is very competitive, with considerable, efficient foreign production, so that just how much growth potential there is in Canada remains an issue.

A review of new essential oil crops in Canada concluded that monarda (*Monarda punctata* L., and hybrids) was the most promising possibility (Ference 1989). Several lines of monarda, each line yielding a distinctive fragrance, were bred at AAFC, Morden, Manitoba (Anon. 1992; Mazza et al. 1993). A similar attempt was made to develop several lines of anise-hyssop (Anon. 1992). Considerable investment in research went into these projects, and although useful cultivars resulted, in retrospect it is unclear whether the efforts were worthwhile economically. For additional information on essential oil crops in Canada, see Small (1997a).

MEDICINAL CROPS

This category of crops overlaps with nutraceuticals. Therapeutic crops primarily marketed as herbal preparations or for medicinal extracts can be considered to be medicinal crops, while those serving as sources for extraction or preparation of nutritional substances can be segregated as nutraceutical crops. There is a limited amount of wildcrafting of medicinal plants in Canada, and cultivation is now predominant. A wide variety of medicinal crops are grown in Canada, but none is very important except ginseng. However, there is considerable enthusiasm for the cultivation of medicinal herbs, and the industry is expanding in Canada. Small and Catling (in press) presents a detailed analysis of medicinal crops in Canada, as well as a guide to Canadian experts who specialize in medicinal crop development.

Ginseng (i.e. "American ginseng," *Panax quinquefolius* L.) has continued to dominate the cultivation of medicinal plants in Canada, with two-thirds grown in Ontario, and most of the remainder in British Columbia. The industry in British Columbia is only two decades old, and demonstrates very well how a lucrative crop can be successfully transplanted to another region. Wild collection of ginseng is no longer authorized in Canada, and ginseng populations have been greatly reduced in this century. Most of the crop is exported to Asia, and in 1996 over 1 million kg of ginseng roots were exported. Ginseng grows well on well-drained, sandy-loam soil, much like that for tobacco, and so most of Canada's ginseng is grown in the traditional tobacco area of southwestern Ontario. There is some interest in reviving woodland cultivation of ginseng. A very limited amount of Asian ginseng (*P. ginseng* C.A. Mey.) is raised in Canada, and there is very strong opposition to this from the ginseng industry, which fears the possible introduction of diseases or other unknown problems.

NUTRACEUTICAL CROPS

"Nutraceuticals" may be any of a wide variety of consumable products, such as foods, pills, powders, and indeed the definition of nutraceuticals varies geographically. In Europe, nutraceuticals are generally restricted to natural ingredients, because of strong consumer interest in natural herbal medicines. In the US, nutraceuticals are not necessarily of natural origin. In Canada, nutraceuticals, "functional foods," and "pharmafoods" are terms frequently used to describe food or products that have physiological benefits or reduce the incidence of chronic disease. Foods recommended by a physician to manage a disease or health condition such as obesity, gluten intolerance, or diabetes are often excluded from the definition of nutraceutical, although this is arbitrary. Various extracted constituents from food crops provide important nutraceuticals. The extracted fiber of oatmeal, wheat bran, barley, lentil, peas, and beans provides health benefits. The insoluble fiber in wheat bran, lentil, and brown rice reduces potential carcinogens in the colon. The soluble fiber in oat bran, dried common bean, and barley fights heart disease by reducing the absorption of cholesterol. Flaxseed can reduce cancer and heart disease, its omega-3 fatty acids decreases blood cholesterol, and its phytoestrogens or lignans are thought to reduce the incidence of breast cancer. Canola oil is healthy because of its relatively low level of saturated fatty acids, but it also contains significant amounts of essential fatty acids, such as oleic acid, which lowers plasma cholesterol levels, and linoleic and linolenic acids, which convert to hormone-like substances (eicosanoids) that affect physiological reactions ranging from blood clotting to immune response. Soybean contain isoflavones, which help reduce the negative effects of estrogen on the human body, ease some symptoms of menopause, reduce hypertension and heart disease, and may be related to a lowered rate of breast and prostate cancer. Chicory (*Cichorium intybus* L.) was recently examined as a source of inulin, which has nutraceutical and feed applications (Lachance 1996). Inulin can also be produced from Jerusalem artichoke (*Helianthus tuberosus* L.).

Nutraceuticals are exploding in importance in the natural health and food products industry, and represent a multi-billion dollar market. The US market has been estimated to be about $1.5 billion, and the Canadian market about $150 million. The growth of nutraceuticals is due to an increasing tendency to fortify foods with disease preventing qualities. This trend is related to an aging population, increasing health care costs, consumer interest in nutrition, and food technology advances. The growing market provides an excellent opportunity for the Canadian agri-food sector to diversify. It is especially significant financially because the products that can be produced have high value added (Spak 1998a).

The chief roadblock at present to the development of both the nutraceutical and medicinal herbal industry in Canada is regulatory uncertainty with respect to the marketing of these products. In Canada, as in many countries, these often fall into a grey and ambiguous area between food and drug. Canada currently does not allow claims of specific health benefits for herbal remedies and nutritional supplements, unless they are elevated to the category of drug, in which case efficacy must be proven by clinical trials. All crops potentially can yield nutraceuticals, but crops that are produced in very large amounts are especially significant. Most of the 60 million tonnes of grains, oilseeds, and special crops produced annually in Canada have the potential to be processed into nutraceuticals for domestic and global markets. This area is attracting a great deal of scientific research and market development in Canada, and is likely to be an important factor in the future regarding the cultivation of new crops in various regions.

Gamma linolenic acid (GLA) is used medicinally, particularly for treating atopic eczema, a common skin rash, and for nutritional deficiencies. It is obtainable from the seeds of many plants, although to date two species have been the chief sources. Borage (*Borago officinalis* L.) is an ancient, essentially undomesticated species, native to the Mediterranean, that has been considered as a diversification crop in many countries, both as a forage and as a medicinal plant. Considerable amounts have been grown in Saskatchewan. Most borage oil is produced in Europe and Asia, but since the crop seems well adapted to the cool growing conditions in parts of the prairies, and possibly also to eastern Canada, it has potential for further development in

Canada. The chief drawback of borage is that most available cultivars produce seeds over a long period, shattering them gradually, so that only about 20% of the seed crop can be collected by conventional means (in Europe, some relatively non-shattering cultivars have been bred). Evening primrose (*Oenothera biennis* L.) seed oil also contains high levels of GLA, and is occasionally grown as a substitute crop for tobacco in southern Ontario. Ference (1989) concluded that evening primrose is a better diversification crop for Canada than borage. Foreign competition, market volatility and lack of processing facilities limit both crops in Canada. The possible use of hemp for GLA production is discussed below. For further information on these species as new Canadian crops, see Small (1997a) and Small and Catling (in press).

Sea buckthorn (*Hippophae rhamnoides* L.) is a shrub that is widely planted on the Canadian prairies as a nitrogen-fixing, soil-conserving, hardy shelterbelt. The berries are used to a small extent to make jams and jellies, but this usage is very limited. However, a venture is underway to process sea buckthorn berries and leaves into a range of health foods and herbal products (Henckes 1998b), with the goal of harvesting sea buckthorn orchards from British Columbia to Manitoba. The fruit is high in vitamins C and E and beta carotene, as well as flavonoids, and the oil is high in essential fatty acids. Essential oils from the fruit can be used in nutritional supplements, skin creams, and other products. In Eurasia the oil has a reputation for relieving pain, reducing inflammation, and fighting bacterial infections. This potential new crop requires a great deal of development, but has succeeded in stimulating considerable interest in Canada.

TEXTILE FIBER CROPS

There is an absence of successful textile fiber plants in Canada, and accordingly there is interest in finding new fiber plants. Oilseed flax is discussed above as a very promising new crop for Canada. In the US flax is extensively employed in papermaking, especially using decorticated straw ("flax tow") from oilseed cultivars grown for linseed oil and meal, obtained from both the US and Canada. Although fiber flax is adapted to Canada's cool-temperate climates, international competition seems to have suppressed the textile crop industry in Canada. Fiber flax cultivars have been grown in Canada, but have only acquired a fraction of the success of oilseed cultivars. As noted below, fiber hemp is likely to be used for fiber production in Canada, but the clothing use will likely remain limited. A shortage of imported fiber during the second world war led to exploration of fluff from common milkweed species (*Asclepias syriaca* L. and *A. speciosa* Torr.) as a substitute. This proved unsuccessful, but with the success of this crop in the US (Witt and Nelson 1992), a further Canadian exploration of the potential for textile, paper, fiberfill, and insulation was initiated by AAFC at Saint-Jean-sur-Richelieu. This has not led to the establishment of a new fiber industry in Canada.

Hemp (*Cannabis sativa* L.) grown under license for fiber as well as oilseed products is the most publicized new crop in Canada, and there are currently thousands of hectares under cultivation by hundreds of authorized farmers. Hemp is being grown both for traditional textile, paper, cordage, and oilseed usages, but also for a variety of new uses (composite building and manufacturing materials; nutritional, cosmetic, and industrial preparations; biomass applications). As detailed information on Canadian hemp is given by Moes (this symposium), and in several publications cited below, the presentation here is limited. Until very recently the prohibition against drug forms of the plant prevented consideration of cultivation of fiber and oilseed forms in Canada. However, in the last 5 years three key developments occurred: (1) much-publicized recent advances in the legal cultivation of hemp in western Europe, especially for new value-added products; (2) enterprising farmers and farm groups became convinced of the agricultural potential of hemp in Canada, and obtained permits to conduct experimental cultivation; and (3) lobby groups convinced the Parliament of Canada that narcotic forms of the hemp plant are distinct and distinguishable from fiber and oilseed forms. There are indeed two categories of plants, formally recognized as subspecies by Small and Cronquist (1976): *C. sativa* subsp. *sativa*, comprising fiber and oilseed forms, as well as similar wild plants; and *C. sativa* subsp. *indica* (Lam.) E. Small & Cronq., comprising drug forms of the plant, as well as similar wild plants. These two subspecies were defined on the basis of their intoxicant potential: subsp. *sativa* with less than 0.3% tetrahydrocannabinol (THC) in the foliage and flowering parts (dry weight basis); and subsp. *indica* with more than 0.3%. This level of 0.3% THC is now used in the European Economic Community and Canada as a criterion for authorization of cultivation. Cultivars with less than 0.3% may be legally cultivated under license, whereas those with levels of 0.3% or greater may not. There are considerable efforts underway in the US to convince the authorities that hemp cultivation should be carried out similarly as in Europe and Canada, but concern over the narcotic types of plant remains the key obstacle, as it did for many years in Europe and Canada. Objective evaluation of hemp as a crop is complicated because of the emotion connected with the drug use of the species. Exaggerated claims of hemp's values are common, and have contributed to the growing bandwagon of authorized cultivation in Canada. The next few years will determine the extent to which hemp can be grown as a new economic crop in Canada, and could influence its future possibilities in the US, where historically it was grown very successfully for a very long period, in much greater quantities than ever occurred in Canada. For further information on hemp in Canada, see Reichert (1994), Small (1979, 1995a,b, 1997b) and Montford and Small (1999).

PULP, BIOMASS, AND ENERGY CROPS

These categories are discussed together because the crops used often serve the three purposes.

Pulp

Kenaf (*Hibiscus cannabinus* L.), a native of east-central Africa, was recently considered to be a candidate for diversification in Ontario, as a newsprint source, although it is much better adapted to the southern US, where it is now grown and has received much favorable publicity. Experimental growth in southern Ontario succeeded only in showing that the crop cannot be grown competitively here. As noted below, poplar and willow lines are available as agroforestry crops in Canada for pulp, and because they can occupy vast areas of marginal Canadian land, they are a much more likely source of pulp than herbaceous crops that require prime agricultural land. However, 40% of Canada is forested, and this vast supply of wood and wood products limits the prospects for all new pulp sources.

Straw-Based Particleboard

As much as 40 million tonnes of cereal straw are produced annually in Canada as a by-product of growing grain. Traditionally, these crop residues have been incorporated back into the soil, used as bedding or a feed additive for livestock or, least desirably, burned on the field, producing air pollution. Rarely, the straw has been used for pulp for paper, ethanol, space heating, and building materials. Very recently, high-grade particleboard (made with the resin methyl-diphenyl-isocyanate, better known as "crazy glue"), using mostly wheat straw, has come into commercial production on the Canadian Prairies. At least here, this venture seems economical. Plywood and oriented strandboard (long strips of wood oriented and blended with adhesives) can be produced relatively cheaply in heavily wooded areas such as the Pacific Northwest, the US Northeast, Quebec, and New Brunswick. Waferboard (wood flakes glued together under pressure) can be cheaply produced from aspen, which is very abundant in Canada, including much of the Northern Prairies, and the long-term success of cereal straw board remains to be demonstrated. Also, fiberboard (produced by fiberizing wood particles as in a semi-pulping process to create a dense board that holds screws better than particleboard) is an economical way of using up the rather large amounts of waste material from the lumber industry. As noted above, hemp is also is under investigation as a new type of biofiber/resin board in Canada. Reichert (1995, 1996) reviews the potential of straw-based particleboard in Canada.

Biodiesel

Biodiesel fuels are derived from renewable biological resources for use in diesel engines. Biodiesel is more viscous than conventional diesel, and therefore less useful at lower temperatures, limiting its use in Canada. Nevertheless, in Canada and similar cold areas, biodiesel can be marketed as an additive in a 5–10% blend with conventional diesel fuel. Canadian biodiesel technology has refined the hydro-treating method using a conventional refining process like that used in the petroleum industry. This produces cetane, a booster for diesel fuel, as well as naphtha and other products. Oilseed crops such as soybean, canola, and sunflower are particularly useful for producing biodiesel fuel (which can also be obtained from other vegetable oils, such as maize and peanut oil, and animal fats). Biodiesel is primarily useful as a means of improving air quality, particularly for lowering sulphur emissions from fossil fuels. Other benefits include decreasing energy dependence on foreign imports, reduction of greenhouse gases, lower toxicity resulting from accidental spills, improved biodegradation, and the creation of employment (as biodiesel production is several times more labor intensive per unit of production than fossil fuels). Another result is the production of high-quality glycerine (used in such products as hand creams, toothpaste, and lubricants) as a by-product (when the transesterification process is used), but this has created a glut of glycerine on the market. The cost of biodiesel production may be two to three times that of petroleum fuels, and biodiesel seems justified only in areas where air pollution is a significant problem. Europe has been the primary area of support of the use of biodiesel, and the European Community has committed large sums of money to this. When one considers the negative environmental impacts of increased fertilizer and pesticide use in using oilseeds to manufacture biodiesel, it is less clear that biodiesel is desirable environmentally. In Canada, biodiesel production seems like a reasonable alternative for using up surplus and frost-damaged oilseeds. Moreover, with increasing industrialization, pollution is likely to increase in large Canadian cities, making subsidization of production more attractive. For an analysis of the biodiesel situation in Canada, see Gowan (1996).

Ethanol

Alcohol can be used as a fuel for vehicle engines, a fuel extender when blended with gasoline, and as an octane enhancer. The two most important alcohols for fuel use are ethanol and methanol. The latter is produced from natural gas, coal and wood, all in high supply in Canada. Alcohol-blended fuels (gasohol) are widely considered desirable for their environmental benefits, reducing polluting vehicle emissions and contributing to the effort to stabilize and reduce emissions of greenhouse gases related to global warming. Crops such as maize, potato, sorghum, and sugarbeet, which produce considerable biomass and a large carbohydrate yield, have the potential for production of alcohol. In North America, about 95% of fuel ethanol is manufactured from maize. Small amounts are made from wheat and barley, and can potentially be obtained from cull potato, Jerusalem artichoke, other cereal crops, lignocellulosic material such as maize hulls, switchgrass (*Panicum virgatum* L.), wood waste, and municipal solid waste (McKeague 1994). Ethanol production also offer opportunities to the grains and oilseeds industries. An integrated ethanol plant that consumes grains or oilseeds can produce edible livestock products in addition to alcohol. Like biodiesel fuel discussed above, current production costs relative to current petroleum is prohibitive, unless there is subsidization. The US has provided considerable subsidization to the production of fuel alcohol from maize. Canada does not have the highly polluted areas which occur in the US, so that there is less incentive to enact environmentally-oriented legislation supporting fuel alcohol production. Maize is a more prominent crop in the US, and maize technology is more advanced. Nevertheless ethanol-enhanced fuels have gained some popularity in Canada, and there is a reasonable prospect that a variety of new and old crops can be adapted with new technology to a more substantial bioethanol industry in Canada. In particular, maize for the production of ethanol is considered to have excellent potential in Canada (McKinnon 1997c), a part of the trend toward use of renewable raw material for industrial processes rather than non-renewable petroleum. Several industrial plants use maize in Western Canada, and new ethanol production plants have been established in Ontario and Quebec. Jerusalem artichoke (*Helianthus tuberosus* L.) has undergone selection in Canada as a new diversification crop (Hergert 1991), but it was found that it could not be grown profitably in Canada as an alcohol source (Baker et al. 1990).

Canadian scientists have developed hybrid poplars (*Populus* spp.) and willows (*Salix* spp.) as biomass crops for ethanol, pulp and the generation of electricity and heat (Mitchell et al. 1992; Abuja and Libby 1993). The trees thrive on marginal soils and cool climates, annually yielding up to 3.7 tonnes of biomass per hectare. It has been estimated that if willows were grown on 10% of Canada's marginal farmland, the fuel could replace ten nuclear generating plants and supply much of Canada's gasoline (Gogerty 1991). The use of trees for fuels, although less efficient than crops, may be a more reasonable alternative for Canada, since the trees can occupy the huge areas of marginal land unsuitable for crops. Switchgrass has acquired a considerable reputation as a potential biomass, ethanol and pulp plant in the US, and has been considered like willows and poplars for energy and fuel alternatives in Canada. All prospective energy crops in Canada must compete with relatively inexpensive supplies of natural gas, hydroelectric power, and wood, so that traditional fuel sources are unlikely to be replaced in the foreseeable future.

LANDSCAPE PLANTS AND CHRISTMAS TREES

The landscape industry in Canada is very large and diverse. However, national statistics are compiled only for sod, which in 1996 had a farm gate value of about 66 million dollars. AAFC at Morden, Manitoba, has been especially active in Canada in developing the landscape industry, generating new and improved winterhardy roses (e.g. 'Morden Blush' and 'Morden Fireglow'), herbaceous perennials, trees, and shrubs. Recent contributions from Morden include two new lilies for the Prairies, a fast-growing disease-resistant male poplar suitable for shelters, a new winter-hardy hedge rose ('Prairie Joy'), and a new fall-bearing raspberry ('Red River'). 'Nicolas' and 'Lambert Closse', new roses in the Explorer series, were recently released by AAFC at Saint-Jean-sur-Richelieu, Quebec.

Christmas trees have become a growth industry in Canada, and according to the 1996 Census there were 51,071 ha cultivated. There are more than 2,000 Christmas tree farms, about a third in Quebec. In 1995, the last year for which data are available, 3.2 million Christmas trees were harvested in Canada, and 2.1 million of these were exported to the US. Traditionally popular species include Scotch pine (*Pinus sylvestris* L.), white pine (*Pinus strobus* L.), and white spruce [*Picea glauca* (Moench) Voss], but there is regional specialization. In British Columbia, Douglas fir [*Pseudotsuga menziesii* (Mirb.) Franco] and Scotch pine are the leaders in Christmas tree sales. By contrast, Fraser and balsam firs [respectively *Abies fraseri* (Pursh) Poir. and *A. balsamea* (L.) Mill.] are the trees of choice in the maritime provinces, where they are easy to grow. Because preferences for the different species change (firs are becoming more popular) and it requires about a decade to grow a Christmas tree (in Canada, a fir requires 9 to 15 years to reach marketable height, while a Scots pine takes 8 to 10 years), consumer trends need to be predicted a decade in advance.

AQUACULTURE

Aquaculture* is one of the fastest growing food production industries in the world. Most aquacultural production occurs in Asia, but the industry is expanding everywhere, including North America. Below, the two principal categories are reviewed.

The Fishery

Animals that can be grown for food in water include finfish, molluscs (notably clams and oysters) and crustaceans. Because these are cold-blooded and do not use energy to produce heat, a very efficient conversion of feed to flesh can be realized (as low as 1.3 has been achieved for fish in the US). Canada has recently experienced catastrophic reductions in the East Coast fishery from overfishing, and the West Coast salmon fishery is experiencing widespread threats to its supply. Increased harvest of the oceans by fishing fleets from around the world has depleted ocean going fish stocks. Pollution and climate change have also been accused of playing a role in reducing the ocean's resources. The protein requirement of about 5% of the world's population is satisfied by fish and other seafood products, and this tradition of eating fish provides a stable market, especially in Japan. In developed western countries such as the US, Canada, and the European Community, it has become common knowledge that fish provide extremely beneficial dietary components, and have been linked in various studies to the reduction of disease. The market for fish is becoming more expensive to satisfy, and the culture of seafoods to meet this need is becoming more widespread. Culture of fish stocks also provides a way of restocking the sport fish industry, and even of providing captive fish for sport. High-gluten wheats such as are commonly produced in Western Canada are particularly well suited to capture a portion of the shrimp feed market, both as feed ingredients and binding agents, and soybean and canola meal also have potential for use in shrimp rations (McKeague 1993). Trout and salmon also provide a possible new outlet for products from grains and oilseeds, both for a new domestic aquaculture industry and for the export market. For a review of aquacultural potential of Canadian feeds, see Kurbis (1996a).

---

*Aquaculture is the controlled cultivation and harvest of aquatic animals and plants (mariculture is the raising of such crops in the sea). Aquiculture is often considered to be a variant spelling, but is also used in a narrower sense for hydroponics, the cultivation of plants in artificial aqueous nutrient media.

Seaweeds

The larger marine algae or seaweeds have a diversity of uses, like many terrestrial plants. Various seaweeds are used as vegetables and condiments, principally in China and Japan. Many brown algae have been extensively used as agricultural fertilizers, especially as a source of potash (which supplies potassium). As a soil amendment, seaweeds also tend to be rich in micronutrients and nitrogen, but are low in phosphate. They have the advantage of being free of terrestrial weeds and fungi. Algin from kelp and other algae is used in the manufacture of more than 300 commercial products, and is particularly valued for its ability to suspend agents in food, cosmetics, and a variety of commercial liquid mixtures. Like other seaweeds, kelp are also harvested as dried fodder for terrestrial livestock in coastal areas, and sometimes grown as forage for cultivated aquatic animals, such as abalone. A special roe on kelp, greatly valued in sushi bars of Japan, has been produced in British Columbia by herring spawning in penned kelp enclosures. The industry is largely managed by native people and the harvest is valued at over 20 million dollars annually. A recent development is the use of kelp as a substrate for biogas (methane) production, a technology that was developed during the OPEC crisis by General Electric of the United States. The value of this technology in reducing the high energy costs in the Canadian Arctic is currently being explored by Canadian companies.

Uncontrolled harvesting has led to reductions of wild supplies of some algae, but seaweed farms are well established in parts of the world, reducing the pressure on natural stands. "Polyculture" (combining fish farming and seaweed culture) is a clever way of using seaweeds to metabolize by-products of the fish culture. China is a major seaweed producer, growing over 2.5 million tonnes of the brown alga *Laminaria japonica* Aresch annually. Japan's primary aquaculture seaweed is nori (mostly the red alga *Porphyra yezoensis* Ueda), with an estimated value of $1.5 billion annually (by contrast, the edible seaweed market in the US is currently valued at only about $30 million annually). Canada's development of a cultivated seaweed industry lags far behind that of several countries with low labor costs and a warm climate that allows year-round cultivation. However, there has been some cultivation of *Laminaria saccharina* (L.) Lamouroux and *L. groenlandica* Rosenvinge (*L. bongardiana* Postels et Ruprecht) on the Pacific coast, for Oriental and health food markets.

Although Canada currently provides less than 2% of the world's seaweed resources, there is considerable potential for increasing commerce, particularly on the species-rich Pacific coast. In fact, with 20 kelp species, coastal British Columbia is one of the world's major regions of kelp diversity, and it has been estimated that 650,000 tonnes of wild kelp currently grow along the B.C. coastline. Current Canadian regulations permit 100,000 tonnes to be harvested annually under strict conservation guidelines, but less than 1% of this amount was harvested in 1996. Canada has the longest coastline of any nation, suggesting that greater use should be made of the cold-water marine plant resources. Canada's marine environment may remain much less polluted than elsewhere, providing an attractive situation for algae and algal products intended for human consumption. For additional information on the economics of seaweeds in Canada, see Small and Catling (in press).

ORGANIC AGRICULTURE

Organic farming employs agronomic practices minimizing or eliminating highly processed chemical inputs, stressing soil building programs, and long term environmental sustainability. Methods include the use of crop rotations, natural insect predators, and organic nutrient sources, as opposed to the use of pesticides, fungicides, insecticides, and chemical fertilizers. In Canada, there are at least 50 different organic certification agencies. Organic grains and oilseeds grown in Canada include wheat (including durum), oats, barley, rye, buckwheat, flax, canola, and sunflower. The majority of Canadian organic grains and oilseeds are exported (as are conventionally produced corresponding crops). The organic grains go primarily to Europe and the US, and the majority of organically-grown oilseeds to the US. Estimating the size of the organic grains and oilseed industry is difficult, but likely it is less than 1% of the conventional crops. Organic crops command higher prices, typically 30% for grains and oilseeds, but the extent to which production can be made competitive with conventional cropping is unclear (Christie 1995b). The organic industry is expected to increase the share of the food retail market in Canada as consumers become increasingly conscious of the issues.

Wise manure usage is related to the subject of organic agriculture. Canada has a huge livestock industry (over 100 million hens and chickens; 111 million hogs; almost 5 million beef cows and over 1 million dairy cows; more than one-quarter of the nation's farms are beef farms), and consequently an associated production of manure. Manure is a valuable resource due to its nutrient content and soil amending properties. On the other hand, manure is expensive to transport, produces an odor that the public finds objectionable, and over-application may pollute surface and ground water. New crops and crop systems that utilize all this manure are extremely desirable, and are a priority in Canada.

TRANSGENIC CROPS

New genetically engineered crops are becoming common in Canada, as in the US and elsewhere. Transgenic crops pose the following risks: (1) escape of the transgenic plants and proliferation in the wild with subsequent displacement of natural vegetation; (2) hybridization with and transgene infiltration into related weedy species, resulting in invigorated weeds; (3) hybridization with and transgene infiltration into native wild species, resulting in alteration of natural gene frequencies; and (4) damage to habitats and ecosystems, resulting from any or all of the above possibilities. Because of the cold winters, the relatively depauperate flora of Canada, and the relatively narrow spectrum of crops grown, transgenic crops pose less of a risk to Canada than they do to more southern countries. However, crops engineered for cold-tolerance pose a special risk for Canada, since temperature is the chief limiting factor for Canadian crops, and indeed for invasive weeds. Increased cold tolerance is a key factor in increasing the area of Canadian crops. Accordingly, genetically transformed plants that have been altered to increase cold tolerance must be given special scrutiny to ensure that escaping genes or plants will not influence native biodiversity. Cold stress tolerance could be expressed in several ways, for example: increased physiological tolerance, adaptation to early or late frosts, adaptation to a short growing season, conversion of a perennial incapable of winter survival to an annual that overwinters as seed, development of compact forms that self-insulate, and development of deep-growing over-wintering roots or rhizomes. The risks of genetically engineered crops in Canada are reviewed by Small (1997c) and Warwick and Small (In press).

SUMMARY ANALYSIS

The principal north-temperate grains (wheat, barley, oats, and maize) have long dominated the crops of Canada. New cultivars are regularly created to meet the challenges of the Canadian climate, diseases, pests, and the international marketplace. Some new grains and pseudograins have been added, most notably canary seed, but remain minor. By contrast, in recent decades the oilseed sector has witnessed huge increases in the cultivation of canola (rapeseed), and soybean, and the introduction of new types of flaxseed, as well as several other species grown in minor amounts. The success of these new crops has been due to intensive Canadian breeding programs. Canadian forages and fodders are dominated by the major legume and grass species imported from temperate Eurasia, although some new Eurasian species have recently been added to the mix. An extensive effort is currently underway to develop native Canadian grass species as new forages and landscape restoration plants. The areas devoted to dry peas and beans have been increasing rapidly in the Canadian prairies, a planned successful response to recent decreases in wheat area. Vegetable production has been increasing in Canada for many years. A considerable variety of vegetable and culinary herb crops is grown in Canada, making it difficult for any new crop to gain much of a market share. Nevertheless, in response to growing ethnic diversity and appreciation for the importance of vegetables to health, many new species and new cultivars are appearing in Canada. By contrast, fruit crops have generally been decreasing in Canada, although new cultivars have appeared. There are numerous possibilities for domesticating native Canadian fruits and vegetables as new food plants, but the marketplace appears to be hostile to these prospects. Greenhouse cultivation is increasing very rapidly in Canada, with emphasis on ornamentals and a few vegetables, most notably tomato. The landscape and mushroom industries are also thriving. The greenhouse, landscape, and mushroom sectors are offering a wide diversity of new crops, although individual species are not likely to capture a large proportion of the markets. New medicinal and nutraceutical crops are attracting considerable attention and research in Canada, and while the farm gate values are limited, value-added processing makes this economic sector very promising. Textile fiber crops are basically absent from Canada, and the prospects for finding new crops for this category appear poor. The legal cultivation of hemp for a variety of purposes, particularly as an oilseed and high-quality pulp fiber source, is undergoing explosive growth in Canada, but the eventual success of this new crop remains to be determined. Herbaceous crops for biomass, energy, fuels, ethanol, and biodiesel have attracted attention in Canada, as in other countries. Subsidization is necessary for these, and is provided in Canada only for a limited amount of fuel ethanol production. Cheap hydroelectric power and abundant natural gas, as well as a huge wild supply of trees, strongly restrict the prospect of these industries in Canada. Among the challenges for the future are the following: utilizing the huge promise of new genetically-engineered crops without harming the environment and the native biota; finding crops and systems to utilize the large amount of manure from the livestock industry; and wisely developing the potentially large coastal seaweed industry.

REFERENCES

Abuja, M.R. and W.J. Libby (eds.). 1993. Clonal forestry. 2 vols. Springer-Verlag, Berlin.

Acton, D.F. 1995. Development and effects of farming in Canada. p. 11–18. In: D.F. Acton and L.J. Gregorich (eds.), The health of our soils: toward sustainable agriculture in Canada. Agriculture and Agri-Food Canada, Research Branch, Ottawa.

Agriculture Canada. 1989. Growing together: a vision for Canada's agri-food industry. Agriculture Canada Publication 5269/E. Ottawa.

Aiken, S.G., P.F. Lee, D. Punter, and J.M. Stewart. 1988. Wild rice in Canada. NC Press, Toronto.

Anon. 1985. Sorghum gets a good look. The Furrow 90(4):4.

Anon. 1992. High value, low volume: a look at the ultraspecialty crops. The Furrow 97(4):16–17.

Anon. 1994. Sunola lights up diversification. Agvance (Agriculture and Agri-Food Canada) 3(2):8–9.

Anon. 1995. Imported matsutake cultivates Japanese taste for autumn. Mushroom World 1995(March):60–61.

Anon. 1998a. New canola cuts the mustard. Agvance (Agriculture and Agri-Food Canada) 7(1):5.

Anon. 1998b. Durum set to grow stronger. Agvance (Agriculture and Agri-Food Canada) 7(1):6.

Baker, L., P.J. Thomassin, and J.C. Henning. 1990. The economic competitiveness of Jerusalem artichoke (*Helianthus tuberosus*) as an agricultural feedstock for ethanol production for transportation fuels. Can. J. Agr. Econ. 38:981–990.

Bateman, J. 1997. The Canadian feed industry. Bi-weekly Bulletin (Agriculture and Agri-Food Canada) 10(21):1–4. (http://www.agr.ca/policy/winn/biweekly/English/biweekly/volume10/v10n21e.htm)

Beckman, C. 1997. Flaxseed: situation and outlook. Bi-weekly Bulletin (Agriculture and Agri-Food Canada) 10(22):1–4. (http://www.agr.ca/policy/winn/biweekly/English/biweekly/volume10/v10n22e.htm)

Beckman, C. 1998. Vegetable oil: situation and outlook. Bi-weekly Bulletin (Agriculture and Agri-Food Canada) 11(10):1–4. (http://www.agr.ca/policy/winn/biweekly/English/biweekly/volume11/v11n10e.htm)

Borie, K.B. 1998. Sweet stevia. National Gardening 1998(Sept.–Oct.):12,14,15.

Canadian Federation of Agriculture. 1995. Agriculture in Canada. The Canadian Federation of Agriculture, Ottawa.

Catling, P.M., A.R. McElroy, and K.W. Spicer. 1994. Potential forage value of some eastern Canadian sedges (Cyperaceae: *Carex*). J. Range Manage. 47:226–230.

Christie, R. 1995a. Sunflower: situation and outlook. Bi-weekly Bulletin (Agriculture and Agri-Food Canada) 8(13):1–4.

Christie, R. 1995b. Organic grains and oilseeds. Bi-weekly Bulletin (Agriculture and Agri-Food Canada) 8(17):1–4.

Chubey, B.B. (Chairman). 1983. Proceedings of the new crops workshop, Winnipeg, Manitoba. Agriculture Canada Research Branch.

Corley, R.H.V. 1989. Assessment of new crops for plantations. p. 53–65. In: G.E. Wickens, N. Haq, and P. Day (eds.), New crops for food and industry. Chapman and Hall, London, UK.

Crop Development Centre. 1991. Special cropportunities (Proceedings Agriculture Canada Workshop on Alternative Crops Aug. 6–7, 1991). Crop Development Centre, Univ. Saskatchewan, Sask.

Dietz, J., R. Henckes, and S. McGill. 1998. Tailoring wheat to the market. The Furrow 103(3):7–8.

Dore, W.G. and J. McNeill. 1980. Grasses of Ontario. Agriculture Canada Research Branch Monograph 26, Ottawa.

Elliott, C.R. and J.L. Bolton. 1979. Licensed varieties of cultivated grasses and legumes. Agr. Can. Publ. 1405, Ottawa.

Environment Bureau. 1997a. Biodiversity in Agriculture—Agriculture and Agri-Food Canada's action plan. Minister of Public Works and Government Services, Ottawa. (http://www.agr.ca/envire.html)

Environment Bureau. 1997b. Biodiversity initiatives—Canadian agricultural producers. Minister of Public Works and Government Services, Ottawa. (http://www.agr.ca/envire.html)

Ference, D. [D. Ference and Associates Ltd.]. 1989. Economic opportunities for Canada in essential oils and medicinal crops. Agriculture Canada Policy Branch Working Paper 10/89, Ottawa.

Gogerty, R. 1991. Farming for factories. The Furrow 96(6):10–13.

Gowan, T. 1996. Biodiesel as an alternative fuel. Bi-weekly Bulletin (Agriculture and Agri-Food Canada) 9(17):1–4. (http://www.agr.ca/policy/winn/biweekly/English/biweekly/volume9/v9n17e.htm)

Gray, K. 1997a. Canada: dry edible beans situation and outlook. Bi-weekly Bulletin (Agriculture and Agri-Food Canada) 10(20, Part 1):1–4. (http://www.agr.ca/policy/winn/biweekly/English/biweekly/volume10/v10n20eb.htm)

Gray, K. 1997b. Canada: canary seed market situation and outlook. Bi-weekly Bulletin (Agriculture and Agri-Food Canada) 10(20, Part 2):1–4. (http://www.agr.ca/policy/winn/biweekly/English/biweekly/volume10/v10n20ea.htm)

Gray, K. 1998. Special crops: situation and outlook. Bi-weekly Bulletin (Agriculture and Agri-Food Canada) 11(7):1–4. (http://www.agr.ca/policy/winn/biweekly/English/biweekly/volume11/v11n07e.htm.)

Hankes, R. 1992. Forages for all reasons: seeds of a new generation. The Furrow 97(4):10–12.

Henkes, R. 1995. The remaking of grasspea. The Furrow 100(3):25–26.

Henkes, R. 1998a. The rise of the omega-3 egg. The Furrow 103(5):20–21.

Henkes, R. 1998b. Commercializing seabuckthorn. The Furrow 103(5):24.

Henkes, R. and J. Dietz. 1997. Breakthrough for buckwheat. The Furrow 102(4):28.

Henkes, R. and L. Leake. 1994. Legendary lupin makes a comeback. The Furrow 99(3):28–29.

Hergert, G.B. 1991. The Jerusalem artichoke situation in Canada. Alternative Crops Notebook 5:16–19.

Jamieson, A.R. 1996. Germplasm resource utilization in small fruits: A Canadian perspective. Five pages (irregularly paginated). In: C.G. Davidson and J. Warner (eds.), Proc. 2nd workshop on clonal genetic resources: "Emerging Issues and New Directions" (Jan. 23–23, 1996, Ottawa). Agriculture and Agri-Food Canada, Ottawa.

Johnson, D.L. 1990. New grains and pseudograins. p. 122–127. In: J. Janick and J.E. Simon (eds.), Advances in new crops. Timber Press, Portland, OR.

Jolliff, G.D. and S.S. Snapp. 1988. New crop development: opportunity and challenges. J. Prod. Agr. 1:83–89.

Joyce, J. 1993. Exploring native grass seed production in Western Canada. Agriculture Canada Prairie Farm Rehabilitation Administration and Ducks Unlimited Canada. (Printed in Saskatoon, Sask.)

Kiehn, F.A. and M. Reimer. 1992. Alternative crops for the prairies. Agriculture Canada Publ. 1887/E, Ottawa.

Kurbis, G. 1996a. Global aquaculture: situation and outlook. Bi-weekly Bulletin (Agriculture and Agri-Food Canada) 9(16):1–4.

Kurbis, G. 1996b. Central and Eastern Canada: regional production and consumption of corn. Bi-weekly Bulletin (Agriculture and Agri-Food Canada) 9(21):1–4.

Lachance, A. 1996. Chicory offers diversification potential. Agvance (Agriculture and Agri-Food Canada) 5(4):8.

Lehmann, J.W. 1991. the potential of grain amaranths in the 1990s and beyond. Alternative Crops Notebook 4:1–3.

Loughton, A., M.J. Columbus, and R.C. Roy. 1991. The search for industrial uses of crops in the diversification of agriculture in Ontario. Alternative Crops Notebook 5:20–27.

Lennox, G. 1996. Ontario wheat: situation and outlook. Bi-weekly Bulletin (Agriculture and Agri-Food Canada) 8(11):1–4.

Lennox, G. 1998. Durum wheat outlook. Bi-weekly Bulletin (Agriculture and Agri-Food Canada) 11(3):1–4. (http://www.agr.ca/policy/winn/biweekly/English/biweekly/volume11/v11n03e.htm)

Mazza, G. and C.G. Davidson. 1993. Sasakatoon berry: A fruit crop for the prairies. p. 516–519. In: J. Janick and J.E. Simon (eds.), New crops. Wiley, New York.

Mazza, G., F.A. Kiehn, and H.H. Marshall. 1993. *Monarda*: A source of geraniol, linalool, thymol and carvacrol-rich essential oils. p. 628–631. In: J. Janick and J.E. Simon (eds.), New crops. Wiley, New York.

McKeague, D. 1993. Aquaculture: Shrimp feed—opportunity for grains and oilseeds. Bi-weekly Bulletin (Agriculture and Agri-Food Canada) 6(15):1–4.

McKeague, D. 1994. Ethanol. Bi-weekly Bulletin (Agriculture and Agri-Food Canada) 7(15):1–4.

McKinnon, D. 1996. Triticale: situation and outlook. Bi-weekly Bulletin (Agriculture and Agri-Food Canada) 9(15):1–4.

McKinnon, D. 1997a. Ontario: Wheat, corn and soybean outlook. Bi-weekly Bulletin (Agriculture and Agri-Food Canada) 10(11):1–4. (http://www.agr.ca/policy/winn/biweekly/English/biweekly/volume10/v10n11e.htm)

McKinnon, D. 1997b. Malting barley: Situation and outlook. Bi-weekly Bulletin (Agriculture and Agri-Food Canada) 10(18):1–4. (http://www.agr.ca/policy/winn/biweekly/English/biweekly/volume10/v10n18e.htm)

McKinnon, D. 1997c. Corn: Situation and outlook. Bi-weekly Bulletin (Agriculture and Agri-Food Canada) 10(23):1–4 (http://www.agr.ca/policy/winn/biweekly/English/biweekly/volume10/v10n23e.htm)

McKinnon, D. 1998. Oat: Situation and outlook for 1998–99. Bi-weekly Bulletin (Agriculture and Agri-Food Canada) 11(11):1–4. (http://www.agr.ca/policy/winn/biweekly/English/biweekly/volume11/v11n11e.htm.)

Mitchell, C.P., J.B. Ford-Robertson, T. Hinckley, and L. Sennerby-Forsse (eds.). 1992. Ecophysiology of short rotation forest crops. Elsevier, London.

Montford, S. and E. Small. 1999. Measuring harm and benefit: the biodiversity friendliness of *Cannabis sativa*. Global Biodiversity 8(4):2–13.

Munro, D.B. and E. Small. 1997. Vegetables of Canada. National Research Council Press, Ottawa.

Prescott-Allen, C. and R. Prescott-Allen. 1986. The first resource. Yale Univ. Press, New Haven, CT.

Quamme, H.A. 1996. Gene resources for future tree fruit breeding. Eight pages (irregularly paginated). In: C.G. Davidson and J. Warner (eds.), Proc. 2nd workshop on clonal genetic resources: "Emerging Issues and New Directions" (Jan. 23–23, 1996, Ottawa). Agriculture and Agri-Food Canada, Ottawa.

Read, C. 1994. What's new in the Canadian grape and wine industry? p. 212–214. In: Canadian agriculture at a glance. Statistics Canada, Agriculture Division, Ottawa.

Redhead, S.A. 1992. An overview of commercial harvesting in Canada. p. 15–19. In: N. De Geus, S.A. Redhead, and B. Callan (eds.), Wild mushroom harvesting session. Integrated Resources Branch, British Columbia Ministry Forests, Victoria.

Redhead, S.A. 1997. The pine mushroom industry in Canada and the United States: A global economy explains why it exists and where it is going. p. 15–54. In: M. Palm, I. Chapela, and H. Ignacio (eds.), Mycology in sustainable development: expanding concepts, vanishing borders: results of a workshop (1995 August 5; San Diego, CA), Parkway, Boone, NC.

Reichert, G. 1994. Hemp (*Cannabis sativa*). Bi-weekly Bulletin (Agriculture and Agri-Food Canada) 7(23):1–4.

Reichert, G. 1995. Straw-based particleboard. Bi-weekly Bulletin (Agriculture and Agri-Food Canada) 8(3):1–4.

Reichert, G. 1996. Pulp and paper production from straw. Bi-weekly Bulletin (Agriculture and Agri-Food Canada) 9(2):1–4.

Reid, I.R. 1995. Country report for Canada to the International Conference and Programme for Plant Genetic Resources. Agriculture and Agri-Food Canada, Research Branch, Ottawa.

Science Council. 1991. It's everybody's business: Submissions to the Science Council's Committee on Sustainable Agriculture. Science Council of Canada, Ottawa.

Skrypetz, S. 1998. Dry peas: situation and outlook. Bi-weekly Bulletin (Agriculture and Agri-Food Canada) 11(13):1–4. (http://www.agr.ca/policy/winn/biweekly/English/biweekly/volume11/v11n13e.htm)

Small, E. 1979. The species problem in *Cannabis*, science and semantics. 2 vols. Corpus, Toronto.

Small, E. 1995a. Crop diversification in Canada with particular reference to genetic resources. Can. J. Plant Sci. 75:33–43.

Small, E. 1995b. Hemp (*Cannabis sativa*). p. 28–32. In: J. Smartt and N.W. Simmonds (eds.), Evolution of crop plants, 2nd. ed. Longman, London, UK.

Small, E. 1996. George Lawson Medal Review: Adaptations to herbivory in alfalfa (*Medicago sativa* L.). Can. J. Bot. 74:807–822.

Small, E. 1997a. Culinary herbs. National Research Council Press, Ottawa.

Small, E. 1997b. Cannabaceae. p. 381–387. In: Flora North America Editorial Committee (ed.), Flora of North America North of Mexico, Vol. 3. Oxford Univ. Press, NY.

Small, E. 1997c. Biodiversity priorities from the perspective of Canadian agriculture: Ten commandments. Can. Field-Nat. 111:487–505.

Small, E. 1999. Why is crop diversification important for Canada? p. 9–18. In: S. Blade (ed.), Special crops conference proceedings, opportunities and profits II—into the 21st century (Edmonton, Nov. 1–3, 1998). Alberta Agriculture, Food and Rural Development, Edmonton, Alberta.

Small, E. and P.M. Catling. 1996. Canadian wild plant germplasm—importance and needs with special reference to clonal crops. Fifteen pages (irregularly paginated). In: C.G. Davidson and J. Warner (eds.), Proc. 2nd workshop on clonal genetic resources: "Emerging Issues and New Directions" (Jan. 23–23, 1996, Ottawa). Agriculture and Agri-Food Canada, Ottawa.

Small, E. and P.M. Catling. 1999. Canadian medicinal crops. National Research Council Press, Ottawa. In press.

Small, E. and A. Cronquist. 1976. A practical and natural taxonomy for *Cannabis*. Taxon 25:405–435.

Small, E., B. Brookes, L.P. Lefkovitch, and D.T. Fairey. 1997. A preliminary analysis of the floral preferences of the alfalfa leafcutting bee (*Megachile rotundata*). Can. Field-Nat. 111:445–453.

Spak, S. 1998a. Nutraceuticals. Bi-weekly Bulletin (Agriculture and Agri-Food Canada) 11(1):1–4. (http://www.agr.ca/policy/winn/biweekly/English/biweekly/volume11/v11n01e.htm)

Spak, S. 1998b. Canada: area seeded for 1998–99. Bi-weekly Bulletin (Agriculture and Agri-Food Canada) 11(5):1–4. (http://www.agr.ca/policy/winn/biweekly/English/biweekly/volume11/v11n05e.htm)

Statistics Canada. 1986. Human activity and the environment. Statistics Canada, catalogue 11509E, Ottawa.

Statistics Canada. 1994. Canadian agriculture at a glance. Statistics Canada, Agriculture Division, Ottawa.

Statistics Canada. 1996. Greenhouse, sod and nursery industries. Statistics Canada, Agriculture Division, Ottawa.

Statistics Canada. 1997a. Historical overview of Canadian agriculture. Statistics Canada, Agriculture Division, Ottawa.

Statistics Canada. 1997b. Agricultural profile of Canada. Statistics Canada, Agriculture Division, Ottawa.

Statistics Canada. 1998a. Agriculture economic statistics, June 1998 [for 1997]. Statistics Canada, Agriculture Division, Farm Income and Prices Section, Ottawa.

Statistics Canada. 1998b. Fruit and vegetable production, February 1998 [for 1997]. Statistics Canada, Agriculture Division, Horticulture Crops Unit, Ottawa.

Turner, N.J. and A.F. Szczawinski. 1979. Edible wild fruits and nuts of Canada. National Museum of Natural Sciences, National Museums of Canada, Ottawa.

Vincent, M. 1995. Dry edible beans: situation and outlook. Bi-weekly Bulletin (Agriculture and Agri-Food Canada) 8(16):1–4.

Vincent, M. and N. Longmuir. 1996. Buckwheat: situation and outlook. Bi-weekly Bulletin (Agriculture and Agri-Food Canada) 9(8):1–4.

Warwick, S.I. and E. Small. Invasive plant species: A case study of evolutionary risk in relation to transgenic crops. Proc. Seventh Int. I.O.P.B. Symposium "Plant Evolution in Man-Made Habitats," Aug. 10–15, 1998, Amsterdam. Univ. Leiden, The Netherlands. In press.

Witt, M.D. and L.A. Nelson. 1992. Milkweed as a new cultivated row crop. J. Prod. Agr. 5:167–171.

The Dynamic Contribution of New Crops to the Agricultural Economy: Is it Predictable?

Stephan R.P. Halloy*

The diversification of agriculture and the development of new crops are closely related. Although diversification is often justified by reference to practical experience, we know little about the theoretical basis for diversification or its causes and effects. Is it theoretically possible to feed the world on a fixed and unchangeable set of species and varieties? Empirical evidence suggests not, but are there fundamental system properties that make this impossible? This paper addresses the following questions:

1. What is the "life cycle" of an economic crop from development to decline?
2. Does "successful" have any objective meaning or threshold in regards to crops?
3. Should we group all crops together or are there distinct functional groups or guilds? Guilds is used as in the ecological literature to signify a group of species having similar ecological (in this case agricultural management) requirements and therefore having similar economic roles.
4. Are there repeatable patterns in the abundance distribution of crops allowing the prediction of future patterns?

The actual abundance distribution patterns of crops in New Zealand is tested against three models: the average, the lognormal, and a resource attraction model (RAM). The first two models are top-down (i.e. inferred from the actual distribution pattern). The average distribution over 150 years approximates an exponential function and represents the null model of a return to the mean. The lognormal is one of the distributions which most closely fits many diverse plant and animal communities (Preston 1948, 1962; Sugihara 1980; May 1981). The lognormal is a middle ground approximation between a variety of distributions ranging from the exponential and power functions to the broken stick model (Magurran 1988; Wilson et al. 1998). Here it represents the null model for competing species. The RAM is a bottom-up model (as is the broken-stick), which leads to distributions in the lognormal to power range based on simple rules of competition (i.e. the pattern emerges from rules affecting the bottom level of system integration such as the individual species, Halloy 1998).

METHODOLOGY

New Zealand was chosen as a valuable case history for crop dynamics as it has a relatively short, but well documented, history of crop introductions and is well known for its development of some high value new crops. In addition, it is physically bounded by the ocean, providing a naturally defined microcosm.

All statistics gathered by the Government of New Zealand from 1842 to 1990 on the area covered by cultivated plants were collected and analyzed. Data were pooled by decades. Minimum sample size considered in the statistics has varied from 0.1 to 4 ha (see Halloy 1994 for details). Species were ranked by abundance and frequency distributions were obtained and standardized by area and species. The change in distribution for 1981–90 was tested against predictions based on an average distribution over 150 years, a fitted lognormal and a resource attraction model. The resource attraction model utilized the following conditions:

Total resources, 1981:	1,445,858 ha
Total resources, 1990:	1,764,922 ha
Number of species:	136
Number of sites:	280 + 20 boundary sites
Arrangement of species:	Random among 280 sites
New resources:	Rain of resource particles of random size, varying 80% around the mean. Particles refer to a finite quantity or portion of resources, in this case for example a number of hectares.

*The present study was made possible through the support of the New Zealand Institute for Crop & Food Research and the New Zealand Foundation for Research, Science and Technology New Crop program.

Mean resource particle: Actual increase 1981–1990/10 years/136 spp.
Minimum species abundance: Maximum particle size (i.e. mean +80%)
Number of links: 20 (10 each side)
Number of steps: 20 (equivalent to 20 half years)
Loss: 0.9% of each species at each step
Distance exponent: 2, fixed. This is the exponent to which the virtual distances between
                                 particles is elevated, see Halloy (1998).

GROWING CULTIVATED AREA, DECREASING DIVERSITY

During the last 150 years the total cultivated area rises steadily, then slowly levels off (Fig. 1). Sown pastures represent developed areas for which species-specific statistics are unavailable. Species in the remaining areas are divided by the government census into three major functional groups or guilds: field crops, forestry, and orchards. Field crops include annual crops (cereals, pulses, tubers, vegetables), fodder, ornamental, and industrial crops.

The number of species (squares) recorded forms a sigmoid curve, which is still rising today although more slowly than in previous decades (Fig. 2). However, a Shannon-Weaver diversity index (triangles) rises steadily to 1951, then falls dramatically. This is largely due to the dominance of a single forestry species, radiata pine (*Pinus radiata*, Pinaceae) (Halloy 1994).

RESULTS

Dynamics

Crop abundance, as expressed by the proportional area covered, has varied dynamically over the 150 years of record (Fig. 3). Crops arise from nothing, stay around or become abundant for some time then, in turn, disappear, becoming "extinct" at the given scale. The dynamics are reminiscent of other dynamic systems such as the fossil record, where stochastic explanations (Raup 1981) and self-organization (Solé et al. 1997) have been invoked. Stochastic explanations argue that these patterns are the result of random variations in abundance and branching patterns between taxa. Self-organization postulates that such patterns arise from simple rules of interaction multiplied by innumerable individuals in a complex system. In our case, external determinants are also apparent at the level of functional groups: first the dominance of potato and wheat relating to the need to feed the new population; then a prolonged period of dominance of feed crops, to feed the working horse; finally the rise and dominance of longer lived timber crops.

Rank Shifting

The shading in Table 1 shows the percentage area covered by each crop every decade. Each of the 10 highest ranking species in 1990 covered more than 1% of the cultivated land area. Only four of these (oat, wheat, turnip, barley) were as

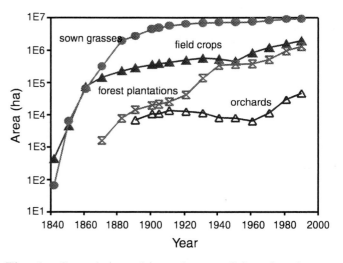

Fig. 1. Growth in cultivated area of functional crop groups.

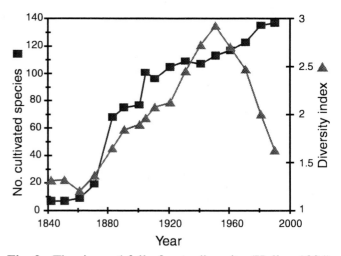

Fig. 2. The rise and fall of crop diversity (Halloy 1994).

54

important in 1891. Of the next 10, each covering areas of 0.1% or more in 1990, only five covered 0.1% or more in 1891. The average life cycle of a species is 10 years for abundances above 1% of all cultivated areas. The longest lasting species has been barley, persisting for 150 years between 1 and 10%. Only one species (radiata pine), for one decade, surpassed 60% abundance. Only five species have ever surpassed the 20% mark (radiata pine for 4 decades, wheat 6, oats 8, potato 1, maize 1).

Abundance Patterns

The abundance-rank distribution of New Zealand crops expressed on a log-log scale repeats similar patterns over time (Fig. 4). A choice of a different cut-off point for the minimum cultivated areas could fit the series to a single straight line, that is, there are long segments within each time cut which are straight. Straight lines on a log-log scale represent power distributions, a signature of self-organized criticality (Bak et al. 1988), and are found to be widespread in natural complex systems (Zipf 1949; Bak 1997). However, when considered in their entirety, these curved log-log representations of abundance distribution correspond to the lognormal distribution on a frequency-abundance representation (Fig. 6).

Each economic grouping of crops shows an abundance-rank distribution similar to the total but varying in slope (Fig. 5). The more diverse and stable groups (fruit and field crops) show the lesser slopes.

The frequency-abundance distribution of crops resembles a lognormal distribution for the 88 "macro-economic" species (Fig. 6). The 49 minor species on the left are separated by a wide gap. Although only 1990 is illustrated, this pattern was maintained with only slight alterations throughout the 150 years studied. Within the macro-economic species irregularities are partly due to historical contingencies and partly to the fact that the system is composite and made up of several functionally distinct groups, as seen above. If there is a regular trend toward a lognormal it should be possible to predict, within some confidence limits, the development of the abundance distribution over time.

The change in abundance distributions between 1981 and 1990 was tested against three predictions: 1) the average of the distributions over 150 years, 2) the lognormal fitted to 1981, and 3) a resource attraction model (RAM). The changes predicted by these three models are compared to the actual changes from 1981 to 1990 in Fig. 7. The resource attraction model approximates the actual change most closely (Chi square of 0.80 as opposed to 1.33 for the lognormal and 3.73 for the average model).

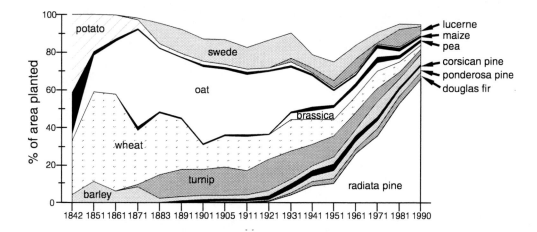

Fig. 3. Abundance dynamics of the 14 highest ranking crops in 1990. Radiata pine, *Pinus radiata* D. Don, Pinaceae; douglas fir, *Pseudotsuga taxifolia* Britt., Pinaceae; ponderosa pine, *Pinus ponderosa* Dougl., Pinaceae; corsican pine, *Pinus nigra* Arnold, Pinaceae; barley, *Hordeum vulgare* L., Gramineae; turnips, *Brassica rapa* L., Brassicaceae; wheat, *Triticum aestivum* L., Gramineae; brassica, *Brassica oleracea* L., Brassicaceae; peas, *Pisum sativum* L., Fabaceae; oats, *Avena sativa* L., Gramineae; maize, *Zea mays* L., Gramineae; potato, *Solanum tuberosum* L., Solanaceae; lucerne, *Medicago sativa* L., Fabaceae; swedes, *Brassica napus* L., Brassicaceae.

Table 1. New Zealand cultivated plants ranked by abundance. Shading indicates percent of cultivated area covered by each crop. Numbers indicate the rank order of the cultivated area of each crop for the given year (1= most abundant).

Shading legend: <0.8% | 0.8–3.2% | 3.2–12.8% | 12.8–51.2% | >51.2%

Rank order by year

Crop	1842	1851	1861	1871	1883	1891	1901	1905	1911	1921	1931	1941	1951	1961	1971	1981	1990
Radiata Pine =Insignis pine											5	5	2	1	1	1	1
Barley, total	4	4		3	6	6	7	7	7	6	8	10	10	10	3	4	2
Lucerne, total						73	73	73	73	73	11	12	11	5	17	2	3
Douglas Fir								16	20	20	14	13	13	13	9	5	4
Ponderosa pine =Western Yellow Pine						13	13	14	14	12	10	9	9	9	8	6	5
Turnip, all types	2	1		6	3	3	3	3	3	2	3	3	5	2	4	7	6
Wheat, all			1	2	2	2	2	2	2	4	2	2		3	2	3	7
Brassica oleracea, total all var. (broccoli, cabbage, cauliflower, etc.)					41	41	41	41	41	53	6	7	7	6	7	10	8
Pea, total all var.				5	9	11	15	12	12	15	21	14	16	15	10	11	9
Oat, all	5	3	2	1	1	1	1	1	1	1	1	1	6	8	6	9	10
Maize, total of maize and sweet corn	3	5	5	8	11	12	10	11	10	14	17	17	22	19	14	12	11
Swede, total				7	4	4	4	4	4	3	4	6	3	7	5	8	12
Rape, total fodder and oil seed				16	5	8	5	5	5	5	13	4	4	4	11	14	13
Kiwifruit													64	46	40	16	14
Corsican pine						10	11	10	11	7	7	11	12	12	12	13	15
Apple		7	7	18	19	22	23	23	26	25	23	25	23	19	19	17	16
Potato	1	2	3	4	7	8	8	8	8	9	12	16	17	22	15	15	17
Grape, total indoor and outdoor					60	20	21	24	25	33	38	38	35	40	26	19	18
Pumpkin (including Squash)							21	22	19	26	31	31	32	30	32	24	19
Tama for fodder																18	20
Asparagus					29	34	34	34	52	55	60	60	73	73	35	23	21
Onion, total (for seed and as vegetables)				13	15	20	22	24	28	27	28	32	36	33	30	21	22
Lentil										44	46	28	27	26		134	23
Lupin, total															31	33	24
Avocado															67	49	25
Carrot, total				12	13	18	17	19	23	29	29	35	44	35	33	32	26
Tomato, total outdoors and indoors					54	54	54	54	54	63	33	34	34	32	36	31	27
Peach					33	33	33	33	33	36	35	29	31	29	29	27	28
Rye, total						17	19	20	19	26	44	30	30	28	24	20	29
Chicory										48	51	53	60	39	37	29	30
Beans, all field (*Phaseolus vulgaris*) = french or kidney beans				11	12	16	16	17	22	21	39	39	50	37	34	25	31
Nectarine										46	53	51	49	52	55	43	32
Pear					55	55	55	55	55	41	40	33	33	36	42	44	33
Apricot					36	36	36	36	36	38	48	41	39	38	43	39	34
Lettuce					57	57	57	57	57	70	70	70	83	83	39	30	35
Orange					68	68	68	68	68	45	49	48	47	49	41	35	36
Kumara = Sweet Potato	6	6	9	19	21	28	29	38	38	35	45	49	43	42	44	37	37
Asian Pear = Nashi																38	38
Blackcurrant							95	95	93	103	103	103	53	58	63	26	39
Mandarin													63	57	48	51	40
Persimmon																77	41

No.	Item	Values (as printed)
42	Tangelo	40 51 67 20
43	Linen flax and Linseed totals	22 54 44 18 57 16 16 58 58 58 58 14
44	Tamarillo	42 49 53 56 40 37
45	Millet, total	45 28 34 40
46	Blueberry	54 76
47	Tobacco	28 25 25 28 27 25 17 6
48	Plum	48 46 41 37 43 39 65
49	Strawberry	53 52 50 75 75 75 75 74
50	Boysenberry	34 59 63
51	Brown-top, for seed	41 27 21 22
52	Raspberry	46 45 45 38 68 68 59 59 59 59
53	Cherry	62 62 55 52 52 50 70 70 70 70
54	Lemon	63 56 48 42 47 49 39 39 39 39
55	Feijoa	52 64 61
56	Watermelon	50 101 101 101 101 101 101 37 37 37 37
57	Marrow and Courgette total	68 61 54 48 41 45 45 45 45 37
58	Grapefruit, total	47 47 47 45 66 66 66 35 18
59	Cucumber (outdoor)	74 72 109 109 109 62 62 62
60	Parsnip, total	56 53 51 59 37 47 31 37 31 25 24 21 16 15
61	Hops	57 50 43 41 30 36 30 28 21 20 19 14
62	Celery	65 60 110 110 101 37
63	Garlic	59
64	Leek	66 120 97 97 97 97 61 31 31 31 29 26
65	Capsicum and Pepper (indoor and outdoor total)	58 57
66	Walnut	72 70 56 46 40 32 32 44 44 44
67	Chestnut	85 91 91 91 91 91 91 91 91
68	Bean, broad	110 110 68 68 62 62 34 62 44 44 44 44 31
69	Rock Melon = Cantaloupe	71 105 105 105 105 105 105 62 62 62 62 31
70	Passionfruit = Passionfruit vine	70 66 62 57 36 50 36 17 13 13 12 10
71	Beet, total (Beet, sugar beet, mangel, mangold, silver beet)	60 58 31 29 23 19 19 17 13 30 28 24 31 15 10
72	Blackberry (includes "other brambles" 1986–1990)	64 79
73	Cymbidium Orchid	79
74	Rhubarb	76 98 98 98 98 98 98 98 98
75	Gherkin	67
76	Sorghum and Broom corn total	36 102 102 102 102 102 102 51 51 51 26 24 34
77	Cherimoya	82
78	Carnation	
79	Rose (indoor in 1990)	
80	Loganberry	81 71 65 66
81	Gypsophila (indoor)	
82	Gooseberry	73 68 60 54 99 99 99 66 66
83	Chrysanthemum	55 65
84	Red Currant	84 77 64 61 77 77 53 53 53 53 34
85	Mushroom	78 74
86	Babaco	92
87	Almond	86 75 66 62 55 47 55 42 42 42 42 42
88	Gerbera (indoor)	75 66 56 47 42 42 42 75 62 66

SUMMARY AND CONCLUSION

The number of crop species in New Zealand has increased over time and seems to be reaching an asymptote while crop diversity (as measured by diversity indices) has increased then decreased in the last four decades. Crop abundances show dynamic trends similar to other complex systems including the fossil record, with emergence of new crops, a period of economic success, and eventual decline. Effective life spans of crop cultivars in many species are around 5 years (4–10 for wheat, Brennan and Byerlee 1991), equivalent to a depreciation of the present value of crop cultivars by 7% per year (Swanson 1996). The New Zealand data suggests that a similar dynamic is operating at the species level, with species having effective life spans (as important species covering more than 1% of cultivated area) of around 10 years. Out of the 20 major economic species present in New Zealand today, which account for more than 98% of the planted area, 11 have had to be developed from "new" crops within the last 100 years. This includes the largest export earners renowned as examples of new crop development such as radiata pine and kiwifruit, which were practically unknown in 1891. Conversely, of the 20 most important crops in 1891, eight are now minor crops. The turnover of the most abundant species is 3%–33% per decade (Halloy 1994). In New Zealand up to 20 new crops need to be developed to become macro-economic species every decade while one at least is likely to become one of the 20 most important species. Both persistence and dynamics are important economically. The aggregation over time of species which have produced relatively short economic booms (e.g. linen flax, kiwifruit) is as important to a vigorous economy as those rarer species which persist for long periods at lower levels (e.g. barley).

Functional economic groups follow similar patterns of abundance distribution as all the species at a range of scales. In every period the abundance distribution shows a grouping of major macro-economic spe-

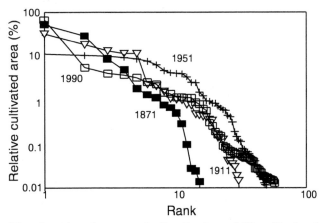

Fig. 4. Abundance-rank distribution of New Zealand crops expressed on a log-log scale for 1871 to 1990.

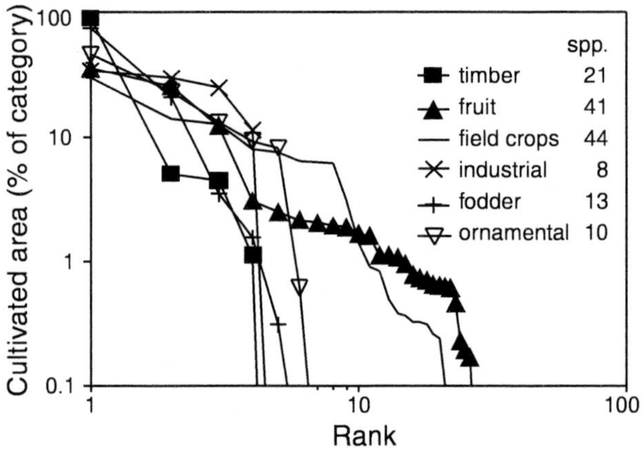

Fig. 5. Abundance-rank by functional crop groups.

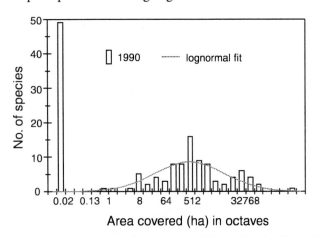

Fig. 6. Frequency-abundance of New Zealand cultivated plants.

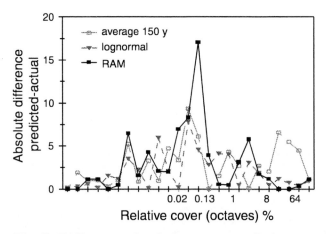

Fig. 7. Difference of real changes to predictions.

cies that is very distinct from that of minor species (Fig. 6). This break or boundary suggests an objective threshold which may be used to define "successful" species. The abundance distribution of crops approximates the well known power and lognormal functions. This regularity provides an objective yardstick against which to interpret and analyze the diversity and dynamics of cropping systems. The comparison of actual abundance patterns with fitted lognormal or power distributions provides interesting diagnostic tools which may have potential for measuring properties such as stability, persistence, sustainability, resilience, or susceptibility to disturbance (Frontier 1985; Halloy 1997; Kevan et al. 1997). Projections based on the lognormal distribution and a resource attraction model show that patterns of crop relative abundances may be predictable. The resource attraction model provided a closer fit than a top-down fitted lognormal model. This may be due to the fact that the RAM model is sensitive to actual resource increase, producing a pattern which emerges from simple rules.

A dynamic flow of crops has been maintained in the past by accessing new germplasm from abroad. All of our economically important species have been obtained from unsung global reservoirs of biodiversity, and return to them when they fall out of favor. Present large scale extinctions imply that we should not be complacent about the permanence of this essential resource.

REFERENCES

Bak, P. 1997. How nature works. The science of self-organized criticality. Oxford Univ. Press, Oxford.

Bak, P., C. Tang, and K. Wiesenfeld. 1988. Self-organized criticality. Physical Rev. A 38(1):364–374.

Brennan, J.P. and D. Byerlee. 1991. The rate of crop varietal replacement on farms, measures and empirical results for wheat. Plant Varieties & Seeds 4:99–106.

Frontier, S. 1985. Diversity and structure in aquatic ecosystems. Oceanography and Marine Biology Annu. Rev. 23:253–312.

Halloy, S. 1994. Long term trends in the relative abundance of New Zealand agricultural plants. Otago Conference Series 2:125–141.

Halloy, S. 1997. Ecosystem disturbance indicators: keystone species and structure of biodiversity. p. 73–81. In: M. Liberman and C. Baied (eds.), Sustainable development of mountain ecosystems: Management of fragile areas in the Andes (in Spanish). UNU and PL-480, La Paz, Bolivia.

Halloy, S.R.P. 1998. A theoretical framework for abundance distributions in complex systems. Complexity International 6. http://life.csu.edu.au/complex/ci/vol6/halloy/halloy.html

Kevan, P.G., C.F. Greco, and S. Belaousoff. 1997. Log-normality of biodiversity and abundance in diagnosis and measuring of ecosystemic health: pesticide stress on pollinators on blueberry heaths. J. Appl. Ecol. 34:1122–1136.

Magurran, A.E. 1988. Ecological diversity and its measurement. Princeton Univ. Press, Princeton, NJ.

May, R.M. 1981. Patterns in multi-species communities. p. 197–227. In: R.M. May (ed.), Theoretical ecology, principles and applications. 2nd ed. Blackwell Scientific Publications, London.

Preston, F.W. 1948. The commonness, and rarity, of species. Ecology 29:254–283.

Preston, F.W. 1962. The canonical distribution of commonness and rarity. Ecology 43:185–215, 410–432.

Raup, D.M. 1981. Probabilistic models in evolutionary paleobiology. p. 51–58. In: B.J. Skinner (ed.), Paleontology and paleoenvironments. William Kaufmann Inc., Los Altos, CA.

Solé, R.V., S.C. Manrubia, M. Benton, and P. Bak. 1997. Self-similarity of extinction statistics in the fossil record. Nature 388:764–767.

Sugihara, G. 1980. Minimal community structure: An explanation of species abundance patterns. Am. Naturalist 116: 770–787.

Swanson, T. 1996. Global values of biological diversity: the public interest in the conservation of plant genetic resources for agriculture. Plant Genetic Resources Newslett. 105:1–7.

Wilson, B.J., H. Gitay, J.B. Steel, and W.M. King. 1998. Relative abundance distributions in plant communities: effects of species richness and of spatial scale. J. Vegetation Sci. 9:213–220.

Zipf, G.K. 1949. Human behavior and the principle of least effort: An introduction to human ecology. Addison-Wesley, Cambridge, MA.

New Forage, Grain, and Energy Crops for Humid Lower South, US

Gordon M. Prine and Edwin C. French

The Humid Lower South (HLS) of the United States (Fig. 1) has a predominantly subtropical climate with high rainfall, a long warm growing season and mild winters. Temperate perennial crops tend to have difficulties surviving the summer and many tropical perennial crops tend to have difficulties surviving the winters.

RHIZOMA PERENNIAL PEANUT, NEW FORAGE

Until the introduction of rhizoma perennial peanut (RPP) (*Arachis glabrata* Benth., Fabaceae) to Florida from Brazil in 1936, there was no persistent forage legume available. Two cultivars of RPP have been released as forage: 'Florigraze' (Prine et al. 1981) and 'Arbrook' (Prine et al. 1986). The RPP must be propagated by rhizomes as they make few seed and these seed do not usually breed true. Bermudagrass sprig harvesters and planters have been adapted to dig and plant RPP. The two forage cultivars of RPP, are now planted on about 8,000 ha in HLS. Most of the hectarage (7,500) is in 'Florigraze' which can grow under colder conditions and wetter soils. Of the two cultivars, 'Arbrook' is the most drought resistence and best adapted to deep droughty sands. RPP is adapted to the climate of the HLS which extends to the northern limit indicated by the dashed line shown on map in Fig. 1. Two other perennial peanuts, *A. pintoi* and *A. kretschmerii*, which can be propagated by seeds, are restricted to warmer portions of peninsular Florida due to their intolerance to winter freezes.

For a more complete review of RRP refer to the proceedings of an international workshop entitled "Biology and Agronomy of Forage *Arachis* held at International Center for Tropical Agriculture at Cali, Colombia in 1993 (Kerridge and Hardy 1996). French et al. (1996) covered RPP as a forage crop in United States at the workshop in much greater detail than will be possible in this report.

RRP has been fed to dairy and beef cattle, horses, dairy and meat goats, sheep, swine, rabbits, and ostrich as hay, silage and pasture. The hay is usually more palatable than alfalfa and good quality hay compares favorably with alfalfa. Rhizoma peanut leaf meal also compared favorably with yellow maize and alfalfa meal as xanthophyll pigment source for egg yolk coloring in laying hens (Janky et al. 1986). Wild hogs, deer, turkey, rabbits and other wildlife readily graze RPP and are attracted to RPP plantings.

The dry matter hay yields of RRP have ranged from 7.0 to 15.7 Mg ha^{-1} yr^{-1} with 9 to 11.2 Mg ha^{-1} yr^{-1} yields most common. The protein content of the forage has reached 22% with a range of 15 to 17% being most common. Three cuttings of hay are possible in good seasons when cut at 7 or 8 weeks intervals. Drought often reduce the number of hay cuttings to two a year.

The rhizoma peanuts have a mass of rhizomes just below the soil surface like bermudagrass and an occasional tap root which can go to great depths in many soils. Fine roots are attached to the rhizomes and tap roots. The RPP have exceptional drought tolerance. They make little growth during severe drought conditions, but often remain green beyond the point that grasses have browned off and resume growth quickly after rainfall begins. RPP is useful for creep-grazing beef calves whose mothers are grazing low quality bahiagrass pastures (Saldivar et al. 1981).

The use of 'Florigraze' and 'Arbrook' as forage crops is well established now. Soon to be released RPP plant introductions, PI 262839, 'Arblick' and PI 262840, 'Ecoturf' are shorter in height and have better flower production than currently used forage types. The flushes of yellow flowers from the RPPs planted as turf or ground cover on lawns, highway medians, and shoulders, parks, playgrounds, off

Fig 1. The Humid Lower South, US.

play areas of golf courses, and any low traffic area brighten up the landscape. Better drought tolerance and freedom from need of nitrogen fertilization and are bonuses for using RPP instead of grasses. The dense rhizoma mat of RPP sod helps control soil loss from water and wind erosion.

Persistence is the attribute which sets RPP apart from other forage legumes in HLS. The first planting of RPP made by first author in 1961 still survives today. Many commercial RPP plantings are now over 18 years old and still performing well. This persistence is due to plants tolerance to disease and insects and high resistance/immunity to root knot nematodes. The RPP appears to be conservative in partitioning nutrients to top growth and maintains food reserves in the rhizome system. The root system is a good procurer of nutrients and water from the soil.

Low temperature which freeze the rhizomes in the upper soil zone and flooding for long periods limit the location where RPP can be planted in HLS. RPP is not adapted to poorly drained soils or soils which flood. Soil pH for RPP should be in range of 4.8 to 7. At pH near 7 and higher, RPP may demonstrate iron deficiency chlorosis. Optimal forage production for Florida released cultivars is at pH 6.0.

PIGEONPEA NEW GRAIN LEGUME

Pigeonpea [*Cajanus cajan* (L.) Millsp., Fabaceae] is an ancient crop in the world but a recent introduction to the Lower South US. Pigeonpea in an important grain legume crop in many parts of the world including India, Asia, Africa, Latin America, and the Caribbean. In these areas, the grain is used as a food component at the mature green and dried stages. Migrants from these regions who reside in the United States still favor this legume in their diet. Except for Hawaii and Puerto Rico, the pigeonpea is not commercially grown in the US. Although a tropical crop, growth of adapted pigeonpea is possible in the Lower South. The pigeonpea is a weak perennial but performs as an annual where it is killed by freezing temperatures. Grain production has to occur between early spring to early fall and requires a long growing season. A pigeonpea cultivar named 'Norman' was released by the North Carolina Experiment Station as a green manure crop, but has disappeared from seed trade. This paper has purpose of introducing a couple of pigeonpea populations adapted for grain production in Florida and Lower South, US which are being increased for release as cultivars by Florida Agricultural Experiment Stations.

The experimental pigeonpeas, 76 WW and 99 WW, are the result of a pigeonpea developmental program started in 1976 to develop an adapted pigeonpea cultivar for grain production. Some 96 lines were obtained from two pigeonpea crosses received from ICRISAT. In 1977, 3 additional pigeonpea populations were obtained from the University of West Indies, Trinidad. Other pigeonpea accessions were obtained from USDA Plant Introductions. Individual plants and lines were selected for earliness, large seed size, and high seed yields, and in later years for light-colored seed coat. Besides selective breeding, various management and seed production trials with various selected pigeonpea lines and populations have been conducted over the years at Gainesville. In recent years, three selected populations, FL 76WW, FL 99WW, and FL DO, were evaluated at Florida A&M University at Tallahassee.

Grain yields of pigeonpea lines at Gainesville have varied from 340 to 3360 kg/ha with most grain yields falling between 1200 and 2000 kg/ha. Pigeonpea yields are not consistent from year to year or location to location. Fall freezes have killed pigeonpea or reduced seed yield and are the reason for our slow development of the crop. In 1996, grain yields at Tallahassee when harvested at the mature green and dried stages were 3220 and 1180 kg/ha for FL 76WW, 2640 and 1470 kg/ha for FL 99WW and 2270 and 1080 kg/ha for FL DO, respectively. Pigeonpeas are a promising home garden and U-pick vegetable harvested at the mature green pod stage. They are shelled by hand or with a commercial pea sheller.

Pigeonpeas are day length sensitive requiring day lengths under 12.5 h for best flowering and seed production. Date of planting has great effect on plant size, earlier planted plants being largest. Planting date may affect grain yields but varies from season to season. June is the best time to plant for combine harvesting for grain but earlier plantings in late April or May are best for home garden and U-pick operations as a vegetable.

Pigeonpeas are attacked by nematodes, insects, and diseases. Rotations and insecticides for controlling insects like corn earworm and stink bug on flowers and seed pods are usually all that is needed for successful production. Pigeonpeas are very drought resistant and produce seed where other grain legumes will fail. Pigeonpeas are resistant to most root-knot nematodes, but are susceptible to the peanut root-knot nematode

Meloidogyne arenaria race 1. It is recommended that pigeonpeas not be planted in rotations with other grain legume crops. Pigeonpea will grow on well-drained soils with a wide range of textures and pH's.

BIOENERGY CROPS

The Humid Lower South has the most suitable climate (warm temperature, high rainfall and longest warm growing season) for biomass crops in the continental US. High biomass yields are attained from full-season growth of adapted cultivars of tallgrasses; elephantgrass (*Pennisetum purpureum* K.Schumach., Gramineae), erianthus (*Erianthus arundenaceum*, Michx. Gramineae), energy and sugar cane (*Saccharum* spp., Gramineae); the tree legume, leucaena (*Leucaena* spp., Mimosacea); and short rotation woody crops, cottonwood (*Populus deltoides* Bartr. ex Marsh., Salicaceae), slash pine (*Pinus elliottii* Engelm., Abietaceae) and *Eucalyptus* sp., Myrtaceae (Prine and Rockwood 1998).

Energy crops must be efficient converters of solar energy to biomass for long periods of time each growing season. Perennial energy crops must also be very persistent and long lived and give dependable production for five or more seasons from the same planting. The tall grasses such as sugarcane, energycane, elephantgrass, erianthus and intermediate grasses such as switchgrass (*Panicum virgatum*) have the C4 metabolic system and stature (strong enough to hold up an entire seasons growth without lodging) to make high biomass yields over a long growing season. The tall grasses have high biomass yields because of linear crop growth rates of 18 to 27 g m^{-2} d^{-1} for long periods, 140 to 196 days and sometimes longer (Woodard and Prine 1993). Oven dry biomass yields of tall grasses have varied from about 20 to 45 mg ha^{-1} yr^{-1} in colder subtropical and warmer temperate locations to over 60 Mg/ha in the lower portion of the Florida peninsular. The results of a 3 year study for several tall grasses in HLS is shown in Table 1. Many tallgrass genotypes are long-lived and resistant to winter kill, the killer of tropical grasses in the colder subtropics.

Most of the perennial tall grasses are propagated vegetatively, which limits their use as energy crops, but once established adapted crops may persist many years. Switchgrass will fit into many bioenergy situations because it can be seeded, has good resistance to winter kill and is persistent and sustainable, and has yielded up to 22 Mg ha^{-1} yr^{-1} (Bransby and Sladden 1995). Overlooked among bioenergy crops is castor (*Ricinus communis* L. Euphorbiaceae). A tall ecotype produced stems seven meters tall that had dry weight of 40 Mg/ha at Gainesville, FL in 1997.

Sewage Effluent Experiments

Lower South cities have sewage sludge and effluent disposal problems plus high energy requirements. Applying this waste on energy crop plantations and using the crop can satisfy some of the cities energy needs. The energy content of 1 Mg of oven dry tall grass and leucaena is equivalent to that of 423 and 466 L (112 and 123 gallons) of number 2 diesel fuel, respectively. Table 2 shows the annual dry matter yields of selected tallgrasses on Tallahassee and Leesburg sewage spray fields. More complete information on these sewage effluent experiments is given by Prine et al. (1997) and Prine and

Table 1. Average annual biomass yield of elephantgrass (eg), energycane (ec), sugarcane (sc), at five locations in southeastern United States over three growing seasons (Prine et al. 1991, 1997).

	Oven dry biomass (Mg ha^{-1} yr^{-1})				
	Florida				Alabama
Crops	Ona	Gainesville	Quincy	Jay	Auburn
N-51 eg	46.7	39.7	33.8	32.1	24.0
PI 300086 eg	41.6	28.6	24.1	24.0	18.6
L79-1002 ec	23.3	32.2	30.1	33.9	24.2
CP72-1210 sc	19.4	10.4	19.2	8.2	6.0

Table 2. Dry matter yields of selected tall grasses grown on the sewage effluent spray field of cities of Tallahassee and Leesburg, FL.

	Average annual dry matter biomass yield (Mg/ha)	
Tallgrass energy crop	Tallahassee, FL (3 seasons)	Leesburg, FL (2 seasons)
'Merkeron' elephant grass	22.3	32.8
N-51 elephant grass	21.0	37.0
PI 300086 elephant grass	21.9	52.0
US 72-1153 energycane	19.3	--
L79-1002 energycane	18.7	21.7
Alamo switchgrass	9.1	--

Table 3. Performance of three cottonwood clones in Florida. Source: Rockwood et al. 1996.

Location	Measurement	Clone		
		Ken 8	S7C1	S13C20
Tallahassee (irrigated with sewage effluent)	7 mo. ht. (m)	1.9	1.7	1.4
	43 mo. ht. (m)	8.3	9.1	8.9
	43 mo. survival (%)	48.0	43.0	34.7
	43 mo. volume (m³/ha)	72.3	71.1	69.4
Quincy	4 mo. ht. (m)	1.4	1.0	1.1
	8 mo. ht. (m)	2.6	2.3	2.2
	3 mo. coppice ht. (m)	1.2	1.0	1.0
Winter Garden	3 mo. ht. (m) with effluent	1.0	0.7	0.5
	3 mo. ht. (m) with effluent, compost and mulch	1.4	1.1	1.2

McConnell (1996). The productivity of cottonwood receiving sewage effluent of Tallahassee was promising (Table 3).

Phosphate Mining Land

The use of phosphatic mining land in central Florida is also favorable for biomass energy crops. The approximately 100,000 ha phosphate mining land base includes about 30,000 ha of phosphatic clays. The phosphate lands have considerable potential to grow sugarcane, energycane, elephantgrass, leucaena, *Eucalyptus*, and slash pine (Table 4). Crop yields for the tropical grasses, generally more than 40 Mg/ha dry matter dry weight (Stricker et al. 1993 and 1996), varied slightly with land type. Dry matter yields for the woody crops exceeded 20 Mg ha⁻¹ yr⁻¹. Leucaena (Cunilio and Prine 1995; Prine et al. 1997) is projected to be the most productive and after an 18 month establishment period, would be harvested annually for at least 10 years. Leucaena can be harvested at longer cycles where top growth is not killed by freezes. Of the several *Eucalyptus* species that may be grown in central Florida (Rockwood 1997), *Eucalyptus grandis* Hill ex Maiden is now showing the greatest potential for the mined lands, with particular clones being most productive. It would be harvested initially after three years and would continue to be harvested in five more three-year cycles. Stricker et al. (1995) discussed the economic development through integration of various biomass systems in phosphatic mining area of Central Florida.

Short Rotation Woody Crops

SRWC hectarege in Southeastern United States, some 12,000 in 1995 was projected to increase to 27,000 in the year 2000 (Land et al. 1996). A considerable portion of the SRWC area in Humid Lower South will be planted to cottonwood. With the assistance of many public and private cooperators in 1997 and 1998, seed were collected from some 200 cottonwood trees distributed across the Southeastern Atlantic, Eastern Gulf, and Central Gulf portions of cottonwood's range (Land et al. 1996). Approximately 1,300 clones are now being produced from these seedlots for field testing and eventual release of improved planting stock for the region. As has been shown for currently available clones (Table 3), cottonwood productivity in the Humid Lower South can be enhanced by sewage effluent, various clones, and is greater in coppice rotations.

Other species suitable for SRWC (Prine and Rockwood 1998) in the Hu-

Table 4. Annual yields (dry Mg/ha)/length of cropping cycle (years)/number of cycles for promising biomass crops by phosphate land type. Source: Segrest et al. (1998).

Biomass Crop	Phosphate land type		
	Clay	Overburden	Cropland
Sugarcane	49.3/1/6	40.3/1/4	40.3/1/4
Energycane	44.8/1/6	35.8/1/5	33.6/1/5
Elephantgrass	40.3/1/6	40.3/1/6	40.3/1/6
Leucaena	35.8/1/10	33.6/1/10	26.9/1/10
Eucalyptus grandis	29.1/3/6	26.9/3/6	24.6/3/6
Slash Pine	--	20.1/8/1	20.1/8/1

mid Lower South include slash pine (*Pinus elliottii*) and, for Florida's more subtropical regions, several *Eucalyptus* species. Slash pine intensively cultured and closely spaced can produce up to 23 dry Mg ha^{-1} yr^{-1}. Under similar culture, *E. amplifolia* can yield as much as 25 dry Mg ha^{-1} yr^{-1} on good sites in northeastern Florida, and *E. grandis* can yield up to 35 dry Mg ha^{-1} yr^{-1} in central and southern Florida. Yields can be enhanced by the addition of wastewater and waste products.

CONCLUSIONS

Perennial peanut is a good persistent forage legume for most soils in Humid Lower South. Pigeonpea is a new warm season grain legume for the area on well-drained soils. The availability of land with an excellent climate and several adapted rapid growing bioenergy crops makes the Humid Lower South a likely area for development of bioenergy industries.

REFERENCES

Bransby, D.I. and S.E. Sladden. 1995. The need and potential to further raise switchgrass yields based on 10 years of research in Alabama. Proc. 2nd Biomass Conference of Americas, Aug. 21–24, 1995, Portland, OR. p. 261–266.

Cunilio, T.V. and G.M. Prine. 1995. Leucaena as a short rotation woody bioenergy crop. Soil Crop Sci. Soc. Fla. Proc. 54:44–48.

French, E.C., G.M. Prine, W.R. Ocumpaugh, and R.W. Rice. 1996. Regional experience with forage *Arachis* in the United States. p. 169–186. In: C.P.C. Kerridge and B. Hardy (eds.), Biology and agronomy of forage *Arachis*. IAT Publ. 240. Cali, Colombia.

Janky, D.M., D.L. Damron, C. Francis, D.L. Fletcher, and G.M. Prine. 1986. Evaluation of Florigraze rhizoma peanut leaf meal (*Arachis glabrata*) as a pigment source for laying hens. Poultry Sci. 65:2253–2257.

Kerridge, P.C. and B. Hardy (eds.). 1996. Biology and agronomy of forage *Arachis*. CIAT Publ. 240. Cali, Colombia.

Land, S.B., Jr., A.W. Ezell, S.H. Schoenholtz, G.A. Tuskan, T.J. Tschaplinski, M. Stine, H.D. Bradshaw, R.C. Kellison, and J. Portwood. 1996. Intensive culture of cottonwood and hybrid poplars. Proc. 35th LSU Forestry Symposium, Baton Rouge. p. 167–189.

Prine, G.M., L.S. Dunavin, J.E. Moore, and R.D. Roush. 1981. 'Florigraze' rhizoma peanut a perennial forage legume. Univ. Fla. Agr. Expt. Sta. Circ. S-275.

Prine, G.M., L.S. Dunavin, R.J. Glennon, and R.D. Roush. 1986. 'Arbrook' rhizoma peanut, a perennial forage legume. Univ. Fla. Agr. Expt. Sta. Cir. S-332.

Prine, G.M., P. Mislevy, R.L. Stanley, Jr., L.S. Dunavin, and D.I. Bransby. 1991. Field production of energycane elephantgrass and sorghum in southeastern United States. Paper 24. In: D.L. Klass (ed.), Proc. Final Program of Conference on Energy from Biomass and Wastes XV. March 25–29, 1991. Washington, DC.

Prine, G.M. and W.V. McConnell. 1996. Growing tall grass energy crops on sewage effluent spray field at Tallahassee, FL. Proc. Seventh National Bioenergy Conf.-Bioenergy '96. p. 770–777.

Prine, G.M. and D.L. Rockwood. 1998. Energy crop opportunities in humid lower south, USA. Proc. Bioenergy 98 Conf. Expanding Bioenergy Partnerships. Oct 4–6, 1998. Madison, WI Vol 2:1192–1199.

Prine, G.M., J.A. Stricker, and W.V. McConnell. 1997. Opportunities for Bioenergy Development in Lower South USA. Proc. 3rd Biomass Conf. of the Americas held in Montreal Canada, Aug. 24–29, 1997. Vol. 1, p. 227–235.

Rockwood, D.L., S.M. Pisano, and W.V. McConnell. 1996. Superior cottonwood and *Eucalyptus* clones for biomass production in wastewater bioremediation systems. Proc. Bioenergy 96, 7th National Bioenergy Conference, Sept. 15–20, 1996, Nashville, TN. p. 254–261.

Rockwood, D.L. 1997. *Eucalyptus*—Pulpwood, mulch or energywood? Florida Cooperative Ext. Ser. Circ. 1194.

Saldivar, A.J., W.R. Ocumpaugh, G.M. Prine, and J.F. Hentges. 1981. Southern Branch of American Society of Agronomy. Madison, WI. (Abstr.).

Segrest, S.A., D.L. Rockwood, J.A. Stricker, A.E.S. Green, and W.H. Smith. 1998. Biomass co-firing with coal at Lakeland, FL, USA, Utilities. Proc. 10th European Biomass for Energy and Industry Conf., June 8–11, 1998, Wurzburg, Germany. p. 1472–1473.

Stricker, J.A., G.M. Prine, K.R. Woodard, D.L. Anderson, D.B. Shibles, and T.C. Riddle. 1993. Production of biomass/energy crop on phosphatic clay soils in Central Florida. Proc.1st Biomass Conf. of the Americas: Energy, Environment, Agriculture and Industry. Vol. 1. p. 254–259.

Stricker, J.A., A.W. Hodges, J.W. Mishoe, G.M. Prine, D.A. Rockwood, and A. Vincent. 1995. Economic development through biomass systems integration in central Florida. Proc. 2nd Biomass Conf. of Americas: Energy, Environment, Agriculture and Industry. p. 1608–1617.

Stricker, J.A., G.M. Prine, D.L. Anderson, D.B. Shibles, and T.C. Riddle. 1996. Biomass/energy crops grown on phosphatic clay in central Florida. Proc. 7th National Bioenergy Conf. Bioenergy '97. Nashville, TN. Sept. 15–20, 1996. p. 822–826.

Woodard, K.R. and G.M. Prine. 1993. Dry matter accumulation of elephantgrass, energycane and elephant millet in a subtropical climate. Crop Sci. 32:818–824.

The North Carolina Specialty Crops Program

E. Barclay Poling

The Specialty Crops Program, formed in July 1997, is a publicly supported inter-agency program in Eastern North Carolina that is designed to accelerate the process of developing new "cash crops" for small and medium size growers who are primarily dependent on tobacco for their livelihood. Substantial new resources have been identified and committed to equipping a research station in Kinston, North Carolina, with state-of-the art facilities and equipment, including a new polycarbonate greenhouse and a refrigerated diesel truck for making deliveries to various test market locations. At the 160 ha Cunningham Research Station in Kinston, University scientists and technicians are working to generate information on the most efficient field and greenhouse growing systems for the specialty crops and varieties in greatest demand. The program is designed to develop markets for new crops, to help farmers sell what they grow. Many of the research station personnel involved in the program are highly motivated to find high value alternatives as their own families have been growing tobacco for several generations.

The idea of weaning tobacco farmers from tobacco onto other crops is not new. Author Wendell Berry, born in tobacco country (Kentucky), has advocated that government and universities need to become more interested in local food economies. He would like to see an agricultural economy of diversified small farms that produce for local markets and local consumers. In his book chapter, *"The Problem of Tobacco"* (1993), he writes:

> *"Tobacco farmers are farmers and among the best of them; their know-how is a great public asset, if only the public recognized it. They are farming some very good land. They should be growing food for people of their region, the people of neighboring cities—or they should have a choice in doing so. The people who condemn them for growing tobacco should be just as eager to help them find alternative crops."*

The initial strategy of the North Carolina Specialty Crops Program is to take advantage of an unprecedented opportunity in the marketplace to exploit the home grown theme and to develop local and regional markets for high-value specialty crops such as strawberry, peach, blueberry, tomato, asparagus, seedless watermelons, and muskmelon! In the future, the proximity of the Global Transpark, an effort to link North Carolina by air with markets throughout the world, is likely to play a role in the eventual development of export markets for specialty fruits, vegetables, cut flowers, and herbs grown in the 13 southeastern coastal plain counties which comprise the Global Transpark zone.

MARKETING AND RISK

The Specialty Crops Program explores new market possibilities. We work in inter-agency networks and advisory teams to understand the production and market feasibility of any new item. But, there is a large measure of risk and a lack of security involved with any new specialty crop. Few growers are comfortable with the ambiguity of not knowing the market situation in precise detail—especially growers of tobacco who have been conditioned to working in a crop where production is controlled and provides grower price supports (Toussaint 1992).

For a number of years, the Cooperative Extension Service has been advising growers to proceed cautiously, and to "avoid planting newer crops until you get a market first." But, relatively few farmers have the interest, time, experience, or financial resources needed to successfully introduce a new crop to the market! For this very reason, a forward-thinking group of tobacco farmers from Eastern North Carolina called the *Alternative Crops Diversification Committee (ACDC)* came to the University and to the Department of Horticultural Science in March 1997 to ask for our leadership and help in developing a program that would provide "visual demonstrations" of potential new crops at the Cunningham Research Station (CRS). It was felt that at the CRS we could grow and evaluate a variety of potentially important new fruit cultivars, vegetables, and herbs for production performance and pest reactions. But, our job would not stop with field and greenhouse production trials. To capture the full potential of this research, these growers wanted marketplace evaluations of each new specialty crop or value-added item!

It was obvious from this initial meeting that if we were going to accelerate the process of introducing promising specialty fruits, vegetables, and herbs to the market, it would be essential to combine the talents and resources of both the North Carolina State University's College of Agriculture and Life Sciences (CALS) with the North Carolina Department of Agriculture's (NCDA) Marketing Division.

INTER-AGENCY TEAMS

A group of CALS faculty in the Department of Horticultural Science, marketing specialists from the NCDA, area extension specialists, research technicians, and front-line research station employees get together monthly at the Cunningham Research Station. These meetings are very important to keeping everyone informed and involved. We always devote the early portion of the meeting to an exploratory discussion of new specialty crops items and marketing ideas. The balance of the meeting is used to review resource requirements, communicate schedule information, delegate tasks, get status updates, and report project details.

Having such a mixed inter-agency group is one of our main strengths. Another strength is our flexibility. That flexibility lets us go to where the opportunities are. Many of those opportunities begin with a visit to another state or country. Trade meetings such as PMA and United Fresh are mandatory events for our group if we are to stay informed of new developments on the food trends superhighway. In 1998, Frieda Caplan, Frieda's Inc., Los Alamitos, California, sponsored two visits from members of the North Carolina Specialty Crops team. But, building bridges to the thought leaders of the specialty foods industry, such as Frieda Caplan, can only help us in our desire to fashion some winning strategies for tobacco farmers in the Carolinas who want to be a part of this nation's food trend superhighway (Caplan 1996).

The Specialty Crops Program must choreograph a number of important transitions in what is basically still a field crop economy in Eastern North Carolina based on tobacco, maize, and soybeans. The main problem or challenge that the Specialty Crops Program must address is to somehow cause a regional demand and a supply to come into existence simultaneously (Berry 1992). While fruits and vegetables are the initial focus of the program, it may eventually include crops such as herbs and cut flowers.

A valuable resource for us has been a "critical mass" of horticultural research and extension specialists resident at North Carolina State University. We feel quite privileged that we are able to directly harness the expertise and research products of our internationally known researchers and extension specialists in fruits, vegetables, and herbs. The fruit and vegetable breeders have a growing warehouse of unconventional cultivars that we believe are going to be "the pick of the produce aisle." But, with each new specialty variety and/ or selection comes the critical need to allocate some funds for cultural research trials. In my view, a new selection or cultivar exists only to the extent that we can demonstrate to a farmer that it is well-adapted and economical to grow. Accordingly, The Specialty Crops Program has a research-arm at the Cunningham Research Station in Kinston that is strictly focused on developing precision field and greenhouse recommendations for unconventional crops (including non-food crops) and specialty varieties.

The Specialty Crops Program market research and grower-assistance component is a key element of the effort. A team of marketing specialists led by Don Thompson, Eastern Marketing Center, and Ross Williams, Assistant Director, Division of Marketing, North Carolina Department of Agriculture and Consumer Services, are working to help identify the product quality, volume and packaging requirements of national and regional supermarkets, and foodservice buyers specialty food products and value-added items. Working in close concert with Extension's Area Fruit and Vegetable Specialist, W.R. "Bill" Jester, Jr., they have already succeeded in helping a group of "early adopters" to form a grower association, Southeastern Growers' Association (SGA), that is now shipping and test-marketing various "Carolina Specialties," including eastern grown specialty melons. Fortunately, in the immediate area of the Cunningham Research Station, a large volume of excess refrigeration capacity exists that is now being effectively utilized by the new grower cooperative that has increased from 12 tractor trailer loads of specialty melons in 1997 to over 70 loads in 1998!

The chainstores have been well satisfied with the eastern North Carolina source of cantaloupes and seedless watermelons, and the Specialty Crops Program is now busily working with SGA and other area farmers to introduce lesser know specialty melons such as 'Emerald Jewel' (Fig. 1) and 'Sprite' (Fig. 2).

FUNDING

Conducting field research and greenhouse trials requires some very deep pockets! In 1998, we were fortunate in that the Research Administration of the College of Agriculture and Life Sciences was able to provide seed money for specialty crops initiatives involving off-season asparagus production (fall cropping), early winter blueberry production with low chill selections (greenhouse), greenhouse strawberries (Fig. 3), and seedless watermelon production (Fig. 4).

In early 1999, we will be aggressively seeking the support of the state legislature. North Carolina is the only one of the 46 states covered by the landmark $206 billion settlement (Nov. 1998) to specifically earmark money to help out tobacco-dependent communities. North Carolina will receive $160 million to $190 million annually over the next 25 years. Half of this amount will go to the foundation set up to assist tobacco communities. This new foundation can help us defray the considerable research expense of developing grower and market-ready specialty crops.

Fig. 2. 'Sprite' is a smaller oriental melon from the Sakata Seed Co. which has a white and yellow exterior with a very sweet and crisp white interior flesh.

Fig. 1. 'Emerald Jewel' is a muskmelon with green flesh and green netted exterior with excellent flavor and handling characteristics. Photo credit: Don Thompson, Marketing Specialist, Eastern Marketing Center, NCDA &CS

Fig. 3. 'Sweet Charlie' is a short day strawberry under evaluation for early winter greenhouse forcing in the program's new polycarbonate greenhouse at CRS. Photo credit: E. Barclay Poling, Coordinator, Specialty Crops Program.

Fig. 4. Seedless watermelon consumer evaluations at Specialty Crops Field Day. The test marketing for specialty melons focuses on a combination of consumer preference surveys, supermarket buyer, foodservice and restaurant evaluations, and market and supply data from various shipping points.

PROGRAM PHILOSOPHY

Our program philosophy is to: "start small and learn big"! (Sturdivant and Blakely 1999) The basic program goals that we have established are to:

1. Demonstrate the "regional market appeal" of new cultivars of fruits, vegetables, and herbs bred by North Carolina State University breeders, other universities, as well as private seed and nursery companies, that are grown, harvested, packed, and shipped from the Cunningham Research Station in Kinston.
2. Cultivate a national as well as international network of breeders, scientists, and private seed companies that can supply the Specialty Crops Program with promising new cultivars of potentially high market appeal.
3. Generate complete packages of timely and reliable production and test-market information that can guide eastern North Carolina farmers in their decisions on the most profitable new markets for specialty fruits, vegetables, and herbs.
4. Cultivate cooperation between small and independent producers to form private cooperatives that can meet the volume and seasonal supply requirements of regional and national chains.
5. Cultivate a regional network of buyers for specialty fruits, vegetables, and herbs grown in Eastern North Carolina.
6. Identify through research and market-testing new opportunities for growing out-of-season crops.
7. Identify through research and market-testing the potential for growing specialty fruits vegetables and herbs for developing ethnic markets throughout the state and region.
8. Develop a regional identity for "Carolina Specialties."

We are excited by our mission of introducing unconventional products into the marketplace and making buyers aware that North Carolina growers have the capability to grow and deliver some of the best tasting specialty fruits and vegetables in the world! The market feedback of consumers and buyers to regionally produced specialties, including more flavorful strawberries, low-acid peaches, specialty lettuces, green-flesh muskmelons, red and yellow seedless watermelons, and specialty peppers and squashes (kabocha), is proving to be very positive. Working together, the state Land Grant University and the Department of Agriculture, can leverage a considerable wealth of resources for the purpose of answering the questions tobacco farmers have about specialty crop production practices, marketing strategies, and profit potential. The Specialty Crops Program Web page is available at the following URL: http://www.cals.ncsu.edu/specialty_crops/

REFERENCES

Berry, W. 1992. Sex, economy & community. Pantheon Books, NY. p. 64–65.
Caplan, F. 1996. Marketing new crops to the American consumer. p. 122–126. In: J. Janick (ed.), Progress in new crops. ASHS Press, Alexandria, VA.
Sturdivant, L. and T. Blakely. 1999. The bootstrap guide to medicinal herbs in the garden, field & marketplace. San Juan Naturals, Friday Harbor, WA. p. 6.
Toussaint, W.D. 1992. The flue-cured tobacco program. North Carolina Cooperative Extension Service, AG-476.

New Uses from Existing Crops

Paul J. Caswell

For manufacturers, researchers, and entrepreneurs interested in processing new agricultural crops, this is exciting (and hopefully profitable) times. In the past two years we have seen major chemical companies switch their focus away from petrochemical-based materials to the genetic engineering of seeds. We have seen staid, old seed companies sell for multiples of ten times sales (not earnings). Major companies, like Monsanto, Dupont, Dow, and Hoescht have selected life sciences and naturally-derived products as their engines of corporate growth for the next century.

This change in focus is being driven by the exciting new technology of genetic engineering. We now have the ability to convert a plant into a miniature chemical production facility. Industrial crops have become a major growth area with new crops on the horizon and existing crops being improved. New products produced from these crops are no longer just "natural." They are now competitively priced and functionally superior. These natural products are also getting a boost in the market place by a consumer preference for natural products in our healthcare, food, clothing and all the areas that closely touch our lives. Natural products are being accepted for more than their renewable source. They are accepted because people like you are making the products better than their petrochemical derived competitors.

ARCHER DANIELS MIDLAND COMPANY

At Archer Daniels Midland I am Director of New Technology. My job is to locate new technologies that can provide us with new markets or improve the existing ones. ADM is first an agricultural processor and second a producer of food and feed products throughout the world. ADM's roots are from industrial crops. We began as a linseed oil manufacturer over 130 years ago and we still process linseed oil today. Our newest industrial product is a chemically modified linseed oil, called Archer 1, where the linoleic acid has been changed to a conjugated double bond. This makes the oil more reactive, like a tung oil, but at a much lower cost and greater availability. It is gaining significant acceptance as an ingredient in adhesives for fiber board. Future application research is planned in inks and polymers. We believe, as the name implies, that this is the beginning of a line of industrial products that will be derived from our existing processed crops.

ADM processes most of the major vegetable oils. The list includes: soybean oil, corn oil, rapeseed, flax, peanut oil, cottonseed oil, palm oil, and canola. We are the world's largest soybean oil processor, wheat miller, and corn processor. We process: 3 ha of maize per minute, 8 ha of soybeans per minute, 5 ha of wheat per minute. We do this in 190 processing plants in the US and 88 foreign plants. We have 37 domestic and 9 foreign oil seed crushing plants processing 84,000 tonnes per day. These processing plants are supported by a network of 142 elevators, 62 domestic terminal and river loading facilities that have a combined storage capacity of 100 million bushels. In addition we operate three corn wet mills and two dry mills in the US that process 1.6 million bushels of day. These facilities create a river of carbohydrates that we use to produce over 100 different products using chemical, physical and fermentation technologies. The maize plants are an integrated production site, similar to major petrochemical facilities. The incoming grains are separated into their major components and purified and modified into products like ethanol, citric acid, xanthan gum, soy proteins, vitamin E, and isoflavones.

This storage and processing capacity, couple with a distribution network that includes 42 ocean ships, 11,000 rail cars, 2000 barges, and 1000 trucks, puts ADM in a unique position to take advantage of the new revolution of genetically engineered crops with improved functional traits. ADM has the capability to identity preserve a new crop through collection, distribution, and processing. ADM is able to recover the modified component and provide value to all the coproducts (not byproducts) produced in the crop. We can do all of this in a big way to help keep cost down.

New, genetically engineered crops are a key source of "new uses for existing crops," but this technology is still in its infancy. Improvements in purification, fermentation, and chemical modification technology is a much more immediate source of "new uses for existing crops." It is these technologies that the new genetic engineered crops must compete with in the market place. Direct production in the seed may be the lowest cost, but the 7 year time lag to develop sufficient seed for a new crop, and the lack of flexibility to quickly change the product's composition, make the technology slow to respond to changes in the market. The same genetic engineering in a fermentation organism can change a product in a few months. In reality we see a combination of all of this sources of new products as the best route. A new genetically engineered trait can first be introduced to a microorganism for immediate production to begin the market development, while a genetically engineered seed is being developed. When the seed is ready production can convert to this lower cost source. The fermentation, or a chemical technology can then be used to make specialty grades of the new product to keep the product flexible to market needs.

DIRECTIONS FOR NEW USES

To determine ADM's direction for obtaining "new uses for existing crops" it is best to look at the past. The products introduced over the last 20 years are a clear guideline to the future. The attached diagram (Fig. 1) shows in the center of the circle the basic corn, wheat, and oilseed businesses as they existed in 1979 at ADM. From these businesses the company expanded each area geographically and into value added products. The concentric circles represent the decades of the 1980s and 1990s. In the 1980s oilseeds expanded into functional 70% soy concentrates and 90% soy isolates, used as a healthy source of vegetable protein in meat substitutes and meat extenders. In corn processing, ADM added ethanol, enzymes, and carbon dioxide from fermentation, high fructose corn sweeteners and dextrose from enzymatic reactions, and industrial starches from chemical reactions.

In the 1990s the soy proteins were formulated into harvest burgers, a healthy hamburger substitute, now going nationwide under the Worthington Foods label. ADM also expanded into lecithin emulsifiers, which are purified from a byproduct of soy oil refining. In addition, we began separation and purification of vitamin E, which has superior health benefits to synthetic material, and sterols, which are a feedstock for steroid pro-

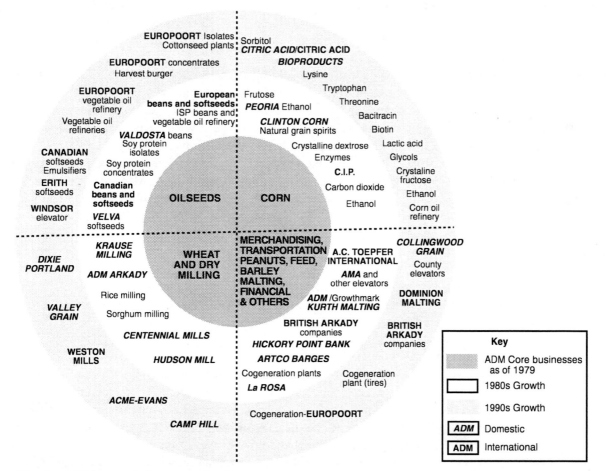

Fig. 1. ADM growth from core businesses.

duction, including androsteinedione, made famous by Mark McGuire. On the corn side, building off of the fermentation technology developed in the 1980s, ADM added the amino acids: lysine, tryptophan, and threonine; the organic acids: citric, lactic and glucono delta lactone; the industrial and food thickener, xanthan gum; and the vitamin, riboflavin. Also in the corn processing side ADM developed crystalline fructose which uses an improved chromatographic separation technique developed originally for the production of 55% high fructose corn syrup.

For the next ten years what products is ADM targeting? To determine this you must begin by looking at ADM's guidelines for new product selection. These are:

1. The raw material source must come from the basic crops already available in ADM. This is our key advantage.
2. The technology for production must build off the basic technologies that we already understand. Key technologies in ADM processing include fermentation, enzyme reactions, basic chemical reactions, like hydrogenation and esterification; separation processes using centrifuges, screens, membranes, and resins; extractions; evaporization and crystallizations; and drying.
3. The products should require a significant capital expenditure that provides a barrier of entry to competitors. To lower this capital cost for ADM we have developed our own fabrication company and do most of our own engineering. We build large facilities using standard sized equipment developed in other projects that the fabricators already know how to build and the operators know how to run.
4. The new process must be able to be integrated into our existing facilities to take advantage of raw material sources and byproduct process synergy's.
5. The new processes must increase the demand for the source crop (grind more corn or beans), or add value to a coproduct stream.
6. Normally, the product must be sold to markets that we already service. This is primarily food and feed markets. Through joint venture and distribution agreements this requirement is changing because we are able to contract for the marketing expertise that we lack. This makes chemical product a possibility in the future as long as corn and soybeans maintain a favorable cost relationship to petroleum and the supplies of these grains remain high.

Using these guidelines some of ADM's "new uses for existing crops" for the next 10 years can be identified. For the oilseeds the priority is more value for the coproducts. The first new product is isoflavones, which are extracted from the waste of soy protein production. Isoflavones structurally are close to estrogen and have been proven to aid in the fight against breast and prostate cancer. We have constructed the world's largest facility for their production in Decatur, Illinois and begun production this summer. The next new product is tocotrienols, which is a strong antioxidant, like vitamin E, that is extracted from palm oil processing byproducts. This product should be available next year in limited quantities. Also from this palm oil stream ADM is researching the production of a natural betacarotene. Additional soy oil byproduct in preparation are phosphytidyl choline and phosphytidyl serine. Both products are in research. These products have health benefits and can be sold into the neutriceutical market. For industrial products ADM is already producing Archer 1, A conjugated linseed oil, and we plan to build off that patented technology to produce other oils and fatty acids that will have applications in lubricants and plastics. We are currently producing biodiesel oil in Europe, but are unsure if the product will ever be cost effective. Many of the new specialty oils will be produced through chemical modification, but in the future, this modification is likely to be accomplished directly in the seed of genetically altered crops.

From the corn processing, "new uses" are:

1. Vitamin C. Produced from a new two stage fermentation based process to 2-keto-gulonic acid, which is then chemically converted to vitamin C. This facility is in construction and will start up in January.
2. Biotin. This is a new fermentation based product from a patented, genetically engineered organism.
3. Astaxanthin. This is a pigment found in natural salmon and trout that gives the fish its pink color. In farmed salmon and trout it must be added to the diet. ADM will produce a yeast that contains the pigment. The yeast will be fed to the salmon direct and provide part of the fishes protein needs also. The competitive astaxanthin is produced synthetically. This production will also start in January.

4. Isoleucine. This is the next critical amino acid for feeds. A new genetically engineered organism is nearing completion.
5. Zeaxanthin. Another pigment used in shrimp farming is in development.
6. Lutein. Another neutriceutical to be produced by fermentation.
7. Ethyl lactate. This is an environmentally friendly solvent that can replace many chlorinate hydrocarbons. It is currently in production at ADM. Significant application work has been completed to determine the industries where it can be cost effective. Other solvents are also being developed from fermentation and chemical processing of our glucose streams.
8. Polyols. In addition to our sorbitol stream we will be marketing mannitol next year from a new facility and are developing new processes for glycerol, propylene glycol, and ethylene glycol.
9. Organic acids. A number of organic acids are in development. These acids currently have small markets now, which are limited by the high cost of the acids. ADM is working on new processes to reduce these costs.

In the oilseed and corn areas significant increase in the grinding of these crops are expected to come from industrial products that compete in the chemical and polymer markets. The nutraceutical and pigment areas will provide higher margin products.

CONCLUSION

I would like to at least mention the key technologies that are driving our new product development. These technologies are:
1. Genetic engineering. New organisms and new crops with good yields that can be easily purified make many of the new product possible
2. Computer control. The new processes can be run as a batch or continuous process with less manpower under tighter limits that give the optimum yields.
3. Membrane separations. Membranes with tightly controlled pore sizes, or that are functionalized to allow separation according to other properties, make purification of natural products much easier.
4. Chromatographic resins. In most cases, if the separation can be done in your lab liquid chromatograph, then it can be done in the plant.
5. Supercritical gas extraction. Allows for specific separation, fractionation, and easy solvent removal.
6. Enzymes and precious metal catalyst. These compounds allow for quick reactions and good yields.

We at ADM are excited about the future of existing crops. We see many new uses and new processes to make existing products from renewable crops.

Corn Stover Potential: Recasting the Corn Sweetener Industry*

David Glassner, James Hettenhaus, and Tom Schechinger

Corn stover is by far the largest single available biomass not being used, representing more than one-third of the total waste, including municipal solids. An estimated 200 million dry tonnes (t) remain each year as aboveground residue.

THE CORN STOVER ISSUE

The corn sweetener industry is based on processing corn grain (maize), creating value-added products such as glucose, dextrose, and fructose. Conversion of corn stover to sugars has been stymied for years due to cost. Environmental benefits, wider adaptation of sustainable farming practices and the relentless improvements in biotechnology are expected to overcome the economic hurdle within the next five years and recast the corn sweetener industry. The corn sweetner industry uses about 8% of the corn crop (20 million t) in 1997; the fuel ethanol industry used about 13 million t (515 million bu). These combined demands equal 13% of domestic demand for corn.

Corn stover consists of the stalks, leaves, and cobs remaining aboveground after the corn kernels are harvested. About 1 kg of stover is produced per kg of grain. In 1997 about 200 million dry t of stover was produced. The mass of stover increases with the yield of the corn—expected to increase 1% to 2% annually

More than 90% of the stover is left in the fields. Less than 1% of corn stover is collected for industrial processing. About 5% is baled for animal feed and bedding. Much of the remaining 90+% must be plowed under for planting to proceed on schedule, ensuring the best yield, eliminating weed seeds, insects and disease harbored by the stover, and reducing the threat of alpha-toxins in the corn. Although some residue is required to protect the soil from erosion, some residue can be safely removed.

Improved management of the residue has the potential to be a win-win for the producer, processor, and the environment. Now, up to 75% of the surface material decays to CO_2, a greenhouse gas (Parr and Papendick 1978; Gale and Cambardella 1999). Excessive residue makes no-till farming more difficult and reduces crop yield; cold soil temperature in the spring slows field drying, retards germination, and reduces growing seasons. It also contributes to problems with disease, weeds, pests, and irrigation. As a result of plowing, a carbon deficit can occur in the soil. The plowing activity exposes soil carbon to oxidation, increasing organic carbon loss with the release of CO_2. The plowing activity also releases soil N to the atmosphere by increasing soil oxidation, similar to the oxidation of soil carbon. This can lead to increases in NO_x and N_2O emissions. Plowing residues back into the soil also increases the amount of fertilizer chemicals that need to be applied. Residues raise the soil carbon content relative to the soil N content. If no N is added, the cash crop harvest is adversely affected while the residue decomposes.

Farmers and the corn processing industry—along with the fuel ethanol producers—are in position to benefit from conversion of excess corn stover to fuels and other products. The farmer wins from the stover sales, reduced cultivation costs and possible carbon credits for the greenhouse gas (GHG) offset. The processors grow to meet fuel market needs and create additional products, some new and others previously produced from petroleum. The environment benefits from improved agricultural practices and fewer GHG emissions.

Drivers for achieving this vision include moving to more sustainable agricultural practices, improved biomass conversion technology, worldwide commitment to reduce GHG emissions, and increases in petroleum cost as supply dwindles. Many envision this scenario to occur within the next five years.

---

*The financial support from the US Department of Energy, Office of Energy Efficiency & Renewable Energy Ethanol Program to perform this work is most appreciated.

Impediments for Corn to Ethanol Expansion

Present annual ethanol production of 5.7 billion liters (1.5 billion gal) is just 1% of the total fuel transportation market. If corn grain is used to double ethanol production, the following disruptive effects ensue according to studies by the Government Accounting Office (USGAO 1990) and the USDA Economic Research Service (Petrulis et al. 1993): (1) higher corn prices, 9% to 15%; (2) increased livestock costs; (3) reduced corn export market, 5%; and (4) reduced soybean price, 6% to 11%.

Both studies show corn price would increase over baseline projections by $7 to $13/t (19 to 32 cents/bu). The higher corn price causes a 6% to 10% increase in feed cost, leading livestock producers to reduce the amount of corn purchased for animal feed and lower the number of cattle by 3% to 4%. Export markets for corn would also shrink, as higher prices would reduce the foreign demand for American-grown corn 4% to 5%.

Soybean processors and producers would face lower demand for their products because ethanol from corn generates protein-rich feed and corn oil by-products that compete with soybean meal and oil. Soybean prices would drop $13 to $426/t (31 to 66 cents/bu), 6% to 11% from the baseline and production would drop 3% to 5%.

Corn Stover Expansion Potential

Innovative corn stover harvest, collection and transportation practices have reduced the corn cost to $30–$35/dry t delivered in Western Iowa where 20,000 ha (50,000 acres) were collected during the 1997 crop year. Planned improvements in productivity and storage stability are expected to reduce costs to less than $30/dry t.

The sustainable amount removed depends on soil, topography, crops, crop rotation, tillage practice, and environmental constraints. Removing just one-third of the stover and hydrolyzing its 38% cellulose content with improved cellulase enzyme systems (currently being developed) results in 29 million t (64 billion lb) of glucose with a targeted cost of $132/t ($6/cwt) or less, twice the amount of sweeteners shipped in 1997. Low sugar cost can spur further market expansion to replace other petroleum-derived products.

The stover also contains 32% hemicellulose. When converted to pentose sugars that have less food value, their most likely future use is a nutrient for fermentation processes, with the largest being alcohol. Taking 80% of the hemicellulose to alcohol employing any of at least three engineered organisms presently under development produces 14 billion liters (3.6 billion gallons) of ethanol from one-third of the corn stover, again more than twice the 5.7 billion liters (1.5 billion gallons) produced annually. The targeted cost is less than $0.25/liter ($1/gal).

The market potential for sugar to supply the ethanol industry and other sugar-based products can be improved significantly by using excess corn stover. This excess may supply more than 5% of US gasoline needs—an additional 20 to 40 billion liters (5 to10 billion gallons) of ethanol. Farm income can be raised significantly without the production of feed coproducts and their adverse effects on livestock and soybeans.

Some surface residue—a minimum of 30% surface coverage—is required to comply with USDA guidelines for erosion protection. Removing 3 t/ha from 9 t/ha produced in many areas can comfortably attain erosion compliance with conservation tillage. With no-till, the quantity removed could likely be doubled.

Today, the excess corn stover decomposes; its potential to offset fossil fuel feedstock is lost. Other environmental benefits result from increased soil organic matter (SOM) formation that occurs by reducing the need to plow under the excess stover.

The stover offers an inherent cost advantage over corn, other grain or energy crops since the crops carry a production cost. The historical net corn cost is about twice the delivered cost of corn stover. The historical cost between 1980 and 1997 for dry mill corn ethanol plants averaged about $60/dry t (Lewis 1997). Delivered cost of corn stover is $32/dry t. Improvements are projected to lower the cost to less than $25/dry t delivered (Glassner et al. 1998).

Improved conversion of the biomass feedstock to sugar is proceeding rapidly. Application of proven biotechnological tools is projected to reduce fermentation sugar costs to less than 6 cents/lb. An early result is the current construction of two biomass conversion plants by BC International (BCI) and a joint venture between Iogen and Petro-Canada. Both are expected to be operating in 2000 (Wald 1998; Canadian Report 1998).

MARKET PULL

There are three drivers for increased ethanol in the transportation fuel market: (1) energy independence; (2) GHG offset of fossil fuel; and (3) cleaner air from vehicle emissions. The dependence of the US on imported fuel continues to increase. Global production of fossil fuel is likely to peak in the next decade. Questions regarding the security and availability of future supply continue to be raised (Scientific American 1998; Campbell 1998). There is increasing likelihood that interruptions and shortages will occur, along with the price hikes, recessions and political struggle unless actions are taken now.

To offset this scenario, domestic production of transportation fuel from biomass is receiving increased attention. For example, the last Congress extended the fuel ethanol subsidy through 2007. Also, the US Department of Energy has recently embarked on efforts to increase use of biodiesel in urban buses and other transportation sectors and "Bridge to the Corn Ethanol Industry" using the infrastructure in the corn ethanol industry.

The fuel ethanol market needs to grow quickly to meet these needs, according to former CIA Director Jim Woolsley (1998). Fuel ethanol represents just 1% of the transportation fuel market. He urges the industry to increase supply to 10% of the market "quickly" or risk being seen as an insignificant factor in energy markets. The California oxygenate market for ethanol is expected to become available in 1999. How ethanol production can be expanded nearly twofold to meet this new demand is currently an open question. Cleaner air from vehicle emissions has been established for a wide range of ethanol blends (Lynd 1996). Increasing the amount of ethanol in the blend reduces improves the quality of the emissions.

Using biomass to replace fossil fuel reduces GHG. The US has signed the Kyoto agreement to reduce GHG emissions, but that agreement faces an uncertain future in the Senate. Nonetheless, large multi-national corporations are faced with compliance in the EU, Canada, Australia, and other parts of the world. They are working to reduce GHG now and are pressing Congress to pass legislation that would give them valuable credits for early actions (Cushman 1999).

SUPPLIER PUSH

Excessive corn stover may significantly reduce crop yields, providing a push for its partial removal from the surface. If the surface residue is not managed well, the following problems can result: (1) germination is delayed due to low soil temperature in northern corn belt; (2) it contributes to weeds, pests and possibly results in toxins in the corn; (3) more chemicals are required for pest and weed control; and (4) excess restricts water in irrigated fields.

As producers have increased corn yields, disposition of the increasing amounts of stover remaining on the surface has become a greater problem for the producer. For example, 9.4 t/ha corn (150 bu/acre, 15.5% moisture) has approximately 9.4 t/ha aboveground residue, whereas a comparable 2.7 t/ha soybean crop (45 bu/acre) on that same land produces just 4.1 t/ha, has less than half the stover amount.

In the northern corn belt, excess stover on the surface can significantly reduce in crop yields, particularly on poorly drained soils and in cooler-than-normal growing seasons due to lower soil temperature in the spring. Lower soil temperature reduces the rate of seed germination and plant growth. The surface cover acts as insulation, causes higher soil moisture, and retards the soil-warming rate. Lower soil temperature also slows nutrient intake by the crops, increasing crop susceptibility to pests and disease.

Throughout the corn belt excess stover is an excellent harborage for weed seeds, insects, and disease. Crops can be seriously damaged if these issues are not addressed. Since herbicides and pesticides are partially absorbed by the residue, control is more expensive and difficult. Where irrigation is practiced, the excess can restrict water flow in the fields. Uneven stands and lower yields result.

SUPPLIER CHOICES

The producer choices for removing excess stover are usually limited to baling or plowing. Just leaving the corn stover on the surface to decompose is often not an effective solution. The rate of stover decomposition is relatively slow due to its carbon to nitrogen ratio. A C:N ratio of 10:1 is near optimum for rapid microbial action. Soybean stubble, with C:N ratios of 20:1, breaks down more quickly. Corn residue with a C:N ratio of 30:1 up to 70:1 is much slower to decompose (Shomberg et al. 1996).

Limited Bale Market

The bale market is limited to less than 10% of the total, almost exclusively for animal feed and bedding. Its use as feed is sub-marginal (Klopfenstein 1996). Although its fiber content suggests papermaking potential, economics remain a hurdle. Currently, the only industrial market—furfural production—uses less than 81,000 ha (200,000 acres) of the annual 32 million ha (80 million acres) of corn grown.

Tillage Chosen for Removal

Conventional tillage is the choice for most corn growers. In 1998, 61% of the land planted in corn was conventional-till, according to the national crop residue survey conducted annually by the Conservation Technology Information Center, CTIC. The CTIC survey data include crop by area and type of tillage practice: no-till, ridge-till, mulch-till, conventional tillage with 15% to 30% residue and 0% to 15% residue.

Table 1 shows virtually no change in conventional tillage practice in the US for corn over the past 7 years. In contrast, soybean conventional tillage has declined 20%, from 61% to 48%. These values for conventional-till are the sum for the land covered with 0%–15% and 15%–30% residue after tilling as reported on the CTIC web site: www.ctic.purdue.edu.

The motivation for tillage choice is not reported. Climate, crop rotation, soil type, previous crop yield and cultural values are all factors. The CTIC data do confirm the impact of climate. Colder weather in the northern corn belt makes producers in states such as Minnesota 28% more prone to plow under the surface residue in contrast to all farmers (Table 2).

Plowing under has many negatives including: (1) estimated $20/ha ($8/acre) or more to bury the stover; (2) additional N required for the stover; (3) loss of SOM; and (4) long-term adverse impact on soil quality from plowing versus no-till.

Tillage Cost. Using the Iowa farm custom rate survey (ISU Extension 1998), conventional tillage cost is $20 to $37/ha ($8 to $15/acre). Disking or soil finishing adds another $12 to $27/ha ($5 to $11/acre).

Additional Nitrogen Required. Plowing under heavy amounts of corn stover requires about another 10 kg N/t of stover to prevent nutrient deficiency. Crop residue that has a high C:N ratio, such as stover, can limit the availability of nutrients, especially nitrogen, for the succeeding crop as microorganisms that decompose the residue are competing for these nutrients.

Organic Matter Loss. Soil organic matter (mostly sequestered C and N in the soil) is also lost when the soil is disrupted by tillage. The loss of SOM is directly related to the amount of disruption. Moldboard plowing has the highest SOM loss, tandem disc is less, and chisel plow is still smaller. No-till soil organic matter (SOM) loss is mostly from the surface residue.

An initial burst of CO_2 occurs and the loss continues over weeks. Aeration simulates microbial activity, SOM is rapidly consumed. Plowing mixes fresh residues into the soil where conditions for decomposition are more favorable than on surface. More than 40 studies have been made showing this effect on SOM (Janzen et al. 1997; Paustian et al. 1997). Fig. 1 shows the initial tillage induced tillage loss; Fig. 2 shows the loss over 19 days (Reicosky et al. 1993; Reicosky 1997).

Long-Term Tillage Impact. Continued tillage reduces SOM and fertility. Other negative effects include increased erosion, water run-off is often contaminated with chemicals, soil pore size is reduced, soil crusting oc-

Table 1. Conventional till area of corn and soybean. Source: CTIC's National Crop Residue Survey.

	Conventional till area (% of total crop)							
Crop	1992	1993	1994	1995	1996	1997	1998	% decline 1992–1998
Corn	61	57	60	59	60	59	61	0
Soybeans	61	53	54	51	51	48	48	20

Table 2. Conventional till area of corn in Minnesota and the US. Source: CTIC's National Crop Residue Survey.

	Conventional till area (% of total corn crop)							
Location	1992	1993	1994	1995	1996	1997	1998	% ave. 1992–1998
Minnesota	75	73	78	80	79	N/A	N/A	77
US	61	57	60	59	60	59	61	60

77

curs. The list goes on, and the CTIC web site offers information and other links. www.ctic.purdue.edu.

EXPAND MARKET FOR CORN STOVER

The market for corn stover can be expanded by low-cost collection and storage of stover, along with meeting the conversion target of $134/t ($6/cwt) of glucose and less than $0.25/liter ($1/gal) ethanol. Collecting stover has many benefits, including: (1) producers bale excess stover to resolve surface residue problems and increase margins; (2) processors have low-cost feedstock for new market opportunities, expanding fuel ethanol, and other fermentation products from sugar; and (3) environmental benefits from GHG offset and cleaner air as a result more no-till, increased SOM, and fossil fuel offset.

Sustainable Collection

If corn stover is baled, removing mainly the stalk portion of the stover and leaving much of the leaf, husk, and cob on the surface, along with the anchored stubble, it is most likely more producers could plant no-till. Properly managed, with adequate residue for erosion compliance, soil tilth and productivity are expected to improve.

The surface residue does have an important role in controlling wind and water erosion, SOM, and soil physical and chemical properties. Factors related to managing its removal include: (1) soil type; (2) topography; (3) crops; (4) crop rotation; (5) environmental constraints; (6) weather history; (7) tillage practice; and (8) value judgments.

Considerable effort has been expended to determine removal effects. As a result of the energy crisis more than 20 years ago, a team of USDA scientists studied residue availability: Woodruff and Siddoway (1965), Bisal and Ferguson (1969), Wischmeier and Smith (1978), Allmaras et al. (1979), Campbell et al. (1979), Gupta et al. (1979), Holt (1979), Larson (1979), Skidmore et al. (1979), and Lindstrom et al. (1979). Generally, they caution that removal should be justified only when the long-term impact on soil productivity is inconsequential. Reduced and no-till emphasis had not yet emerged.

Corn stover removal impact on erosion for reduced tillage, no-till, and conventional till was reported by Lindstrom (1986) for two sites. The experimental results show the excess stover is greater than can be removed with baling equipment. For 9.4 t/ha (150 bu/acre), the stover to be removed, 76% and 82%, is above the normal 70% limit of commercial balers. These site-specific results are summarized below.

The Universal Soil Loss Equation was used to determine the retained surface residue mass, Y, required to control soil loss to a tolerable level, T. The slope was 6% in both cases. The Y for Swan Lake was 2,240 kg/ha. For Madison, Y was 1,680 kg/ha. In both locations, leaving the amount of residue results in less actual soil loss than the tolerance level. Table 3 shows the results.

The effect of residue removal on SOM and sequestered C is less known. The relative contribution of roots and surface residue to SOM is receiving increased attention. Campbell et al. (1997) shows no adverse

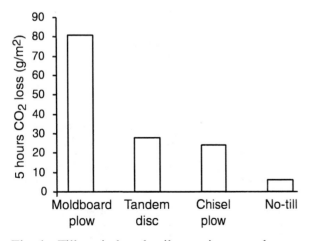

Fig. 1. Tillage induced soil organic matter loss as CO_2, Source: Reicosky (1993).

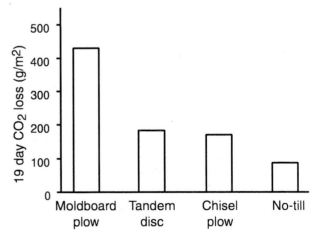

Fig. 2. Tillage induced soil organic material loss as CO_2, Source: Reicosky (1993).

effect on soil C and N since 1990 by removing straw from no-till fields. They address the use of several SOM measurement methods, and indicate the methods vary considerably. A search of the USDA CRIS database indicates much additional research is underway in this area, but not yet published.

Low-Cost Collection

Recent collection of corn stover show delivered cost of $30/dry t or less is achievable (Glassner et al. 1998). An industrial processor contracted with 440 producers for 20,000 ha (50,000 acres) of stover near Harlan, IA last year. Custom operators performed all baling, collecting, and hauling.

Table 4 summarizes corn stover pricing used for the 1997–1998 crop year. The delivered price for the stover was $34.76/dry t if about 8 t/ha (3 tons/acre) were collected. If only 4 t/ha (1.5 tons/acre) were harvested, the price increased to $39.30/dry t. Most producers chose higher collection, Case 2. Producers with sloped land mostly chose Case 1. The two prices were based on perceived quality differences in bale composition. No difference was seen and two-tier pricing was discontinued this year. The baler received $16.06/dry t, regardless of the density. The producer and the hauler shared the remainder based on distance. Table 4 summarizes corn stover pricing used for the 1997–1998 crop year.

The combine left a windrow by shutting off the spreader. No raking was required. Both round and square balers were used. Bale density was about 550 dry kg/bale. Round bales were triple wrapped with plastic net that permitted outside storage.

Table 3. Effect of residue harvest and till practice on soil loss.

Location Tillage	Total[z] residue (kg/ha)	Residue retained & surface cover (kg/ha)	(%)	Residue removed (kg/ha)	(%)	Soil loss tolerance (kg/ha)	Actual loss (kg/ha)
Swan Lake, Minnesota							
Conventional-till	9,430	9,430	9	0	0	11,200	7,100
Reduce-till, 0.5Y	9,430	1,120	16	8,310	88	11,200	11,800
Reduce-till, Y	9,430	2,240	29	7,190	76	11,200	6,500
Reduce-till, 2Y	9,430	4,480	33	3,950	42	11,200	1,600
Madison, South Dakota[y]							
Conventional-till	9,430	9,430	10	0	0	11,200	32,800
No-till, 0.5Y	9,430	840	33	8,590	80	11,200	42,700
No-till, Y	9,430	1,680	46	7,750	82	11,200	10,400
No-till, 2Y	9,430	3,360	56	6,070	64	11,200	5,500

[z]Based on 9,430 kg/ha (150 bu/acre) corn harvest.
[y]Two exceptional rainstorms resulted in high erosion at Madison, SD.

Table 4. Corn stover pricing summary (1997–1998).

	Corn stover payments ($/dry t)			
	Hauling radius			
Revenue	0–25 km	26–49 km	50–80 km	81–164 km
Producer				
4 dry t/ha (Case 1)	16.50	13.56	10.63	7.70
>5.5 dry t/ha (Case 2)	12.00	9.05	6.12	3.19
Baler	16.06	16.06	16.06	16.06
Hauler	6.71	9.65	12.58	15.51
Total, Case 1	39.30	39.30	39.30	39.30
Total, Case 2	34.76	34.76	34.76	34.76

Table 5. Projected sugar cost based on improved conversion technology.

Process improvement	Technology availability date	Total sugar costs ($/MT)	($/cwt)
Current industrial cellulase expression with mesophilic fermentation system	1998	154	(7.00)
3× specific activity improvement, thermophilic expression system	2003	117	(5.30)
10× specific activity improvement, increased sugar fermentation	2008	106	(4.80)
Higher carbohydrate feedstock infrastructure to support larger plants	2013	84	(3.80)

High-speed tractors were used with "load and go" wagons. One operator could load 17 round bales, about 12 to 14 t, in less than 20 min. Safe tractor-wagon speeds, up to 70 km/hr, permitted economic transport over a 50-km radius. Less than 10 min was required to weigh, sample and unload at the processing location.

Cost reduction to $25/dry t or less is projected from improved baling productivity, transportation efficiency and regulatory changes that include the custom operators in the same category as producers for transporting the stover to the processor. This activity furthers infrastructure development for biomass collection. It also improves the base for accelerating entry of energy crop production.

Corn Stover Conversion

The corn stover is primarily composed of cellulose, hemicellulose, and lignin. The cellulose and hemicellulose are hydrolyzed to hexose and pentose sugars using thermochemical and enzymatic unit operations. Lignin has high value and is used as a fuel for steam and electrical production used in biomass conversion process. Removing just one-third of the stover and hydrolyzing its 38% cellulose content with improved cellulase enzyme systems currently being developed results in 29 million t (64 billion lb.) of glucose with a targeted cost of $132 t ($6/cwt) or less, twice the 58 million t of sweeteners (32 billion lb.) shipped by the corn refiners in 1997.

The stover also contains 32% hemicellulose. When converted to pentose sugars that have less food value, their most likely future use is a nutrient for fermentation processes, with the largest being alcohol. Taking 80% of the hemicellulose to alcohol employing any of at least three genetically engineered organisms (GEOs) presently under development produces 14 billion liters (3.6 billion gallons) of ethanol from one-third of the corn stover, again more than twice the 27 billion liters (1.5 billion gallons) produced annually. The targeted cost is less than $0.25/liter ($1/gal).

Application of proven biotechnological tools is expected to reduce fermentation sugar costs to less than 6 cents/lb. Table 5 summarizes the expectations for process improvement.

BC International is currently starting up a 76 million liter (20 million gallon) per year plant in Jennings, LA for the conversion of hemicellulose using a GEO. BCI will employ one of the three GEO for pentose conversion to ethanol. Also, Iogen and Petro-Canada announced a partnership for construction of a 2 million liter (0.5 million gallon) cellulose and hemicellulose demonstration plant. They will use two other GEO for cellulose conversion to glucose and pentose conversion to ethanol. Both plants are expected to be operating in 2000.

CONCLUSIONS

Corn stover has the potential to produce more than 50 million t of sugar annually, more than doubling the current sweetener industry production. The projected sugar conversion cost can open up new market possibilities by replacing petroleum-derived products. Transportation fuels and other petroleum derived products can be produced from the stover-derived glucose and pentose sugars, reducing the future need for fossil fuels. Realization of this scenario depends of the following: (1) sustainable harvest of corn stover in sufficient quantities; (2) attainment of improved corn stover conversion technology; (3) a worldwide commitment to reduce GHG emissions; and (4) increase in the price of petroleum.

REFERENCES

Bruce, J.P., M. Frome, E. Haites, H. Janzen, R. Lal, and K. Paustian. 1998. Carbon sequestration in soils, Discussion Paper, Carbon Sequestration in Soils Workshop, Calgary, AB., Soil Water Conserv. Soc., Silver Spring, MD.

Campbell, C. 1998. How secure is our oil supply: Dealing with limited reserves. Science Spectra 12:18–24.

Campbell, C.A., F. Selles, G.P. Lafond, B.G. McConkey, and D. Hahn. 1997. Effect of crop management on C and N in long-term crop rotations after adopting no-tillage management: Comparison of soil sampling strategies. Can. J. Soil Sci. 78:1.

Campbell, R.B., T.A. Matheny, P.G. Hunt, and S.C. Gupta. 1979. Crop residue requirements for water erosion control in six southern states. J. Soil Water Conserv. 34:83–85.

Cushman, J.H. 1999. Industries press plan for credits in waste control. New York Times, Jan 3, p. 1.

Gale, W. J. and C.A. Cambardella. 1999. Carbon dynamics of surface residue and root derived organic matter under simulated no-till. Soil Sci. Soc. Am. J. (in press).

Glassner, D., J.R. Hettenhaus, and T.M. Schechinger. 1998. Corn stover collection project. BioEnergy98, Madison, WI p. 1100–1110.

Gupta, S.C., C.A. Onstad, and W.E. Larson. 1979. Prediction the effects of tillage and crop residue management on soil erosion. J. Soil Water Conserv. 34:77–79.

Holt, R.F. 1979. Crop residue, soil erosion and plant nutrient relationships. J. Soil Water Conserv. 34:96–98.

Iogen/Petro. 1998. Can boost fuel-ethanols spirits. p. 1–2. In: Canadian Report on Fuel Ethanol, D. Nixon (publ.), 7-309 Frank St., Ottawa, Ont. K2P 0X7.

Iowa State Univ. Extension Service. 1998. Iowa farm custom rate survey. FM-1698, Rev.

Janzen, H.H., C.A. Campbell, E.G. Gregorich, and B.H. Elert. 1997. Soil carbon dynamics in Canadian agrosystems, p. 57–80. In: R. Lal, J. Kimble, R. Follett, and B. Stewart (eds.), Soil processes and carbon cycles. CRC Press, Boca Raton, FL.

Klopfenstein, T. 1996. Crop residue as an animal feed. p. 315–340. In: P. Unger (ed.), Managing agricultural residues. Lewis Publ., Boca Raton, FL.

Larson, W.E. 1979. Crop residues: Energy production or erosion control? J. Soil Water Conserv. 34:74–76.

Lewis, S. 1997. Value added enzyme applications for fuel ethanol production. 1997 Fuel Ethanol Workshop, Omaha, NE.

Liegois, W.A. 1997. Wet milling operating cost study for Southern Illinois Univ., Edwardsville.

Lindstrom, M.J., E.L. Skidmore, S.C. Gupta, and C.A. Onstad. 1979. Soil conservation limitations on removal of crop residues for energy production. J. Envir. Qual. 8: 533–537.

Lindstrom, M.J., S.C. Gupta, C.A. Onstad, R.F. Holt, and W.E. Larson. 1981. Crop residue removal and tillage-effects on soil erosion and nutrient loss in the corn belt. USDA Agr. Info. Bul. 442.

Lindstrom, M.J. 1986. Effects of residue harvesting on water runoff, soil erosion and nutrient loss. Agr. Ecosyst. Environ. 16:103–112.

Lynd, L.R. 1996. Overview and evaluation of fuel ethanol from cellulosic biomass. Annu. Rev. Energy Environment. 21:403–464.

Parr, J.F. and R.I. Papendick. 1978. Factors affecting the decomposition of crop residues by microorganisms. Chap. 6. In: W.R. Orschwald (ed.), Crop residue management systems. Am. Soc. Agron. Spec. Publ. 31, Madison, WI.

Paustian, K., O. Andren, H. Janzen, R. Lal, G. Tian, H. Tiessen, M. van Noordwijk, and P. Woomer. 1997. Agricultural soil as a carbon sink to offset CO_2 emissions. Soil Use Manage. 13:230–244.

Petrulis, M., J. Sommer, and F. Hines. 1993. Ethanol production and employment. Agr. Inform. Bul. 678. USDA, Washington, DC.

Reicosky, D.C. and M.J. Lindstrom. 1993. Impact of fall tillage method on short term carbon dioxide flux from soil. Agronomy J. 85:1237–1243.

Reicosky, D.C. 1997. Technologies for improved soil carbon management and environmental quality. p. 127–136. In: Incorporating climate change into crop business strategies. Intl. Climate Change Conf. Proc., Baltimore, MD.

Schomberg, H., P.B. Ford, and W.L. Hargrove. 1996. Influence of crop residues on nutrient cycling and soil chemical properties. Chapt. 6. In: P. Unger (ed.), Managing agricultural residues. Lewis Publ., Boca Raton, FL.

Scientific American Special Report. 1998. The end of cheap oil. 278:77–95.

Skidmore, E.L., M. Kumar, and W.E. Larson. 1979. Crop residue management for wind erosion control in the Great Plains. J. Soil Water Conserv. 34:90–94.

Stumborg, M., L. Townley-Smith, and E. Coxworth. 1995. Crop residue export-issues and concerns. Innovations in Straw Utilization Symp., Oct 23–25. 1995, Winnipeg, Manitoba. Agriculture and Agri-Food Canada Research, Swift Current, Sask.

Wald, M.L. 1998. New technology turns useless agricultural byproducts into fuel for autos. New York Times. p. 20.

Wischmeier, W.H. and D.D. Smith. 1978. Predicting rainfall-erosion losses, A guide to Conservation Planning. Agr. Handb. 537, USDA, Washington, DC.

Woodruff, N.P. and F.H. Siddoway. 1965. A wind erosion equation. Soil Sci. Soc. Am. Proc. 29:602–608.

Woolsley, R.J. 1997. Alcohol and driving can mix. Wall Street J. Oct 24.

U.S. General Accounting Office. 1990. Report to the Chairman, Subcommittee on Energy and Power, Committee on Energy and Commerce, House of Representatives. Alcohol Fuels Impacts From Increased Use of Ethanol Blended Fuels. GAO/RCED-90-156.

POLICY

Policy Considerations in New Crops Development

Gary D. Jolliff *

Profitable new crop alternatives have been needed by US farmers for most of the peacetime history of the nation but evidence of comprehensive strategic planning is lacking. Lack of new crop options has been very costly in economic, environmental, social, and political terms and the US trails several countries in strategic planning for new crops development. The long-term, high-risk nature of new crops development is a common barrier to mobilizing private sector interest. Thus, voice, public funding, and leadership, are essential. But competition for resources and the attention of Congress is fierce. A key opportunity and challenge is to develop and present compelling evidence to convince policy makers of their responsibility to legislate for R&D incentives to stimulate new crops development. The European Community (EC) is advanced in doing this. Australia and New Zealand are intermediate in their activities. Global alliances for sharing the risks and the benefits of such developments would advance a strategic approach. United States leadership would strengthen US and world agriculture sustainability.

Economic theory predicts under-investment in agricultural research because it is too easy for "free-riders" to benefit from investments while the investor may not realize acceptable rates of return (Alston et al. 1994). This concept can apply to individuals, companies, states, nations, or groups of nations. The situation is even worse for investment with the higher risks and the longer-term nature of new crops development. Yet, new crops development will be needed even more as widespread adoption of the GATT (General Agreement on Tariffs and Trade) and NAFTA (North American Free Trade Agreement) increases exposure of traditional agricultural crops to free-market forces.

Two key questions are pertinent: (1) is it wise public policy to invest public resources in new crops development? and (2) if the investment is wise, what should be the mechanism for policy development? The objectives of this paper are: (1) to provide reasons why new crops development is needed, (2) to provide evidence that new crops development policy changes are needed, (3) to identify selected barriers to change, and (4) to make policy recommendations to bring about needed changes.

HISTORICAL PERSPECTIVE

More than a century ago, writers were noting the needs and opportunities for profitable new crop alternatives for US farmers. In 1775, George Washington, in a letter, said: "..neither my overseers nor manager will attend properly to anything but the crops they have usually cultivated; and, in spite of all I can say, if there is the smallest discretionary power allowed them, they will fill the land with corn, although even to themselves there are the most obvious traces of its baneful effects. I am resolved, however, as soon as it shall be in my power to attend a little more closely to my own concerns, to make this crop yield in a degree to other grains, to pulses, and to grasses" (Rasmussen 1975). He then continued with a discussion of his own attempts to grow new crops.

Low prices for US farm commodities became a problem soon after the farming sector had recovered from the Civil War in the 1860s (Taylor and Taylor 1952). During the 1890s, the Peffer Report (US Senate 1894) and the George Report (US Senate 1895), from a subcommittee of the US Senate Committee on Agriculture recommended that cotton growers diversify their cropping system and introduce new crops (Taylor and Taylor 1952). Both reports pointed to the problems of low prices for crops produced in surplus, the lack of markets, and the high costs of production.

Farm commodity surpluses again became a problem in the US after the recovery from the disruption caused by World War I. Some of the most useful history on crop surplus problems is in the 1920s writings concerning causes and consequences (for example, Schmidt and Ross 1925; McMillen 1929). The writers understood and correctly predicted the complexity and long-term nature of the problem of agricultural surpluses and the stress they impose on farmers. Yet there were forces to blunt their message for decades while millions of US farmers went bankrupt.

---

*Adapted from G.D. Jolliff (1997).

In 1928, Mead and Ostrolenk stated: "If American farmers were generally to adopt the proven methods of scientific agriculture, the farming industry would go quickly and in grand manner into bankruptcy. Farm prices would dwindle to the vanishing point. ... We have here an amazing paradox. Scientific agriculture is the salvation of the individual, but the ruin of the masses" (Mead and Ostrolenk 1928).

In 1929, McMillen warned: "To the end that the increasing farm output shall not have to struggle for the further expansion of already overstuffed American stomachs nor compete in foreign food markets, greatly increased appropriations to develop new non-food uses [including new crops] are needed. The new national wealth eventually created will pay the bill, and probably with the usual 500% annual interest" (McMillen 1929).

In the 1930s, the Secretary of Agriculture, Henry Wallace, apparently had at least some sympathy for the efforts of the National Farm Chemurgic Movement in promoting the production of industrial raw materials from agriculture, as alternatives to surpluses. However, his heart, and USDA funding, was with corn (Hughes 1935; Wallace 1935). History shows that corn was a big winner, but the farmers' recommendations for new crops development apparently fell on deaf ears, or at least received very limited long-term support.

By 1936, Wallace had observed: "Farm adjustments necessitated by the loss of foreign trade emphasize the importance of developing new uses for crops and crop byproducts. The Department has appointed a committee of scientists from several of its bureaus to study the research now being done in this field and to indicate promising new lines. At present the farmers have profitable uses for only about half of what they grow. The other half they must either throw away or turn to some low-yielding use."

"...Both hope and fears cluster about the possibilities of chemical research and its bearing upon new uses for the products and byproducts of the farm... Both the hopes and the fears should be discounted. There is no possibility either that chemistry will solve all the farmers difficulties overnight or that it will do away with the need for farms" (Wallace 1936).

In 1937, Dr. Karl Compton, President of the Massachusetts Institute of Technology, addressed the Third Dearborn Conference of the Farm Chemurgy movement. His remarks give some evidence of frustration with the forces which resisted the objectives of the Chemurgic Movement: "To my mind the most significant of all encouraging signs is the phenomenal growth of this farm chemurgic movement which is sweeping the country despite opposition from those who misunderstand it or who believe that their personal interests will be served by its failure. But it will not fail because it is pointed right in the direction of progress; it is based on the new philosophy of creating wealth and opportunity for all rather than the age-old instinct of taking wealth from others; it is essentially cooperative between agriculture, industry and the general public rather than competitive between them" (Compton 1937). Apparently, significant opposition was coming from non-farm and petroleum-producing states (Wright 1996).

A letter from the Secretary of Agriculture reporting a survey made by the USDA, relative to four regional research laboratories, in 1939, stated: "Perhaps the most frequent recommendation made to the Department of Agriculture has been that work should be done in the new laboratories to encourage new or replacement crops which may be developed by agriculture" (US Senate 1939).

Although chemurgists and farmers during the 1930s had been discussing the need for development of new crops, political forces seem to have erected a legislated barrier. Wolff and Jones (1958) reported that the Agricultural Adjustment Act of 1938 did not permit any substantial part of the work of the four USDA regional research laboratories to be directed to new crops development. "According to the original mandate, therefore, practically all USDA utilization research in succeeding years was directed toward finding new, improved, or expanded uses of existing commodities" (Wolff and Jones 1958).

This is one of several cases of internal competition within agriculture for resources, resulting in the lack of resources for new crops development. It begs the question: Is the USDA the most favorable place to locate the leadership for new crops development, because of its vulnerability to political pressure? At the outset, the Farm Chemurgic Council was outside the USDA and financed by the Chemical Foundation, Inc. from funds used for the advancement of science (Benson 1937). "There are no strings on chemurgy, no political affiliations, no loyalties to organizations or obligations of any kind" (Benson 1937). It seemed to flourish until the funding mechanism became political. Modern-day farm bill debates have revealed the fierce political pressures from vested interests to which USDA can be subjected.

World War II and the Korean War (1942–53) disrupted world agriculture and created enough export demand that there were few problems with the US excess agricultural production capacity. Several federal investments in new crops development during World War II were justified on the basis of strategic raw material needs, such as guayule for rubber, kenaf for fiber, and pyrethrum for insecticide. The ravages of war in Europe were a significant factor in creating demand for US crop production for several years until the European agricultural capacity was restored.

In the meantime, US agricultural crop production capacity had been expanded by technological developments such as mechanization, nitrogen fertilization, chemical weed control, improved crop cultivars and production management, and irrigation projects. Expanded markets developed during the war years provided the outlets to encourage full use of this improved production capacity.

Thus, in the 1950s, an ever larger US agricultural excess production capacity revisited the US farmer with surpluses and low prices, combined with mounting outlays by the federal government. President Eisenhower appointed a Bipartisan Commission on the Increased Industrial Use of Agricultural Products. Some of the Farm Chemurgic Council efforts may have had some influence on the policy-makers asking for this report. A focus was placed within USDA on increased industrial use of agricultural products (Task Group on New and Special Crops 1957).

Recommendations from the Task Group included the discovery, domestication, and commercialization of new crops. Several potential new crops were discovered, including wild plants from within the borders of the US. Unfortunately, only meager funds, if any, were provided for developing cultivars, production practices, and markets for these crops. Research chemists in USDA had major control of this program. Funding of research in crop improvement, production management, and marketing was low to nil. And, nationally, new crops development again was considered secondary to research aimed at finding new uses for existing commodities.

At the same time, policy makers were convinced of the threat of widespread starvation, and the need for the US to be competitive in the global marketplace. As a result, resources were focused on producing the highest possible yields of existing crops. At that time, the soybean was catching on as a new crop after languishing throughout the late 1800s and early 1900s. Investment in soybean had been stimulated by the crisis of a vegetable oil embargo after World War I. Soybean development had advanced enough to begin carrying sufficient political voice to sustain at least a moderate level of federal funding for R&D. But, a strategic plan for new crops development was still missing from US agriculture policy.

Between 1956 and 1958, Wolff and Jones were participants in the renewed, but short-lived, USDA efforts to find industrial crops as alternatives to the post-war surpluses (Wolff and Jones 1958). In 1957, "The Commission strongly recommends that an adequate annual investment in research and development for new crops be favorably considered, along with suitable authority to the administrator of the program to provide incentives where essential to bridge over the awkward stage of establishment" (Task Group on New and Special Crops 1957. Report to the President's Bipartisan Commission). In 1956, the Federal Budget Office appropriated $100 million per year for R&D work on increased industrial use of agricultural products. It might be instructional to know where those funds were used.

There seems to be little evidence that adequate annual investments were made after 1957 to "bridge over the awkward stage" of development for few, if any, new crops. Some wild plants discovered by USDA in that effort have been domesticated and commercialized to some degree by various piece-meal efforts, but most potential new crops still languish in their development because of a lack of strategic planning, a lack of focused responsibility, and a lack of public resources for research, development, and commercialization.

Frustration with commodity policies, which distorted or destroyed incentives to develop new crops, seemed to be expressed in 1958 by the following statement published on the front page of the April issue of the Chemurgic Digest: "As long as the artificial incentives for overproduction keep pouring on surplus, as Assistant Secretary Peterson indicated in his Conference address, research can hardly be expected to overtake the policy errors. Perhaps we should say to inquirers, Write your senators and congressman and demand that the recommendations of the President's Commission be put into action. Then both public and private industrial research would be stimulated." (Note that the 1995 Farm Bill legislation passed as the Federal Agriculture Improvement and Reform Act of 1996, eliminated several crop price support subsidies for the first time in more than 60 years. Based on the preceding quote, some would view such a policy change as a major advance to help level the playing field for new crops development but it may take a few years for the consequences of the change to be observed and understood.) The oil embargo of 1974 drove up world prices for most commodities, including agricultural products. Additionally, agricultural production systems were faltering in some of the world's large, centrally-planned economies. Consequently, markets developed for US agricultural commodities, again prompting the US agricultural production machine to be unleashed, from fence-row to fence-row. This again resulted in severe economic stress for rural Americans as excess production returned with increased force. Federal government programs designed to take US cropland out of production and control supplies were used more than ever before in history. As a consequence of these market gyrations, lack of profitable alternative crops, lack of relevant policies, and bad decisions by farmers, large numbers of farmers went bankrupt in the 1980s. Very low commodity prices in 1998–1999 again caused severe stress.

Recommendations for new crops development were made again to the USDA and to Congress in Task Force Reports in 1984 (CAST 1984) and in 1987 (Sampson et al. 1987). Congress responded by creating an institutional innovation in the 1990 Farm Bill. It was the Alternative Agricultural Research and Commercialization (AARC) Center (now Corporation) AARC Center mentioned below.

The negligible amount of funding for new crops research within the AARC Center underscores the need for a national commitment to new crops-specific development, with direct funding sources within budgets or within organizations. This would short-circuit the redirection of resources away from new crops to new uses of existing crops as has been accomplished so easily, and so consistently in the past. The creation of at least one separate, narrowly targeted program, is necessary.

NEW CROPS POLICY

There is evidence that it is wise United States public policy to invest public resources in new crops development (Janick et al. 1996; Jolliff 1996). Benefits of past new crops development need to be estimated and documented in the literature, to counter the political barriers to the advancement of new crops development. For example, soybean developed as a new crop in the US from 1920 to 1970 has contributed to: (1) farm-gate wealth generation for the US, (2) rural prosperity, (3) improved US balance of trade, (4) multiplier effects in the US economy (including the advancing of meat animal industries—domestic and foreign, (5) decades of reduced government commodity program payments by providing an alternative crop to surplus crops such as maize, (6) bio-control of pests through interruptions in the maize monoculture, (7) modeling a transition to more environmentally friendly and renewable resource use, (8) the increase in no-till practices to reduce soil erosion. In 1996, soybean was no longer considered a new crop. Public-funded strategic policy is needed for developing new-crops that are economically productive, socially equitable, and ecologically benign (Kloppenburg 1988).

New crops introduced from foreign lands form the foundation of agriculture in the US (Janick et al. 1996) and Australia (Wood et al. 1994a). In fact, very few US or Australian crops have been domesticated from indigenous species.

For more than a century, US farmers have had insufficient numbers of profitable new crops as alternatives to surpluses. The needs for, benefits of, and calls for, new crops development have been well documented (US Senate 1894, 1895, 1939, 1957; Holmes 1924; Schmidt and Ross 1925; National Industrial Conference Board 1926; Mead and Ostrolenk 1928; McMillan 1929; Taylor and Taylor 1952; Task Group on New and Special Crops 1957; Wolff and Jones 1958; Sampson et al. 1987; Jolliff and Snapp 1988; Jolliff 1989; Rexen 1992; Wright 1993, 1995, 1996; Jolliff 1994, 1996; Janick et al. 1996).

However, the US has few, if any, examples of comprehensive tactical operations in new crops development as a result of any non-crisis-based strategic plan. Why is this? Is it that new crops have no voice to compete for resources? Have the command-and-control politics of US agriculture been simply too powerful (Browne 1990)? Are the past benefits of new crops not recognized? Are the costs of not having new crop options not recognized? Do agriculturists at all levels, including scientists, have insufficient understanding of new crops and their development to be advocates of public resource allocation to this topic?

In the 1990 Farm Bill, the US Congress approved the creation of the Alternative Agricultural Research and Commercialization Center (AARC) within the USDA. The Center was established in 1992. There have been many budget battles for funding the Center, partly because of differing views of what the Center should do. During the first three years of operation, most of the Center's resources were devoted to joint-venture commercialization projects related to new uses of existing agricultural commodities. Funded projects reflected personal, corporate, academic, or political affiliations of several of the AARC Board members. Projects involving corn, wheat, cotton, alfalfa, and peanuts received the majority of the resources. Only a few new crops (e.g. kenaf, lesquerella, and milkweed) received funding, and only one project was a research grant.

This is typical of US agricultural politics of the past 200 years. Congressional fights during the debates on the 1995 Farm Bill (which was passed in 1996) confirm that US agricultural policy is heavily influenced by powerful special interests (O'Connell 1994; Kilman 1995). Historically, US agricultural policy has lacked a strategic plan to develop profitable new crop options for US farmers and the national good. There have been several abortive attempts to develop new crops, but these have been crisis-based and piecemeal. They lacked the organizational structure, support, and commitment within a Congressional mandated strategic plan. The USDA needs to rely on strategic planning to attain its goals (O'Connell 1994) because of the swing in political power away from production agriculture. This could be a good sign for new crops development.

Europe

The EC leads the world in strategic planning for new crops development. The First Framework Programme (1984–88) introduced medium-term planning of agricultural research on a European-wide level for the first time. Research on industrial crops and non-food production was the focus because of the costly surpluses of products like sugar, cereals, and animal products and deficits of cellulosic fibers and proteins. Consumer preferences also were changing towards more "natural" and "environmentally-friendly" products (Rexen 1992). A breakthrough was achieved with the Second Framework Programme which ran from 1988 to 1993 (Rexen 1992). The Third Framework Programme (1990–94) overlapped with the Second Framework Programme by two years, and further developed the strategic multidisciplinary, trans-national approach, involving both public and private sectors.

These Framework Programmes included 429 Agriculture, Agro-Industry and Fisheries proposals involving 3,332 participants with a total research budget of 780 MECU (Rexen 1992; Mangan et al. 1995).* Non-food components were expanded in the Fourth Framework Programme, launched at the end of 1994. Mangan et al. (1995) reported on 133 of the projects from 18 countries.

New crops activities in Europe are mainly focused on industrial crops for non-food use or renewable resources (Capelle 1996). The crops activities are placed into three categories: (1) existing crops for new markets; (2) new crops for existing markets; and (3) new crops for new markets. "The third category is the most fundamental, strategic, and is highly risky. Therefore, governments and EC authorities see a clear function for themselves in stimulating activities within this category. Agricultural income, overproduction, rural policy, and environment all play a role in this government support. In the UK, France, and Germany the main topic is alternative energy (biomass, bioethanol, biodiesel), while in other countries, especially the Netherlands, a niche approach is preferred," with an emphasis on vegetable oils (*Crambe abyssinica, Calendula officinalis, Limnanthes alba*), specialty fibers (*Cannabis sativa*), and secondary metabolites (*Carum carvi*).

The UK is participating in the EC Framework Programmes. The Alternative Crops Unit was set up within the UK Ministry of Agriculture, Fisheries & Food (MAFF) to act as a catalyst to carry forward the Government's initiative (and strategy) on developing new crops for industry and energy (MAFF 1995). In recent decades, food production had been the overriding priority for the UK agricultural industry. Now there is widespread interest in new crops for industrial feedstocks, energy, manufacturing, rural development, and use of land no longer needed for food production (MAFF 1995). This effort apparently was launched in July 1994 with the release of a document entitled "Alternative Crops, New Markets" and was advanced by the

---

*One ECU has been referenced as equivalent to approximately 1.20 to 1.33 US dollars (Capelle 1996).

Technology Foresight Programme which emphasized wealth creation and the enhancement of quality of life, by making recommendations for development.

Australia

Australia has an institutional innovation which aids new crop development efforts; the Rural Industries Research and Development Corporation (RIRDC) is jointly funded from public and private sources (Wood and Fletcher 1994). A five-year new crops project jointly funded by this Corporation and the Grains Research and Development Corporation was initiated in July 1993, providing resources for several activities, including support for the First Australian New Crops Conference. This project followed a previous investment in market based methodology to choose new crops for development (Russell et al. 1992) and is supplemented by several small projects on a range of species. In these projects there must be commercial investment to qualify for RIRDC support. Even though five years is a short term for new crops work, the concept is a good start, and one of the more innovative approaches. Time will tell how much political control of programs is exercised by entities tied primarily to traditional crops.

The two-volume report by Wood et al. (1994a,b), also funded and published by RIRDC, is perhaps the most thoughtful, comprehensive, and well-written treatment of new crops development experiences available. They analyzed the development of about twenty species, some of which have become well established crops in Australia and, at the other end of the spectrum, some which have not realized their expected potential. These volumes illustrate the kind of publication which is needed to develop a strategic plan and to design operational tactics. There is little or no economic incentive for private industry to be the sole source of funds for this kind of work.

DEFINING THE PROBLEM

Agriculture is complex and new crops development is not a panacea. But new crops development has a place, and the merits and potential of new crops have shown their worth in the US through the development of crops such as maize, wheat, and soybean. The problem is the over-concentration of efforts and resources on a few major crops during the past century or more. At the same time, there has been under-investment in the development of economically viable alternative crop choices for farmers in sustainable farming systems.

There are positive consequences from increased crop diversity. Profitable new crop options expand the opportunities for farmers to use their ingenuity in developing new cropping systems which can also improve soil and water quality (National Research Council 1993). The social and environmental benefits from new crops development need to be quantified and publicized. "It is a challenge to link the social and economic factors that determine producer behavior with the physical, chemical, and biological factors that determine the effects of that behavior on soil and water quality" (Batie 1993).

Many calls, reports, and recommendations have been made for the development of profitable alternatives to surplus crops (Janick et al. 1996). The literature strongly supports the notion that new crops development could provide opportunities for enhanced agricultural sustainability (US Senate 1894, 1895, 1939, 1957; Holmes 1924; Schmidt and Ross 1925; National Industrial Conference Board 1926; Mead and Ostrolenk 1928; McMillan 1929; Taylor and Taylor 1952; Task Group on New and Special Crops 1957; Wolff and Jones 1958; Sampson et al. 1987; Jolliff and Snapp 1988; Jolliff 1989; Busch et al. 1994; Jolliff 1994; Wright 1995, 1996; Jolliff 1996; Busch 1997). Yet there continues to be serious under-investment in this area.

It is a high-risk, long-term activity and benefits are diffused across producer, processor, and consumer sectors, often without regard to who made the initial investments (Alston et al. 1994; Janick et al. 1996). Alston et al. (1994) explained the economics of public-sector agricultural R&D, and the basis for government intervention in agricultural R&D. New crops research is a prime example of the necessity of public funding to establish and sustain a core national program to identify and develop profit potentials in new crops. This in turn could attract private participation and risk-taking which is necessary for market forces to function as the development of a new crop advances through the commercialization process.

However, new crops development, by definition, has no established support constituency to secure the needed public funding. Therefore, this important area of research and development receives minuscule public investment compared to the need and the opportunity for national return on that investment (Jolliff and Snapp 1988; Jolliff 1989, 1994, 1996).

The Need for More New Crop Options

Farmers lack adequate profitable new crop alternatives to make their cropland productive in a sustainable manner. This is evidenced by data on government-financed idled cropland (i.e. acreage set-aside, or cropland reduction), costly government subsidy programs, and farmer displacement from the land (Jolliff 1996). Data for idled cropland are available from the USDA, since 1933. Approximately 30 to 60 million acres, or 10% to 20% of US cropland, has been taken out of production each year during the majority of the peacetime years since 1933. Had there been profitable new crops available (while meeting newly developed market demands), this land could have been producing new wealth each year. That wealth could have been multiplying throughout the nation indefinitely, as it was invested in various enterprises. Instead, taxpayer funds have been used to take the wealth-generating resource out of production. Opportunities exist to partner internationally to share the costs, risks, and benefits of new crops development. This should be done in the same manner as the international agricultural research centers.

Financial Consequences

Since 1933, the federal government has employed at least six different programs to take cropland out of production to control surplus crop production (Crosswhite and Sandretto 1991). Large financial, social, environmental, and political costs have accrued to the nation as a consequence of idled cropland. Jolliff (1996) estimated the costs and lost opportunities for 1978 through 1994 of certain components of agricultural crop commodity surplus-related programs to be $932 billion, in 1995 dollars. These programs were created (and costs/lost opportunities incurred) substantially because of insufficient profitable new crop alternatives for farmers to remain economically viable, and operate at sustainable risk levels.

The financial cost estimate included: (1) government outlays for farm income stabilization; (2) certificate programs which were used in place of cash; (3) annually-compounded interest on the outlays and certificates (because the government continues to pay interest on borrowed money spent for these programs); (4) lost revenue opportunities on idled land that was available to produce new crops (had they been developed); and (5) the lost multiplier effect (for economic activity) on the lost revenue opportunities. No financial cost estimates were made related to the surpluses during the eras of 1956 to 1973, 1933 to 1942, or earlier eras such as 1919 to 1935, or 1870 to 1914. Further, no estimates were made of the social, environmental, or political costs of farm policy in promoting surplus crop production for any period of US history.

Loss of Farmers

The consequences of the lack of adequate profitable new crop alternatives for farmers has been serious for farmers themselves. Since 1933, more than 4 million US farms (>70%) have been lost as economic units (National Research Council 1995), resulting in approximately 20 million people being displaced from farms. Many farmers have had little or no profitable choice but to produce crops which are commonly in surplus, and part of government programs, and to use the available tools to try to remain economically viable. The improved tools, such as cultivars, technologies, and inputs associated with those government program crops are increasingly controlled by large agribusinesses. This further limits the choices available to farmers (Smith 1992a,b, 1993).

"Some farm groups are concerned that some of the technological innovations that will be necessary to develop new agricultural products, such as genetic engineering, will be made and patented by large agribusinesses. Such a development could contribute to the growing trend of economic concentration in agriculture, these critics contend, and could mean a shift in economic power from individual farmers to private industry, whose interests may not represent what is best for American agriculture" (Congressional Research Service 1988).

Availability of an array of profitable new crop options could mobilize farmer ingenuity at the local level. However, farmers are not well organized outside of their association with specific commodity-based entities.

Therefore, avenues for farmer voice in Washington, DC are generally tied to existing enterprises and industries and innovative areas such as new crops lack representation and action.

Environmental Consequences

The lack of profitable crop and management options during the past sixty years has forced many farmers to accelerate industrialization and specialization in fewer crops in accord with government programs. The results often have been unfriendly to the environment because of increased soil erosion and water quality deterioration. Consequences of government crop production programs are seen, for example, in the changing level of annual average nitrate concentration in ground water (National Research Council 1993). The effects of government programs have been described in a National Research Council Report as follows: "Incentives are perverse. Price support, deficiency payment, and supply control policies should be reformed to remove barriers to voluntary adoption of improved farming systems. The structure of US farm programs induces a bias toward intensive farming practices to boost yields and to expand the base acreage of the cropland that can be enrolled in the price support program" (National Research Council 1993).

Social Consequences

Farm policies have promoted an extractive type of agriculture, producing undifferentiated bulk commodities which are substantially under government control and regulation. Large supplies of low-cost bulk commodities can mean economic prosperity for the many trade industries dealing with marketing, and those providing production inputs. But, increasing portions of that business by-passes rural communities. A small, and declining, share of the retail food dollar is going to the farmer (Smith 1992a,b, 1993). Smith (1992b) reported that "In real terms from 1910 to 1990, the value of the marketing sector grew from $34.5 billion to $216.8 billion, the input sector from $12.6 billion to $57.9 billion, while the farm sector shrank from $24.2 billion to $22.6 billion. The absolute values of the market sector and input sector increased 627% and 460%, respectively, while the value of the farm sector declined over the same period." Thus, rural America is participating in less of the nation's economic activity associated with the agricultural industry (Smith 1992b). Off-farm corporate industries may flourish, even as farmers go bankrupt, rural communities disappear, and rural poverty spreads.

Political Consequences

Government-subsidized export of surplus US crop commodities on the world market, sometimes at prices below the costs of production, place serious negative economic pressure on the agricultural sectors in some developing countries. Development of more crop options would allow free-market forces to operate more widely.

ESTABLISHMENT OF FARM POLICY AND RESEARCH FOR NEW CROPS DEVELOPMENT

General Policy-making Forces and Tendencies in National Politics

Just and Huffman (1992) described how research funding in the land grant system has oriented public research toward private good when it should be more oriented toward public good. The history of new crops research and development bears out the difficulty of retaining focus on the public good.

"The loss of public support has tended to exacerbate political-economic problems of the land-grant system by (1) increasing emphasis on serving the [narrow] clientele that provides political support, (2) increasing efforts by interest groups and administrators to control and prioritize research agendas, and (3) increasing privatization of land-grant activities" (Just and Huffman 1992).

Federal farm policy development has been substantially driven by short-term profits and crisis-based factors. Such forces run counter to the long-term strategic planning needed to give priority to new crops development. Zulauf and Tweeten (1993) see the need to re-order the mission of agricultural research at land-grant universities by providing a new vision for using farm products for non-food and non-feed uses. But, it is likely that Congress would need well-documented, strong justification to take needed action in the absence of strong lobbying forces. It may be necessary for Congress to fund the development of those justifications because there is a lack of resources for such activity. An alternative is to wait until the next crisis of agricultural surpluses, low prices, and farm bankruptcies, and hope that enough "corporate" memory is developing to amass the political pressure needed to pass legislation for sufficient funding. At the end of 1999 it appears that "the next" crisis of agricultural surpluses and low prices arrived in 1998, and continues in 1999. There appears to be little evidence of the amassing of corporate memory to develop needed political influence to fund the Thomas Jefferson Institute for Crop Diversification that Congress authorized in June 1998.

Just and Rausser (1993) argued for the necessity of strategies to alter the landscape of special versus public interests in agriculture. They believe the current trend for scientists within land grant universities to seek private funding, produce privately appropriable products for organized interest groups, and to produce private goods and patents that compete with private sector products will undermine the very foundation of the land grant institutions' existence. Alternatively, they recommended an "active role of organizing political support for public good research activities, and of restructuring incentives to enhance the public-good productivity of research and outreach." They suggest institution building that lowers transaction costs of public interest group formation. These principles apply to the needs for new crops development.

Mtika et al. (1994) studied the perceptions that external interest groups have influence on land-grant university agricultural and natural resource research agenda-setting. Their concern was that the sustainability of the American agricultural system is being threatened by some of the very practices that have contributed to its success. They found that the priority concern of commodity and agricultural industry support groups is to maximize production and profits, while consumer and environmental organizations place priority on consumer and environmental welfare. They cited long-standing calls to redirect land grant university emphasis in agriculture and natural resources research. The call has been to shift from production-oriented agricultural research to that which emphasizes a broad food systems approach which includes linkages between agricultural production, community viability, food safety and security, public health, and environmental protection. The paper recommended that land grant universities focus on reconciling interests of established constituents and potential constituents by developing partnerships in the funding, design, implementation, and evaluation of research. This new paradigm could be tested in developing a national, and global, new crops initiative at a level needed by society to improve agricultural sustainability. This would provide a win-win situation even as the larger agricultural industrial complex sees the opportunities for no-win situations to develop; united voices could encourage Congress to pass needed legislation for new crops development.

Agrarian Myths and Agricultural Policy

Agrarian myths have had a major influence on US farm policy. Browne et al. (1992) demonstrated how specific social and economic interests have used agrarian myths to gain legitimacy in the eyes of the general public. Since World War II, the notion that the world will soon be starving gained a foothold, on the basis of fear founded in experiences from the Great Depression and World War II. Higher yields of the major US field crops was touted as the answer to these frightening problems. This led to the development of the production ideology.

For many years this ideology espoused maximizing production, while little attention was given to the consequences (economic, social, environmental, and political) of such a strategy. Technology was viewed as the answer to all ills. This ideology has been around for the past half century and has become deeply-held dogma. Several elements of the agricultural community placed great trust and focus on yield increases of major field crops as the answer to the needs of rural America and to the alleviation of starvation in developing nations. Curricula in agricultural universities followed suit, eroding attention to topics which explore the potential of, and the prospects for, new crops development and related issues.

The Paradox of Success

The production ideology has led to the paradox of success, in which the consequences of technologies to promote high crop yield served to undermine public support for productivity research (Weaver 1993). Agricultural research has provided technological advances that have dramatically increased agricultural production. However, major economic, social, and environmental consequences have been externalized. Society has reacted. Subsequently, a paradox of success has emerged (Weaver 1993). The result has been waning societal support for an agricultural research system which had successfully generated solutions to quantitative productivity problems for specific sectors of agriculture. Little attention was given to the economic, social, and environmental consequences of the science and technology which led to crop yield increases. Improved understanding, within the agricultural industrial sectors, of these concerns may improve the prospects for broad-based support of longer-term investments in federally-funded research.

Society is forcing a paradigm shift in agricultural research through the reduced funding for research in production agriculture. Because of budget cuts, scientist staffing is being reduced in the land grant university research system—dramatically in some instances. The disciplinary infrastructure critical to a sustainable science and technology base for crop production is in jeopardy for some disciplines at some universities, e.g., funding sources are forcing major shifts of soil scientist efforts from agriculture to environmental soil science at some universities. A need exists for broad-based expertise to address new crops development. A national focus on new crops development would serve as a worthy place to apply and revitalize that disciplinary expertise while producing valuable new annually renewable wealth.

The history of agriculture has not been emphasized in the land grant university curricula in recent decades, so past mistakes may be repeated. Recent generations of graduates in agriculture have been focused sharply on technology and the desire for well-paying jobs. Students were given relatively little, if any, exposure to the factors, thought processes, belief systems, and social interaction skills which are critical for effectively working where agricultural science and technology interface with society. Congress could provide funds to remedy this situation. A well-funded Thomas Jefferson Institute for Crop Diversification (Janick et al. 1996), including an educational objective with World Wide Web resources, would be one good way to start.

Farmer Creativity Compromised

Farmers are among the most innovative workers in the United States. However, when it comes to long-term plant breeding, agronomy, chemistry of utilization, and marketing of new crops, the task is beyond the realm of the farm. Farmers need profitable new crops as the tools to use in developing innovative and sustainable farming systems. Therefore, the national long-term agricultural research agenda is critical to supporting sustainable agriculture. The production ideology, crop subsidies, and current research, focused on a few major crops, have, by default, formed the basis of neutralizing or destroying significant amount of the creative capacity of many farmers who could contribute to the development of profitable new crop alternatives—and, in new farming systems. For several decades, public research funding provided farmer technologies to produce greater excesses when, in reality, they also required profitable new crop alternatives.

Farmers can do much to facilitate the development of profitable new crop alternatives. To do so, they need mechanisms through which to collaborate. They need freedom from the seriously distorting effects which government programs for farm income stabilization and research programs have had on the investment of resources in crops research, development, and marketing. Profitable new crops are tools which many farmers would learn how to use wisely, if they had a chance. Farmers have done this in the US in the past.

Recognition of Reality

Only in recent years has it become widely recognized that frequently, world hunger is the result of economic, social, religious, and political problems, rather than the result of global incapacity to produce food. Hunger in Somalia was a recent compelling illustration of this point, which convinced more of the general public that simple production of more food in the US will not necessarily reduce the number of starving people in the world. It is increasingly recognized that the dumping of US grain on world markets at prices which have been below the costs of production can have serious negative effects on the food production systems of developing countries. Thus, there are far-reaching justifications for the US to increase development efforts for profitable new crop alternatives for farmers globally to facilitate strong agricultural and food production systems world-wide.

Special Interests of Off-farm Agricultural Trade Industries

Numerous off-farm trade groups stand to gain by promoting the production ideology. This leads to the funding of research projects of direct benefit to them in the short run, in preference to the funding of new crops development. Industrial trade suppliers of farm inputs, and buyers of farm outputs can benefit from increased acreages of government program crops. Unfortunately, the low crop prices force farmers into bankruptcy, even though, after changing hands, the land would most likely remain in production (Smith 1992a,b). The costly negative consequences of the production ideology have usually been externalized; that is, the costs are borne by people who are different from those who reap the rewards of the practices (National Research Council 1993). Most of these external costs are not included in the economic assessments of the "fruit" of research investments in the production ideology. However, calls for accountability in agriculture-related industries are increasing (National Research Council 1993; Jolliff 1994).

Browne et al. (1992) have supported the argument that US agricultural policies contribute little to the public well-being. The burden of proof for making agricultural policy changes needs to be shifted from those who wish to change it, to the agricultural interests who defend the status-quo (Browne et al. 1992). Martin (1996) has suggested that there is a need for more comprehensive analysis of the distributional impacts of agricultural research and new technologies; who wins, who loses?

Reichelderfer (1991) discussed how "economic input has often been linked with the perceived, pure objectivity of its dollars-and-cents conclusion...yet even the most objective economic scientist must make subjective judgments in order to conduct an analysis...thus any number of held beliefs, including one's view or definition of "sustainability," can influence the outcome of an analysis of the agriculture and sustainability issue."

Cultures of the Land Grant University and the Federal Agricultural Research Systems

Land grant universities have been a primary source of graduates employed as research scientists and agricultural policy-makers within its own ranks, and within the USDA and other government agencies. Thus, there has been a tendency for certain mind-sets and paradigms to be perpetuated. Some of these paradigms may be so difficult to change that new institutional innovations will be necessary (Meyer 1993, 1995). In such an environment, the forces of organizational inertia may prevail. Systematic distortion of information can take place for the preservation of the organization (Bella 1987, 1996, 1997). Change may be impossible. It is important to consider how the strengths of the existing organizations might be used for new crops development while avoiding those characteristics which may hinder progress.

A national focus on new crops development would need to center its research with the scientific expertise of the land grant university system. Collaboration with scientists within the USDA and other federal agencies would be vital for an effective and efficient program. That type of collaboration was conducted in the 1930s when the Farm Chemurgic Council, which was funded by The Chemical Foundation, led new crop development efforts and promoted USDA and land grant university/agricultural experiment station involvement (Barnard 1937).

Danbom (1991) suggested that the tension in the American agricultural research system has become a crisis of purpose and direction. If that is true, then national leadership, long-term funding, and institutional structure for new crops development becomes all the more important. Otherwise, it may be very difficult for public scientists to dedicate themselves to the complex, and poorly-funded, problem-solving aspects of new crops development. It may be necessary, for example, for non-profit organizations to be associated with selected land-grant universities to facilitate the wise use of federal, and possibly state, funding to get efforts focused in new directions which do not have strong political voices, but which are in the public interest.

The land grant university system, which includes the state agricultural experiment stations and the cooperative extension service, is one of the most successful social experiments in history (Meyer 1995; Campbell 1995). However, the US agricultural research system is under stress, as evidenced by the waning public support for traditional productivity goals (Meyer 1995; Campbell 1995; Weaver 1993). There is increasing public pressure for the focus of research to shift away from purely sectoral goals, such as quantitative productivity, which are usually associated with corporate agriculture, to more broadly-defined societal goals of maximum social welfare, or the greatest good for the most people (Weaver 1993). Such a change of focus in the land grant university research system would represent a quantum leap for some scientists in departments such as crop science and agronomy. With appropriate funding, such changes are feasible.

New crops development is consistent with society's desires to shift research focus to emphasize the diversity of farming systems approaches which include linkages with environmental, social, food safety and security, and public health concerns. However, environmentalists are not in favor of developing new crops with the prospect of simply shifting a traditional extractive type of agricultural system to different plant species. Given the opportunity for funding support, a shift of research focus would provide an opportunity to enlist substantial scientific expertise to advance new crops development. As stated earlier, resources allocated in past efforts to initiate new crops development research regularly have been diverted to new-uses research on existing commodities. Therefore, to get the job done, it is essential that Congress design an innovative institutional structure and provide financial resources (Janick et al. 1996) to mobilize sufficient expertise in problem-oriented research teams, and coordinate efforts to realize successes within much shorter time frames than experienced in the past. Such successes would satisfy and encourage Congress, the public, the scientists, farmers, and the agriculture trade industries so that investment in the national effort could grow, and gain widespread commitment as a long-term investment strategy.

Scientist Culture

Many scientists within agriculture at land grant universities are second and third generation students of the land grant system. That professional environment in recent decades was dominated by the production ideology which led to the "paradox of success."

As a result of the integration of the myths of agriculture and the production ideology, a culture has developed within the community of land grant university agricultural research scientists (Castle 1993; Meyer 1993, 1995). It has been fueled and guided by research granting systems (Huffman and Just 1994) and professional reward structures which promote individual faculty achievement in selected, narrow discipline-focused topic areas. Some of the resulting incremental advances in crop yield served sectoral needs of vested interest groups that play major roles in gaining research funding. Professional scientific societies catered to those trends, but are now seeing the need to help institutional change by taking risks, promoting lifelong learning, advancing diversity, and finding balance, for example, through a common vision (CAST 1996).

Research scientists have had little choice but to work on investigations for which funding was made available by the powers-that-be within the system. There are problems in the funding systems. For example, the competitive grants in the federal peer-review system have been found to have conflicts of interest that subvert the process and make it political (Huffman and Just 1994). Yet numerous scientists with vested interests in the competitive grant system exude an attitude of elitism with that system once they have established their financial foothold. Competitive grants have been a major source of research support in selected areas which commonly serve special interest groups who lobby directly and indirectly to set the research agenda for use of federal and state funds. The competitive grant system is subject to political pressure at all levels.

Just and Huffman (1992) have discussed several facets of the fundamental challenge facing the land-grant system, which are applicable to the topic of new crops development. "The fundamental challenge of the land-grant system is how to achieve effective disciplinary integration, and appropriate balance among basic, pretechnology, and applied research and resident teaching activities while effectively utilizing the individual creativity and broad base of intelligence in the scientific community to respond to emerging public needs in both the short and long run" (Just and Huffman 1992).

Need to Democratize Agricultural Research Agenda-Setting

Busch (1984) argued that democratization of the research formation process, through the broadening of the range of interests that have access to it, would go a long way toward ensuring that the research agenda was not controlled by special interests. For example, if the general public knew the true costs of not having developed profitable new crop options for farmers, new crops development would probably gain significant support in a democratized system of research funding. Scientists, especially those in the public sector, have a responsibility to aid in developing remedies for an apparently faltering system, even if the remedy requires involvement in national policy development.

The Challenge to Change

The process of change in the scientist culture is difficult. "The major problem facing the Land Grant Colleges of Agriculture is that they are part of an intractable, even unruly organization called a university, encumbered with unique characteristics or principles that should not be compromised" (Meyer 1995). Meyer (1993) suggested that the land grant colleges needed help in escaping from old ideas, which meant escaping from old organizations built on the past. "Changing the mindset of personnel throughout the organization is the most difficult task of all, but most essential if renewal is to occur" (Meyer 1995). National agriculture policy, through priority setting and research funding, with a clear priority for new crops development would provide one more incentive for change within the land grant system. Adequate funding in new priority areas would effect such a change quickly.

The Peer-Review System

Peer-reviewing has been used to develop credibility in scientific publications, research agenda-setting and research resource allocation. It has served society well in the past, and may continue to do so in the future. However, there is concern about the system. In the broader university community, Goodstein (1995) sees the peer review system in academia threatened. As referees, scientists perform a professional service and are not required to justify what they write in reviews. Thus, they can "with relative impunity...delay or deny funding or publication of their rivals. When misconduct of this kind happens, it is the referee who is guilty, but it is the editors and program officers who are responsible for propagating the system that make misconduct almost inevitable...purely intellectual competition has now become an intense competition for scarce resources."

Competition for scarce resources has long been a challenge at all levels of funding when working with new crops development. The competition occurs with established crops, between new crops, and among investigators on a single new crop. Significant competition occurs on Capitol Hill where resources are allocated.

The result of this scientist culture has been discipline-oriented or scientist-oriented research which provides results that bring rewards to the scientist, but often do not provide a good solution to the problem because of unanswered questions (Hatfield 1991). A contrasting organizational research structure is one in which the problem, and the development of the research questions to solve the problem, drive the process.

New crops development ought to be an example of a problem-oriented research structure which addresses the needs of society. Problem-oriented research provides for a better understanding of the knowledge gaps and necessary solutions (Hatfield 1991). Teaching problem-based issues requires enough funding to put teams to work. Funding of new crops development has seldom, if ever, had that kind of approach in the US.

Seitz (1991) discussed evidence for dissatisfaction with higher education reward structures, a need for universities to develop clearer statements of mission which are more relevant to social problems, and a need to tighten the linkage of mission and faculty rewards. Again, a focused institutional innovation on new crops development would give a clear mission to which many scientists could contribute, and would have clear social relevance and environmental and economic benefits.

RECOMMENDATIONS FOR NATIONAL AGRICULTURE POLICY CHANGE

Education

Selling policy ideas based on rational judgment requires education. Education about new crops development can serve as motivation, but the topic has not been included in core secondary or higher education. Scientists seldom write about new crops development policy. Little is written about the history of the issue. Incentives to do so are meager, if existent. There is a large education gap, and consequently most present-day leaders need the benefit of exposure to this information. If a new crop is sufficiently profitable, it will be grown. In fact, it may be difficult to prevent it from being grown, as is the case with illicit drug crops. Therefore, profit development is the key to successful commercialization. Profit development requires research investment; and, this has been evident historically with most industries.

Policy-makers respond to their clientele, who would also benefit from information on the benefits of new crops development. The public is readily receptive to the concept of new crops development.

Unfortunately, agricultural policy decisions are controlled by special interest groups who have particular programs to protect. Such groups often have short time scales for problem solving, compared with that required for strategic planning with new crops. So, these special interest groups are important targets for information. They need to be shown the benefits likely to accrue from new crops development.

Three types of information are especially pertinent: (1) benefits of past new crops development, (2) costs and lost opportunities from the lack of sufficient new crops development to meet farmers' needs, (3) evidence of the sufficiency of genetic materials, technologies, expertise, and financial resources to develop profitable new crops, (4) progress with new crops development efforts in various countries of the world (e.g. the EC), and (5) the positive economic, environmental, social, and political impact possible from new crops development.

Credible and convincing data need to be prepared and venues identified for educating those persons who otherwise view new crops development as a threat to their interests.

Incentive Barriers

New crops development involves a complex production system. The diversity of genetic materials, soils, climates, pests, crop husbandry, and human cultures (e.g. economic, education, research, agrarian, industrial, political, religious, military) throughout the world provides every imaginable kind of incentive and counter-incentive for new crops development.

For example, for-profit organizations "live and die" by the economic gains of their endeavors. Therefore, such organizations should not be expected to invest in the early stages of new crops research and development (R&D) which are very long-term and high-risk. When profit potential is discernible, there should be policies to remove barriers preventing private industry involvement. On the other hand, such private organizations, including established crop commodity organizations, should be made aware of the merit of providing political support for public investment in new crops R&D.

Research scientists at public land grant universities in the US over the past 40 years have been rewarded by a system based heavily on publishing in refereed scientific journals. So long as scholarly work was conducted, little scrutiny was placed on the economic, environmental or social relevance of the scientific investigations conducted. Very few resources have been available to university research scientists to conduct investigations on new crops, when compared with the scale and stability of funding available for the established crops. Thus, to the credit of the researchers, new crops research work often was accomplished as the result of curiosity and scholarly inquiry even if funding was scarce. But, that situation has not led to the commercial developments needed by farmers.

There has also been little incentive for a scientist to risk investing time in advancing strategic planning for new crops development, for the good of society. For the past 30 years, there has been an increasing focus in the academic reward system on the acquisition of extramural research grants. Scientists have become entrepreneurs, brokers of science and influence, and business operatives, in order to sustain the economic base of academic research. This has seriously chilled relationships for free information exchange.

Scientists' energies and investigations have been increasingly dictated by the sources of funding. Therefore, the control of the funding sources is substantially the control of the research agenda for public scientists. And, public research funding is highly directed by vested interests through the political process. Thus, long-term high-risk research on new crops receives little support. This topic is discussed further below.

In a similar manner, public scientists working for the United States Department of Agriculture (USDA) invest their time in response to the funding and political pressures applied. Very little incentive, and significant disincentives exist for working on new crops development.

Therefore, policy considerations for new crops development should address the incentive system currently in place.

Making New Crops Development a National Priority.

New crops development needs to be a national priority to attract resources in adequate amounts with sufficient stability to meet strategic plans. Such an approach has been adopted in some countries, as discussed above.

New crops development has no voice or organized industry. Therefore, a substitute force is needed for Congress to use as a basis for action. A comprehensive documentation of the past, present, and future benefits of new crops development could be a powerful force for Congress to use as justification for new innovative legislation. Articulation of the national justification for developing profitable new crop alternatives for farmers should become a focused priority as one aspect of national agriculture policy development.

Public resources should be invested to acquire independent and objective assessments by rural sociologists, crop scientists, agronomists, horticulturists, historians, economists, economic botanists, ethicists, and others to provide the necessary data to accurately characterize and justify needed investment in solving this long-standing national need. In economic terms, the financial, environmental, social and possibly political costs of not having developed sufficient profitable new crop options for US farmers during the past century need to be calculated and added to benefit estimates. Congress would be well served by supporting comprehensive studies of this nature to justify public investments on a long-term basis for new crops development for US farmers. These kinds of studies could also evaluate the institutional innovation needs and opportunities for accomplishing the new crops development task.

Using New Crops-specific Legislative Language

Specific language is required to distinguish new crops development from work on new uses of existing commodities, and to guard against resource redirection or diversion from new crops, as has happened repeatedly in the US. Work on new uses of existing crops has a well-established voice which can speak for itself and should be distinguished from new crops R&D.

National agriculture policy language should be specific for new crops development. This type of support is necessary to bring new crops to a profitability level which will then attract private sector participation in joint ventures for commercialization as now promoted by the AARC Center. Crop domestication, development, and commercialization time frames of 40 to 80 years need to be shortened to 20 to 30 years, via improved funding, to better serve the needs of US farmers and to simultaneously be serving the national good.

Minimum funding of $100 million per year is proposed. This level of funding was appropriated by The US Budget Bureau in 1956 to launch an effort to increase industrial use of agricultural products (Task Group on New and Special Crops 1957). Unfortunately, a political barrier had been written into the Agricultural Adjustment Act of 1938 and did not "permit any substantial part of the work of the four USDA regional research laboratories be directed to new crops development" (Wolff and Jones 1958). Current US agricultural policy should be purged of any similar types of language which may be counterproductive for the good of farmers, rural America, and the nation.

Creating an Institutional Innovation

Such an innovation would require an organizational structure, and voice, targeting specific objectives and known barriers to new crops development. This has been recommended many times in the US, and attempted more than once. As explained above, new crops development has faltered without a Congress-sanctioned and mandated new crops-specific voice (US Senate 1939; Wolff and Jones 1958; I. Wolff pers. commun. 1984; Janick et al. 1996).

Unfortunately, there has not been adequate attention paid to the four policy considerations, above. The outcome has been piece-meal, short-lived programs that have yielded very little commercial product. Several countries of the EC and Australia have some institutional innovations which are likely to be more successful with strategic planning and implementation programs.

New crops development should be kept distinct in mandate from programs designed to develop new uses for existing commodities commonly produced in surplus. This is because farmers still lack sufficient profitable new crop alternatives, which they have needed for more than a century. The remedy is Congressional action.

SUMMATION

United States farmers have needed profitable new crop alternatives to grow in place of surplus crops throughout the entire peacetime history of the nation. Despite many efforts during the past century to encourage change (Janick et al. 1996), agriculture policy initiatives have been inadequate to meet the needs. Why is this?

New crops development is long-term, high-risk, and the benefits often accrue to persons other than those making the initial investments. Thus, it lacks attraction for private investment. Since new crops development benefits the nation in general, public resources are appropriate for support. However, public resources commonly are allocated to areas where there are special interests with forceful representation—and new crops development is not one of those areas.

Therefore, it is necessary to create an environment in which Congress has the incentive to press for the funding of such work. The political climate of the mid 1990s suggested that budget deficits and the national debt provided an important mechanism for connecting with the consciousness of Congress and the American public.

More than four million US farmers (>70% of the total number), plus their families, have been displaced from the land since 1935 (National Research Council 1995). This is substantially because of insufficient economic opportunities, i.e., the lack of sufficiently profitable alternatives to the few major crops commonly produced in surplus. Public resources have been concentrated in traditional crops, and have not been invested in new crops development. Resultant distortions in research investment and markets have been large, widespread, and costly. Economic costs and lost opportunities from not having developed profitable new crop options for farmers has been estimated to total as much as $932 billion, in 1995 dollars, for the brief period between 1978 and 1994 (Jolliff 1996). This problem has been in place for more than a century.

Barriers to policy development include the nature of the policy development process, agrarian myths, special interests which compete with public interests for resources, and organizational cultures in the public-funded research communities.

Society is insisting that the agricultural research system become more responsive to the broader needs of society as contrasted with the narrower, more traditional focus on the ideology of production agriculture. One of the most easily quantified evidences of pressure for change is the reduction in public support for the public agricultural production research system.

Therefore, in the public interest, it is recommended that the United States Congress:
1. Make new crops development a national priority to improve US farmers" sustainability and to enhance rural development.
2. Use legislative language that is new crops-development-specific to: (a) distinguish its intent from work on new uses of existing commodities, (b) to assure intended results, and (c) to guard against resource redirection or diversion—as has happened repeatedly in the past.
3. Create, or modify, an organizational entity to ensure the focused capability, responsibility, and accountability to: (a) respond to the unique needs, opportunities, and challenges of new crops development, and (b) compete with powerful established interest groups for public resources.

REFERENCES

Barnard, H.E. 1937. The background of the Farm Chemurgic Council and The Chemical Foundation. In: Proc. Pacific Northwest Chemurgic Conference with Washington State Planning Council. March 22–23, 1937. Spokane, Washington. Published by Ernest N. Hutchinson, Secretary of State. State of Washington. S.D.P.W. Program 101A-W.P.A. A5101:54–55.

Batie, S.R. 1993. Preface. p. ix. In: Soil and water quality: An agenda for agriculture. National Research Council (U.S.). Committee on Long-Range Soil and Water Conservation, Board on Agriculture. National Academy Press.

Bella, D.A. 1987. Organizations and the systematic distortion of information. J. Prof. Issues Engineer. 113:360–370.

Bella, D.A. 1996. The pressures of organizations and the responsibilities of university professors. BioScience 46:772–778.

Bella, D.A. 1997. Organized complexity in human affairs: The tobacco industry. J. Business Ethics 16:977–999.

Benson, H.K. 1937. Closing remarks and adjournment. In: Proc. Pacific Northwest Chemurgic Conference with Washington State Planning Council. March 22–23, 1937. Spokane, Washington. Published by Ernest N. Hutchinson, Secretary of State. State of Washington. S.D.P.W. Program 101A-W.P.A. A5101:120–121.

Browne, W.P. 1990. The fragmented and meandering politics of agriculture. In: U.S. agriculture in a global setting: An agenda for the future. Resources For the Future: Washington, DC.

Browne, W.P., J.R. Skees, L.E. Swanson, P.B. Thompson, and L.J. Unnevehr. 1992. Sacred cows and hot potatoes: Agrarian myths in agricultural policy. Westview Press, Boulder, CO.

Busch, L. 1984. Science, technology, agriculture, and everyday life. Research in rural sociology and development 1:289–314. JAI Press, Stamford, CT.

Busch, L., V. Gunther, T. Mentele, M. Tachikawa, and K. Tanaka. 1994. Socializing nature: Technoscience and the transformation of rapeseed into Canola. Crop Sci. 34:607–614.

Campbell, J.R. 1995. Reclaiming a lost heritage: Land Grant and other education initiatives for the Twenty-first Century. Iowa State Univ. Press, Ames.

Capelle, A. 1996. New industrial crops for Europe. p. 19–21. In: J. Janick (ed.), Progress in new crops. ASHS Press, Alexandria, VA.

Castle, E.N. 1993. Public policy, credibility and land grant universities. Land Grant Days' Program. Address given Sept. 27, 1993. Utah State Univ., Logan.

Chemurgic Digest. 1963. Special board meeting in New York. Chemurgic Dig. 31:1.

Compton, K.T. 1937. Farm chemurgy: A forecast. Chemurgic Dig. 1(2):1. 1942.

CAST. 1984. Development of new crops: Needs, procedures, strategies and options. Report 102. Council for Agricultural Science and Technology, Ames, IA.

CAST. 1996. Scientific societies: Conversations on change. Special Publication 20. Council for Agricultural Science and Technology, Ames, IA.

Congressional Research Service (U.S.). 1988. New crops and new farm products: A briefing. Dec. 19, 1988. The Library of Congress, Washington, DC.

Crosswhite, W.M. and C.L. Sandretto. 1991. Trends in resource protection policies in agriculture. p. 42–49. In: Agricultural Resources: Cropland, water, and conservation. Situation and Outlook Report AR-23. Economic Research Service, U.S. Dept. Agr. Washington, DC.

Danbom, D. 1991. Challenges to the agricultural research system from a historical perspective. p. 7. In: Proc. Conference on Innovative Policies for Agricultural Research. Nov. 21–22, 1991. Tufts Univ. School of Nutrition, Boston, MA.

Goodstein, D. 1995. Peer review after the big crunch. American Scientist 83:401–402.

Hatfield, J.L. 1991. Problem-oriented vs. scientist-oriented research structures. p. 9. In: Proc. Conference on Innovative Policies for Agricultural Research. Nov. 21–22, 1991. Tufts University School of Nutrition, Boston, MA.

Holmes, C.L. 1924. The economic future of our agriculture. J. Political Econ. 32(5):505–525. Agriculture in reconstruction. The present crisis and the probable future of our agriculture. p. 529–546. In: L.B. Schmidt and E.D. Ross (eds.), Readings in the economic history of American agriculture, Macmillan, New York.

Huffman, W.E. and R.E. Just. 1994. Funding, structure, and management of public agricultural research in the United States. Am. J. Agr. Econ. 76:744–759.

Hughes, H.D. 1935. The future of corn production. Contributions from the Iowa Corn Res. Inst. 1(1):151–152.

Janick, J., M.G. Blase, D.L. Johnson, G.D. Jolliff, and R.L. Myers. 1996. Diversifying U.S. crop production. p. 98–109. In: J. Janick (ed.), Progress in new crops. ASHS Press, Alexandria, VA [CAST Issue Paper 6].

Jolliff, G.D. 1989. Strategic planning for new crop development. J. Prod. Agr. 2:6–13.

Jolliff, G.D. 1994. New crop development as part of sustainable agriculture. p. 93–131. In: Proc. policy discussion workshop held Dec. 8–9, 1993. USDA AARC Center, Washington, DC.

Jolliff, G.D. 1996. New crops R&D: Necessity for increased public investment. p. 115–118. In: J. Janick (ed.), Progress in new crops. ASHS Press, Alexandria, VA.

Jolliff, G.D. 1997. Policy considerations in new-crops development. Invited plenary session lead paper. p. 1–28. In: New crops, new products: New opportunities for Australian agriculture. Vol. 1 (Principles and case studies). Proc. First Australian New Crops Conference held 8–11 July 1996 at The Univ. Queensland Gatton College. Gatton, QLD, Australia. RIRDC Research Paper 97/21.

Jolliff, G.D. and S.S. Snapp. 1988. New crop development: Opportunities and challenges. J. Prod. Agr. 1:83–89.

Just, R.E. and W.E. Huffman. 1992. Economic principles and incentives: Structure, management, and funding of agricultural research in the United States. Am. J. Agr. Econ. 74(Dec. 1992):1101–1108.

Just, R.E. and G.C. Rausser. 1993. The governance structure of agricultural science and agricultural economics: A call to arms. Am. J. Agr. Econ. 75(Oct. 1993):69–83.

Kilman, S. 1995. Risk averse: How Dwayne Andreas rules Archer-Daniels by hedging his bets. The Wall Street Journal CXXXIII No. 83. Western Edition. Oct. 27. Front Page.

Kloppenburg, J.R. Jr. 1988. First the seed: the political economy of plant biotechnology, 1492–2000. Cambridge Univ. Press.

MAFF. 1995. Crops for industry and energy: The government strategy for renewable raw materials for industry and energy. Ministry of Agriculture, Fisheries & Food, Alternative Crops Unit, London.

Mangan, C., B. Kerckow, and M. Flanagan (eds.). 1995. AIR: Agriculture, agro-industries, fisheries: Non-food, bio-energy and forestry. Catalogue of contracts for research and demonstration projects concerning alternative land use, production and processing of biological raw materials for bio-energy, chemical, polymers and forestry products from renewable resources. European Commission Publication EUR 16206EN. European Commission, Luxembourg.

Martin, M.V. 1996. Users of research assessment: Land Grant and industry perspectives. p. 11–12. In: Agricultural research assessment: A symposium summary. Council on Food, Agricultural, and Resource Economics: Washington, DC.

McMillan, W. 1929. Too many farmers: The story of what is here and ahead in agriculture. William Morrow & Company.

Mead, E.S. and B. Ostrolenk. 1928. Harvey Baum: A study of the agricultural revolution. Univ. Pennsylvania Press, Philadelphia. Humphrey Milford, Oxford Univ. Press, London.

Meyer, J.H. 1993. The stalemate in food and agricultural research, teaching, and extension. Science 260:181.

Meyer, J.H. 1995. Transforming the land grant college of agriculture for the Twenty-first Century. Dept. Animal Science, Univ. California, Davis.

Mtika, M., J. Wilkins, L.M. Butler, L.S. Lev, L.J. Gaines, H. Murray, R. Carkner, and R.P. Dick. 1994. Land grant university agricultural and natural resources research: Perceptions and influence of external interest groups. Research report supported by the USDA Western Regional Sustainable Agriculture Research and Education Program. Washington State Univ., Puyallup, and Oregon State Univ., Corvallis.

National Industrial Conference Board, Inc. 1926. The agricultural problem in the United States. National Industrial Conference Board, Inc., New York.

National Research Council (U.S.). 1993. Soil and water quality: An agenda for agriculture. Committee on Long-Range Soil and Water Conservation, Board on Agriculture. National Academy Press, Washington, DC.

National Research Council (U.S.). 1995. Colleges of agriculture at the land grant universities: a profile. Committee on the Future of Land Grant Colleges of Agriculture. Board on Agriculture. National Research Council. National Academy of Sciences. National Academy Press, Washington, DC.

O'Connell, P.F. 1994. Consensus building. p. 77–82. In: Enviro/Economic Sustainability Workshop. Proc. policy discussion workshop held Dec. 8–9, 1993, Chicago, IL. USDA Alternative Agricultural Research and Commercialization Center, Washington, DC.

Rasmussen, W.D. 1975. Agriculture in the United States: A documentary history. Vol. 1. A Letter from Washington to Thomas Jefferson (Oct. 4, 1775). Random House, New York.

Reichelderfer, K.H. 1991. Agriculture and resource sustainability: Can economics help? p. 53–63. In: Understanding the true cost of food: Considerations for a sustainable food system. Symposium Proc. March 1991. Institute for Alternative Agriculture, Inc., Greenbelt, MD

Rexen, F. 1992. The non-food dimension in the EEC research programmes. Ind. Crops Prod. 1:1–3.

Russell, J.S., F.G. Pollock, and D.B. Prestwidge. 1992. A global market analysis of FAO data as a guide to selection of potential new crops for Australia. CSIRO Tropical Agronomy Technical Memorandum 75. CSIRO Division of Tropical Crops and Pastures, Brisbane.

Sampson, R.L. (chairman, and ed.), et al. 1987. New farm and forest products: Responses to the challenges and opportunities facing American agriculture. A Report from the New Farm and Forest Products Task Force to The Secretary, U.S. Department of Agriculture. U.S. Department of Agriculture, Washington, DC.

Schmidt, L.B. and E.D. Ross (eds.). 1925. Agriculture in reconstruction, the present crisis and the probable future of our agriculture. Readings in the economic history of American agriculture. Macmillan, New York.

Seitz, W.D. 1991. Changing professional reward structures in higher education. p. 15. In: Proc. Conference on Innovative Policies for Agricultural Research. Nov. 21–22, 1991. Tufts Univ. School of Nutrition, Boston, MA.

Smith, S. 1992a. Farming activities and family farms: Getting the concepts right. Paper presented at the symposium: Agricultural Industrialization and Family Farms: The Role of Federal Policy. October 21, 1992. Joint Economic Committee, United States Congress, Washington, DC.

Smith, S. 1992b. "Farming:" It's declining in the U.S. Choices. First Quarter:8–10.

Smith, S. 1993. Sustainable agriculture and public policy. Maine Policy Rev. 2(1):68–78. Margaret Chase Smith Center for Public Policy: Univ. Maine, Orono.

Task Group on New and Special Crops. 1957. Report to [U.S.] President's Appointed Bipartisan Commission on Increased Industrial Use of Agricultural Products. R.D. Lewis, Chairman. Texas Ag. Expt. Sta., College Station.

Taylor, H.C. and Taylor, A.D. 1952. The story of agricultural economics in the United States, 1840–1932. Iowa State College Press, Ames.

Tweeten, L. 1995. The twelve best reasons for commodity programs: Why none stands scrutiny. Choices 2:4–7, 43–44.

United States Senate. 1894 and 1895. Committee on Agriculture and Forestry. Agricultural depression; Causes and remedies, Report by Senator W.A. Peffer, submitted to the Senate Committee on Agriculture and Forestry, Feb. 15, 1894 (53rd Congress, 3rd Session, Senate Report 787, serial 3288, v. 1, Washington, 1895). In: H.C. and A.D. Taylor, The story of agricultural economics in the United States, 1840–1932. Iowa State College Press, Ames. p. 19–29.

United States Senate. 76th Congress, 1st Session. 1939. Senate Document 65. p. 16. Regional Research Laboratories, Department of Agriculture. Letter from the Secretary of Agriculture transmitting A Report of a Survey made by the Department of Agriculture Relative to Four Regional Research Laboratories, One in Each Major Farm Producing Area.

United States Senate. 1957. Report to the 85th Congress, 1st Session. p. 61–62. Senate Document No. 45. Pursuant to Public Law 540. Commission on Increased Industrial Use of Agricultural Products. Presented by Mr. Curtis. New crops to prevent surpluses.

Wallace, H.A. 1935. Six decades of corn improvement and the future outlook. Contr. Iowa Corn Research Inst. 1(1):153–158.

Wallace, H.A. 1936. The Secretary's report to the President. p. 1–117. In: Yearbook of Agriculture. United States Department of Agriculture, Washington, DC.

Weaver, R.D. 1993. Strategic issues facing the U.S. agricultural research system. p. 333–353. In: R.D. Weaver (ed.), U.S. Agricultural Research: Strategic challenges and options. Agricultural Research Institute, Bethesda, MD.

Wolff, I. and Q. Jones. 1958. Cooperative new crops research: What the program has to involve. Chemurgic Dig. 17:4–8.

Wood, I., P. Chudleigh, and K. Bond. 1994a. Developing agricultural industries. Lessons from the past. Rural Industries Research and Development Corporation. Vol. 1. R1RDC Research Paper Series 94/1..

Wood, I., P. Chudleigh, and K. Bond. 1994b. Developing agricultural industries. Lessons from the past. Rural Industries Research and Development Corporation. Vol. 2. R1RDC Research Paper Series 94/1.

Wood, I. and R.J. Fletcher. 1994. Editorial. The Australian New Crops Newsletter 1:1.

Wright, D.E. 1993. Alcohol wrecks a marriage: The farm chemurgic movement and the USDA in the alcohol fuels campaign in the spring of 1933. Agr. Hist. 67(1):36–66.

Wright, D.E. 1996. Agricultural Editors Wheeler McMillen and Clifford V. Gregory and the Farm Chemurgic Movement. Agricultural History 69:272–287.

Zulaf, C.R. and L.G. Tweeten. 1993. Reordering the mission of agricultural research at land grant universities. Choices 1993(2):31–33.

New Crops and the Search for New Food Resources*

Jules Janick

Plants are the basis for the human food supply, either consumed directly or fed to animal intermediaries. In prehistory, in various parts of the world, our forbears brought into cultivation a few hundred species from the hundreds of thousands available and in the process of domestication, transformed them to crop plants through genetic alteration by conscious and unconscious selection. Through a long sequence of trial and error, a relatively few plant species have become the mainstay of present day agriculture. The 30 most important crops consumed directly by humans (in order of production by weight of agricultural product) include sugarcane, rice, wheat, maize, potato, sugar beet, cassava, barley, sweet potato, soybean, banana/plantain, tomato, cottonseed, orange, grape, sorghum, apple, coconut, cabbage, watermelon, onion, rape, yam, oat, peanut, millet, sunflower, rye, mango, and bean. Our sustenance as a species is now based on the production of these species. There are three options available for increasing future crop resources: (1) emphasize genetic improvement and more efficient production of the major crops; (2) reinvestigate little known and underutilized crops; or (3) explore plant biodiversity to discover completely new crops. The first option continues to receive the most attention because of political support from vested interests such as growers and processors so that traditional crops have received the bulk of research support by the public sector and practically all of the private sector support, while their agricultural production has been reinforced by expensive subsidies or tax advantages. Furthermore, new advances in biotechnology have focused on the concept of altering major crops rather than minor ones because it offers the best way to increase returns on investment. Present experience indicates that improvement of major crop yields per unit or area of the major crops continues although the research cost per unit of yield increase has also risen. The consequence of this emphasis on major crops results in a continuing erosion of agricultural biodiversity. The expansion of underutilized or completely new crops offers many potential benefits including production diversification providing a hedge for financial and biological risks, national economic advantages by increasing exports and decreasing imports, improvement of human and livestock diets, creation of new industries based on renewable agricultural resources and substitutions for petroleum-based products, and the spur of economic development in rural areas by creating local, rural-based industries. Although interest in underutilized crops has increased as a result of increasing world globalization because new immigrants continue to prefer their traditional foods, there is no world strategic plan for new crop research, which is presently curtailed by lack of long term support. Similarly, the investigation of completely new crops is virtually ignored and is confined at present to the ornamental and pharmaceutical industries. The long term nature and high risk of exploring, developing, and commercializing completely new crops make it unlikely that the private sector can be successful so that government support and leadership is essential. An optimum strategy for expansion of future food resources will require a balance of effort between the three options described above.

The story of humankind is intimately connected with the search for sustenance and nourishment. An analysis of the food habits of other primates indicates that humans were originally scavengers and collectors of food while our dental structure and digestive biochemistry confirms that humans are omnivorous and well adapted to a varied diet. Our history as a species, from inception of the hominid line a million or so years ago to the present, can be viewed in light of changing technology for obtaining food and to increases in our population, both numerically and spatially, as a consequence of these technological changes. If population is indeed a measure of fitness, we are an extremely successful species, multiplying at an ever increasing and now alarming rate (Table 1).

WORLD POPULATION

The growth in human population, although increasing inexorably, has not been uniform over time, and when plotted on a logarithmic scale (Fig. 1) appears as three surges reflecting stages in our cultural and technological evolution (Deevey 1960). The first surge, from one million to about 10,000 years ago, represents

*Also published as *The Search for New Crop Resources* in Plant Biotechnology 10(1), 1999.

technological advances such as progress in tool making, the discovery of fire, and the development of social organization reflecting a change from gathering and scavenging to successful group hunting. This shift in technology caused a rapid increase in the human population and our species dispersed over the entire earth forming associations into communal tribes. However, the expansion of gathering and hunting populations is limited by the fact that human populations must be kept in equilibrium with the carrying capacity of the land. This

Table 1. World population growth, 1990–2100 (Bongaarts 1995).

Countries	Population (billions)			Increase (%)
	1990	2025	2100	1990–2100
Developing	4.08	7.07	10.20	150
Developed	1.21	1.40	1.50	24
World total	5.30	8.47	11.70	121

has been accomplished by a number of adaptive strategies including sexual codes to delay conception and restrict population, constant warfare to maintain territoriality, or even drastic measures such as infanticide for the young or euthanasia for the old. This long phase of human existence as members of hunting societies has had a tremendous influence on our collective psyche. Its influence is felt today in various ways such as the appeal of the chase, the division of labor between men and women, and the social tensions in human interaction ranging from cooperation and community to our predilection for actual or ritualized warfare.

The second great change affecting human history is the invention of agriculture, a series of technologies involving plants and animals used for food (Harlan 1992). Cultivated plants and domesticated animals substituted for the bounty of wild species previously harvested by gathering or hunting. About 10,000 years ago, agriculture first appears as a sweeping and sudden change, at least in the time frame of archeologists who have named it the Neolithic Revolution. The precise origins of agriculture are unknown but earliest evidence for it is found in the highlands of Tigris-Euphrates River complex. It led to another momentous population change in the history of humankind, and our destiny as a species again altered irrevocably.

The third surge in population, brought about by the scientific-industrial revolution, is barely 200 years old, and is with us now. This scientific and technological revolution enormously increased food productivity and efficiency, but the increase in population arose as a consequence of advances in sanitation and medical care that reduced mortality rates, especially in the young. The birth rate fell as populations, no longer needed in a mechanized and more efficient agriculture, exited the rural economy for an urban existence, but not fast enough to compensate for the decline in death rate. The birth rate decline lag in the demographic transformation from high birth and death rate to low birth and death rates has resulted in a huge increase in the growth of human population in poor areas of the world. Equilibrium has been achieved in North America, Europe, and Japan, but not in the rest of Asia or Africa with important consequences for the human condition in the next 100 years (Bongaarts 1995).

THE DISCOVERY OF AGRICULTURE

The ubiquitous association of agriculture and humans makes it tempting to ascribe a single locus and a diffusion pattern. But the evidence suggests that agriculture has resulted from independent but similar discoveries throughout many parts of the world. For example, we find each great ancient civilization based on grain, a nutritious, compact, and versatile source of food (wheat in the Near East, rice in Asia, maize in the Americas, and sorghum and millets in Africa) a demonstration of the technological brotherhood of humans. Although widespread over the earth, the discovery of agriculture was by no means universal. For example, the aborigines of Australia or the

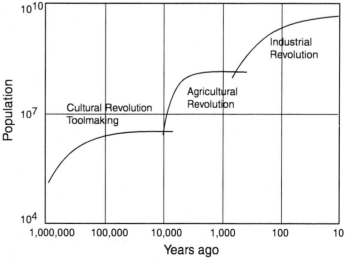

Fig. 1. The growth of human populations reveals population surges based on advances in technology (Deevey 1960).

105

Inuit cultures of the far north never entered this phase and remained as true gathering and hunting societies. What is remarkable about agriculture is the perspicacity of each population in ferreting out desirable food species and transforming them into new entities: crops and domestic animals, a process known as domestication.

Domestication involves two distinct events. One is to identify potentially useful species and the other is to actually transform them into dependable servants. The choice of appropriate species seems obvious when it is completed but so are all acts of genius. The virtue of the original unimproved selected species may not have been so obvious. Cassava, for example, is poisonous, and many crop are unpalatable or inedible without the cooking process. The change from wild plant to crop is accomplished by no less than a genetic transformation achieved through selection of genetic variants that intensify desirable traits and eliminate undesirable characteristics. Selection (differential reproduction) led inexorably to evolutionary changes as some weedy food plants were converted to domestic crops dependent upon humans to complete their life cycles. The traits desirable from a human perspective, such as nonshattering and loss of seed dormancy, are often those which limit survival of the plant. Cultivated plants, unlike weeds, are usually unadapted to exist without the benefit of human interference. The development of crops resulted in a loss of independence of both humans and plants. As in the case of the dairy farmer and his herd, it is not clear who serves whom the most. Many crops, maize, for example, have been so altered that they no longer exist outside of cultivation, and a direct connection to their progenitors has been all but obliterated.

The success of domestication assured the expansion of agriculture. Examples of fundamental alterations in crops are changes that ensure dependable cultivation and increase harvestability and alterations that increase productivity, usually by altering the proportion of the plant that is economically useful (harvest index) rather than an increase in true biological efficiency.

The end result of the agricultural revolution has been a fundamental change in the human condition. The interaction of humans, crops, and domestic animals has resulted in fused genetic destinies. An abundance of food causes changes in selection pressure and alterations of human evolution equivalent to those wrought by the domestication of plant and animal species. Agriculture, by creating not only a dependable food supply but a surplus to be stored, permitted civilization to develop. In the process this new system pushed out the hunter and the nomad and rapidly expanded to all usable land, filling it with people even beyond its capacity! As agriculture produced more food, it instilled the quest for fertility—of corn, of cattle, of soil, of women. The present population explosion has its roots in that phenomenon. The social ramifications of the Neolithic Revolution remain. They include the implication of territoriality and land ownership, our feelings regarding fertility and population, and our attitude regarding community.

FOOD RESOURCES

Vital to our agricultural systems is the choice of servant species to sustain us. The options are prodigious. Thus, there may be 350,000 plant species of which it is estimated about 80,000 are edible. However, at present only about 150 species are actively cultivated, and of these, 30 produce 95% of human calories and proteins (Menini 1998). About half of our food derives from only four plant species (rice, maize, wheat, and potato) and three animal species (cattle, swine, and poultry). Most marine food is still largely harvested from the sea, but this technology is now changing. It needs to be stressed that anonymous and unsung farmers and herdsmen in prehistory, not agricultural scientists, made the choices of most of our current agricultural species. Despite the tremendous advance made by the scientific revolution, the discovery or creation of new crops is a rare and unusual event. We are, in fact, dependent upon Stone Age crops and animals.

It is reasonable to pose the question whether the resource of food species that we now depend upon is sufficient and adequate for the future. Have our forebears made, in fact, the best choices of servant species? Are we hostage to the solutions of the past or can we begin anew? One is awed by the conservatism of the human species seemingly held captive by the resource base of the past. One might intuitively expect, in light of increasing population pressure, that we would be expanding the number of species to sustain and nourish us. The fact of the matter is that the trend for our food economy has been the other way, with fewer and fewer species accounting for more and more of our food. The agricultural history of the United States chronicles the rise and fall of introduced species, but through a process of introduction, trial, and error, it is now based on a very narrow group of food crops with almost 80% of annual row crop area planted to maize, soybean, and wheat.

Many crops have been developed over time in various parts of the world. Food crops may be classified on the basis of their economic importance as follows:

Major crops are cultivated worldwide in adapted areas with high economic value and are associated with high genetic input. They include grains, forages, oilseeds and grain legumes, tuber crops, fruits, vegetables, and sugar crops (Table 2).

Specialty crops are niche crops that, while economically important, have small markets that can be filled by a relatively few growers. Included are a number of horticultural species including fruit, vegetable, and spice crops.

Underutilized crops were once more widely grown but are now falling into disuse for various agronomic, genetic, economic, or cultural factors. In general, they are characterized by much less genetic improvement than the major crops but they are being lost because they are less competitive. Examples include cereals such as emmer and spelt; pseudocereals such as buckwheat; and oilseeds such as sesame and safflower.

Neglected crops, traditionally grown in their centers of origin and where they are important for the subsistence of local communities, are maintained by socio-cultural preferences and traditional uses. These crops remain inadequately characterized and, until recently, have been largely ignored by agricultural researchers and genetic conservation. Yet they may represent our most valuable potential resource for the future. In some cases, their lack of exploitation is an historical accident. Examples include the Andean root and tuber crops, and the minor millets such as *Panicum*, *Paspalum*, and *Digitaria* species.

New crops include those recently developed from wild species whose virtues are newly discovered, formerly collected or wild-crafted species, or synthesized crops created from interspecific or intergeneric crosses. They represent only a handful of cultivated species and very few are included as new foods. Totally new crops from wild species are mainly associated with industrial crops such as *Limnanthes alba* (meadowfoam), a source of unique seed oils, or *Taxus brevifolia*, a source of Taxol, a valuable anticarcinogen. Kiwifruit (*Actinidia deliciosa*), now an important world fruit, is an example of a new crop developed in New Zealand from a crop only previously gathered in China. Newly synthesized crops include triticale, developed from intergeneric crosses between wheat and rye, and two crops derived from interspecific crosses in *Brassica*: harukan, a heading crucifer, and oo, a fodder rape.

Genetically transformed crops include those modified by recombinant DNA technology. Gene splicing is now an established technique with over 50 transgenic crops field tested in the United States. Rates of adoption by farmers for transgenic cotton, soybean, and maize have been very high from the first releases in 1996. In 1998 there were about 2.8 million hectares of transgenic cotton, mostly *Bt* (54% of the total), 8.0 million hectares of transgenic soybean, all herbicide resistant (28% of the total), and 6.9 million hectares of transgenic maize, mostly *Bt* (21% of the total).

What is the reason for this diminution of diversity in our food crops? One would expect that there would be many species among the 350,000 available to us, to have equal or better attributes than those we now consume. I propose four explanations:

1. The crops chosen were not random ones but represent thousands of years of trial and error. They have

Table 2. The 30 major food crops, 1995 (megatonnes).

Cereals	wheat (554), rice (551), maize (515), barley (143), sorghum (54), oat (29), millet (27), rye (23)
Oilseeds and Legumes	soybean (126), cottonseed (58), coconut (47), rapeseed/canola (35), peanut (29), sunflower (27)
Vegetables	tomato (84), cabbage (46), watermelon (40), onion (37), bean (18)
Fruits	banana/plantain (85), orange (57), grape (55), apple (50), mango (19)
Tubers	potato (285), cassava (164), sweetpotato (136), yam (33)
Sugar crops	sugarcane (1168), sugarbeet (265)

survived because of unique attributes that cannot be denied. Wheat, an ancient crop of Southwest Asia, is a complex interspecific hybrid, adapted to bright sunny weather and cool climates for early growth. Its unique properties are based on a combination of seed proteins (gliadin and glutenin) that make possible varied bakery products such as bread, pastry, and pasta. Maize, an ancient crop of Central America, is a C-4 plant that is amazingly productive. Its nutritional deficiencies (low lysine) can be overcome by complementing animal rations with the protein of grain legumes. Rice, a native of China, is especially adapted to grow in wet climates. Potato, adapted to cool climates, has very high potential yields, well-balanced protein, and high versatility in storage and processing.

2. Our major crops have received an increasing amount of grower and scientific attention that has overcome or compensated for many of their deficiencies and increased their adaptation. Value-added processing has increased their economic importance. Thus, maize, grown principally as a source of poultry and swine feed, is now widely used as a source of starch, a sweetener replacing cane or beet sugar, and as a source of ethanol. Soybean oil is used to produce many products including margarines, shortening, and salad dressing, and the resulting protein-rich meal is used in animal feed. Soybean is also the source of many food products including miso (soy paste), shoyu (soy sauce), tofu (soy curd), soy milk (extracted fluid), tempeh (fermented cake-like product), cooked immature beans (edamame), sprouts, and is the source of textured protein for meat substitutes.

3. Our important crops have become part of our social fabric as well as our religious and cultural heritage. We have become addicted to them and, in various culinary forms, they have become mainstays of our diet. A meal without rice is unacceptable in Asia (and much of South America), as is a meal without bread or potato in Europe or North America. It is very difficult to change basic food habits.

4. Finally, the political influence of the growers of basic food grains has encouraged governments to protect them with subsidies and to support them indirectly with basic research funds and marketing assistance. This is especially true in Japan where rice cultivation is even found in urban areas, an indefensible practice from an economic standpoint. It is true in the United States, where maize growers had long been protected by subsidy, and which now, even in light of a planned phased elimination, continues still in the form of support for the ethanol industry.

A STRATEGY FOR EXPANDING NEW FOOD RESOURCES

A legitimate case can be made for expanding crop diversity and for reversing the trend toward monocultures in many parts of world agriculture (Janick et al. 1996). There are, of course, extremely successful examples of new food crops developed from underutilized species of which soybean and canola are the best examples. The soybean has contributed more than $500 billion to the US economy from 1925 to 1985 and canola (low erucic acid rapeseed) has become a major crop of Canada, recently valued at a billion dollars per year by virtue of its healthfulness as a cooking oil based on a significant fraction of long-chain monoenoic fatty acids. New crops advocates suggest that successful new introductions offer alternative means to increase farm income by diversifying products, hedging risks, expanding markets, increasing exports, decreasing imports, improving human and livestock diets, and creating new industries based on renewable agricultural resources. Diversification could spur economic development in rural areas by creating local, rural based industries such as processing and packaging and by providing general economic stability. Furthermore, an expansion of alternate crops could serve the strategic interests of nations by providing domestic sources for imported materials and by providing substitutes for petroleum based products. Diversification would also serve as a form of world food security and would make agronomic sense because reliance on few species poses special hazards and risks due to biotic hazards. The southern maize leaf blight epidemic of 1970 arose because the common male sterile (T) cytoplasm of practically all hybrids grown in the United States was susceptible to an outbreak of a new strain of *Helminthosporium maydis*, a fungal pathogen, that caused a billion dollar loss in a single year. Finally, the use of new species are also important for potential sources of new industrial products of industrial compounds, new foods, and new medicinals.

The long time required for the genetic improvement of wild species, and the high risk involved, makes it unlikely that a rescreening of wild germplasm would be a profitable activity for uncovering new food crops.

In general, a search through wild species only makes sense for medicinal crops. Problems in this area relate to tensions between the governments in the countries where these plants are found and private drug companies, who in any case have shown little incentive to explore botanicals because they are not patentable. It seems clear that this effort will require cooperation between the public and private sector and the countries which claim these untapped germplasm resources.

The reinvestigation of neglected and underutilized crops is a better strategy to obtain new food crops. Recent work with pearl millet (*Pennisetum glaucum*) in the United States suggests that a number of grains could have wider appeal worldwide, especially for special situations such as for arid areas, or for double cropping (e.g. following summer wheat where the season may not be long enough for soybeans). Furthermore, the globalization of our economy has increased interest in ethnic foods, opening up an expanded market for new products.

Many neglected and underutilized crops are locally well adapted and constitute an important part of the local diet, culture, and economy; require relatively low inputs; and contribute to high agricultural sustainability. However, traditional agricultural research in developed countries has hitherto paid little attention to or ignored these crops and, consequently, they have attracted little research funding despite the fact that they are adapted to a wide range of growing conditions, contribute to food security, especially under stress conditions, and are important for a nutritional well-balance diet. Although these traditional crops often are low yielding and cannot compete economically with improved cultivars of major crops, many of these crop species have the potential of becoming economically viable.

A major factor hampering the development of these traditional crops is the lack of genetic improvement and narrow genetic diversity for important agronomic traits. Further constraints are the lack of knowledge on the taxonomy, reproductive biology, and the genetics of agronomic and quality traits. However, because these crops represent the greatest resource for meeting new food needs in the next century, publicly funded research is required. The development of the Consultative Group on International Agricultural Research (CGIAR) Centers and other associated groups have carried out research efforts in this area. These include the International Potato Center (CIP) that supports work in tuber crops; the International Crops Research Institute for the Semi-Arid Tropics (ICRISAT) and the International Centre for Agricultural Research in Dry Areas (ICARDA) which carries out work on crops of the dry semiarid tropics such as sorghum, pearl millet, pigeon peas, chickpeas, and lentils; and the Asian Vegetable Research and Development Center (AVRDC) which carries out work in tropical vegetable improvement. Unfortunately, funding for publicly supported research, both nationally and internationally, is no longer increasing, and in many cases is declining at the same time that the cost of doing research is soaring. This has prevented a serious, long term, world strategic plan for the type of research effort that would increase biodiversity. With a shortage of funds, international research efforts have understandably emphasized only those few major food crops grown in the tropics and subtropics that lead to food security: rice, wheat, maize, sorghum, banana and plantain.

At the present time, there are two competing strategies for meeting the food needs of the future. One is to increase food diversity by exploiting the potential in underexploited and neglected crops. However, genetic improvement requires a long-term, sustained effort. Unfortunately, there are no financial incentives either in the public or private sector to accomplish this feat. Only emphasis on world cooperation will be able to maximize this effort.

The other competing strategy is to seek further improvement of our present major crops emphasizing the new technology of molecular biology now fortified by genomics. For example, it has been successfully demonstrated that oil quantity and quality is amenable to change. The proponents of molecular biology stress the likelihood of altering our present oil crops (soybean or canola) to duplicate other oils. It should be possible, for example, to genetically engineer soybeans to produce oils very close to olive oil, sunflower oil, or canola, and vice versa. Clearly the incentives to do this are powerful. The present protection of intellectual property rights through patents will encourage the private sector to pursue this goal.

The current success of genetically transformed crops in the United States (*Bt* maize and cotton, and herbicide resistant soybean) provides a rationale for this approach. However, because of the enormous expense of this endeavor, the multinational research companies are reluctant to move outside of any but the most important crops. Thus, the trend toward reducing genetic diversity in agriculture is constantly being reinforced.

The coming controversy will be to decide which strategy leads to a more productive and sustainable agriculture. It should not be overlooked that molecular biology may also contribute to the genetic improvement of underutilized and neglected species by overcoming bottlenecks, but the problem is that many of these crops are not inherently productive. Thus, traditional plant breeding is still essential. I expect all these avenues to be pursued, but if I were a betting person, I would not wager against the molecular biological approach because the tide of history is in its favor. In my opinion, a way must be found to pursue both options. The only way to do this is to foster true cooperation between the public and private sector, between national and international research organizations, and among universities and other researchers. The challenge of increasing food resources to meet a doubling of the population before the end of the next century depends on such an approach.

REFERENCES

Bongaarts, J. 1995. Global and regional populations projections to 2025. p. 7–22. In: N. Islam (ed.), Population and food in the early twenty-first century. Meeting future food demand of an increasing population. International Food Policy Research Institute (IFPRI). Washington, DC.

Deevey, E.S. Jr. 1960. The human population. Scientific American Sept. 1960.

Harlan, J.R. 1992. Crops and man. Am. Soc. Agron., Madison, Wisconsin.

Janick, J., M.G. Blase, D.L. Johnson, G.D. Joliff, R.L. Myers. 1996. Diversifying U.S. crop production. CAST Issue Paper 6. Council of Agricultural Science and Technology, Ames, Iowa.

Menini, U.G. 1998. Introductory remarks. World Conference on Horticultural Research. June 17–20. Rome, Italy.

Policy Challenges in New Crop Development

Robert L. Myers

THE PRESENT AGRICULTURAL CONTEXT FOR DIVERSIFICATION

Agricultural systems throughout human history have tended to be much more diverse than present US farms, and many non-western types of agricultural still include many plant species. Even early in this century, most farms had a diversified crop base and kept some native vegetation in place. There are probably several reasons why we have evolved to the present system. One factor has been the increased specialization and cost of farm machinery, although it is worth noting that many alternative crops can be grown with present farm equipment. More significant factors in the move towards monoculture agriculture include the lack of marketing options, the extensive research support for traditional crops, and the emphasis in USDA support programs on commodity crops.

As the agricultural marketplace has become more dominated by large companies, there has been increasing pressure on grain elevators and processors to handle relatively few crops. New processing plants are often built to serve just a single crop, despite the fact that many seed crops can be processed in a very similar fashion. For example, soybean processing facilities can often be modified at reasonable cost to handle other oilseeds. The buying demands of the big grain marketers also connects to the unwillingness of many grain elevator managers to deal with cleaning and storing separately another grain, especially one they are not used to handling, such as canola. Agronomic researchers have tended to concentrate their efforts on the major crops: maize, and to a lesser extent, wheat, soybeans, and cotton. This gives a competitive advantage to the traditional commodities. Within the policy arena, the emergence of politically potent commodity organizations for crops such as maize, soybeans, cotton, and wheat, has assured that these crops get special support from Congress and USDA. This extends through direct commodity payments, export assistance programs, crop insurance offerings, and even rules written for conservation programs.

WHY THE LACK OF SUPPORT FOR NEW CROPS?

New crop development has clearly suffered from a lack of political and policy support, for several reasons. The most obvious is the lack of support for new crops by well-positioned agricultural companies and organizations, especially the commodity groups. However, there have been other challenges, such as explaining to policy makers the complex and time-consuming nature of new crop development. Although new crops can help in the long term with problems like commodity surpluses or weather-induced crop losses, they are not the type of short term fix favored by federal policy makers. It is much more politically potent to announce a $100 million wheat export or hog purchase, that will take effect within a few weeks, rather than launch a long-term effort with new crops that may take decades to reach maximum impact.

For the more research minded policy makers, new crops fail to fit into the push towards basic research, especially genomics and biotechnology. At a national policy level, we have been eager to put hundreds of millions into more basic research on crops we produce in surplus, yet doing applied plant breeding to develop a new crop gets no attention. There is the often expressed attitude, too, that new crops must not have much merit or farmers would already be growing them.

Getting policy support for new crops will depend in part on providing anecdotes of success stories, rather than a heavy dose of statistics on new crop need or potential. Elected officials need to be able to relate to how new crops can help individual farmers or communities that the represent.

SUPPORTING NEW CROPS THROUGH POLICY CHANGES

A number of specific policy steps are needed to spur further new crop development. Foremost is gaining additional funding for research, education, and marketing programs on new crops. A key step in doing this has been getting Congress to authorize a national Thomas Jefferson Initiative for Crop Diversification, which provides for a coordinated program on new crop development.

Appropriations support is still need to make the Jefferson Initiative a reality. Funding for the Initiative needs to be long-term, especially to allow germplasm development and plant breeding to take place. The current trend of making USDA research funds go into short-term competitive grants, is not suited to supporting long-term breeding programs, although such grants can work for certain aspects of new crop development, such as utilization studies.

Jefferson Initiative

In addressing the issue of developing more national for new crops, a Council of Agricultural Science and Technology (CAST) task force recommended a new national initiative (Janick et al. 1996). They proposed that Congress authorize and fund a Thomas Jefferson Initiative for Crop Diversification. Specifically, CAST suggested that the Jefferson Initiative consist of three parts: a set of regional centers, a national coordinating center, and a pool of grant funds available to open competition.

In action on the Agricultural Research, Education, and Extension Reauthorization Act of 1997 (P.L. 105-185), Congress included the recommendations of CAST. Congress authorized appropriations for a national program "conducting research and development, in cooperation with other public and private entities, on the production and marketing of new and nontraditional crops needed to strengthen and diversify the agricultural production base of the United States." The Congress designated that at least half of any appropriated funds should be for "regional efforts centered at colleges and universities." Competitive grants may be awarded to colleges or universities, non-profit organizations, public agencies, or individuals (such as producers). The national coordinating center was authorized to work with the regional centers and help support marketing efforts and other national needs, including development of appropriate federal guidelines affecting new crop development and use. If Congress provides appropriations for the Jefferson Initiative during 1999 action, funds would be made available through USDA-CSREES sometime in the year 2000. At this point there is good support for the Jefferson Initiative among relevant policy makers in Washington, but there are also many competing demands for new funding. The current low prices of the major commodities will help focus attention on alternative approaches, however, such as crop diversification.

Other Areas of Needed National Policy Change

In addition to obtaining federal funding for research and development, there are several other areas where national policy change is needed, including crop insurance. Most alternative crops are not eligible for federal crop insurance. This barrier is amplified because many banks will not give a farm operating loan to plant a crop for which there is no insurance available. The producer who needs an operating loan may have no choice but to grow more conventional crops.

Grain grading standards or other market class guidelines are needed for most alternative crops. Without these in place, farmers who grow a premium product are penalized, and worse, potential buyers may lack confidence in the quality of the crop because of lack of standards.

Obtaining pesticide registration for alternative crops is usually very difficult, given their small acreage. Agrichemical companies have no incentive to spend millions of dollars to register a chemical product for a crop grown on only a few thousand acres. This issue could be partially addressed if EPA were more flexible in extending existing pesticide registrations to similar crops; pesticides registered for canola, for example could reasonably be applied to crambe, which is a member of the mustard family like canola. The lack of current pesticide registrations for many alternative crops means that organic or pesticide-free production methods much be developed, and those production guidelines accepted by organic certification programs.

Support for new and alternative crops can and should be included in other government programs as well, such as conservation programs and export programs. In situations where acreage reduction of commodity crops is a goal, such acreage diversion programs could allow a selected amount of alternative crops to be "demonstrated" on the set aside ground. Taking this approach eliminates the land cost as part of the production, which may amount to 1/3 to 1/2 of the regular cost of growing the crop. This would be a substantial incentive to try a new crop. When conservation programs provide direct payments to farmers, they should allow for consideration of diversified crop rotations, since these often help with erosion control and other environmental goals. New crops can blend with other conservation approaches such as agroforestry or buffer strips as well, and guidelines need to provide enough flexibility to innovate in cropping system design.

Billions of dollars have been spent to export the major commodities of maize, wheat, soybeans, and cotton. Yet not one penny has been spent by the USDA marketing division to offset imports of crops such as canola, sesame, and guar. There is no question that the commodity crops should receive the bulk of our trade assistance dollars, but at the same time it would be highly appropriate and very beneficial to spend a small proportion of these funds to develop crops that can be grown domestically, rather than imported, or to develop crops with good export potential, such as buckwheat, adzuki bean, and kenaf. State departments of agriculture could also help with trade and export issues surrounding alternative crops, rather than focusing so heavily on commodities.

At the local level, institutional and company policies directly affect the potential for new crop success. Farm lenders must be flexible in supporting new crops. Farm machinery dealers must be knowledgeable about how to adapt equipment to work with new crops. Grain elevator managers need to be willing to handle grain other than corn, soybeans, or wheat. In the case of non-grain crops, there needs to be delivery points and arrangements made for marketing the crops through food/crop brokers.

CONCLUSIONS

To be successful in reducing policy barriers to new crop development, a variety of approaches will be needed. These changes will require sustained contact with policy makers over a significant period of time, providing appropriate new crop success stories that can gradually develop a support base for new crops. Immediate needs include obtaining more federal support for new crop research and development, such as through the national Jefferson Initiative. Longer term policy needs involve changes in federal crop insurance, market class guidelines, trade programs, and conservation programs. In addition to change at the federal level, education and outreach is need to reduce attitudinal and policy barriers in local institutions and companies. Over time, the local and national barriers to new crops can be reduced, making diversification more feasible, and providing long term benefit to both the agricultural community and taxpayers as a whole.

REFERENCES

Janick, J., M.G. Blase, D.L. Johnson, G.D. Jolliff, and R.L. Myers. 1996. Diversifying U.S. crop production. CAST Issue Paper 6, Council for Agricultural Science and Technology, Ames IA.

A Strategy for Returning Agriculture and Rural America to Long-Term Full Employment Using Biomass Refineries

Donald L. Van Dyne, Melvin G. Blase, and L. Davis Clements

Production agriculture in Rural America has two major problems. The first is that *it has long-term excess productive capacity at acceptable prices.* The second, accentuated by the recent change in Federal farm program policy, is *widely fluctuating prices with their resulting "boom and bust" cycles.* These severely damage the economies of communities in which production agriculture provides the majority of the economic stimulus. These two problems are closely linked and the foci of this conceptual analysis.

HISTORICAL

Within the context of the two major long-term problems mentioned above, two major events during the last two years will have significant change for Rural America. The first was passage of the Federal Agriculture Improvement and Reform Act of 1996 (PL 104-127, FAIR legislation) that effectively shifted price risks for major grains from the Federal Government to individual farmers. No longer will land diversion be part of the annual absorption of the sector's excess capacity (recognizing that the longer-term Conservation Reserve Program (CRP) remains for up to 10 additional years under current law). Many believe that elimination of Federal farm price supports will increase price instability, thus resulting in wider price swings in the future than have existed previously.

The second major result was the development of a potential negating force for some of the price instability problem that resulted from recently completed research by this projects principal investigators (PI's). It was the discovery of a process to produce a set of important industrial products including ethanol that is *economically viable* without government subsidies as is now the case with fuel ethanol. By simulating the production of ethanol as a coproduct with a higher value chemical, furfural, documentation now exists that huge volumes of grain crop residues and other lignocellulosic (LCF) materials can be converted in biomass refineries to multiple industrial products. In addition to crop residues LCF resources also can include woody biomass, energy crops, wood construction debris, the paper and organics portions of municipal solid waste, and other residues. For purposes of this chapter LCF resources will be described as crop residues when, in fact, other types of residues might be used depending on availability and price. Since the conversion process is fermentation, biomass refineries can select the mix of feedstocks—including grain and LCF resources—that is most profitable. *This concept is extremely important since it would be market driven, not the result of year-to-year Federal agricultural policy.* Finally, not only can the biomass refinery of the future use multiple feedstocks but also it will be able to shift output from the production of one chemical to another in response to market demands. Given that the markets for these chemicals, especially ethanol, are very large, a national chain of biomass refineries can redefine the agricultural/industrial interface. Obviously, these recent findings, described by some as a breakthrough, have tremendous implications for helping to stabilize the price "roller coaster ride" that is beginning to characterize agricultural markets.

An excellent indicator of the long-term excess productive capacity at acceptable prices in production agriculture is the number of cropland acres that have been idled by various types of Federal farm programs. The number of cropland acres that have been idled annually for all Federal farm programs from their beginning in 1933 through 1996 is identified in Fig. 1. Of particular note is that during three multi-year periods large acreages of cropland have been idled. The first was during the 1930s and early 1940s just prior to World War II. The second was from the late 1950s through the early 1970s when more than 50 million acres were idled during nine of those years. Over 60 million cropland acres were idled during three years of this same period. The last multi-year period was from the mid-1980s to today when almost 80 million acres of cropland have been idled during three of the years. During these three multi-year periods there was sufficient excess productive capacity to justify major expenditures by the Federal government to take land out of production, thus helping to reduce production and the cost of storing excess crop supplies. However, when large acreages of cropland were not retired, full production was needed to produce adequate volumes of traditional food, feed and fiber crops to supply both domestic and foreign demand.

Another indication of farm gate price instability is shown for corn (maize) in Fig. 2. The average annual farm gate price in the US is compared to carryover stocks from the previous year. Note the inverse relationship between the two series. For instance, at the end of the 1995 crop year corn stocks were estimated to have dropped to 426 million bushels and the corresponding average farm gate corn price was $3.24 per bushel. This low carryover level resulted primarily because of devastating floods of 1993 and 1995, even though a record production of 10.1 billion bushels was achieved in 1994 (USDA 1998). This degree of instability occurred in spite of the fact that Federal farm programs had protected farmers against low prices in previous years. Now even that protection is gone unless some type of emergency aid is granted such as the $5.9 billion disaster relief aid granted by Congress in 1998, an election year. A more likely scenario is that of the last three years during which Midwest corn prices gyrated from about $5.25 per bushel at the farm gate ($5.54 futures price, Rudel) to about $1.50 per bushel at some Midwest elevators ($1.87 futures price, Rudel). Clearly, such instability does not bode well for Rural America.

DIAGNOSIS OF A RECENT CYCLE

A brief discussion of the events surrounding the period from the late 1970s to date will provide an indication of forces facing agriculture and rural communities on a perpetual and unpredictable basis. Using corn as an example, they are as follows:

1978–82   Farmers urged to plant "fence row-to-fence row" to produce for both domestic and very strong international grain demands.

1980–81   No cropland idled by Federal farm programs as the result of strong demand.

1983   Federal PIC (payment in kind) program where farmers idled cropland and accepted grain from government storage to: (1) reduce government storage costs, and (2) entice farmers to take land out of production.

1984   Reduction in idled acres via conventional programs to restore grain stocks to a "comfortable level."

1985–86   Large crop production years with carryover corn stocks in the 1986–87 crop year of 4.882 billion bushels (USDA 1998, p. 60)

1987–88   Idled 76.2 million and 77.7 million acres, respectively in an attempt to reduce carry over grain stocks and maintain prices at acceptable levels.

1989–95   Idled acreage of 60.9, 61.6, 64.5, 54.9, 59.2, 49.2, 54.8 million acres, respectively. Again, large acreage reductions were used in an attempt to reduce excess supply and the high costs of carryover stocks.

1993, 95   Major flooding throughout much of the corn belt resulted in low levels of corn production on a

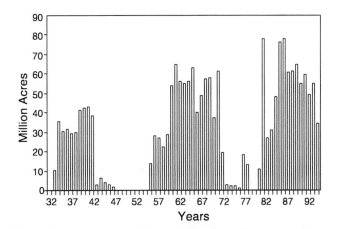

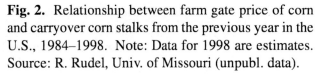

Fig. 1. Cropland idled by Federal farm programs in the US, 1932–1996. Note: Federal farm program legislation was implemented in 1933. Source: Anderson and Magleby (1997).

Fig. 2. Relationship between farm gate price of corn and carryover corn stalks from the previous year in the U.S., 1984–1998. Note: Data for 1998 are estimates. Source: R. Rudel, Univ. of Missouri (unpubl. data).

reduced number of harvested acres, with production of 6.3 and 7.4 billion bushels, respectively in the two years (USDA 1998).

1994   Historic record corn production of 10.1 billion bushel (USDA 1998).

1996   Elimination of most Federal crop price support programs, except the CRP program.

1998   Emergency farm "bailout program" for agriculture with Federal expenditure estimated at $5.9 billion, with grain prices plummeting as illustrated by corn at $1.50 per bushel in some local Midwest cash markets.

While idled farm acreage provides a measure of excess agricultural productive capacity it also suggests the excess capacity that occurs in supporting rural businesses. For example, for each acre of cropland idled, there are corresponding reductions in the purchases of fertilizer, seed, chemicals and other inputs; reductions in the volumes of grain to be stored and marketed; plus reductions in all other supporting rural businesses such as banking services, health care, restaurants, automotive purchases and repairs, etc. Thus, idling of farm crop acreage actually result in a "spiraling down" of virtually all economic activity in rural communities—especially in those areas where agriculture is the primary industry.

From the beginning of Federal farm programs in 1933 until today there have been only 10 years when no cropland was idled. Thus, US agriculture has tremendous excess productive capabilities at acceptable prices in most—but not all—years. *This excess capacity calls for efforts to diversify agriculture beyond traditional food, feed, and fiber to include the production of industrial products.* Markets for traditional commodities are mature. However, markets for industrial products are large. For instance, approximately 115 billion gallons of gasoline and 50 billion gallons of number 2 distillate fuel (about one-half for transportation) are used in the US annually. Large volumes of a multitude of chemicals also exist, as evidenced in Fig. 5. Although intuitively appealing, there are two problems. First, profitable methods must be found for producing large volumes of industrial products from agricultural feedstocks. Potential progress in this area, to be reported in the next section, has been made recently that suggests a breakthrough in this area. Second, what would processing plants producing industrial products from agricultural feedstocks do in those years when all cropland is needed (bid away) to produce traditional food, feed, and fiber? Building upon the recent findings concerning profitable uses of agricultural feedstocks for industrial production, this problem will be addressed with a non-conventional design for the production of ethanol and higher value chemicals.

PREVIOUS AND UNDERLYING RESEARCH

The study that was alluded to above is highly relevant. This analysis, completed by the PI's, evaluated the technical and economic feasibility of converting lignocellulosic feedstocks (LCF) resources into ethanol and higher value chemicals. It was based on technology that could be commercialized today. Part of this technology is being implemented by BC International (BCI) in a plant under construction in Jennings, Louisiana (SERBEP). The most important results include the following. First, the production of ethanol as a *sole* product from LCF resources is technically feasible, but not economically viable when priced at $1.25 per gallon with no subsidies. Second, *co-production of ethanol and higher value chemicals (using furfural as an example) can be highly profitable.* Finally, the optimum size processing plant—with a given volume and location of LCF resources—was estimated for an area in West Central Missouri (See Fig. 3). This chart shows the discounted present value earnings over a 15-year lifetime versus processing plant size represented by the tons of LCF resources that would be processed

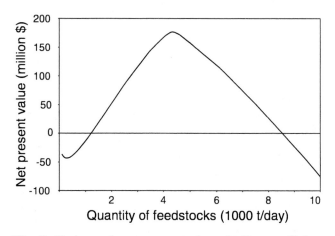

Fig. 3. Estimated net present value of a ligno-cellulose-to-ethanol and furfural plant with varying levels of feedstock use, Carroll County, MO, 1998. Source: Van Dyne et al. (1998)

116

daily. An important finding of the research was that economies-of-scale for processing plant size is extremely important. A plant of less than about 1,300 tons of feedstock used daily is not profitable when producing both ethanol and furfural. Optimum profitability would be reached with a plant that uses about 4,360 tons of feedstocks daily. However, the most likely plant size probably would range from about 2,000 to 2,500 tons of feedstock used daily. Summary results for a plant that would process 4,360 tons of feedstocks daily include:

1. dilute acid processing technology was assumed to be used;
2. potential feedstocks included crop residues, woody biomass and the paper portion of municipal solid waste (MSW);
3. prices assumed for ethanol and furfural were $1.25 per gallon and $0.32 per pound, respectively;
4. plant produces 47.5 million gallons of ethanol and 323 thousand tons of furfural annually;
5. processing plant would cost an estimated $455 million to construct;
6. annual income, costs and net profit would be $281 million, $173 million, and $108 million, respectively;
7. increases employment by an estimated 6,095 jobs (additional temporary jobs would be created during plant construction);
8. increases personal income by $155 million and total economic activity by an estimated $624 million annually;
9. increases local tax base, increases revenues for state and federal income taxes, real and personal property taxes, sales taxes, plus state and federal excise taxes on fuel sales; and
10. reduces imports of petroleum and chemicals.

SUMMARY OF RECENT RESEARCH FINDINGS

This research is monumental because it demonstrates that ethanol can be produced profitably without subsidy if it is coproduced with other higher value chemicals! Further, it begins to suggest how the boom-and-bust cycles can be moderated. In periods of surplus capacity this BioRefinery approach can use both grain and LCF materials to produce ethanol and other higher value chemicals. During periods of high grain prices this approach can utilize LCF materials alone, such as corn stalks or wheat straw or forages (that can be safely produced on highly erodible land like that in the Conservation Reserve Program).

Based on the above, the specific purpose of this conceptual analysis is to provide a method that could help agriculture produce and operate at full capacity on a long-term sustainable basis by including the production of industrial products in addition to traditional food, feed and fiber yet continue the production of industrial products when all the land resources are needed to produce traditional products.

CONCEPTUAL ANALYSIS

This concept proposes to help improve the long-term sustainability of agriculture and rural communities by: (1) more fully utilizing the long-term excess capacity in rural communities, and (2) help stabilize agricultural prices, thus reducing the wide fluctuations in agricultural prices that create "boom and bust" cycles throughout rural America. This will be accomplished by diversifying agriculture beyond traditional food, feed, and fiber products to also include profitable production of industrial products. *It includes the need to further evaluate the technical and economic feasibility of using BioRefineries to convert agricultural feedstocks—both grain and LCF materials—into industrial products.*

Attention now will be turned to the description of various types of BioRefineries. Subsequently, a more thorough description will describe the characteristics of the one studied in this conceptual analysis.

Phase I BioRefinery

An example of this type of processing plant is a dry mill ethanol plant. It uses grain as a feedstock, has a fixed processing capability, and produces fixed outputs of ethanol, feed coproducts, and carbon dioxide. It has almost no flexibility in processing. This type will be used for comparison purposes only.

Phase II BioRefinery

An example of this is the current wet milling technology. This technology uses grain feedstocks, yet has the capability of producing various end products depending on product demand, prices, and contract obligations. Such products include starch, high fructose corn syrup, ethanol, corn oil, plus corn gluten feed and meal. This type begins to suggest the chain of industrial plants needed to moderate the boom-and-bust economy in agriculture today.

Phase III BioRefinery

A Phase III BioRefinery is the type evaluated in this conceptual analysis. A Phase III BioRefinery can not only *produce a variety of chemicals, fuels and intermediate or end products*, but also can *use various types of feedstocks* and *processing methods* to produce the products for the industrial marketplace (see Fig. 4). The *flexibility of feedstock use* is the factor of primary importance for accommodating the changes in the demand for and the supply of feed, food, fiber, and industrial commodities. In brief, this plant of the future will: (1) accommodate a mix of agricultural feedstocks; (2) have the ability to use various types of processing methods, and (3) have the capability to produce a mix of higher value chemicals while coproducing ethanol.

Of extreme importance in the success of long-term production of industrial products from agricultural resources is markets for those products. Fig. 5 provides a schematic description of some of the more important industrial products that can be developed from fossil resources (natural gas, petroleum, and coal). Note that information in Fig. 5 includes not only the path of derivation of various industrial products but also included in parenthesis under each chemical is the volume that is consumed in the US annually (Morris and Ahmed 1992). Most of the consumer end-products identified in the right hand column—as well as many others—can also be derived from biobased resources. In fact, most were derived from biomass prior to the petroleum era which began after the turn of the century. *This conceptual analysis establishes the methodology for evaluating the technological feasibility and the profitability of a multi-input, multi-output plant, i.e., a Phase III BioRefinery.*

Of primary importance in stabilizing agricultural production—and thus commodity prices—over time can be seen by evaluating the significance of Fig. 1, 2, and 5. First, the industrial products such as chemicals and fuels compete mostly with fossil resources for market share, not other agricultural commodities and products. Second, the mix of feedstocks used by the BioRefinery depends on their relative prices. For instance, when large cropland acreages need to be retired and large volumes of grain are in storage, grain prices are typically depressed. In this situation the BioRefinery will use considerable volumes of grain to produce industrial products. However, when little or no cropland is retired, carryover stocks will be low and grain prices will be relatively high. Under this scenario the BioRefinery will use most/all lignocellulosic feedstocks and no grain. *This substitution is possible with the front end equipment complement in the BioRefinery because both types of feedstocks yield sugars as intermediate products in a BioRefinery. Once the glucose is available as a fermentable sugar, essentially the same process equipment can be used to produce ethanol, acetic acid, acetone, butanol, succinic acid, or other products of fermentation.* Thus, the flexibility of different feedstock use in the production of industrial products by the BioRefinery is important for the successful development of the conceptual analysis. The flexible feedstock use is essential in developing the most profitable mix of higher

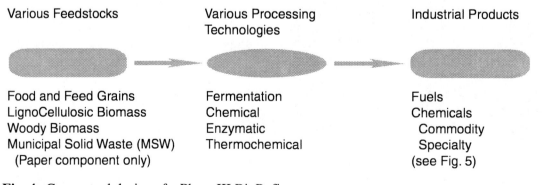

Various Feedstocks	Various Processing Technologies	Industrial Products
Food and Feed Grains	Fermentation	Fuels
LignoCellulosic Biomass	Chemical	Chemicals
Woody Biomass	Enzymatic	Commodity
Municipal Solid Waste (MSW)	Thermochemical	Specialty
(Paper component only)		(see Fig. 5)

Fig. 4. Conceptual design of a Phase III BioRefinery.

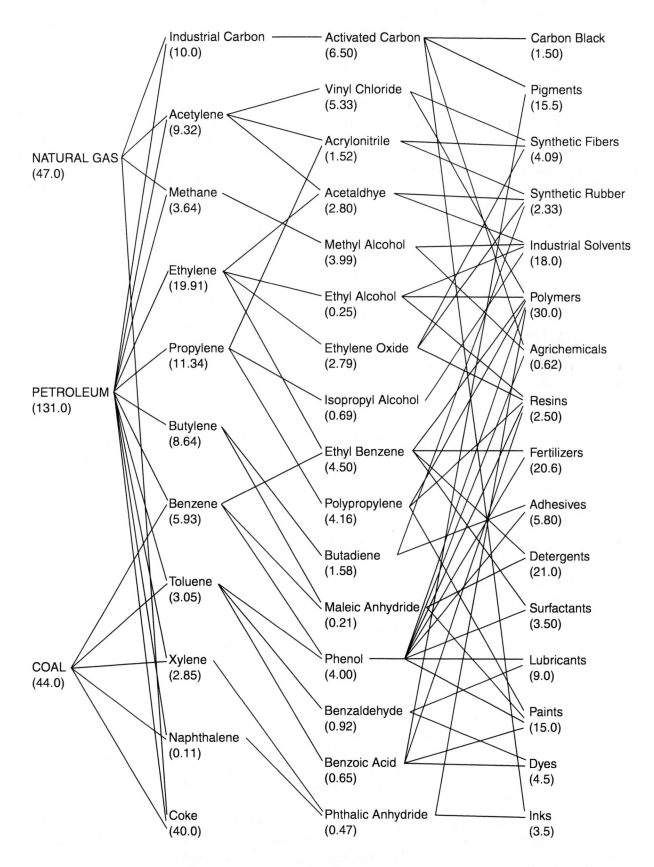

Fig. 5. Interrelationship of major precursor chemicals derived from fossil fuels for the manufacture of consumer end-products in the US (millions of tons). Source: Morris and Ahmed (1992)

value chemical outputs. Moreover, "smoothing" the year-to-year changes also helps stabilize prices in the long-term and increases the economic sustainability of Rural America.

Three important factors should be reemphasized. *First,* the stabilizing of production, use and prices will be possible as the result of relative prices in the marketplace—without government intervention. *Second,* it also assures that the production of industrial products will not be at the expense of having inadequate supplies of food, feed, and fiber commodities because the supply-demand conditions are based on market demand, not inflexible Federal farm policies. *Finally,* the BioRefinery will help to fulfill the large void anticipated as the result of the Federal government transferring risks to farmers created in the "Freedom-to-Farm" legislation.

Example of a BioRefinery Approach

Lignocellulosic materials are made up of three primary chemical fractions, hemicellulose (a polymer of five-carbon sugars), cellulose (a polymer of glucose, a six-carbon sugar), and lignin (a polymer of phenol). The sugar polymers, hemicellulose and cellulose, can be converted to their component sugars through the chemical addition of water. This process, called hydrolysis, is relatively easy in the case of the hemicellulose and rather more difficult for the cellulose fraction.

The glucose fraction from cellulose can be fermented to produce materials as described above. The xylose fraction from hemicellulose can be fermented by some organisms, but these fermentations are notably low in yield. The remaining material, the lignin, is a natural adhesive which has some commercial value, but because it has a heating value approaching that of sub-bituminous coal, but with no sulfur content, it is a premium quality solid fuel. An attractive option for the use of the xylose fraction, however, is its conversion into furfural, and as described below, into furfural derivatives, including nylon.

The biomass-to-nylon process takes advantage of the fact that the hemicellulose fraction is readily hydrolyzed to xyloses. The reaction conditions most commonly used are reactions with dilute sulfuric acid at a temperature of about 160°C. The same reaction conditions that hydrolyze the hemicellulose also can further convert the xyloses to furfural, an industrial chemical used in refining of motor oils, for making certain plastics, and use in new "clean" liquid fuels.

$$\text{Lignocellulose} + H_2O \rightarrow \text{Lignin} + \text{Cellulose} + \text{Hemicellulose}$$

$$\text{Hemicellulose} + H_2O \rightarrow \text{Xylose}$$

$$\text{Xylose } (C_5H_{10}O_5) + \text{Acid Catalyst} \rightarrow \text{Furfural } (C_5H_4O_2) + 3\ H_2O$$

$$\text{Cellulose } (C_6H_{10}O_6) + H_2O \rightarrow \text{Glucose } (C_6H_{12}O_6)$$

Furfural in fact has many uses, but important to this discussion is that furfural can be converted into both of the precursors of nylon 6,6 or into the raw material for nylon 6. The original process for making nylon 6,6 was based on furfural. The last of these plants closed in 1961 because of the artificially low price of petroleum. Nevertheless, the size of the market for nylon 6 is huge.

The production of the nylon precursors requires hydrogen which is a product of the gasification reactions used to destroy waste materials in the waste management/power production part of the plant. Similarly, the nylon plant produces carbon monoxide, carbon dioxide, and methane which are products in the fuel gas made in the gasifier section of the waste management plant.

In the case of the biomass conversion, there are additional opportunities. The hydrolysis of the hemicellulose takes about one-third of the total biomass fraction. The remaining two-thirds, cellulose and lignin, can be used for power production, but this discards an additional valuable resource, the cellulose.

The hydrolysis of cellulose to glucose can be carried out through chemical processing or by enzymatic processing. The rotting of wood in the forest is the natural version of enzymatic hydrolysis that can be used for the preparation of glucose. The cost of commercial enzymes is high, and the process is rather slow. Chemical methods, particularly the use of mineral acids and higher temperatures than those required for hemicellulose hydrolysis, lead to commercially viable production of glucose from cellulose. The original acid hydrolysis process for cellulose, called the Schöller process, was used extensively before and during World War II to produce ethanol for fuel.

An attractive addition to the biomass-to-nylon process is the hydrolysis of cellulose to glucose. Once the glucose is available as a fermentable sugar, essentially the same process equipment can be used to produce ethanol, acetic acid, acetone, butanol, succinic acid, or other fermentation products.

In sum, the process strategy is to create a highly integrated waste management, power production, and chemicals and fiber production complex. The three aspects of the complex complement each other in terms of exchange of inputs and products, making the sum much more profitable than the parts. *In essence, the effluents from one segment of the operation are the inputs for another segment. The result is an extremely efficient, near-zero discharge facility. The several technologies involved are all proven—the competitive advantage lies in the integration of the parts.*

Implications for Economic Analyses

Economic analyses—based on the assumption that the processing capabilities are technically feasible—are absolutely essential for the BioRefinery concept to work. That is, industrial products made from agriculture—relative to those made from other feedstocks including petroleum, coal, and natural gas must be price competitive. Also, they must be at least as high quality, have similar or better performance characteristics, and have as dependable supply over time as those derived from traditional resources.

A tool that is useful in estimating the economic feasibility of a BioRefinery is a non-linear optimization model, General Algebraic Modeling System (GAMS). Initially, GAMS can be useful in estimating the optimum size BioRefinery by evaluating the tradeoffs between economies-of-size of the processing plant versus increasing feedstock costs that result from delivery of biomass resources from more distant locations for ever larger plant sizes. The optimum size BioRefinery plant will vary among locations because of differing feedstock densities, as well as feedstock characteristics (relative composition of cellulose, hemicellulose, and lignin) that are used in making industrial products. The GAMS model of a Phase III BioRefinery allows a variety of feedstocks—both grains and crop residues—in a variety of locations to be used as inputs, while also allowing a variety of industrial products, depending on: (1) the feedstock composition (cellulose, hemicellulose, lignin, starch, etc.), and (2) economic parameters such as feedstock price, transportation costs, product prices, processing plant costs, and other relevant factors. Hence, the model will show not only the minimum cost feedstock mix, but also the location from which it will be drawn and, of course, the location of the BioRefinery. Moreover, the optimum size plant also will identify the mix of industrial products to be produced, given the product prices. The sensitivity of the optimum profitability of the plant can be evaluated by varying the most important and relevant prices, costs, and conversion efficiencies.

The methodology to evaluate the economic feasibility of a BioRefinery described above is similar to that used in the petroleum refinery industry. While their feedstock is crude oil, the mix of products to be produced for the day or week varies depending on relative prices, product demand, and contract obligations. The determination of the volume of each type of product, i.e., gasoline, distillate, residual, etc., to be produced is determined by some type of mathematical program such as linear programming (LP) or nonlinear program such as the GAMS model described previously. The frequency of using LP or GAMS in determining product output depends on various factors such as volatility of relative product prices and changing contract status for various products.

Implications for Policy Analyses

Development of a series of Phase III BioRefineries throughout the US could address the two major problems in Rural America identified in the opening paragraph of this paper: (1) long-term excess productive capacity at acceptable prices, and (2) widely fluctuating prices with their resulting "boom-and-bust" cycles. Producing industrial products in addition to traditional food, feed, and fiber feedstocks would help to more fully utilize all rural resources on a long-term sustainable basis. Also, the feedstock flexibility of Phase III BioRefineries would result in firms responding to relative market prices to choose between using grains and

LCF resources, thus helping reduce wide price swings. This would provide much more efficient and reasonable use of agricultural resources, than relying on prices influenced by Federal government programs as we have since their beginning in 1933.

An extremely important issue is how to go from a Federal farm program mentality to an agriculture that would include the production of industrial products as an important long-term mission for rural America. While considerable detail is necessary to move toward biobased production of fuels and chemicals, the single most important factor is *the reduction of risk to the investors and producers*. Fortunately, the risk reduction issue is not new to agriculture and rural communities. Risk sharing is commonly done through loans and loan guarantees from various Federal agencies. For example, USDA provides various types of loans and loan guarantees. The Farm Service Agency offers, "direct and guaranteed farm ownership and operating loan programs to farmers who are temporarily unable to obtain private, commercial credit." Also, loan guarantees can be made for housing in rural areas. Additionally, various types of support are available through the Rural Business and Cooperatives Service, the Rural Community Development Service, the Rural Housing Service and the Rural Utilities Service. Services provided by these USDA agencies range from loans and loan guarantees to helping to assess business opportunities. Other Federal agencies also provide various types of loans and loan guarantees to help share risk.

CONCLUSIONS

The potential for enhancing rural development and economic viability as well as improve environmental benefits throughout rural America represents an overwhelming opportunity for seriously studying and developing a strategy for moving forth with the BioRefinery concept. This nation-wide effort would effectively help to more fully utilize rural resources while also reducing the year-to-year fluctuations in commodity prices and thus helping to provide stability in rural economies.

The primary reasons that the BioRefinery can be a valuable component in helping to stabilize and enhance rural economies include the capabilities of: (1) using a variety of grain and lignocellulosic feedstocks; (2) using a variety of processing capabilities; and (3) producing a wide variety of intermediate and end consumer industrial products, including fuels plus commodity and specialty chemicals. The flexibility in feedstock use provides the opportunity to vary input use based on relative feedstock prices that are market driven, not the result of Federal farm policy. Secondly, a Phase III BioRefinery would have various types of processing capabilities which should result in more fully utilizing feedstocks, thus reducing the volume of low value or "waste" products from the processing stream. Finally, the flexibility of being able to produce various types of products and coproducts based on relative prices helps to assure maximum returns to the BioRefinery.

Diversifying beyond traditional feed, food, and fiber crops to also include production of feedstocks for industry has immense potential. A major question at this time is with the potential positive opportunities that such an effort would result, why didn't this effort occur long ago? Basically, while the effort has been attempted numerous times previously—usually in periods of surplus commodities—the technological feasibility of a Phase III BioRefinery now makes the potential for success much more promising in the future of rural American economies. Moreover, with the potential for much larger swings in product prices resulting from elimination of Federal farm programs, it becomes much more important now than ever before. The time to act is now!

REFERENCES

Anderson, M. and R. Magleby (ed.). 1997. Agricultural resources and environmental indicators, 1996–97, USDA, Natural Resource and Environmental Division, Agricultural Handb. 712.

Clements, L.D. and J.R. Halvorsen. 1990 Production of polyethylene and polyvinyl chloride from ethanol using super critical processing. Paper presented at the 1990 AIChE National Meeting.

Clements, L.D., S.R. Beck, and C. Heintz. 1983. Chemicals from biomass feedstocks. Chem. Engr. Prog. 79:59–62,

Landucci, R., B. Goodman, and C. Wyman. 1994. Methodology for evaluating the economics of biologically producing chemicals and materials from alternative feedstocks. Appl. Biochem. Biotech. 45–46.

Morris, D. and I. Ahmed. 1992. The carbohydrate economy: Making chemicals and industrial materials from plant matter, The Inst. for Local Self Reliance, Washington, DC.

SERBEP. 1998. World's first commercial biomass-to ethanol plant. SERBEP Update, Southeastern Regional Boipmass Energy Program, Tennessee Valley Authority, Muscle Shoals, AL.

USDA. 1998. Feed: Situation and outlook report. Economic Research Service, FDS-1998.

Van Dyne, D.L., M.S. Kaylen, and M.G. Blase. 1998. The economic feasibility of converting ligno-cellulosic feedstocks into ethanol and higher value chemicals. Draft Report prepared for the National Ethanol Research Institute, administered by the Consortium for Plant Biotechnology Research, Inc. and the Missouri Division of Energy, University of Missouri, Dept. Agricultural Economics.

Wei, J., T.W.F. Russell, and M.W. Swartzlander. 1979. The structure of the chemical processing industries. McGraw-Hill, New York.

Legal and Technological Measures to Prevent Farmers from Saving Seed and Breeding Their Own Plant Varieties

Hope Shand

The Rural Advancement Foundation International (RAFI) is an international civil society organization, headquartered in Canada, with an affiliate office in Pittsboro, North Carolina (USA). We work primarily at the international policy level. RAFI is concerned about the loss of biological diversity—especially in agriculture—and about the impact of intellectual property on farmers and world food security. We also conduct research on the social and economic impacts of biotechnology on agriculture and farm communities. Our staff has campaigned against Plant Breeders Rights and other forms of plant intellectual property since 1980. We believe that monopoly control over plants, animals, and other life forms jeopardizes world food security, undermines conservation and use of biological diversity, and threatens to further marginalize the world's poor.

CHALLENGES TO AGRICULTURL BIODIVERSITY

I'd like to mention just three trends that we see as major challenges facing conservation and sustainable use of agricultural biodiversity. The first trend, and probably the most obvious one, is *genetic erosion*. We're losing genetic resources for agriculture at an unprecedented rate. What's at stake is the biological basis for world food security. Forests are falling, fisheries are collapsing, plant and animal genetic diversity is eroding all over the world. The statistics are sobering:

- Crop genetic resources are disappearing at the rate of 1–2% per annum.
- Domestic animal breeds are disappearing at an annual rate of 5%, or 6 breeds per month. FAO considers 30% of all livestock breeds endangered or critical.
- Tropical forests are falling at a rate of just under 1% per annum, or 29 hectares per minute (CGIAR 1996). From 1980–1990, this is equivalent to an area the size of Ecuador and Peru combined.
- All of the world's main fishing grounds are being fished at or beyond their limits. About 70% of the world's conventional marine species are fully exploited, overexploited, depleted or in the process of recovering from over fishing (FAO 1995). During this century, about 980 fish species have become threatened (CGIAR 1998).

The loss of biodiversity threatens global food security, especially for the poor, who rely on biodiversity for 85 to 90% of their livelihood needs.

The second trend I'd like to mention is privatization of plant breeding and seed sales. The first half of 1998 witnessed a dramatic consolidation of power over plant genetic resources worldwide, a trend that began over three decades ago. The global seed trade is now dominated by "life industry" corporations whose vast economic power has effectively marginalized the role of public sector plant breeding and research. Of course, the consolidation trend is not just in seeds but in all sectors of the life industry.

- 20 years ago there were thousands of seed companies, most of which were small and family owned. Today, the top 10 global seed companies control 30% of the $23 billion commercial seed trade.
- 20 years ago RAFI was monitoring about 65 agrochemical companies involved in the development of new crop chemicals. Today the top 10 pesticide manufacturers account for 82% of the $30 billion dollar global market.
- Two decades ago, the top 20 global pharmaceutical companies controlled roughly 5% of the global pharmaceutical trade. Today, the top 10 pharmaceutical companies account for 36% of the $251 billion global pharmaceutical market.
- Today, the top 10 firms held 61% of the animal veterinary market valued at (US) $16 billion.

If you look at the dominant companies in all of these sectors—in plant breeding, pesticides, veterinary medicines, and pharmaceuticals—you'll find that many of the same companies are dominant in all of these sectors. Companies such as Monsanto, Novartis, Dow, and Dupont are everywhere.

The bottom line is that fewer and fewer companies are making critical decisions about the agricultural research agenda, and the future of agriculture worldwide. With the advent of genetic engineering life industry corporations are staking far-reaching claims of ownership over a vast array of living organisms and biological processes. The power of exclusive monopoly patents is giving these companies the legal right to determine who gets access to proprietary science and at what price. This has serious implications for the future of world food security and for conservation of genetic resources.

The third trend I want to mention has to do with farmers and their role in conserving and using agricultural biodiversity. The future of world food security depends not just on stored crop genes, but on the people who use and maintain crop genetic diversity on a daily basis. In the long run, the conservation of plant genetic diversity depends not so much on a small number of institutional plant breeders, but on the vast number of farmers who select, improve, and use crop diversity, especially in marginal farming environments. That's why we are particularly alarmed by the current trend to eliminate the right of farmers to save and exchange seed, and breed their own crops. Both public and private sector institutions are using, developing, and promoting a variety of legal and technological tools that are designed to give the seed industry greater control over plant genetics and eliminate the right of farmers to save and re-plant seed from their harvest.

Let me back up a minute to explain, what's so important about the right of farmers to save seeds? Two points: Farmers are not just saving seeds, they are plant breeders who are constantly adapting their crops to specific farming conditions and needs. For over 200 generations, farmers have been selecting seeds and adapting their plants for local use. This genetic diversity is key to maintaining and improving the world's food supply. The second point I want to make is this: No plant breeder or genetic engineer starts from scratch when they develop a new cultivar of maize, or tomato, or barley. They are building on the accumulated success of generations of farmers, who have selected and improved seeds for thousands of years. Companies like Dupont and Monsanto and Novartis tell us that they 'invented' their genetically engineered plants and that they should be rewarded with exclusive monopoly patents. In reality, corporate plant breeders are fine-tuning and modifying plants that were developed by anonymous farmers and the more recent contributions of institutional plant breeders.

Intellectual property regimes have evolved very rapidly. Eighteen years ago, RAFI testified at Congressional hearings in Washington, DC on US plant variety protection laws—a type of plant patenting. Gary Nabhan also testified at those hearings. We were there to oppose the expansion of these laws, and we argued that, if the seed industry got its way, farmers would *lose* the right to save seeds and that genetic diversity would be threatened. I'll never forget the response we got from the Chairman of the House Agriculture Committee, Kika de la Garza from Texas, he laughed out loud and said, "No one's going to take away the right of farmers and backyard gardeners to save seeds!"

Well, here it is, 1998, and I'm sorry to say that Kika de la Garza was dead wrong. Under US patent law, it's now *illegal* for farmers to save patented seed and re-use it. Today, Monsanto requires farmer's—its customers—to sign a gene licensing agreement before they buy the company's patented, genetically engineered seeds. If farmers are caught infringing the patent, Monsanto is "vigorously prosecuting" them in court. The penalties may include criminal charges, on-farm field inspections, and damages that could exceed $1 million dollars. In some areas, the company has hired Pinkerton investigators to root-out farmers who are saving Monsanto's patented seed. In other words, farmers are being turned into criminals and rural communities are becoming corporate police states. Monsanto has spent over $8 billion acquiring seed and biotech companies since 1996, and it's now the second largest seed corporation in the world.

Historically, plant variety protection regimes guaranteed the right of farmers to save seed from their harvest. But the farmers' right or farmers' exemption is disappearing from international plant patenting regimes. Earlier this year, the 1991 Act of UPOV entered into force, replacing the 1978 Act. UPOV is the international body that coordinates a common legal regime for plant variety protection. Under the new Act, the farmers' right to save seed is no longer an automatic feature of the new Convention—it is recognized only if member states make special provision for it in national legislation. This is a big deal because many nations of the South are under intense pressure to join UPOV to satisfy their obligation under GATT-TRIPs.

TERMINATOR TECHNOLOGY

The latest assault on the right of farmers to save seed is a technology developed by USDA and the seed industry. In March of this year the USDA and Delta & Pine Land Co., a Monsanto subsidiary, proudly announced that they received a patent on a technique that genetically alters seed so that it will not germinate if re-planted the following season. In other words, scientists have developed plants that are genetically engineered to kill their own seeds. The technology is designed to prevent farmers from saving seed from their harvest, thus forcing them to return to the commercial seed market every year. This technology is still in the early stages of development. But, if it works as advertised, it's a dream-come-true for the multinational seed industry. Because we view it as a potentially "lethal" technology, RAFI calls it the "Terminator Technology."

Probably some of you know that RAFI and many other NGOs have adopted a campaign to ban the Terminator. In RAFI's opinion, the Terminator technology is the neutron bomb of agriculture. It's a dead-end for farmers, and for biological diversity. This is a product that will bring *no* agronomic benefit to farmers or consumers—it's designed simply to increase seed industry profits. The technology is aimed primarily at seed markets in Africa, Asia, and Latin America, where over 1.4 billion people depend on farm-saved seed and on-farm plant breeding. Patents on the Terminator technology are pending in 87 foreign countries. The president of Delta & Pine Land, Murray Robinson, says that his company's seed sterilizing technology could be used on up to 1 billion acres (400 million ha) worldwide (an area the size of South Asia), and that it could generate revenues for his company in excess of $1 billion per annum.

If farmers lose the right to save seed—they lose the ability to select seed and adapt crops to their unique farming conditions. If farmers eat or abandon their traditional seeds in the process of adopting this new technology, centuries of crop genetic diversity will be lost forever.

But that's not all. Scientists warn that, under certain conditions, the trait for seed sterility will flow, via pollen, from Terminator crops to surrounding plants, making the seeds of neighboring plants sterile. Simply put, the Terminator technology will restrict the food-producing capacity of farmers and it will destroy diversity.

The good news is that the highly unpopular Terminator technology has catalyzed international debate on these issues and there's an avalanche of public opposition. Over 2,500 people have written to US Secretary of Agriculture Glickman (from RAFI's web site alone) urging him to cease negotiations with Monsanto on the licensing of the technology, and to abandon development of the technology. Two weeks ago, the world's largest international agricultural research network—the Consultative Group on International Agricultural Research—adopted a policy banning the use of Terminator technology (and other genetic seed sterilization technologies) in its plant breeding programs.

- African delegates to the United Nations FAO have said that they don't want Terminator genes used on African soil.
- In May, the Conference of the Parties to the Convention on Biological Diversity (COP IV) directed its scientific body to examine the technology's impact on farmers and biodiversity.
- India's agriculture minister has taken action to ban the import of seeds containing the terminator gene because of the potential harm to Indian agriculture.

I realize that we're not here to debate the Terminator technology—but I believe the subject is entirely relevant to our discussion tonight. I also know that there are many dedicated and well-meaning scientists at USDA who are doing very important work to support conservation and sustainable use of biodiversity. If the US government is serious about supporting the conservation and sustainable use of biodiversity, then it's sending precisely the *wrong* message by developing a technology that aims to restrict the right of farmers to save seed and select crops. The right to save seed is fundamental to agriculture and global food security. That's the bottom line. If farmers lose the right to save seed, we all lose.

REFERENCES

CGIAR. 1996. Press Release, "Poor Farmers Could Destroy Half of Remaining Tropical Forest," Washington, DC, 4 August 1998.

CGIAR. 1998. Statement of the CGIAR to the 4th Meeting of the Conference of Parties to the Convention on Biological Diversity, Bratislava, Slovakia, May, 1998.

FAO. 1995. The State of World Fisheries and Aquaculture, FAO, Rome, p. 8.

Gene Protection Technologies:
A Monsanto Background Statement

In recent months, there has been considerable publicity in a number of countries about a potential new plant technology dubbed the "terminator." This technology is one of a class of so-called gene protection or gene control technologies, still in research and development, that may be used to control the germination of seeds produced by plants modified by biotechnology.

Monsanto has been the subject of much of that publicity because of our announced intention to purchase Delta and Pine Land Company (D&PL). D&PL, along with the U.S. Department of Agriculture, developed this particular gene protection technology.

The news stories and numerous interested parties have raised questions about the impact of such technologies on traditional farming methods and on the production of adequate food supplies to meet the anticipated increases in the world's population.

Many seed companies around the world, as well as government and independent research institutions, are in some stage of research and development on gene protection technologies. We know of none, however, that have moved beyond the research and development phase. The securing of a patent related to these technologies is one part of any research and development effort, and does not predict commercial viability or acceptability.

Companies are researching these technologies because they believe they may provide a number of benefits, the primary benefit being protection of the investment required to develop the seeds. Such protection encourages more research and investment in future agricultural improvements and thereby would expedite access to the benefits of biotech seeds by farmers who want them.

At the same time, however, the fact that there is so much concern being expressed about this type of technology indicates that there are many who have serious misgivings about them and their potential impact on food production.

We believe that the concerns about gene protection technologies should be heard and carefully considered before any decisions are made to commercialize them.

We have conferred in the last several months with a number of international scientific and agricultural leaders about this situation including Ismail Serageldin of the World Bank, Professor Swaminathan of India, Calestous Juma, former head of the Secretariat for the Convention on Biodiversity, Jose Sarukhan, Director of CONABIO, the National Biodiversity Council in Mexico, and Jim Moody, President and CEO of InterAction, an umbrella organization representing 160 development and humanitarian aid organizations.

These individuals have been generous in their counsel and have recommended that thorough, independent and comprehensive consideration be given to the concerns raised about the impact of new gene protection technologies. We agree, and we are calling for just such consideration and public discussion, covering the full range of questions and issues that relate to the impact of these technologies on farming practices around the world.

One important issue, especially in developing countries, is that of small holder farmers who rely on saved seed to provide growing stock for the next year. We also hope that serious study and consideration will be given to the potential environmental, economic and social impacts of gene protection systems; how they should be developed; under what conditions or circumstances they should be utilized; and who should own them.

There is time for all aspects of this situation to be considered carefully, and in an open, comprehensive and consultative fashion. Experts agree that these systems are not expected to be ready for potential commercial use for at least five years.

It is important that the interests and perspectives of the food security and development communities be fully considered throughout the study and consultative process. In this connection, we are pleased that InterAction has agreed to work with its members and other groups to help achieve this objective. It is our hope that the many organizations interested in food security and development will contribute their expertise and perspectives.

Until a thorough, independent examination of gene protection systems has been conducted and all points of view considered, we will not attempt to commercialize these technologies. Moreover, in considering whether to commercialize such technologies, we will respond publicly and fully to the conclusions, opinions and arguments that are raised.

We believe in biotechnology and its potential to help address the very real and serious food and environmental demands facing the world, but we know that the success of biotechnology depends on its acceptance by farmers and the broader public. Many parties, both private and public, have an important stake in the issues surrounding gene protection technologies. We hope they will participate in a careful examination of those issues.

SOURCES OF PUBLIC FUNDING

Sources of New Funding for New Crops and New Uses

Carmela A. Bailey

The Federal government offers a wide variety of programs to support research, development, and commercialization of products from agricultural, biobased materials. These programs promote new sources of energy, and new industrial and consumer products. They provide funding to expand the nation's capabilities to take advantage of new and exciting technologies. As a result, business opportunities are created, value is added to conventional crops and ag residues, agriculture is diversified through new crops development, and sustainable agricultural practices are expanded. Many of the products and processes developed under these programs compete with traditional counterparts, while showcasing environmental stewardship which can ultimately stimulate increased funding for research and development.

Biobased Products Alternative Agricultural Research and Commercialization Corporation (AARC)

Robert E. Armstrong

The AARC Corporation is a wholly-owned corporation of the US Department of Agriculture. Operating under the guidance of a largely private-sector board of directors, the Corporation functions as a venture capital firm that invests in companies commercializing non-food/non-feed products derived from agricultural raw materials. In many instances, these products replace petroleum, or make use of an agricultural waste material. Preference is given to projects that benefit rural communities and are environmentally friendly. AARCC currently has $35 million invested in companies throughout the US.

In summary, the economy of the 21st century will be biobased and venture capital provides active assistance to build and grow companies to meet the challenge. Agriculture has always been recognized as the foundation of manufacture and commerce. Improvements in plants and animals provide greater diversity of raw materials for manufacturing and commerce, and constant improvements in biological processing make biobased raw materials more affordable. An economy based on biology is more environmentally friendly than one based on geology, i.e., petroleum. Agricultural activities serve as the engine of job creation in rural America. Industry categories of biobased products currently widely used include the following: absorbents, adhesive and bonding agents, biocontrol agents, soil media and fertilizer, cleaning agents, detergents, solvents and surfactants, coatings and paints, construction materials and composites, cosmetics and personal care products, degradable polymers, fillers, yarn and insulation, fuels, inks, lubricants, paper and packaging, pharmaceuticals and veterinary products.

Venture capital is an organized pool of high risk capital, invested in businesses with high growth potential. Venture capital provides active assistance to build management teams and to stimulate company growth. The goal is creation of wealth (with the ancillary benefit of job creation). Providing risk capital is not the role of lenders. Funding sources are limited because corporations are limited to strategic interests, the high net worth investor market is not organized, high risk activities are too small and time consuming for pension funds and other institutional investors, corporations and institutional investors may have difficulty evaluating the opportunity, e.g. lack of time and skills, and are not as likely to be able to provide assistance.

AARCC has demonstrated the ability to identify high-potential biobased investment opportunities. AARCC's portfolio balances three variables as performance measures: (1) return on investment, (2) job creation, and (3) increased use of agricultural land.

The areas of greatest potential growth in biobased products are in adhesives and bonding agents, cleaning agents/detergents/solvents/surfactants, coatings and paints, construction materials and composites, fuels, lubricants, pharmaceuticals, and fibers. Today's situation and future projections for these potential growth areas are as follows:

Adhesives and Bonding Agents

Current status: annual sales $13.3 billion

1/3 of all adhesives products are of natural or renewable origin

Emerging trends: adhesives are projected to grow 3.3% annually through year 2000 and natural adhesives are expected to grow 3.8% annually though year 2000

AARCC investment: $0

Cleaning Agents, Detergents, Solvents, and Surfactants

Current status: annual sales $44.9 billion

Emerging trends: concerns about food safety and sanitation prompting continued trend towards antibacterial products

AARCC investment: $1.74 million

Coatings and Paints

Current status: annual sales $43.3 billion

Emerging trends: coatings and paints are expected to grow at an average rate of 2% and continued strong commitment to research and development of new products is projected to increase 25% by year 2005

AARCC investment: $0

Construction Materials and Composites

Current status: annual sales $80.4 billion

Emerging trends: prefabricated and manufactured housing is predicted to grow 20% of new housing starts by the year 2000 and agricultural based building materials will develop a growing niche if existing environmental standards are enforced

AARC investment: $11.8 million

Fuels

Current status: annual sales $587.8 billion

Emerging trends: National Research Council calls for eventually satisfying up to 50% of liquid fuel needs with biobased products

AARCC investment: $0.8 million

Lubricants

Current status: annual sales $5.1 billion

Emerging trends: projected growth rate is 15% through year 2000; consumer demand for "green" products in the automotive sector and "natural" plant-based materials in the personal care sector will enhance opportunities for biobased end-use products

AARCC investment: $1.89 million

Pharmaceuticals and Veterinary Products

Current status: annual sales $169.7 billion

Emerging trends: projected growth rate is 7% through year 2000, with a likely increase as baby boom generation retires; biotechnology and agricultural-based raw materials will create new opportunities for companies in this field.

AARCC investment: $0.66 million

Fibers

Current status: annual sales $170 billion

Emerging trends: agricultural fibers will replace and /or supplement wood and synthetic fibers ad demand increases

AARCC investment: $1.19 million

Contact Information

USDA/AARCC

1400 Independence Ave. SW   202-690-1533

Room 0156 South Building   202-690-1655 (fax)

Washington, DC 20250-0401   www.usda.gov/aarc

US Department of Commerce Advanced Technology Program

Robert Bolksberg-Fireovid

The Advanced Technology Program (ATP), a program within the US Department of Commerce, provides funding on a cost-shared basis to industry for high-risk/high payoff R&D on emerging and enabling technologies. The ATP program concentrates on those technologies that offer significant, broad-based benefits to the nation's economy, but that are unlikely because of the high technical risks, to be developed in a timely fashion without ATP's support. The subjects of ATP research projects are proposed by industry, and awards are made on the basis of announced competitions that consider both the technical and business merits of the proposed work. Since its inception, ATP and industry have partnered to do 400 projects, representing over $2 billion of total industry/government investment. Excluding 1998 awards, which are yet to be announced, industry/government funding for ATP projects involving agriculture and bioprocessing industries has totaled $55 M. General information about the ATP, including policy, procedures and projects funded to date, can be found on the ATP website at http://www.atp.nist.gov

Details about the ATP portfolio from 1990–1998 include the following:
- $2.78 billion has been invested and half of the funding has been provided by industry.
- 431 projects with 1,010 participants have been funded.
- 3,585 proposals have been submitted by industry
- There is strong participation by small businesses. Over 50% of the projects are led by small businesses and joint ventures typically include small businesses
- Universities play a significant role. More than 130 different universities have been involved with more that 400 instances of participation

ATP-funded projects are high risk, high payoff that present a substantial technical challenge, offer innovative solutions, are on a sound scientific basis, present a credible commercialization plan, and offer a potential high payoff to the US. Companies typically are interested in research that is less generic, lower risk, shorter term, and offers large benefits to the company. ATP funds research that is more generic, higher risk, longer term, and offers large benefits to the Nation. The basic characteristics of ATP are unique mission focus, partnership with industry, published selection criteria, built-in sunset provisions, extensive rigorous peer review, rigorous project and program impact assessment. Funding is awarded under general competitions and focused programs.

Intellectual property provisions include the following:

Companies incorporated in the US keep the intellectual property rights and they can license. The government reserves the right to royalty-free non-exclusive license for government use. Trade secrets are protected with a non-disclosure agreement, and government rights are rarely invoked. Universities and non-profit research organizations may receive royalties but cannot own title to patentable inventions.

Project selection criteria are based on scientific and technical merit (30%), broad-based benefits to the US economy (20%), commercialization planning (20%), level of commitment and organizational structure (20%), and experience and qualifications (10%). The project must have a business plan (65%) as well as a R&D plan (35%).

There are two ways to participate in ATP. The first is as a single proposer. In this case, projects are no longer than 3 years, total NIST funds are limited to $2 million, NIST pays only direct costs, for large companies the cost share must be at least 60% of project cost, and no direct funding is provided to universities, governmental labs or non-profit research organizations. The second way to participate in ATP is a joint venture. Projects are limited to 5 years, there is no limit on award amount, NIST cost share is less than 50%, and the joint venture must involve at least two for-profit companies (universities and other non-profits may also participate). Non-profits, but not universities, may administer.

Of the 431 awards made between 1990–1998, they can be broken out by technology area as follows: computing, information and communications—29%; electronics—19%; biotechnology—17%; materials—13%; manufacturing—9%; chemicals and chemical processing—7%; energy and environment—6%.

The benefits of participating in ATP are broad. The program facilitates and accelerates high risk research, stimulates collaboration and formation of strategic alliances, shortens the R&D cycle, accelerates commercialization of the ATP-related technology, attracts additional funding, improves the company's competitive standing, stimulates discovery of new applications for ATP technology, and facilitates changes in corporate philosophy. ATP also offers the R&D Alliance Network (www.atp.nist.gov/alliance). This network promotes formation of ATP joint ventures and provides a sense of the resources required for a successful ATP joint venture. The Collaboration Bulletin Board offers a site where potential proposers can seek partners. The R&D Alliance Forum provides the opportunity for users to exchange ideas and questions.

Examples of agriculture-related R&D funded by ATP include:

AgriDyne: "US Self-Sufficiency in High-Quality Pyrethrin Production"

Agracetus: "Transgenic Cotton Fiber with Polyester Qualities via Biopolymer Genes"

Aquatic Systems/Kent SeaFarms: "Development of New Technologies for Treating and Recycling Wastewater from Aquaculture Facilities"

Mycogen: "Oleaginous Yeast Fermentation as a Production Method for Squalene and other Isoprenoids"

Genencor: "Continuous Biocatalytic Systems for the Production of Chemicals from Renewable Resource"

General Electric: " Synthesis of Monomers"

Agritope: "Using Biotechnology to Control Fruit Ripening"

CropTech Development: "Enhanced Manufacturing Technologies for Bioactive Proteins and Peptides in Transgenic Tobacco"

Aquatic Systems/Kent SeaFarms: "Superfingerlings: Animal Husbandry Techniques for Use in Aquaculture"

AviGenics: "Development of Hen Oviducts as Bioreactors via Promoter-less Minigene Insertion"

Maxygen: "Whole Genome Shuffling: Rapid Improvement of Industrial Micro-organisms"

Henkel/GE: "Phage Display-Based Platform Technology for Engineering Selective Catalysts"

Contact Information

Advanced Technology Program	Phone: 1-800-ATP-FUND (287-3863)
National Institute of Standards and Technology	Fax: 301-926-9524 or 301-590-3053
100 Bureau Drive, Stop 4701	e-mail: atp@nist.gov
Gaithersburg, MD 20899-4701	internet: www.atp.nist.gov

US Department of Energy

Phillip C. Badger

The US Department of Energy (DOE) has several avenues for funding activities that have an energy-related component. These avenues include a variety of solicitations issued by various departments throughout the year, including the Regional Biomass Energy Program, grants through the Small Business and Innovation Research (SBIR) and Inventions and Innovations programs, Cooperative Research and Development Agreements (CRADAs), and other avenues. DOE funds are also frequently available through programs conducted by state energy offices. Most DOE funding requires cost sharing from non-federal sources. Examples are described below.

The Inventions and Innovations Program provides funds to establish technical performance and conduct a project in the early development phase. Financial assistance is offered at two levels: up to $40,000 or up to $100,000. One solicitation per year is issued, however submissions can be made anytime. Other assistance can come from technical partners, commercial sponsors, or business plan resources.

The NICE3 (National Industrial Competitiveness Through Energy, Environment, Economics) program focuses on simultaneous reduction of industrial energy use, reduction of pollution generated by industrial processes, and improvement of the economics of energy production. The proposer must justify the use of government funds and awardees are expected to finance continuation after initial government funding. Awards are made through the states.

The SBIR (Small Business Innovation Research) program requires the participation of all federal agencies with budgets over $100,000 and 2.5% is set aside to implement this program. SBIR contains specific topical areas that vary yearly and consists of 3 phases: phase 1—up to $100,000 may be awarded for a 6 month feasibility study; phase 2—up to $75,000 may be awarded over a 2 year period to conduct principal R&D; phase 3—non-federal capital is used to pursue commercial applications. One solicitation is issued per year. The anticipated FY 1999 budget is $75 million.

Request for Proposals (RFP's) and Program Opportunity Notices (PONs) for energy-related R&D are offered by various offices within DOE and include the Office of Industrial Technologies, the Office of Transportation Technologies, the Office of Utilities Technology, and the Regional Biomass Energy Program. Announcements are made in the *Commerce Business Daily* (CBD). The amount of awards, the timing of RFPs and the award processes vary greatly.

Another mechanism to promote R&D is the CRADA (Cooperative Research & Development Agreement). Government and a company cooperate on R&D through a formal agreement to determine how much each party puts into R&D. The company keeps the intellectual property and the government keeps the information for a period of time. This mechanism is heavy on bureaucracy and red tape.

Unsolicited proposals have little likelihood of being funded. The numerous solicitations that are offered throughout the year are usually general enough to cover most subjects. For more information about funding opportunities that have been described, check the following websites:

www.eren.doe.gov

www.eren.doe.gov/golden

and check the Commerce Business Daily (CBD)

US Department of Agriculture Agricultural Research Service

Dwayne R. Buxton

The Agricultural Research Service (ARS) is charged with conducting USDA's in-house research. In meeting this mission, ARS generally does not maintain extramural programs unless directed by Congress and, thus, provides little direct funding to researchers outside ARS. Still, ARS collaborates with university, industry, and other scientists and thus often enhances the ability of these groups to conduct research. With a budget of approximately $800 million and a cadre 2,000 permanent scientists, ARS conducts its research at more than 100 locations throughout the United States and some foreign countries. ARS's research is organized into 23 National Programs with a focus on regional and national priorities or in support of the action and regulatory agencies within USDA. ARS does provide extramural funding for plant explorations and for those activities identified and supported by the various Crop Germplasm Committees.

ARS devotes about $16 million to support research on new uses, products, and materials. Much of this work is conducted at the four regional research centers in Wyndmoor, PA; Peoria, IL; Albany, CA; and New Orleans, LA. Additional work is conducted at Madison, WI; Hilo, HI; and Winter Haven, FL. Congress in 1938 established the four regional centers to conduct basic and applied research on agricultural commodities to promote better utilization of farm products. This research seeks to transform raw agricultural materials into commercially valuable products such as biodegradable plastics, adhesives, industrial lubricants, cosmetics, insect and weed control agents, and natural rubber from domestic plants.

Kenaf and flax have potential as fiber crops to supply niche markets and needs. Additionally, huge US trade deficits with petroleum imports have characterized the last decade. For transportation purposes alone, $50 billion of petroleum was imported in 1996. America's dependency on foreign oil (now at 50 percent and rising) is not only an economic issue, but is one of the national security, particularly in times of global unrest. These factors, coupled with environmental concerns regarding the combustion of fossil fuels and production of atmospheric carbon dioxide, fostered the expansion of the fuel ethanol industry. The capacity of the US industry now exceeds 1.5 billion gallons per year. ARS devotes about $5 million to support research on alternate fiber and energy crops at the four regional laboratories and at Athens, GA, and Stoneville and Mississippi State, MS.

ARS dedicates about $2 million to develop products from novel crops at Oxford, MS; Stoneville, MS; Wyndmoor, PA; Peoria, IL; and Albany, CA. The Oxford laboratory is new, and its mission is to discover natural products for use in agricultural pest management, with emphasis on pest management agents derived from plants. Discovery efforts are focused on products for agricultural sectors that are of little interest to the agrichemical industry, such as some horticultural crops and aquaculture. A secondary mission is to support development of medicinal plants as alternative crops.

ARS commits about $2 millions to support research on industrial crops. This includes research on guayule at Albany, Phoenix, and Oxford; crambe at Peoria; cuphea at Peoria and Ames, IA; safflower at Sidney, MT; flax at Athens, GA; and meadowfoam at Peoria. The mission of the New Crops Research Unit at Peoria is to develop new industrial products from meadowfoam, lesquerella, jojoba, crambe, milkweed, cuphea, *Euphorbia lagascae*, vernonia, or other potential new crops; to solve seed processing problems; and to evaluate germplasm and breeding lines for the proposed crops in collaborative programs with plant scientists. At Phoenix, the New Crops research program seeks to develop agricultural diversification in semiarid and arid regions through the commercialization of new crops adapted to low water requirements. The focus is on guayule, lesquerella, and vernonia. Basic and applied research is conducted to improve available germplasm through selection and breeding techniques. In addition, appropriate cultural management practices are being developed for the eventual economical commercialization of the new crops. Research at Albany, CA is centered on genetically engineering guayule with rubber biosynthesis genes to obtain enhanced yields of high quality rubber and reduce or eliminate the requirement for cold induction of rubber biosynthesis in this species. Additionally, work is being conducted to develop transformation protocols for temperate-zone, rubber-producing annual plants such as milkweed, maize, cotton, soybean, rice, and tobacco.

ARS also conducts research on fruit and nut crops and on floral and nursery crops. A major emphasis is on improved variety development. Complementary programs are conducted in plant pathology and entomology. This includes research on tree fruits at Byron, GA; Kearneysville, WV; Geneva, NY; East Lansing, MI; Yakima, WA; Wenatchee, WA; Corvallis, OR; and Fresno, CA. Research on small fruits is carried out at Fresno, CA; Corvallis, OR; Davis, CA; Poplarville, MS; Orlando/Ft. Pierce, FL; and Beltsville, MD. Research on nuts is conducted at Byron, GA; Davis, CA; and College Station, TX. Research on floral and nursery crops is ongoing at Beltsville, MD; Washington, DC; and McMinnville, TN.

More specific information about ARS research programs at these and other locations can be obtained through the Internet by logging onto ARS's Homepage at http://www.ars.usda.gov. Links to homepages at specific ARS locations can be obtained from this source.

Biobased Products: US Department of Agriculture and the Cooperative State Research, Education and Extension Service

Daniel E. Kugler

Biobased industrial products made from renewable agricultural and forestry materials are an area of priority for the US Department of Agriculture (USDA). Research, development, and commercialization of these new uses products can lead to new and expanded markets, thereby calling for increased, sustainable farm production of the raw materials. The intended outcome is diversification and expansion of the agricultural economy, frequently resulting in substituting for imports of critical and strategic materials such as petroleum. USDA funding for biobased industrial (nonfood and fuels) products was $76 million in each of fiscal years 1997 and 1998, compared to about $50 million for new/improved food products. (Table 1)

Leadership for USDA is vested in the Biobased Products Coordination Council, chaired by the Under Secretary for Research, Education and Economics, and including ten agencies: Alternative Agricultural Research and Commercialization Corporation (AARCC); Agricultural Marketing Service; Agricultural Research Service (ARS); Cooperative State Research, Education, and Extension Service (CSREES); Foreign Agricultural Service; Forest Service (FS); Natural Resources Conservation Service; Office of the Assistant Secretary of Administration; Office of Energy and New Uses; and Rural Business-Cooperative Service.

AARCC, ARS, CSREES, and FS are major agencies either conducting or sponsoring new uses research, development, and commercialization. AARCC and FS work only on nonfood industrial products, while ARS and CSREES work on both food and nonfood products. ARS and FS conduct fundamental and applied research in USDA laboratories and facilities. CSREES supports mainly fundamental and applied research through its university partners in the land grant system. AARCC makes equity investments in businesses to commercialize biobased industrial products.

While ARS ($82 million) and CSREES ($26 million) dominate the funding and support for new uses (Table 2), it is the combination of all agencies and their various functions that characterize biobased products in USDA. There exists a capability to work with a promising new or improved material or process from its discovery in the laboratory, through product design and specification, to a precommercial technology demonstration project, and finally venture equity funding for market entry. Each stage is linked to the next and, as a new/improved product a process moves closer to being marketable and bankable, the involvement and interest of the private sector heightens.

The CSREES partnership with the land grant college and university system for new uses comes in three main forms: competitive programs, formula fund programs and noncompetitive special grants. Referring to Table 3, in fiscal year 1998 the competitive programs (NRI, SBIR, and SARE) accounted for $18.6 million of the CSREES $26 million total. Noncompetitive or directed special research grants added $3 million. The remaining approximately $4 million of the total comes from Hatch (research) and Smith-Lever (extension) funds distributed to the States and territories by formula.

For further information about biobased industrial products/new uses, access the USDA Homepage (http://www.usda.gov) for individual agencies. For biobased products information in CSREES, contact Carmela Bailey at USDA-CSREES, Mail Stop 2220, Washington, DC, 20250-2220.

Table 1. Funding by category for new uses for agricultural commodities, US Department of Agriculture.

Category	Funding ($ million)	
	FY 1997	FY 1998
Food	50	49
Nonfood	68	67
Fuels	8	9
Total	126	125

Table 2. Funding by agencies for new uses for agricultural commodities, US Department of Agriculture.

Agency	Funding ($ million)	
	FY 1997	FY 1998
Agricultural Research Service (ARS)	82	82
Cooperative State, Research, Education, and Extension Service (CSREES)	26	26
Forest Service (FS)	8	9
Alternative Agricultural Research and Commercialization Corporation (AARC)	8	7

Table 3. Funding by program for new uses for agricultural commodities.

CSREES program	FY 1998 funding ($ million)
National Research Initiative (NRI)	2 (Nonfood)
	2 (Wood)
Small Business Innovation Research (SBIR)	1.6
Sustainable Agriculture Research and Education (SARE)	13
Hatch Act, Smith-Lever (HA, S-L)	formula
Special Research Grants (KYC)	3

Crop Diversification Projects

Robert L. Myers

TThere are a number of programs that can potentially support work on crop diversification or adding value to crops. Many of these are described in the publication "A resource guide to federal programs in sustainable agriculture and forestry," published by the USDA Sustainable Agriculture Research and Education (SARE) program and the Michael Fields Institute in 1998. This publication describes over 50 federal programs that can provide funding or technical assistance for projects in value-added agriculture or other aspects of sustainable agriculture. The publication is available free from ATTRA (phone 800-346-9140), or can be viewed on their website (www.attra.org).

Among the programs to consider for funding support of crop diversification is the USDA SARE program, administered through the CSREES. This program, which has been funded at about $11.5 million annually in recent years offers three types of grants. The majority of the funds are available through the SARE research and education grants, which are typically 2–3 year projects of $30,000 to $200,000. Universities, nonprofit organizations, and government agencies are eligible for both these grants, and the SARE professional development grants, which are somewhat smaller awards (<$100,000) aimed at providing professional development to extension staff and NRCS personnel. The third SARE program is designed to directly fund individual farmers or teams of farmers. They can apply, using a simple form, for up to $5000 in grant aid to carry out an on-farm study, education program, or marketing project. All of the SARE programs are reviewed and administered on a regional basis. Projects on new crop development, on marketing, or on agroforestry would all be appropriate for proposals or ideas to SARE. Additional information can be obtained from the national SARE office (phone 202-720-5203) or website (www.sare.org).

Another federal grants program that can potentially provide competitive funds for crop diversification is the USDA-CSREES National Research Initiative (NRI). The NRI encompasses grant programs on several topics, the most relevant being the agricultural systems grant program. Awards in the ag systems NRI grant program have typically been in the $100,000 to$250,000 range, with awards being made once per year. Universities, non-profit organizations, and government researchers are eligible to apply. Guidelines for this program can be obtained from the USDA-CSREES website (www.reeusda.gov). That same website also covers other competitive grant programs, including the Small Business Innovation Research (SBIR) program, which can help fund certain types of value-adding projects. Information on funding programs available from other USDA agencies, including the Alternative Agriculture Research Commercialization (AARC) program, can be obtained from the resource guide described above, or through links from the main USDA website (www.usda.gov).

A few private foundations have funded diversification projects as part of programs on sustainable agriculture, but unfortunately, support from this sector has been declining in the last few years. For some states, a better source of support has been through the state department of agriculture. Several states now fund grant programs, either for individual farmers, or for organizations and universities to apply to, that include the opportunity to fund new crops or value-added agriculture projects.

Beginning in 2000, a new source of funding for new crop development may be available through USDA. A new national initiative, the Thomas Jefferson Initiative for Crop Diversification, was authorized by Congress as part of the 1998 Agricultural Research, Education, and Extension Reauthorization Act. Appropriations are currently being sought, and if obtained, would first be available as part of the fiscal year 2000 budget. This funding would provide support for regional crop diversification centers at universities, for competitive grants on new crops, and for a national coordinating center. More information on the status of this program can be obtained from the Jefferson Institute (phone 573-449-3518).

Technical information on new crops is available from several sources. The best on-line source is the Purdue University NewCROP website (www.hort.purdue.edu/newcrop). Other information resources include the Alternative Farming Systems Information Center at the National Agricultural Library (phone 301-504-6559, website www.nal.usda.gov/afsic), and the Appropriate Technology Transfer for Rural Areas (ATTRA) center, (phone 800-346-9140, website www.attra.org).

PART II
CONSERVATION & INFORMATION

GERMPLASM CONSERVATION

Challenges and Strategies in Promoting Conservation and Use of Neglected and Underutilized Crop Species

Stefano Padulosi, Pablo Eyzaquirre, and Toby Hodgkin

There is an increasing interest in neglected and underutilized crop species (NUS) throughout the world, reflecting a growing trend within agriculture to identify and develop new crops for export and domestic markets. The FAO Global Plan of Action for the Conservation and Sustainable Use of Plant Genetic Resources for Food and Agriculture, which was adopted in 1996 by approximately 150 countries, identified the improved conservation and use of NUS as one of its 20 main activities. Interest in NUS stems from a variety of factors, including their contribution to agricultural diversification and better use of land, their economic potential and the opportunities they provide for diet diversification. NUS are being often presented as new species though they have been used by local populations in traditional ways for many centuries. Their novelty is thus not related to their introduction to new areas but rather to the ways in which old and new uses are being re-addressed to meet today's needs.

The International Plant Genetic Resources Institute (IPGRI) has been concerned to improve conservation and use of NUS, and has spearheaded, over the last few years, specific activities at national and international level for the better conservation and use of these species. Collaborative works on a number of NUS-oriented Networks have been supported together with the production of a series of monographs on selected NUS. IPGRI's strategy for meeting the challenges of NUS promotion is based on the premise that the deployment of plant genetic diversity in agriculture will lead to more balanced and sustainable patterns of development. Because NUS may never command a large percentage of national resources and are often characterized by local specificity they require however an approach different from that used for other crops. Key elements of the IPGRI's strategy for NUS conservation consists in enhancing conservation through use, identifying specific indicators to measure threats of genetic erosion, promoting the interest of national programs and other agencies and facilitating exchange of information and collaboration among interested parties. IPGRI is particularly concerned to develop general tools and methods for promoting conservation and use of NUS which can be adapted to different contexts and enable national programs and local institutions to use them for the species to which they give priority.

TRENDS IN THE USE OF PLANT GENETIC RESOURCES

Today, only 30 plant species are used to meet 95% of the world's food energy needs (FAO 1996). These crops are widely and intensively cultivated and have been selected from a large agrobiodiversity basket containing more than 7,000 food species (Wilson 1992), which is approximately 1/10 of the estimated number of edible species present in nature (Myers 1983). Although an analyses of the data on a country-by-country basis, indicates that food supply is provided on average by 103 species, yet the exploitation of our plant genetic diversity wealth remains still far lower than what existing potentials would in fact allow (Prescott–Allen and Prescott–Allen 1990).

The reasons behind such a narrow focus of agriculture are well understood. Such factors as physical appearance, taste, nutritional properties, cultivating techniques, processing qualities, environmental adaptability, range of possible uses, and storability have fuelled the promotion of these "major crops" and ensured their success across continents and their acceptance by so many different cultures. The success of the breeding programs of these crops depends on the genetic diversity collected through thousands of germplasm collecting missions. Today, more than 6 millions of accessions of genetic resources for food and agriculture are stored in some 1,300 germplasm collections around the world (FAO 1996). Most of this ex situ conserved diversity (about 80%) belongs to major crops and their close relatives (Padulosi 1998a). The presence of a limited amount of germplasm of the so called "minor crops" in gene banks and its poor representation in terms of genetic diversity (Padulosi 1998a) represents a great challenge for the successful improvement and promotion of this group of species. Yet the narrow agricultural portfolio of today's agriculture raises serious questions on how effectively major crops alone can contribute towards food security, poverty alleviation, and ecosystem conservation as we become more and more aware of the fact that diversification of crops at all levels and in all types of agro-systems is the most crucial element for sustainability (Collins and Hawtin 1998). Emerging opportunities over the last few years for "minor" crops (particularly those underutilized or neglected) signal a new attention of the public opinion on biodiversity and its sustainable use along with an increasing interest of the public and private sector towards "new" crops, "new" uses and new markets.

CHALLENGES AND OPPORTUNITIES

The True Novelty of Neglected and Underutilized Crop Species

Interest on NUS stems from a variety of concerns and needs, including their contribution to agricultural diversification, better use of marginal land and changing environments, food security and a more balanced diet, better safeguard of our agrobio-diversity and associated cultural heritage, self-reliance of agricultural systems, additional source of income to farmers, and employment opportunities (Padulosi 1998a). Although some of these aspects might have a "new" connotation, often NUS are being presented as new crops in spite of the fact that they have been used by local populations in traditional ways for generations. This point is fundamental to underscore the importance of these species in our life (Bhag Mal 1994). Indeed the true novelty of underutilized and neglected species and their potentials is not related to their introduction to new areas but rather to the ways in which old and new uses are being re-addressed to meet today's needs. Any work in support to underutilized and neglected species must be consistent with this aspect to achieve their sustainable promotion and safeguard. From an international perspective, emerging global and national attention on neglected and underutilized species originates basically from three major areas of interest and concern: new markets and uses, the environment, and food security and nutrition.

Development of New Markets and New Uses

Modern technologies have the ability to transform crops and other plants into diverse products, from plastics to surgical tissue, to extend shelf life of fresh vegetables, and to enhance commercialization and marketing systems. These benefits undergird the growing demands for "new" foods and plant products.

Concern on Environmental Change and Ecosystem Stability

Climate change, degradation of land and water resources have led to growing interest with crops and species that are adapted to stresses and difficult environments. Many of these species are neglected and underutilized and their use is restricted to the niches where they are maintained by poor farming communities in marginal environments (high mountains, desert margins, poor soils, etc.). The deployment of a wider species diversity within agro-ecosystems enhance greater stability and ecosystem health.

Concern Over Food Security and Nutrition

Many neglected and underutilized species are nutritionally rich and are adapted to low input agriculture. The erosion of these species can have immediate consequences on the nutritional status and food security of the poor. Also, growing market opportunities for such species may generate additional income to those poor farmers in less favored environments.

A STRATEGY FOR THE PROMOTION OF NEGLECTED AND UNDERUTILIZED CROP SPECIES AT IPGRI

IPGRI is an autonomous international scientific organization operating under the aegis of the Consultative Group on International Agricultural Research (CGIAR). IPGRI's Headquarters are based in Rome, Italy, with regional Offices located in various locations around the world. IPGRI's mandate is to advance the conservation and use of plant genetic resources for the benefit of present and future generations. IPGRI works in partnership with other organizations, undertaking research, training, and the provision of scientific and technical advice and information.

Table 1. List of constraints per type of crop indicated by the participants to the conference on neglected and underutilized crop species priority settings for the central and west Asia and north Africa region held in Aleppo on 9–11 Feb. 1998. Highest numbers indicate highest weight of correspondent constraints.

Constraints (ranked by degree of importance)	Medicinals, aromatics	Forest trees	Fruit trees, nuts	Vegetables	Forages, browses	Industrial crops	Ornamentals	Legumes	Cereals
Low competitiveness	3	3	3	3	3	3	3	3	3
Lack of knowledge on uses	3	3	3	3	3	3	3	3	3
Lack of research on genetic diversity assessment and use	3	3	3	3	3	3	3	2	2
Policy & legislation	3	3	3	3	3	3	1	1	1
Loss of traditional knowledge	3	3	2	3	2	2	2	1	1
Lack of market/poor commercialization	3	2	2	2	3	3	3	1	1
Low income	2	3	2	2	3	3	1	2	2
Lack of propagation techniques	3	3	2	1	2	2	3	1	1
Scarce knowledge on cultural practices	3	2	2	2	2	1	3	2	1
Lack of attractive traits	1	2	3	3	1	2	1	1	1

Table 2. List of constraints in national programs to promote conservation and use of neglected and underutilized crop species, areas of action, activities required and comparative advantage.

Constraints (ranked by degree of importance)	Area of action	Activities	Comparative advantage
Low competitiveness	Enhance competitiveness	Promote better links between growers and marketers Investigate new areas of cultivation	Study environmental adaptation Ecogeographic distribution
Lack of knowledge on uses	Enhance documentation, information, public awareness	Carry out surveys on use Ethnobotanic studies Public awareness material	Study uses of genetic resources and gather ethnobotanic information Disseminate the information and raise public awareness on usefulness
Lack of research on genetic diversity assessment & use	Enhance research on genetic diversity	Characterization Evaluation Descriptors lists	Descriptors lists Characterization Evaluation

Constraint	Strategy	Activities	Specific actions
Policies & legislation	Promote better policies/legislation Carry out public awareness	Seminars, meetings, public awareness material etc. Investigate on restrictions affecting use Study ways to support NUS Public awareness to stimulate interest from public and private sector	Technically valid recommendations on sustainable use of wild/gathered species Genetic erosion assessment surveys/ monitoring Study various land use systems and recommend better ways to use genetic resources of NUS Tackle the distortions resulting from subsides/ policies in favor of narrow range of species Carry out impact studies for value-added strategies Survey conservation status and recommend/ carry out conservation actions accordingly
Loss of traditional knowledge	Enhance documentation & information on traditional knowledge Carry out public awareness Promote participation of farmers/users	Gathering info. on traditional knowledge Documenting traditional knowledge Public awareness on role of traditional knowledge	Ethnobotany work Community participation Promote local knowledge systems
Lack of market/poor commercialization	Carry out marketing and commercialization studies	User definition Promotion campaign Market niche studies Studies on price, processing presentation to user	Strategies for diversity-rich markets
Low income	Support NUS income Investigate on better agricultural systems Promote new uses	Study alternative ways to support NUS income	Public awareness Study crop compatibility for enhancing cultivation possibilities
Lack of propagation techniques	Enhance research on propagation techniques	Carry our research on most suitable technologies	Carry our research on most suitable technologies
Scarce knowledge on cultural practices	Enhance research on agronomic aspects	Research on yield, pests, harvesting, growth rate	Research on yield, pests, harvesting, growth rate.
Low quality	Enhance breeding work	Research on taste, flavor, appearance etc.	Research on taste, flavor, appearance etc.

IPGRI has been concerned to improve conservation and use of underutilized and neglected species since its establishment in 1974, and has spearheaded, over the last few years, specific activities at national and international level for the better conservation and use of these species. Collaborative works on a number of Networks have been supported (Padulosi 1998b) together with the production of a series of 23 monographs on selected species.

IPGRI's strategy for meeting the challenges of the promotion of underutilized and neglected species is based on the premise that the deployment of plant genetic diversity in agriculture will lead to more balanced and sustainable patterns of development.

The IPGRI strategy on underutilized and neglected species aims at four main goals:

1. **Enhance the conservation** through use of plant genetic resources of a wider range of useful species
2. **Strengthen the work of other actors** who are working on the documentation, evaluation, domestication of neglected or underutilized species
3. **Strengthen research on the choice of species** based on strategic factors for conservation, development, and food security
4. **Identify criteria** for research, development, and conservation actions on neglected and underutilized species that place the conservation and use of these genetic resources in the context of national and global strategies for sustainable agriculture, to improve the livelihoods of the rural poor, and to broaden the bases of food security.

Selecting Priority Species

In view of the fact that there are so many NUS species for which greater attention is being called for around the world, IPGRI has developed a specific approach for guiding their selection process. IPGRI will select those species that will allow to work with a wide spectrum of partners for the development and test of new approaches on plant genetic resources conservation and use that are of a wide-scale significance. These selected species will serve as models for the conservation and use of genetic resources of other NUS of local/regional significance, that are key to the livelihood of farmers in similar environments or the same region.

As part of this selection process across the regions of its Institutional mandate, IPGRI has organized in 1998 an International Conference in Aleppo, Syria, focusing specifically on the Setting of Priorities on NUS for the Central and West Asia and North Africa Region (CWANA). The Conference output consisted of a number of key presentations addressing the status of NUS across CWANA and of the identification of those limiting factors that halt their full exploitation along with a list of priority actions needed for their sustainable promotion. In addition, a number of species (divided into 9 categories viz. medicinal and aromatic species; forest species; fruit trees and nuts; vegetables species; forage and browses; industrial species; ornamental

Table 3. Recommended species selected by the participants attending the conference on neglected and underutilized crop species. Species have been selected on the basis of their contribution to: 1. Food security, 2. Ecosystem conservation and 3. Poverty alleviation in the central and west Asia and north Africa region.

Groups of species	Recommended species
Cereals	*Secale cereale;* hulled wheat (einkorn, emmer, spelt); *Stipa lagascae*
Forages & browses	*Atriplex halymus; Salsola* spp.; *Lathyrus* spp.; *Hedisarum* spp.; *Dactylis glomerata*
Forest trees	*Juniper* spp.; *Pistacia* spp.; *Quercus* spp.; *Acacia* spp.; *Abies* spp.
Fruit trees & nuts	*Pistacia vera*; *Ceratonia siliqua; Cydonia oblonga; Ziziphus* spp.; *Prunus* spp. (wild relatives of fruit species)
Industrial	*Carthamus* spp.; *Rhus* spp.; *Crocus* spp.; *Laurus nobilis; Stipa tenacissima*
Medicinal & aromatic	*Origanum* spp.; *Artemisia* spp.; *Thymus* spp.; *Rosmarinus* spp.; *Coriandrum* spp.
Ornamental	*Tulipa* spp.; *Nerium* spp.; *Iris* spp.; *Limonium* spp.; *Cercis siliquastrum*
Pulses	*Trigonella foenum-graecum; Lupinus* spp.
Vegetables	*Cichorium* spp.; *Capparis* spp.; *Brassica* spp.; *Malva* spp.; *Scolymus* spp.

species; pulses and cereals) assessed by participants as particularly valuable for the whole region were also recommended as priority in future initiatives in this domain (Tables 1, 2, and 3).

Monographs Produced by IPGRI

A series of monographs have been produced by IPGRI on neglected and underutilized species. Crops covered include: aibika/bele, *Abelmoschus manihot* (L.) Medik.; Andean roots and tubers: ahipa, arracacha, maca and yacon; bambara groundnut, *Vigna subterranea* (L.) Verdc.; black nightshades, *Solanum nigrum* L. and related species; breadfruit, *Artocarpus altilis* (Parkinson) Fosberg; buckwheat, *Fagopyrum esculentum* Moench; carob tree, *Ceratonia siliqua* L.; cat's whiskers, *Cleome gynandra* L.; chayote, *Sechium edule* (Jacq.) SW.; chenopods, *Chenopodium* spp.; coriander, *Coriandrum sativum* L.; grass pea, *Lathyrus sativus* L.; hulled wheats, (*Triticum monococcum, T. dicoccum,* and *T. spelta*); lupin, *Lupinus* spp.; niger, *Guizotia abyssinica* (L.f.) Cass.; oregano, *Origanum* spp.; peach palm, *Bactris gasipaes* Kunth; physic nut, *Jatropha curcas* L.; pili nut, *Canarium ovatum* Engl.; safflower, *Carthamus tinctorius* L.; sago palm, *Metroxylon sagu* Rottb.; tef, *Eragrostis tef* (Zucc.) Trotter; traditional African vegetables; yam bean, *Pachyrhizus* DC.

REFERENCES

Bhag Mal. 1994. Underutilized grain legumes and pseudocereals: Their potentials in Asia. RAPA/FAO, Bangkok.

Collins W. and G. Hawtin. 1998. Conserving and using crop plant biodiversity in agroecosystems. In: W.W. Collins and C. Qualset (eds.), Biodiversity in agroecosystems. CRC Press, Washington, DC.

FAO. 1996. Report on the state of the world's plant genetic resources for food and agriculture, prepared for the International Technical Conference on Plant Genetic Resources, Leipzig, Germany, 17–23 June 1996. Food and Agriculture Organization of the United Nations, Rome, Italy.

Myers, N. 1983. A wealth of wild species: Storehouse for human welfare. Westview Press, Boulder, CO.

Prescott–Allen R. and C. Prescott–Allen. 1990. How many plants feed the world. Conservation Biol. 4:365.

Padulosi, S. 1998a. Criteria for priority setting in initiatives dealing with underutilized crops in Europe. Paper presented at the European Symposium on Plant Genetic Resources for Food and Agriculture, Braunschweig, Germany, 29 June–5 July 1998. (in press).

Padulosi, S. 1998b. The underutilized Mediterranean species project (UMS): An example of IPGRI's involvement in the area of underutilized and neglected species Third Regional Workshop of MEDUSA, Coimbra, Portugal 27–28 April 1998. CIHEAM-MAICh, Crete, Greece. (in press).

Wilson, E.O. 1992. The diversity of life. Penguin, London.

Conservation of the Wild Relatives of Native European Crops

Vernon Heywood

Interest in the conservation of wild relatives of cultivated plants has increased considerably in recent years and is recognized as one of the priority activities of the Leipzig Global Plan of Action. Wild relatives have contributed to the improvement of most crop plants and are used mostly as sources of desirable genes as well as in research relating to crop improvement.

Although not often thought of as a major center of crop diversity, the European continent harbors rich wild gene pools of many crop species. These include: cereals, particularly oats (*Avena*) and rye (*Secale*); food legumes such as pea (*Pisum*) and lupins (*Lupinus*); fruit crops, such as apple (*Malus*), pear (*Pyrus*), plums and cherries (*Prunus*), grape vine (*Vitis*), raspberries and blackberries (*Rubus*), olive (*Olea*) and fig (*Ficus*); vegetables—including lettuce (*Lactuca*), carrot (*Daucus*), parsnip (*Pastinaca*), cabbage and other brassicas (*Brassica*), beet (*Beta*), celery, celeriac (*Apium*), leek (*Allium*), asparagus (*Asparagus*), salsify (*Tragopogon*), and artichoke (*Cynara*). The wild inventory is also very rich in the assemblage of pot herbs, condiments, and aromatic plants such as: caper (*Capparis*), mints (*Mentha*), marjoram (*Origanum*), lavender (*Lavandula*), thyme (*Thymus*), sage (*Salvia*), rosemary (*Rosmarinus*), mustards (*Sinapis, Brassica*), horseradish (*Armoracia*), water cress (*Nasturtium*), chives and leek (*Allium*), fennel (*Foeniculum*), caraway (*Carum*), that have their close wild relatives in Europe.

There is also a very large number of ornamentals, many of which have been taken into cultivation in Europe itself and represent a significant part of its cultural heritage. These include: candytuft (*Iberis*), stock (*Matthiola*), wallflower (*Erysimum*), mountain ash (*Sorbus*), sweet william and pinks (*Dianthus*), delphinium (*Delphinium*), larkspur (*Consolida*), columbine (*Aquilegia*), pansy and violets (*Viola*), sweet pea (*Lathyrus*), marigold (*Calendula*), snapdragon (*Antirrhinum*), bluebell (*Hyacinthoides*), snowdrop (*Galanthus*), cyclamen (*Cyclamen*), gladiolus (*Gladiolus*), narcissus, jonquil, daffodil (*Narcissus*), crocus (*Crocus*), lily of the valley (*Convallaria*).

Europe is also rich in forestry resources such as: pine (*Pinus*), fir (*Abies*), spruce (*Picea*), oak (*Quercus*), and poplar (*Populus*) and in fodder plants: rye grass (*Lolium*), cock's foot grass (*Dactylis*), clover (*Trifolium*), alfalfa (*Medicago*), etc. Many of the cultivars of these trees, grasses, and legumes have been derived from wild forms that are native to this continent.

THE INVENTORY

The Council of Europe organized a Colloquy on the "Conservation of Wild Progenitors of Cultivated Plants" in 1989 in cooperation with the Israel Nature Reserve Authorities and this reported to the Council of Ministers of Member States on the need to convene a Group of Experts to address the problems involved in conserving the wild relatives of native European cultivated plants.

The Group of Experts on Biodiversity and Biosubsistence was created in 1990 and one of its outputs was A Catalogue of the Wild Relatives of Cultivated Plants Native to Europe (Heywood and Zohary 1995) that provides an initial survey of the wild genetic resources of European cultivated plants. The area covered is Europe as circumscribed by Flora Europaea with the addition of Cyprus and the Canary Islands. For Turkey only species occurring in European Turkey are included.

The catalogue enumerates species and subspecies belonging to the following main groups of cultivated plants that are grown in Europe and also have their wild relatives on this continent, including Grain crops, Oil plants, Fruits & vegetables, Fibres, Herbs, Condiments, Medicinal plants, Fodder, Timber, Ornamentals, Nuts and Others. The survey of the food crops is more exhaustive than for the other groups and only the principal timber trees, fodder crops, medicinals, and ornamentals have been included. Even so, the total number of wild relatives of cultivated plants of economic importance in Europe is larger that many might suspect: over two hundred taxa are included in the Catalogue.

The Catalogue is only a first step and detailed surveys are now needed on a country-by-country basis. In addition, the Group of Experts organized a series of three workshops whose aim was: "to combine the skills of conservation biologists, crop plant evolutionists, population biologists, biochemists, cytogeneticists, and molecular biologists, with those of conservationists, managers of protected areas, and gene bank managers and apply it to the surveying, conservation, and management of the genetic diversity of the wild gene pools of the cultivated plants of Europe. The workshops will identify key elements for the elaboration of management and survival strategies for these plants."

WORKSHOP

The main themes covered by the workshops were: (1) Linking ecological and genetic variation; (2) Variation and demography of populations; (3) Genetic systems; (4) Environmental stress and survival strategies; (5) Conservation and management; (6) Protection of genetic variability in forest tree populations; Case studies. The proceedings of the workshops were combined in a single volume edited by Valdés et al. (1997).

The Group of Experts is now associated with the DIVERSITAS programme of biodiversity science as a European node of the Programme Theme "Conservation of the genetic diversity of wild species, especially those used in human activities" (Convenor V.H. Heywood).

Wild relatives was one of the topics covered by an "ad hoc Inter-agency Consultation on Promoting the Conservation and Sustainable Use of Wild Plants of Importance for Food and Agriculture" that was held at UNESCO, Paris, 11–13 February 1998, convened by DIVERSITAS, in association with UNESCO, FAO, IPGRI, and the Convention on Biodiversity (CBD). The consultation brought together different organizations concerned with the conservation and sustainable use of wild plants of importance for food and agriculture, including forestry, medicine, and other groups of interest to humans, with a view to developing a draft framework for collaboration amongst organizations. The report was been submitted to the Fourth meeting of the Conference of the Parties to the Convention on Biological Diversity held at Bratislava in May 1998 as an Information paper (UNEP/CBD/COP/4/Inf.17). Copies may be obtained from the DIVERSITAS Secretariat, c/o UNESCO-M.AB, 1 rue Miollis, 75015 Paris, France. E-mail: diversitas@unesco.org

A second inter-agency Consultation is planned take place at FAO in May 1999 at FAO, Rome, Italy, at which collaborative action will be planned in detail, coordination mechanisms proposed and funding sources identified.

The Wild Relatives Group is planning to (1) build on the existing catalogue by commissioning in-depth studies of the species concerned on a country-by-country basis; (2) extend the catalogue so as to cover the whole of the Mediterranean Region; (3) prepare a set of guidelines for the conservation and sustainable use of European wild relatives.

For further information contact:

Professor Vernon Heywood
Centre for Plant Diversity and Systematics
School of Plant Sciences
The University of Reading
PO Box 221, Whiteknights
Reading RG6 6AS, UK
Tel: +44 (0)118 9318160 / 9780185
Fax: +44 (0)118 9891745 / 9753676
Email: v.h.heywood@reading.ac.uk

REFERENCES

Heywood, V.H. and D. Zohary. 1995. A catalogue of the wild relatives of cultivated plants native to Europe. Flora Mediterranea 5:375–415.

Valdés, B., V.H. Heywood, F. Raimondo, and D. Zohary. (eds). 1997. Conservation of the wild relatives of European cultivated plants. Bocconea 7. Palermo.

The MEDUSA Network: Conservation and Sustainable Use of Wild Plants of the Mediterranean Region

Vernon Heywood and Melpo Skoula

The Mediterranean region is one of the world's major centers of plant diversity, housing approximately 25,000 species, about half of which are endemic to the region. It is one of the most important of the eight centers of cultivated plant origin and diversity identified by Vavilov who listed over 80 crops from the region, the most important of which are the cereal crops, pulses, fruit trees, and vegetables. Also found are many native species that are economically less important including notably medicinal and aromatic plants, herbs and spice-producing plants, neglected horticultural crops such as *Eruca sativa, Lepidium sativum, Portulaca oleracea, Smyrnium olusatrum, Scolymus hispanicus*, and ornamentals, all of which play an important role in local cultures. Some of these may well be worth consideration for further development and improvement as crops suitable for marginal areas. Many crop relatives occur in the Mediterranean basin.

The traditional use of wild plants assumes that such plant resources will continue to be available without any specific action to ensure this. However, as noted in the FAO Global Plan of Action, current programs for conservation research and development tend to neglect these species and no concerted effort has been made to ensure their continued availability in the face of the threats posed by over-exploitation caused by increasing demand, increasing human population and extensive destruction of the plant-rich habitats of the Mediterranean ecosystems.

To address these issues, a network on the "Identification, Conservation and Use of Wild Plants in the Mediterranean Region" called MEDUSA, was formally established in June 1996, by CIHEAM (Centre International des Hautes Etudes Agronomiques Méditerranéennes) and its constituent organ MAICh (Mediterranean Agronomic Institute of Chania). The Network is financially supported partly by the Directorate General I of the European Union and partly by CIHEAM.

The eventual aim of the Network is to propose methods for the economic and social development of rural areas of the Mediterranean Region, using ecologically-based management systems that will ensure the sustainable use and conservation of plant resources of the area. These plant genetic resources are of actual or potential importance to agriculture, various industries, and human health, and consequently will improve the quality of life. The particular goal of the Network, is the exploration of possibilities for the sustainable utilization of such resources as alternative crops for the diversification of agricultural production for improved product quality. This will involve a considerable amount of prior survey work regarding those wild species that are currently used in the region, whether through wild harvesting or through small-scale cultivation of semi-domesticated material.

The Network will contribute to the implementation of the FAO Global Plan of Action agreed at Leipzig in 1996, especially with regard to the conservation and sustainable use of underexploited wild species, and to the Decision XII/1 on Agricultural Biodiversity of the Conference of the Parties to the Convention on Biological Diversity.

ORGANIZATION AND OBJECTIVES

The objectives of the Network are:

1. The identification of native and naturalized plants of the Mediterranean Region, according to use categories such as food, food additives, animal food, bee plants, invertebrate foods, materials, fuels, social uses, vertebrate poisons, non-vertebrate poisons, medicines, perfumery and cosmetics, environmental uses, and gene sources (Table 1).

2. The creation of a Regional Information System that will include: scientific plant name and authority, vernacular names, plant description, chemical data, distribution, habitat description, uses, conservation status, present and past ways of trading, marketing and dispensing, and indigenous knowledge (ethnobiology and ethnopharmacology), including references to literature sources.

3. Preliminary evaluation of the conservation status and potential utilization in agriculture of these plants as alternative minor crops.

Table 1. The most useful plants of the Mediterranean region as determined by the MEDUSA survey

Food (Food, including beverages, for humans only)
Arbutus unedo L. (Ericaceae)
Castanea sativa Miller (Fagaceae)
Ceratonia siliqua L. (Leguminosae) Fabaceae
Cichorium intybus L. (Compositae) Asteraceae
Cynara cardunculus L. (Compositae) Asteraceae
Ficus carica L. (Moraceae)
Foeniculum vulgare Mill. (Umbelliferae) Apiaceae
Fragaria vesca L. (Rosaceae)
Malva sylvestris L. (Malvaceae)
Pinus pinea L. (Pinaceae)
Portulaca oleracea L. (Portulacaceae)

Food Additives (Processing agents and other additive ingredients used in food preparation)
Apium graveolens L. (Umbelliferae) Apiaceae
Carum carvi L. (Umbelliferae) Apiaceae
Coriandrum sativum L. (Umbelliferae) Apiaceae
Coridothymus capitatus (L.) Rechb.f. (Labiatae) Lamiaceae
Cuminum cyminum L. (Umbelliferae) Apiaceae
Foeniculum vulgare Mill. (Umbelliferae) Apiaceae
Humulus lupulus L.(Cannabinaceae)
Laurus nobilis L. (Lauraceae)
Mentha pulegium L. (Labiatae) Lamiaceae
Origanum vulgare L. (Labiatae) Lamiaceae
Rosmarinus officinalis L. (Labiatae) Lamiaceae

Animal Food (Forage and fodder for vertebrate animals only)
Ceratonia siliqua L. (Leguminosae) Fabaceae
Lathyrus cicera L. (Leguminosae) Fabaceae
Lupinus angustifolius L. (Leguminosae) Fabaceae
Stipa tenacissima L. (Gramineae) Poaceae
Trifolium pratense L. (Leguminosae) Fabaceae

Bee Plants (Pollen or nectar sources for honey production)
Coridothymus capitatus (L.) Rechb. f. (Labiatae) Lamiaceae
Melissa officinalis L. (Labiatae) Lamiaceae
Rosmarimus officinalis L. (Labiatae) Lamiaceae

Materials (Woods, fibers, cork, cane, tannins, latex, resins, gums, waxes, oils, lipids etc. and their derived products)
Cupressus sempervirens L. (Cupressaceae)
Linum usitatissimum L. (Linaceae)
Olea europaea L. (Oleaceae)
Pistacia lentiscus L. (Anacardiaceae)
Quercus suber L. (Fagaceae)

Fuels (Wood, charcoal, petrol substitutes, etc.)
Juniperus oxycedrus L. (Cupressaceae)
Quercus ilex L. (Fagaceae)
Quercus suber L. (Fagaceae)

Social Uses (Plants used for social purposes, not definable as food or medicines, such as masticatories, smoking materials, narcotics, hallucinogens and psychoactive drugs, contraceptives and abortifacients and plants with ritual or religious significance)

Datura stramonium L. (Solanaceae)
Hyoscyamus albus L. (Solanaceae)
Laurus nobilis L. (Lauraceae)
Lawsonia inermis L. (Lythraceae)
Olea europaea L. (Oleaceae)
Peganum harmala L. (Zygophyllaceae)
Ruta montana L. (Rutaceae)

Vertebrate Poisons (Plants poisonous to vertebrates, both accidentally and usefully, e.g. for hunting and fishing)

Citrullus colocynthis (L.) Schrad. (Cucurbitaceae)
Hyoscyamus niger L. (Solanaceae)
Taxus baccata L. (Taxaceae)
Urginea maritima L. (Liliaceae)

Non-Vertebrate Poisons (Both accidental and useful poisons e.g. molluscicides, herbicides, insecticides to non-vertebrate animals, plants, bacteria and fungi)

Lavandula stoechas L. (Labiatae) Lamiaceae
Lawsonia inermis L. (Lythraceae)
Mentha pulegium L. (Labiatae) Lamiaceae

Medicines (Both human and veterinary)

Coridothymus capitatus (L.) Rechb. f. (Labiatae) Lamiaceae
Crataegus monogyna Jacq. (Rosaceae)
Datura stramonium L. (Solanaceae)
Marrubium vulgare L. (Labiatae) Lamiaceae
Mentha pulegium L. (Labiatae) Lamiaceae
Rosmarinus officinalis L. (Labiatae) Lamiaceae
Taxus baccata L. (Taxaceae)
Teucrium polium L. (Labiatae) Lamiaceae
Urginea maritima L. (Liliaceae)

Environmental Uses (Ornamentals, hedges, shade plants, windbreaks, soil improvers, plants for revegetation and erosion control, waste water purifiers, indicators off metals, pollution, or underground water)

Cupressus sempervirens L. (Cupressaceae)
Myrtus communis L. (Myrtaceae)
Nerium oleander L. (Apocynaceae)

The Network is coordinated by the Mediterranean Agronomic Institute of Chania (MAICh) and includes members who are representatives of International Organizations (CIHEAM, IUCN, IUBS ICMAP, FAO, IPGRI-WANA, LEAD) and form the Steering Committee, and representatives of Institutions from countries of the Mediterranean basin, acting as the Focal Point Coordinators. The participating countries are: Algeria, Cyprus, Egypt, France, Greece, Italy, Morocco, Portugal, Spain, Syria, Tunisia and Turkey. It is envisaged that the Network will include eventually members from all the Mediterranean countries and from relevant National Institutions and other International Organizations.

MAJOR ACTIVITIES

Regional Workshops

The first workshop, organized at the Mediterranean Agronomic Institute of Chania, Crete, Greece, was entitled "Identification of Wild Food and Non-Food plants of the Mediterranean Region" on June 28th-29th 1996. development. The Proceedings of this Workshop have been published in the Cahiers Options Méditeranéennes. The second MEDUSA Workshop, was held in Port El Kantaoui, Tunisia on May 1st–3rd, 1997. General and Country Reports on the Governmental and Non-Governmental Organizations involved in any aspects of the study, cultivation, sustainable use, conservation of plant genetic resources used or of potential use in agriculture, and habitat conservation and restoration, were presented. The Proceedings were published in December 1998.

The third MEDUSA Workshop was held at the University of Coimbra, 27–28 April 1998. In addition to general papers and progress reports on activities of the Network, a series of case studies on particular medicinal and/or aromatic species was presented. The Proceedings will be published in 1999.

Newsletter

To facilitate operation of the Network, a newsletter is published, with the financial support of FAO. The first, was issued in the summer of 1997 and the second will appear in early 1999.

Priority Species List And Database

A specific questionnaire aimed at establishing a list of the priority species in the region, following the use categories as defined in the objectives, was been distributed and completed by most of the current Country Focal Point Coordinators of the Network. The results of the questionnaires have been entered in a database held at MAICh with a view to compiling a database of the mostly widely used species. This is currently being revised and expanded. So far, 1335 records have been received and correspond to 684 taxa and 361 genera that belong to 104 families. Half of the records as well as half of the taxa and 40% of the genera refer only to seven plant families which are in order of importance: Labiatae, Leguminosae (Fabaceae), Compositae (Asteraceae), Gramineae (Poaceae), Umbelliferae (Apiaceae), Rosaceae and Solanaceae. Work is also in progress, in collaboration with SEPASAL, ILDIS, Euro+Med PlantBase and FAO, on revising the use categories and on defining fields for use in an enhanced database.

Conservation of Medicinal and Aromatic Plants in Brazil

Roberto F. Vieira*

THE BRAZILIAN VEGETATION

Approximately two thirds of the biological diversity of the world is found in tropical zones, mainly in developing countries. Brazil is considered the country with the greatest biodiversity on the planet, with nearly 55,000 native species distributed over six major biomes (Fig 1): Amazon (30,000); Cerrado (10,000); Caatinga (4,000); Atlantic rainforest (10,000), Pantanal (10,000) and the subtropical forest (3,000).

The Brazilian Amazon Forest (tropical rainforest) covers nearly 40% of all national territory, with about 20% legally preserved. This ecosystem is rather fragile, and its productivity and stability depend on the recycling of nutrients, whose efficiency is directly related to the biological diversity and the structural complexity of the forest (Anon. 1995). Giacometti (1990) estimated that there are about 800 plant species of economic or social value in the Amazon. Of these, 190 are fruit-bearing plants, 20 are oil plants, and there are hundreds of medicinal plants (Berg 1982).

The "Cerrado" is the second largest ecological dominion of Brazil, where a continuous herbaceous stratum is joined to an arboreal stratum, with variable density of woody species. The cerrados cover a surface area of approximately 25% of Brazilian territory and around 220 species from cerrado are reported as used in the traditional medicine (Vieira and Martins 1998).

The "Caatinga" extends over areas of the states of the Brazilian Northeast and is characterized by the xerophitic vegetation typical of a semi-arid climate. The soils that are fertile, due to the nature of their original materials and the low level of rainfall, experience minor runoff (Anon. 1995). Various fruit species and medicinal plants have their centers of genetic diversity in this region, and the use of local folk medicines is common. Several important aromatic species are reported for this region (Craveiro et al. 1994), such as *Lippia* spp. and *Vanillosmopsis arborea*.

The Atlantic Forest extends over nearly the whole Brazilian coastline, and is one of the most endangered ecosystems of the world, with less than 10% of the original vegetation remaining. The climate is predominantly hot and tropical, and precipitation ranges from 1,000 to 1,750 mm. The land is composed of hills and coastal plains, accompanied by a mountain range (Anon. 1995). Several important medicinal species are found in this region, such as *Mikania glomerata, Bauhinia forficata, Psychotria ipecacuanha*, and *Ocotea odorifera*.

The territory of the Meridional Forests and Grasslands includes the mesophytic tropical forests, the subtropical forests, and the meridional grasslands of the states of southern Brazil. The climate is tropical and subtropical, humid, with some areas of temperate climate. The naturally fertile soils, associated with the mild climate, allowed a rapid colonization during the last century, mainly by European and, more recently, by Japanese immigrants (Anon. 1995). Several medicinal plants, such as chamomile (*Matricaria recutita*), calendula (*Calendula officinalis*), lemon balm (*Melissa officinalis*), rosemary (*Rosmarinus officinalis*), basil (*Ocimum basilicum*), and oregano (*Origanum vulgare*), were introduced and adapted by immigrants.

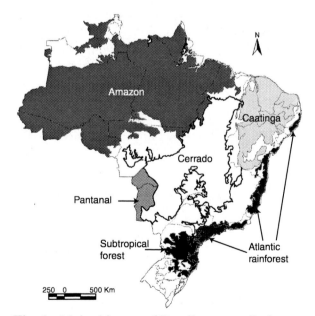

Fig. 1. Major biomes of Brazil, source: Embrapa, Cenargen.

---

*I thank Sergio Eustaquio and all germplasm curators for their suggestions and collaboration and Dr. Jules Janick and James Simon for their review.

The Pantanal is a geologically lowered area filled with sediments which have settled in the basin of the Paraguay River. Pantanal flora is formed by species from both Cerrado and Amazon vegetation. More than 200 species useful for human and animal consumption as well as for industrial use have been recorded in this region (Anon. 1995).

GERMPLASM CONSERVATION

In the last decade, serious efforts to collect and preserve the genetic variability of medicinal plants have been initiated in Brazil. The National Center for Genetic Resources and Biotechnology—Cenargen, in collaboration with other research centers of Embrapa (Brazilian Agricultural Research Corporation), and several universities, has a program to establish germplasm banks for medicinal and aromatic species (Table 1).

The first step is to establish criteria to define a species priority, based on economic and social importance, markets, and potential genetic erosion. Vieira and Skorupa (1993) proposed the following criteria to define priority, as follows: (1) species with proven medicinal value including those containing known active substance(s) or precursor(s) used in the chemical–pharmaceutical industry with proven pharmacological action, or at least demonstrating pre-clinical and toxicological results; (2) species with ethnopharmacological information widely used in traditional medicine; and which are threatened or vulnerable to extinction; (3) species with chemotaxonomical affinity to botanical groups which produce specific natural products.

Conservation of threatened germplasm includes seed banks, field preservation, tissue culture, and cryopreservation. Seed storage is considered the ideal method; seeds considered orthodox can be dried and are able to be preserved at sub-zero temperatures (–20°C), while recalcitrant seeds, including most tropical species, lose their seed viability when subjected to the same conditions. Maintenance of the germplasm in field collections is costly, requires large areas, and can be affected by adverse environmental conditions. Tissue culture or cryopreservation techniques can be also considered in some cases.

The next step is to decide which germplasm conservation method will be applied: ex situ or in situ. In an ex situ procedure, the germplasm is collected from fields, markets, small farms, and other sites, in form of seeds, cuttings, underground systems, and sprouts. The collected samples should represent the original population with passport data and herbarium vouchers. In a long term, mutation can take place over the years in a cold chamber or in vitro conservation. In contrast, in situ conservation maintains population in its preserved natural area, allowing the evolutionary process to continue, although genetic reserves are subject to anthropogenic action and environmental effects. Most in situ conservation has focused in forest species, with some medicinal species included, such as *Pilocarpus microphyllus* and *Aniba roseodora*. The establishment of genetic reserves in Brazil has relied on National Parks and conservation areas established by the environmental protection agency of Brazil, Ibama.

There are now five forest genetic reserves in Brazil: one in the Amazon Tropical Rainforest, state of Para; one in the Caatinga, state of Minas Gerais; two in the Cerrado in the Federal District, and one in the Meridional Forest (Subtropical) in the state of Santa Catarina. Four other genetic reserves are being created; two in the Atlantic Forest in the states of Rio de Janeiro and Espirito Santo, one in the Caatinga in the state of Piaui, and another in the Tropical Humid Forest in transition with Cerrado in the state of Minas Gerais (Anon. 1995). These reserves aim to conserve the most endangered species and those of greatest economic interest, including medicinal and aromatic plants.

The Brazilian program on medicinal germplasm conservation has three foci: (1) ethnobotanical studies; (2) germplasm collection and characterization; and (3) in situ conservation. Ethnobotanic and phytogeographic studies on the medicinal flora of Cerrado have been able to identify and collect genetic material for conservation. About 110 species used in traditional medicine were reported in the Cerrado region (Vieira et al. 1998). Bibliography review and a herbaria search were carried out allowing an estimation of the medicinal potential of each species studied, their geographic distribution, and period of fruit maturation.

In 1994, a cooperative project between the Brasília Botanical Garden and Embrapa/Cenargen was established. An in vivo collection of medicinal plants from Cerrado, now contains 161 accessions (Dias et al. 1995). The collection has facilitated phytochemical and pharmacological studies of this plant materials, and an anti-inflamatory agent has been identified on *Lychnophora salicifolia* (Miguel et al. 1997).

PRIORITY SPECIES

A few germplasm collections of medicinal and aromatic plants have been established in Brazil (Table 2). The following species, listed alphabetically, have been recognized as priority for germplasm conservation.

Maytenus ilicifolia Martius ex Reiss., Celastraceae (Espinheira Santa)

Espinheira santa is a small shrub evergreen tree reaching up to 5 m height. It is native to many parts of southern Brazil, mainly in Paraná and Santa Catarina states.

Leaves of *Maytenus* species are used in the popular medicine of Brazil for their reported antiacid and antiulcerogenic activity. The effects of a boiling water extract of equal parts of *M. aquifolium* and *M. ilicifolia* leaves have been tested in rats and mice. Attempts to detect general depressant, hypnotic, anticonvulsant, and analgesic effects were reported by Oliveira et al. (1991). The potent antiulcerogenic effect of espinheira santa

Table 1. List of institutions, germplasm collections of medicinal and aromatic plants, accessions, curator, and contact.

Institution	Major germplasm collections	No. accessions	Curator	Contact
Agronomic Institute of Paraná	*Pfaffia*	Unknown	P. Guilherme	www.pr.gov.br/iapar
Brasília Botanical Garden	Medicinal plants from cerrado	165	A. Lucia	Jardim Botanico de Brasilia, Lago Sul, Brasilia, DF
Embrapa–Genetic Resources and Biotechnology	*Phyllanthus, Pilocarpus, Stevia, Solanum*, plants from cerrado	335	T. Dias R. Vieira	www.cenargen.embrapa.br
Embrapa–Ocidental Amazon	*Croton cajucara*, general collection of medicinal and aromatic plants	Unknown	A. Franco	www.embrapa.br/cpaa, Rodovia Am - 010, km 24, CP 319, 69048-660, Manaus
Embrapa–Oriental Amazon	*Psychotria, Pilocarpus*	109	I. Rodrigues	Trav. Dr. Enéas Pinheiro s/n, Marco, CP 48 66095-100, Belém, PA
Maranhão State Univ.	*Pilocarpus*	27	G. Silva	Campus Universitário Paulo VI, CP 09, São Luis-MA
São Paulo State Univ., Botucatu	*Lippia, Ocimum*	Unknown	L. Ming	Unesp, Faculdade de Ciencias Agrarias, Departamento de Agronomia, Botucatu, SP
Univ. of Brasilia	*Pfaffia, Mentha* spp., *Labiatae, Phitolacca dodecandra*	Unknown	J. Kleber	Universidade de Brasília, Departamento de Agronomia, CP 04364, 70000, Brasília, DF
Univ. of Campinas, Cpqba	*Maytenus, Artemisia, Phyllanthus, Pfaffia, Cordia, Stevia*	330	P. Melillo	Unicamp, Cpqba, C.P. 6171, Campinas, SP
Univ. of Ceará	Aromatic plants: *Lippia, Croton, Cymbopogon*	224	F. Mattos	Lab. de Produtos Naturais/UFC, Campus do Pici, 60021-970, Fortaleza,CE
Univ. of North Fluminense	*Psychotria*	10	E. Martins	Universidade Estadual do Norte Fluminense, Lab. de Melhoramento Genetico Vegetal, Av. Alberto Lamego, 2000, Horto, 28015-620, Campos dos Goytacazes, RJ
Univ. of Paraná	*Maytenus*	78	M. Scheffer	Universidade Federal do Parana, Escola de Florestas, Departamento de Silvicultura e Manejo, Rua Bom Jesus, 650, Juveve, 80035-010, Curitiba, PR

Table 2. List of medicinal and aromatic species with high priority for germplasm collection and conservation in Brazil.

Species	Common name	Habit	Active substance/ pharmacological action	Region	Conservation form
Achyrocline satureioides L.	Macela	Herb	Hypotensive, spasmolytic	Cerrado	Field collection
Ageratum conyzoides L.	Mentrasto	Herb	Anti-inflammatory	Ruderal	Field collection
Aniba roseodora Ducke	Pau rosa	Tree	Linalool	Amazon forest	In situ
Astronium urundeuva (Fr. All.) Engl.	Aroeira	Tree	Anti-inflammatory, anti-ulceric	Cerrado chamber	In situ, cold
Baccharis trimera DC.	Carqueja	Herb	Hepatic disturbs	Ruderal	Field collection
Bauhinia forficata L.	Pata de Vaca	Tree	Diabetes	Atlantic forest	Cold chamber
Caryocar brasiliensis Camb.	Pequi	Tree	Anti-inflamatory	Cerrado	In situ
Copaifera langsdorffi Desf.	Copaiba	Tree	Oil, anti-inflamatory	Cerrado	In situ, cold chamber
Croton cajucara Benth.	Sacaca	Herb	Linalool	Amazon	Field collection
Croton zehntneri Pax et Hoff.	Cunha	Shrub	Anetol, eugenol	Caatinga	Field collection
Datura insignis B. Rodr.	Toe	Shrub	Escopolamina	Amazon forest	Cold chamber
Dimorphandra mollis Benth.	Faveiro	Tree	Rutin, anti-hemorragic	Cerrado	Cold chamber
Echinodorus macrophyllus (Kunth.) Mich	Chapeu de Couro	Herb	Diuretic	Cerrado	Field collection, cold chamber
Jatropha elliptica (Pohl) Baill.	Batat de Tiu	Shrub	Jatrophone	Cerrado	In situ, field collection
Lippia spp.	Alecrim pimenta	Shrub	Source of volatile oils, anti-microbial	Caatinga	Field collection
Lychnophora ericoides Mart.; *L. salicifolia* Mart.	Arnica do Cerrado	Shrub	Volatile oils	Cerrado	Field collection, in situ
Mandevilla vellutina Mart.		Shrub	Anti-inflamatory, bradykynin antagonist	Cerrado	In situ, field collection
Maytenus ilicifolia Mart. ex. Reiss; *M. aquifolium* Mart.	Espinheira Santa	Tree	Anti-ulceric	Meridional forest	Cold chamber, in situ
Mikania glomerata Spreng.	Guaco	Herb	Bronchitis, coughs	Atlantic forest	Field collection
Ocotea odorifera (Vell.) Rohwer	Canela Sassafraz	Tree	Safrol, metileugenol	Atlantic forest	In situ
Operculina macrocarpa (L.) Farwel	Batata de Purga	Herb	Purgative	Caatinga	Cold chamber
Piper hispidinervum DC.	Pimenta longa	Herb	Safrol	Amazon	Cold chamber, field collection
Pfaffia paniculata (Martius) Kuntze	Ginseng brasileiro	Herb	Antitumor compounds	Margins of Parana river	Cold chamber, field collection
Phyllanthus niruri L.	Quebra pedra	Herb	Hepatitis B, renal calculus	Ruderal	Cold chamber
Pilocarpus microphyllus Stapf.	Jaborandi	Shrub	Pilocarpine	Amazon forest	Cold chamber, in situ
Psychotria ipecacuanha (Brot.) Stokes	Ipecac	Herb	Emetin, cefaline	Amazon and Atlantic forest	Cold chamber, in situ
Pterodon emarginatus Vogel	Sucupira	Tree	Analgesic, antinoceptive, cercaricide	Cerrado	In situ, cold chamber
Solanum mauritianum Scopoli	Cuvitinga	Shrub	Solasodine	Ruderal, southeast and southern Brazil	Cold chamber
Stryphnodendron adstringens (Mart.) Coville	Barbatimao	Tree	Tannin, anti-inflamatory	Cerrado	In situ, cold chamber
Tabebuia avellanedae (Lor.) ex. Griseb.	Ipe roxo	Tree	Lapachol	Cerrado	In situ
Vanillosmopsis arborea (Aguiar) Ducke	Candeia	Shrub	Bisabolol	Caatinga	In situ, field collection

leaves was demonstrated effective compared to two leading anti-ulcer drugs, Ranitidine and Cimetidine (Souza-Formigoni et al., 1991). Toxicological studies demonstrated the plant's safety.

Seeds of *Maytenus ilicifolia* can be classified as orthodox and stored at –20°C in long-term cold chambers (Eira et al. 1995). The Forestry Department of the University of Paraná began a project in 1995 to study the genetic variability of natural populations of *Maytenus ilicifolia* and 78 accession were collected in the states of Parana, Santa Catarina, and Rio Grande do Sul. Field collections are maintained at the university campus (Scheffer et al. 1998). Although cultivation of *M. ilicifolia* is the object of several studies in Brazil, a research focus on in situ conservation and sustainable systems of harvesting are required.

Pfaffia paniculata Martius, Amaranthaceae (Brazilian Ginseng)

Pfaffia is a large, shrubby ground vine, which has a deep root system. *Pfaffia* is well known in Central and South America with over 50 species growing in the warmer tropical regions of the area and has been exploited for more than 15 years. The species grow in the borders of Paraná river, but predatory collection has greatly reduced the natural populations.

In Brazil, *Pfaffia* is known as *para tudo*, which means "for all things" and also as Brazilian ginseng, since it is widely used like American and Asian ginseng (*Panax* spp.). The active substances are found in the roots.

This action is attributed to the anabolic agent, beta-ecdysterone as well as three novel ecdysteroid glycosides which are found in high amounts in *Pfaffia*. This species is such a rich source of beta-ecdysterone. The extraction methods employed to obtain it from this root is protected by a Japanese patent (Nishimoto et al. 1988).

The root of *Pfaffia* contains about 11% saponins. These saponins include a group of novel chemicals called pfaffosides as well as pfaffic acids, glycosides, and nortriperpenes. These saponins have clinically demonstrated the ability to inhibit cultured tumor cell melanomas and help to regulate blood sugar levels (Takumoto et al. 1983; Nishimoto 1984). The pfaffosides and pfaffic acid derivatives in *Pfaffia* have been patented as antitumor compounds in two Japanese patents (Japanese Patent 84184198, Oct. 19, 1984 by Rohto Pharmaceutical Co., Ltd.).

Few accessions of *Pfaffia* are available in any of the present field collections. This species requires an immediate recollection to preserve the plant. Due to its economic importance a germplasm collection and characterization of its chemical constituents, is fully warranted.

Phyllanthus niruri L., Euphorbiaceae (Quebra Pedra)

Quebra pedra is a small erect annual herb growing up to 30 to 40 cm. height. Although several species are recognized by this common name, *P. niruri* and *P. sellovianus* are the most scientifically studied. The antispasmodic activity of alkaloids in *Phyllanthus sellovianus* explained the popular use of the plant for kidney and bladder stones. The alkaloid extract demonstrated smooth muscle relaxation specific to the urinary and biliary tract which facilitates the expulsion of kidney or bladder calculi (Calixto 1984; Santos 1994, 1995)

Quebra pedra has gained world-wide attention due to its effects against Hepatitis B (Thyagarajan 1982; Mehrotra 1990; Yeh, et al. 1993; Wang 1995). Recent research on quebra pedra reveals that its antiviral activity extends to the human immunodeficiency virus (HIV). The HIV-1 reverse transcriptase inhibition properties of *P. niruri* can be obtained with a simple water extract of the plant (Qian-Cutrone 1996). There have been no side effects or toxicity reported in any of the clinical studies or in its many years of reported use in herbal medicine.

Several species, called quebra pedra, contain the same or similar active compounds. A germplasm collection to study the genetic and chemical variation, as well as the seed physiology of this species is necessary and warranted.

Pilocarpus microphyllus Stapf., Rutaceae (Jaborandi)

Jaborandi is an indigenous name (*ia-mbor-end*) of this species. *Pilocarpus microphyllus* contain the highest pilocarpine content in the leaves. The plant is an understory species, 6 to 8 m in height, of the pre-Amazonian rain forest in the states of Pará, Maranhão, and Piauí.

Pilocarpine is an imidazolic alkaloid that stimulates the secretions of the respiratory tract, the salivary, lachrymal, gastric and other glands, weakens the heart action, accelerates the pulse rate, increases intestinal peristalsis and promotes uterine contractions (Morton 1977). In the treatment of glaucoma, the alkaloid pilocarpine acts directly on cholinergic receptor sites, thus mimicing the action of acetylcholine. Intraocular pressure is thereby reduced, and despite its short-term action, pilocarpine is the standard drug used for initial and maintenance therapy in certain types of primary glaucoma (Lewis and Elvin-Lewis 1977). Recently, the US Food and Drug Administration approved pilocarpine for use to treat post-irradiation xerostomia (dry mouth) in patients with head and neck cancer (Pinheiro 1997).

The exploration of this product, due to its high economic value, has led to great scientific interest in research and development effects for domestication and conservation. Pinheiro (1997) reports that the price of jaborandi leaves has reached US$4.00/kg. The wild harvest or collection of leaves from wild *P. microphyllus* has been carrying out to such an extent that it has significantly reduced the natural populations, and this species is included in the official list of endangered plants from Brazilian flora (Anon. 1992).

In 1991, the Cenargen initiated a project for recollecting and conservation of the genetic variability of *Pilocarpus microphyllus* and related species. From 1991 to 1993, two collection expeditions were undertaken, covering the states of Pará and Maranhão. A total of 27 accessions were collected in form of seeds and seedlings (Vieira, 1993). A germplasm bank of Jaborandi was established at Maranhão State University, São Luis, and at Embrapa—Ocidental Amazon, Belém, Pará State.

Studies on the methodology of *P. microphyllus* conservation led to the conclusion that seeds of this species are considered orthodox. Seeds can be dried down to 6–8% moisture content and be conserved for a long period at –18°C and 5% relative humidity (Eira et al. 1993). A seed sample of all collected accessions is being maintained at Embrapa, Cenargen.

Native populations of *P. microphyllus* have suffered from anthropogenic activity, with plants of shorter size than normal due to intensive harvesting of leaves. It will be challenge to stimulate the management and cultivation of this species in its native habitat. Although seeds can be preserved for long periods, in situ conservation must be initiated and natural reserves established. This species can be only found in indigenous areas, and some private lands.

Psychotria ipecacuanha (Brot.) Stokes, Rubiaceae (Ipecac)

Psychotria ipecacuanha (Brot.)Stokes [= *Cephaelis ipecacuanha* (Brot.) A.Rich.] is a shrub, whose medicinal value relates to the production of emetine in the roots. Ipecac is found in the humid forests of Central America, Colombia, southern part of the Amazon Forest in the States of Rondônia, Mato Grosso, and Atlantic forest, in the States of Bahia, Espírito Santo, Minas Gerais, and Rio de Janeiro (Skorupa and Assis 1998).

Ipecac as a powerful emetic, is used in gastrointestinal diseases, diarrhea, and intermitent fevers. It is employed as an expectorant, in bronchitis, broncopneumonia, asthma and mumps, and also as a vasoconstrictor. In 1959, dihydroxi-emetine, an emetine analogue, was presented as an amoebicide due to its reduced toxic effect on cardiac muscle (Lewis and Elvin-Lewis 1977).

The global production of ipecac averages 100 t a year, originated mainly from Nicarágua, Brazil and India (Husain 1991). Considering the economic and medicinal values of ipecac, the deforestation of the areas of occurrence and the extrativist nature of its production, in 1988, Cenargen has began a program for the recollecting and conservation of the genetic variability of this species. From 1988 to 1991, five collecting expeditions were undertaken, covering the States of Rondonia, Mato Grosso, Pernambuco, Bahia, Espirito Santo, Rio de Janeiro, and Minas Gerais, and a total of 86 accessions were collected (Skorupa and Assis 1998) and now maintained in field germplasm banks at Embrapa—Ocidental Amazon, Belém, Para, and at Florestas Rio doce, Linhares, Espirito Santo. Recently, other germplasm collections was established at the University of North Fluminense, which contains 10 accessions originated from the Atlantic Forest area (states of Rio de Janeiro and São Paulo).

Solanum mauritianum Scop., Solanaceae (Cuvitinga)

The steroidal alkaloids of the Solanaceae are compounds of considerable pharmaceutical interest as starting materials for the synthesis of steroid compounds such as anticontraceptive steroids and corticosteroids. The world demand for steroid precursors continues to increase while some of the traditional sources of steroidal raw material, such as yams (*Dioscorea* spp.) of Mexico and Central America, are becoming rapidly depleted (Roddick 1986). Solasodine is a chemical analog of diosgenin, and may be a substitute for this drug.

There are around 1,100 species of *Solanum* in South America, and *S. mauritianum* is among the species with the highest solasodine content (Vieira and Carvalho 1993). *Solanum mauritianum* is a subtropical shrub which grow all over southern Brazil. The solasodine content of *S. mauritianum* was evaluated in green fruits of natural populations growing on two different soils. High contents of solasodine were found in both population of *S. mauritianum* (from 2% to 3.5% of total dry weight) (Vieira 1989). Germplasm collections are needed to continue the study of genetic and environmental variation of solasodine, and to provide foundation study for future development programs.

Exotic Species

Although the major focus of germplasm conservation is on native species, several exotic, introduced and adapted species have been widely used and cultivated in Brazil. Many of them, such as lemongrass [*Cymbopogon citratus* (D.C.) Stapf.] and aloe (*Aloe* spp.), are cultivated in backyard gardens. Others, such as picão-preto (*Bidens pilosum* L.), mastruço (*Chenopodium ambrosioides* L.), and mentrasto (*Ageratum conyzoides* L.), whose adaptation through the years, has allowed a spontaneous wide distribution throughout the country, have had their use well disseminated (Dias 1995). In southern Brazil, due to favorable cultural and environmental conditions, several exotic species are cultivated in large areas. These include chamomile (*Matricaria recutita*), calendula (*Calendula officinalis*), rosemary (*Rosmarinus officinalis*), *Duboisia* sp., and Japanese mint (*Mentha arvensis*), all of which are deserving of collection and preservation due to the use of their natural products and the agricultural-based industries that produce these crops. The germplasm collection of exotic species also needs to be expanded to provide genetic resource for species adapted in Brazil. Although Brazil is not their genetic center of origin, different chemotypes have been naturalized (Mattos, pers. commun. 1994) and need to be conserved. One example is *Coleus barbatus*, which was introduced from Africa, and is clonally propagated in Brazil. However, several volatile oils chemotypes are found in Brazil for this species, probably due to different introductions from Africa in the past.

REFERENCES

Anon. 1992. Instituto Brasileiro do Meio ambiente e dos recursos naturais renováveis. Portaria no. 06N, de 15.01.1992. Diário Oficial, Brasília, 23, Jan., 1992. p. 870–872.

Anon. 1995. International Conference and Programme for Plant Genetic Resources (ICPPGR). Country Report. http://www.cenargen.embrapa.br/rec_gen/country/country.

Berg, M.E. van den. 1982. Plantas medicinais na Amazônia: Contribuição ao seu estudo sistemático. Belém, CNPq/PTU.

Calixto, J.B. 1984. Antispasmodic effects of an alkaloid extracted from *Phyllanthus sellowianus*: Comparative study with papaverine. Braz. J. Med. Biol. Res. 17(3–4):313–321.

Craveiro, A.A.; M.I.L. Machado, J.W. Alencar, and F.J.A. Matos. 1994. Natural product chemistry in northeastern Brazil. p. 95–102. In: G.T. Prance, D.J Chadwick, and J. March (eds.), Symposium on ethnobotany and search for new drugs. Ciba Foundation Symposium 185. Fortaleza, Brasil.

Dias, T.A.B., R.F. Vieira, M.V.M. Martins, C.M.C. Mello, M.C. Boaventura, A.E. Ramos, M.C. Assis, F.A. Ramos, P.P. Monteiro, and G.M.C.L. Reis. 1995. Conservacao ex-situ de recursos geneticos do cerrado: plantas medicinais, ornamentais e meliponideos. In: Proc. Int. Savanna Simposium, Brasilia, DF, Embrapa/CPAC. p. 195–197.

Eira, M.T.S., T.A.B. Dias, and C.M.C. Mello. 1995. Physiological behaviour of *Maytenus ilicifolia* seeds during storage. Hort. Bras. 13(1):32–34.

Eira, M.T.S.; R.F. Vieira, C.M.C. Mello, and R.W.A. Freitas. 1992. Conservação de sementes de Jaborandi (*Pilocarpus microphyllus* Stapf.). Rev. Brasileira de Sementes 14(1):37–39.

Giacometti, D.C. 1990 Estrategias de coleta e conservacao de germoplasma horticola da America tropical. In: Proc. Simposio Latinoamericano sobre recursos geneticos de especies horticolas, 1. Campinas/SP. Fundacao Cargill. p. 91–110.

Husain, A. 1991. Economic Aspects of Exploitation of Medicinal Plants. In: O. Akerele, V. Heywood, and H. Synge (eds.), Conservation of medicinal plants. Cambridge Univ. Press, Cambridge.

Lewis, W.H. and P.F. Elvin-Lewis. 1977. Medical botany. Wiley, New York.

Mehrotra, R. 1990. In vitro studies on the effect of certain natural products against hepatitis B virus. Indian J. Med. Res. 92:133–138.

Miguel, O.G., E.O. Lima, V.M.F. Morais, S.T.A. Gomes, F.D. Monache, A.B. Cruz, R.C.B. Cruz, and V. Cechinel Filho. 1996. Antimicrobial activity of constituents isolated from *Lychnophora salicifolia* (Asteraceae). Phytotherapy Res. 10:694–696.

Morton, J.F. 1977. Major medicinal plants. Charles C. Thomas, Illinois. p. 187–189.

Nishimoto, N., S. Nakai, N. Takagi, S. Hayashi, T. Takemoto, S. Odashima, H. Kizu, and Y. Wada. 1984. Pfaffosides and nortriterpenoid saponins from *Pfaffia paniculata*. Phytochemistry 23:139–42.

Nishimoto, N, Y. Shiobara, S.S. Inoue, M. Fujino, T. Takemoto, C.L. Yeoh, F.D. Oliveira, G. Akisue, M.K. Akisue, and G. Hashimoto. 1988. Three ecdysteroid glycosides from *Pfaffia iresinoides*. Phytochemistry 27:1665–1668.

Oliveira, M.G.M., M.G. Monteiro, C. Macaubas, V.P. Barbosa, and E.A. Carlini. 1991. Pharmacologic and toxicologic effects of two *Maytenus* species in laboratory animals. J. Ethno-Pharmacology 34:29–41.

Pinheiro, C.U.B. 1997. Jaborandi (*Pilocarpus* sp., Rutaceae): a wild species and its rapid transformation into a crop. Econ. Bot. 51:49–58.

Qian-Cutrone, J. 1996. Niruriside, a new HIV REV/RRE binding inhibitor from *Phyllanthus niruri*. J. Nat. Prod. 59:196–199.

Roddick, J.G. 1986. Solanaceae: Biology and systematics. Columbia Univ. Press, New York.

Santos, A.R. 1994. Analgesic effects of callus culture extracts from selected species of *Phyllanthus* in mice. J. Pharm. Pharmacol. 46:755–759.

Santos, A.R. 1995. Analysis of the mechanisms underlying the antinociceptive effect of the extracts of plants from the genus *Phyllanthus*. Gen. Pharmacol. 26:1499–1506.

Scheffer, M.C., L.C. Ming, and A.J. Araujo. 1998. Conservacao de recursos geneticos de plantas medicinais. In: Simposio sobre Recursos Geneticos do Semi-Arido. Embrapa -Semi- Arido, Petrolina, PE (in press).

Skorupa, L.A. and M.C. Assis. 1998. Collecting and conserving Ipecac (*Psychotria ipecacuanha*, Rubiaceae) germplasm in Brazil. Econ. Bot. 52:209–210.

Souza-Formigoni, M.L.O., M.G.M. Oliveira, M.G. Monteiro, N.G. Silveira-Filho, S. Braz, and E.A. Carlini. 1991. Antiulcerogenic effects of two *Maytenus* species in laboratory animals. J. Ethno-Pharmacology 34 (1):21-27.

Takemoto, T., N. Nishimoto, S. Nakai, N. Takagi, S. Hayashi, S. Odashima, and Y. Wada. 1983. Pfaffic acid, a novel nortriterpene from *Pfaffia paniculata* Kuntze. Tetrahedron Lett. 24, 1057–1060.

Thyagarajan, S.P. 1982. *In vitro* inactivation of HBsAg by *Eclipta alba* Hassk and *Phyllanthus niruri* Linn. Indian J. Med. Res. 76:124–130.

Vieira, R.F. 1989. Avaliação do teor de solasodina em frutos verdes de *Solanum mauritianum* Scop. sob dois solos no estado do Paraná, Brasil. MS theses. Curitiba, Universidade Federal do Paraná.

Vieira, R.F. 1993. *Pilocarpus microphyllus* Stapf. G15 Gene Bank Medicinal and Aromatic Plants Newslett. 3–4:4–5.

Vieira, R.F. and L.A. Skorupa. 1993. Brazilian medicinal plants gene bank. Acta Hort. 330:51–58.

Vieira, R.F. and L.D.A. de Carvalho. 1993. Espécies medicinais do gênero *Solanum* produtoras de alcalóides esteroidais. Rev. Brasileira Farmácia 74:97–111.

Vieira, R.F. and M.V.M. Martins. 1998. Estudos etnobotanicos de especies medicinais de uso popular no Cerrado. In: Proc. Int. Savanna Simposium, Brasilia, DF, Embrapa/CPAC. p. 169–171.

Wang, M. 1995. Herbs of the genus *Phyllanthus* in the treatment of chronic hepatitis B: observations with three preparations from different geographic sites. J. Lab. Clin. Med. 126:350–352.

Yeh, S.F. et al. 1993. Effect of an extract from *Phyllanthus amarus* on hepatitis B surface antigen gene expression in human hepatoma cells. Antiviral Res. 20:185–92.

Indigenous Plant Genetic Resources of South Africa

Cobus Coetzee, Elton Jefthas, and Emmy Reinten

South Africa is considered to be a "hotspot" for biodiversity and more than 22,000 plant species occur within its boundaries. This represents 10% of the world's species, although the land surface of South Africa is less than 1% of the earth. The country is divided into seven biomes and into 68 vegetation types (Low and Rebelo 1996). The Savanna biome covers 33% of the surface of the country, but it is especially the Flora Capensis that is unique. This, the Cape Floral Kingdom, is the smallest of the world's six Floral Kingdoms. It contains 8,700 species of which 68% are endemic.

Despite the enormous richness in plant species, relatively few of these plants are economically utilized. Business ventures that have developed from the use of indigenous plants is the trade in medicinal and cultural plants, food crops, and ornamental plants. Although indigenous wood has been previously used, the source is almost depleted and today these wood types are utilized on a limited scale. Dekriet (*Chondropetalum tectorum* Pillans), and dekgras (*Schizachyrium semiberbe* Nees) is still largely used to thatch vernacular buildings.

South Africa beholds her indigenous plants as a valuable natural resource and accepts responsibility to conserve the unique flora. Attempts are also made to utilize the plant kingdom economically for the nation, but with legal acknowledgement of the legal owners.

MEDICINAL AND CULTURAL USE OF INDIGENOUS PLANTS

Indigenous medicinal plants are used by more than 60% of South Africans in their health care needs or cultural practices (Table 1). Approximately 3,000 species are used by an estimated 200,000 indigenous traditional healers (Van Wyk et al. 1997).

Due to urbanization, a large informal trade business has been established with medicinal plants. Unfortunately, utilization of the plants has depleted the wild populations, resulting in many plant species being considered vulnerable, and being lost from their natural habitat. If raw materials of medicinal plants can be delivered in sustainable quantities (Mander et al. 1995), indigenous plants will continue to form an important component of the primary health care in Southern Africa.

Few medicinal plants are cultivated and only *Warburgia salutaris* (Bertol.f.) Chiov. (pepperbark tree) and *Siphonochilus aethiopicus* (Schweinf.) B.l. Birtt (African ginger) are known to be propagated for cultivation. Phytomedicine plants in South Africa with a position in the international trade are Cape aloes (*Aloe ferox* Miller), buchu (*Agathosma* spp.), devil's claw (*Harpagophytum procumbens* DC), umkcaloabo (*Pelargonium sidoides* DC) and uzara (*Xysmalobium undulatum* R.Br) (N. Gericke 1998 unpubl. report). Buchu is cultivated commercially, but is also harvested in the wild, although under supplied. This can lead to over utilization of the natural habitat. *Aloe ferox* Miller is sustainably harvested from the wild. In the case of devil's claw the natural habitat can be over utilized. From the diversity of medicinal plants, a large number of species containing chemical components have the potential to play a role in the medicinal market on a global scale. At present bioprospecting is done on all plants in South Africa to determine among other things its pharmaceutical potential.

FOOD INDUSTRY

Although Southern Africa is very rich in the diversity of plant species, only a few are used as edible food material. The leaves and roots of edible plants have a high nutritional value and can play an important role in the prevention of malnutrition in rural areas. (Venter and Van den Heever 1998). With urbanization there is a movement away from traditional crops and more Western eating habits are developing.

Some of the indigenous food types (Table 2) such as rooibos tea [*Aspalathus linearis* (Burm.f.) R. Dahlgren] and honeybush tea (*Cyclopia* spp.) have developed as an agricultural industry with export potential. Buchu (*Agathosma* spp.), one of the traditional medicinal plants, is exported in large volumes. This however, is not for medicinal uses but mainly as a fixative in the food industry. Buchu is also used on a small scale in the ornamental industry.

Table 1. Selection of indigenous medicinal plants used in South Africa.

Species	Family	Popular name
Agathosma betulina (Bergius) Pill.	Rutaceae	Buchu
Agathosma crenulata (L.) Pill.	Rutaceae	Buchu
Aloe ferox Miller	Asphodelaceae	
Artemisia afra Jacq. ex Willd.	Asteraceae	Wormwood
Balanite maughamii Delile	Balanitaceae	Torchwood
Bersama tysoniana Oliv	Melianthaceae	White ash
Boophane disticha (L.f.) Herbert.	Amaryllidaceae	Tumbleweed
Bowiea volubilis Harv.	Hyacinthaceae	Climbing lily
Cassine papillosa (Hochst.) Kuntze	Celastraceae	Common saffron
Clivia miniata Regel.	Amaryllidaceae	Bush lily
Cryptocarya latifolia Sond.	Lauraceae	Broad leaved quince
Curtisia dentata (Burm.f.) C.A. Smith	Cornaceae	Assegaai
Dioscorea sylvatica (Kunth) Ecklon	Dioscoreaceae	Elephant's foot
Eucomis autumnalis (Mill.) Chitt.	Hyacinthaceae	Wild pineapple
Gunnera perpensa L.	Gunneraceae	Wild rhubarb
Harpagophytum procumbens DC.	Pedaliaceae	Devil's claw
Ocotea bullata (Burchell) Baillon	Lauraceae	Stinkwood
Pelargonium sidoides DC.	Geraniaceae	Umkcaloabo
Pittosporum viridiflorum Sims	Pittosporaceae	Cheesewood
Rapanea melanophloeos (L.) Mez	Myrsinaceae	Cape beech
Scilla natalensis Planch.	Hyacinthaceae	Blue hyacinth
Siphonochilus aethiopicus (Schweinf.) B.l. Birtt	Zingiberaceae	African ginger
Stangeria eriopus Nash	Stangeriaceae	Natal grass cycad
Warburgia salutaris (Bertol.f.) Chiov.	Canellaceae	Pepperbark tree
Xysmalobium undulatum R.Br	Asclepiadaceae	Uzara

Table 2. Indigenous edible wild plants used in South Africa.

Species	Family	Popular name
Agathosma betulina (Bergius) Pill.	Rutaceae	Buchu
Agathosma crenulata (L.) Pill.	Rutaceae	Buchu
Aponogeton distachyos L. f.	Aponogetonaceae	Waterblommetjies
Amaranthus hybridus (L.f)	Amaranthaceae	Marog
Amaranthus tricolor L.	Amaranthaceae	Marog
Aspalathus linearis (Burm.f.) R. Dahlgren	Fabaceae	Rooibos tea
Cajanus cajan Mill sp.	Leguminosae	Pigeon pea
Carpobrotus edulis (L) N.E.Br.	Mesembryanthemaceae	Sour fig
Cleome gynandra L.	Capparidaceae	Leafy vegetable
Colocasia antiquorum var. *esculenta* Schott	Araceae	Amadumbie
Cyclopia genistoides (L.) Vent.	Fabaceae	Honeybush tea
Dovyalis caffra Warb.	Flacourtiaceae	Kei apple
Plectranthus escullentus N.E.Br.	Lamiaceae	
Sclerocarya birrea subsp. *caffra* Sond.	Anacardiaceae	Marula
Solanum retroflexum Dun	Solanaceae	
Vigna subterranea (L.) Werdc.	Fabaceae	Bambara ground nut
Vigna unguiculata (L.) Walp.	Fabaceae	Cowpea

ORNAMENTAL INDUSTRY

The Flora Capensis, a plant kingdom contained within the boundaries of South Africa, comprises 8600 species (Bond and Goldblatt 1984). From this floricultural wealth, European explorers collected plant material and a range of new horticultural products were developed over the last two hundred years

Gladiolus and *Freesia* (Table 3), important fresh cut flowers on the world market, are originally from this region. An important component of the agricultural sector in the Western Cape is based on Flora Capensis, namely the indigenous flower industry (Wessels et al. 1997). The indigenous flower industry supports 20,000 people and is an important job creator in the Western Cape. In the past, indigenous flowers were harvested in the natural habitat—today the industry is in a transformation process and flowers are cultivated. If the process is not successful, South Africa stands to lose its protea industry and countries like Australia, New Zealand and Zimbabwe will take a leading role in production.

ECONOMIC EXPLOITATION

Economic exploitation of South Africa's rich natural plant resources is limited. At present only the indigenous flower industry has relatively successfully established small and medium scale entrepreneurs. The indigenous medicinal plant industry is large, but fully based on harvesting from the wild. This is not sustainable and will have to be supplemented with cultivation. The commercial utilization of food types is limited, with the exception of the buchu industry which has already been established as a cultivated industry with an export market. Only aloe and devil's claw (*Harpagophytum procumbens* DC.) are exported for medicinal use.

The utilization of South African indigenous flora can only be successfully explored if the existing indigenous knowledge of the inhabitants is made available to science. By forming associations between natural healers and scientists, medicinal plants can be investigated. From these associations, industries can be formed to commercialize the products.

Table 3. Some indigenous ornamental plants originally from South Africa, cultivated worldwide.

Genus	Family
Agapanthus spp.	Alliaceae
Clivia spp.	Amaryllidaceae
Chlorophytum spp.	Asphodelaceae
Erica spp.	Ericaceae
Freesia spp.	Iridaceae
Gerbera spp.	Asteraceae
Gladiolus spp.	Iridaceae
Leucadendron spp.	Proteaceae
Leucospermum spp.	Proteaceae
Nerine sarniensis	Amaryllidaceae
Osteospermum spp.	Asteraceae
Pelargonium spp.	Geraniaceae
Plumbago spp.	Plumbaginaceae
Protea spp.	Proteaceae
Strelitzia spp.	Strelitziaceae
Ornithogalum spp.	Hyacinthaceae
Thamnochortus spp.	Restionaceae
Veltheimia spp.	Hyacinthaceae
Zantedeschia spp.	Araceae

Commercial utilization will promote the creation and development of rural SMME's (small micro and medium entrepreneurs). This will enable communities to create wealth from indigenous technologies and plants (M. Deliwe 1998 unpubl. report), and will ensure that natural habitats are protected. The problem is to prohibit illicit exploitation of plant material, as well as other prejudicial actions. At present no official legislation exists, but a proposed law known as the "Protection of Indigenous Knowledge Act" is being prepared to advance the promotion and protection of indigenous knowledge. This act does not prohibit the exploitation of indigenous plants. On the contrary, the act attempts to promote and develop the use of indigenous genetic material. The primary aim is to ensure that the lawful owner is recognized in the development.

The proposed legislature will contribute to documenting indigenous knowledge. The act makes provision for manners and customs related to food, production of traditional medicine from herbs or other sources, and fermentation techniques to be documented without giving away ownership. This aspect is invaluable in ensuring that the cultural heritage is conserved for generations to come.

Communities fear the illicit use and exploitation of indigenous knowledge by outsiders, with the result that most knowledge and especially indigenous medicinal plant knowledge is being kept secret. The proposed act will lay ghost to this fear as the law will now protect the individuals and communities. A protocol and guidance is available to assist communities in negotiations on use of indigenous knowledge. These developments will enhance the maintenance and availability of indigenous knowledge and will contribute to research and development, contributing to the African Renaissance, not only in South Africa but in the whole of Africa.

REFERENCES

Bond, P. and P. Goldblatt. 1984. Plants of the Cape Flora: A descriptive catalogue. J. South African Botany. Vol. 13. (Suppl.).

Low, A.B. and A.G. Rebelo. 1996. Vegetation of South Africa, Lesotho and Swaziland. Published by the Department of Environmental Affairs & Tourism, Pretoria.

Mander, M., J. Mander, N. Crouch, S. Mckean, and G. Nichols. 1995. Catchment action: Growing and knowing muthi plants. Share Net Resource, Howick, South Africa.

Van Wyk, B-E., B. Van Oudtshoorn, and N. Gericke. 1997. Medicinal plants of South Africa. Briza Publ., South Africa.

Venter, S.L. and E. Van Den Heever. 1998. Vegetable and medicinal uses of traditional edible seed and leafy vegetable crops. Abstracts SANCRA (Southern African New Crop Research Association). Symposium 28–30 Sept. 1998.

Wessels J., P. Anandajaysekeram, G. Littlejohn, D. Martella, C. Marasas, and C. Coetzee. 1997. Socio-economic impact of the Proteaceae Development and Transfer Program. SACCAR. (Southern African Center for Cooperation in Agricultural and Natural Resources Research and Training).

CROP INFORMATION SYSTEMS

A Site-Specific Retrieval System for Crop Information

Jules Janick, James E. Simon, and Anna Whipkey

NewCROP (New Crops Resource On-line Program) is a web-based internet resource developed by the Purdue University Center for New Crops and Plant Products emphasizing new and specialty crops (**www.hort.purdue.edu/newcrop**). Designed to deliver instant topical information on crop plants, NewCROP first came on-line in 1995, and since that time has continued to grow significantly in content information and represents the most extensively used virtual library of new crop information on the web (Simon et al. 1996). Since that initial report, advances in web-based technology, coupled with the addition of new databases have enabled NewCROP to continue to address the needs of the new crop community. Linkage to crops is based on either a comprehensive index using common and scientific names (CropINDEX) or via a search engine (CropSEARCH). The information on crops is based on in-depth information prepared by crops experts (FactSHEETS), information from the proceedings of national new crop symposia, crop monographs, and links to outside sources. A portion of NewCROP now concentrates specifically on herbs, spices, aromatic, and medicinal plants (Aromatic-MedicinalPLANTS). Other useful databases include information on import permits phytosanitary certificates, quarantine, and inspection information (IMPORT-EXPORT), information on species consumed in times of food scarcity (Famine Foods), directories of researchers and experts (CropEXPERT), and announcements (CropEVENTS). In 1998, the web site received and responded to over 1.7 million requests for 650,623 pages, serving 218,836 distinct hosts. An historical graph of the site traffic from July, 1996 to April, 1999 shows a continuous upward trend (Fig. 1). The site is accessed nationally and internationally by many users (Table 1). (See http://www.hort.purdue.edu/stats/98newcrop.html)

At the heart of this virtual library, NewCROP incorporates the three proceedings organized by the Center for New Crops and Plant Products (*Advances in New Crops*, 1990; *New Crops*, 1993; and *Progress in New Crops*, 1996) and will include this present volume, *Perspectives on New Crops and New Uses*, 1999, organized in conjunction with the Association for the Advancement of Industrial Crops and the New Uses Council. These four volumes contain over 400 articles with references to over 7,000 crops (scientific and common names) and crop products. All texts are hyperlinked.

Table 1. NewCROP web server domain report for 1998.

User domain	Pages % of total
Commercial	24.4
Network	18.6
Unresolved	21.2
US government	0.7
Education	6.6
Organizations	0.9
Foreign*	25.6

*Includes Canada, Australia, United Kingdom, Germany, New Zealand, Belgium, Netherlands, Denmark, Mexico, Italy, Brazil, Switzerland, Spain, Japan, France, Malaysia, Sweden, South Africa, Portugal, Singapore, Argentina, Israel, Finland, India, Greece, Thailand, Hungary, South Korea, Indonesia, Colombia, Chile, Norway, Austria, Peru, Poland, Turkey, Costa Rica, Czech Republic, Philippines, Ireland, Yugoslavia, Slovenia, Hong Kong, Croatia, Russia, Venezuela, Nicaragua, Taiwan, Ecuador, Jamaica, Uruguay, Egypt, Estonia, Dominican Republic, Slovak Republic, United Arab Emirates, China, Guatemala, Romania, Botswana, Iceland, Lithuania, Ukraine, Trinidad and Tobago, Cyprus, Zambia, Bahrain, Bulgaria, Namibia, Oman, Paraguay, Bolivia, Kenya, Lebanon, Jordan, Malta, Macedonia, Mauritius, Guyana, Luxembourg, El Salvador, Pakistan, Bosnia-Herzegovina, Panama, Former USSR, New Caledonia, Ethiopia.

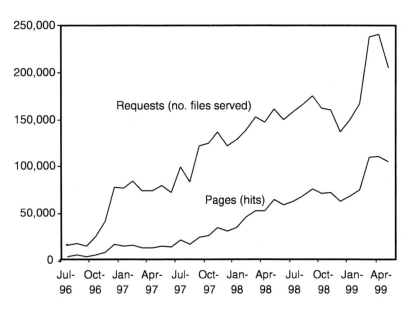

Fig. 1. NewCROP web server statistics.

INDIANA CROPMAP

A site-specific retrieval system (CropMAP) to identify current and alternative crops was developed using the state of Indiana as a prototype (Fig. 2). CropMAP was conceptualized to enhance the usefulness of the NewCROP database to growers, marketers, and extension personnel. CropMAP is based on three concepts: (1) the development of an interactive state map; (2) the incorporation of county crop statistics from the US Agricultural Census; and (3) the development of crop lists by state experts subdivided into traditional crops, recommended crops, experimental crops, and non recommended crops. Each crop is linked to detailed information available in our databases and enriched by extension information. The prototype, Indiana CropMAP, is described below.

The interactive map of Indiana is divided into it's 92 counties. Each county can be accessed by clicking the image map or county name. The county page brings up information and options as follows: (1) a highway map in color; (2) a link to extension resources; (3) crop statistics from the most recent agricultural census; (4) a list of crop commodity groups (cereals and pseudocereals, forage legumes and grasses, fiber and industrial crops, vegetables, fruits and nuts, aromatic, herb, spices, and medicinal crops). Selecting a crop group leads to a list of crops under four headings: traditional, recommended, experimental, or not recommended. Each crop is followed by a statement briefly describing general suitability and adaptation throughout the state. The crop name is also linked to general information contained in NewCROP enriched by specific extension information from the state or region.

The prototype for Indiana does not yet include ornamental, greenhouse, or forest crops. Information on crop statistics was obtained from the 1992 US agricultural census but is being updated by inclusion of data from the 1997 census. The list of crops was suggested by different extension personnel at Purdue University. The list is not static, and new information can be added or deleted as needed.

FUTURE DEVELOPMENT

The expansion of CropMAP to other states is best handled on a state to state basis through local experts. We are proposing a Coordinator for each state to choose the specific Crop Specialists for each of the crop classes. The State Coordinator needs to define the specific crop regions for each state. The Crop Specialists will complete the list of traditional, recommended, experimental, and not recommended crops for the following classes (1) Cereals, pseudocereals, and oilseeds, (2) Forage grasses & legumes, (3) Vegetables, (4) Fruits & nuts, (5) Herb, aromatic, medicinal, & bioactives, (6) Industrial & fiber (7) Nursery & ornamental (8) forest crops. The crop lists can be modified to reflect the needs of specific regions. We plan to initially expand CropMAP to five states in 1999 and to include the entire US by 2001. Should this system prove useful it could then be expanded into other countries.

One of the problems we forsee is to develop agricultural zones of the United States similar to the hardiness zones. A system has been developed by Sunset Books, Inc. and these maps which are copyrighted are found in Sunset National Garden Book (1997).

REFERENCES

Janick, J. and J.E. Simon (eds.). 1990. Advances in new crops. Timber Press, Portland, OR.

Janick, J. and J.E. Simon (eds.). 1993. New crops. John Wiley and Sons, Inc., New York.

Janick, J. (ed.). 1996. Progress in new crops. ASHS Press, Alexandria, VA.

Simon, J.E., J. Janick, and A. Hetzroni. 1996. The NewCROP electronic network. p. 142–147. In: J. Janick (ed.), Progress in new crops. ASHS Press, Alexandria, VA.

Sunset Books, Inc. 1997. Sunset National Garden Book, Menlo Park, CA

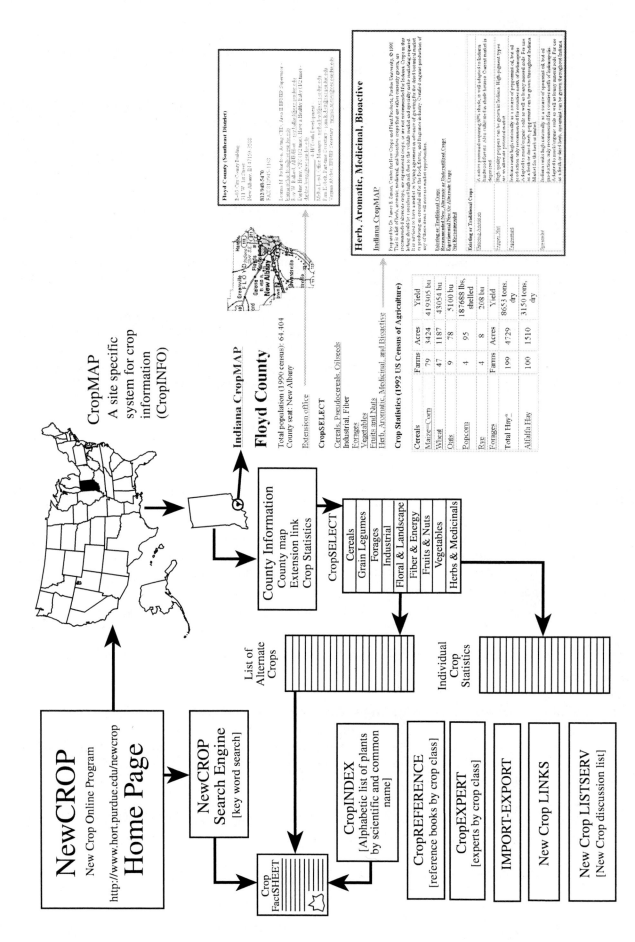

Fig. 2. Indiana CropMAP

PLANT INFO: A Component of FAO's Global Plant and Pest Information System

P. Griffee, T. Putter, M. Marsella, H. Pfeiffer, and P. Diemer

In co-operation with a range of government and non-government institutions, FAO has established an international shared knowledge resource on technical information about plants and pests. This Global Plant & Pest Information System (GPPIS), is an Internet-based service for information access but also a process for harnessing global knowledge within a framework of standard procedures and protocols for data collection, verification, validation, and distribution, that are endorsed by participants. Implemented under the auspices of FAO, the GPPIS forum creates a dynamic framework for continuous collective knowledge processing within the sharing community. The Internet makes this possible because it distributes the task required to create a global resource, while concentrating and multiplying the benefits of collaboration. GPPIS can be implemented in a variety of cross-platform environments on the Internet, in local Intranets, but also as a complete Internet-equivalent system that runs on CD-ROM.

Plant Info, as integrated part of GPPIS, aims at developing and maintaining a platform (interactive compendium) for access to reliable information on crop production, designed to guide technical decisions within different agroecozones and production systems, for increased food security and sustainable agricultural development. Following the usual pattern of compendia, wherein a few scientists with expertise in a discrete set of specialized fields voluntarily devote their time to writing and illustrating crop data sheets or profiles, GPPIS is an Internet adapted dynamic, interactive, digital implementation of the same traditional approach. Specifically, individuals are invited and encouraged to participate in the GPPIS community by "adopting" a crop species, a discrete "layer" (topic) of information about plants, e.g. the "medicinal uses" layer of information associated with separate plant records, or simply by serving as public referees or scrutineers of information in the GPPIS knowledge commons. The main difference it is not a single, one-time, printed product that requires regular reprinting of new editions, but a permanently updated set of data managed by a set of methods and protocols for data collection and maintenance under password-protection.

Data quality in GPPIS is further managed in a variety of ways. Each record has a single primary editor: this editor receives a password which is required to edit or add data. Each record makes provision for a few "topic editors" to function as a small editorial committee to support and advice the primary editor. The information in GPPIS is displayed under the name and logo of the contributing author or institution, resulting in a public, transparent display of assumption of responsibility through volunteering expertise. Coupled to the continuous public scrutiny that characterizes the Internet, as well to GPPIS' procedures for immediate feedback by public viewers and users, data quality in GPPIS is controlled dynamically and in real-time. Finally, partnership program with key institutions who have a recognized role of knowledgeable authority in a particular group of records which are then "clustered" and presented under the banner and logo of that institution.

GPPIS data are in the public domain and not copy righted because its content is created and maintained by the community of individuals who choose to participate in the building of GPPIS. Data contribution by sponsors and supporters is voluntary and motivated by the realization that individuals sharing their individual expertise enables all participants to have access to the sum of their collective knowledge. Information is considered not as product but as advertising the expertise which the institution hopes to sell through service provision, in order to sustain its continued financial viability.

Apart from their individual significance, the main elements of GPPIS also produce interaction effects especially when considering supporting or ancillary information in GPPIS such as glossary of terms, bibliographic references, the picture databank and GPPIS-related methodologies. This leads to a quantitative synergy between separate data elements—an interaction that produces knowledge that is more than the sum of its individual information elements. GPPIS proposes unique and original procedures whereby users at the periphery of the GPPIS network and community can control the process of establishing hypertext links between discrete units of information that convey a message or "*memes*." Each editor controls the hypertext linking process explicitly and thus GPPIS is a knowledge creation system as opposed to an information broadcasting system. This original GPPIS' procedures that allow users to create "reverse" hypertext links led to participatory methods of knowledge building whereby users update and own the data.

FAO is fundamentally concerned about developing countries where Internet access is currently unavailable. Therefore, GPPIS was designed to be packaged also on CD-ROM (periodical snapshot) for independent local stand-alone or Intranet use. GPPIS will be distributed as regular, free CD-ROMS, and user connected to the Internet can, at any time, "switch" from local CD-ROM to current, real-time information on the Internet. Procedures are being developed to enable individuals to submit personal interest profiles that would filter and dispatch by E-mail information according to their specific needs. GPPIS is also a meeting point where specific subjects could be discussed among resource person groups (discussion forum), a location where specialized resource persons could be identified and where rapid information circulation and specialised advice on specific alert situations could be sourced.

The descriptions above provide the framework and values that inspire and sustain GPPIS. However, a summary like this cannot anticipate and address all the questions and misgivings that this example will stimulate. However, GPPIS exists and does answer them and readers are invited to visit the website at http://pppis.fao.org. A manual for GPPIS is available at this URL (Fig 1.). You are invited to join the GPPIS community by contributing information that makes you an individual expert—no matter how modest the contribution, because a snowstorm is made of many small flakes and we are changing the world of plant production and protection information, one person at a time.

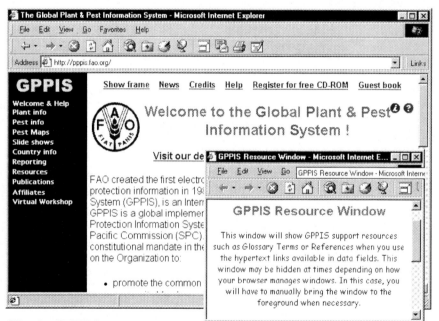

Fig. 1. GPPIS homepage.

Appendix : Data areas & free form text guides in Plant Info for the Global Plant and Pest Information System (GPPIS).

Plant menu headings / Information headings / Sub-headings	Description of entries	Menu's with given values (all menu's under Ecocrop but "selected value" appear under the heading as shown)
Identity	Taxonomy and introduction of the species and information on Editor and Sponsor	
Logo	The Sponsor Institution logo is usually an image about 20×20 pixels in .gif format. It can send to the GPPIS Supervisor as an e-mail attachment.	
Plant illustration	The Plant illustration is usually a black and white drawing, 100 pixels wide and about 5-13 Kbytes. It can be .gif or .jpg format and can be sent to the GPPIS Supervisor as an e-mail attachment.	
Preferred name	Scientific name (edited by the GPPIS Supervisor)	
GPPIS Code	Code used by the programme (generated by GPPIS)	
Authority	Taxonomic authority of the species	
Family	Taxonomic family of the species. If there is a synonym for the family name mention this after a ";", e.g. Leguminosae, Fabaceae	
Synonyms	Taxonomic synonyms of the preferred name of the species	
Common names	Common names of the species in English first and there after in other languages. Note if possible languages (e.g. Fr.; French, Sp.; Spanish, Ge; German etc) or the country	
Editor	Editor name (edited by the GPPIS Supervisor)	
Sponsor	Sponsor name (edited by the GPPIS Supervisor)	
Topic editors	Topic editor names (edited by the GPPIS Supervisor)	
Life form	Life form of the species, value(s) selected under EcoCrop	Grass, herb, vine, sub-shrub, shrub, tree erect, prostrate/procumbent/semi-erect, climber/scramblet/scandent, rosette plant
Habit	Growth habit of the species, value(s) selected under EcoCrop	
Life span	Life span of the species, value(s) selected under EcoCrop	Ephemeral, annual, biennial, perennial
Category	Category of the species, value(s) selected under EcoCrop	Cereals & pseudocereals, pulses (grain legumes), roots/tubers, forage/ pasture, fruit & nuts, vegetables, industrial, ornamentals/turf, medicinal & aromatic, forest/wood, cover
Plant attributes	Main attribute(s) of the species, value(s) selected under EcoCrop	Grown on large scale, grown on small scale, harvested from wild, previously widely grown, weed, parasite
Status		
Overview	Present economical and social description of the species	
Role	General importance of the species as text information and as value(s) selected under EcoCrop; General and specific roles of the species (and its comparative advantages) in agricultural and overall socio-economic development at farm and macro-economic level	

Production	Area cropped. Total and local production (primary and secondary products). World trade. Collaboration FAOSTAT, country or world production, economics etc. Production in main and secondary products. Cost/benefit analysis
Economics	General economics of crop production within specific production systems (at farm level)
Ethno-botany	Social and cultural importance of the species, beliefs, customs indigenous knowledge (IK)
Description	
General	Physical and physiological description of the species. See also under Ecocrop Short overview and description of appearance of the species. Including height & width (habit) of species at maturity or at any important state of development (or growth stages)
Morphology	Description of each item (Roots, Stems, etc.). If relevant, including special adaptations
Roots	Root system of the plant
Stems	Stem(s) of the plant
Leaves	Leaves of the plant
Flowers	Flowers of the plant
Fruits	Fruits of the plant
Seeds	Seeds of the plant
Physiology	Overview of major characteristics such as photosynthetic pathway (C3/C4), any important state of the plant development/growth, DM accumulation, N fixing ability and other main attributes
Environment	The environmental requirements of the species including specific tolerances, susceptibilities and limitations. See also under Ecocrop
Habitat	Geographical distribution. Overview of the distribution of the species by: - agro-ecological or agro-climatic zones, by vegetation type: phytogeographic zones, phytosociology, and by terrain and climatic seasons requirements
Latitude	Latitudinal range where in the species is found
Altitude	Altitudinal range, if possible describe the range within specific climatic zones
Temperature	Temperature range for growth, killing temperature, vernalisation requirements, frost tolerance, etc.
Water	Rainfall and/or irrigation range for growth.
Radiation	
Range & intensity	Radiation range and intensity for growth described in flux or as bright light, clear skies, overcast, light shade, etc.
Photoperiodism	Short day, day neutral or long day plant
Soil	
Physical	Texture, depth, OM content, drainage, erosion susceptibility, etc.
Chemical	pH, fertility (CEC, exchangeable cations, Al/Mn), salinity, etc.
Notes	Other specific abilities, susceptibilities or tolerances such as response drought, trampling, flooding, fire, ability to compete with companion plants etc.
Crop management	Main cultivation practices are described within major production systems. For each item, factors such as requirements: methods, timing, equipment, material and labour input are considered, if relevant. Also regional/local practices are mentioned, if relevant

172

Overview — Cultivation status (harvested from wild, grown on small scale, large scale) and short description of the crop cycle including its variability

Cropping systems — Main cropping patterns at farm level and their interaction with available resources and applied production technologies, within major farming systems

Land management — Site and crop selection (terrain, rotation, association) and soil conservation. Land preparation operations (clearing, leveling, anti-erosion measures, ploughing, harrowing, ridging). Erosion sustainability in relation to the cultivation practice of the crop. Alternatively, eg. in case of no tillage, indicate the real crop soil requirements such as well structured, no compactions, no clods or stones. Shaping of the surface —leveling, ridging, bedding, depending upon soil type, type of mechanization and water regime of the crop.

Propagation material — Type, size, quality, production, selection, preparation and storage of planting material. Plant nursery operations for the production of planting material.

Planting — General guidelines on planting dates, densities, planting methods within different cropping and production systems

Water management — Principles for crop irrigation (method, quantity and timing). Drainage and water conservation

Fertility management — Principles for fertility management (organic, mineral) related to crop production systems and environments. Ranges of nutritive elements requirements and exports, toxicity levels and symptoms

Weed control — Guidelines for integrated weed control practices including tolerance to herbicides. Mechanical, thermic, chemical; cultural practices, plant competition; influencing factors like way of planting —broadcast, drilled, row, split row.

Grazing management — Guidelines for the management of the species under grazing, including reaction to grazing pressure, persistence under grazing, any special grazing requirements.

Compatibility with other species — Ability of the species to combine with other pasture species to form persistent pasture under grazing, especially grasses with legumes and legumes with grasses.

Other crop management practices — General cultivation including: plant staking, pruning, training, thinning, fruit thinning defoliation, etc.

Pollination — Guidelines for the pollination of the species in different cultivation situations, pollination agents, etc.

Harvest — Harvest operations at field level: one stage harvest, two stage with in field drying or conditioning phase field drying or conditioning phase —depending on the mechanization level. Main harvest operations: picking/stripping, cutting/pulling, threshing, chopping, digging/uprooting —combinations possible dependence on level of mechanization. Yields (average and variability). Crop pre-treatment. Maturity and number of harvests, time range. Moisture content, Harvest index. Principal, secondary part to be harvested & residues (%, residues, use). Methodology and output per yield unit. Machinery type, setting, maintenance, energy type & units per yield. Separation

	of main produce and by-produce components (for example grain/straw) in the field. Transport: fresh produce, chopped produce, dried produce, compressed produce. Pre-cleaning type and output. Hazards.
R&D Notes	On-going research and development programmes for improved crop management and production technologies (ICM, others). Existing experiences with 0-tillage.
Crop improvement	Description of conservation status and improvement work for the species including cultivars
Genetic resources	Conservation in-situ, ex-situ, IUCN status and where to find genetic material
Variability & cultivars	Short description of major varieties and cultivars of the species. Sources for further information
R&D Notes	On-going breeding programmes including biotechnology for growth and productivity, resistance to pest and diseases and quality
Pest & Diseases	Description of Pest and Diseases associated with the species
Pest notes	Crop-specific pest issues such as economic importance /impact, specific crop management, interaction between crop ecology and local management. Crop-specific plant-pest relations including symptoms, dispersal and control measures within different agro-ecological environments, their importance/impact for each economically important pest. Technical options for the application of plant protection agents on different mechanization levels: manual, tractor, (airplane); accessibility, penetration, parts to be controlled with different pests. Note that any general information applicable across crops should be added to the concerned Pest records in Pest Info (send your contribution the pest Editor)
Pest list	List of pests and diseases associated with the species as presented under Pest Info
Products & Uses	Processing, storage, product characteristics, use, nutrition and research work in these fields for the species
Processing & storage	Methodologies and outputs for principal part processed and residues. Principal part processed and %, Residues% and use, Methodology and output per yield unit, Machinery type, setting, maintenance, Energy type & units per yield, Pre-packaging, Packaging, label standards, Quality standards, Transport, Shelf life, Hazards. Type of produce to be stored: "dead" (staple grain, straw, hay, leaves) or "living" (oxygen requirements) (seed grain, potatoes, fruit). Physical properties of the produce: grain, stalk, leaves, units, bulk. Method of conservation: controlled atmosphere, airtight, cooling, drying. On farm processing and cleaning operations, drying, conditioning, grading.
Product characteristics	Principal ingredients/composition of main and secondary products. Quality standards, labeling and major recorded hazards of fresh or transformed products/by-products
Wood properties	
Uses	Principal and secondary uses of primary or transformed products/by-products as text information. See also under EcoCrop
Nutrition	Major nutritional aspects
Medicine	Major medicinal aspects

174

Nutritional quality & animal production	The nutritive value of the species in terms of digestibility and intake, proximate analysis, and the ability of the species to produce animal product in terms of meat, milk or fiber.
Toxicity	Anti-nutritive factors which may be present in the species due to secondary compounds, fungal toxins, mineral disorders, or other anti-palatability factors.
R&D Notes	On-going research and developing programmes for improved transformation, storage technologies and crop alternative uses and product quality
Pictures	Picture, Author, Caption and Keywords (either selected among previously entered references or entered through: Resources—Picture Databank)
History	
Overview	Origin, history and development of the species
Event sequence	Origin, history and development of the species as an overview
Key references	Development of the species as a sequence of events
	Author(s), title, year and abstract of references of key importance (either selected among previously entered references or entered through: Resources—Bibliographic References)
Notes	Free form notes
Ecocrop	Data sheet description (environmental requirements, physical appearance, use and cultivation) of the species through selected values and numeric inputs.

A Geographic Information System to Identify Areas for Alternative Crops in Northwestern Wyoming

J.A. Young, B.M. Christensen, M.S. Schaad, M.E. Herdendorf, G.F. Vance, and L.C. Munn*

Agriculture is the third largest industry in Wyoming after mining and tourism (R. Micheli, Wyoming Department of Agriculture Director, pers. commun.). The Bighorn Basin, located in northwestern Wyoming, is one of the largest agricultural production areas of the state. This area accounts for 27% of the value of crops produced in Wyoming (Wyoming Agricultural Statistics Service 1998). The Bighorn Basin was developed for agricultural use when irrigation was introduced into the area in 1905 with the completion of the Buffalo Bill Dam. Major crops currently grown in the Bighorn Basin include: spring wheat, barley, oats, dry beans, sugar beets, alfalfa hay, and corn.

One alternative for increasing profit margins of the Wyoming agricultural industry is through the introduction of new crops. Wallis et al. (1989) defines new or alternative crops as either crops new to a particular county, region or state, or as a crop that has been, or is being developed, from a plant that has never been cultivated for commercial production. Alternative crops such as cabbage, mint, pumpkins, squash, and grass seed have been cultivated in the Bighorn Basin, but are most often marketed for local consumption rather than being commercially produced (J. Jenkins, Park County Cooperative Extension Agent, pers. commun.).

The study area is noted for producing 80% of Wyoming's sugar beet and barley crops. Intense cultivation, however, has resulted in reduced production and increased use of pesticides, tillage, and other management practices. The sugar beet nematode (*Heterodera schachtii*) has greatly reduced the yield of sugar beets, the highest value crop grown in the area. Crop rotation is one of the most effective cultivation practices used to help control pests, including the sugar beet nematode (Gardner, 1994). Introduction of alternative crops to the Bighorn Basin would aid in providing crop rotations, thus breaking disease cycles and pest infestation.

A tool that can be used to predict alternative crop growth is a Geographic Information System (GIS). GIS is "an organized collection of computer hardware, software, geographic data, and personnel designed to efficiently capture, store, update, manipulate, analyze, and display all forms of geographically referenced information" (ESRI 1995). A GIS stores spatial data as data layers. These layers are defined as a set of features, each having both location and attributes. The advantage of GIS is attributed to its ability to spatially analyze multiple data layers. This powerful tool allows decision-makers to simulate effects of management and policy alternatives within a geographic area prior to implementation (Congalton and Green 1992).

OBJECTIVES

The Bighorn Basin of Wyoming is an agricultural community that relies heavily on the production of staple crops. Production of alternative crops such as canola, buckwheat, and mint could provide new opportunities for agriculturalists to increase income possibilities and to break herbicide and pest cycles. This study used a GIS to investigate the potential for cultivating new crops in the Bighorn Basin. Environmental traits of the basin (e.g., rainfall, temperature, and soil) influence where new or alternative crops could be produced. Maps developed in this project provide producers with possible alternatives to current practices in the Bighorn Basin. The objectives of this study were to: (1) develop a spatially-correlated database that includes environmental and climatic data; (2) develop a continuous soils layer for the Bighorn Basin study area; (3) determine crop growth parameters for 28 agricultural crops; and (4) combine growth parameters with environmental data to display and describe areas of potential alternative crop production.

*We express our appreciation to Dr. Renduo Zhang for his assistance in the geostatistical analysis and development of the climactic data layers used in this study.

METHODOLOGY

Site Description

The Bighorn Basin is located in the northwestern corner of Wyoming (Fig. 1). The study area includes parts of three different counties and is comprised of 760,000 ha. The area elevation ranges from 1220 to 1525 m; this low elevation, relative to the rest of the state, results in the region having an average of 90–120 frost-free days (Western Regional Climate Center 1998).

The study area was delineated in the GRID module of ARC/INFO. The 1:24,000 USGS topographic quadrangle maps and the Wyoming State Land Cover Classification were overlayed onto the Park, Big Horn and Washakie counties. These coverages were clipped to the study area. Due to irrigation constraints, topographic maps having a majority of slopes greater than 5% were eliminated. This resulted in the removal of only a portion of the topographic maps for the three counties. From these topographic maps, the quadrangles with greater than 25% agricultural land were selected. A buffer of one quadrangle was placed around the entire study area to include smaller areas of agricultural land. This analysis resulted in the inclusion of 81 quadrangles in our study area. Upon ground-truthing in May 1998, 22 of the quadrangles were excluded from the study area because they lacked either agricultural production potential due to topography or had insufficient irrigation capabilities.

Soils Description

Published soil survey data for the majority of the study area is unavailable; Washakie county soils have been delineated, while Park and Big Horn counties have not been completed. In order to obtain soils information, a GIS simulation model was created to predict soil occurrence in the Bighorn Basin (Fig. 2). Bedrock and surficial geology covering the state of Wyoming was obtained from the Spatial Data and Visualization Center at the University of Wyoming (1998). These coverages were clipped to the study area. Soils series in the Washakie County comprehensive soil survey were examined to determine bedrock and surficial geology combinations upon which they occurred (Iiams 1983). These combinations, and combinations predicted from other unpublished work in the region, were used to predict soils for the entire study area. The model was applied utilizing decision rules in Arc Macro Language (Munn and Arneson 1998). The predicted soils map was field checked and updated accordingly.

Weather Records

Eighteen weather stations exist both in and around the study area (Fig. 3). These stations have collected weather data for 30 years or longer (Western Regional Climate Center, 1998). The weather station attribute data were used to create continuous weather patterns for the basin; contour lines were extrapolated from the point data utilizing geostatistics. Environmental layers developed from the weather station data include frost-free days (80% probability), growing degree-days (base 5°C), annual precipitation, and August mean temperature. Geostatistics has proven to be advantageous when compared to other methods of extrapolation (Kravchenko et al. 1996). Semi-variograms of each environmental attribute were plotted and cross-validation and kriging techniques were applied. Kriging resulted in a grid of each attribute with values placed at the nodes or intersections of the grid. This information was imported into ARC/INFO and used to create continuous grid coverages of the four environmental layers shown in Fig. 3.

Fig. 1. Study area location.

177

Crop Growth Parameters

The crop growth parameters of 28 alternative crops were derived through documented sources; many parameters were taken from Purdue University's NewCrop Resource Online Program (Purdue University 1998). Other sources of information were obtained through the Cooperative Extension Publications from Montana State University, University of Nebraska–Lincoln, and Colorado State University. These parameters were compiled into one database for analysis. Table 1 illustrates crop growth parameters that were analyzed in the GIS database for buckwheat and canola potential production areas.

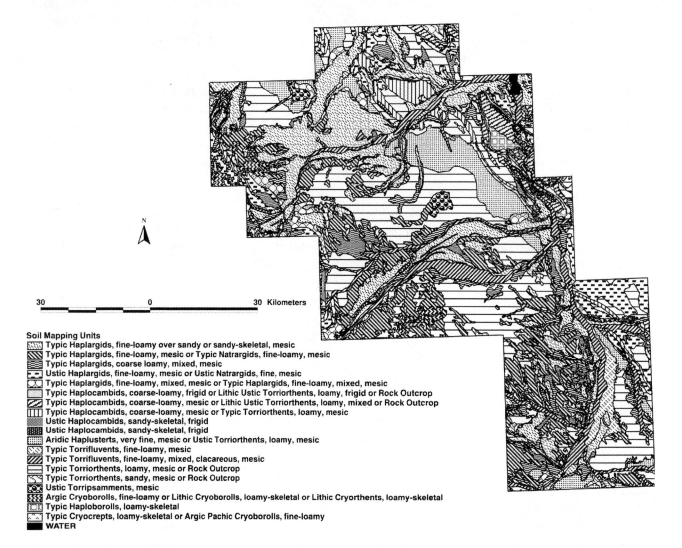

Soil Mapping Units
Typic Haplargids, fine-loamy over sandy or sandy-skeletal, mesic
Typic Haplargids, fine-loamy, mesic or Typic Natrargids, fine-loamy, mesic
Typic Haplargids, coarse loamy, mixed, mesic
Ustic Haplargids, fine-loamy, mesic or Ustic Natrargids, fine, mesic
Typic Haplargids, fine-loamy, mixed, mesic or Typic Haplargids, fine-loamy, mixed, mesic
Typic Haplocambids, coarse-loamy, frigid or Lithic Ustic Torriorthents, loamy, frigid or Rock Outcrop
Typic Haplocambids, coarse-loamy, mesic or Lithic Ustic Torriorthents, loamy, mixed or Rock Outcrop
Typic Haplocambids, coarse-loamy, mesic or Typic Torriorthents, loamy, mesic
Ustic Haplocambids, sandy-skeletal, frigid
Ustic Haplocambids, sandy-skeletal, frigid
Aridic Haplusterts, very fine, mesic or Ustic Torriorthents, loamy, mesic
Typic Torrifluvents, fine-loamy, mesic
Typic Torrifluvents, fine-loamy, mixed, clacareous, mesic
Typic Torriorthents, loamy, mesic or Rock Outcrop
Typic Torriorthents, sandy, mesic or Rock Outcrop
Ustic Torripsamments, mesic
Argic Cryoborolls, fine-loamy or Lithic Cryoborolls, loamy-skeletal or Lithic Cryorthents, loamy-skeletal
Typic Haploborolls, loamy-skeletal
Typic Cryocrepts, loamy-skeletal or Argic Pachic Cryoborolls, fine-loamy
WATER

Fig. 2. Soils of the Bighorn Basin, WY.

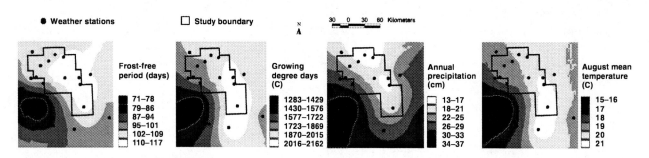

Fig. 3. Environmental layers for the Bighorn Basin, WY.

Table 1. Crop growth parameters for buckwheat and canola.

	Buckwheat (*Fagopyrum esculentum*) Warm season broadleaf	Canola (*Brassica campestris*) Cool season broadleaf
Optimal temperature	< 25°C	12–30°C, needs temp. below 25°C for flowering
Soil attributes	Wide range of soils, no crusting, prefers medium textured soils	Non-crusting clay loam soils, medium texture
Precipitation	Sensitive to drought stress	Dryland >30.5 cm annually
Frost-free season	Matures in 75–90 days, easily killed by frost	> 90 days
Minimum temp.	0°C	5°C
Maximum temp.	Unavailable	< 11 days >32°C in July
Growing degree-days	1200 days (base 5°C)	860–920 (base 5°C)

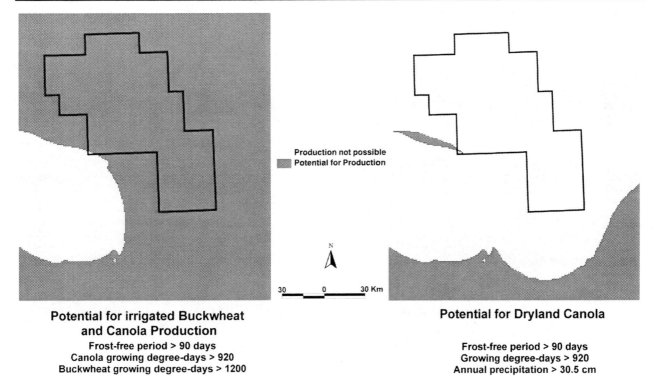

Potential for irrigated Buckwheat and Canola Production
Frost-free period > 90 days
Canola growing degree-days > 920
Buckwheat growing degree-days > 1200

Production not possible
Potential for Production

Potential for Dryland Canola
Frost-free period > 90 days
Growing degree-days > 920
Annual precipitation > 30.5 cm

Fig. 4. Production potential for buckwheat and canola in the Bighorn Basin, WY.

Model Formation

Environmental data were compiled into a GIS and analysis using map algebra was performed. The parameters queried included length of frost-free season, growing degree-days, annual precipitation, and average August temperature. Areas with selected crop growth parameter combinations were displayed. A separate map representing two alternative crops was created showing potential areas for its production. Each map utilized the specific requirements of the individual crop.

RESULTS

Soils of the study area were a mosaic of 19 mapping units. Ground-truthing conducted in May 1998 confirmed that the soils map created in the simulation model was suitably accurate. Classification of only one soil mapping unit required change. The creation of the soils data allows for the use of textural classes, pH, and water-holding capacities in our alternative crop analysis. A large proportion of the study area consists of

Torrifluvents, which is where a significant portion of the Bighorn Basin irrigated agriculture currently occurs. In an arid region, crop development commonly occurs in flood-plain soils.

Contour maps created from the kriging process are shown in Fig. 3. These contour maps are related to the topography of the surrounding study area, i.e., with an increase in elevation there is a decrease in frost-free period, a decrease in growing degree-days, an increase in precipitation, and a decrease in temperature. These maps display frost-free period (80% probability), growing degree-days (base 5°C), annual precipitation, and August mean temperature.

Preliminary findings utilizing frost-free period and growing degree-days indicate that canola and buckwheat can be grown in virtually all of the study area under irrigated conditions (Fig. 4). Canola production requires a frost-free period of 90 days or greater and a value of 920 or greater growing degree-days (base 5°C) or heat units. Buckwheat also requires a frost-free period exceeding 90 days, but needs 1200 growing degree-days. Figure 4 also displays the areas of potential dryland production of canola; the factors for frost-free period and growing degree-days were used in addition to greater than 30.5 cm annual precipitation. There is less than 5% dryland agriculture production currently in the study area.

CONCLUSION

The Bighorn Basin in Wyoming has potential for the introduction of many new or alternative crops, though only two examples are described in this paper. As more of the environmental variables are created through the use of geostatistical methods, a more accurate picture of potential alternative crops will emerge. A GIS provides an excellent means for exploring the possibilities of cultivating new crops. The process of locating areas suitable for growth is rapid, once the necessary information is entered into the database.

The study area has a short growing season that lends itself to the cultivation of cool-season crops. While many of the crops being investigated can be grown in the Bighorn Basin, the commercial yield of these crops is uncertain. A crop yield model should be employed to predict if alternative crop production could compete in the marketplace. An economic analysis of the feasibility of the growth of these new crops is also an area that requires further analysis.

REFERENCES

Congalton, R.G. and K. Green. 1992. The ABCs of GIS. J. Forestry 90(11):13–20.

ESRI. 1995. Understanding GIS: The ARC/INFO Method. GeoInformation International, United Kingdom and Wiley, New York.

Gardner, J.C. 1994. Choosing among alternative crops. North Dakota Farm Research 50(2):3–5.

Iiams, J.E. 1983. Soil survey of Washakie County, Wyoming. National Cooperative Soil Survey. U.S. Government Printing Office, Washington, DC.

Kravechenko, A., R. Zhang, and Y.K. Tung. 1995. Estimation of mean annual precipitation in Wyoming using geostatistical analysis. p. 271–282. In: H.J. Morel-Seytoux (ed.), Proc. 16th Annual American Geophysical Union Hydrology Days. Hydrology Days Press: Atherton, CA.

Munn, L.C. and C.S. Arneson. 1998. Soils of Wyoming: A digital statewide map at 1:500,000-Scale. Agr. Expt. Sta. Rpt. B-1069. Univ. Wyoming, College of Agriculture, Laramie, Wyoming.

Purdue University—Center for New Crops and Plant Products. New Crop Resource Online Program. Online. Available: http://www.hort.purdue.edu/newcrop/. July 1998.

Spatial Data and Visualization Center. Wyoming Natural Resources Data Clearinghouse. Online. Available: http://www.sdvc.uwyo.edu/clearinghouse/. March 20, 1998.

Wallis, E.S., I.M. Wood, and D.E. Byth. 1989. New crops: A suggested framework for their selection, evaluation and commercial development. p. 36–52. In: G.E. Wickens, N. Haq, and P. Day (eds.), New crops for food and industry. Chapman and Hill, London and New York.

Western Regional Climate Center. Wyoming climate summaries. Online. Available: http://www.wrcc.dri.edu/summary/climsmwy.html. July 15, 1998.

Wyoming Agricultural Statistics Service. 1998. Wyoming Agricultural Statistics 1998. Wyoming Department of Agriculture, Univ. Wyoming, College of Agriculture.

PART III
STATUS OF NEW CROPS & NEW USES

CEREALS & PSEUDOCEREALS

Kamut®: Ancient Grain, New Cereal

Robert M. Quinn

Kamut® is a registered trademark of Kamut International, Ltd., used in marketing products made with a remarkable grain. The new cereal is an ancient relative of modern durum wheat, two to three times the size of common wheat with 20–40% more protein, higher in lipids, amino acids, vitamins and minerals, and a "sweet" alternative for all products that now use common wheat (Fig.1). Nutritionally superior, it can be substituted for common wheat with great success. Kamut brand wheat has a rich, buttery flavor, and is easily digested. A hard amber spring type wheat with a huge humped back kernel, this grain is "untouched" by modern plant breeding programs which appear to have sacrificed flavor and nutrition for higher yields dependent upon large amounts of synthetic agricultural inputs.

Although the Kamut brand wheat is thousands of years old, it is a new addition to North American grain productions. It's origins are intriguing. Following WWII, a US airman claimed to have taken a handful of this grain from a stone box in a tomb near Dashare, Egypt. Thirty-six kernels of the grain were given to a friend who mailed them to his father, a Montana wheat farmer. The farmer planted and harvested a small crop and displayed the grain as a novelty at the local fair. Believing the legend that the giant grain kernels were taken from an Egyptian tomb, the grain was dubbed "King Tut's Wheat." But soon the novelty wore off and this ancient grain was all but forgotten. In 1977, one remaining jar of "King Tut's Wheat" was obtained by T. Mack Quinn, another Montana wheat farmer, who with his son Bob, an agricultural scientist and plant bio-chemist soon perceived the value of this unique grain. They spent the next decade propagating the humped-backed kernels originally selected from the small jar. Their research revealed that wheats of this type originated in the fertile crescent area which runs from Egypt to the Tigris-Euphrates valley. The Quinns coined the trade name "Kamut" an ancient Egyptian word for wheat. Egyptologists claim the root meaning of Kamut is "Soul of the Earth."

In 1990, the USDA recognized the grain as a protected variety officially named 'QK-77'. The Quinns also registered Kamut as a trademark. Perhaps the most significant aspect of the introduction and cultivation of Kamut brand wheat is that it is an important new crop for sustainable agriculture. This grain's ability to produce high quality without artificial fertilizers and pesticides make it an excellent crop for organic farming.

The real history of the Kamut brand grain has been as elusive as its taxonomic classification. Although not thought to have been in commercial production anywhere in the world in the recent past, most scientists believe it probably survived the years as an obscure grain kept alive by the diversity of crops common to small peasant farmers perhaps in Egypt or Asia Minor*. It is thought to have evolved contemporary with the free-threshing tetraploid wheats. Scientists from the United States, Canada, Italy, Israel, and Russia have all examined the grain and have reached different conclusions regarding its identification. All agree that it is a *Triticum turgidum* (AABB) which also includes the closely related durum wheat. The correct subspecies is in dispute. It was originally identified as *polonicum*. Some now believe it is *turanicum*, while others claim it is *durum*. One Russian scientist believes it is a durum cultivar called 'Egiptianka' or "the durum of Egypt." Still others believe it is may evolve from a mixture of many types which would be consistent with its supposed descent from an ancient landrace originally gathered by primitive farmers from the wild. The majority now identify the grain as *turanicum* commonly called Khorasan wheat. Although its true history and taxonomy may be disputed, what is not disputed is its great taste, texture, and nutritional qualities as well as its hypo-allergenic properties.

KAMUT BRAND WHEAT

Kamut Brand Wheat can be found in cereals, breads, cookies, snacks, waffles, pancakes, bread mixes, baked goods, and prepared and frozen meals. Because of the inherent sweetness of this grain (referred to by some as "the sweet wheat"), no sugar is required to hide the subtle bitterness associated with most wheats and whole wheat products. Many are utilizing the natural firmness of the kernels to produce tasty pilafs, cold

---

*This belief has now been substantiated by collections made in the upper Nile area of Egypt in mid 1998 under the direction of the author.

salads, soups, or a substitute for beans in chili. Kamut brand bulgur and couscous are also popular in Europe. Kamut brand wheat also makes an outstanding pasta which is superior to all other whole grain pastas in texture and flavor. Because of the strong gluten in the protein, Kamut pasta can be frozen and reheated without losing its firmness. Recently, Green Kamut was introduced. It is becoming the rage of the green foods market because of its concentrated health benefits and mild, fresh taste when compared to other wheat grass juices. Thus the leaves as well as the grain of this remarkable plant are proving to be valuable.

Fig 1. Kamut® wheat.

The complete nutritional analysis of Kamut brand grain substantiates that it is higher in energy than other wheats. Compared to common wheat, it is higher in eight out of nine minerals; contains up to 65% more amino acids; and boasts more lipids and fatty acids. The most striking superiority of Kamut brand wheat is found in its protein level—up to 40% higher than the national average for wheat. Because of its higher percentage of lipids, which produce more energy than carbohydrates, Kamut brand can be described as a "high energy grain." Athletes, people with busy lives and anyone looking for quality nutrition will find Kamut brand products a valuable addition to their diet. A bowl of hot Kamut cereal in the morning, or a delicious serving of Kamut pasta at noon will satisfy between meal hunger pangs as well as urges for snacking.

For those suffering wheat sensitivities, Kamut brand products also play a unique role. Recent research by the International Food Allergy Association (IFAA) concluded "For most wheat sensitive people, Kamut grain can be an excellent substitute for common wheat." Dr. Ellen Yoder, President of IFAA and a team of independent scientists and physicians reached this conclusion through their work with two different wheat sensitive populations—those who have immediate immune responses and those with delayed immune responses. In the delayed immune response group, a remarkable 70% showed greater sensitivity to common wheat than Kamut brand grain. In the immediate immune response group—the severely allergic—70% had no, or minor, reaction to Kamut brand wheat. However, those with severe allergies should always seek the advice of a physician. Research is now underway in Austria to study gluten intolerance but is yet unfinished so no recommendations can be made for those suffering this affliction. For many wheat sensitive people, however, Kamut brand grain has become "the wheat you can eat."

Yield comparison between Kamut and hard red spring wheats are similar to results observed in comparison between "the covered wheats" and other free threshing wheats (Stallknecht et al. 1996). Kamut will outyield spring wheats under drought stress during the growing season, but yields equal or lower in ideal seasons. Plant height is 127 cm with good to excellent straw strength.

THE KAMUT ASSOCIATIONS

The Kamut Association of North America (KANA) was formed to promote the use of the Kamut brand, provide consumer education and to encourage the expansion of organic agriculture. KANA members include food manufacturers and distributors of Kamut brand products. KANA provides information about available Kamut products and their manufacturers, results of research and nutritional studies, general history and background of the grain, and also information and other assistance in available to the general public as well as retailers and manufacturers by request from the offices of KANA. A similar association has been formed in Europe (KAE). These associations maintain a homepage at kamut.com.

REFERENCE
Stallknecht, G.F., K.M. Gilbertson, and J.E. Ramey. 1996. Alternate wheat cereals as food grains: Einkorn emmer, spelt, kamut, and triticale. p. 156–170. In: J. Janick (ed.), Progress in new crops. ASHS Press, Alexandria VA.

Variability in 'Plainsman' Grain Amaranth

F.R. Guillen-Portal, D.D. Baltensperger, L.A. Nelson, and N. D'Croz-Mason*

Grain amaranth is a pseudo-cereal that played an important role as human food in the ancient civilizations of America. Current interest in amaranth resides in the fact that it exhibits a high nutritional value, a C_4 photosynthetic pathway, a great amount of genetic diversity, and phenotypic plasticity (Dowton 1972; Hauptli 1977; Jain et al. 1979; Kauffman 1981; National Research Center 1984). In 1992, a cooperative amaranth breeding program between the Rodale Research Center and the University of Nebraska resulted in the release of the cultivar 'Plainsman', an interspecific hybrid between *Amaranthus hypochondriacus* L., a gold seeded selection from Mexico, and *Amaranthus hybridus* L., a black seeded selection from Pakistan. Breeding objectives for 'Plainsman' development were early maturity, light seed color, and short plant height. Single plant selection in earlier generations (F_2 to F_5) and mass selection in advanced generations (F_6 to F_7) was used (Baltensperger et al. 1992). 'Plainsman' maturity is very early in Western Nebraska, reaching maturity 110 days after planting. Evaluation tests in Nebraska, Colorado, Missouri, Minnesota, and South Dakota indicated that 'Plainsman' was one of the most promising amaranth cultivars for the United States. In 1994, approximately 1200 ha of 'Plainsman' were grown (Baltensperger 1992; Myers 1994) but field observations indicated a great deal of variation. The pollination mechanism of amaranth is complex in nature, varying from low to high outcrossing rates and furthermore being strongly affected by the environment (Jain et al. 1982; Hauptli and Jain 1985). The complexity of its pollination mechanism along with the breeding strategy employed in its development provided some evidence to suspect that some residual genetic variability is still present in 'Plainsman'. However, the large phenotypic plasticity observed in the amaranths might be the cause of such variability. Phenotypic plasticity has been defined as the physiological and/or morphological alteration by an organism in response to environmental differences (Schilichting 1986) and plastic variance as the amount of variation due to the environment and due to the genotype by environment interaction ($\sigma^2_{Pl} = \sigma_E^2 + \sigma_{GxE}^2$). Plasticity has been defined as the ratio of plastic variance to total phenotypic variance (Scheiner and Goodnight 1984). Further improvement by selection within 'Plainsman' requires more knowledge about the cause of variability in seed production traits. The objectives of this study were to investigate the amount of morphological variation present in 'Plainsman', to estimate the genetic and environmental components of variance in agronomic traits in a random sample of 'Plainsman', and to predict the response to selection within 'Plainsman' under a selfing-selection scheme.

METHODOLOGY

'Plainsman' foundation seed from the Foundation Seed Division of the University of Nebraska was planted on 6 Jan 1995 in 140 pots. Sixty days after planting, plants were self pollinated by covering the panicles with pollinating bags. The panicles were hand-harvested 120 days after planting, allowed to dry at 32°C for three days, threshed, and cleaned by hand. Morphological traits measured included color of the main stem during the seed-filling period, inflorescence color during seed-filling period, branching index based on presence or absence of primary, secondary, or tertiary branches developed from the main stem, and inflorescence compactness just before harvesting.

In 1995, the 140 self pollinated families were planted at the High Plains Agricultural Laboratory near Sidney, Nebraska, and the Panhandle Research and Extension Center at Scottsbluff. At Sidney, a non-irrigated trial was planted on 16 June in a Duroc loam (Pachic Haplustol) soil and an irrigated trial on 17 June in a Keith silt loam (Aridic Arguiustoll) soil. The irrigated trial at Scottsbluff was planted on 18 July in an Otero loam (Fluventic Haplustol) soil using a Wintersteiger air seeder. Transplanting and hand thinning was necessary to achieve uniform populations at each experiment. Distance between plants was 0.15 m for the Sidney non-irrigated plots (51,000 plants/ha), 0.3 m for the Sidney irrigated plot (25,500 plants/ha), and 0.3 m at Scottsbluff (25,500 plants/ha). For all experiments, nitrogen was applied before planting at a rate of 100 kg/

*Univ. of Nebraska Agr. Res. Div. J. Series no. 12496. The authors express their gratitude to Dr. Kent Eskridge for assistance in the statistical analysis of the results, and to Dr. Stephen Mason for manuscript review.

ha and 85 kg/ha was applied at panicle emergence at Scottsbluff. The plots were hand-weeded.

A replications-in blocks experimental design with two replications at each location was used. At each location, the experiment included ten blocks each containing a set of 14 randomly chosen families. The sets were kept together between replications and locations with a different randomization of families within each set. Single-row plots 5 m long with 0.76 m row spacing were used. Morphological traits measured were the same as those measured in the greenhouse. In addition, days to panicle emergence, flowering, and maturity, plant height, stem diameter, panicle length, grain yield per plant (the last four averaged over 5 plants randomly chosen per plot), and 1000-seed weight were determined. Morphologic data was analyzed by using a Chi-square goodness-of fit test between observations collected in the greenhouse (G_0 generation) and those collected in the field (G_1 generation). The least square mean method was used for the analysis of the agronomic data using Statistical Analysis System (SAS Institute, Inc. 1989). The analysis of data combined over locations was performed using a random linear model where the hypothesis test for genetic differences among families within blocks was H: $\sigma_f^2 = 0$. Components of variance were estimated by equating the observed mean squares to the expected mean squares from the combined analysis of variance. The standard errors associated with these estimates were calculated as described by Anderson and Bancroft (1952). Broad sense heritability (H) was calculated as the ratio of the genetic variance to the phenotypic variance, $H = \sigma_f^2 / \sigma_p^2$ where σ_f^2 = genetic variance among families, and σ_p^2 = phenotypic variance. Plasticity (P_t) was calculated as the ratio of the plastic variance to the sum of the variance components, $P_t = \sigma_{Pt}^2 / \sigma_{Pt}^2 + \sigma_f^2 + \sigma_c^2$ where σ_{Pt}^2 = plastic variance, σ_f^2 = genetic variance among families, and σ_c^2 = residual variance. Approximate standard errors estimated for H and P_t were computed as SE(H) = SE(σ_f^2)/ σ_p^2 and SE(P_t) = SE(σ_{Pt}^2)/ $\sigma_{Pt}^2 + \sigma_f^2 + \sigma_c^2$. The predicted genetic gain from selection Gp was calculated as Gp = k σ_f^2/σ_p where k = standardized selection differential, σ_f^2 = genetic variance among families, σ_p = estimated phenotypic standard deviation.

EXPERIMENTAL RESULTS

Frequency distributions of morphological traits of selfed plants grown in the greenhouse (G_0) and of resulting progeny grown in the field in Scottsbluff (G_1) had three or four classes for all traits in both G_0 and G_1

Table 1. Chi-square test of differences of classes within four morphological traits in two generations of 140 selfed families from the cultivar 'Plainsman'.

Trait	Class[z]	Generation[y] G_0	G_1	χ^2
Branching index	No branches	56	49	21**
	Few branches at the bottom	37	26	
	Few branches at the top	41	39	
	Branches all along the stem	6	18	
Stem color	Green	13	13	18**
	Pink base	110	119	
	Red or darker base	17	0	
Inflorescence color	Green	1	2	13**
	Dark amaranthine	125	129	
	Light amaranthine	14	1	
Inflorescence density	Lax	92	51	127**
	Near lax	15	47	
	Near dense	15	32	
	Dense	18	2	

[z] Classes within traits were established following Brenner (1994).

[y] G_0 based on 140 individuals grown in the greenhouse, G_1, Based on 132 individuals grown in the field (Scottsbluff).

** significant at 1% level.

generations, with one class being larger than the others (Table 1). No branches was the most frequent class for branching index with 39% of the total across generations. Pink-base stem color was present in 85%, while dark amaranthine inflorescence color was present in 93% of the plants. Lax density was the most prevalent class for inflorescence density with 52% of the plants. A Chi-square test of goodness-of fit to a G_0:G_1 ratio was highly significant for all the characteristics studied (Table 1), indicating that expression of these traits was greatly affected by the environment. It is evident that 'Plainsman' is characterized by a large amount of morphological variation, which is higher for branching index and inflorescence density than for stem and inflorescence color.

Panicle emergence and flowering days, and days to maturity were uniform among families (data not shown). Substantial variation over locations for all the agronomic traits was found (Table 2). Weather conditions contributed to this variation during the growing season. In late June, a hailstorm occurred at the Sidney non-irrigated location decreasing the plant population at this experiment to 10,000 plants/ha.

Analysis of variance showed variation among locations for all of the traits ($p < 0.01$), suggesting that the growing conditions at each location were different (Table 3). Family differences were found for plant height and stem diameter ($p < 0.01$) and 1000-seed weight ($p < 0.05$). Family by location interactions were found for plant height ($p < 0.01$) and stem diameter ($p < 0.05$), suggesting these traits were more susceptible to environmental changes. It appears that 'Plainsman' possesses a genetic structure composed mainly by homozygous-heterogeneous lines. The analysis of variance components indicated that the genetic components were small compared to the other variance estimates (Table 4), which reinforces the idea about the homozygous structure

Table 2. Mean, maximum, minimum values and their corresponding standard errors (se) of five agronomic traits measured in 140 selfed families from the cultivar 'Plainsman'.

Parameter	Plant height (cm)	Stem diameter (cm)	Panicle length (cm)	Grain yield per plant (g)	1000-seed weight (g)
Sidney (non irrigated)					
Max.	152	3.0	66	15.7	1.1360
Min.	69	1.0	25	0.7	0.5480
Mean	118	1.8	46	9.1	0.6642
SE	13.4	0.35	10.05	4.45	0.0324
Sidney (irrigated)					
Max.	174	6.8	75	--	--
Min.	104	1.6	29	--	--
Mean	155	2.3	50	--	--
SE	8.52	0.3	6.81	--	--
Scottsbluff (irrigated)					
Max.	228	3.8	93	52.3	1.0797
Min.	120	1.6	34	11.5	0.6130
Mean	170	2.6	59	26.8	0.7133
SE	10.54	0.25	6.08	7.90	0.0364
Combined over locations[z]					
Max.	228	6.8	93	52.3	1.1360
Min.	69	1.0	25	0.7	0.5480
Mean	147	2.2	51	18.3	0.6900
SE	11.02	0.32	7.94	6.57	0.0346

[z]Plant height, stem diameter and panicle length averaged over the three locations. Grain yield per plant and 1000-seed weight averaged over locations Sidney (non irrigated) and Scottsbluff.

186

of 'Plainsman'. The family by location (genotype by environment) interaction variance component was greater than the genetic variance component, suggesting that most of the variability in 'Plainsman' was plastic. A small negative estimate for grain yield per plant was obtained, but since it is small, it can be ignored. The phenotypic variance was greater than the genetic and the family by location components of variance for all the traits. Residual error variance estimates showed larger values than the rest of the components, except for plastic variance. The estimated error variances among locations were relatively uniform for 1000-seed weight, stem diameter, and plant height but not for panicle length and grain yield per plant where a 1:3 error variance

Table 3. Combined analysis of variance of five agronomic traits measured in 140 selfed families from the cultivar 'Plainsman' tested at Sidney (irrigated and non irrigated conditions) and Scottsbluff, Nebraska, in 1995.

			Mean squares				
Source	df	Plant height (cm)	Stem diameter (cm)	Panicle length (cm)	df	Grain yield/plant (g)	1000-seed weight (g)
Locations (L)	2	161592.4**	36.34**	9201.7**	1	29171.3**	0.2321**
Blocks (B)	9	552.6**	0.58**	136.7*	9	160.2**	0.0025*
B × L	18	707.3**	0.26**	242.9**	9	65.5 NS	0.0006 NS
Replications/B/L	30	110.9 NS	0.17*	59.6 NS	20	63.7 NS	0.0023 NS
Families/B	130	227.5**	0.16**	72.4 NS	129	43.7 NS	0.0016*
F/B × L	251	169.5**	0.13*	71.4 NS	115	37.5 NS	0.0014 NS
Error	326	121.5	0.10	63	170	43.1	0.0012
C.V. (%)		7	15	15		36	5

*, ** significant at 5% and 1% levels.

Table 4. Estimates of components of variance and their associated standard errors (SE) of five agronomic traits in 140 selfed families from the cultivar 'Plainsman'.

Variance component[z]	Plant height (cm)	Stem diameter (cm)	Panicle length (cm)	Grain yield/plant (g)	1000-seed weight (g)
Family (σ_f^2)	11.04 ± 6.05	0.006 ± 0.004	0.19 ± 2.08	1.99 ± 2.17	0.0000 ± 0.0001
Family by location (σ_{fxl}^2)	26.21 ± 9.74	0.015 ± 0.008	4.57 ± 4.39	-3.47 ± 3.88	0.0001 ± 0.0000
Phenotypic (σ_p^2)	43.31 ± 5.33	0.031 ± 0.004	13.78 ± 1.69	14.03 ± 1.61	0.0005 ± 0.0001
Plastic (σ_{Pt}^2)	709.62 ± 485.49	0.168 ± 0.109	42.63 ± 27.98	157.06 ± 127.60	0.0013 ± 0.0010
Residual (σ_e^2)	121.54 ± 9.49	0.108 ± 0.008	63.05 ± 4.92	43.15 ± 4.65	0.0012 ± 0.0000

[z]Plant height, stem diameter, and seed-head length estimates based on a combined analysis over three locations: Sidney (irrigated and non-irrigated) and Scottsbluff. Grain yield per plant and 1000-seed weight estimates based on a combined analysis over two locations: Sidney (non-irrigated) and Scottsbluff.

Table 5. Broad sense heritability (H) and plasticity (Pt) estimates and their associated standard errors.

Parameter	Plant height (cm)	Stem diam. (cm)	Panicle length (cm)	Grain yield/plant (g)	1000-seed wt. (g)
Broad sense heritability (H)	0.25±0.14	0.19±0.13	0.01±0.15	0.14±0.15	0.00±0.20
Plasticity (P_t)	0.84±0.58	0.60±0.39	0.40±0.26	0.77±0.63	0.52±0.40

Table 6. Predicted gain (G_p) from selection per year using three different standardized selection differentials (k) in five agronomic traits from the cultivar 'Plainsman'. Numbers in parenthesis correspond to genetic gain expressed as a percentage of the mean.

Standardized selection differential (k)	Expected genetic gain				
	Plant height (cm)	Stem diameter (cm)	Panicle length (cm)	Grain yield/plant (g)	1000-seed weight (g)
2.64 (1%)	4.43 (3)	0.09 (4)	0.13 (0)	1.41 (8)	0.00 (0)
2.06 (5%)	3.46 (2)	0.07 (3)	0.10 (0)	1.10 (6)	0.00 (0)
1.75 (10%)	2.94 (2)	0.06 (3)	0.09 (0)	0.93 (5)	0.00 (0)

proportion among locations was observed (data not shown). Plastic variance components as defined by Scheiner and Goodnight (1984) had the greatest value for all five agronomic characters, and plastic variance was at least 30 fold greater than the genetic variance. It is evident that separation of genetic variation from phenotypic variation is not easy in amaranth cultivars (Kauffman 1981).

Estimates of broad sense heritability showed the largest heritability for plant height, followed by stem diameter, grain yield per plant, and panicle length. A zero heritability was observed for 1000-seed weight (Table 5). These estimates do not agree with those reported in the literature. Espitia (1994), in a population of amaranth races, found very high heritability of 0.92 for plant height and a moderately high heritability of 0.43 for grain yield. Joshi (1986), studying Indian landraces of amaranth, found high heritability of 0.77 for 1000-seed weight, 0.63 for inflorescence length, and 0.61 for plant height. Since the heritability estimates were small for the five agronomic characteristics studied, it is concluded that improvement for these traits through selection would be limited within this cultivar. Estimates of plasticity were high for all of the agronomic traits studied, indicating that they were greatly affected by the environment.

Estimates of the predicted gain from selection per year using three standardized selection differentials corresponding to a selection of the upper 10%, 5%, and 1% families based on estimates over locations show a relatively high genetic gain for plant height, a low genetic gain for yield per plant and stem diameter, and no genetic gain for 1000-seed weight (Table 6).

CONCLUSION

The breeding biology of amaranth is complex in nature being strongly affected by the environment (Jain et al. 1982; Hauptli and Jain 1985). Based on theoretical considerations (Allard 1960; Simmonds 1979; Fehr 1987) the breeding method employed in the development of 'Plainsman' seems appropriate, although self-pollination was assumed to be the prevalent reproductive system in the developing populations (Weber and Kauffman 1990; Kauffman 1981; Schulz-Schaeffer et al. 1991). Under a single plant selection scheme, it is generally accepted that at the F_5 generation near homozygosity is reached, so preliminary yield trials may begin at F_6 generation (Allard 1960; Simmonds 1979). However, some residual genetic variability is retained, and it might be present indefinitely. Evidence of this has been found in cotton (*Gossypium hirsutum*) and sorghum (*Sorghum* spp.). Homologous pairing, new mutations, and recombination of linkage blocks promoted by homozygozity have been suggested as possible causes (Simmonds 1979). Amaranths in general are considered to have an intrinsic ability to attenuate the effects of strong environmental variations (Kauffman 1981).

It has been hypothesized that plasticity and heterozygosity act in an opposite way. According to one hypothesis, phenotypic plasticity should increase as heterozygosity decreases due to the increase in developmental instability caused by deleterious homozygous recessive genes. Another hypothesis considers plasticity and heterozygosity as two antagonistic conditions in the sense that they represent alternative methods to deal with environmental heterogeneity. Thus, a population in which a consistent plastic response is observed has no need for genetic variation, and vice versa (Schlichting 1986). Yet under another hypothesis, plasticity and heterozygosity might well be expressed together so that a population could respond to an extremely variable environment by becoming both more plastic and more genetically variable (Scheiner and Goodnight 1984). The study showed the large extent to which environmental conditions affect the expression of morphological and agronomic traits in a population exhibiting a small degree of genetic variability, and that genetic improvement through a selfing–selection scheme would be limited.

REFERENCES

Allard, R.W. 1960. Principles of plant breeding. Wiley, New York. p. 115–125.

Anderson, R.L. and T.A. Bancroft. 1952. Statistical theory in research. McGraw-Hill, New York. p. 313–335.

Baltensperger, D.D., L.E. Weber, and L.A. Nelson. 1992. Registration of 'Plainsman' grain amaranth. Crop Sci. 32: 1510–1511.

Brenner, D. 1994. Characterizing system for *Amaranthus* germplasm. NCRPIS. USDA/ARS. Iowa State Univ, Ames. p. 2–7.

Dowton, W.J.S. 1972. *Amaranthus edulis*: A high lysine grain amaranth. World Crops J. 20. p. 25.

Espitia-Rangel, E. 1994. Breeding of grain amaranth. p. 23–38. In: O. Paredez-Lopez (ed.), Amaranth: Biology, chemistry, and technology. CRC Press, Boca Raton, FL.

Fehr, W. 1987. Principles of cultivar development: Theory and technique. Vol. I. McGraw-Hill, New York. p. 332–337.

Hauptli, H. 1977. Agronomic potential and breeding strategy for grain amaranths. p. 71–78. In: Proc. 1st. Amaranth Sem. Rodale Press, Emmaus, PA.

Hauptli, H. and S. Jain. 1985. Genetic variation in outcrossing rate and correlated floral traits in a population of grain amaranth (*Amaranthus cruentus* L.). Genetica 66:21–27.

Jain, S.K., K.R. Vaidya, and B.D. Joshi. 1979. Collection and evaluation of Indian grain amaranths. p. 123–128. In: Proc. 2nd. Amaranth Conf. Rodale Press, Emmaus, PA.

Jain, S.K., H. Hauplti, and K.R. Vaidya. 1982. Outcrossing rate in grain amaranths. J. Hered. 73: 71–72.

Joshi, B.D. 1986. Genetic variability in grain amaranth. Indian J. Agr. Sci. 56:574–576.

Kauffman, C. 1981. Grain amaranth varietal improvement: Breeding program. OGFRC. Rodale Press, Emmaus PA.

Myers, R.L. 1994. Regional amaranth variety test. Legacy 7(1). The Amaranth Institute, Bricelyn, MN. p. 5–8.

National Research Council. NRC. 1984. Amaranth: Modern prospects for an ancient crop. National Academy Press, Washington, DC.

SAS Institute Inc., SAS/STAT. 1989. User's guide version 6, 4th. ed. Vol. 2. SAS Institute Inc., Cary, NC.

Scheiner, S.M. and C.J. Goodnight. 1984. The comparison of phenotypic plasticity and genetic variation in populations of the grass *Danthonia spicata*. Evolution 38:845–855.

Schlichting, C.D. 1986. The evolution of phenotypic plasticity in plants. Annu. Rev. Ecol. Syst. 17:667–693.

Schulz-Schaefer, J., D.E. Baldridge, H.F. Bowman, G.F. Salknecht, and R.A. Larson. 1991. Registration of 'Amont' grain amaranth. Crop. Sci. 31:482–483.

Simmonds, N.W. 1979. Principles of crop improvement. Longman, London.

Weber, L.E. and C. Kauffman. 1990. Plant breeding and seed production. p. 115. In: Amaranth: Perspectives on production, processing, and marketing. Minnesota Ext. Service, St. Paul.

Plant Population Influence on Yield and Agronomic Traits in 'Plainsman' Grain Amaranth*

F.R. Guillen-Portal, D.D. Baltensperger, and L.A. Nelson

Grain amaranth, a pseudo-cereal with a rich history as a source of food in Meso America, is an attractive alternative crop suitable to be produced in semi-arid conditions. Grain amaranth production in the United States was 1,800 ha in 1991 (Stallknecht and Schulz-Schaeffer 1991). Most amaranth production is located in the Great Plains, especially in Western Nebraska (Williams and Brenner 1995), a semi-arid region with an average annual precipitation of less than 400 mm. Under these conditions, 'Plainsman' is the most widely grown grain amaranth cultivar. Grain yields range from 800 to 1100 kg/ha (Baltensperger et al. 1991; Myers 1994).

Currently, there is no commercially available planting equipment for amaranth because of its unusually small seed size (two million seeds/kg). Adapting existing planters for other crops has become a practical solution for the growers. The lowest seeding rate at present is about 1 kg seed/ha. Observations in the field indicates that 2 million plants/ha may be too high. In other crops, reduced grain yield, increased lodging, and lack of plant vigor are problems associated with excessively high plant populations.

Previous research conducted on grain amaranth suggests that yield is maximized at low plant populations (60,000 to 75,000 plants/ha) (Duncan and Volak 1979; Mnzava and Ntimbwa 1985; Henderson et al. 1991). However, most of these studies dealt with non-uniform landrace populations or with evaluations under wider row spacings. Identification of plant populations that maximize grain yields in improved cultivars is still an aspect that requires attention. Information obtained from these studies might be useful for the design of equipment capable of seeding more appropriate plant populations. The objective of this study was to evaluate the effect of plant population on grain yield and some agronomic components of 'Plainsman' grain amaranth.

METHODOLOGY

A dryland study was conducted at the University of Nebraska Panhandle Research and Extension Center at Scottsbluff, NE and at the High Plains Agricultural Laboratory near Sidney, NE in 1991 and 1992. 'Plainsman' grain amaranth (*A. hypochondriacus* × *A. hybridus*) (Baltensperger et al. 1992) was planted at Scottsbluff on 16 June 1991 and 21 June 1992 in an Otero loam (Tripp very fine sandy loam) soil. At Sidney, planting was on 24 June 1991 and 22 June 1992 on a Duroc loam (fine-silty, mixed, mesic, Pachic Haplustoll) soil. Ammonium nitrate was applied pre-plant at a rate of 100 kg N/ha. Plots were hand-weeded. The experiment was arranged as a randomized complete block design with plant populations of 2 M (million) (control), 1.4 M, 0.7 M, 0.35 M, 0.17 M, 0.085 M, and 0.043 M plants/ha with four replications at each location and year. Populations were established by over-planting and hand thinning the seedlings three weeks after planting to the appropriate plant spacing. Plots were four 76 cm rows wide and 4 m long in the experiments at Scottsbluff and 6 m long at Sidney. Measurements taken at maturity (100 days after planting) from 10 plants randomly chosen in each plot were plant height (measured from the ground to the top of the inflorescence head), inflorescence length (measured from the base to the top), and stem diameter (measured in the main stem at two thirds of the height above). Lodging percentage was determined on a plot basis before harvesting. Grain yield was obtained from the center two rows in each plot. Data from the experiments were analyzed using Statistical Analysis System (SAS Institute, Inc. 1989). For the analyses of variance, plant population was considered as fixed and blocks as random effects. For the combined analysis of variance, all effects except plant population were considered random. Consequently, the environment × plant population interaction effect was used as an error term for testing differences among plant populations. Plant population effect was partitioned using linear and quadratic orthogonal contrasts to identify the nature of the response. Additionally, linear orthogonal contrasts were used to compare means among treatments.

---

*Univ. of Nebraska Agr. Res. Div. J. Series no. 12130.

EXPERIMENTAL RESULTS

Conditions were extremely dry at Scottsbluff in 1991 but precipitation was near normal in 1992. Above-average moisture conditions were prevalent in the Sidney area in 1992 but precipitation was below normal in 1991. Growing season temperature at both locations was warm in 1991 and cooler in 1992 (data not shown). As a result of an early freeze at Scottsbluff in 1991, grain yield at maturity was substantially low (34 kg/ha) compared to normal yields (> 400 kg/ha). Also, grain yield at the 1992 Scottsbluff experiment was not collected because of a hailstorm which destroyed the field plots during the last week of August. Therefore, grain yield for these experiments were not included in the analysis.

At Sidney in 1991, significant differences in grain yield among plant populations were found (data not shown). Populations of 0.043 M, 1.4 M and 2 M plants/ha were different for grain yield but populations between 0.085 M and 0.7 M plants/ha produced similar grain yield. In the 1992 Sidney experiment, no significant yield differences among plant populations were observed.

A combined analysis of grain yield over environments showed significant differences between environments and plant populations. Differences in grain yield might be attributable in part to changes in the environmental conditions during the growing seasons of 1991 and 1992. For the range of populations studied, there was a significant quadratic response for grain yield (data not shown). A regression analysis of plant population on grain yield (Fig. 1) showed that grain yield reached its maximum (1050 kg/ha) at a population of about 0.47 M plants/ha, although the increase in grain yield at this population is only 2% compared to the overall mean (1023 kg/ha). The response of grain yield at populations in the range between 0.043 M and 0.95 M plants/ha was above the overall mean. Increase in grain yield at populations from 0.043 M to 0.47 M plants/ha suggests that little interplant competition occurs in this range. Decrease in grain yield at populations higher than 0.47 M plants/ha might be attributable to interplant competition. A linear contrast analysis showed the response of grain yield was higher at the lowest populations (0.043 M and 0.085 M plants/ha) and it was lower at the highest populations (1.4 M and 2 M plants/ha).

No significant differences in plant height among plant populations were observed in the experiment at Scottsbluff in 1991. Contrastingly, it was significant at Sidney. In 1991 at Sidney, plants at the lower populations were significantly taller than at the higher populations, and plants at the control population were significantly shorter than the other populations. No differences in plant height at populations between 0.085 M and 0.7 M plants/ha were observed. However, at Sidney in 1992 the highest population was significantly taller

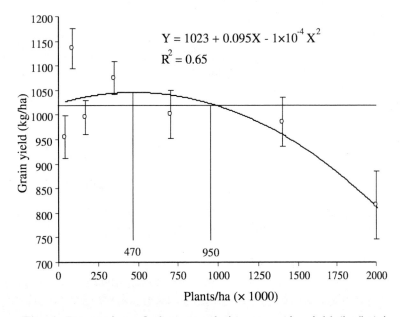

$$Y = 1023 + 0.095X - 1 \times 10^{-4} X^2$$
$$R^2 = 0.65$$

Fig. 1. Regression of plant population on grain yield (kg/ha) in 'Plainsman' grain amaranth. Sidney, Nebraska, 1991, 1992.

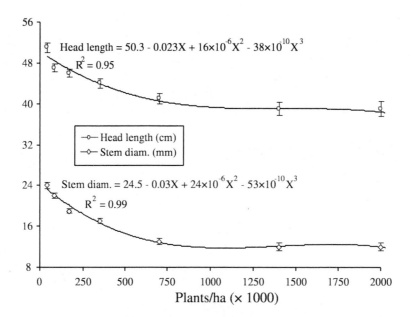

Fig. 2. Regression of plant population on seed-head length (cm) and stem diameter (mm) in 'Plainsman' grain amaranth. Sidney and Scottsbluff, Nebraska, 1991, 1992.

than the other populations. In general, plants at the low populations were significantly shorter (data not shown). Since conditions at Sidney in 1991 and 1992 were highly variable in precipitation, it can be inferred that the effect of plant population on plant height is substantially dependent on soil water availability. Combined over years and locations, plant height was unaffected by plant population (data not shown). The lack of response of plant height to a wide range of plant populations over environments may be the result of the ability of the crop to compensate for environmental factors by minimizing competition for water and light.

Seed head length response to plant population was significant at Scottsbluff in 1991. At this location seed head length reached its maximum at the lowest population. No significance for seed-head length among plant populations was found at Sidney. Combined over years and locations, the effect of plant population on seed head length was highly significant showing a cubic response and the same pattern at all environments (no environment × plant population interaction) (data not shown). Seed head length reached its maximum at the lowest plant population and then declined as plant population increased (Fig. 2). It is evident that grain yield was maintained at lower populations by an increase in seed head length and was decreased at higher populations as a result of interplant competition.

Stem diameter was greatly affected by plant population (data not shown). Consistently among locations and combined over locations stem diameter showed a cubic response to plant population (Fig. 2). In general, plants exhibited a robust main stem at low populations decreasing at higher populations. The decrease in stem diameter with increasing plant population may be the result of water and light interplant competition. The lack of response of plant height, the large change in stem diameter, and the modest response of grain yield to plant population suggests that grain amaranth compensates for a high plant population by translocating reserve assimilates from the stem to the reproductive organs. This is in agreement with the results obtained by Hauptli (1977), who found in a cultivated form of *Amaranthus* spp. a negative correlation between seed yield per plant and allocation of energy to stem.

Total lodging was unaffected by plant population at any of the locations and years. The highest lodging rates corresponded to populations between 0.085 M and 0.35 M plants/ha (data not shown). It is important to note, however, that at low populations peduncle lodging was observed, whereas at high densities stalk lodging was prevalent.

CONCLUSION

The modest response of grain yield to a wide range of plant populations may be interpreted as the ability of the amaranth to compensate for environmental variations through a) allocation of more energy to reproductive organs in the plant at the expense of a restriction in stem diameter, and b) minimizing the effects of water and solar radiation competition. Grain yield above the overall mean was obtained at a plant population in the range between 0.043 M and 0.95 M plants/ha. Considering this study as an initial step, further studies including other improved grain amaranth cultivars need to be conducted with the ultimate goal of providing information for the design and development of equipment and techniques for planting grain amaranth for rain-fed areas.

REFERENCES

Baltensperger, D.D., D.J. Lyon, L.A. Nelson, and A. Corr. 1991. Amaranth grain production in Nebraska. NF 91-35. Univ. Nebraska–Lincoln.

Baltensperger, D.D., L.E. Weber, and L.A. Nelson. 1992. Registration of 'Plainsman' grain amaranth. Crop Sci. 32:1510–1511.

Duncan, A. and B. Volak. 1979. Grain amaranth: Optimization of field population density. In: Proc. 2nd. Amaranth Seminar; July 29, 1979. Maxatawny, PA.

Hauptli, H. 1977. Agronomic potential and breeding strategy for grain amaranths. p. 71–78. In: Proc. 1st. Amaranth Sem., Rodale Press, Emmaus, PA.

Henderson, T.L., A.A. Schneiter, and N. Riveland. 1991. Row spacing and population effects on yield of grain amaranth in North Dakota. p. 219–221. In: J. Janick and J.E. Simon (eds.), New crops. Wiley, New York.

Mnzava, N.A. and T. Ntimbwa. 1985. Influence of plant density on edible leaf and seed yields of vegetable amaranth following repeated leaf harvest. Acta Hort. 158:127–132.

Myers, R.L. 1994. Regional amaranth variety test. Legacy. The Official Newsletter of the Amaranth Institute. Vol. VII. No. 1

SAS Institute Inc., SAS/STAT. 1989. User's guide version 6, 4th ed. Vol. 2, SAS Institute Inc., Cary, NC.

Stallknecht, G.F. and J.R. Schulz-Schaeffer. 1991. Amaranth rediscovered. p. 211–218. In: J. Janick and J.E. Simon (eds.), New crops. Wiley, New York.

Williams, J.T. and D. Brenner. 1995. Grain amaranth (*Amaranthus* species). p 129–186. In: J.T. Williams (ed.), Cereals and pseudocereals. Chapman and Hall, London.

LEGUMES

Legume Genetic Resources with Novel "Value Added" Industrial and Pharmaceutical Use*

J. Bradley Morris

Consumer preferences and scientific developments are changing and this is leading to a significant adjustment for US agriculture. During the last century, most agronomic research and production were to increase yields of food and fiber (Abelson 1994). However, during the last decade more attention is being focused on the production of new and alternative crops and their by-products for industrial, and pharmaceutical use. The legume family (Fabaceae) is the third largest family of flowering plants, with approximately 650 genera and nearly 20,000 species (Doyle 1994). Its species range from large tropical canopy trees to small herbs found in temperate zones, humid tropics, arid zones, highlands, savannas, and lowlands (NPGS 1995).

The Fabaceae contains many taxa of industrial, or pharmaceutical importance. Legume seeds are the second most important plant source of human and animal food (Vietmeyer 1986). Other new products would include new food sources, but the majority would provide industrial products such as dyes from Indigo, fiber pulps, vegetable, and pharmaceutical products. Many legumes contain organic chemicals in sufficient quantity to be economically useful as feedstocks or raw materials for many scientific, technological, and commercial applications. Legumes can biologically fix nitrogen, adding annually up to 500 kg N/ha/year to the soil (NAS/NRC 1979). Not only do other legume species provide hope for combating food shortages in developing countries, but they also can provide many specialty products such as rotenoids (Balandrin et al. 1985) for use as pesticides in developed countries.

Genetic variation in legume species and their wild relatives is of prime importance to the successful breeding of improved crop cultivars with added value and durable resistance to pests. The collection and preservation of legume germplasm has been established to ensure that scientists have access to as many genes as possible. The USDA, ARS Plant Genetic Resources Conservation Unit (PGRCU) is dedicated to acquiring, conserving, characterizing, evaluating, documenting, and distributing the genetic resources of crops, including special-purpose legumes. More than 4,000 accessions of special-purpose legumes are stored as seed at –18°C at the USDA, ARS, PGRCU in Griffin, Georgia. The purpose of this article is to highlight some outstanding new uses where some underexploited legumes seem notably promising.

LEGUMES WITH USEFUL PHYTOCHEMICALS

The USDA, ARS, PGRCU is dedicated to conserving 17 leguminous species with potentially useful phytochemicals (Table 1). Examples of commercially useful phytochemicals are rotenone, tephrosin, and deguelin, which are used in limited quantities as pesticides (Beckstrom–Sternberg and Duke 1994; Gaskins et al. 1972; Minton and Adamson 1979; Tyler et al. 1976). The use of pesticidal plants is widespread in the developing countries (Balandrin et al. 1985). The legume tephrosia (*Tephrosia purpurea*) contains insecticidal properties and the antitumor compound, lupeol (Beckstrom-Sternberg and Duke 1994). Rotenoid compounds derived from fish poison bean (*Tephrosia vogelii*) (Lambert et al. 1993) are also used as insecticides and rotenone has been reported to have antitumor potential (Beckstrom–Sternberg and Duke 1994).

Some legumes are potential sources of glycosides, biologics, antibiotics, and alkaloids which are used in drug manufacturing by the pharmaceutical industry (Tyler et al. 1976). The glycosides include aloe-emodin, chrysophanol, emodin, and rhein. Another phytochemical with potential use as an antibiotic is prodelphinidin derived from snout bean (*Rhynchosia minima*) (Beckstrom–Sternberg and Duke 1994). The alkaloid genistein, derived from kudzu has been found to retard cancer growth (Brink 1995). Trigonelline, an anticancer agent is derived from jackbean (*Canavalia ensiformis*) (Beckstrom-Sternberg and Duke 1994). Canavanine, extracted from jackbean has been found to be cytotoxic to human pancreatic cancer cells (Swaffar et al. 1994; Swaffar et al. 1995). Jackbean is also cultivated throughout the tropics as a cover crop, forage and green manure (Oropeza et al. 1993).

---

*The author wishes to acknowledge the S9 regional project and the University of Georgia for partial support of this research.

196

Both cowitch (*Mucuna pruriens*) and kudzu have been reported to contain multiple useful phytochemicals (Beckstrom–Sternberg and Duke 1994). Cell suspension cultures of cowitch accumulated the anti-Parkinson drug L-Dopa (Pras et al. 1993). The chemical, daidzin found in kudzu (*Pueraria montana* var. *lobata*) not only is cancer preventive, estrogenic, and spasmolytic but has also been effective in reducing alcohol consumption in hamsters (Braddock 1995). Kudzu starch extracted from the tuberous roots in Japan is sold as a health food worldwide (NAS/NRC 1979). Not only does winged bean (*Psophocarpus tetragonolobus*) provide useful phytochemicals such as polyunsaturated fatty acids used as an antipolyneuritic (Beckstrom–Sternberg and Duke 1994), but it also produces edible leaves, shoots, flowers, pods, and tubers as well as seeds whose composition duplicates that of soybeans. The most interesting feature of winged bean tubers is their protein content. Winged bean tubers average 20% protein as compared to 1% for cassava (*Manihot esculenta* Crantz, Euphorbiaceae) and 3–7% for potato (*Solanum tuberosum* L., Solanaceae) (NAS/NRC 1979). To date, winged bean has not yet met expectations due in part to its intolerance of cold temperatures during the fall and winter.

Lablab bean (*Lablab purpureus*) has a myriad of uses (NAS/NRC 1979). The young pods, dried seeds, leaves, and flowers can be eaten. Tyrosinase found in lablab bean has potential use for antihypertensive treatment (Beckstrom–Sternberg and Duke 1994). Lablab bean occurs in two botanical types. The garden type is twining, late maturing, and used mainly as a vegetable. The field type is erect, bushy, early maturing, and used as forage, cover crop, and an ornamental (NAS/NRC 1979).

Common indigo (*Indigofera tinctoria*) has been found to contain indirubin which is useful for the treatment of chronic myelocytic leukemia (Han 1994). Butterfly pea (*Clitoria ternatea*) contains antifungal proteins and has been shown to be homologous to plant defensins (Osborn et al. 1995). *Desmodium gangeticum* is used in Nigerian traditional medicine and has been evaluated for possible antileishmanial acitivity (Iwu et al. 1992). Tick clover (*Desmodium adscendens*), a medicinal herb used in Ghana has been evaluated for three active components including dehydrosoyasaponin I, soyasaponin I, and soyasaponin III for potential use as antiasthma (Mcmanus et al. 1993).

INDUSTRIAL LEGUMES

Indigo dye derived from indigo (*Indigofera arrecta*) (Purseglove 1981) is just one example of the usefulness and unfortunate obscurity of several potential industrial legume species (Table 2). The crop has been cultivated in India, but its importance has declined due to synthetic dye production. There were more than 600,000 ha cultivated in India in 1896, however by 1956 this had declined to 4000 ha. Indigo is still used for local dyeing in tropical Africa, but seldom enters into international trade (Purseglove 1981).

Guar (*Cyamopsis tetragonoloba*) is native to tropical Africa and Asia. The young pods are eaten as a vegetable and seeds are used as cattle feed in India and Pakistan, where it is also used as forage and green manure. Both guar and dhaincha (*Sesbania bispinosa*) contain galactomannan gum. This gum is water soluble, produces a smooth, light-colored, coherent, and elastic film useful for sizing textiles and paper, as well as for stabilizing the mud used in oil drilling (Vietmeyer 1986). Galactomannan gum is also used as a stabilizer and thickener in food products such as ice cream, bakery mixes, and salad dressings. Guar is grown for gum production in India and the southwestern United States. The plant is hardy and very drought resistant and grows well on alluvial and sandy loams (Purseglove 1981). Dhaincha can be grown in a rotation scheme for soil improvement, to provide fiber for paper pulp, for fodder, and has ornamental value (Vietmeyer 1986). Dhaincha appears to produce well on a large scale with little care or investment, and survives well on saline or wet soils (NAS/NRC 1979).

Leadtree (*Leucaena leucocephala*) is a multi-purpose legume tree providing fiber for paper products and is also beneficial as a cover crop, fodder, green manure, and ornamental (Mureithi et al. 1994; National Research Council 1984). In North America, the best-known nitrogen fixing trees are black locust (*Robinia pseudoacacia* L.) and honey locust (*Gleditsia triacanthos* L.). In the tropics, leadtree is so productive that on the most suitable areas it has reached heights of almost 6 m in its first year and 20 m thereafter in 6 years (Vietmeyer 1986). Leadtree produces vigorous sprouts after cutting and when young trees are grazed, the plants vigor reappears as lush shoots making useful livestock feed. High weight gains have been measured on cattle browsing a mixture of leadtree and grass in northern Australia (Vietmeyer 1986).

Table 1. Legume genetic resources with useful phytochemicals conserved at the PGRCU.

Name		Use		
Scientific	Common	Agricultural	Bioactive	Phytochemical (Pharmacological)
Canavalia ensiformis (L.) DC.	Jackbean	Forage, green manure, pulse	Pesticide, bactericide fungicide	Betonicine (Hemostat) Canavanine (Antiflu, anti-viral) Trigonelline (Anticancer: cervix, liver, hypocholesterolemic, hypoglycemic)
Clitoria ternatea (L.)	Butterfly pea	Cover crop, forage, ornamental	Antifungal	Antifungal proteins
Crotalaria juncea L.	Sunn hemp	Fiber, green manure		Senecionine (Antitumor hypotensive) Seneciphylline (Antitumor)
Crotalaria retusa L.			Pesticide	Monocrotaline (Antileukemic, antitumor, cardiodepressant, hypotensive Retusin (Antitumor)
Desmodium adscendens (Sw.) DC.	Tick clover			Dehydrosoyasaponin I, Soyasaponin I, Soyasaponin III (Antiasthma)
Desmodium gangeticum (L.) DC.				Antileishmanial
Indigofera tinctoria (L.	Common indigo	Pesticide Dye		Indigotin (Antiseptic, astringent) Indirubin (Antileukemic)
Lablab purpureus (L.) Sweet	Hyacinth bean	Browse, forage, ornamental, pulse		Tyrosinase (Antihypertensive)
Mucuna pruriens (L.) DC.	Cowitch	Pulse	Pesticide	Bufotenine (Cholinesterase inhibitor) Dopa (Anti-parkinsonian) Mucunain (Anthelminthic) Serotonin (Antiaggregant, antigastric, cholinesterase inhibitor, coagulant, myo-relaxant, myo-stimulant)
Psophocarpus tetragonolobus (L.) DC.	Wingbean	Vegetable		Erucic acid (Antitumor) PUFA (AntiMS, antiacne, antieczemic, antipolyneuritic)
Pueraria montana var. *lobata* (Willd.) Maesen & S. Almeida	Kudzu	Forage, human food		Daidzein (Anti-inflammatory, antimicrobial, coronary dilator, estrogenic, spasmolytic) Daidzin (Cancer preventive, estrogenic, spasmolytic) Genistein (Antileukemic, antimicrobial, cancer preventive, estrogenic)

Species	Common name	Use	Chemical constituents (activity)
Rhynchosia minima (L.) DC.			Puerarin (Antimyocarditis, hypoglycemic, hypotensive); Robinin (Cancer preventive); Tectoridin (Anti-inflammatory)
Senna alata (L.) Roxb.	Snout bean / Ringworm bush	Forage / Pesticide	Prodelphinidin (Antibiotic); Aloe-emodin (Antileukemic, antiseptic, antitubercular, antitumor); Chrysophanol (Antiseptic, hemostat); Emodin (Antiaggregant, anti-inflammatory, antimutagenic, antiseptic, antitumor (breast), antiulcer, spasmolytic); Rhein (Anticarcinomic, antiseptic, anti-tumor)
Senna occidentalis (L.) Link	Coffee senna	Coffee substitute / Pesticide	Aloe-emodin (Antileukemic, antiseptic, anti-tubercular, anti-tumor); Anthraquinone (Laxative); Chrysophanol (Antiseptic, hemo-stat); Emodin (Antiaggregant, anti-inflammatory, anti-mutagenic, anti-septic, antitumor (breast), antiulcer, spasmolytic); Physcion (Antiseptic); Rhein (Anticarcinomic, antiseptic, anti-tumor)
Tephrosia candida DC.	White tephrosia	Cover crop	Rotenone (Antitumor) — Acaricide, antifeedant, pesticide, piscicide; Tephrosin — Antifeedant, cytotoxic, pesticide, piscicide
Tephrosia purpurea (L.) Pers.			Lupeol (Antirheumatic, antitumor) — Cytotoxic; Rotenone (Antitumor) — Acaricide antifeedant, pesticide, piscicide; Tephrosin — Antifeedant, cytotoxic, pesticide, piscicide
Tephrosia vogelii Hook. f.	Fish poison bean		Deguelin — Cytotoxic, insecticide, pesticide, piscicide; Rotenone (Antitumor) — Acaricide, antifeedant, pesticide, piscicide; Tephrosin — Antifeedant, cytotoxic, pesticide, piscicide

Table 2. Industrial legume genetic resources conserved at the PGRCU

Name		
Scientific	Common	Use
Cyamopsis tetragonoloba (L.) Taub.	Guar	Vegetable gum, forage, green manure, source of galactomannan gum
Indigofera arrecta Hochst. ex A. Rich	Indigo	Dye
Leucaena leucocephala (Lam.) de Wit	Leadtree	Fiber, forage, fuel, fodder, human food, green manure
Sesbania bispinosa (Jacq.) W. Wight	Dhaincha	Fiber, pulp, cover crop, fodder, green manure, ornamental, gum, source of galactomannan gum

CONCLUSIONS

Leguminous genetic resources have hardly been explored and sampled for their offerings. Intensive efforts should be expanded for maximizing the use of potentially useful leguminous species. Biotechnology can accentuate the use of these leguminous genetic resources by facilitating the isolation of unusually promising genetic characteristics. Advances in biotechnology have increased the value of legume genetic resources for industry and the pharmaceutical industry. The cost of manipulating new genetic material or identifying and isolating new phytochemicals is declining swiftly (Reid et al. 1995). Leguminous plant natural products have been and will continue to be important sources and models of forage, gums, insecticides, phytochemicals, and other industrial, medicinal, and agricultural raw materials. Since most of these legume species have not been examined for chemical or biologically active components, it is logical to expect that new sources of valuable substances remain to be discovered. The Plant Genetic Resources Conservation Unit is dedicated to quality conservation of legume genetic resources for distribution to scientists worldwide.

REFERENCES

Abelson, P.H. 1994. Continuing evolution of U.S. Agron. Sci. 264:1383.

Balandrin, M.F., J.A. Klocke, E.S. Wurtele, and W.H. Bollinger. 1985. Natural plant chemicals: Sources of industrial and medicinal materials. Science 228:1154–1160.

Beckstrom-Sternberg, S.M. and J.A. Duke. 1994. The phytochemical database. http://probe.nalusda.gov:8300/cgi-bin/query?dbgroup=phytochemdb (ACEDB version 4.0 - data version July 1994).

Braddock, P. 1995. Kudzu: It's sobering. The Atlanta Journal, The Atlanta Constitution, Atlanta, GA, Oct. 7. G1.

Brink, S. 1995. Looking beyond beta carotene. U.S. News and World Report, Washington DC. p. 92–93.

Doyle, J.J. 1994. Phylogeny of the legume family: An approach to understanding the origins of nodulation. Annu. Rev. Ecol. Systemat. 25:325–349.

Gaskins, M.H., G.A. White, F.W. Martin, N.E. Delfel, E.G. Ruppel, and D.K. Barnes. 1972. *Tephrosia vogelii*: a source of rotenoids for insecticidal and piscicidal use. U.S. Dept. Agr., Tech. Bul. 1445.

Han, R. 1994. Highlight on the studies of anticancer drugs derived from plants in China. Stem Cells 12:53–63.

Iwu, M.M., J.E. Jackson, J.D. Tally, and D.L. Klayman. 1992. Evaluation of Plant-extracts for antileishmanial activity using a mechanism based radiorespirometric microtechnique (RAM). Planta Med. 58:436–441.

Lambert, N., M.-F. Trouslot, C. Nef-Campa, and H. Chrestin. 1993. Production of rotenoids by heterotrophic and photomixotrophic cell cultures of *Tephrosia vogelii*. Phytochemistry 34:1515–1520.

Mcmanus, O.B., G.H. Harris, K.M. Giangiaacomo, P. Feigenbaum, J.P. Reuben, M.E. Addy, J.F. Burka, G.J. Kaczorowski, and M.L. Garcia. 1993. An activator of calcium dependent potassium channels isolated from a medicinal herb. Biochemistry 32:6128–6133.

Minton, N.A. and W.C. Adamson. 1979. Response of *Tephrosia vogelii* to four species of root-knot nematodes. Plant Dis. Reptr. 63:514.

Mureithi, J.G., W. Thorpe, R.S. Tayler, and L. Reynolds. 1994. Evaluation of *Leucaena* accessions for the semi-humid lowland tropics of East Africa. Trop. Agr. (Trinidad) 71:83–87.

NAS/NRC. 1979. Tropical legumes: resources for the future. Report by an ad hoc advisory panel of the Advisory Committee on Technology Innovation, Board on Science and Technology for International Development, Commission on International Relations, National Academy of Sciences and the National Research Council, Washington, DC.

National Research Council. 1984. *Leucaena*: promising forage and tree crop for the tropics. Report of an ad hoc panel of the Advisory Committee on Technology Innovation, Board on Science and Technology for International Development, Office of International Affairs, second edition, National Academy Press, Washington, DC.

NPGS. 1995. Germplasm resources information network (GRIN). Database Management Unit (DBMU), National Plant Germplasm System, U.S. Dept. Agr., Beltsville, MD.

Oropeza, C., G. Godoy, J. Quiroz, and V.M. Loyola-Vargas. 1993. *Canavalia ensiformis* L. DC (Jackbean): in vitro culture and the production of canavanine. p. 34–50. In: Y.P.S. Bajaj (ed.), Biotechnology in agriculture and forestry, medicinal and aromatic plants IV. Springer-Verlag, Berlin, Heidelberg.

Osborn, R.W., G.W. De Samblanx, K. Thevissen, I. Goderis, S. Torrekens, F. Van Leuven, S. Attenborough, S.B. Rees, and W.F. Broekaert. 1995. Isolation and characterisation of plant defensins from seeds of Asteraceae, Fabaceae, Hippocastanaceae and Saxifragaceae. FEBS Lett. (Netherlands) 368:257–262.

Pras, N., H.J. Woerdenbag, S. Batterman, J.F. Visser, and W. Vanuden. 1993. *Mucuna pruriens*: Improvement of the biotechnological production of the anti-parkinson drug L-dopa by plant cell selection. Pharm. World Sci. 15:263–268.

Purseglove, J.W. 1981. Leguminosae. p. 199–332. In: Tropical crops dicotyledons. Longman Group LTD, Essex, UK.

Reid, W.V., C.V. Barber, and A. La Vina. 1995. Translating genetic resource rights into sustainable development: gene cooperatives, the biotrade and lessons from the Philippines. Plant Gen. Res. Newslett. 102:1–17.

Swaffer, D.S., C.Y. Ang, P.B. Desai, and G.A. Rosenthal. 1994. Inhibition of the growth of human pancreatic cancer cells by the arginine antimetabolite L-canavanine. Cancer Res. 54:6045–6048.

Swaffer, D.S., C.Y. Ang, P.B. Desai, G.A. Rosenthal, D.A. Thomas, P.A. Crooks, and W.J. John. 1995. Combination therapy with 5-fluorouracil and L-canavanine: In vitro and in vivo studies. Anti-cancer Drugs 6:586–593.

Tyler, V.E., L.R. Brady, and J.E. Robbers. 1976. Glycosides. p. 76–103. In: Pharmacognosy. Lea and Febiger, Philadelphia.

Vietmeyer, N.D. 1986. Lesser-known plants of potential use in agriculture and forestry. Science 232:1379–1384.

Chickpea, Faba Bean, Lupin, Mungbean, and Pigeonpea: Potential New Crops for the Mid-Atlantic Region of the United States

Harbans L. Bhardwaj, Muddappa Rangappa, and Anwar A. Hamama*

The New Crops Program of Virginia State University, established in 1991, has evaluated the production feasibility of a wide array of leguminous crops including chickpea, faba bean, mungbean, and pigeonpea under Virginia's agro-climatic conditions (Bhardwaj et al. 1996). Such crops could provide alternatives to farmers in Virginia and adjoining Mid-Atlantic States. These farmers, in general, rely on a limited number of crops and are interested in diversification. The close proximity of these farmers to the Washington, DC metropolitan area where the international community is familiar with these crops can provide a market for these crops.

The evaluations of chickpea, pigeonpea, and mungbean were conducted as replicated field experiments. The evaluations of faba bean and lupin germplasm were conducted by planting single row plots of each accession. All field experiments were conducted at the Randolph Farm of Virginia State University which is located approximately 37° 15' N and 077° 30.8' W.

CHICKPEA

Cicer arietinum L., an ancient crop, was probably grown in Turkey 7400 years ago. Most chickpea world production is in India. The mature chickpea seed are used as a dry bean and green immature seed are used as a vegetable. In chickpea, two seed types exist: *kabuli* or garbanzo (large seeded) and *desi* (small seeded). Chickpea is an annual plant generally requiring a cool season. However, it can be planted in spring in Virginia. The chickpea plant is 20–100 cm tall. Chickpea has a deep tap root and is considered drought tolerant.

The results of chickpea evaluations are presented in Table 1. The mean yield of *desi* type chickpea lines (1153 kg/ha) was significantly higher than that of *kabuli* type lines (719 kg/ha). However, the larger kabuli-type chickpea are known to be sold at premier prices at the green-immature stage for use as a vegetable. Recent research has indicated that 'Sanford' and 'Dwelly' (*kabuli* type cultivars) and 'Myles' (*desi* type cultivar) are adaptable and high yielding in Virginia.

FABA BEAN

Vicia faba L. is known to be an efficient nitrogen fixer and there is interest among farmers to grow faba bean as a vegetable crop to market the green beans in the Washington, DC metropolitan area. The faba bean is generally a cool season crop but can be planted in Virginia during spring. A diverse germplasm collection of faba bean germplasm has been evaluated for production potential. This collection has included lines from ICARDA (Syria); US collection at Pullman, Washington; and lines from Dr. Al Slinkard (University of Saskatchewan, Saskatoon, Canada). The seedling and foliar diseases have been a major hindrance in faba bean production under Virginia conditions. Although our results with faba bean have been disappointing, two cultivars, 'Fatima' and 'Chinese', seem to have promise under Virginia conditions.

LUPIN

White lupin (*Lupinus albus* L.) is making a comeback in the southern United States due to its high potential in both conventional and sustainable production systems. Since 1997, white lupin is being evaluated in

*We thank Dr. I.W. Budenhagen (University of California, Davis, California), Dr. L.H. Edwards (Oklahoma State University, Stillwater, Oklahoma), Dr. R.M. Hannan (USDA-ARS, NPGS, Pullman, Washington), Dr. R.S. Malhotra (ICARDA, Syria), Dr. L.C. Merrick (University of Maine, Orono, Maine), Dr. F.J. Muehlbauer (USDA-ARS, Pullman, Washington), Dr. S.C. Phatak (University of Georgia, Tifton, Georgia), Dr. B. Schatz (North Dakota State University, Carrington, North Dakota), Dr. C. Simon (USDA-ARS, NPGS, Pullman, Washington), and Dr. A. Slinkard, University of Saskatchewan, Saskatoon, Sask., Canada) for providing seeds of various crops.

Virginia as a winter grain legume crop and as a green manure crop to meet nitrogen needs of following summer crops. Lupin can potentially fix 150 to 200 kg/ha nitrogen for the use of a succeeding crop (Reeves et al. 1990). It has been estimated that if lupin replaced a quarter of wheat area in the southeastern United States, 95000 t of nitrogen fertilizer worth $50 to $60 million per year could be saved (Reeves et al. 1990).

The fiber-rich lupin flour is also gaining attention as a food source for humans. The nutritionally-rich lupin flour, due to its high content of potassium, calcium, carotenes, and protein, can be used to enrich pastas, cake mixes, cereals, and other baked goods (Birk 1993). Sweet lupin have been observed to be good sources of macro- and micro-nutrients, protein, fat, carbohydrates, minerals, and vitamins (Yanez 1996) for normal growth and development of humans and other animal species. Sweet lupin seeds lack trypsin inhibitors and can make a valuable contribution to dairy, beef, swine, sheep, and poultry rations at the farm since high temperature cooking to eliminate anti-nutritional factors is not needed. A survey of historic weather data for Virginia (1961–1990), has indicated that successful lupin production in Virginia and the mid-Atlantic region would get a boost from development of cold-tolerant lines.

During fall of 1997, a lupin collection of 284 lines representing four species: *Lupinus albus, L. angustifolius, L. luteus,* and *L. mutabilis*, were evaluated for cold tolerance and 148 selected lines are now being evaluated for yield potential and cold-tolerance. Greenhouse experiments conducted during 1997 have indicated that nodulation effectiveness was dependent upon specific *Bradyrhizobium* strain and lupin genotype combination. Lupin yields have been unstable in the mid-Atlantic region. Field experiments conducted at Orange, Virginia during 1995–96 season resulted in an average yield of 3480 kg/ha as compared to average Alabama yields of less than 1740 kg/ha (Noffsinger et al. 1998).

A comparison of lupin seed produced in Maine to that produced in Virginia (Bhardwaj et al. 1999) indicated that growing environment significantly affected total sugar, amino acids, oil, fatty acids, and minerals but not protein. The results indicated that lupin seed has potential as human food. The lupin seed produced in Virginia contained approximately 3 percent ash, 37 percent protein, 5 percent oil, and 7 percent sugar.

MUNGBEAN

Vigna radiata (L.) Wilczek. is native to northeastern India–Burma (Myanmar) region of Asia. It is primarily grown in Asia, Africa, South and North America, and Australia principally for its protein-rich edible seeds. Mungbean is also known as mung, *moong, mungo*, green gram, golden gram, chop-suey bean. Human consumption of mungbean is as dry seeds or sprouts. Mungbean also has potential as a green manure and a forage crop. In the United States, mungbean was grown as early as 1835. Oklahoma, California, and Texas account for about 90% of the US production (about 50,000 ha). Approximately, 7 to 9 million kg of mungbean are consumed annually in the United States and nearly 75% of this amount is imported (Oplinger et al. 1990). Enhanced domestic production can help offset annual imports of approximately 5 to 7 million kg of mungbean.

Seven mungbean lines were evaluated during 1993 and 1994 with encouraging results (Table 2). Mungbean planted in June or July, may be a suitable crop in rotation with winter wheat. The soybean farm machinery and production technology are generally suitable for mungbean culture. During 1997 and 1998, mungbean was commercially produced in Virginia on a small scale for sale to businesses intending to use it as a dry pulse.

Table 1. Performance of *desi* and *kabuli* chickpea lines during 1993 when planted in March and harvested in July.

Line	Type	Seed yield (kg/ha)
Aztec	*Desi*	1400
ICC 4948	*Desi*	1360
ICC 10136	*Desi*	1343
C 235	*Desi*	1183
ICC 4	*Desi*	1097
NEC 1163	*Desi*	1003
Garnet	*Desi*	964
PI 12074	*Desi*	876
Mean		1153
LSD(.05)		453
UC 8532	*Kabuli*	1083
UC 85150	*Kabuli*	1047
UC 27	*Kabuli*	996
UC 15	*Kabuli*	929
UC8624	*Kabuli*	925
UC85183	*Kabuli*	811
UC5	*Kabuli*	620
SR20I	*Kabuli*	576
UC8554	*Kabuli*	559
Surutato 77	*Kabuli*	431
Surutato	*Kabuli*	349
UC8536	*Kabuli*	307
Mean		719
LSD(.05)		432

PIGEONPEA

Cajanus cajan (L.) Millsp. is one of the oldest food crops of the world and ranks 5th among edible legumes in worldwide production. Pigeonpea is known to produce more nitrogen per unit of plant biomass than most other legumes and can nodulate in most soils. It is also considered to be tolerant to low and high temperatures. There is considerable variation among pigeonpea germplasm for crop duration which may vary from 80 to 250 days. Pigeonpea is useful as a grain, forage, or a green manure crop. Both determinate and indeterminate genotypes of pigeonpea exist. Seeds of pigeonpea are known to be a rich source of proteins, carbohydrates, and minerals with protein content generally varying from 18 to 25% and as high as 32%. Pigeonpea seeds are rich in sulfur-containing amino acids, methionine, and cystine. In pigeonpea, green immature seeds are used as a vegetable and could be important income for small and part-time farmers. A market for green pods of pigeonpea is known to exist in the Washington, DC metropolitan area.

During 1992, the seed yield varied from 349 to 2042 kg/ha with a mean yield of 1236 kg/ha (Table 3). The mean yield of determinate lines (1751 kg/ha) was significantly superior to that of indeterminate lines

Table 2. Seed yield of mungbean during 1993 and 1994 (Source: Bhardwaj et al. 1997).

	Yield (kg/ha)				
	1993		1994		
Entry	June 9–Oct. 6[z]	July 7–Oct. 6	May 17–Oct. 21	June 16–Oct. 21	July 2–Nov. 29
LSB 8205	2068	1799	1955	1257	1242
Johnston's California	1758	1651	2794	2516	888
TexSprout	1535	1338	2362	3025	936
Lincoln	1522	1737	2663	1892	927
Berken	1516	1265	3263	991	805
M 12	1382	1469	2621	1885	669
OK 12	1189	1065	3287	2258	848
Mean	1567	1475	2706	1975	902
LSD(.05)	414	298	745	1118	ns

[z]Planting and harvest dates.

Table 3. Performance of pigeonpea during 1992 in Virginia.

					Green bean		
Line	Type	Seed yield (kg/ha)	No. seeds/pod	Seed wt (g/100)	Harvest index (%)	Yield (kg/ha)	Moisture (%)	Shelling (%)
VXPP-I1	Determinate	1925	4.9	11.8	20.2	13184	78.9	52.1
VXPP-I2	Determinate	2042	4.2	9.2	28.4	15696	82.4	53.7
VXPP-I3	Determinate	1287	3.7	7.6	21.9	11888	84.1	55.1
VXPP-I4	Indeterminate	597	4.3	10.0	8.9	--	--	--
VXPP-I5	Indeterminate	1217	4.5	9.6	11.7	--	--	--
VXPP-I6	Indeterminate	349	4.4	10.2	4.3	--	--	--
Mean		1236	4.3	9.8	15.9	13589	81.8	53.6
LSD (.05)		494	0.5	1.4	5.1	ns	ns	ns

(721 kg/ha). A mean green bean yield of 13589 kg/ha was obtained in this study. These results indicated that pigeonpea can be successfully grown in Virginia and the mid-Atlantic region. During 1998 summer, a Virginia farmer grew about 0.4 ha of pigeonpea for marketing of green pods.

CONCLUSIONS

Based on these results, we consider chickpea, mungbean, and pigeonpea to be potential crops for the mid-Atlantic region of the United States. The possibility of producing lupin in this region seems encouraging. Faba bean evaluation indicates that seedling and foliar diseases are a major hindrance to successful production.

REFERENCES

Bhardwaj, H.L., A.A. Hamama, and L.C. Merrick. 1999. Genotypic and environmental effects on lupin seed composition. J. Plant Foods Human Nutrition (in press).

Bhardwaj, H.L., M. Rangappa, and A.A. Hamama. 1997. Potential of mungbean as a new summer crop in Virginia. Virginia J. Sci. 48:243–250.

Bhardwaj, H.L., A. Hankins, T. Mebrahtu, J. Mullins, M. Rangappa, O. Abaye, and G.E. Welbaum. 1996. Alternative crops research in Virginia. p. 87–96. In: J. Janick (ed.), Progress in new crops. ASHS, Alexandria, VA.

Birk, Y. 1993. Anti-nutritional factors (ANFs) in lupin and other legume seeds: Pros and cons. p. 424–429. In: Martins and da Costa (eds.), Advances in lupin research. Proc. 7th Int. Lupin Conf. held at Evora, Portugal, 18–23 April, 1993.

Noffsinger, S.L., D.E. Starner, and E. van Santen. 1998. Planting date and seeding rate effects on white lupin yield in Alabama and Virginia. J. Prod. Agr. 11:100–107.

Oplinger E.S., L.L. Hardman, A.R. Kaminski, S.M. Combs, and J.D. Doll. 1990. Mungbean. In: Alternative field crops manual, Univ. Wisconsin, Cooperative Extension Service, Madison.

Reeves, D.W., J.T. Touchton, and R.C. Kingery. 1990. The use of lupin in sustainable agriculture systems in the Southern Coastal Plain. p. 9. In: Abstracts of Technical Papers, 17, Southern Branch ASA, Feb 3–7, 1990, Little Rock, AR.

Yanez, E. 1996. Sweet lupin as a source of macro and micro nutrients in human diets. Proc. 8th Int. Lupin Conf., Pacific Grove, California. 11–16 May, 1996. Univ. California, Davis.

Chickpea: A Potential Crop for Southwestern Colorado

Abdel Berrada, Mark W. Stack, Bruce Riddell, Mark A. Brick, and Duane L. Johnson

Chickpeas (*Cicer arietinum* L.) better known in the United States as garbanzo beans are an important source of protein in human diets. It is believed that chickpeas originated in southeastern Turkey. They have been cultivated in the Middle East, India, the Mediterranean, and Ethiopia since antiquity and were brought to the New World through trade and conquests (Duke 1981). Chickpea is an important crop in the US, Mexico, and Australia. Commercial production of chickpea in the US is concentrated in the central and coastal valleys of California and the Palouse region in eastern Washington and northern Idaho (Brick et al. 1998).

Chickpea is a deep-rooted (up to 2 m), self-pollinated, annual legume crop. Its stems are branched, erect or spreading, and leaves are either pinnately compound with 3 to 8 leaflet pairs or less commonly unifoliate. Leaf color is olive, dark green, or bluish. There are two groups of chickpea, depending on seed size, shape, and color. The large-seeded chickpeas (in excess of 26 g/100 seeds) are called *kabuli* and the smaller ones are called *desi*. Almost all of the chickpeas grown in the US are of the kabuli type. Kabuli chickpea often have rounded and pale-cream seeds. Their plants are tall (up to 1 m), have white flowers and no anthocyanin pigmentation. Desi plants are shorter, more prostrate, and have smaller leaflets. Their flowers and stems usually have anthocyanin pigmentation. Flowers can be white, pink, purplish, or blue and seeds can be irregularly shaped and yellow, brown, black, or green. Desi types are traditionally grown in India, other parts of Asia, and in Ethiopia and account for more than 80% of the world production of chickpea (Muehlbauer et al. 1982).

Chickpea seed contains 13% to 33% protein, 40% to 55% carbohydrate, and 4% to 10% oil (Stallknecht et al. 1995). Fatty acid composition varies with chickpea type but is approximately 50% oleic and 40% linoleic (Duke 1981). Chickpea seeds are consumed "fresh as green vegetable, parched, fried, roasted, and boiled; as snack bar, sweet, and condiments" (Muehlbauer and Tullu 1997). They are also ground into flour and used to make soup, bread, and sweetmeats. Chickpeas are commonly found in salad bars in the US and are extensively used for making dishes such as "couscous," "hummus," and "falafel" in the Mediterranean region, the Middle East, and parts of Asia.

Chickpea production was introduced to southwestern Colorado in the early 1980s but was short-lived due to agronomic, processing, and marketing constraints. Renewed interest in chickpea in recent years has been prompted by the release of more adapted cultivars and the need for alternative crops. Research on chickpea at the Southwestern Colorado Research Center (SWCRC) in 1994–1998 included cultivar yield trials, gradient irrigation experiments, and planting date trials. Elevation at the research center is 2128 m. The number of days with minimum temperature > –2.2°C is 143 in 8 out of 10 years (Colorado Climate Center). The average annual precipitation is 40.5 cm with June being the driest month. The predominant soil series is Wetherill silty clay loam (fine-silty, mixed, superactive, mesic Aridic Haplustalf).

The main objectives of the studies conducted at the SWCRC have been to evaluate the adaptation and yield potential of new cultivars of chickpea in southwestern Colorado; to determine the response to irrigation and N rate; and to identify challenges and opportunities that may arise from chickpea production.

YIELD POTENTIAL OF CHICKPEA IN SOUTHWESTERN COLORADO

Cultivar Differences

In 1995, five chickpea cultivars and eight advanced lines from the International Crops Research Institute for the Semi-Arid Tropics (ICRISAT) were tested at the SWCRC (Tables 1 and 2). The Kabuli-type chickpeas 'UC27', 'UC15', 'Dwelley', and 'Sanford' were planted with a White 3407 air planter. 'Myles' (desi-type) was planted with a hand-push Precision Garden Seeder due to its smaller seeds. Row spacing was 76.2 cm and planting rate 6.6 seeds/m of row. The plate used with the White planter was not adequate for planting large-seeded chickpea such as 'Dwelley', which resulted in numerous skips. 'UC15' and 'UC27' are cultivars released by the University of California-Davis for production in the coastal areas and central valleys of California, respectively (Helms et al. 1992a,b). Both cultivars produce seed with desirable canning quality but are susceptible to Ascochyta blight caused by *Ascochyta rabiei* (Pass.*)*, a serious chickpea disease that spreads rapidly under cool moist conditions and is difficult to control (Wiese et al. 1995). 'Dwelley' and 'Sanford' were developed by the USDA-ARS in cooperation with Washington State University, the University of Idaho, and Oregon State University and released in 1994 (Muehlbauer et al. 1998a; Muehlbauer et al. 1998b) and both have good resistance to Ascochyta blight. 'Sanford' matures 3 to 4 days earlier, has slightly smaller seed, and is more productive than 'Dwelley' (Muehlbauer et al. 1982). 'Myles' is a desi-type chickpea that was developed by USDA-ARS and released in 1994 based on its resistance to Ascochyta blight (Muehlbauer et al. 1998c).

Dryland cultivar Trial 1 averaged 2238 kg/ha in above ground dry matter (DM) and 1088 kg/ha in seed yield (Table 1). 'Dwelley' had the lowest plant population, dry matter, and seed yield at harvest. 'Myles' was the first to mature. It started to flower in late June and was ready to harvest by the third week of August. 'Myles' is easily recognizable from the kabuli-type seed due to its purple flowers and smaller leaflets (fern-type). There were no significant differences in DM or seed yield among the ICRISAT lines (Table 2). Yield

Table 1. Results of the 1995 chickpea cultivar trial 1 at Yellow Jacket, Colorado[z].

Cultivar	Type	No. plants/m² at harvest	Dry matter[y] (kg/ha)	Seed yield (kg/ha)	100-seed weight (g)	Diseased plants (%)
Myles	Desi	7.5	2088	1265	18.6	6.4
UC27	Kabuli	5.8	2293	1211	50.3	11.9
UC15	Kabuli	4.8	2382	1210	47.6	8.0
Sanford	Kabuli	8.4	2633	1178	42.8	8.8
Dwelley	Kabuli	2.9	1796	578	51.5	11.3
Mean		5.9	2238	1088	42.2	9.3
LSD (5%)		1.5	335	201	1.5	--
CV (%)		16.6	10	12	2.3	--

[z]Planting date: May 3, 1995; Harvest: Aug. 23, 1995 (Dwelley Aug. 29)
[y]Total above ground dry matter (DM)

Table 2. Results of the 1995 chickpea cultivar trial 2 at Yellow Jacket, Colorado[z].

Entry (Kabuli)	No. plants/m² at harvest	Dry matter[y] (kg/ha)	Seed yield (kg/ha)	100-seed weight (g)	Diseased plants (%)
ICCV95301	5.6	2213	1278	39.2	6.5
ICCV94305	6.1	1987	1119	42.7	24.0
ICCV95401	5.6	2108	1115	39.3	11.8
ICCV94304	5.5	2216	1081	38.5	12.5
ICCV92328	5.7	2126	1074	39.7	10.6
ICCV92310	5.8	2053	1073	40.4	12.1
ICCV95402	7.3	2056	1065	51.0	4.6
ICCV95501	5.4	1992	1064	37.8	14.5
Mean	5.9	2094	1109	41.1	12.1
LSD (5%)	NS[x]	NS	NS	2.5	--
CV (%)	8.1	8	7	3.4	--

[z]Planting date: May 3, 1995; Harvest: Aug. 23, 1995
[y]Total above ground dry matter (DM)
[x]Not significant

levels in cultivar Trials 1 and 2 were similar with the exception of 'Dwelley'. All the entries in the two trials were infested with Pea Enation Mosaic Virus (PEMV) and possibly other unidentified diseases. Infestation was 6% to 12% in Trial 1 and 5% to 24% in Trial 2. Leafminer (*Liriomyza sativa* Blanchard) was observed on a small number of chickpea plants. The effect of PEMV or leafminer on DM and seed yield was not quantified.

'Myles' had the lowest seed weight as was expected and dark-cream to brownish seeds. Among the kabuli-type cultivars, 'UC27' had the most marketable seeds. It had very few cracked or stained seeds, good seed size (50.3 g/100 seeds), and attractive seed shape (round) and color (light cream). Both 'UC27' and 'UC15' had round seeds with smooth edges compared to the seeds of 'Dwelley' and 'Sanford', which were somewhat wrinkled. 'Dwelley' and 'UC15' had similar seed size to that of 'UC27' but had more stained seeds. 'Sanford' had smaller seeds as did most of the ICRISAT lines. The only ICRISAT line with acceptable seed size for canning was 'ICCV 95402' but it had more stained seeds than 'UC27' (data not shown).

Irrigation and Nitrogen Management

Most of the chickpea crop in the world is produced on residual moisture but supplemental irrigation can enhance production. Irrigation during the pre-flowering period and at early pod fill resulted in increased yield at several locations in India (Saxena 1980). Irrigation prolonged the reproductive period of chickpea and produced higher total biomass and more pods per plant. Conversely, 100-seed weight and harvest index were reduced (ICRISAT 1987).

Line-source irrigation experiments were conducted at Yellow Jacket to test the response of 'Dwelley' (1994) and 'Sanford' (1994–1996) to irrigation amount and N rate. The irrigation system used was similar to the one described by Hanks et al. (1976). Planting dates were 2 June 1994, 5 May 1995, and 21 May 1996. Metolachlor (Dual 8E) was applied pre-plant at 2.2 to 2.8 kg a.i. ha^{-1} and incorporated with a field cultivator. The granular rhizobial inoculant (Implant Plus, LiphaTech) was broadcast at 20 kg ha^{-1} shortly before planting. Row spacing was 76.2 cm and seeding rate 13.2 seeds/m of row. Phosphorus and/or K rates were based on soil tests.

Table 3. Seed yield and size of 'Dwelley' and 'Sanford' chickpea as influenced by irrigation and nitrogen in 1994 at Yellow Jacket, Colorado.

| | Dwelley | | | | Sanford | | | |
| | Check | | 56 kg N/ha | | Check | | 56 kg N/ha | |
Irrigation amount[z] (cm)	Seed yield (kg/ha)	100 seeds (g)	Seed yield (kg/ha)	100 seeds (g)	Seed yield (kg/ha)	100 seeds (g)	Seed yield (kg/ha)	100 seeds (g)
0.0	888	48.4	1042	56.6	990	45.5	889	44.8
3.6	857	--	1260	--	1085	--	1096	--
7.1	879	46.1	1734	59.8	1146	45.2	1432	48.3
10.7	904	--	1994	--	1335	--	1889	--
14.2	803	39.6	2373	58.5	1264	39.8	2289	48.9
17.7	917	--	2388	--	1374	--	2473	--
21.3	892	38.2	2467	52.4	1445	37.3	2488	42.4
24.8	936	--	2362	--	1580	--	2498	--
28.3	972	38.5	2051	46.6	1624	36.0	2500	37.2
Mean	894	42.1	1963	54.8	1316	40.8	1951	44.3
LSD (5%)	271	2.0	271	2.0	277	1.9	277	1.9

[z]Derived from the regression equation, irrigation amount (cm) = 31.88 – 1.93X, where X = distance from the sprinkler line in m. R^2 = 0.99

Irrigation, nitrogen fertilization, and their interaction all had a highly significant effect on seed yield and size of 'Dwelley' and 'Sanford' in 1994 (Table 3). The seed yield of 'Dwelley' more than doubled with the application of 56 kg N/ha and that of 'Sanford' increased by about one-half compared to the zero N treatment. There was no significant increase in seed yield of either 'Dwelley' or 'Sanford' in the N treatment beyond 14.2 cm of cumulative irrigation. In addition, a uniform irrigation of 5.0 cm before planting and 2.5 cm after planting was applied to ensure adequate stand establishment. Total rainfall from planting to harvest was 5.8 cm. The increase in yield due to irrigation was more pronounced where nitrogen was applied, particularly in the case of 'Dwelley'. Nitrogen and/or less irrigation water applied resulted in larger and/or heavier seeds of 'Dwelley' and 'Sanford'. A composite sample of 'Sanford' seeds from the 1994 harvest had few defects, good seed color, and excellent imbibing quality (Stan Murray, Klein-Berger Co., pers. commun.).

'Sanford' seed yield increased significantly with increasing amounts of irrigation in 1995 (Table 4). A maximum of approximately 2000 kg/ha was reached with 28.7 cm of water (irrigation plus precipitation). Hundred-seed weight averaged 48.2 g with no significant differences among irrigation levels. Nitrogen fertilization did not influence seed yield or size despite low soil test levels in 1995 (data not shown). Much cooler conditions prevailed during and following planting in 1995 than in 1994 or 1996 (Colorado Climate Center). Consequently, 50% emergence did not occur until at least four weeks after planting in 1995 compared to about two weeks in 1994 and 1996. The cooler conditions in 1995 may have reduced soil nitrogen mineralization and plant uptake. Saxena (1980) reported a positive response to 15 to 25 kg N/ha of starter fertilizer in soils low in organic matter. Murray et al. (1987) recommended the application of 22 to 34 kg N/ha at planting on soils with less than 22 kg N/ha of residual nitrogen. Starter nitrogen may be beneficial for stand establishment or until chickpea plants are able to fix their own nitrogen.

The period from Oct. 1995 through May 1996 was extremely dry. Consequently, the pre-irrigation in 1996 was barely enough to ensure adequate seed germination and plant emergence. The soil profile below 20 cm was at or near the wilting point at planting. Subsequent irrigations and rainfall were not enough to fill the root zone or meet the crop water demand, which would explain the relatively low yields in 1996 (Table 5). Seed yield and 100-seed weight increased significantly with 56 kg N/ha and greater amounts of irrigation water.

Table 4. Seed yield and size of 'Sanford' as influenced by irrigation in 1995 at Yellow Jacket, Colorado.

Irrigation[z] amount (cm)	Seed yield (kg/ha)	100 seeds (g)
0.0	1248	46.6
2.0	1248	--
4.2	1394	47.8
6.2	1567	--
8.6	1583	49.2
10.8	1780	--
13.0	1877	48.8
15.2	2054	--
17.3	2027	48.7
Mean	1642	48.2
LSD (5%)	125	NS

[z]Derived from the regression equation, Irrigation amount (cm) = 19.54 − 1.19 X, where X = distance from the sprinkler line in m. $R^2 = 0.99$

Table 5. Seed yield and size of 'Sanford' as influenced by irrigation and nitrogen in 1996 at Yellow Jacket, Colorado.

Irrigation amount[z] (cm)	Check Seed yield (kg/ha)	Check 100 seeds (g)	56 kg N/ha Seed yield (kg/ha)	56 kg N/ha 100 seeds (g)
0.0	595	42.7	680	43.0
2.9	570	--	743	--
6.6	682	44.0	870	44.3
10.3	802	--	1114	--
14.0	853	44.2	1296	47.8
17.7	1079	--	1557	--
21.5	1066	43.8	1674	46.2
25.2	1122	--	1772	--
28.9	1179	44.6	1756	45.5
Mean	884	43.8	1273	45.4
LSD (5%)	177	1.3	177	1.3

[z]Derived from the regression equation, Irrigation amount (cm) = 32.64 − 2.03 X, where X = distance from the sprinkler line in m. $R^2 = 0.97$

Planting Date

Chickpea should be planted when soil temperature is above 5°C although "some cultivars can tolerate temperatures as low as –9.5°C in early stages or under snow cover" (Muehlbauer et al. 1982; Muehlbauer and Tullu 1997). Chickpea is a quantitative long-day plant but flowers in every photoperiod (Smithson et al. 1985). A 1984 chickpea planting date study at SWCRC showed no significant difference in seed yield between 1 May, 15 May, and 1 June planting dates (unpublished data). Additional studies (Table 6) were conducted in 1997 and 1998 to determine the effects of chickpea cultivar, planting date, and their interaction on seed yield and quality. A new release 'Evans' from the USDA-ARS, Washington State University, and the University of Idaho was included in this study. 'Evans' is a kabuli-type chickpea with good resistance to Ascochyta blight. Its seed size is larger than 'Sanford' but smaller than 'Dwelley' and matures 2 to 3 days earlier than 'Sanford' or 'Dwelley'.

The effect of planting date and cultivar on seed yield and 100-seed weight was highly significant in 1997 and 1998. Seed yield increased substantially when planting was delayed until 20 or 22 May. Further yield increase occurred in 1997 when chickpea was planted as late as 16 June. Seed yields were generally higher in 1997 than in 1998 due to more favorable moisture conditions. Most chickpea cultivars evaluated at SWCRC exhibited a somewhat indeterminate growth habit. For example, 'UC27' and 'Evans' planted in April or early May were ready to harvest by mid- to late-Aug. in 1997 and 1998. However, frequent rain events in late July and Aug. induced new vegetative growth leading to a flush of flowering and pod formation. 'Sanford' and 'Dwelley' showed similar characteristics but their maturity appeared to be more uniform for the late May and early June plantings than at the other planting dates.

'UC27' had the highest average seed yield in 1997 and 1998. 'Dwelley' outperformed 'Evans' and 'Sanford' in 1997 due to the higher yield at 2 June and 16 June planting dates, but had similar yield in 1998. The effect of planting date by chickpea cultivar on seed yield was significant in 1997 at $\alpha = 0.05$. Seed weight was influenced by both the cultivar and planting date. 'Dwelley' produced the largest and/or heaviest seeds at all planting dates followed by 'UC27' in 1997 and 1998. Based on 100-seed weight, the seeds of 'Dwelley' and 'UC27' would be suitable for canning. 'Evans' and 'Sanford' did not produce seeds of canning quality,

Table 6. Planting date effect on seed yield of four chickpea cultivars in 1997 and 1998 at Yellow Jacket, Colorado.

Planting date	Dwelley		Evans		Sanford		UC27		Mean	
	Seed yield (kg/ha)	100-seed weight (g)	Seed yield (kg/ha)	100-seed weight (g)	Seed yield (kg/ha)	100-seed weight (g)	Seed yield (kg/ha)	100-seed weight (g)	Seed yield (kg/ha)	100-seed weight (g)
1997										
22 Apr.	902c[z]	54.81c	683b	44.98d	843c	46.78c	832c	50.51b	815	49.26
5 May	1019c	59.23a	731b	46.61c	1008c	50.00b	885c	51.22b	911	51.76
20 May	1462b	57.08ab	1407a	53.02a	1482ab	53.53a	1897b	56.01a	1562	54.91
2 June	1671b	55.47bc	1504a	51.40ab	1256b	49.02bc	2228b	54.91a	1665	52.70
16 June	2127a	54.12c	1529a	49.67b	1844a	47.46bc	2596a	56.23a	2024	51.88
Mean	1436	56.14	1171	49.14	1287	49.36	1687	53.78	1395	52.10
1998										
1 May	999	56.14ab	1028	46.26b	1158	49.30ab	1140	49.33d	1081b	50.26
8 May	982	56.44a	1071	47.99a	1076	50.51a	1370	51.09c	1125b	51.51
22 May	1346	54.68bc	1033	48.45a	1112	49.14ab	1769	53.66ab	1315a	51.48
3 June	1398	53.51c	1197	46.89ab	1325	48.53b	1585	52.14bc	1376a	50.27
16 June	1137	57.31a	1191	47.45ab	1244	47.76b	1867	54.94a	1360a	51.86
Mean	1172b	55.62	1104b	47.41	1183b	49.05	1546a	52.23	1251	51.10

[z]Mean separation within cultivars and years by Duncan's Multiple Range Test, 5% level.

Table 7. United States chickpea production 1992 to 1997. Source: NASS, USDA

State	Production (t)						
	1992	1993	1994	1995	1996	1997	Average
California	8,573	10,070	12,519	13,063	15,966	6,169	11,060
Idaho	680	771	408	2,858	1,678	4,627	1,837
Oregon	862	454	590	907	154	2,994	993
Washington	1,860	2,359	1,996	4,627	3,674	3,493	3,001
Total	11,975	13,654	15,513	21,455	21,472	17,283	16,892

except at 5 May ('Sanford') and 20 May/2 June ('Evans') in 1997. Only mature seeds were used for 100-seed weight.

When marketing chickpea, seed lots with a high percentage of stained or green seeds may be rejected or sold as animal feed. On this basis, all the production from 2 June 1997 would be discarded since it had approximately 49% immature seeds (data not shown). Above average precipitation and frequent rains prolonged the vegetative and reproductive stages of the chickpea cultivars planted on or after 20 May 1997. There were less immature (green) seeds in 1998 than in 1997 due to drier conditions in 1998. Visual observations showed that with the exception of the early harvested chickpea cultivars, the early June plantings had the most uniform maturity and the least percentage of green and stained seeds. When seed maturity is delayed due to irrigation and/or weather conditions, chickpeas should be swathed and left to dry in the field for several days before they are threshed (Muehlbauer et al. 1982). Desiccants such as paraquat (Gramoxone) have also been used to hasten chickpea maturity.

MARKET CONSIDERATIONS

The US imports approximately $12 million chickpea each year, compared to domestic production valued at $2 million (Yarris 1984). World chickpea production from 1992 to 1996 averaged 7.3 million t, 90% of which was produced in Asia. India, alone, produces 4 to 5 million t of chickpea annually (Bean Market News, 1997 Summary). Other major chickpea producing countries are Pakistan, Turkey, Iran, Australia, Mexico, and Ethiopia. US production represents less than 1% of world production with California leading the way (Table 7). Almost all the chickpeas produced in the US are of the kabuli type, while over 80% of world production is of the desi type (Muehlbauer et al. 1982).

Chickpeas are graded as a miscellaneous bean under US standards in which damaged and defective seeds and foreign material are considered in determining grades (Muehlbauer et al. 1982). Most chickpeas grown in the US are sold for canning. The rest, about 10%, is sold for dry packaging or as animal feed. Desirable qualities for canning are medium seed size (50.6 to 52.5 g/100 seeds), golden color, rough texture, high water intake, and a seed coat that does not fracture easily (Stan Murray, Klein-Berger Co., pers. commun.).

Chickpea can be processed with the same equipment used for pinto beans. Elevator legs, bins, conveyors, screw augers, air cleaners, and secondary equipment should provide sufficient separation to process a market grade product. Commercial processing includes seed conditioning, sizing, and packaging or canning of the product (Brick et al. 1998). A minimum of 100 t is required to make processing of chickpea in southwestern Colorado economically feasible. In 1998, at least one-half of the chickpea production in southwestern Colorado was shipped in bulk to a bean dealer in Kansas.

Table 8. Dealer pinto bean and chickpea prices. Source: USDA Bean Market News, 1996 and 1997 summaries.

Year/ commodity	Price ($/t)	
	Pinto[z]	Chickpea[y]
1992–1993	524.77	916.86
1993–1994	733.49	714.98
1994–1995	470.11	932.95
1995–1996	567.31	1,018.03
1996–1997	607.42	711.01
1997–1998	563.56	726.66
Mean	577.78	836.75

[z]Northern Colorado
[y]California

Available markets for chickpea include dry packers, canners, dehydrated products, and livestock feed (inferior quality seeds). Southwestern Colorado would be ideal for a dry packer or canner. The lack of railroad or waterways may reflect higher freight rates, depending on product destination.

ADVANTAGES OF CHICKPEA OVER PINTOS IN SOUTHWESTERN COLORADO

Southwestern Colorado is traditionally a dry bean (primarily pinto) producing area. From 1990 to 1997, the harvested crop area of dry bean averaged 12500 ha of which 87% was dryland. Seed yields averaged 1750 kg/ha with irrigation and 360 kg/ha dryland (without irrigation) during the same period (Colorado Agricultural Statistics). Evaluation of chickpea at the SWCRC in the early 1980s and in 1994–1998 showed that average yields of at least 2000 kg/ha with irrigation and 500 to 1000 kg/ha dryland are feasible in southwestern Colorado. Limited on-farm testing and commercial production has confirmed these results (Bruce Riddell pers. commun.).

In addition to seed yield, chickpea prices have been consistently higher than those of pinto beans (Table 8) although seed quality appears to have more effect on chickpea prices. Acceptable seed quality for dry packaging with 'Sanford' and for canning with 'UC27' and 'Dwelley' has been produced in southwestern Colorado. Marketing may be the biggest obstacle for chickpea production in southwestern Colorado.

Chickpea planted early i.e. before mid-May at high elevation seem to produce smaller seeds than later planted chickpea. If chickpeas are planted late i.e. after the first week of June, the incidence of immature seeds increases. Above normal precipitation, particularly in Aug. and Sept., tends to prolong the vegetative and reproductive growth of chickpea in southwestern Colorado, which could lead to lower seed quality. Ideally, chickpea should be planted in mid-May (or earlier at lower elevations) and harvested by early Sept. Pinto beans are more sensitive to frost damage than chickpea and are usually not planted in southwestern Colorado until early June. An early frost will damage the seeds of both chickpea and pinto beans, although chickpea plants will continue to grow if the frost is not too severe.

Unlike pintos, chickpea can be direct-combined which would speed up the harvest operation and reduce costs. Swathing should be considered if maturity is delayed due to late planting or wet conditions, to allow sufficient time for the seeds to dry. Dry beans are usually undercut with knives in early to mid-Sept., windrowed and left to dry in the field for at least 10 days depending on the weather. They are then threshed with a combine equipped with a pickup attachment that lifts the windrow into the header (Berrada et al. 1995a). This practice often delays winter wheat planting that was shown to reduce wheat yield (Brengle et al. 1976). Other production practices for chickpeas are similar to pinto beans, reducing the need for new equipment investment (Berrada et al. 1995b).

CONCLUSION

The results of the chickpea trials at Yellow Jacket from 1994 to 1998 show good yield potential and adaptability of several cultivars. Comparable seed yields to pinto bean under irrigation and higher yields under dryland conditions were achieved. Both crops can be grown and processed using similar equipment and management practices. Chickpea can be planted and harvested earlier than pintos, which would allow for a more optimum planting date of winter wheat. Southwestern Colorado is well suited to dry packers and canners. The freight availability may create some problems in timely deliveries and transportation expenses. Good seed quality has been produced in southwestern Colorado but late planting and/or frequent rains can delay maturity and increase the incidence of stained seeds. More research is needed to identify cultivars and management practices that will enhance the marketability of chickpea produced in southwestern Colorado.

REFERENCES

Bean Market News. 1997. Summary. USDA Agr. Marketing Service, Greeley, CO.

Berrada, A., D.L. Johnson, D.V. Sanford, and M.W. Stack. 1995a. The feasibility of garbanzo bean production in Southwest Colorado. Agron. Abstr., Am. Soc. Agron., Madison, WI. p. 120.

Berrada, A., M.W. Stack, D.V. Sanford, and A.G. Fisher. 1995b. Management systems for dryland wheat and bean production in southwestern Colorado-Conservation tillage project, 1989–93. Agr. Expt. Sta. Tech. Bul. TB95-2, Colorado State Univ., Fort Collins.

Brengle, K.G., H.D. Moore, and A.G. Fisher. 1976. Comparison of management systems for dryland wheat production in southwestern Colorado. Agr. Expt. Sta. Gen. Ser. Bul. 958, Colorado State Univ., Fort Collins.

Brick, M.A., A. Berrada, H.F. Schwartz, and J. Krall. 1998. Garbanzo bean production trials in Colorado and Wyoming. Colorado and Wyoming Agr. Expt. Sta. Tech. Bul. TB 98-2, Colorado State Univ., Fort Collins and Univ. of Wyoming, Laramie.

Colorado Agricultural Statistics. 1991 to 1997 Reports. Colorado Dept. of Agr., Englewood.

Duke, J.A. 1981. Handbook of legumes of world economic importance. Plenum Press, New York. p. 52–57.

Hanks, R.J., J. Keller, V.P. Rasmussen, and G.D. Wilson. 1976. Line source sprinkler for continuous variable irrigation-crop production studies. Soil Sci. Soc. Am. J. 40:426–429.

Helms, D., L. Panella, I.W. Buddenhagen, F. Workneh, C.L. Tucker, K.W. Foster, and P.L. Gepts. 1992a. Registration of 'UC27' chickpea. Crop Sci. 32:500.

Helms, D., L. Panella, I.W. Buddenhagen, F. Workneh, C.L. Tucker, K.W. Foster, and P.L. Gepts. 1992b. Registration of 'UC15' chickpea. Crop Sci. 32:500.

ICRISAT. 1987. 1986 Annual Report. International Crops Research Institute for the Semi-Arid Tropics. Patancheru, A.P. 502 324, India.

Muehlbauer, F.J., W.J. Kaiser, and I. Kusmenoglu. 1998a. Registration of 'Dwelley' chickpea. Crop Sci. 38:282–283.

Muehlbauer, F.J., W.J. Kaiser, and I. Kusmenoglu. 1998b. Registration of 'Sanford' chickpea. Crop Sci. 38:282.

Muehlbauer, F.J., R.W. Short, and W.J. Kaiser. 1982. Description and culture of garbanzo beans. Coop. Ext. Publ. EB 1112, Washington State Univ., Pullman.

Muehlbauer, F.J., and A. Tullu. 1997. *Cicer arietinum* L. NewCROP FactSHEET. Center for New Crops and Plant Products. Purdue University, West Lafayette, IN. http://www.hort.purdue.edu/newcrop/cropfactsheets/Chickpea.html

Muehlbauer, F.J., H.A. Van Rheenen, and W.J. Kaiser. 1998c. Registration of 'Myles' chickpea. Crop Sci. 38:283.

Murray, G.A., K.D. Kephart, L.E. O'Keeffe, D.L. Auld, and R.H. Callihan. 1987. Dry pea, lentil and chickpea production in Northern Idaho. Agr. Expt. Sta. Bul. 664, Univ. of Idaho, Moscow.

Saxena, M.C. 1980. Recent advances in chickpea agronomy. p. 89–105. In: Proc. Int. Workshop on Chickpea Improvement, 28 Feb–2 Mar 1979, Hyderabad, India.

Smithson, J.B., J.A. Thompson, and R.J. Summerfield. 1985. Chickpea (*Cicer arietinum* L.). p. 312–390. In: R.J. Summerfield and E.H. Roberts (eds.), Grain legume crops. Collins, London, UK.

Stallknecht, G., K.M. Gilberston, G.R. Carlson, J.L. Eckhoff, G.D. Kushnak, J.R. Sims, M.P. Wescott, and D.M. Wichman. 1995. Production of chickpeas in Montana. Montana Agr. Res. 12:46–50.

Wiese, M.V., W.J. Kaiser, L.J. Smith, and F.J. Muehlbauer. 1995. Ascochyta blight of chickpea. CIS 886. Univ. of Idaho, College of Agriculture, Coop. Ext. System and Agr. Expt. Sta., Moscow.

Yarris, L. 1984. Garbanzos on the Palouse. Agr. Res. Feb., p. 6.

OILSEED & INDUSTRIAL CROPS

The Future of New and Genetically Modified Oil Crops

Denis J. Murphy

Oil crops are one of the most valuable traded agricultural commodities and are probably worth over $100 billion/yr. The current volume of traded vegetable oils is over 70 Mt per year and is predicted to increased to over 100 Mt per year by the year 2010 (Murphy 1996). This is probably a considerable underestimate of the total volume of oil production since, particularly in developing countries, most vegetable oils are consumed locally, rather than entering into trading networks. Despite the large volume of globally traded vegetable oil, only four major crops, soybean, oil palm, rapeseed, and sunflower contribute about 75% of this production. The domination of the oilseed market by the "big four" oil crops is set to continue for the foreseeable future. Indeed, the establishment of new plantations in South East Asia is predicted to double oil palm production in the next 15 years, making it, by then, the most important source of vegetable oils in the world.

The vast majority of vegetable oils are currently used for edible commodities, such as margarines, cooking oils, and processed foods. Only about 15% of production goes towards the manufacture of oleochemicals, i.e. industrial products derived from oil crops. Over the next few decades two important factors will contribute to an expansion in the markets for both edible and industrial oil crops. Firstly, according to current demographic trends, the global human population will at least double before stabilizing at some time in the mid 21st century. This, coupled with rising levels of affluence in many developing countries, will lead to increased demands for both raw and refined edible oil products. Secondly, it is well known that global hydrocarbon reserves are a non-renewable resource. The most recent estimates from six different national and international agencies predict that world oil production from petroleum reserves will peak at some time between the years 2000 and 2020, and will decline thereafter (Kerr 1998). In addition to being an important source of energy, e.g. electricity generation and vehicle fuel, petroleum is also the source of a huge range of petrochemicals. These petrochemicals are the raw materials for products such as plastics, textiles, lubricants, paints, varnishes. Once non-renewable hydrocarbon resources such as petroleum and coal are exhausted, there will be no other source of such products, other than oleochemicals derived from oil crops.

The challenge for researchers in the coming years will be to produce oil crops with higher yields to satisfy increased demands and also to increase the spectrum of useful products, whether for edible or industrial use, that can be derived from these crops. To date, the vast majority of research and development activities has focused on improving existing crops and most notably the "big four" oil crops as above. More recently, there has been considerable interest in using recombinant DNA technology to transfer genes from other oil-producing species into the major oil crops, in order to extend the range of fatty acids produced by such crops. Until now, relatively little effort has gone into the domestication of completely new oil-producing species. Partly, this has been due to the considerable agronomic problems faced in adapting a wild or semi-wild species for large scale agricultural production. Nevertheless, recent advances in plant science, which have been considerably assisted by technical developments related to the biotechnology industry, now make it a realistic option to consider domestication as an alternative to transgenic technology, in order to produce novel oils for future generations (Murphy 1998).

POTENTIAL AND LIMITATIONS OF TRANSGENIC TECHNOLOGY

The manipulation of seed oil content via transgene insertion has been one of the earliest successful applications of modern biotechnology in agriculture. For example, the first transgenic crop with a modified seed composition to be approved for unrestricted commercial cultivation in the US was a lauric oil, rapeseed, grown in 1995 (Murphy 1996). Many of the major agricultural companies have considerable investments in transgene technology, particularly as applied to oil crops. One of the major drivers for such investment was a perception that oil quality is a relatively plastic phenotype and that substantial changes could be effected by the insertion of one or two key genes. In addition, one of the major oil crops, rapeseed, turned out to be particularly amenable to tissue culture and regeneration and could therefore be transformed relatively easily. More recently, soybean transformation has also become relatively routine (at least within companies) albeit much more labor intensive than rapeseed transformation (Krebbers et al. 1997). There were also some significant early successes, mostly notably the achievement of 40% to 60% lauric content in rapeseed oil, which normally accumulates little or no lauric acid (Murphy, 1996). Nevertheless, attempts to achieve high levels of other novel fatty acids in seed oils have met with much less success and there have been several reports that the presence of novel fatty acids in transgenic plants can sometimes lead to the induction of catabolic pathways which break down the novel fatty acid, i.e. the plant recognizes the "strange" fatty acid and, far from tolerating it, may even actively eliminate it from the seed oil (Ecclestone and Ohlrogge 1998; Murphy et al. 1999).

During the past two years there has also been an increasing recognition that the metabolic pathways involved in seed oil biosynthesis are considerably more complex than was first appreciated. For example, several new enzymes have been discovered which may be involved in remodeling oil, even after it has been deposited into storage oil bodies (Stobart et al. 1997; Mancha and Stymne 1997). It has proved extremely difficult to achieve the high levels (70 to 90%) of a particular fatty acid in the seed oil that is often important for its commercial viability. One of the key problems faced in the production of novel fatty acids in transgenic plants is that the major oil crops, such as rapeseed, may lack mechanisms which are able to channel such fatty acids towards storage oil and away from accumulation in membrane lipids. It is likely that the presence of even relatively small amounts of certain unusual fatty acids in membrane lipids could cause serious disruption to important cellular processes. It is possible that such plants seek to avoid these problems by identifying and breaking down novel fatty acids, such as petroselinic acid as has recently been shown in our laboratory (Murphy et al. 1999).

In addition to the technical problems associated with producing novel oils in transgenic crops, it has recently been pointed out that there are considerable challenges involved in the management of such crops (Murphy 1999). For example, at present there are at least 10 to 20 different transgenic cultivars of rapeseed at various stages of development, both in the laboratory and in field trials. All of these cultivars appear identical and the only differences are in their seed oil compositions. It is these different seed oil profiles that determines that one cultivar may be used for detergent manufacture, while other cultivars may be used for margarines, cosmetics, therapeutic agents, or lubricants. Clearly, the segregation and identity preservation of such mutually incompatible commodity streams raises formidable challenges at all levels of production, ranging from sowing, harvesting, storage, crushing, and processing. The expensive and sophisticated analytical equipment required to differentiate between the different transgenic cultivars will not be available to all growers, crushers or processors and therefore there is serious potential for mixing and cross-contamination of different seed lots.

In many countries, particularly in Europe, there is also a considerable consumer resistance to the cultivation of genetically modified crops for food use. For example, France has already imposed a three year moratorium on the release of genetically modified crops and the UK is considering a similar moratorium at present. Although this situation may well change with better public education about genetic research and with more thorough risk assessment programs, consumer resistance is likely to be an important factor which may limit the application of transgene technology in the near future.

A final argument against an ever increasing reliance on a very small number of major crops is the concern that such large scale monocultures may be more prone to opportunistic infection by pests and diseases, as well as reducing biodiversity at the farm level. It is official policy of the European Union to encourage greater crop diversity and therefore to favor the introduction of new crops rather than the continued increase in cultivation of existing major crop species.

DOMESTICATION OF NEW CROP SPECIES

All of our existing major crop species have been through a continual process of domestication and improvement since the beginnings of agriculture more than ten thousand years ago. However, research aimed at the domestication of new species rarely finds favor with the agribusiness industry or with Government funding agencies. The process of domestication is perceived to be extremely slow. It is also obviously limited by the climatic range of the candidate crop. For example, some potential novel oil crops are found in tropical regions and therefore are unlikely ever to be domesticated for cultivation in temperate regions. Nevertheless, given the extraordinary diversity of seed oil contents found in the natural world, it is likely that a species producing useful quantities of a particular seed oil can be found in several different climatic zones. Another problem is that the novel crops may have a different growth habit to existing crops and may therefore not be suitable for harvesting using existing equipment. These problems are added to the, often serious, agronomic difficulties exhibited by many candidate oilseed plants. Nevertheless, we should not be discouraged by such challenges. The application of modern biotechnological methods such as genome mapping and molecular marker assisted selection, now make it feasible to consider domestication of such species within one or two decades (Martin 1998; Murphy 1998). This time horizon is well within the normal development lifetime of most new pharmaceutical products or large civil engineering projects.

There are many agronomic problems faced by new oil crops. Among the most common of these are asynchronous flowering, premature pod shattering, allogamy, low seed oil content, poor germination rates, and low seed yield. Until recently, the solution to such problems was to look for natural variation, e.g. reduced pod shattering or increased oil yield and to attempt to produce reasonably defined lines in which these characters were well expressed. The production of more uniform inbred lines has been assisted by techniques such as double haploids and by the development molecular maps for at least some of the major species. In the future, new kinds of molecular markers, such as microsatellites, promise to make the process of producing a detailed genetic map for any species much more rapid and much less expensive than in the past.

The recent advances in genomics and in gene function studies has allowed us to understand the detailed genetic basis of many complex traits, such as flowering time, height, and disease resistance (Murphy 1998). Many of these complex traits had been regarded as being controlled by large numbers of genes, which made them difficult to manipulate by simple Mendelian genetics. However, there are now several striking examples where the vast majority of the variation underlying such complex characters has been mapped to only a very small number of quantitative trait loci (QTL) (Doebley 1993; Martin 1998). Once such genes have been mapped in a model species such as *Arabidopsis*, techniques such as positional cloning can be used to isolate the gene of interest and to verify its function in the laboratory. Within the next few years, more and more important agronomic traits will be explicable in terms of a relatively small number of key genes which account for most of the observed variation. Such information can then be used for the selection of plants expressing such genes. For example, in our own laboratories we are currently studying genes regulating characters such as pod shattering, oil yield, oil quality, flowering and canopy architecture. Such research has the potential to provide tools for the much more rapid domestication of new oil crops within the next few years.

CONCLUSIONS

It is likely that, in the future, transgenic oil crops and newly domesticated oil crops will both be developed in order to provide the increased amount and diversity of oils which will be required for both edible and industrial use. It is important that we recognize that both approaches have both positive and negative points. It will be a combination of these two strategies that is most likely to supply the increasing demands for plant oils in the 21st century and beyond.

REFERENCES

Doebley, J. 1993. Genetics, development and plant evolution. Current Opinion in Genetic Development 3:865–872

Eccleston, V.S. and J.B. Ohlrogge. 1998. Expression of lauroyl-acyl carrier protein thioesterase in *Brassica napus* seeds induces pathways for both fatty acid oxidation and biosynthesis and implies a set point for triacylglycerol accumulation. Plant Cell. 10:613–621.

Kerr, R.A. 1998. The next oil crisis looms large—and perhaps close. Science 281:1128–1131.

Krebbers, et al. 1997. Biotechnological approaches to altering seed composition. p. 595–633. In: B.A. Larkins and I.K. Vasil (eds.), Cellular and molecular biology of plant seed development. Kluwer Academic Publ., The Netherlands.

Mancha, M. and S. Stymne. 1997. Remodeling of triacylglycerols in microsomal preparations from developing castor bean (*Ricinus communis* L.) endosperm. Planta 203:51–57.

Martin, G.B. 1998. Gene discovery for crop improvement. Current Opinion in Biotechnology 9:220–226.

Murphy, D.J. 1996. Engineering oil production in rapeseed and other oil crops. TIBTECH 14:206–213.

Murphy, D.J. 1998. Impact of genomics on improving the quality of agricultural products. p. 199–210. In: G.K. Dixon, L.G. Copping, and D. Livingstone (eds.), Genomics: Commercial opportunities from a scientific revolution. Society of Chemical Industry.

Murphy, D.J. 1999. Production of novel oils in plants. Current Opinion Biotech. 10:175–180.

Murphy, D.J., D.J. Fairbairn, and S. Bowra. 1999. Expression of unusual fatty acids in transgenic rapeseed causes induction of glyoxylate cycle genes. John Innes Centre Annual Report. In press.

Stobart, K., M. Mancha, M. Lenman, A. Dahlquist, and S. Stymne. 1997. Triacylglycerols are synthesised and utilised by transacylation reactions in microsomal preparations of developing safflower (*Carthamus tinctorius* L.) seeds. Planta 203:58–66.

Producing Wax Esters in Transgenic Plants by Expression of Genes Derived from Jojoba

Michael W. Lassner, Kathryn Lardizabal, and James G. Metz*

Jojoba [*Simmondsia chinensis*, Simmondsiaceae, (Link) Schnieder], a native of deserts of the American Southwest, is unusual and perhaps unique amongst higher plants in that its seed storage lipids are liquid waxes rather than triglycerides. These waxes are esters of long chain (mostly; C20, C22, and C24), monounsaturated, fatty acids and alcohols. The waxes are produced in developing embryos during seed formation. Jojoba oil is primarily used as an ingredient in cosmetics. The stability of the oil under high temperatures and pressures suggests it has potential as a component of industrial lubricants, however, the high price of jojoba oil has served as a barrier to its widespread use for these types of applications. The production of liquid waxes in an oilseed crop more suited to large scale agriculture could allow the production of wax esters at a price more acceptable to industrial users.

As in other plants, jojoba fatty acids up to 18 carbons in length are synthesized in plastids. Oleic acid (18:1) is exported from plastids, and its CoA esters serve as substrates for microsomal membrane enzymes involved in wax formation. Three enzyme activities are involved in the formation of waxes (Fig.1, Pollard et al. 1979):

1. Fatty acid elongase (FAE) is a complex of four enzymes that converts 18:1 to eicosenoic acid (20:1), erucic acid (22:1), and nervonic acid (24:1).
2. An acyl-CoA reductase catalyzes the formation of fatty alcohols from 20:1-, 22:1-, and 24:1-CoAs.
3. Wax synthase is an acyltransferase that forms wax esters from fatty alcohol and acyl-CoA substrates.

In this manuscript we describe the isolation of genes encoding the enzymes involved in wax synthesis and the use of these genes to produce wax esters in transgenic arabidopsis plants.

FATTY ACID ELONGASE

The membrane-associated nature of the elongation enzymes has hindered investigation of their biochemistry. As in animal systems (Bernert and Sprecher 1979), fatty acid elongation in plants is believed to occur by way of a four-step mechanism that is similar to fatty acid synthesis, except that CoA, rather than ACP, is the acyl carrier (Cassagne et al. 1994b; Fehling and Mukherjee 1991; Stumpf and Pollard 1983). The first step in the cycle involves condensation of malonyl-CoA with a long-chain acyl-CoA to yield carbon dioxide and a ß-ketoacyl-CoA in which the acyl moiety has been elongated by two carbons. The enzyme that catalyzes this reaction is ß-ketoacyl-CoA synthase (KCS). Subsequent reactions are: reduction to ß-hydroxyacyl-CoA, dehydration to an enoyl-CoA, and a second reduction to yield the elongated acyl-CoA. In both mammalian and plant systems where the relative activities of the four enzymes have been studied, the initial condensation reaction is the rate-limiting step (Suneja et al. 1991; Cassagne et al. 1994a). Thus, the overexpression of KCS can increase the quantity of very long chain fatty acids and increase the substrates available for wax formation.

We previously described the isolation of a jojoba gene encoding the KCS associated with fatty acid elongation in developing seeds, and showed that expression of this gene complimented the canola (*Brassica napus*) mutation which knocked out the production of very long chain fatty acids in the seed oil. Additionally, the influence of the jojoba KCS substrate preferences was evident from the appearance of up to 5% nervonic acid in the transgenic oil—nervonic acid is present in jojoba oil but is found at very low levels in rapeseed oil (Lassner et al. 1996). More recently, we isolated the corresponding gene from seeds of *Lunaria annua*, Brassicaceae. *Lunaria* seed oil is rich in nervonic acid. Expression of the *Lunaria* KCS in transgenic *Brassica* resulted in rapeseed oils with greater than 25% nervonic acid (Lassner 1997; M.W. Lassner unpubl. data).

*The authors thank Bill Hutton for analyzing the *Arabidopsis* seed by ^{13}C NMR. Brenda Reed and Theresa Grunder grew the transgenic plants. JoAnn Turner was responsible for *Brassica* transformation. Tom Hayes and Jaques Fayet-Faber helped with GC analysis of seed oils. Bill Schreckengost and Jason Fenner helped with DNA sequencing.

ACYL-COA REDUCTASE

The jojoba fatty acyl-CoA reductase is also associated with membranes in developing embryos. Based on both the wax composition and in vitro assays, the enzyme has a strong preference for very long chain acyl-CoA substrates (24:1 > 22:1 > 20:1) and has very little activity towards 18:1-CoA. The enzyme was purified from a microsomal membrane fraction isolated from developing embryos. We used the detergent, 3-[(3-cholamidopropyl)-dimethylammonio]-1-propane-sulfonate (CHAPS), to solubilize the reductase. Although the presence of high levels of CHAPS in assay solutions resulted in an apparent loss of enzyme activity, this inhibition could be completely reversed by dilution of the detergent to below its CMC. We routinely diluted CHAPS to 0.3% (w/v) in our assays.

The reductase enzyme was purified using three chromatography steps. Blue A agarose chromatography was used to separate the reductase activity from a majority of the proteins. Size exclusion chromatography on Sephacryl S-100 HR did not result in an increase in specific activity, but did improve subsequent chromatography. The final purification step was affinity chromatography in which the reductase was bound to palmitoyl-CoA agarose and eluted using NADPH. Two polypeptides with apparent molecular masses of 54 and 56 kDa were present in the NADPH eluted fractions from the palmitoyl-CoA column. Several peptides, generated using either trypsin or cyanogen bromide, were isolated from the two proteins and N-terminal sequences were determined. All of the sequences obtained from peptides of the 54 kD protein were found in peptides isolated from the 56 kD polypeptide. Subsequent immunoblot analysis of a jojoba embryo cell free extract showed that only the 56 kD polypeptide was present. Thus, the smaller 54 kDa polypeptide appeared to be an artifact of protein purification.

The peptide sequences were used to design oligonucleotide primers, and this enabled PCR amplification of a partial cDNA from jojoba embryo tissue. The PCR product was used to isolate a full-length cDNA clone. The cDNA clone was expressed in *E. coli*. Extracts from the transformed *E. coli* gained acyl-CoA reductase enzyme activity when compared to the control extracts, and analysis of the lipids revealed that the transformed *E. coli* had increased levels of fatty alcohols (Fig.2). This demonstrated that we had indeed isolated the gene encoding the acyl-CoA reductase responsible for fatty alcohol production.

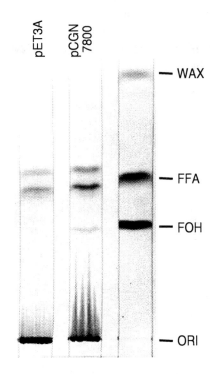

Fig. 2. TLC analysis of *E. Coli* expressing the jojoba acyl-CoA reductase. Lipids were extracted from *E. coli* in hexane:isopropanol (3:2) and separated by normal phase TLC. The plasmid pCGN7800 contains the acyl-CoA reductase under control of a T7 polymerase promoter. Plasmid pET3A is the vector used for expression of the reductase. The lipids extracted from *E. coli* containing pCGN8500 contain fatty alcohols not found in the lipids from *E. coli* containing the empty vector. Abreviations: FFA, free fatty acid; FOH, fatty alcohol; and ORI, origin.

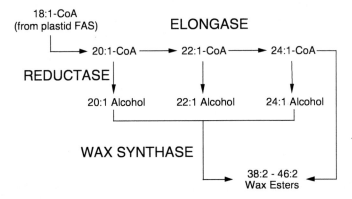

Fig. 1. The wax biosynthetic pathway in jojoba. Three enzyme activities compose the wax synthesis pathway: fatty acid elongase, acyl-CoA reductase, and wax synthase. The relative activities of the three enzymes determines the chain lengths of the waxes found in jojoba oil.

221

The reductase cDNA was placed under the control of oleosin regulatory sequences. Expression of the acyl-CoA reductase gene in transgenic high erucic acid rapeseed (HEAR) directed the synthesis of low levels of fatty alcohol in the seed oil. The best transgenic line had 0.4% alcohol. Analysis of the nucleotide composition of the reductase gene showed that the A+T content of the gene exceeded 60% in many portions of the gene (determined by examining the A+T content of windows of 25 nucleotides). The gene was resynthesized to reduce the A+T composition of the gene to between 40% and 50% in all 25 nucleotide segments. Seed oils extracted from individual seeds from HEAR transformed with the resynthesized reductase gene now contained up to 4.4% (by weight) fatty alcohol. Interestingly, a portion of these fatty alcohols were esterified to fatty acids, forming wax esters. The wax content of the seed oil ranged up to 8.5% weight percent of the seed oil. Apparently an endogenous wax synthase in the *Brassica* embryos could convert some of the alcohol to wax esters. Enzyme assays showed that low levels of wax synthase activity could be detected in cell free extracts of *Brassica* embryos. The wax content was variable and ranged from less than the alcohol content to twice the alcohol content (full conversion of alcohol to wax). Seeds with the highest alcohol content germinated poorly. Since high concentrations of fatty alcohols are likely to be toxic to cells, we postulated that the addition of an efficient wax synthase activity would allow the generation of transgenic plants with higher alcohol and wax content.

WAX SYNTHASE AND PRODUCTION OF WAX ESTERS IN TRANSGENIC *ARABIDOPSIS*

Initial attempts to solubilize the jojoba embryo wax synthase using conditions developed for the fatty acyl-CoA reductase were unsuccessful. CHAPS was used as the detergent, however a higher detergent to protein ratio was required to achieve solubilization and higher concentrations of CHAPS in the chromatography buffers was needed to prevent aggregation. Additionally, after treatment with the detergent, enzyme ac-

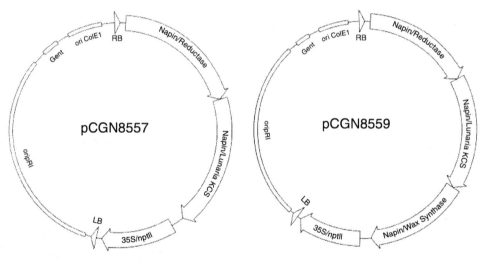

Fig. 3. Plasmids used to express wax ester biosynthetic genes into *Arabidopsis*. The plasmids pCGN8557 and pCGN8559 were introduced into *Agrobacterium tumefaciens*. These *Agrobacterium* strains were used to transform *Arabidopsis thaliana*.

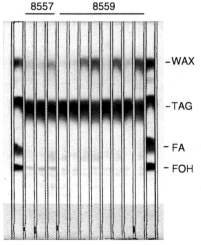

Fig. 4. TLC analysis of *Arabidopsis* seed oil from plants transformed with jojoba wax synthesis genes. Lipids were extracted from transgenic *Arabidopsis* seeds in hexane and separated by normal phase TLC. Plasmid pCGN8557, the control plasmid, contains KCS and acyl-CoA reductase genes. Plasmid pCGN8559 contains the same two genes with the addition of a wax synthase gene. Seed oils from plants transformed with pCGN8559 contain higher quantities of wax esters than. untransformed plants and plants transformed with pCGN8557. Abbreviations: TAG, triglyceride, FA, free fatty acid; and FOH, fatty alcohol.

tivity was dependent upon reconstitution into phospholipid vesicles. The wax synthase was partially purified by chromatography on Blue A agarose, size exclusion, and hydroxyapatite columns. A 33 kDA protein was identified whose abundance correlated with the amount of enzyme activity present in fractions from multiple chromatographic separations. This protein was isolated by SDS-PAGE, and digested with trypsin. The sequences of several peptides were determined. The peptide sequences were used to design oligonucleotide primers, and a PCR product representing a partial cDNA was isolated. DNA sequence of the PCR product enabled the cloning of the 5' and 3' ends of the cDNA via RACE (Frohman et al. 1988).

Two plasmids were constructed for plant transformation (Fig. 3). One construct, pCGN8557, contained the *Lunaria* KCS and the jojoba acyl-CoA reductase under control of napin regulatory sequences. The second construct contained the *Lunaria* KCS, the jojoba acyl-CoA reductase and wax synthase candidate under control of napin regulatory sequences. Napin is a *Brassica* seed storage protein and the napin regulatory sequences drive high level expression of the associated transgenes in embryos of transgenic plants (Kridl et al. 1991). The two plasmids were introduced in *Agrobacterium tumefaciens*, and subsequently transformed into *Arabidopsis thaliana*.

Immature seeds were dissected from the transgenic plants, and wax synthase enzyme assays were performed. Seed extracts from most of the plants transformed with pCGN8559 exhibited wax synthase activity, while very little activity was detected in the seed extracts from pCGN8557 plants. This demonstrated that we had isolated the gene encoding wax synthase.

TLC analysis showed much higher levels of wax esters in the oil derived from pCGN8559 plants than present in oil derived from pCGN8557 plants (Fig. 4). Gas chromatography analysis of transmethylated oil showed that up to 16 weight percent of the oil from the transgenic pCGN8559 plants was fatty alcohols. If the wax synthase was efficient at converting fatty alcohol to wax esters, this fatty alcohol composition data suggested that the oil was comprised of 30 weight percent wax. Another method of analyzing the wax content is ^{13}C NMR. Fig. 5 shows the region of the NMR spectrum that indicates the glycerol backbone carbons of triglyceride and the #1 carbon of the alcohol moiety of wax esters. NMR analysis suggested that the oil was comprised of approximately 50 mole percent (approximately 40% by weight) of wax.

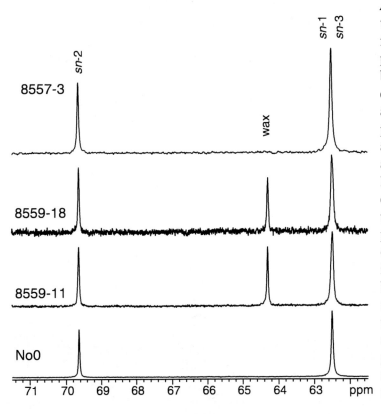

Fig. 5. ^{13}C NMR analysis of seeds from *Arabidopsis* plants transformed with jojoba wax synthesis genes. NMR analysis was performed on whole seeds from *Arabidopsis* plants. The peak at 62.4 ppm represents carbons at the *sn*-1 and *sn*-3 positions of the glycerol backbone of triglycerides, while the peak at 69.6 ppm represents carbons in the sn-2 position of the glycerol backbone. The peak at 64.3 ppm represents the #1 carbon of the alcohol moiety of wax esters. Since each triglyceride has one *sn*-2 glycerol carbon, and each wax molecule has one alcohol molecule, the relative areas of the 64.3 and 69.6 ppm peaks provide an estimate of the molar ratios of wax esters and triglycerides, respectively, in the oil samples. The chromatograms labeled 8559-11 and 8559-18 show the analysis of seeds from two independent pCGN8559 transformants, the chromatogram labeled 8557 shows the analysis of seeds from a pCGN8557 plant, and the chromatogram labeled NoO shows the analysis of seeds from an untransformed control plant.

CONCLUSIONS

The work described in this manuscript shows that we have isolated a number of key genes involved in jojoba wax biosynthesis. Data from plant genome projects which unveil orthologous genes in other plant species may lead to understanding how jojoba evolved to use wax esters as their seed storage lipids. The levels of wax achieved in transgenic *Arabidopsis* are quite promising—50 mole % of the seed oil is wax in the transgenic pCGN8559 plant with the best phenoytpe. The analyzed seeds of this plant were segregating for the presence of the transgene and the wax content of homozygous plants would be expected to be higher. Currently, the transgenes are being introduced into high erucic acid varieties of rapeseed. The development of transgenic rapeseed varieties could provide a new source of wax esters with a lower price than the liquid wax produced in jojoba.

REFERENCES

Bernert, J.T. and H. Sprecher. 1979. The isolation of acyl-CoA derivatives as products of partial reactions in the microsomal chain elongation of fatty acids. Biochim. Biophys. Acta 573:436–442.

Cassagne. C., J.-J. Bessoule, F. Schneider, R. Lessire, B. Sturbois, P. Moreau, and C. Spinner. 1994a. Modulation of the very-long-chain fatty acid (VLCFA) formation in leek. p. 111–114. In: J.K Kader and P. Mazliak (eds.), Plant lipid metabolism. Kluwer Academic Publ. Dordecht, the Netherlands.

Cassagne, C., R. Lessire, J.-J. Bessoule, P. Moreau, A. Creach, F. Schneider, and B. Sturbois. 1994b. Biosynthesis of very long chain fatty acids in higher plants. Prog. Lipid Res. 33:55–69.

Fehling, E. and K.D. Mukherjee. 1991. Acyl-CoA elongase from a higher plant (*Lunaria annua*): metabolic intermediates of very-long-chain acyl-CoA products and substrate specificity. Biochim. Biophys. Acta 1082:239–246.

Frohman, M.A., M.K. Dush, and G.R. Martin. 1988. Rapid production of full length cDNAs from rare transcripts: Amplification using a single gene-specific oligonucleotide primer. Proc. Nat. Acad. Sci. (USA) 85:8998–9002.

Kridl J.C., D.W. McCarter, R.E. Rose, D.E. Scherer, D.S. Knutzon, S.E. Radke, and V.C. Knauf. 1991. Isolation and characterization of an expressed napin gene from *Brassica rapa*. Seed Sci. Res. 1:209–219.

Lassner M.W., K. Lardizabal, and J.G. Metz. 1996. A jojoba ß-ketoacyl-CoA synthase cDNA complements the canola fatty acid elongation mutation in transgenic plants. Plant Cell 8:281–292.

Lassner, M. 1997. Transgenic oilseed crops: A transition from basic research to product development. Lipid Technol. 9:5–9.

Pollard, M.R., T. McKeon, L.M. Gupta, and P.K. Stumpf. 1979. Studies on biosynthesis of waxes by developing jojoba seed. II. The demonstration of wax biosynthesis by cell-free homogenates. Lipids 7:651–662.

Stumpf, P.K. and M.R. Pollard. 1983. Pathways of fatty acid biosynthesis in higher plants with particular reference to developing rapeseed. p. 131–141. In: J.K. Kramer, F. Sauer, and W.J. Pigden (eds.), High and low erucic acid rapeseed oils. Academic Press, New York.

Suneja, S.K., M.N. Nagi, L. Cook, and D.L. Cinti. 1991. Decreased long-chain fatty acyl CoA elongation activity in quaking and jumpy mouse brain: Deficiency in one enzyme or multiple enyme activities? J. Neurochem. 57:140–146.

Breeding Advances and Germplasm Resources in Meadowfoam: A Novel Very Long Chain Oilseed

Steven J. Knapp and Jimmie M. Crane*

Meadowfoam (*Limnanthes alba* Benth., Limnanthaceae) produces a very-long chain seed oil with novel physical and chemical characteristics (Earle et al. 1959; Smith et al. 1960; Bagby et al. 1961; Miller et al. 1964; Isbell 1997). These characteristics have propelled the development of meadowfoam as a specialty oilseed crop and the development of industrial markets for meadowfoam oil and oil derivatives.

Meadowfoam has been produced on a limited scale for more than 20 years in Oregon (Fig. 1). Markets for the oil began emerging in 1992. Production has steadily increased with progress in marketing the oil and the development and marketing of a variety of specialty chemicals from meadowfoam fatty acids and triglycerides (Isbell 1997). Oregon growers have ensured a supply of oil for the meadowfoam industry.

Meadowfoam has an important role as a rotation crop in the Willamette Valley of Oregon and is primarily grown by grass seed producers. The rotational fit of meadowfoam is excellent, particularly for growers with weed or other pest problems in grass seed production fields. Weed control and other crop rotation benefits have stimulated production in Oregon. Although miniscule on a global scale, meadowfoam has had a significant economic impact in Oregon (seed production), California (seed and oil processing), and Illinois (marketing and product development). The number of hectares of meadowfoam produced outside Oregon is not publicly recorded. Meadowfoam has been produced on a pilot scale in New Zealand and Virginia (US) and field tests have been carried out in the United Kingdom, the Netherlands, France, and elsewhere.

The development of the meadowfoam industry has been impeded by low and erratic seed yields. Although meadowfoam production has steadily increased since 1993 (Fig. 1), meadowfoam seed yields have not increased and are not much greater today than they were 10 to 15 years ago in Oregon. Significant seed yield increases are needed to solidify and sustain the growth of the meadowfoam industry.

The goal of our research has been to increase the productivity of meadowfoam by developing superior cultivars, discovering and developing novel phenotypes, and advancing our understanding of the genetics of economically important traits. Our breeding work has concentrated on increasing seed yield, seed oil concentration, lodging resistance, and *Scaptomyza* resistance, developing self-pollinated lines and cultivars, and developing novel oils. We review recent progress in meadowfoam breeding, describe meadowfoam germplasm resources, and review breeding and genetics research needs for meadowfoam.

EARLY GERMPLASM SCREENING AND SPECIES SELECTION

The development of meadowfoam as a crop began with the discovery of the novel fatty acid profile of the seed oil (Earle et al. 1959; Smith et al. 1960). The oil has four key characteristics that have stimulated the development of the meadowfoam industry. The oil has (1) very high concentrations of very-long-chain (C_{20} and C_{22}) fatty acids (~980 g/kg), (2) very low concentrations of saturated fatty acids (~10 g/kg), (3) very high concentrations of $\Delta 5$ double bonds (~840 g/kg), and (4) excellent oxidative stability (Earle et al. 1959; Smith et al. 1960;

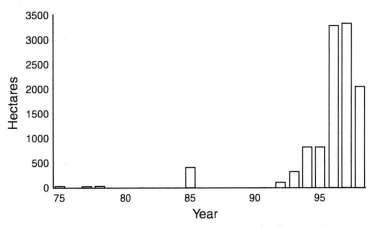

Fig. 1. Hectares of meadowfoam produced in Oregon between 1974–75 and 1998–99.

*This research was funded by grants from the USDA (#58-5114-8-1021 and #58-3620-8-107) and the Fanning Corporation. Oregon Agr. Exp. Stat. Tech. Paper No. 11,445.

Bagby et al. 1961; Miller et al. 1964; Isbell 1997).

The early germplasm screening and species selection work produced several key findings that have since dictated the course of meadowfoam research and development. *First*, the novel fatty acids of meadowfoam are found throughout the genus in similar proportions and concentrations (Bagby et al. 1961; Earle et al. 1959; Miller et al. 1964; Smith et al. 1960). This discovery showed that novel oils were produced by all of the known species and that species could be selected for commercial development on the basis of agronomic performance alone. *Second*, *L. alba* and *L. douglasii* were shown to produce greater biomass and seed yields than other meadowfoam species (Gentry and Miller 1965; Higgins et al. 1971; Pierce and Jain 1977; Jain et al. 1977; Brown and Jain 1979; Brown et al. 1979; Jain and Abeulgasim 1981; Krebs and Jain 1985; Jain 1989). This discovery narrowed the focus of crop development research to these two species. *Third*, the lack of non-shattering phenotypes in *L. douglasii* and presence of non-shattering phenotypes in *L. alba* (Jain 1989; Dole and Jain 1992) narrowed the focus of cultivar development research in Oregon to *L. alba* and led to the development of Foamore (PI 543894), the first non-shattering cultivar (Calhoun and Crane 1974).

GERMPLASM RESOURCES

Limnanthes (meadowfoam) is a genus of 17 species and subspecies belonging to two Sections (Inflexae and Reflexae) of the Limnanthaceae (Mason 1952) (Table 1). The USDA maintains germplasm for several of these species. Our laboratory maintains an *L. alba* germplasm collection and concentrates on the development of *L. alba* germplasm and cultivars. *L. alba* is a predominantly allogamous, self-compatible, diploid (x = 5) in Section Inflexae (Mason 1952; Jain 1978).

Two inter-fertile *L. alba* subspecies have been described (*L. alba* ssp. *alba* and *L. alba* ssp. *versicolor*). Partially fertile inter-specific hybrids can be made between *L. alba* and secondary gene pool species of Section Inflexae (e.g., *L. gracilis* and *L. floccosa*) (Table 1). Hybrids between *L. alba* and tertiary gene pool species of Section Reflexae species (e.g., *L. douglasii*) are sterile. The genetic diversity of several related species can be accessed through inter-specific crosses. We assigned species and subspecies to primary, secondary, and tertiary gene (germplasm) pools on the basis of the fertility of inter-subspecific and inter-specific hybrids with *L. alba* (Table 1).

The *primary gene pool* of *L. alba* is comprised of *L. alba* ssp. *alba* and *L. alba* ssp. *versicolor*. Crosses between these subspecies are fertile and seem to undergo normal meioses (Mason 1952; Knapp and Crane 1998, unpubl. data). Partially fertile progenies have been produced from crosses between *L. alba* and all of the species we assigned to the secondary gene pool (Table 1). Fertile progenies have not been produced from crosses between *L. alba* and the species we assigned to the tertiary gene pool. One backcross to *L. alba* greatly increases the fertility of progeny from crosses between primary and secondary gene pool species. Secondary gene

Table 1. Meadowfoam germplasm resources held by the United State Department of Agriculture (USDA) National Plant Germplasm System (NPGS) and the Oregon State University (OSU) Center for Oilseed Research (CORE).

Limnanthes species	No. accessions	Source
Primary gene pool		
L. alba	18	USDA NPGS
L. alba	~ 30	OSU
L. alba ssp. *alba*	4	USDA NPGS
L. alba ssp. *versicolor*	1	USDA NPGS
L. alba ssp. *versicolor*	5	OSU
Secondary gene pool		
L. gracilis	2	USDA NPGS
L. gracilis ssp. *parishii*	1	USDA NPGS
L. gracilis ssp. *gracilis*	1	USDA NPGS
L. floccosa	5	USDA NPGS
L. floccosa ssp. *bellingeriana*	2	USDA NPGS
L. floccosa ssp. *pumila*	2	USDA NPGS
L. floccosa ssp. *grandiflora*	1	USDA NPGS
L. floccosa ssp. *floccosa*	1	OSU
L. floccosa ssp. *californica*	1	OSU
L. montana	1	USDA NPGS
Tertiary gene pool		
L. bakeri	2	USDA NPGS
L. douglasii	3	USDA NPGS
L. douglasii ssp. *douglasii*	1	USDA NPGS
L. douglasii ssp. *nivea*	5	USDA NPGS
L. douglasii ssp. *rosea*	2	USDA NPGS
L. macounii	1	USDA NPGS
L. striata	2	USDA NPGS

pool species, however, are vastly inferior to *L. alba* for agronomically important traits and have limited utility for enhancing meadowfoam seed and oil yield in the short run.

The USDA National Plant Germplasm System (NPGS) holds seed of 23 *L. alba* and 15 secondary gene pool accessions (Table 1). The working collection is maintained by USDA-ARS Plant Introduction Station (Pullman, WA). Passport data on the working collection can be accessed through the Germplasm Resource Information Network or GRIN (http//:www.ars-grin.gov). The OSU Center for Oilseed Research (CORE) presently holds 30 or more *L. alba* accessions, in addition to families and lines from numerous segregating populations and mapping populations (Table 2). The CORE accessions are comprised of unreleased germplasm populations (selected and unselected), recollected wild populations, newly discovered and collected wild populations, unreleased inbred lines, novel genetic stocks, unreleased experimental cultivars, and released cultivars.

The *L. alba* germplasm collection held by the USDA is primarily comprised of wild populations, most of which were collected between 1962 and 1972. These accessions have not been regenerated since being deposited and have slowly and steadily degraded. Although meadowfoam seed can be kept alive in low humidity cold storage for ~15 years, much of the seed in the collection is more than 25 years old. These accessions will eventually be completely lost unless the collection is regenerated or the accessions are recollected and routinely regenerated.

Table 2. Meadowfoam germplasm accessions held by the Oregon State University (OSU) Center for Oilseed Research (CORE).

Accession	Description
Knowles (OMF69)	Released insect-pollinated cultivar
OMF78	Experimental insect-pollinated cultivar
OMF86	Experimental insect-pollinated cultivar
OMF103	Experimental insect-pollinated cultivar
OMF156	Experimental insect-pollinated cultivar
OMF78-C$_1$	Experimental insect-pollinated cultivar
OMF58	Unselected insect-pollinated germplasm population (diverse pedigree)
OMF59	Unselected insect-pollinated germplasm population (diverse pedigree)
OMF157	Unselected insect-pollinated germplasm population (OMF62 × OMF154)
OMF62	High oil insect-pollinated germplasm population
OMF87	High oil insect-pollinated germplasm population
OMF154	Large seeded insect-pollinated germplasm population
OMF-Redding (OMF66)	Wild *L. alba* ssp. *versicolor* population
OMF158	Wild *L. alba* ssp. *versicolor* population (recollected PI 283705)
OMF159	Wild *L. alba* ssp. *versicolor* population (recollected PI 374791)
OMF160	Wild *L. alba* ssp. *versicolor* population (recollected PI 374801)
OMF161	Wild *L. alba* ssp. *versicolor* Population (recollected PI 374802)
OMF63 S$_5$	Self-pollinated inbred line
OMF64 S$_5$	Self-pollinated inbred line
OMF66 S$_5$	Self-pollinated inbred line
OMF104-Bulk S$_1$	Bulk of self-pollinated S$_1$ *L. alba* ssp. *versicolor* progeny
OMF104-Bulk S$_2$	Bulk of self-pollinated S$_2$ *L. alba* ssp. *versicolor* progeny
OMF104-Bulk S$_3$	Bulk of Self-Pollinated S$_3$ *L. alba* ssp. *versicolor* progeny
Mermaid S$_5$	Insect-pollinated inbred line
Globus	Fused sepal mutant
LAG64-1 F$_4$	Self-pollinated *L. alba* × *L. gracilis* ssp. *parishii* inbred line
LAG64-2 F$_4$	Self-pollinated *L. alba* × *L. gracilis* ssp. *parishii* inbred line
LAG64-3 F$_4$	Self-pollinated *L. alba* × *L. gracilis* ssp. *parishii* inbred line

There are three cultivars in the USDA and CORE collections: two *L. alba* cultivars (Foamore and Mermaid) and one *L. alba* ssp. *alba* × *L. floccosa* ssp. *grandiflora* cultivar (Floral). 'Foamore' (PI 543894) (Calhoun and Crane 1975) and 'Mermaid' (PI 601232) (Calhoun and Crane 1984; Jolliff 1986) have been released to the USDA working collection. 'Floral' (PI 562386) (Jolliff 1994) has been deposited in the USDA NPGS National Seed Storage Laboratory (Fort Collins, Colorado), but has not been released to the USDA working collection.

Many wild *L. alba* populations could be extinct. Seed of the wild *L. alba* populations collected by Dr. S.K. Jain (University of California, Davis) and his students over the years (Pierce and Jain 1977) has been lost and many of the original collection sites for *L. alba* have been disturbed by human development. *L. alba* cannot be safely preserved in situ. As noted by Knapp and Crane (1997), the site of the OMF-Redding population was destroyed by human development soon after seed was collected. Several *L. alba* populations have undoubtedly had similar fates over the last 20 years in California.

Six wild *L. alba* populations from the University of California, Davis meadowfoam collection have been used in our breeding work. We received seed samples of four wild *L. alba* ssp. *alba* populations (UC-302, UC-305, UC-308, UC-312) from Dr. Jain in 1987. These were bulked and intermated under cages at Corvallis, Oregon in 1988–89. We used the bulk population (UC-Bulk) to develop OMF58. We received seed of two additional *L. alba* accessions (UC-Sonoma and UC-Calaveras) from Dr. Jain in 1988. These accessions were intermated en masse with 21 *L. alba* accessions in an isolated field at Corvallis, OR in 1992 to develop OMF59 (Table 2). Pierce and Jain (1977) described the origin of some of the UC accessions. The geographical origins of several accessions are not known. Most of the collected diversity of *L. alba* was pooled in OMF58 and OMF59.

A census of wild populations has not been taken for many years, the number and distribution of present day wild populations is not known, and germplasm has not been systematically collected from the wild for many years. Historical records (Mason 1952; Pierce and Jain 1977; www.ars-grin.gov) show that germplasm has apparently not been collected from parts of the natural range of *L. alba*, particularly along the easterly margins of the range along the Sierra Nevada foothills.

Several *L. alba* ssp. *versicolor* populations were collected by Jimmie Crane and Daryl Ehrensing in May of 1993 from sites near Redding, California. They collected seed from sites (populations) of four previously collected populations (PI 283705, PI 374791, PI 374801, and PI 374802) and from a new *L. alba* var. *versicolor* population (OMF-Redding or OMF66) near the other four sites. Seed of the recollected populations (OMF158, OMF159, OMF160, and OMF161) is held in the CORE meadowfoam collection (Table 2). OMF-Redding was found to be a source of self-pollinated phenotypes (Knapp and Crane 1997). Seed of OMF-Redding, OMF158, OMF159, OMF160, and OMF161 has not yet been regenerated or deposited in the NPGS working collection (Table 2).

Self-pollinated inbred lines have been developed from OMF66 and from two other wild *L. alba* ssp. *versicolor* populations (PI374791 and PI374801) (Knapp and Crane 1997). These inbred lines (OMF63-S_5, OMF64-S_5, and OMF66-S_5) are being increased for release as enhanced germplasm. Seed of OMF63-S_5, OMF64-S_5, and OMF66-S_5 will be sent to the working collection in July of 1999. We are presently developing a diverse self-pollinated population (OMF104-Bulk) from a bulk of wild *L. alba* ssp. *versicolor* populations. Two cycles of inbreeding and mass selection for self-pollination have been completed in this population. The third cycle will be completed in 1998–99. OMF104-Bulk S_3 seed will be released in July of 1999.

INSECT-POLLINATED GERMPLASM AND CULTIVAR DEVELOPMENT

The development of enhanced insect-pollinated germplasm is the core of our breeding program. Aside from partially self-pollinated wild populations and strongly self-pollinated lines (Knapp and Crane 1997), most *L. alba* germplasm is allogamous and insect-pollinated and most of the genetic diversity of this species resides in insect-pollinated populations. We have primarily used recurrent mass, half-sib family, and S_1 family selection in these populations with a strong focus on increasing seed yield and lodging resistance.

Meadowfoam germplasm screening began in 1966 and cultivar development began in 1971 at Oregon State University. 'Foamore', the first cultivar developed for this crop, was released in 1974 (Calhoun and

Crane 1975). 'Foamore' was eclipsed by 'Mermaid' (Calhoun and Crane 1984; Jolliff 1985) and 'Mermaid' was eclipsed by 'Floral' (Jolliff 1994). 'Foamore' and 'Mermaid' were developed by mass selection for seed yield in wild *L. alba* ssp. *alba* populations, while 'Floral' was developed by mass selection for seed yield in an interspecific population (*L. alba* ssp. *alba* × *L. floccosa* ssp. *grandiflora*). All three cultivars are insect-pollinated. 'Floral' is difficult to distinguish from *L. alba* and has few if any obvious *L. floccosa* ssp. *grandiflora* traits. The latter is a diminutive species with very low biomass. 'Floral' (Jolliff 1994) was the primary cultivar grown in commercial fields in Oregon between 1993 and 1998. Over 3000 ha of 'Floral' were grown in Oregon in 1997–98. 'Foamore' and 'Mermaid' have not been grown in Oregon for several years.

Several new experimental insect-pollinated cultivars have been developed from the breeding program we initiated in 1991. 'Knowles', OMF78, and OMF86 were the first of these. 'Knowles' (tested as OSU-EXP-OMF69), a newly released insect-pollinated cultivar (Crane and Knapp 1999; Knapp and Crane 1999), was developed by one cycle of recurrent half-sib family selection for increased seed yield, seed oil concentration, and lodging resistance in OMF58. OMF78 was developed by one cycle of recurrent half-sib family selection for increased seed yield and seed oil concentration in OMF59. OMF86 was developed by two cycles of recurrent half-sib family selection for increased seed yield, seed oil concentration, and lodging resistance in OMF58. OMF78 and OMF86 have not yet been officially released.

'Knowles', OMF78, OMF8, 'Mermaid', and 'Floral' were grown in replicated yield trials at Hyslop Farm (Corvallis, OR) in 1996–97 and 1997–98. The mean seed yield of 'Floral' was 783 kg/ha versus 688 kg/ha for 'Mermaid'. The mean seed yield for 'Knowles' and OMF78 were 1,014 kg/ha and 1,016 kg/ha, respectively. The seed yields of 'Knowles' and OMF78 were significantly greater than the seed yields of 'Mermaid' and 'Floral' ($LSD_{0.05}$ = 83 kg/ha), but still may not be great enough to dramatically affect the price of meadowfoam oil. Over 1000 ha of 'Knowles' were planted in Oregon in October of 1998. 'Knowles' and OMF78 are presently being tested at four sites in the western US (Davis, California; Medford and Corvallis, Oregon; and Mt. Vernon, Washington).

OMF86 has the strongest upright growth habit and the greatest lodging resistance of any cultivar we have developed or tested thus far. The seed yield of OMF86 was not significantly different from Knowles in 1996–97 (1,222 versus 1,200 kg/ha), but was significantly less than Knowles in 1997–98 (713 versus 829 kg/ha) ($LSD_{0.05}$ = 83 kg/ha). Because of this, OMF86 may not be released as a new cultivar.

The first wave of new cultivars will undoubtedly have short commercial lives and be rapidly superceded by other new cultivars (OMF156 and others). The cultivar replacement process is especially dynamic in meadowfoam because of the short breeding history of this crop. Even though self-pollinated cultivars have tremendous appeal and commercial promise, insect-pollinated germplasm and cultivars are the cornerstone of our breeding program and the development of insect-pollinated cultivars is presently more advanced than the development of self-pollinated cultivars.

SELF-POLLINATED CULTIVAR DEVELOPMENT

The self-pollinated germplasm described earlier lacks many of the traits necessary for commercial production (Knapp and Crane 1997; Crane and Knapp 1999), e.g., $OMF63-S_5$, $OMF64-S_5$, and $OMF66-S_5$ grow flat and produce less seed than elite insect-pollinated *L. alba* cultivars. Several new self-pollinated inbred lines have been developed from crosses between insect-pollinated *L. alba* ssp. *alba* cultivars (e.g., 'Knowles') and self-pollinated *L. alba* ssp. *versicolor* lines (e.g., OMF64). Some of these lines may be released as enhanced germplasm or cultivars. With further development, self-pollinated cultivars might significantly impact the meadowfoam industry by reducing production costs (honeybee hive rentals) and increasing the stability of seed yields across years.

Meadowfoam seed yields tend to be volatile across years. Poor honeybee pollination, rightly or wrongly, is routinely blamed for low seed yields. Oregon growers experienced some of this volatility the last two years. The mean seed yield of all cultivars in our yield trial was significantly greater ($LSD_{0.05}$ = 45 kg/ha) in 1996–97 than 1997–98 (1,094 kg/ha versus 654 kg/ha). Similar seed yields were produced in commercial fields in those years. The yield decline in 1997–98 was partly blamed on poor pollination. Although pollination undoubtedly plays a role in seed yield volatility across years, there are no experimental data that directly incriminate pollination as the primary cause of low seed yields. Yield tests of insect- and self-pollinated cultivars across years should shed light on this. If poor pollination is the primary problem, then self-pollinated cultivars should experience less seed yield volatility across years than insect-pollinated cultivars.

BREEDING BASICS

As note earlier, *L. alba* is self-compatible and predominantly allogamous (Jain 1978; Brown and Jain 1979). Although naturally self-pollinated progeny are found in certain wild *L. alba* ssp. *versicolor* populations (Knapp and Crane 1997), most populations lack self-pollinated progeny and strongly self-pollinated progeny have not been found in wild (unselected) populations. The frequency of self-pollinated progeny among F_2 progeny from crosses between self- and insect-pollinated lines is very low (unpublished data) because the self-pollinated phenotype (seeds per flower) is a function of two threshold traits (protandry and heterostyly). The genetics of self-pollination are poorly understood in meadowfoam. Molecular breeding experiments are underway in our laboratory to dissect the genetics of self-pollination. We suspect that the genetics are complex and that two or more quantitative trait loci (QTL) underlie phenotypic differences for self-pollination, heterostyly, and protandry.

The versatility of the mating system of meadowfoam means that diverse breeding methods and schemes can be employed. We have used several breeding methods to develop the germplasm described in this review: recurrent mass and family selection and backcross breeding in insect-pollinated populations and backcross, pedigree, single seed descent, and bulk breeding in crosses between self- and cross-pollinated lines. We used recurrent half-sib family selection for early breeding experiments so that families could be grown in large plots (2 × 7 m) for direct combining. Half-sib seed can be easily produced on field grown plants spaced 1 m or more apart in isolated nurseries. Such plants typically yield 15 to 45 g of seed (~1,500 to 4,500 seeds per family). This is sufficient to plant one or two replications per family in 2 × 7 m plots. Individual plants do not produce enough seed to test families in more than one year or location in 2 × 7 m plots, so one cycle of selection is typically completed with families tested in one year and one location. We test families in non-isolated fields and produce selected populations by bulking remnant seed of selected families.

We have begun using S_1 rather than half-sib family recurrent selection in insect-pollinated populations. This switch was made for a few reasons. First, half-sib family seed must be produced on spaced plants in field isolations, whereas S_1 family seed can be produced in the greenhouse. Selected S_1 families are intermated under cages in the field, thus circumventing the need for a field isolation. Second, S_1 family selection should produce greater gains from selection than half-sib family selection. The S_1 family selection cycle is longer than the half-sib family selection cycle (three versus two years), but the theoretical gains are greater for the former. Third, the process of developing S_1 families is a natural starting point for discovering and selecting novel phenotypes and developing inbred lines. We produce S_1 family seed in screened greenhouses by manually pollinating individual flowers within a plant with cotton swabs. Numerous flowers on a single plant can be pollinated in one to two minutes. The process of selfing a plant typically spans eight to 12 days with repeat pollinations every two to three days. We typically produce 100 to 200 S_1 seeds per plant. This is enough seed to plant two rows 2 m long. We bulk harvest S_1 plots with a forage harvester and thresh the seed with a stationary thresher.

Developing inbred lines is straightforward in meadowfoam. We have developed and are developing several inbreds for breeding and genetics experiments by manually selfing greenhouse grown plants. Many meadowfoam populations are depressed by inbreeding and lines are frequently lost in inbreeding experiments (Jain 1976, unpubl. data). We have observed inbreeding depression in several *L. alba* ssp. *alba* populations. Some *L. alba* ssp. *versicolor* populations, however, seem to have no genetic load (Knapp and Crane 1997), e.g., inbreeding coupled with selection for increased self-pollination led to increased fecundity in OMF64. The effects of inbreeding and heterosis needs to be systematically and intensively studied in the self- and cross-pollinated gene pools of meadowfoam. Meadowfoam is an excellent candidate for hybrid seed production; however, a system for producing hybrid seed has not been discovered or developed for this crop. With such a system, single-cross hybrids could be produced between self-pollinated inbred lines, thereby short-circuiting inbreeding depression and producing simultaneous gains from self-pollination and heterosis. The development of a hybrid seed production system would have a revolutionary impact on meadowfoam.

Crosses are easily produced in meadowfoam. We emasculate flowers one day before buds break and typically pollinate emasculated flowers one to two days later. Anthers are excised using forceps. The seed to seed generation time for greenhouse grown plants ranges from 120 to 150 days. We typically complete three generations per year in the greenhouse. One of the keys to speeding up the greenhouse cycle is rapidly germinating freshly harvested seed. Meadowfoam seed typically must be after-ripened for one month before a new generation can be started; however, we discovered that freshly harvested seed, when heat treated, can be immediately germinated. Physiologically mature seeds are held at 50°C for 48 h to break dormancy. This trick reduces the seed to seed generation times by three months over the course of one year.

We have had situations where we needed to field test segregating backcross progeny (e.g., BC_1S_1 progeny) for pre-anthesis traits and backcross to the recurrent parent. Because producing crosses in the field is difficult and impractical, we grow the recurrent parent in the greenhouse so that the field and greenhouse generations will nick. Stems of flowering plants are collected from field grown plants. The cut stems are placed in water in the greenhouse and pollen is harvested from flowers as they open and the anthers dehisce. We have produced crosses from flowering stems held this way for one to five days in the greenhouse. The females, of course, must planted ahead so that they nick with field grown male plants. We have not developed or tested a protocol for storing meadowfoam pollen.

KEY AGRONOMIC PROBLEMS AND PRODUCTION BARRIERS

The most serious pest of meadowfoam in Oregon is *Scaptomyza*, a fruit fly that bores into developing buds and crowns, thereby destroying seed or whole plants. The taxonomy of this pest is unclear. Glenn Fischer and Daryl Ehrensing at OSU are studying the life cycle, control, and integrated management of *Scaptomyza* and have developed promising control measures. Floral, Knowles, and several other cultivars are susceptible to *Scaptomyza*. We attribute some of the seed yield decline for all cultivars and the very low seed yield of 'Floral' in our 1997–98 yield trial to *Scaptomyza* damage. *Scaptomyza* damage was severe in our 1997–98 yield trial (Hyslop Farm, Corvallis, Oregon) and tends to be more severe on our experiment farm than elsewhere. This situation undoubtedly exists because meadowfoam, although rotated, is continuously grown within the confines of the experiment farm.

As stated earlier, resistance or tolerance to *Scaptomyza* has been found in OMF78. We developed an experimental insect-pollinated cultivar (OMF156) from *Scaptomyza* tolerant half-sib families and are increasing this cultivar for further tests. The genetic basis for resistance is not known. The biology, control, and genetics of resistance to *Scaptomyza* must be further elucidated to help avert losses to this pest.

As noted earlier, poor pollination of insect-pollinated cultivars is frequently cited as one of the key factors underlying low seed yields in meadowfoam. The perception is that seeds yields are diminished when the weather for pollination is unsatisfactory or when honeybees are not optimally managed. This has been the driving force behind the development of self-pollinated cultivars. The first self-pollinated cultivars are being increased for testing in 1998–99. The cost of producing meadowfoam should be significantly lower for self- versus insect-pollinated cultivars (honeybee rentals typically comprise 30% to 35% of the cost of production). If seed yields are comparable for self- and insect-pollinated cultivars, then the release and commercial production of self-pollinated cultivars should decrease production costs and meadowfoam oil prices.

The commercial development of meadowfoam has been limited by the growing conditions under which this crop thrives. The Willamette Valley of Oregon (45°N and 123°W) seems to be ideal for producing meadowfoam. Several traits and factors listed below underlie the narrow ecological niche of this crop.

1. Meadowfoam (specifically *L. alba*) grows naturally in riparian habitats (vernal pools) in northern California (Mason 1952). The crop tolerates and thrives in saturated soils. Many Willamette Valley soils are saturated from late November to early April or later. Soil drying is typically coupled with increased ambient air and soil temperatures in the Willamette Valley. These factors accelerate flower and fruit develop and senescence. The long wet growing season and dry harvest season in the Willamette Valley are ideal for meadowfoam.

2. Meadowfoam thrives in cool weather (growing temperatures between 4° and 16°C), but is typically killed by ambient air temperatures below 5°C for prolonged periods (more than four days). We speculate that meadowfoam lacks frost tolerant and winter hardy types, but concede that the germplasm

collections have not been systematically screened for either trait. The development of winter hardy cultivars would dramatically expand the range of production for this crop.

3. Meadowfoam can be sown as a spring annual to circumvent winter damage, but the growing season is dramatically compressed and seed yields are typically very low. There is certainly no harm in experimenting with spring or summer planting, but be prepared for failure.

4. Seed dormancy restricts planting to soil temperatures below 15°C (Toy and Willingham 1966, 1967). Secondary dormancy is induced when fully imbided seeds are planted in soils warmer than 16°C (Toy and Willingham 1967). Selection for non-dormancy has not been done in meadowfoam, the genetic basis for seed dormancy is not known, and germplasm has not been screened for non-dormant phenotypes. We speculate that nothing can be gained by planting earlier (via the development of cultivars lacking secondary dormancy) without developing heat tolerant cultivars. Oregon farmers typically have no difficulty planting meadowfoam in early October after soils temperatures have cooled to 15°C or lower.

REFERENCES

Bagby, M.O., C.R. Smith, T.K. Miwa, R.L. Lohmar, and I.A. Wolff. 1961. A unique fatty acid from *Limnanthes douglasii* seed oil: the C_{22} diene. J. Am. Oil Chem. Soc. 26:1261–1265.

Brown, C.R. and S.K. Jain. 1979. Reproductive system and pattern of genetic variation in two *Limnanthes* species. Theor. Appl. Genet. 54:181–190.

Brown, C.R., H. Hauptli, and S.K. Jain. 1979. Variation in *Limnanthes alba*: a biosystematic survey of germplasm resources. Econ. Bot. 33:267–274.

Calhoun, W. and J.M. Crane. 1975. Registration of 'Foamore' meadowfoam. Oregon Agr. Exp. Sta., Corvallis.

Calhoun, W. and J.M. Crane. 1984. Registration of 'Mermaid' meadowfoam. Oregon Agr. Exp. Sta., Corvallis.

Crane, J.M. and S.J. Knapp. 1999. Registration of 'Knowles' meadowfoam. Crop Sci. (in press).

Dole, J.A. and S.K. Jain. 1992. Genetics of the new crop genus *Limnanthes*. I. Five morphological marker loci in *L. alba × L. gracilis* progenies. Plant Breed. 109:198–202.

Earle, F.R., E.H. Melvin, L.H. Mason, C.H. van Ettan, and I.A. Wolff. 1959. Search for new industrial oils. I. Selected oils from 24 plant families. J. Am. Oil Chem. Soc. 36:304–307.

Gentry, H.S. and R.W. Miller. 1965. The search for new industrial crops. IV. Prospectus of *Limnanthes*. Econ. Bot. 19:25–32.

Higgins, J.J., W. Calhoun, B.C. Willingham, D.H. Dinkel, W.L. Raisler, and G.A. White. 1971. Agronomic evaluation of prospective new crop species. II. The American *Limnanthes*. Econ. Bot. 25:44–54.

Isbell, T.A. 1997. Development of meadowfoam as an industrial crop through novel fatty acid derivatives. Lipid Technol. 9:140–144.

Jain, S.K. 1976. Evolutionary studies in the meadowfoam genus *Limnanthes*: an overview. p. 50–57. In: S.K. Jain (ed.), Vernal pools: Their ecology and conservation. Inst. Ecol., Univ. CA, Davis.

Jain, S.K. 1989. Domestication of *Limnanthes* (meadowfoam) as a new oil crop. p. 121–134. In: A. Micke (ed.), IAEA Monograph, Vienna, Austria.

Jain, S.K. 1978. Breeding systems in *Limnanthes alba*: several alternative measures. Am. J. Bot. 65:272–275.

Jain, S.K. and E.H. Abuelgasim. 1981. Some yield components and ideotype traits in meadowfoam, a new industrial crop. Euphytica 30:437–443.

Jain, S.K., R.O. Pierce, and H. Hauptli. 1977. Meadowfoam: Potential new oil crop. Calif. Agr. 31:18–20.

Jolliff, G.D. 1986. Plant variety protection certificate (#8500166) for Mermaid meadowfoam. U.S.D.A., Washington, D.C.

Jolliff, G.D. 1994. Plant variety protection certificate (#9200257) for Floral meadowfoam. U.S.D.A., Washington, D.C.

Knapp, S.J. and J.M. Crane. 1995. Fatty acid diversity of Section Inflexae *Limnanthes* (meadowfoam). Industrial Crops Prod. 4:219–227.

Knapp, S.J. and J.M. Crane. 1997. The development of self-pollinated inbred lines of meadowfoam by direct selection in open-pollinated populations. Crop Sci. 37:1770–1775.

Knapp, S.J. and J.M. Crane. 1998. A dominant gene decreases erucic acid and increases dienoic acid in meadowfoam. Crop Sci. 38:1541–1544.

Knapp, S.J. and J.M. Crane. 1999. Plant variety protection certificate (pending) for Knowles meadowfoam. U.S.D.A., Washington, D.C.

Krebs, S. and S.K. Jain. 1985. Variation in morphological and physiological traits associated with yield in *Limnanthes* spp. New Phytol. 101:717–729.

Mason, C.T. 1952. A systematic study of the genus *Limnanthes*. Univ. Calif. Berkeley Bot. Publ. 25:455–512.

Miller, R.W., M.E. Daxenbichler, and F.R. Earle. 1964. Search for new industrial oils. VIII. The genus *Limnanthes*. J. Am. Oil Chem. Soc. 41:167–169.

Pierce, R.O. and S.K. Jain. 1977. Variation in some plant and seed oil characteristics of meadowfoam. Crop Sci. 17:521–526.

Smith, C.R., M.O. Bagby, T.K. Miwa, R.L. Lohmar, and I.A. Wolff. 1960. Unique fatty acids from *Limnanthes douglasii* seed oil: The C_{20}- and C_{22}-monoenes. J. Am. Oil Chem. Soc. 25:1770–1774.

Toy, S.J. and B.C. Willingham. 1966. Effects of temperature on seed germination of ten species and varieties of *Limnanthes*. Econ. Bot. 20:71–75.

Toy, S.J. and B.C. Willingham. 1967. Some studies on secondary dormancy in *Limnanthes* seed. Econ. Bot. 21:363–366.

Oil Content Distribution of Meadowfoam Seeds by Near-Infrared Transmission Spectroscopy*

Brett E. Patrick and Gary D. Jolliff

Meadowfoam (*Limnanthes* spp., Limnanthaceae) is a recently domesticated herbaceous winter–spring industrial oilseed crop (Jolliff 1989). The virtually pure raw source of long-chain fatty acids has unique composition and very high oxidative stability (Isbell et al. 1999; Muuse et al. 1992). Initial commercial sales were substantially for high-value personal care products. Other market applications may develop with the evolution of such things as price, utilization experience (Isbell et al. 1999), derivative development, and supply logistics.

Increased oil yield per hectare remains a top priority for advancing meadowfoam profitability for farmers and commercialization into broader industrial markets. Many variables interact to influence oil yield of meadowfoam; environment, management, genetics, and pests can have major effects. Improved understanding of the effects of these variables may contribute to improving oil yield. Analysis of bulk seed samples can mask the cause and effects of some variables, and does not allow for the partitioning of variance for measuring important cause and effect relationships. Thus, single-seed analysis provides a means of characterizing variation.

Single-seed oil determination of meadowfoam by near-infrared transmission spectroscopy (NITS) is a fast, efficient, nondestructive procedure, amenable to substantial automation compared to traditional chemical analysis (Patrick and Jolliff 1997). This technology has potential applications evaluating oil content and variability within individual flowers, plants, populations, or bulk samples. It provides the prospect to quantify the effects of numerous kinds of management, environmental, and genetic variables on the oil content of seeds.

The objective of this paper is to determine the distribution of oil content in seed from meadowfoam plants by NITS single-seed oil determinations. This information could be useful in selection and management research programs designed to increase oil yield.

METHODOLOGY

Single-seed oil determinations were made with a Tecator Infratec 1255 scanning monochronometer (Foss NIRSystems, Inc.), with a custom 23 single-seed scanning tray, and an improved NITS (850–1050 nm) calibration with a 3.0% standard error of cross validation (SECV). The calibration was developed from 966 meadowfoam seeds representing 21 individual sources, representing a variety of genetic origins from our breeding program over 7 growing seasons (Patrick and Jolliff 1997).

From our elite breeding materials an additional 78 plants were surveyed for high mean and high variance of oil content by NITS single-seed oil determinations of a random sample of 23 seeds from each plant. Hereafter, a survey refers to a 23-seed random sample from a plant. The 10 plants having surveys of highest mean oil content and highest oil content variance were selected for evaluation. The mean oil contents of these plants were determined by NITS single-seed oil determinations of all seeds from each plant. These selected plants were surveyed an additional three times, to compare plant mean oil content and survey mean oil content. In addition, single-seed mass determinations were made for all seeds from five of these plants.

RESULTS AND DISCUSSION

The NITS single-seed oil determinations of surveys from 78 meadowfoam plants indicated a wide range of oil content and variance. Survey mean oil content varied from 19.5% to 38.8% with a mean of 29.0%. The standard deviation varied from 2.8% to 13.4% with a mean of 6.1%. The oil content distributions of survey results from three plants are compared in Fig. 1.

NITS single-seed oil determinations were performed on 3,685 seeds from the 10 selected plants. The mean oil contents of these plants ranged from 29.2% to 36.2%, and the oil content standard deviations ranged from 6.2% to 9.1%. For the 3,685 seeds, the oil content mean was 31.6%, the oil content range was 2.3% to

---

*Oregon Agricultural Experiment Station Technical Paper Number 11444

55.2%, and the oil content standard deviation was 7.2%. The oil content distribution of the 3,685 seeds is illustrated in Fig. 2. The additional surveys of these 10 selected plants indicated the mean difference between survey mean oil content and plant mean oil content was 3.1%.

Mass determinations of 2,939 individual seeds from 5 of the selected plants yielded a mean mass of 15.0 mg, a mass range of 1.8 to 27.3 mg, and a mass standard deviation of 4.4 mg. The mean seed mass of the plants ranged from 14.0 to 19.1 mg. A plot of oil content vs. seed mass is shown in Fig. 3. The correlation was 0.68.

The highest meadowfoam seed properties found by this analysis were a seed of mass 27.3 mg and a seed of 55.2% oil content. These exceed the highest values of the calibration set, which are 21.9 mg and 45.7% (Patrick and Jolliff 1997). An improved NITS calibration is needed for more accurate oil determinations of seed with very high oil content (over 45.7%) and high mass (over 21.9 mg).

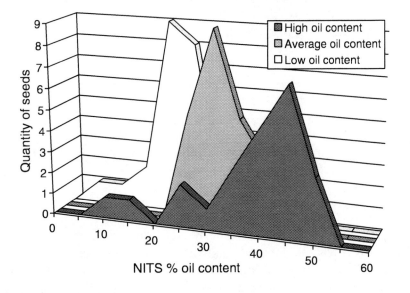

Fig. 1. Oil content distributions of 23 randomly selected seeds from 3 meadowfoam plants are compared, by NITS single-seed oil determinations.

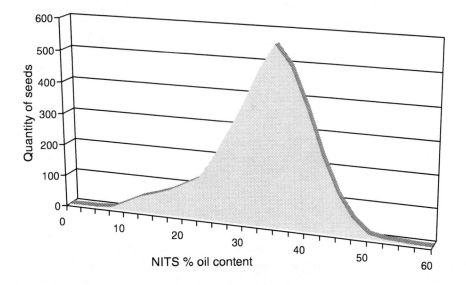

Fig. 2. Oil content distribution of 3,685 individual meadowfoam seeds from 10 plants, by NITS single-seed oil determinations.

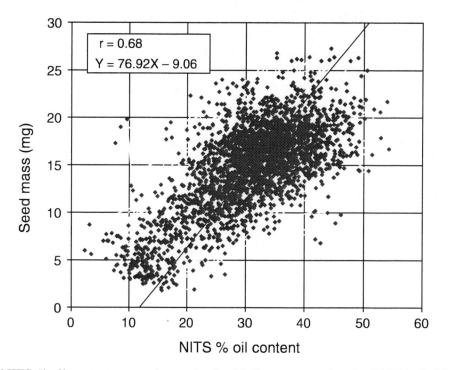

Fig. 3. Plot of NITS % oil content vs. seed mass (mg) with linear regression for 2,939 individual meadowfoam seeds from 5 plants, by mass and NITS single-seed oil determinations.

CONCLUSION

Single-seed oil determinations of meadowfoam by NITS are quick, economical, and non-destructive. Selection of individual seeds is possible for use in cultivar improvement, breeding programs, and physiology studies where further use of live seeds is necessary. In maize (*Zea mays* L.), single-seed selection was found to be more effective in improving oil content than composite sample selection (Silvela et al. 1989). The identification of seeds with high oil content in plants of high oil content suggests a single-seed selection program to improve oil content.

The positive correlation between oil content and seed mass suggests an opportunity to use mechanical separations of high mass (i.e. high weight) seeds from bulk samples as a first phase process of selecting quantities of seeds for high oil content. NITS scanning would be used as the second phase to separate the highest oil content seeds from the high mass seeds. However, an improved NITS calibration should be developed for accurate oil determinations of seeds exceeding the characteristics of the existing calibration.

REFERENCES

Isbell, T.A., T.P. Abbott, and K.D. Carlson. 1999. Oxidative stability index of vegetable oils in binary mixtures with meadowfoam oil. Ind. Crops Prod. 9:115–123.

Jolliff, G.D. 1989. Meadowfoam domestication in Oregon: A chronological history. p. 53–65. In: L.L. Hardman and L. Waters, Jr. (eds.), Strategies for alternative crop development: Case histories. Proc. national symposium, Nov. 29, 1988, Anaheim, CA. Center for Alternative Plant and Animal Products. Univ. Minn., St. Paul.

Muuse, B.G., F.P. Cuperus, and J.T.P. Dereksen. 1992. Composition and physical properties of oils from new oilseed crops. Ind. Crops Prod. 1:57–65.

Patrick, B.E. and G.D. Jolliff. 1997. Nondestructive single-seed oil determination of meadowfoam by near-infrared transmission spectroscopy. J. Am. Oil Chem. Soc. 74:273–276.

Silvela, L., R. Rodgers, A. Barrera, and D.E. Alexander. 1989. Effect of selection intensity and population size on percent oil in maize, *Zea mays* L. Theor. Appl. Genet. 78:298–304.

Establishment of Meadowfoam as a New Crop in Virginia

Harbans L. Bhardwaj, Muddappa Rangappa, and Anwar A. Hamama

Meadowfoam (*Limnanthes alba* Bentham, Limnanthaceae), a low growing winter annual native to northern California and Southern Oregon (Ehrensing et al. 1996), was domesticated at Oregon State University (Franz et al. 1992). The natural habitat of meadowfoam is 35° to 45° N latitude. Meadowfoam seed oil contains long-chain fatty acids (20- and 22-carbon). These fatty acids are unique due to very high levels of mono-unsaturation and very low levels of poly-unsaturation. These characteristics make meadowfoam oil very stable, even when heated or exposed to air. Meadowfoam oil is useful in manufacture of high quality waxes, lubricants, detergents, and plasticizers. The unique fatty acid profile and special properties of meadowfoam oil and its derivatives have also stimulated interest in its use in personal care products such as cosmetics and toiletries. Oil from meadowfoam was first tested in the late 1950s as part of a US Department of Agriculture search for plants to provide new raw materials. Meadowfoam is well adapted to the relatively mild winters and warm dry winters of the Pacific Northwest and is known to grow well in many types of soil including poorly drained areas.

Currently, meadowfoam is produced on about 3000 ha in the Willamette valley of Oregon (Ehrensing et al. 1996) by Oregon Meadowfoam Growers Association (OMGA). However, this production is insufficient to meet the demands of chemical and cosmetic industry. Due to the fact that meadowfoam in Oregon is produced in rotation with grass seed production, an increase in meadowfoam area in Oregon is not possible. OMGA has been interested in expanding the production of meadowfoam due to increasing demand for oil and to stabilize meadowfoam oil's availability. The objectives of this research were to determine feasibility of meadowfoam production in Virginia; develop an agronomic system for meadowfoam production; and facilitate meadowfoam production in Virginia.

MEADOWFOAM RESEARCH IN VIRGINIA

The meadowfoam research was initiated by the New Crops Program of Virginia State University in 1992. Previous research conducted in Virginia had indicated that meadowfoam production in Virginia may not be possible. However, because of increasing demand for meadowfoam oil, it was decided to plant a preliminary, observation-type plot (36 m^2) with 'Mermaid' cultivar during 1992–93. The meadowfoam plants grew well and produced seed. A similar observation plot was again planted during 1993–94 using the seed produced during 1992–93. The meadowfoam in these observation plots matured and produced seed (Bhardwaj and Wohlman 1997; Bhardwaj and Parrish 1998).

Based on these results, a replicated experiment was conducted during 1995–96 season, using 'Mermaid'. In this experiment, five nitrogen rates were evaluated in a randomized complete block design with three replication. This experiment was planted on Nov. 25, 1995. A second experiment was planted on Feb. 26, 1996. Both these experiments were successful. During 1995–96 season, the seed yield varied from 494–606 kg/ha with a mean of 494 kg/ha (Table 1). The mean oil content of this crop was 22.8% whereas the mean long chain fatty acid content was 92.4%. Based on this success, Fanning Corporation (Chicago, Illinois) supported commercial production of four ha of meadowfoam in cooperation with two Virginia farmers during 1996–97. This crop was planted on Dec 12, 1996 with 25 kg seed per ha. This crop received 45 to 70 kg/ha of nitrogen. The meadowfoam was in full bloom by May 9, 1997. Eight beehives were placed in each 2 ha field. The crop started to mature during last week of May, 1997. We tried using various harvesting techniques without much

Table 1. Performance of 'Mermaid' meadowfoam in Virginia.

Year	Seed yield (kg/ha)	Oil content (%)	Long-chain fatty acids (%)
1995–96	494 (251–606)	22.8 (21.0–25.0)	92.4 (92.3–92.5)
1996–97	408	26.0	97.0

success. The meadowfoam was eventually direct combined from June 9–11, 1997. The seed moisture content at this time was 14–18%. The direct combining caused approximately 38% loss due to shattering. The average yield from direct combining was 284 kg/ha. At the time of direct combine harvesting, we harvested four 1 m² plots from one farmers field, by hand, to estimate the actual seed yield. In these plots, the yield level was 408 kg/ha (Table 1). The oil content in this crop was 26% and average long-chain fatty acid content was 97%. Additional research indicated that reduction in seed moisture from 60% to 32% (Table 2) did not significantly reduce the oil content and the long chain fatty acid content. However, based on recommendations from Oregon State University, it might be desirable to swath meadowfoam when the seed moisture is about 42% until substantial data for optimal seed moisture content in Virginia are available.

Table 2. Effect of harvest date on 'Mermaid' meadowfoam quality.

Date of 1997 harvest	Moisture (%)	Oil (%)	Long-chain fatty acids (%)
May 28	79.5aᶻ	17.6d	87.3c
May 30	74.7b	21.9c	93.7b
June 4	60.3c	24.3b	96.1ab
June 9	31.8d	24.7ab	96.4a
June 12	15.9e	26.0a	97.0a

ᶻMean separation in columns by Duncan's Multiple Range Test, 5% level.

During 1997–98, eight Virginia farmers planted approximately 50 ha of meadowfoam in cooperation with Oregon Meadowfoam Growers Association. This crop was generally unsuccessful. Possible reasons for this failure were judged to include: use of an unadapted cultivar, late application of Select herbicide and liquid nitrogen fertilizer, and unsuitable weather for bee foraging at flowering. For the 1998–99 season, approximately 40 ha of meadowfoam has been planted.

CONCLUSIONS

Experimental results indicated that meadowfoam can be grown in Virginia as a winter crop with acceptable seed oil content and quality. Row spacing of 15–30 cm appears optimal and nitrogen needs may be up to 100 kg/ha. The commercial meadowfoam production efforts in Virginia during 1996–97 and 1997–98 seasons suffered from inadequate weed control and lack of harvest technology. The research conducted by Oregon State University scientists has identified herbicides for weed control. We plan to evaluate these herbicides in Virginia to determine optimal rates. In the meantime, plans include to obtain state registrations of two herbicides (Select for grass weed control and Stinger for broadleaf weed control) to meet a special local need [Pursuant to Section 24(C) of the FIFRA] for use in meadowfoam production fields during 1998–99 season. Virginia State University has purchased a swather and a belt pickup attachment for a combine for harvesting of meadowfoam. This equipment will be provided to the farmers. One farmer, who planted approximately 16 ha during 1997–98 season, purchased his own swather and belt pick-up for his combine and is expected to help other farmers with their meadowfoam harvesting operation. It is planned that the meadowfoam crop during 1998–99 will be harvested using swathers at about 42% moisture in the seed, left in the field in windrows to dry, and then combined using the belt pickup attachment.

Our experience with meadowfoam indicates that meadowfoam can be grown in Virginia as a winter crop with acceptable content of oil in the seed, and content of long-chain fatty acids in the oil. However, the successful introduction of meadowfoam into Virginia will depend upon development of adapted cultivars, successful weed control strategies, and efficient harvesting practices.

REFERENCES

Bhardwaj, H.L. and A. Wohlman. 1997. Expansion of meadowfoam production into the Mid-Atlantic area of the United States. Presented at the Int. Conference of Association for the Advancement of Industrial Crops. Sept. 14–18, 1997. Saltillo, Mexico.

Bhardwaj, H.L. and M. Parrish. 1998. Meadowfoam: A new winter oilseed crop for Virginia. http://www.ext.vt.edu/news/periodicals/cses/1998-03/1998-03-02.html

Ehrensing, D.T., R.S. Karow, and J.M. Crane. 1996. Growing meadowfoam in the Willamette Valley. Oregon State Univ. Extension Service, EM 8567. Corvallis, OR.

Franz, R.E., M. Seddigh, G.D. Jolliff, and E.L. Alba. 1992. Meadowfoam seed yield in response to temperature during flower development. Crop Sci. 32:1284–1286.

Salinity Effects on Growth, Shoot-ion Relations, and Seed Production of *Lesquerella fendleri*

Catherine M. Grieve, Michael C. Shannon, and David A. Dierig*

Lesquerella fendleri (Gray) S. Wats., a native of the arid and semiarid regions of southwestern US, has been proposed as an important industrial seed oil crop (Thompson and Dierig 1988). Intensive breeding and agronomic research efforts at the US Water Conservation Laboratory have been directed towards domesticating and commercializing the crop (Dierig et al. 1993; Coates 1994; Brahim et al. 1996; Dierig et al. 1996; Nelson et al. 1996).

Information on the management of lesquerella under saline conditions is limited to a two-season (1993–4 and 1994–5) field trial conducted by Leland E. Francois at the Irrigated Desert Research Station located at Brawley, CA (Grieve et al. 1997). Six salinity treatments were imposed by adding NaCl and $CaCl_2$ (2:1 molar ratio) to Colorado River water. Electrical conductivities of the irrigation waters (EC_i) ranged from 1.4 to 10.0 dS m^{-1}. Both biomass production and seed yield were significantly reduced by salinity, although salt stress had no effect on either the content or the composition of the oil. Based on the combined seed yield data for both years, the salt tolerance threshold (the maximum allowable EC_e without a yield decline) was 6.1 dS m^{-1} and the C_{50} (EC_e value that would result in a 50% yield reduction) was 9 dS m^{-1}. [As an approximation: EC_e = 1.5 EC_i]. Each unit increase in salinity above the threshold resulted in a 19% decrease in yield.

Under nonsaline control conditions in the Brawley plots (EC_i = 1.4 dS m^{-1}), lesquerella was a strong calcium accumulator, and as external calcium increased in concert with salinity, leaf-Ca^{2+} significantly increased. In response to Cl$^-$ salinity, leaf-Cl$^-$ increased three-fold as EC_i increased from 1.4 to 10 dS m^{-1}. Large increases in external Na$^+$ had no effect on leaf-Na$^+$. Sodium levels in lesquerella leaves were at least an order of magnitude lower than leaves of other crucifers grown at the same location with the same compositions of Cl$^-$-dominated salinity (Francois and Kleiman 1990; Francois 1994). Lesquerella apparently possesses an unique and effective exclusion mechanism that limits the accumulation of potentially toxic levels of Na$^+$ in the leaves.

One objective of the current study was to determine the response of lesquerella to sodium sulfate-dominated irrigation waters that are typical of the saline drainage waters commonly encountered in the San Joaquin Valley of California. Lesquerella was evaluated as a crop that could fill a niche in the drainage water reuse system where only moderate tolerance is required. A second objective was to identify and isolate individuals that exhibited salt tolerance and use this germplasm for establishing a base population for breeding more salt tolerant lesquerella.

METHODOLOGY

On 1 Oct 1997, lesquerella seeds, collected from 1993–1994 seed-increase plot located at the US Water Conservation Laboratory, were planted in a greenhouse in Speedling[1] trays containing a peat-sand-perlite mix. Two months after planting, the seedlings were transferred to the US Salinity Laboratory and transplanted to 24 outdoor sand tanks. Each tank contained 12 rows with six seedlings per row. The tanks (1.5 × 3.0 × 2.0 m deep) contained washed sand having an average bulk density of 1.2 Mg m^{-3}. At saturation, the sand had an average volumetric water content of 0.34 m^3 m^{-3}. Plants were irrigated once daily with a nutrient solution consisting of (in mol m^{-3}): 3.5 Ca^{2+}, 2.5 Mg^{2+}, 21.5 Na$^+$, 6.0 K$^+$, 10.9 SO_4^{2-}, 7.0 Cl$^-$, 5.0 NO_3^-, 0.17 KH_2PO_4, 0.050 Fe (as sodium ferric diethyleneamine pentaacetate), 0.023 H_3BO_3, 0.005 $MnSO_4$, 0.0004 $ZnSO_4$, 0.002 $CuSO_4$, and 0.0001 H_2MoO_4 made up with city of Riverside municipal water. This solution, with an electrical conductivity (EC_i) of 3 dS m^{-1}, served as the control treatment. Each sand tank was irrigated from a 3700-L reservoir. Irrigations were of 15 min duration, which allowed the sand to become completely saturated, after which the solution drained to the reservoir for reuse in the next irrigation. Water lost by evapotranspiration

---

*Mineral ion analyses were performed by Donald A. Layfield. Technical assistance was provided by: Terence Donovan, John Draper, Aaron Kaiser, Greg Leake, and James Poss.

[1]Trade names and company names mentioned are included for the benefit of the reader and do not constitute an endorsement by the US Department of Agriculture.

was replenished automatically to maintain constant electrical conductivites in the solutions.

On 15 Dec. 1997, two weeks after the seedlings were transplanted, eight salinity treatments were imposed with irrigation waters designed to simulate saline drainage waters commonly present in the San Joaquin Valley of California, and compositions of increased salinity which would result from further concentration of these drainage waters (Table 1). Electrical conductivities of the saline treatments were increased to the desired levels by incremental additions of the salts over a 9-day period to avoid osmotic shock to the seedlings. Targeted EC_i values of the solutions were 3 (nutrient solution only), 6, 9, 12, 15, 18, 21, and 24 dS m^{-1}. The experiment was a randomized complete block with 8 salinity treatments and three replications. The pH was uncontrolled and varied between 7.5 and 8.5. Initially the tanks were irrigated on a daily basis, and later in the season, on alternate days.

Irrigation waters were analyzed at weekly intervals for Ca^{2+}, Mg^{2+}, Na^+, K^+, and total-S by inductively coupled plasma optical emission spectroscopy (ICPOES), and for Cl^- by coulometric-amperometric titration. Nutrients and salinizing salts were replenished as required.

Plant survival and time of flowering were recorded on a regular basis. On 20 Feb., 15 Mar. and 15 Apr. 1998, plants were sampled and the following measurements were recorded: fresh weight, height of the main axis, number of secondary branches. Leaves and stems were separated. Leaf area was measured. Shoot tissues were washed with deionized water, dried at 70°C, reweighed, and ground in a Wiley mill. Total-S, -P, Ca^{2+}, Mg^{2+}, Na^+, and K^+ were determined on nitric-perchloric acid digests of the plant material by ICPOES. Chloride was determined on nitric-acetic acid extracts by titration. Irrigation was discontinued and plants were sprayed with a desiccant 1 wk before final harvest on 9 Jun. 1998. Individual shoots were weighed, and seeds of each plant were separated and weighed.

Statistical analyses of the ion data were performed using SAS release version 6.12 (SAS Institute Inc. 1997). The effects of salinity on shoot biomass and seed yields were determined by the procedure described by van Genuchten and Hoffman (1984).

EXPERIMENTAL RESULTS

Plant Growth

Within two weeks after imposition of full treatments, transplant survival in the sand cultures was reduced in treatments in which salinity exceeded 15 dS m^{-1} (Fig. 1). Survival continued to decrease with salt treatment over time. In contrast, salinity had no effect on plant survival in the Brawley field plots. The difference in response may be due to plant age at initiation of salinization, e.g. 12 weeks in the field compared with 10 weeks in sand cultures. In addition, root growth of the transplants may not have been well established in the sand tanks prior to salt treatment.

On 10 Apr. 1998, survival was so poor at 24 dS m^{-1} that this treatment was discontinued; the few survivors were rescued and grown in crossing blocks under nonsaline conditions. During the last month of crop growth, stem cracking and stem rot became evident in all treatments. Plants grown at the lower salinity levels were particularly susceptible. *Sclerotinia sclerotinia*, which was isolated from affected stems, appeared to be the causal agent. Because of this infection, survival of plants grown at 3 dS m^{-1} was reduced to about 40% at final harvest. Regardless of salinity level, root morphology, as observed at final harvest, was invariably distorted and taproots were either atypical or absent.

Biomass production, measured at final harvest, decreased significantly in response to salinity (Fig. 2). The salinity level that resulted in a 50% reduction in final shoot weight was

Table 1. Composition of salinizing salts in irrigation waters used for determining salt tolerance of lesquerella in sand cultures.

EC_I (dS m^{-1})	Composition (mol m^{-3})					
	[Ca]	[Mg]	[Na]	[K]	[SO4]	[Cl]
3	3.5	2.4	21.5	6.0	10.9	7.0
6	6.3	4.9	43.6	6.0	22.2	21.1
9	10.0	8.3	73.3	6.0	37.2	35.5
12	12.6	10.0	88.5	6.0	44.9	42.8
15	13.0	13.9	123.0	6.0	58.2	59.6
18	13.4	17.8	157.5	6.0	71.4	73.4
21	13.6	22.2	196.2	6.0	85.7	90.8
24	13.7	26.5	234.8	6.0	100.0	105.3

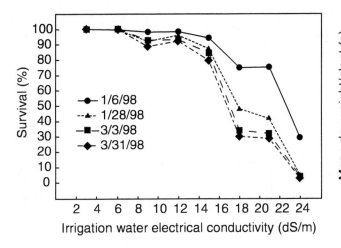

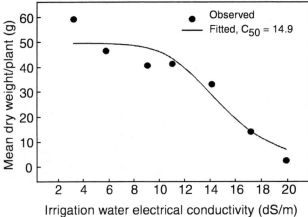

Fig. 1. Survival of lesquerella plants (%) recorded on three dates as a function of increasing irrigation water salinity.

Fig. 2. Lesquerella shoot biomass production as a function of increasing irrigation water salinity.

14.9 dS m⁻¹. Average seed yield per plant was about 2 g in the 3, 6, 9, and 18 dS m⁻¹ treatments and 3 g at 12 and 15 dS m⁻¹ (Fig. 3). Leaf area, determined at a midseason harvest (15 Apr. 1998), decreased consistently and significantly from a mean of 950 to 65 cm² per plant as salinity increased from 3 to 21 dS m⁻¹.

Ion Relations

The divalent cations, Ca^{2+} and Mg^{2+}, were strongly accumulated by lesquerella shoots and both cations were preferentially accumulated in the leaves rather than the stems (Figs. 4A and 4B). Leaf-Ca^{2+} was reduced by salinity when EC_i exceeded 15 dS m⁻¹, although over this salinity range, Ca^{2+} concentration in the irrigation waters increased nearly four-

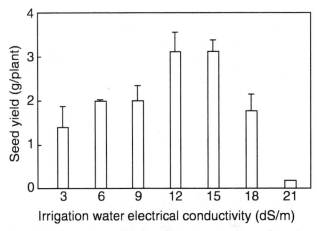

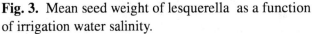

Fig. 3. Mean seed weight of lesquerella as a function of irrigation water salinity.

fold. At higher salinities, leaf-Ca^{2+} decreased significantly as Ca^{2+} in the substrate continued to increase. The Ca^{2+} status of salt-stressed plants is strongly influence by the ionic composition of the external medium. The presence of salinizing ions such as Na^+ or Mg^{2+} in the substrate may reduce Ca^{2+} activity and limit the availability of Ca^{2+} to the plant. High Na^+/Ca^{2+} may affect vital physiological and nutritional processes. External Mg^{2+}/Ca^{2+} ratios greater than 1 are known to reduce growth and yield of several crop species (Grattan and Grieve 1994). In the present experiment, external Mg^{2+}/Ca^{2+} increased from 1.1 in the 15 dS m⁻¹ treatment to 2 at 18 and 21 dS m⁻¹, and at the same time, Na^+/Ca^{2+} increased from 9 to 14 and 17 as salinity increased from 15 to 18 to 21 dS m⁻¹, respectively (Table 1). These imbalances, as well as ion interactions that occurred within the plant, may have contributed to reduced Ca^{2+} in lesquerella leaves and stems.

The *L. fendleri* ecotype evaluated in this salt tolerance study was found growing on calcareous soils of limestone origin (Dierig et al. 1996). These neutral or alkaline soils contain high amounts of calcium and bicarbonate and are generally warmer, drier, and more permeable to water than, for example, silicaceous soils (Larcher 1972). Species or ecotypes differ in their management of Ca^{2+}. Those plants that are adapted to calcareous soils and preferentially take up large amounts of Ca^{2+} for storage in the plant sap are termed calcicoles. Many members of the Brassicaceae are considered calcicoles by virtue of their ability to tolerate high concentrations of external Ca^{2+} and to actively accumulate Ca^{2+}. For example, even under nonsaline conditions, shoot-Ca^{2+} concentrations in *Eruca sativa* (Ashraf and Noor 1993), *Brassica juncea* (Ashraf 1992), as well as *L. fendleri* (Grieve et al. 1997), range from 1200 to 1500 mmol kg dry wt⁻¹. These levels are twice as

241

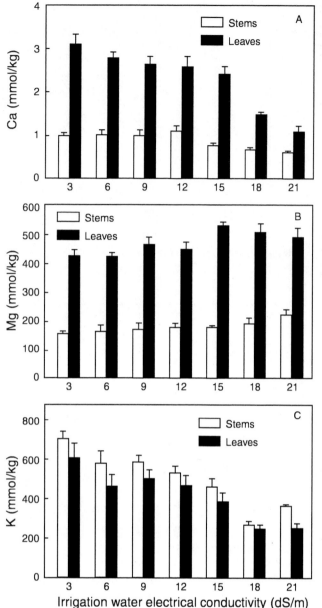

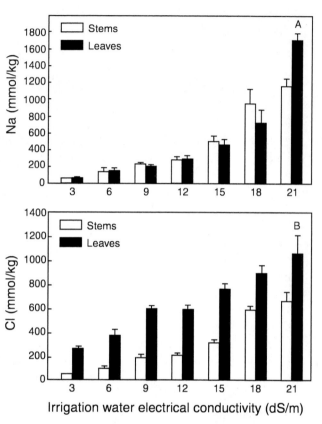

Fig. 4. Concentrations of (4A) Ca^{2+}, (4B) Mg^{2+} and (4C) K$^+$ in leaves and stems of lesquerella grown at seven salinity levels. Values are the means of three replications ± SD.

Fig. 5. Concentrations of (5A) Na$^+$ and (5B) Cl$^-$ in leaves and stems of lesquerella grown at seven salinity levels. Values are the means of three replications ± SD.

high as in the leaves of other crucifers, e.g. *Crambe abyssinica,* (Francois and Kleiman 1990), *B. napus* (Francois 1994), and leafy *Brassica* vegetables (USSL 1997, in preparation).

With increasing salinity, Mg^{2+} in the irrigation water increased from 2.4 to 26.5 mol m^{-3} (Table 1). However, this ten-fold increase had little effect on the Mg^{2+} content of either the leaves or stems over the range of treatments. Regardless of salinity level, both Ca^{2+} and Mg^{2+} were higher in the leaves than in the stems.

With few exceptions, both K$^+$ and Na$^+$ were uniformly partitioned between leaf and stem tissues. Potassium in both leaves and stems decreased with salinity (Fig. 4C), while K$^+$ concentration in the irrigation waters remained constant at 6 mol m^{-3}. Numerous studies with a wide variety of plants have shown that K$^+$ concentration in plant tissues declines with increases in Na$^+$/K$^+$ ratio in the root media. Results from the field study differ from those of the present experiment in that leaf-K$^+$ in field-grown lesquerella was about 400 mmol kg dry wt^{-1} regardless of salinity, whereas in the sand culture study, leaf-K$^+$ was high (700 mmol kg dry wt^{-1}) at the lower salinity levels and did not fall below 400 mmol kg^{-1} even at the highest salinity. Perhaps the external K$^+$ concentration (6 mol m^{-3}) used for the latter experiment was in the luxury range.

Differences between the Brawley field study and the sand culture experiment were also apparent in leaf-

Na$^+$accumulation. In both cases external-Na$^+$ increased with increasing salinity. In the SO$_4$$^{2-}$-system used for the sand cultures, leaf-Na$^+$ rose significantly (Fig 5A), whereas in the field-grown plants, the mean leaf-Na$^+$ concentration across salinity levels was low (< 25 mmol kg dry wt^{-1}) and unaffected by treatment. The absence of what appears to be a very effective Na$^+$ exclusion mechanism in the plants grown in sand cultures may have resulted from factors associated with abnormal root morphology observed in the transplants.

Both leaf- and stem-Cl$^-$ increased as Cl$^-$ concentration in the irrigation water increased. Leaves were stronger Cl$^-$-accumulators than the stems (Fig. 5B). Comparison of leaf-Cl$^-$ levels in response to the lowest salinity at the two experimental sites showed that the field-grown leaves accumulated more than twice as much Cl$^-$ (690 mmol kg^{-1}) as those in sand cultures (270 mmol kg^{-1}) despite a large difference in Cl$^-$ content in the irrigation waters. The control plants in the Brawley plots were irrigated with Colorado River water (average Cl$^-$ = 2.8 mol m^{-3}; EC$_i$ = 1.4 dS m^{-1}), whereas the control solution in the sand cultures contained 7 mol m^{-3} Cl$^-$ (EC$_i$ = 3.0 dS m^{-1}).

This preliminary sand culture experiment serves as a valuable management guide for our continuing studies of the salt tolerance of lesquerella germplasm in outdoor sand cultures. Changes in management practices for the determination of crop response to sodium sulfate-dominated salinity will include: (1) direct seeding to encourage normal root development, (2) altering the irrigation system to permit more rapid delivery to the plants and to prevent excessive flooding, and (3) changing irrigation scheduling to avoid plant injury associated with stem cracking and subsequent pathogen attack. These modifications are currently under evaluation in greenhouse sand culture experiments.

REFERENCES

Ashraf, M. 1992. Effect of NaCl on growth, seed protein and oil contents of some cultivars/lines of brown mustard (*Brassica juncea* (L.) Czern & Coss). Agrochimica 36:137–147.

Ashraf, M. and R. Noor. 1993. Growth and pattern of ion uptake in *Eruca sativa* Mill. under salt stress. Angew. Bot. 67:17–21.

Brahim, K., D.K. Stumpf, D.T. Ray, and D.A. Dierig. 1996. *Lesquerella fendleri* seed oil content and composition: Harvest date and plant population effects. Indust. Crops Prod. 5:245–252.

Coates, W.E. 1994. Mechanical harvesting of lesquerella. Indust. Crops Prod. 2:245–250.

Dierig, D.A., A.E. Thompson, and F.S. Nakayama. 1993. Lesquerella commercialization efforts in the United States. Indust. Crops Prod. 1:289–293.

Dierig, D.A., A.E. Thompson, J.P. Rebman, R. Kleiman, and B.S. Phillips. 1996. Collection and evaluation of new *Lesquerella* and *Physaria* germplasm. Ind. Crops Prod. 5:53–63.

Francois, L.E. 1994. Growth, seed yield, and oil content of canola grown under saline conditions. Agron. J. 86:233–237.

Francois, L.E. and R. Kleiman. 1990. Salinity effects on vegetative growth, seed yield, and fatty acid composition of crambe. Agron. J. 82:1110–1114.

Grattan, S.R. and C.M. Grieve. 1994. Mineral nutrient acquisition and response by plants grown in saline environments. p. 203–226. In: M. Pessarakli (ed.), Handbook of plant and crop stress. Marcel Dekker, New York.

Grieve, C.M., J.A. Poss, T.J. Donovan, and L.E. Francois. 1997. Salinity effects on growth, leaf-ion content and seed production of *Lesquerella fendleri* (Gray) S. Wats. Ind. Crops Prod. 7:69–76.

Larcher, W. 1972. Physiological plant ecology. Springer-Verlag, New York.

Nelson, J.M., D.A. Dierig, and F.S. Nakayama. 1996. Planting date and nitrogen fertilization effects on lesquerella production. Indust. Crops Prod. 5:217–222.

SAS Institute, Inc. 1997. SAS/STAT Software: Changes and enhancements through release 6.12. Cary, NC.

Thompson, A.E. and D.A. Dierig. 1988. Lesquerella: A new arid land industrial oil seed crop. El Guayulero 10:16–18.

van Genuchten, M.Th. and G.J. Hoffman. 1984. Analysis of crop salt tolerance data. p. 258–271. In: I. Shainberg and J. Shalhevet (eds.), Soil salinity under irrigation: Process and management. Ecological Studies 51. Springer-Verlag, New York.

Growth Analysis of Lesquerella in Response to Moisture Stress

Naveen Puppala and James L. Fowler

Lesquerella fendleri (Gray S. Wats.) native to the arid south-western United States has been cited as prime candidate for domestication as a new source of hydroxy fatty acids. Lesquerella oil can replace castor oil and produces a seed containing gum and unique oil, which can be used in a variety of industrial applications and cosmetics. The manner in which plants partition products of photosynthesis into various plant parts is important in determining growth rate and yield. The adaptation of a crop to a particular environment depends upon its efficiency in using resources to produce biomass and its ability to partition the biomass into seed yield. Adequate soil moisture is essential for maximum crop production, but different stages of plant development possess varying sensitivity to moisture stress. Therefore, with limited irrigation, it is essential to distribute the water according to a specific development stage. The objective of this study was to study the effect of moisture stress at different phenological stages on photosynthetic rate, leaf water potential, and growth analysis.

METHODOLOGY

Two field experiments were conducted at the New Mexico State University, Plant Science Center during 1994–95 and 1995–96 growing seasons. The experimental design was a randomized complete block with six replications. Treatments consisted of (1) continuous favorable soil moisture [irrigated at 50% available water content (AWC)], (2) moisture stress (irrigated at 25% AWC) from establishment to final harvest, (3) moisture stress (irrigated at 25% AWC) from establishment to flowering with no stress afterwards (50% AWC), and (4) no stress imposed from establishment to flowering (50% AWC) followed by stress (25% AWC). Lesquerella seeds obtained from Dr. D.A. Dierig, ARS, US Water Conservation Laboratory, Phoenix, AZ, were sown using brillion planter at the rate of 8 kg/ha. The data presented is the mean of two years.

The mean Crop Growth Rate (CGR), Relative Growth Rate (RGR) and Net Assimilation Rate (NAR) were calculated as suggested by Buttery (1970) and Enyi (1962). A random sample was taken from the leaf fraction for leaf area measurements with a LI-3000 automatic leaf area meter (LI-COR, Inc., Lincoln, NE). The growth analysis were conducted in each plot on plants harvested from a ground area of 0.25 m^2 at 15 days intervals from the onset of stress treatments. These components were oven-dried at 65°C for 48 h.

Top most fully expanded leaf of the marked plants was selected from each plot to measure photosynthetic rate (PN) using LICOR 6100 (LICOR, USA). The xylem water potential was measured during mid-day between 12 noon to 1:00 p.m. using a model 3005-1422 Plant Water Status Console (Soil Moisture Corp., Santa Barbara, CA) from one stem of each plant. To evaluate leaf water potential leaf samples were collected during the afternoon in 10 cc syringes and stored in the freezer. The samples were thawed and analyzed using a Model 5130C vapor pressure osmometer (Wescor, Inc., Logan, UT).

EXPERIMENTAL RESULTS

Mean pod yield of two seasons is presented in Table 1. The consumptive use of water (CUW) for the 50% AWC treatment was 662 mm. The seed yield was 925 kg/ha. The water use efficiency was 1.13. Stress prior to and after 50% flowering resulted in 12% and 21% reduction in pod yield compared to the control. The WUE efficiency was 4% percent higher than the control treatment because of saving in one irrigation. Stress after 50% flowering resulted in 5% reduction in WUE compared to control. Irrigating the crop at 25% AWC drastically reduced the yield by 45%. Thus irrigating the crop at 50% after flowering is most beneficial resulting in maximum growth and yield and was similar to that reported by Hunsaker et al. (1998).

Data for the growth analysis during the maximum vegetative stage is presented in Table 2. The maximum CGR occurred when leaf cover was complete and it represented the maximum potential dry matter production (219 DAS) and solar energy conversion rate. The lower CGR rate for treatment 25:50% was due to the release of stress during this period. The RGR decreases with the age of the crop and this decrease are due to the fact that an increasing part of the plant is structural rather than metabolically active tissue and as such does not contribute to growth. The decrease is also due to shading and increase in age of lower leaves (Brown

Table 1. Seed yield, consumptive use of water (CUW) and water use efficiency (WUE) of *Lesquerella fendleri*.

Available water content	Seed yield (kg/ha)	CUW (mm)	WUE
50%	925	662	1.13
25:50%	825	588	1.17
50:25%	725	550	1.09
25%	525	519	1.00

Table 2. Mean growth analysis as influenced by irrigation treatments (207–219 DAS).

Available water content	Ps (mmol CO_2 $m^{-2} s^{-1}$)	CGR ($g m^{-2} d^{-1}$)	RGR ($g g^{-1} d^{-1}$)	NAR ($g m^{-2} d^{-1}$)
50%	27.2	8.47	0.026	0.011
25:50%	28.5	5.57	0.021	0.001
50:25%	20.4	8.80	0.032	0.012
25%	17.5	6.04	0.022	0.005

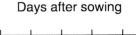

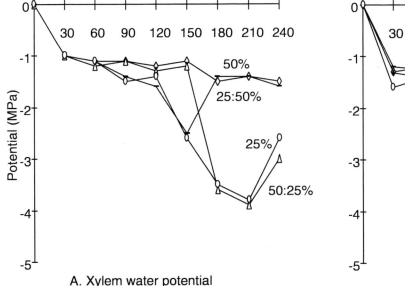

A. Xylem water potential   B. Leaf water potential

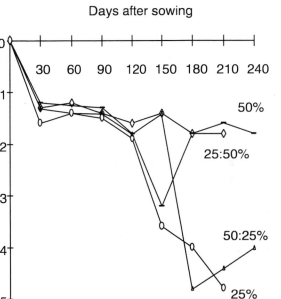

Fig. 1. Effect of water deficit on xylem and leaf water potential of lesquerella.

1984). The NAR is a measure of the average efficiency of leaves on a plant. The high efficiency of NAR in treatments 50% and 50:25% was due to high CGR rates during this period. A disadvantage of using leaf area as a basis for growth expression is that only the average efficiency of leaves in producing dry matter is known.

The effect of stress on leaf water potential was more drastic than xylem water potential (Fig. 1). Stress after 50% flowering and irrigating the crop at 25% resulted in a leaf water potential of 0.5 MPa. The low leaf water potential had a severe impact on the net photosynthetic rate (PN) resulting in lowering the oil synthesis due to lack of photosynthates as most of the fats are synthesized from phosphates by gluconeogensis pathway. The effect of water deficit on oil yield and returns are presented in Fig. 2. Irrigating the crop at 25% AWC resulted in almost 50% reduction in returns compared to control.

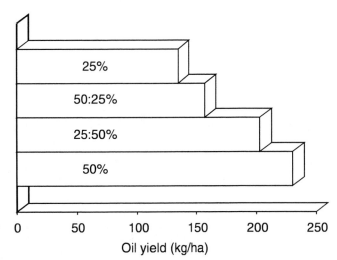

Fig. 2. Effect of water deficit on oil yield of lesquerella.

245

CONCLUSIONS

This study confirms that lesquerella can withstand mild water stress prior to flowering. Lesquerella should not be water-stressed from flowering to early pod formation and is considered to be the most critical stage for irrigation. Irrigating the crop at 75% depletion of the available soil water prior to flowering and 50% depletion after flowering can result in maximum growth and yield.

REFERENCES

Brown, R.H. 1984. Growth of the green plants. p. 153–174. In: M.B. Tesar (ed.), Physiological basis of crop growth and development. Am. Soc. Agron. Madison, WI.

Buttery, B.R. 1970. Effect of variation on leaf area index on maize and soybean. Crop Sci. 10:9–13.

Enyi, B.A.C. 1962. Comparative growth rates of upland and swamp rice varieties. Ann. Bot. 26:467–487.

Hunsaker, D.J., F.S. Nakayama, D.A. Dierig, and W.L. Alexander. 1998. Lesquerella seed production: Water requirements and management. Ind. Crops Prod. 8:167–182.

High Performance 4-Cycle Lubricants from Canola

Duane L. Johnson

IDENTIFICATION OF A MARKET

Developing a new market for an established international crop like canola (*Brassica napus* L., *Brassica rapa* L., Brassicaceae) requires significant care and study. Canola has been identified by several authors as a highly desirable food oil. Canola oils are relatively rich in monounsaturated fatty acids and with small percentages of both saturated and polyunsaturated fats. Canola can be differentiated from rapeseed by Canadian standards of low erucic acid and low glucosinalate content. Canola's relatively high percentage of monounsaturated fats, natural antioxidants, and lubricity make it ideal for lubricant applications

To develop a canola oil industry which does not compete with existing canola markets requires identification of new applications. Three objectives for a canola-based lubricant were identified: first, an oil functionality equal to or better than comparable petroleum oils; second, a cost which is acceptable to the general consumer; and third a product which is environmentally benign, allowing for some premium in cost and market.

Currently, numerous vegetable oils are used in lubricant applications. Current lubricant applications include additives in synthetic oils, transmission fluids, two-cycle motor oils, chain oils, hydraulic oils, greases and biodiesels. US consumption of these oils requires about 8 million kilograms of vegetable oil annually. This represents about 9% of all industrial applications for vegetable oils (USDA-ERS, 1995). To identify a market not currently served by a vegetable oil, a product was selected which is not typically vegetable oil based: four-cycle motor oils. This market consumes 9.66 billion liters of refined petroleum oil annually within the United States for lubrication purposes. Approximately 3.9 billion liters are consumed solely for four cycle engine oil (Fig. 1).

DEVELOPMENT OF A FOUR CYCLE MOTOR OIL

The internal combustion engine is an extreme environment for vegetable oil. The combination of extreme pressure and heat frequently cause a polymerization of the oil, changing the oil into a plastic within a short period of time. This polymerization has been observed since the development of the internal combustion engine. Prior attempts after the turn of the 20th century found the lifespan of vegetable oils to be unacceptable for modern engine applications. As engine compression increased with increasing requirements of power and performance, vegetable oils were found to be severely lacking in performance. Vegetable oils frequently would polymerize within hours of engine startup. Attempts to use fatty acid esters derived from vegetable oils were successful but production costs made them unacceptable to consumers.

The blending of any of several common base oils with hydroxy oils and wax esters, however, provided the initial functional motor oil. Modifications and variations of these oils have been made to adjust for various applications ranging from small, air cooled engine applications to water cooled, gasoline and diesel engines. Using a base oil derived from essentially any vegetable oil, initially being canola, with a relatively high concentration of monounsaturated fatty acids, was key to providing for an oil that met the objectives of the project.

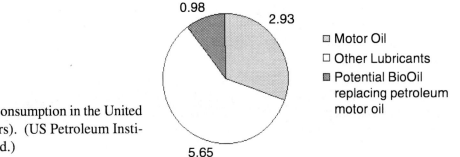

Fig. 1. Lubricant consumption in the United States (Billion liters). (US Petroleum Institute, 1990 modified.)

247

Evaluation of Vegetable Oil Functionality

A canola-based 4-cycle motor oil was constructed for use in a multitude of engine applications. Production cost-estimates were considered in formulations used to simulate actual production and performance expectations.

To test the oil formulation, a series of bench tests and small engine tests were used to confirm the functionality of the oil. Completion of these tests led to evaluations of the oil in everyday automotive applications. Bench tests included boundary friction analysis using a Timpken bearing test to measure pressure, time to seizure and scar size at seizure. Oxidative stability was measured using a modified AOCS CD-12B-92 hydraulic stability test (72 hrs @177°C). The oxidative stability of the oil was measured with a viscometer (cSt). An oil functionality index (b) was developed from the summed data of a commercial 10W-30 petroleum-based oil and various canola- and canola-soy oil formulations. Conventional and high oleic base oils were used in addition to various added oils based on commercial sources of hydroxy and wax ester sources. Base vegetable oils from either commercial canola, high oleic canola or soybean oils were combined with hydroxy oils from castor and wax esters from jojoba. Functionality indexes are given in Table 1. It became obvious that that the oleic content of the oil increases, index values increase. The addition of a hydroxy fatty acid source provides a motor oil superior to conventional petroleum. The further addition of a vegetable liquid wax ester more than doubles the functionality index (Table 1). The general conclusion was that a low cost base oil, combined with a source of hydroxy fatty acids and wax esters would provide adequate quality for use in engine applications. For all engine tests conducted a modified commercial canola-based motor oil formulation [commercial canola oil (can) + hydroxy fatty acid oil (hy) + liquid wax ester enriched oil (wx)] was used. Modification was done with commercially available vegetable-based oil additive packages.

Small Engines

Small engine trials were conducted on 5 and 6 hp air-cooled engines which were run under full load for 300 to 500 hours at regular oil changes at 35 hours. Consistent observations showed a 20% reduction in wear at friction points and a reduction of oil consumption of 63% over the petroleum lubricated check engine. Engine operating temperatures were also reduced by 15%–23% over the check in replicated trials. Observational data also show reductions of exhaust emissions from hydrocarbons and carbon monoxide (D. Peiper pers. commun. Univ. of Wisconsin, 1997).

Automotive Applications

Initially, three pre-1975 automobiles not equipped with catalytic converters or engine computers were selected. More recently, six newer automobiles manufactured from 1990 to 1998 have been included in the field trials. The field trials include daily commuting within municipalities as well as long distance, highway kilometers. A total of 10,000 hours in small engine trials and an additional 20,200 hours (173,000 kilometers) in automotive trials. Automotive trials were conducted in Colorado environments ranging from altitudes of 1,525 meters (5,000 feet) to 3,660 meters (12,000 feet). Ambient air temperatures ranged from –28.9°C (–20°F) to 43.3°C (110°F) and at humidities ranging from 8% to 95%.

Automotive engine trials were conducted using replicated samples nested within engines. This was done to minimize variation effects from engines and was considered a better test than a randomized block test. Prior small engine trials have indicated biasing of emissions data following use of the vegetable motor oil (reducing emissions

Table 1. Functionality indexes for vegetable oils.

Oil	Functionality index
10W-30	12.3
Commercial soybean oil	6.7
Commercial canola oil	8.7
Commercial high oleic canola oil	10.9
Commercial canola oil + hydroxy fatty acid rich oil	11.5
Commercial high oleic canola oil + hydroxy fatty acid rich oil	19.8
Commercial canola oil + hydroxy fatty acid rich oil + liquid wax ester enriched oil	24.6
Commercial high oleic canola oil + hydroxy fatty acid rich oil + liquid wax ester enriched oil	31.2

in the petroleum cycle) and so petroleum cycles required additional run time (10 to 20 hours) and purging with mineral-based oils. Petroleum-based 10W-30 motor oil was then used in the petroleum cycle. A similar purge system was used with the vegetable cycle with each replication.

Pre-1975 automotive applications included an air-cooled 1600 cc 1971 Volkswagen, a 1970 5 liter Ford Mustang, and a 1966 6.4 liter Ford Thunderbird. These older vehicles were selected since they lacked emission control devices such as onboard computers and catalytic converters. More recent additions include: a 1990 2.2 liter Chrysler Cirrus, a 1996 8 liter Dodge Ram truck, a 1998 5.2 liter Dodge Ram truck, a 1998 Jeep Wrangler, and a 1990 4.2 liter Chevrolet APV.

Observations in Automotive Applications

Results of the automotive applications were similar to observations in the small engine tests. All vehicles tested to date show consistent reductions in automotive emissions of hydrocarbons, carbon monoxide and carbon dioxide. Fuel economy has likewise shown consistent increases from 3–6% in all vehicles tested. Across vehicles monitored, a 4.67% increase in fuel economy has been noted (Table 2).

Exhaust emissions have likewise been reduced by conversion to a vegetable-based motor oil. In the example of the 1966 Ford Thunderbird, not only did fuel economy increase but exhaust emissions were significantly reduced. In the case of the Ford, oil consumption was significantly reduced and this may be partially responsible for reductions in hydrocarbons and carbon monoxide (Table 3).

The Environmental Effects of a Canola-Based Motor Oil

Several physical and environmental concerns are addressed by the vegetable motor oil. These data are summarized in Table 4. Table 4 illustrates the "bench test" properties of the canola-based motor oil. The oil has superior properties for flash point, fire point, viscosity index, 4 ball wear, acute toxicity (trout toxicity test), biodegradability, and coefficient of friction. The vegetable oils were inferior to petroleum for pour point, oxidative stability, and low temperature cranking power. More recent studies have addressed these problems and preliminary data indicate they are solvable concerns.

The biodegradability test shows the oil will meet EPA standards for biodegradability. The oil when tested for marine toxicity was found to be over 210,000 times less toxic than its petroleum counterpart. Additional ICP analysis of used automotive oil show no significant metals in either the new or used canola-based motor oil. Heavy metals are an integral part of petroleum-based lubricants but are not added to the vegetable-based oil. No metal contaminants, with the exception of iron, were accumulated by the canola-based oil.

SUMMARY

A canola-based motor oil was found to be a feasible alternative to conventional and synthetic petroleum motor oils. Higher oleic fatty acid content was important to functional properties but due to current costs, a conventional fatty acid profile was used in the test.

Table 2. Sample fuel economy increases for replicated automotive trials using vegetable oil.

Automobile	Fuel economy increase (%)
'71 VW[z]	3.0
'70 Mustang	4.2
'66 Thunderbird	4.3
'90 Chevrolet	4.5
'98 Dodge	6.0

[z]Results of two replications using nested data

Table 3. Summary of emissions and oil consumption data for a 1966 Ford Thunderbird. Emission data from State of Colorado Emissions Testing Station, Ft. Collins, Colorado.

Variable	Canola-based oil	10W-30 petroleum
Hydrocarbons ppm (idle: 1230 rpm)	138	148
Hydrocarbons ppm (high speed: 2500 rpm)	75	89
Carbon monoxide ppm (idle: 1230 rpm)	3.14	4.94
Carbon monoxide ppm (high speed: 2500 rpm)	2.12	3.58
Carbon dioxide ppm (idle: 1230 rpm)	7.7	11.8
Carbon dioxide ppm (high speed: 2500 rpm)	10.7	12.9
Oil consumption (liters/1000 kilometers)	0.65	1.68

Table 4. Comparative functional and environmental properties of a canola-based motor oil. Data summary provided by US Naval Facilities Engineering Service Center.

Physical properties	ASTM method	Canola results	Petroleum 10W-30
Flash point (°C)	D 92	274	221
Fire point (°C)	D 92	320	243
Pour point (°C)	D 97	-30	-35
Viscosity index (cSt)	D 2270	193	93
Oxidative stability (min)	D 2272	164	225
4 ball wear (mm)	D 2266 (mod.)	0.61	0.69
Coeff. friction	D 2266 (mod.)	0.053	0.119
Cold cranking cSt@−20°C	D 5293	1,470	3,500
Biodegradability (%)	D 212	97.7	38.6
Acute toxicity, LC50 (ppm)	EPA/600/4-90/027F, 1993	7,320	0.03

Canola-based motor oils were evaluated in bench trials, small engine trials, and in automotive applications. In general, reductions in oil consumption, fuel consumption, engine operating temperatures and engine wear were universal in all engine trials.

Physically, the canola motor oil was superior to a comparable petroleum oil in five of seven categories. In terms of environmental concerns, the canola-based oil met EPA standards for biodegradability, marine safety, exhaust emissions, and fuel economy.

Additional improvements in the canola-based motor oil are possible and solutions are expected to improve deficiencies to make the oil more functionally superior to current petroleum products. Cost concerns may limit functional improvements but environmental concerns may overcome some cost constraints. Improvements in functionality and environmental safety appear viable within cost constraints.

REFERENCES

U.S. Petroleum Institute. 1990. Annual report. Washington, DC.

USDA-ERS. 1995. Vegetable oil marketing report. U.S. Printing Office. Washington, DC.

Evaluation of Salinity Tolerance of Canola Germination

Naveen Puppala, James L. Fowler, Linnette Poindexter, and Harbans L. Bhardwaj

Canola (*Brassica napus* L., *B. rapa* L., Brassicaceae) is a genetically altered form of rapeseed with low erucic acid, a 22-carbon chain fatty acid that is used in a variety of polymer and lubricant products. Interest in canola is increasing steadily among health-conscious consumers due to its lowest content of saturated fatty acids (<70 g/kg) among major oil seeds. Domestic production of canola would reduce the import costs, enhance the productivity of American farms, and diversify agriculture (Starner et al. 1995). Canola oil is now the third largest source of edible oil following soybean and palm oil (Nowlin 1991). This increased demand, and the need for crop diversification, will undoubtedly promote increased acreage of canola in the western US, where some soils are prone to become saline (Francois 1994). Sims et al. (1993) reported that canola yields in Montana increased greatly with increased availability of water under normal conditions with lowered mean oil content. The average yield of 2.2 to 2.7 t/ha with oil content of 32% to 49% has been reported in Virginia during 1992 to 1995 (Virginia Agricultural Statistics 1995).

The traditional approach to the cropping of arid lands has been to use conventional cultivars and modify the soil and water to meet the needs of the crop, or to make genetic selections from established cultivars for improved performance under arid conditions. Saline soils and saline irrigation waters present potential hazards to canola production. Germination is one of the most critical periods for a crop subjected to salinity. Germination failures on saline soils are often the results of high salt concentrations in the seed planting zone because of upward movement of soil solution and subsequent evaporation at the soil surface (Bernstein 1974). These salts interfere with seed germination and crop establishment (Fowler 1991).

In an effort to develop the low erucic acid cultivars, the plant breeders are at the same time attempting to look for seedlings, which are tolerant to salinity. The two species of canola *B. napus* and *B. campestris* are classified as tolerant to salinity as per Maas and Hoffman (1977) salt tolerance classification table. Maas (1990) reported that even though both the species exhibit high salinity thresholds, the rate of yield decline above the thresholds was much greater than most other crops in the tolerant category. Shafii et al. (1992) reported that winter canola cultivars grown Pacific northwest had significantly higher oil content than the same ones grown in the Southeastern US.

The tolerance of canola to salinity during germination however has not been reported. The objective of this study was to evaluate the germination response of canola to a wide range of cultivars, salinity levels, temperatures, and to determine their interactions.

METHODOLOGY

Evaluation for salinity tolerance during germination was accomplished by placing 50-seed samples in 90 by 15 mm plastic petri dishes containing one blotter paper to which 5 mL of distilled water or various solutions of NaCl and $CaCl_2$ (2:1 molar ratio of NaCl and $CaCl_2$) were added. Germination responses to concentrations of 0, 5.4, 10.1, 16.2, 21.6, and 26.4 dS/m of the combined salts were determined. Electrical conductivities and osmotic potentials were measured with a Model PM-70CB conductivity bridge (The Barnstead Co., Boston, MA) and a Model 5130C vapor pressure osmometer (Wescor, Inc., Logon, UT), respectively. The covered petri dishes were arranged in an incubator in a randomized complete block experimental design with one block per shelf over four shelves. Germination response to salinity at six temperatures (10, 15, 20, 25, 30, and 35°C) was evaluated by replicating the temperatures twice in the same incubator over time. Temperatures were maintained within ±1°C of target levels. Germination counts were made at 3, 6, 9, and 12 days after initiation of germination (DAI). One blank petri dish with a blotter paper per shelf was randomized with the treatment dishes. Distilled water (representing the mean loss of water from the blanks) was added to each petri dish on Day 3, 6, and 9 to maintain salt concentrations near the target levels throughout the germination period. Five cultivars (Falcon, Jetton, HNO31-91, ST9194 and W4689E) were tested for salinity, temperature, salinity × temperature, and salinity × temperature × cultivar responses. Analyses of variance and orthogonal contrasts were used to analyze the data (SAS Institute 1995). Data for each counting date were analyzed independently.

EXPERIMENTAL RESULTS

The germination response of canola seed to a wide range of salinity averaged over cultivars and temperatures are presented in Table 1. Germination generally declined linearly with salinity levels. As the salinity level increased from 10.1 to 16.2 dS/m there was almost 40% reduction in seed germination compared to control. Soils that are higher than 16.2 dS/m salinity can reduce the canola yield due to reduction in plant population.

Among the five canola cultivars used in this study, ST9194 showed significantly higher percent germination on all counting dates and was significantly higher than the other cultivars. This probably represents genetic resistance to salinity and might be exploited as a source of salinity resistance for breeding (Table 2).

Germination was severely limited at 5°C with the control germinating only 23% after 12 days after initiation of germination (DAI). The low germination response at 5°C suggests that this temperature be near the minimum germination temperature for high-quality canola seed. The optimum germination temperature for the control occurred in the 15 to 25°C ranges (Table 3). This range is similar for most of the oilseeds crops. Germination of canola at 0 salt stress was close to 100% after 3 days at 25°C. Significant difference at 0.01 levels were seen among cultivars, salinity levels, and different temperatures were seen at each counting dates. The interactions between cultivar × salinity, cultivar × temperature, and cultivar × salinity × temperature were also highly significant at each counting date at 0.01 levels.

Table 1. Canola germination during a 12 days period in response to different salinity levels averaged over cultivar and temperature.

Salinity levels (dS/m)	Germination (%)			
	Days after initiation			
	3	6	9	12
0	61.0	76.7	84.2	87.2
5.4	53.3	70.6	79.1	82.3
10.1	40.5	62.7	69.6	72.3
16.2	24.5	46.7	51.9	53.8
21.6	8.3	22.6	26.9	27.8
26.4	0.4	0.6	0.8	0.8
LSD (0.05)	3.1	2.6	3.0	3.0

Table 2. Germination of five canola cultivars as a response to salinity and temperature during a 12 d germination period.

Cultivars	Germination (%)			
	Days after initiation			
	3	6	9	12
Falcon	24.5	40.4	45.5	47.8
Jetton	29.6	46.8	52.4	55.6
HNO31-91	35.9	49.0	52.8	54.2
ST9194	38.0	55.8	63.0	64.6
W4689E	28.8	41.4	46.8	47.8
LSD (0.05)	2.8	2.4	2.8	2.8

Table 3. Cumulative germination percentage of canola as a response to cultivar and salinity during a 12 days germination period.

Temperature (°C)	Germination (%)			
	Days after initiation			
	3	6	9	12
5	0	1.0	15.4	22.7
10	0.8	40.6	53.0	57.7
15	39.1	62.4	66.4	67.5
20	48.2	67.9	69.6	69.8
25	53.3	61.1	63.4	63.4
30	45.0	56.2	58.6	58.6
35	33.1	37.7	38.3	38.4
LSD (0.05)	19.5	12.0	13.2	13.6

CONCLUSIONS

Based on the above results, the salinity tolerance of canola seed during germination should be classified as moderately tolerant (Maas and Hoffman 1977) over the 10 to 30°C temperature range. This tolerance classification was determined based on the salinity levels resulting in a 50% reduction in final germination at 12 days. Germination was most tolerant between 15 and 25°C, slightly less tolerant at 10 and 30°C and less tolerant at 35°C, but all falling within the moderately tolerant range.

REFERENCES

Bernstein, L. 1974. Crop growth and salinity. p. 39–54. In: J. van Schiffgaarde (ed.), Drainage for agriculture. Agron. Monogr. 17. ASA, Madison, WI.

Fowler, J.L. 1991. Interaction of salinity and temperature on the germination of Crambe. Agron. J. 83:169–172.

Francois, L.E. 1994. Growth, seed yield, and oil content of canola grown under saline conditions. Agron. J. 86:233–237.

Mass, E.V. 1990. Crop salt tolerance. p. 262–304. In: K.K. Tanji (ed.), ASCE manuals and reports on engineering 71. ASCE, New York.

Mass, E.V. and G.J. Hoffman. 1977. Crop salt tolerance: Current assessment. J. Irrig. Drainage Div., Am. Soc. Civ. Eng. 103:115–134.

Nowlin, D. 1991. Winter canola. Agr. Consultant 47(4):8.

SAS Institute. 1995. SAS user's guide: Statistics. Version 7. SAS Institute, Cary, NC.

Shafii, B., K.A. Mahler, W.J. Price, and D.L. Auld. 1992. Genotype × environment interaction effects on winter rapeseed yield and oil content. Crop Sci. 32:922–927.

Sims, J.R., D.J. Solum, D.M. Wichman, G.D. Kushnak, L.E. Welty, G.D. Jackson, G.F. Stallknecht, M.P. Westcott, and G.R. Carlson. 1993. Canola variety yield trials. Montana State Univ. Ag. Expt. Stat., Bozeman, Montana Ag. Research 10:15–20.

Starner E.D., H.L. Bhardwaj, A. Hamama, and M. Rangappa. 1996. Canola production in Virginia. p. 287–290. In: J. Janick (ed.), Progress in new crops. ASHS Press, Alexandria, VA.

Virginia Agricultural Statistics. 1995. Virginia Department of Agricultural and Consumer Services. Richmond.

Canola Oil Yield and Quality as Affected by Production Practices in Virginia*

David E. Starner, Anwar A. Hamama, and Harbans L. Bhardwaj

Rapeseed (*Brassica napus* and *B. rapa* L., Brassicaceae), is now the third most important source of vegetable oil in the world. Canola (CANada Oil-Low Acid, an international registered trademark owned by the Canola Council of Canada) is the name given to a group of rapeseed cultivars that are low in erucic acid (22:1) and low in glucosinolates. Canola oil is considered healthy for human nutrition due to its lowest content of saturated fatty acids among vegetable oils and moderate content of poly-unsaturated fatty acids. The annual demand for canola oil by US consumers has increased from about 45 million kg to over 635 million kg, worth over $400 million (USCA 1997). The US production of canola increased from virtually zero in 1986 to over 150,000 ha in 1996 but at this level, meets less than 10% of domestic demand.

The two Land Grant Universities in Virginia have attempted to develop canola as an alternate cash crop to substitute for winter wheat. Field research conducted from 1993–1995 in Virginia indicated that canola could yield about 2000 kg/ha which compares well with yield from other US and foreign locations (Starner et al. 1996). The objective of the present studies was to characterize the effects of production practices (nitrogen fertilizer rates, planting dates, and seeding rates) on yield and quality of canola oil in Virginia.

METHODOLOGY

Five canola cultivars (Cascade, Ceres, Cobra, Doublol, and Jetton) were planted in replicated field experiments during 1994–95 crop season to determine the effects of three planting dates (Sept. 13, 28, and Oct. 7) on yield and quality of canola oil. Other experiments evaluated seeding rates (5.4, 3.6, and 1.8 kg seed/ha) and nitrogen application (0, 50, 100, 150, and 200 kg N/ha) with 'Jetton' cultivar. The seeding rate and nitrogen rate experiments were planted on Sept. 28, 1994. All plots were harvested at maturity, approximately during the first week of June, 1995.

Lipids were extracted from 20 g of ground seed three times at room temperature by homogenization with hexane/isopropanol (3:2, v/v) (St. John and Bell 1990). The fatty acid methyl esters (FAME) of the lipid were prepared (Dahmer et al. 1989) and analyzed in a Varian model Vista 6000 Gas Chromatograph equipped with a fused silica capillary column (SP-wax10, 25 m × 0.25 mm i.d.), a flame ionization detector and a Spectra Physics model 4290 integrator. The carrier gas was He at a column flow rate 0.8 ml/min with a split ratio of 1:80. Oven, injector, and detector temperatures were maintained at 210, 240 and 260°C, respectively. Peaks were identified by comparison to retention of FAME standards and quantified by the aid of 17:0 as an internal standard. The data were analyzed using GLM procedure in version 6.11 of SAS (SAS 1996).

EXPERIMENTAL RESULTS

Significant planting date effects existed for seed yield, oil content, oil yield, and the content of 16:1 fatty acid (Table 1) but the interaction between entries and planting dates was, generally, non-significant. The mean oil yield during 1994–95 season was 922 kg/ha (Sept. 13 planting), 799 kg/ha (Sept. 28 planting), and 543 kg/ha (Oct. 7 planting). Delayed planting reduced both seed yield and oil content. The planting date did not effect contents of 14:0, 16:0, 18:0, 20:0, 18:1, 18:2, 18:3, 20:1, and 22:1 fatty acids but planting delay from either Sept. 13 or 28 to Oct. 7 increased the content of 22:0 by almost 3 and 7 times, respectively. The highest content of 16:1 fatty acid (1.05%) was observed from the Sept. 28 planting which was significantly higher than in canola planted on Sept. 13 (0.77%) and Oct. 7 (0.71%). Planting dates did not effect the content of saturated or unsaturated fatty acids. In general, planting date did not affect the quality of canola oil. Our results indicate that canola in the Northern Virginia should be planted from Sept. 13 to 28.

---

*This research was supported by funds allocated to New Crops Program of Virginia State University by National Canola Research Program (Mid-Western Region) of U.S. Department of Agriculture through University of Southern Illinois, Carbondale.

Table 1. Effects of production practices on canola in Virginia.

Production practice	Seed yield (kg/ha)	Oil yield (kg/ha)	Oil content (%)	Fatty acid profile (Mol %)						
				Saturated			Unsaturated			
				14:0	16:0	18:0	16:1	18:1	18:2	18:3
Planting date[z]										
Sept. 13	2530a	922a	35.8a	0.1	4.6	0.7	0.8b	51.4	32.4	9.4
Sept. 28	2433a	799a	32.8b	0.1	4.5	0.8	1.1a	52.0	33.0	8.2
Oct. 7	1651b	543b	32.7b	0.1	4.7	0.7	0.7b	50.2	33.1	9.8
Nitrogen rate[y]										
0	2448	871	35.5	0.1	3.9	0.6	1.0	54.0	31.2	7.2
50	2673	883	33.2	0.1	3.9	0.7	0.7	61.2	26.1	6.9
100	2516	943	37.4	0.1	3.4	0.5	0.6	57.6	24.4	9.8
150	2202	827	37.8	0.1	4.9	0.7	1.0	49.6	34.2	8.5
200	2259	827	36.1	0.1	4.2	0.7	0.7	59.0	27.5	7.3
Seeding rate[y]										
1.8	3031c	1098c	36.2	0.1	5.3b	0.9	0.8a	56.7	27.4	7.9
3.6	3436b	1246b	36.2	0.1	6.0a	0.9	0.4b	53.4	30.5	8.1
5.4	3823a	1402a	36.7	0.1	5.6ab	0.9	0.6ab	54.4	28.8	8.6

[z]All means are from Cascade, Ceres, Cobra, Doublol, and Jetton cultivars. The Entry × Planting Date Interaction was non-significant. Mean separation by Duncan's Multiple Range Test, 5% level.

[y]All data from Jetton cultivar planted on Sept. 28. Mean separation by Duncan's Multiple Range Test, 5% level.

The effects of nitrogen rates on canola oil yield and quality were not significant (Table 1) but there was a trend towards increasing seed yield for nitrogen rate up to about 100 kg/ha. We consider the previously established rate of approximately 100 kg N/ha ideal for most Virginia locations.

Seeding rate significantly affected the seed and oil yield but not oil content or fatty acid profile. Oil yield increased from 1098 to 1402 kg/ha when seeding rate increased from 1.8 to 5.4 kg/ha (Table 1). Seeding rate of 5.4 kg/ha resulted in the highest seed yield of 3823 kg/ha. An increase in seeding rate from 1.8 to 3.6 kg/ha resulted in higher contents of 16:0 fatty acid.

CONCLUSIONS

Canola oil quality was, generally, unaffected by production practices investigated. The contents of saturated fatty acids in the oil of Virginia-grown canola varied from 4.0% to 7.0% indicating that the quality of oil from canola produced in Virginia is comparable to that from other locations. The mean saturated fatty acid in oil from canola produced in Virginia (5.5%) was in fact lower than the mean content of 7.1% reported in the United States (USDA 1998). In spite of these positive results, and our experience indicating profitability of canola in Virginia, locally available crushing facilities are needed before canola can be commercially grown in Virginia. Small crushers, capable of crushing canola produced on a few hundred hectares, offer an opportunity in this situation. The recent low market prices of winter wheat have encouraged Virginia farmers to look for alternative winter crops such as canola.

REFERENCES

Dahmer, M.L., P.D. Fleming, G.B. Collins, and D.F. Hildebrand. 1989. A rapid screening technique for determining the lipid composition of soybean seeds. J. Am. Oil Chem. Soc. 66:543–549.

SAS. 1996. SAS system for Windows. SAS Institute, Inc., Cary, NC.

St. John, L.C. and F.P. Bell. 1989. Extraction and fractionation of lipids from biological tissues, cell, organelles, and fluids. Biotechniques 7:476–481.

Starner, D.E., H.L. Bhardwaj, A.A. Hamama, and M. Rangappa. 1996. Canola production in Virginia. p. 300–303. In J. Janick (ed.), Progress in new crops. ASHS, Alexandria, VA.

USCA. 1997. U.S. Canola Association. Washington, DC.

USDA. 1998. Nutrient database for standard reference. http://www.nal.usda.gov/fnic/.

Colocynth: Potential Arid Land Oilseed from an Ancient Cucurbit

Zohara Yaniv, Ella Shabelsky, and Dan Schafferman

Citrullus colocynthis (L.) Schrad., Cucurbitaceae (colocynth or wild-gourd or bitter-apple), is a non hardy, herbaceous perennial vine, branched from the base. Originally from Tropical Asia and Africa, it is now widely distributed in the Saharo-Arabian phytogeographic region in Africa and the Mediteranean region. The stems are angular and rough; the leaves are rough, 5–10 cm in length, deeply 3–7 lobed; solitary pale yellow blooms. Each plant produces 15–30 round fruits, about 7–10 cm in diameter, green with undulate yellow stripes, becoming yellow all over when dry. Seeds are small (~6 mm in length), smooth and brownish when ripe. *C. colocynthis* occurs in many places in Israel, from the north to the hot desert, in sandy soils and wadis. It flowers between May and August (Feinbrun–Dothan 1978).

During biblical times, fruits were gathered and considered as a deadly poison (II Kings 4:39–40). The fruits are widely used medicinally, especially for stomach pains. The pulp, because of its content of glucosides, such as colocynthin, is a drastic hydragogue, cathartic, and laxative (Dafni et al. 1984; Burkill 1985). The fruits were exported as a laxative from the Gaza Strip to Europe in the early 20th century (Palevich and Yaniv 1991). The seeds are edible and when ground provide a rude bread for the desert Bedouins (Zohary 1982). The seeds have a high oil content (17–19%); in ancient times it was among the oils permitted to be used for candle light (Palevitch and Yaniv 1991)

In recent years there has been much interest in developing new oilseed crops which could be used in food, and for medicinal and industrial purposes (Yaniv et al. 1994). Many melon seeds (*Cucurbita* spp., *Citrullus* spp.) are rich in oil and protein (Al-Khalifa 1996) and although none of these oils has been utilized on an industrial scale, many are used as cooking oils in some African and Middle Eastern countries (El-Magoli et al. 1979). Melon seeds are utilized for oil production, especially in Nigeria (Girgis and Said 1968). Melon seed oil contains a large amount of linoleic acid (C18:2) which is important for human nutrition and an essential fatty acid and very little linolenic acid (C18:3) (Akoh and Nwosu 1992; Huang et al. 1994; Udayasekhara Rao 1994). Such oil composition resembles safflower oil (Yaniv et al. 1996) and is very beneficial for human diets.

Accessions of wild *Citrullus colocynthis* have been collected in arid zones in Israel and kept in the Israel Gene Bank. This gene pool was evaluated for chemical and agronomical characters, in order to test the seeds as candidates for a potential new oilseed crop in Israel—a crop adapted to arid zones.

METHODOLOGY

Plant Material

Seeds of wild *Citrullus colocynthis* were collected during 1983/84 throughout the Negev, Arava, and Sinai Deserts. The seeds were generally collected after fruit ripening, between October and December. Seeds of each accession are stored at the seed storage facilities of the Israel Gene Bank, in sealed aluminum cans, until needed for maintenance and evaluation.

Field Observation Trial of 28 Accessions of *C. colocynthis*

In 1995, 28 accessions of *Citrullus colocynthis* were studied at the Bet Dagan Experimental Farm, The Volcani Center, located in the coastal plain at latitude 32°01'N and an elevation of 50 m. An adequately

*Contribution from the Agricultural Research Organization, The Volcani Center, Bet Dagan 50250, Israel. The authors thank Mrs. Myra Manoah, Seed Exchange Officer, The Israeli Gene Bank for Agricultural Crops, for her assistance and co-operation in supplying seeds and information. We thank Prof. Y. Gutterman, Ben-Gurion Univ. of the Negev, Inst. for Desert Research, for his advice and guidance and Dr. Z. Amar for his contribution to the search for old citations regarding the use of *C. colocynthis* in ancient times.

fertilized and irrigated field, kept free of weeds and diseases, was used. The soil type at the farm is a deep fertile vertisol.

Seeds were sown in germination trays on 2 May 1995. The soil used was a mix of equal parts of tuff, peat, and vermiculite. Full germination (100%) was observed within one week. 23-day old seedlings were transplanted into the field on 25 May 1995 on raised beds, spaced at 2 m from center to center. There was one row per bed with five plants of each accession at a spacing of 50 cm between plants and 3 m between accessions. Each accession occupied an area of 10 m². Basic fertilization was done at the time of soil preparation at rates of 2N–2P–1K. "Trifluralin" (2.5 kg/ha) was applied as a herbicide. At the time of planting, the seedlings received 300 m³/ha water. A drip irrigation system was used throughout the growing period (June to August). The plants were irrigated weekly with 150 m³/ha water. Irrigation was terminated at the onset of fruit ripening. The total amount of water given to the plants was 2,100 m³/ha. Fruits were harvested according to maturation of each accession, and the seeds were analyzed for their fatty acid composition.

Field Experiment on Four Selected Lines

Based on the field-observation trial conducted in the 1995 season, four accessions representing various oleic and linoleic acid values were selected and tested at the Bet Dagan Experimental Farm during the 1996 season. Each accession was replicated four times in a random block design. Soil preparation and the size of the plots (10 m²) were the same as during the 1995 trial. Seeds were sown in germination trays on 6 May 1996 and transplanted into the field on 6 June 1996. All the fruits were harvested from each plot at maturation (when the fruit changed its color from green to yellow and the seed coat was dark brown), from 1 Sept. until 14 Nov., and divided according to five maturation periods: Sept. 1 (1–15 Sept.), Sept. 15 (15–30 Sept.), Oct. 1 (1–15 Oct.), Oct. 15 (15–31 Oct.), and Nov. 1 (1–15 Nov.). In addition, fruits from each maturation period were divided according to size into five sub-groups.

Data regarding the following parameters were collected during the growing season: earliness in flowering and ripening; fruit yield, fruit diameter, number of seeds per fruit, 1000-seed weight, seed yield, oil content (% of dry weight), and fatty acid composition. Oil content was determined on samples taken from each plot. The data were subjected to a two-factorial variance analysis and arranged in a split-plot design (the main plots were accessions and the sub-plots were the various periods) having four replications.

Lipid Extraction

Seeds were dried overnight at 50°C and ground into powder in a Moulinex coffee grinder. Five grams of powder were mixed with 100 ml petroleum ether (40–60°C), and the lipid fraction was extracted in a Soxhlet apparatus for 16 h at 60°C. The solvent was evaporated, and the lipid fraction residues were weighed (Yaniv et al. 1991).

Direct Transesterification from Seeds

Seeds (200 mg) were dried overnight at 50°C and ground into powder with a mortar and pestle, after which 0.3 ml of dichloromethane and 2.0 ml of 0.5N sodium methoxide (MeONa) were added. The tube was shaken and heated for 30 min at 50°C. The reaction was stopped by adding 5.0 ml of water containing 0.1 ml of glacial acetic acid. The esterified fatty acids were extracted with 2.0 ml petroleum ether (40–60°C). The clear fraction was kept at –20°C until further analysis. Samples of 2.0 ml were injected into the gas-chromatograph for fatty acid analysis (Yaniv et al. 1991).

Gas Chromatography of Methylated Fatty Acids

A Megabore column (DB-23, 0.5 mm film thickness, 30 m × 0.54 mm, J&W Scientific) was used in a gas-chromatograph equipped with a flame ionization detector (Varian 3700 GC) and an automatic area integrator (3390A-HP). The flow rate of N_2 was 30 ml/min and the oven temperature range was 135–200°C, programmed at a rate of 4°C/min.

The following fatty acids were identified by comparison with known standards (Supelco): C16:0, palmitic; C18:0, stearic; C18:1, oleic; and C18:2, linoleic acid (Yaniv et al. 1991).

Statistical Analysis

All data were analyzed by means of procedures of the SAS package (SAS Institute 1985). Statistical differences were calculated as Least Significance Difference (LSD) and evaluated with Duncan's Multiple Range Test (DMRT) at P=0.05.

EXPERIMENTAL RESULTS

Evaluation of *C. colocynthis* Wild Accessions

The results of the chemical analyses of the 28 accessions are summarized in Table 1. The predominant fatty acid of the seed oil was linoleic acid, C18:2, ranging from 67.0% to 73.0% of total fatty acids. The oleic acid (C18:1) content ranged from 10.1 to 16.0%. A similar pattern has been observed previously (Akoh and Nwosu 1992; Huang et al. 1994; Udayasekhara Rao 1994; Al-Khalifa 1996) but, the contents of both linoleic acid and oleic acid were generally higher in our study than those reported in these studies (i.e., 77.1% to 89% in our studies as compared with 75.8% to 82.3% in the literature). The fatty acid composition of the seed oil of *C. colocynthis* is very similar to that of safflower oil (Yaniv et al. 1996).

Field Evaluation of Four Selected Lines

Tables 2 and 3 present agronomical and chemical data collected during the cultivation of four selected lines of *C. colocynthis* in the 1996 season. The number of fruits harvested per plot (10 m²) varied within lines from 112 to 129, with Line 4 being the highest. Fruit size was determined by its diameter and varied from 5.7 cm (Line 1) to 6.3 cm (Line 2). Fruits of *C. colocynthis* contain a large number of seeds. Numbers varied, in lines, between 291 and 404 seeds per fruit; 1000-seed weight was the same in Lines 2, 3, and 4 (about 44 g) while Line 1 had significantly smaller seeds (40 g). Lines 2 and 4 excelled in seed yield (2.1 kg/10 m²) while Line 1 yielded only 1.5 kg/10 m².

Table 1. Chemical variability in seed oil of 28 accessions of *Citrullus colocynthis* collected in Israel and grown at Bet Dagan experimental farm during 1995.

Fatty acids		Mean	Range
		% of total oil	
Palmitic	(C16:0)	10.1	8.6–12.0
Stearic	(C18:0)	6.7	5.2–8.2
Oleic	(C18:1)	13.1	10.1–16.0
Linoleic	(C18:2)	70.1	67.0–73.0

Table 2. Agronomic data of *Citrullus colocynthis* lines evaluated in Bet Dagan during the 1996 season. (The observations are averages of four replications.)

Line	No. fruit/ 10 m²	Fruit diam. (cm)	No. seed/ fruit	1000-seed weight (g)	Seed yield (g/10 m²)
IN 34250	127.3a[z]	5.70b	291b	40.0b	1,479c
IN 34256	120.6b	6.32a	396a	43.4a	2,107a
IN 34262	112.5c	6.18a	404a	44.0a	1,893b
IN 34267	128.8a	6.12a	370a	43.6a	2,143a

[z]Mean separation by Duncan's multiple range test, 5% level.

Table 3. Chemical data of *Citrullus colocynthis* lines evaluated in Bet Dagan during the 1996 season. (The observations are averages of four replications.)

Line	Oil content[z] (%)	Calculated oil yield (L/ha)	Oleic FA content (%)	Linoleic FA content (%)	Oleic+linoleic FA yield (L/ha)
IN 34250	17.1c[y]	253b	11.7c	70.9a	209b
IN 34256	19.0ab	400a	15.0a	66.7c	327a
IN 34262	19.5a	369a	14.2b	68.2b	304a
IN 34267	18.5b	396a	14.3b	67.9b	325a

[z]Oil content based on seed dry weight.
[y]Mean separation by Duncan's multiple range test, 5% level.

Evaluation of Chemical Characteristics

Oil content in seeds varied from 17 to 19.5%. Based on this data and seed yield, oil yield of 250 to 400 L/ha was calculated. The two major fatty acids in seed oil of *C. colocynthis* are C18:1 oleic (11.7–15%) and C18:2 linoleic (66–70%). The total yield of unsaturated fatty acids varied from 209 to 327 L/ha. This oil contains only traces of C18:3 linolenic acid and thus could provide a good source of edible oil. Comparison between the lines, which were collected from different locations in Israel, points to the fact that Line 1 was significantly lower in almost all parameters evaluated, including seed and oil yield.

Yield Parameters

Seed yield (Fig. 1) was maximal in the first month of ripening and declined drastically until the end of the ripening period. The differences among lines indicated biodiversity. Early-ripening fruits (first month) contained heavier seeds than later-ripening fruits (Fig. 2), as measured by 1000-seed weight, perhaps late ripening seeds contain less water then early ripening seeds.

Larger fruits were obtained mostly during the first month of ripening (Fig. 3). Later in the season, ripe fruits became smaller and smaller. The best harvest time, in terms of fruit yield, fruit size, seed yield, and total oil yield was during the first four weeks of ripening.

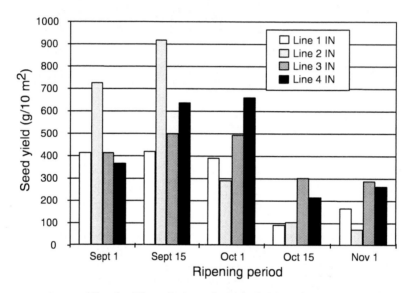

Fig. 1. The relation of seed yield to ripening.

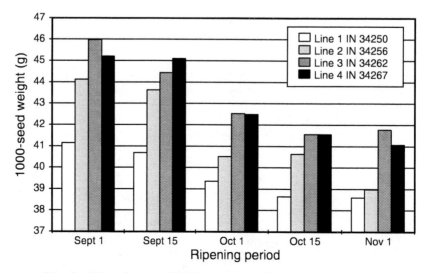

Fig. 2. The relation of 1000-seed weight to ripening period.

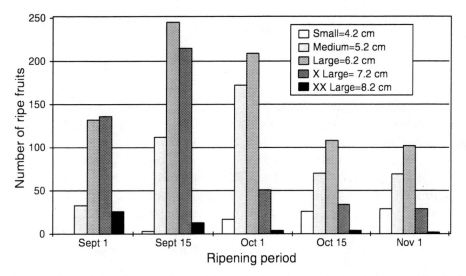

Fig. 3. Number of ripe fruits collected in each ripening period and arranged by fruit size.

CONCLUSIONS

C. colocynthis is a plant of dry climate and its cultivation should now be evaluated under arid conditions. To improve its agronomical properties it is desired to select lines with uniform harvest time and with higher oil content. A special attention should be given to evaluate its medicinal and diuretic potential as a source of high value by-products.

REFERENCES

Akoh, C.C. and C.V. Nwosu. 1992. Fatty acid composition of melon seed oil lipids and phospholipids. J. Am. Oil Chem. Soc. 69:314–316.

Al-Khalifa, A.S. 1996. Physicochemical characteristics, fatty acid composition, and lipoxygenase activity of crude pumpkin and melon seed oils. J. Agr. Food Chem. 44:964–966.

Burkill, H.M. 1985. The useful plants of west tropical Africa. Vol. 1, Families A-D. Royal Botanic Garden. Kew, UK.

Dafni, A., Z. Yaniv, and D. Palevitch. 1984. Ethnobotanical survey of medicinal plants in Northern Israel. J. Ethnopharmacol. 10:295–310.

El-Magoli, S.B., M.M. Morad, and A.A. El-Fara. 1979. Evaluation of some Egyptian melon seed oils. Fette Seifen Anstrichmittel 81(5):201.

Feinbrun-Dothan, N. 1978. Flora Palaestina—Part III. The Israeli Academy of Sciences and Humanities, Jerusalem.

Girgis, P. and F. Said. 1968. Characteristics of melon seed oil. J. Sci. Food Agr. 19:615–616.

Huang, K., C.C. Akoh, and M.C. Erickson. 1994. Enzymatic modification of melon seed oil: Incorporation of eicosapentaenoic acid. J. Agr. Food Chem. 42:2646–2648.

Palevitch, D. and Z. Yaniv. 1991. Medicinal plants of the Holyland. (in Hebrew) Tamus Modan Press, Tel Aviv. p. 56–58.

SAS Institute. 1985. SAS Users' Guide: Statistics, version 5. Cary NC.

Udayasekhara Rao, P. 1994. Nutrient composition of some less-familiar oil seeds. Food Chem. 50: 379–382.

Yaniv, Z., Y. Elber, M. Zur, and D. Schafferman. 1991. Differences in fatty acids composition of oils of wild Cruciferae seeds. Phytochemistry 30:841–843.

Yaniv, Z., D. Schafferman, Y. Elber, E. Ben-Moshe, and M. Zur. 1994. Evaluation of *Sinapis alba*, native to Israel, as a rich source of erucic acid in seed oil. Indust. Crops Prod. 2:137–142.

Yaniv, Z., D. Schafferman, M. Zur, and I. Shamir. 1996. *Matthiola incana*: Source of omega-3-linolenic acid. p. 368–372. In: J. Janick (ed.), Progress in new crops. ASHS Press, Alexandria, VA.

Zohary, M. 1982. Plants of the Bible. Cambridge Univ. Press. Cambridge.

Cropping Systems for Stokes Aster*

Elizabeth J. Callan and Charles W. Kennedy

Stokes aster (*Stokesia laevis* L., Compositae) has the potential to become an industrial oilseed crop for epoxy acid, a compound widely used in the chemical industry. Its achenes contain vernolic acid (12, 13-epoxy-*cis*-9-octadecenoic acid), which is easily converted to epoxy acid (Campbell 1981). One obstacle to the development of Stokes aster into a crop is the fact that the perennial does not flower during its first year of growth (Campbell 1981, 1984). Spring seeded plants will not flower that summer and fall seeded plants will not flower the following summer. To overcome this economically unproductive first season, Stokes aster could be intercropped with a summer annual such as soybean [*Glycine max* (L.) Merr., Fabaceae]. Balasbramanian and Sekayange (1990) indicated that intercropping prolongs the exploitation of resources due to longer combined leaf area duration. Thus, intercropping can make better use of land area by overlapping the time needed in the field by the two crops. The main benefit of a soybean–Stokes aster intercrop would be the initial soybean yield during the first year of Stokes aster growth. Soybean would be the overstory crop in this system, substantially reducing the amount of light available for Stokes aster seedling growth. Larcher (1983) has indicated that plants do adapt to changes in light intensity over time as new tissue and organs form. In growth chamber and greenhouse studies, we have found that Stokes aster growth can be reduced by low light intensity, but adjustments in photosynthesis occurred and plants began recovering after shade was removed (Callan and Kennedy 1995a, 1996). Once recovered, Stokes aster should grow and produce normally through the length of the production cycle although no field research has been conducted to corroborate this. Moreover, the length of the production cycle is somewhat vague, but estimated at 3–5 years (Campbell 1981).

Because of this lack of extensive field data, our objectives were twofold. The first was to determine vegetative growth and seed yield of Stokes aster under three cropping systems (a spring-planted monocrop, a fall-planted monocrop, and a spring-planted intercrop with soybean). The second was to determine the change in yield over a several year period to identify a viable production cycle.

METHODOLOGY

Plantings were initiated in 1992 through 1994 at Baton Rouge, Louisiana (30°N Lat.) on a Mhoon silty clay loam (fine-silty, mixed, nonacid, thermic, typic Fluvaquent). 'Pioneer 9501' soybean was planted May 1, 1992, April 13, 1993, and April 23, 1994 on a 76.2 cm row spacing. Due to a limited seed supply coupled with variability in germination and emergence of the seed, Stokes aster seed of USDA accession BSLE2 and an unknown parent (BSLE2, BSLE1, BSLL1, or BSLL2) were initially germinated in germination paper, transferred to 5 cm-wide peat pots filled with Jiffy mix and transplanted to the field about 1 month later. The entire process was initiated at the time of soybean planting so the growth of Stokes aster would be on the same time frame as soybean, i.e., it simulated a field planting of Stokes aster. Each intercrop plot consisted of 4 soybean rows. Three 19 cm-wide rows of Stokes aster were planted in each of the 3 middles between each row of soybean. The spring-planted monocrop contained 5 rows of Stokes aster. Row width was 19 cm. The fall-planted monocrop of Stokes aster used the same procedures as the spring monocrop but was initiated in early October each year. All Stokes aster plots were 2.1 m long. Plot widths were 0.95 m for spring- and fall-planted monocrops and 2.28 m for the intercrop. Nutrient fertilization consisted of 2.9 mM N, 0.06 mM P, 0.86 mM K, 1 mM Ca, 1mM Mg, 1 mg/L Fe, and 1 ml/L stock micronutrients used to wet the Jiffy mix during seedling transfer. For 1992, a field application of 67 kg N/ha was applied to all spring-planted Stokes aster plots in the late summer and again in May of 1993. The late summer application was repeated for the 1993 planting, but the spring application occurred in late March and was 224 kg N/ha. Thereafter, only an early spring application of this amount of nitrogen was done for each cropping system. Soybean harvest occurred in late September of each year 1992–1994. Stokes aster harvest, beginning in 1993, occurred in mid August of each year.

---

*Approved for publication by the director of the Louisiana Agricultural Experiment Station as manuscript 98-09-0521.

Weed control was a combination of the herbicides vernolate (s-propyl dipropylcarbamothioate) and tri-fluralin (2,6-dinitro-N,N-dipropyl-4-(trifluoromethyl)benzenamine) along with hand weeding. The use of benomyl (methyl-1-[(butylamino)carbonyl]-H-benzimidazol-2-ylcarbamate) was used for *Phomopsis* blight control. Two to three applications of 1.12 kg/ha usually provided adequate control. Plants per m^2 and leaf number per m^2 were determined for each cropping system during the first year of growth. Light interception by Stokes aster under the soybean canopy was determined with a Li-Cor 1-m line quantum sensor. The sensor was placed parallel to the soybean row and light intensity was averaged over 9 equidistant positions across the row middle at the aster leaf level. This was done 3 times at about 30 day intervals during the growing season beginning about 70 days after planting. Monocropped and intercropped soybean yields were determined on 2, 2-m lengths of row (3.05 m^2). Stokes aster yield was determined from a 1 m^2 harvested area in the center part of each plot. The experimental design was a randomized complete block with 4 replications in 1992 and 1993 and 3 replications in 1994. Statistical analysis was conducted using the general linear model technique and mean separation used the least square means method (SAS 1985).

RESULTS AND DISCUSSION

Vegetative Growth

In a perennial-annual intercrop, the goal is an acceptable growth rate for the perennial and a good yield for the annual (Vandermeer 1989). Initial Stokes aster growth varied between planting year and cropping system. In 1992, seedling mortality was a factor in growth per unit land area as measured by leaf production. The soybean provided a shaded understory that undoubtedly reduced soil evaporation and surface temperatures as well as light intensity. This condition resulted in better aster seedling survival than in exposed monocrop plots (Fig. 1) which kept leaf production per m^2 higher in the intercropping system (Fig. 2). However, leaf development per plant was lower under the intercropped soybean canopy (data not shown). In 1993 and 1994, the negative effects of being the understory component of an intercrop was apparent for Stokes aster. Seedling mortality was less of a problem in the monocrop but tended to decline in the intercrop as did leaf production (Fig. 1, 2). Previous work (Callan and Kennedy 1995a) has shown that Stokes aster can tolerate low light intensity of at least 160 μmol/m^2/s photosynthetic photon flux density (PPFD) and continue to grow, although slowly. At PPFD less than 40 μmol/m^2/s Stokes aster leaf development becomes static and recovery is slow (Callan and Kennedy 1996). Although average mid-day PPFD reaching Stokes aster was usually at or above about 160 μmol/m^2/s in the field (Fig. 3), the distribution of light would vary with location of Stokes aster plants relative to the soybean row and also time of day. Thus, the total amount of light received was probably lower than the average would indicate. Moreover, the greatest decline in growth of Stokes aster during the intercropping period in 1993 was late in soybean development and corresponded to the time of soybean leaf drop. These senesced leaves covered the Stokes aster resulting in increased mor-

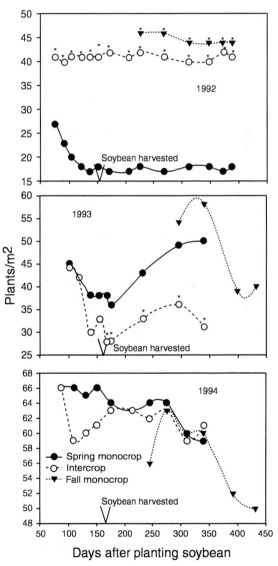

Fig. 1. Plant population of Stokes aster under different cropping systems. Symbols subtended by "*" at a given time are significantly different from the spring monocrop.

263

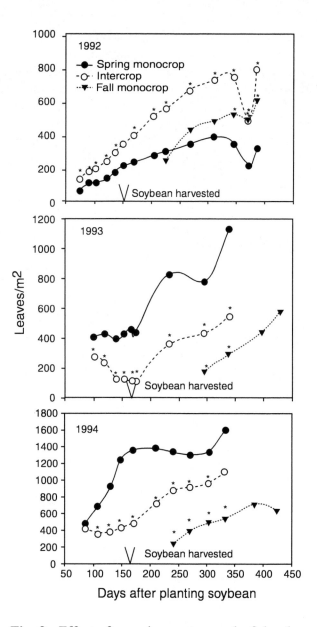

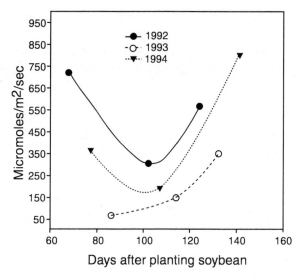

Fig. 3. Photosynthetic photon flux density reaching stokes aster intercropped under a soybean canopy.

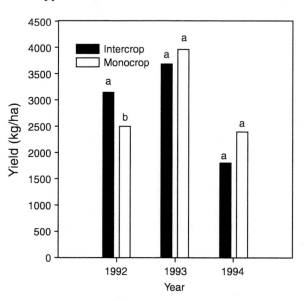

Fig. 2. Effect of cropping system on leaf development of Stokes aster. Symbols subtended by "*" at a given time are significantly different from the spring monocrop.

Fig. 4. Yields of monocropped soybean and soybean intercropped with Stokes aster. Bars having the same letter for a given year are not significantly different.

tality and leaf die-back. Because the population rebounded after a decline, the mortality applied only to the above ground portion of the plant; the growing point remained viable in many 'dead' plants. Regardless of performance under the soybean canopy, Stokes aster increased in growth after the canopy was removed (Fig. 2). This growth, coupled with plant age was enough to allow reproductive development the following spring. The fall-planted monocrop grew throughout its first year, but was too young and/or too small to be generally receptive to stimuli that caused a shift to reproductive growth during the first spring subsequent to planting.

Seed Yield

Soybean yields were generally unaffected by Stokes aster as an intercrop (Fig. 4). Since Stokes aster growth was small during the first 6 months, its effect on soybean was negligible. The effect of environment among years was not significant on seed yield of Stokes aster, but years in production from initial establishment did have a significant impact on seed yield. For this reason, data from each establishment (planting) year was pooled across production year for analysis. The effect of intercropping Stokes aster and the vegetative growth decline during the period in which it was under the soybean canopy did tend to depress aster seed yields in the first year following establishment. The second year harvest always had the greatest average seed yield for the Stokes aster originally intercropped (Fig. 5). Differences between spring-planted monocrop and intercropped Stokes aster yields (Fig. 5) may be for the reasons already alluded and also because 25% of the land area was not planted to Stokes aster in the intercrop. On an area-planted basis, the intercropped Stokes aster averaged higher yields than the monocrop (data not shown). Establishing Stokes aster in the fall resulted in an establishment duration of an additional seven months prior to blooming compared to the other cropping systems. This prolonged duration may have attributed to the slightly lower yields of that system (Fig. 5).

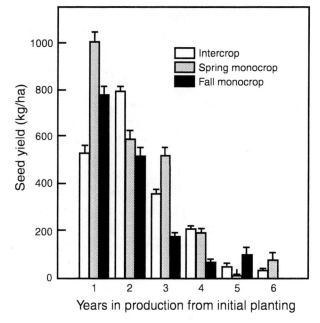

Fig. 5. The effect of cropping system and years in production from the initial planting on seed yield of Stokes aster. Average of three planting years. Error bars represent standard error of the mean. Yields of intercropped Stokes aster reflect the 25% of land not actually planted, i.e., the old soybean rows.

Although seed yield of Stokes aster has been estimated at about 2000 kg/ha and a production cycle of 3 to 5 years, (Campbell 1981), we did not find either to be the case in our study. Maximum yields were generally less than 1000 kg/ha. Moreover, maximum yields did not extend past one year of production (Fig. 5). The yield potential of Stokes aster at this location may have been lower due to soil type or some other limiting unknown factor. The limitation of the production cycle of this perennial crop was primarily a problem of sustained weed control, *Phomopsis* blight, and fire ant mound building. Although some research has been conducted on tolerance of Stokes aster to various herbicides (Campbell 1981; Callan and Kennedy 1995b) the ability to control some weed species, especially perennial clover species (*Trifolium* spp.) within an established Stokes aster planting is lacking. This weed encroachment was probably amplified due to the small plot sizes in this study. *Phomopsis* blight and possibly other diseases reduced the plant stand over time, but these diseases were usually kept in check with fungicide applications. Mound building by fire ants (*Solenopsis saevissima*) smothered established plants. We perceived a greater number of large fire ant mounds in Stokes aster plots than in the surrounding area, but have not substantiated this with data and the use of small plots may have amplified the negative effect. Regardless, application of insecticide can reduce this problem.

CONCLUSIONS

The use of an intercropping system to supplement land productivity during Stokes aster's establishment year appears viable. However, extremely dense overstory canopies and a subsequent large amount of leaf drop might limit the success of intercropped Stokes aster. Maximum yield potential at our location was less than 1000 kg/ha and a viable production cycle could last no more than two years. If Stokes aster is to become a viable industrial oilseed crop, additional efforts in breeding, and herbicide-biotechnology need to be undertaken.

REFERENCES

Balasurbramanian, V. and L. Sekayange. 1990. Area harvests equivalency ratio for measuring efficiency in multiseason intercropping. Agron. J. 82:519–522.

Callan, E.J. and C.W. Kennedy. 1996. Intercropping Stokes aster: Seedling growth under a soybean canopy. p. 363–367. In J. Janick (ed.), Progress in new crops. ASHS Press, Alexandria, VA.

Callan, E. J. and C.W. Kennedy. 1995a. Intercropping Stokes aster: Effect of shade on photosynthesis and plant morphology. Crop Sci. 35:1110–1115.

Callan, E.J. and C.W. Kennedy. 1995b. Tolerance of Stokes aster to selected herbicides. Indust. Crops Prod. 4:285–290.

Campbell, T.A. 1984. Responses of Stokes aster achenes to chilling. J. Am. Soc. Hort. Sci. 1009:736–741.

Campbell, T.A. 1981. Agronomic potential of Stokes aster. Am. Oil Chem. Soc. 9:287–295.

Larcher, W. 1983. Physiological plant ecology. Springer Verlag, New York.

SAS Institute, Inc. 1985. SAS User's guide: Statistics. SAS Institute, Cary, NC.

Vandermeer, J. 1989. The ecology of intercropping. Cambridge Press, New York.

Applications of Vernonia Oil in Coatings

D.L. Trumbo, J.C. Rudelich, and B.E. Mote

One of the more serious issues facing today's global community is preservation of the environment. A major part of this preservation effort is limiting the amount of contaminates introduced into the air and water of the planet. The paint industry shares these concerns and over the past two decades has made much progress in bringing more environmentally friendly products to the marketplace (Chatta 1980; Padget 1993; Nabauurs et al. 1996; Bhabe and Athawale 1997). Of prime concern has been the removal or reduction of volatile organic compounds (VOC) in paint and coatings formulations, as these substances can contribute to air pollution.

Approaches to limiting VOC content have included the introduction of very high solids (>75%) liquid coatings, powder or "solventless" coatings, and waterborne (latex) paints and coatings. In this report we will concentrate on very high solids types of coatings. Very high solids coatings can be achieved by using low molecular weight polymers which are soluble in high concentration in the chosen solvent. The resulting solution must have a viscosity in the appropriate range for the desired end use of the coating. Another approach to very high solids coating formulations is the use of reactive materials as solvents. Such materials become part of the polymeric film left behind when the coating dries or cures and does not contribute to VOC. Utilization of this approach yields formulations that are nearly solvent free.

In order to be used as a reactive solvent a molecule must have at least one and preferably two or more functional groups which are capable of reacting with moieties present in the main polymeric binder of the coating. The molecule must also be a reasonable solvent for the polymeric binder, producing solutions with appropriate viscosities. These are not the only criteria that must be met, particularly in regards to specific end uses, but they are the most important.

An examination of the structure of vernonia oil, Fig. 1, obtained from *Vernonia galamensis* (Cass.) Less., Asteraceae, shows that there are three epoxy groups/molecule of oil. Epoxy groups, even hindered ones such as those in vernonia oil, posses a relatively high degree of reactivity when compared with many other moieties. Additionally vernonia oil itself has a viscosity at ambient temperature which is in the useful range. Vernonia oil then meets two of the requirements of a material being considered for use as a reactive solvent. What was not known at the beginning of this work was vernonia oil's solvent power for various types of synthetic polymers. Thus, the first part of our investigation involved synthesizing styrene containing copolymers using vernonia oil as the solvent for the polymerization. The second part of our investigation involved the use of vernonia oil as a crosslinking agent for thermally curable coatings. We believe this two phase investigation will yield information as to the suitability of using vernonia oil in reactive solvent applications.

EXPERIMENTAL

Refined vernonia oil was obtained from Vertech Inc. of Plano, Texas and was used as received. All other chemicals were commercial grade materials and were used without further purification. Molecular weight measurements were made with a GPC equipped with a Waters 510 pump, 410 refractive index detector and two 30 mm Polymer Labs Linear microstyragel columns. Numerical values for the molecular weights were obtained by comparison to a polystyrene calibration curve solvent resistance was assessed as the number of

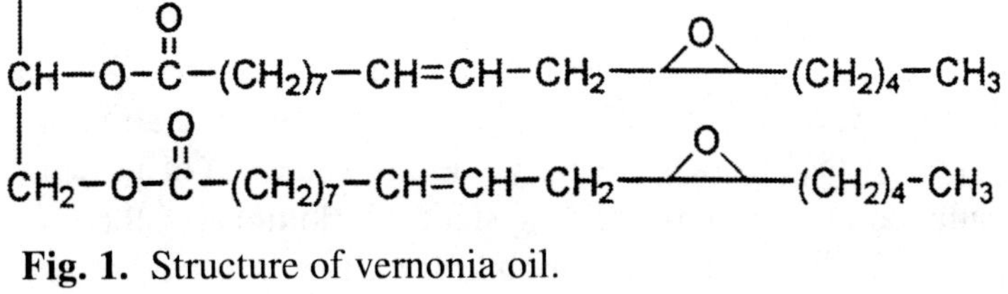

Fig. 1. Structure of vernonia oil.

methyl ethyl ketone double rubs required to break through a paint or coating film to the substrate below. The number of double rubs was measured by using an ATLAS AATCC crockmeter. Gloss measurements were made with a BYK-Gardner Microtrigloss meter. Impact measurements were made with a Gardner Impact tester employing a 4 lb weight. Pencil hardness was estimated with ASTM standard pencils.

Polymer Synthesis, Vernonia Oil Solvent

Vernonia oil (370 g) was charged to a 1 liter round bottom flask equipped with a mechanical stirrer, reflux condenser, thermometer, and pressure equalizing addition funnel. The vernonia oil (after sparging with N_2) was heated to 80°C with stirring and a mixture consisting of styrene (40 g), glycidyl methacrylate (24 g), n-butyl acrylate (136 g) and Vazo 64 (2.0 g) was added drop wise from the addition funnel over the course of 2 hr. After the addition was complete, another 0.3 g of Vazo 64 was added and the reaction mixture was stirred at 80°C for another 1.5 h to ensure complete conversion of monomer to polymer. Solution viscosity = 330 centipoise, Mn = 8000, mol. wt. = 17000.

Maleinized Linseed Oil

This material was synthesized from linseed oil and maleic anhydride via the procedure of Warth et al. (1997). The chemical analysis and physical properties matched these given by Warth for this material.

Coatings Formulation (UV Cures)

Coatings were formulated by mixing 25 g of the vernonia oil acrylic copolymer solution with the desired type and level of cure catalyst. Films were made by drawing the coating formulations over glass or aluminum panels with a #3 Bird bar (dry film thickness ~ 2 mils). The films were cured by passing the panels through a Fusion P-300 UV Cure instrument operating at a wavelength of 285–350 nm with a energy output of 1040 J/ cm^2. The exposure time for each panel was 80 seconds.

Coatings Formulations (Thermal Cure)

Coatings were formulated by mixing stoichometric (based on active functional groups) amounts of polymeric resin with vernonia oil. Because some of the polymeric resins employed had limited solubility in vernonia oil, it was necessary to add 25 wt % of an organic solvent (methyl ethyl ketone) to insure a homogeneous coating formula. Films were made by drawing the coating formulations over aluminum panels, as described above. The films were cured by baking in a forced air oven at 130°C for varying lengths of time. The panels were removed from the oven and allowed to cool ambient temperature before testing was begun.

RESULTS AND DISCUSSION

A summary of the results obtained from the UV cure study are presented in Table 1. Controls for these experiments were films made with vernonia oil alone, the styrene-acrylic alone and the vernonia oil/styrene acrylic copolymer solution without any added cure catalyst. Both the vernonia oil alone and the styrene-acrylic alone had cure catalysts added; 6974 for the vernonia oil and 6990 for the styrene-acrylic polymers.

The vernonia oil alone, with 6 wt % catalyst added, achieved 8–12 MEK double rubs before breakthrough to the substrate. However, the films were somewhat tacky and very soft. The styrene-acrylic resin itself gave films with 30–35 MEK double rub resistance before breakthrough to the substrate. The vernonia oil acrylic achieved 2–4 MEK double rubs.

The gloss readings for the films are very low. This is because the cured films had a significant amount of surface wrinkling. However, this part of our study did show that cured films with reasonable solvent resistance and hardness could be obtained. In addition, we showed that vernonia oil is a reasonable solvent for certain types of styrene-acrylic polymers.

Because most coating applications require films without surface wrinkling, and to test the reactivity of vernonia oil as a crosslinking agent, we decided to investigate the incorporation of vernonia oil into thermoset types of coatings. This part of the study will help guide the choice(s) of polymer resins(s) to be used with vernonia oil in the formulation of very high solids systems. Additionally, the heat applied to thermoset systems allows the polymers in the film to be more mobile for a longer period of time before the onset of cure.

Table 1. UV cure results.

Sample	Substrate	Cure catalyst	Cure catalyst[z] level (wt%)	Methyl ethyl[y] ketone double rubs	Pencil[x] hardness	60°[w] gloss
1	Al	6990	2.0	4	--	14
2	Al	6990	3.0	5	--	7
3	Al	6990	4.0	7	--	6
4	Al	6990	6.0	6	H	6
5	Glass	6990	2.0	7	--	--
6	Glass	6990	3.0	11	B	--
7	Glass	6990	4.0	22	B	--
8	Glass	6990	6.0	35	B	--
9	Al	6974	2.0	6	--	16
10	Al	6974	3.0	8	F	21
11	Al	6974	4.0	12	F	25
12	Al	6974	6.0	16	HB	28
13	Glass	6974	2.0	25	2H	--
14	Glass	6974	3.0	28	2H	--
15	Glass	6974	4.0	33	4H	--
16	Glass	6974	6.0	40	4H	--

[z]Curing agents were Cyracure 6974 and Cyracure 6990 from Union Carbide.
[y]The number of methyl ethyl ketone double rubs required to break through a film is a measure of cure. The greater the number of rubs the higher the degree of cure is judged to be.
[x]Pencil hardness is ranked from softest to hardest as follows: 3B, 2B, B, HB, F and H-8H.
[w]Gloss measurements are measurements of reflected light and are therefore not made on films drawn on glass.

Table 2. Polymer binder resin data.

Polymer	Composition[z]	Mn	mol. wt.	Tg (°C)[y]
Acid functional styrene-acrylic	Styrene/butylacrylate/acrylic acid	6000	8100	70
Amine functional styrene acrylic	Styrene/laurel methacrylate/amine monomer	6400	10300	50
Maleinized linseed oil	Linseed oil/maleic anhydride	1100	1200	--

[z]Exact monomer percentages of the acrylic copolymers is proprietary information.
[y]Tg = glass transition temperatures, the temperature at which a molten or fluid material becomes an amorpous solid.

Accordingly, we blended vernonia oil with an acid functional styrene-acrylic copolymer, an amine functional styrene acrylic copolymer, and maleinized linseed oil. As previously stated a solvent, methyl ethyl ketone, was added at the 25 wt % level to obtain homogeneous formulations. Table 2 summarizes the acrylic copolymer and malenized linseed oil data.

The film properties obtained from these resins cured with vernonia oil are summarized in Table 3. In these experiments two controls were used: vernonia oil alone and each resin alone. The vernonia does not cure under these conditions, even after heating at 130°C for 20 h. Only one or two methyl ethyl ketone double rubs were required to break through to bare metal. The maleinized linseed oil by itself attained 5–6 double rubs after 10 h at 130°C. The AFSA resin has 15–20 double rubs after 10 h at 130°C. The AMSA resins achieves 24 double rubs after 6 h at 130°C.

Table 3. Thermal cure results.

Starting resins[z]	Bake temp (°C)	Bake time (h)	Methyl ethyl ketone double rubs	Pencil hardness	60° gloss
MLO	130	0.50	6	3B	80
MLO	130	0.75	15	B	72
MLO	130	1.00	20	HB	78
MLO	130	3.00	40	H	75
AFSA	130	0.50	24	B	60
AFSA	130	1.00	42	F	72
AFSA	130	2.00	70	2H	81
AFSA	130	4.00	180	6H	88
AMSA	130	0.50	120	HB	96
AMSA	130	1.00	135	H	95
AMSA	30	2.00	145	2H	94
AMSA	130	12.00	260	4H	92

[z]MLO = Malenized linseed oil: AFSA = acid functional styrene-acrylic; AMSA = amine functional styene-acrylic

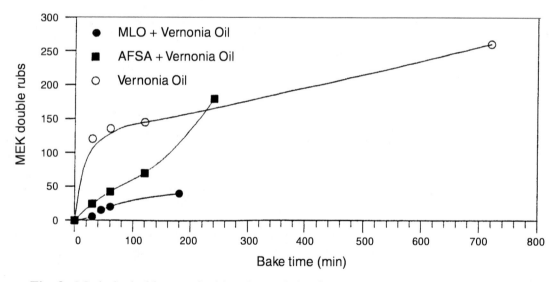

Fig. 2. Methyl ethyl ketone double rubs vs. bake time.

These results demonstrate that the vernonia oil is capable of crosslinking acid, anhydride, and amine functional resins. Not surprisingly, the reaction with the amine functional resins is the most rapid giving a relatively large number of methyl ethyl ketone double rubs in a shorter bake time. The AFSA resin also yields films with high degrees of cure if the films are allowed to bake long enough. The MLO resin attains the lowest double rub resistance, but this is not unexpected given that the degree of functionality. The MLO is less than the other two resins. However, as the plot in Fig. 2 shows, the cure response of the MLO is the same as that of the other resins; i.e. the degree of cure is increasing with increasing bake time.

The gloss numbers do show much higher values for the thermally cured films than for the UV films. This is despite the greater degree of crosslinking in the thermal films. The increased temperature does indeed allow for more polymer flow hence more uniform films with higher glosses.

CONCLUSIONS

The results of this study have shown that vernonia oil is a reasonable solvent for styrene acrylic copolymers (Dirlikov et al. 1990; Dirlikov and Friechinger 1991). These copolymer plus vernonia oil systems can be cured cationically using photoinitiators. However, the films are wrinkled and low in gloss (Crivello and Carlson 1996). In order to further assess the reacticity of vernonia oil and to produce films with higher gloss, vernonia oil was used as a crosslinker for several polymers containing moieties known to react with epoxy groups. In all cases, the vernonia did indeed act as a crosslinker, producing films with high methyl ethyl ketone double rub resistance if baked for a sufficient time. The films cured thermally also had higher glosses. This study showed that vernonia oil has three important attributes of a reactive solvent, reasonable solvating power, good reactivity and suitable viscosity. Work is in progress concerning optimization of vernonia oil based coatings with regard to specific applications requirements.

REFERENCES

Bhabe, M.D. and V.D. Athawale. 1997. Chemoenzymatic synthesis of oil modified monomers as reactive diluents for high solids coatings. Prog. Org. Coat. 30:207.

Chatta, M.S. 1980. High solids coatings. J. Coat Technol. 52:43.

Crivello, J.V. and K.D. Carlson. 1996. Photoinitiated cationic polymerization of natrally occurring epoxidized trigryerides. Macromol. Reports A33:251.

Dirlikov, S. and I. Friechinger. 1991. Low VOC fast air-drying coatings. Polym. Mater. Sci. Eng. Preprints 65:178.

Dirlikov, S., M.S. Islam, and P. Muturi. 1990. Vernonia oil: A new reactive diluent. Mod. Paint Coatings 80(8):48.

Nabauurs, T.R., A. Baijards, and A.L. German. 1996. Alkyd-acrylic hybrid systems for use as binders in waterborne paints. Prog. Org. Coat. 27:163.

Padget, J.C. 1993. Polymers for water based coatings: A systematic overview. Proc. 19th Int. Conf. Org. Coat. Sci. and Technol. Athens, Greece. p. 387–416.

Warth, H., R. Mulhaupt, B. Hoffmann, and S. Lawson. 1997. Polyester networks based upon epoxidized and maleinated natural oils. Angew Makromol. Chem. 249:79.

Variability in Oil and Vernolic Acid Contents in the New *Vernonia galamensis* Collection from East Africa

Ali I. Mohamed, Tadesse Mebrahtu, and Teklu Andebrhan*

Epoxy oils are important in industry for the manufacture of plastic formulations, protective coatings, lubricants, and other products. Current industrial techniques are expensive, generate large amounts of chemical waste, and produce high viscosity oil. A natural low-viscosity, epoxy oil is now available from the seeds of *Vernonia galamensis* (Cass.) Less, a herbaceous member of the sunflower family (Asteraceae). The low viscosity and polymerizing characteristics of this oil make it especially valuable as a solvent in industrial coatings and paints, for environments where fumes from traditional solvents are hazardous or polluting (Kaplan 1989). Some of the products that are being developed from *Vernonia* oil are degradable lubricants and lubricant additives, epoxy resins, adhesives, insecticides and insect repellants, crop-oil concentrates, and the formulation of carriers for slow-release pesticides.

The development of alternative crops is receiving increased recognition as an answer to some of the problems facing today's agriculture. New industrial crops could significantly diversify American agriculture and create markets that are essentially noncompetitive with existing crops. They would also provide a reliable domestic source of essential industrial feedstocks such as unique oils, many of which are currently imported (Aziz et al. 1984; Cunningham 1987; Kaplan 1989; Perdue 1989; Perdue et al. 1986; Thompson et al. 1994a,b,c). Establishment of a new industrial crop such as *Vernonia* can be an answer to problems facing farmers today who need a "high cash" crop as a primary source of income. This is crucial in states where farms consist of relatively small acreages and who are dependent upon a single cash crop

Vernonia galamensis is an annual herb and native of Africa (Perdue et al. 1989). It grows in areas with as little as 20 cm of seasonal rainfall. Plantings in Virginia, Arizona, and other states showed that *Vernonia* is extremely resistant to insects and diseases. *Vernonia* seeds contain up to 40% epoxy oil and this oil has up to 80% vernolic acid (cis-12,13-epoxyoleic acid). Plantings in Eritrea, Kenya, and Zimbabwe confirmed that *V. galamensis* has an excellent seed retention compared to *V. althemantica*.

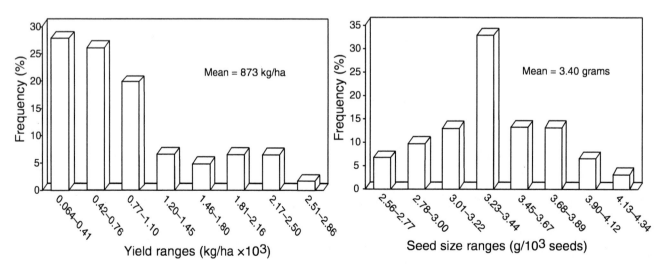

Fig. 1. Frequency of seed yield (kg/ha × 10³) distribution in *Vernonia galamensis* accessions collected in Eritrea.

Fig. 2. Frequency of seed size (grams/10³ seeds) distribution in *Vernonia galamensis* accessions collected in Eritrea.

*The authors are grateful for HBCU/USAID for funding the project and the collaboration of the research and field staff of the Ministry of Agriculture, The State of Eritrea. The authors also extend their appreciation to Drs. David Dierig and Terry Coffelt of the Water Conservation Lab, ARS/USDA for their advice and support.

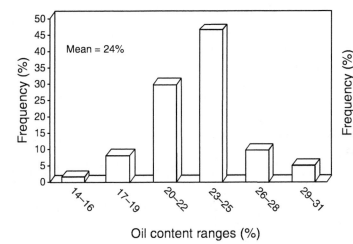

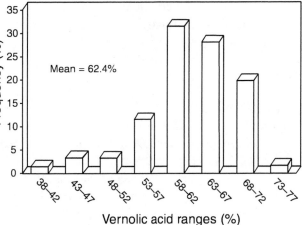

Fig. 3. Frequency of oil content (%) distribution in *Vernonia galamensis* accessions collected in Eritrea.

Fig. 4. Frequency of vernolic acid (%) distribution in *Vernonia galamensis* accessions collected in Eritrea

Despite the highly successful progress in domestication of *Vernonia* (Thompson et al. 1994a,b,c), substantial research is needed to evaluate *Vernonia* accessions, to make new selections for vernolic acid quantity and quality, and to determine the effect of environment, cultural practices, and processing on *Vernonia* oil. The main objectives of this research were to determine seed yield and yield components of the newly collected accessions and to determine oil content and fatty acid pattern

GERMPLASM COLLECTION

The existing *V. galamensis* germplasm collection at ARS/USDA is limited to 63 accessions. During the first year of the project new accessions were collected from Eritrea and Ethiopia through a USAID/HBCU grant. Collection was done by selecting individual matured inflorescences and planted for seed multiplication in Eritrea. A total of 61 accessions was collected and planted each in a single row for seed multiplication at Halhale research station in Eritrea. Accessions with adequate amount of seeds including breeding lines received from the Water Conservation Lab at ARS/USDA, Phoenix, AZ were planted in a four-row plot at the same location. At maturity each accession was evaluated for seed yield, agronomic traits, seed oil content, and fatty acid pattern. Oil

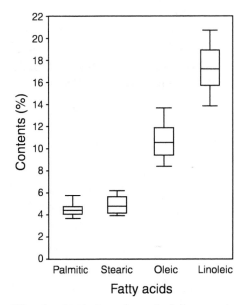

Fig. 5. Variations in palmitic, stearic, oleic and linoleic in *Vernonia galamensis* accessions collected in Eritrea.

and vernolic acid were analyzed using the methods of Ayorinde et al. (1990) and Mohamed et al. (1995a,b).

EVALUATION

During the rainy season of 1996 in Eritrea, the 61 accessions evaluated for agronomic and chemical parameters demonstrated the existence of wide genetic variability that could give the possibilities for genetic improvement of the crop. Significant differences for seed yield, oil content and vernolic acid were observed among the 61 accessions evaluated. The mean seed yield was 873 kg/ha and ranged from 60 to 2800 kg/ha (Fig. 1). The variation in yield was also reflected in seed size and the mean size was 3.4 grams/10^3 seeds (Fig. 2). The mean of the total oil was 24% and ranged from 14% to 31% (Fig. 3) where the majority (46%) of the accessions fall within the overall mean. The vernolic acid mean of the accessions was 62% and ranged from 38% to 77% (Fig. 4) and 49% of the accessions had vernolic acid content which exceeded the mean. The fatty acid profile of the accessions is given in Fig. 5. Emphasis will be given to the accessions that gave

273

highest seed yield, vernolic acid and oil.

A positive correlation (r = 0.28**) between oil percentage and vernolic acid was found. This indicates that breeding *Vernonia* for higher oil content will increase vernolic acid percentage. A highly significant and negative correlation (r = –0.90, –0.82, –0.95, and –0.96) were found between vernolic acid and palmitic, stearic, oleic, and linoleic acid, respectively.

REFERENCES

Ayorinde, F.O., K.D. Carlson, R.P. Palvic, and J. Mcvety. 1990. Pilot plant extraction of oil from *Vernonia galamensis* seed. J. Am. Oil Chem. Soc. 67:512–519.

Aziz, R., S.A. Khan, and A.W. Sabin. 1984. Experimental cultivation of *Vernonia pauciflora*—a rich source of vernolic acid. Pakistan J. Sci Ind. Res. 27:215–219.

Cunningham, I. 1987. Zimbabwe and U.S. develop vernonia as a potentially valuable new industrial crop. Diversity 10:18–19.

Kaplan, K.C. 1989. Vernonia new industrial oil crop. Agr. Res. 37(4):10–11.

Mohamed, A.I., H.L. Bhardwaj, C. Paul, and A.E. Thompson. 1995a. Vernonia lipase and its inactivation by microwave heating. Paper presented at 86th American Oil Chemist Society Annual Meeting, May 7–11, San Antonio, TX.

Mohamed, A.I., H.L. Bhardwaj, A.A. Hamama, and C. Webber. 1995b. Chemical composition of kenaf oil. Indust. Crops Prod. 4:157–165

Perdue, R.E. Jr. 1989. *Vernonia galamensis*, a new industrial oil seed crop for the semiarid tropics and subtropics. US Department of Agriculture (Mimeo)

Perdue, R.E. Jr., K.D. Carlson, and M.G. Gilfert. 1986. *Vernonia galamensis*, potential new crop source of epoxy acid. Econ. Bot. 40:54–68.

Thompson, A.E., D.A. Dierig, E.R. Johnson, G.H. Dahlquit, and R. Kleiman. 1994a. Germplasm development of *Vernonia galamensis* as a new industrial oilseed crop. Indust. Crops Prod. 2:185–200.

Thompson, A.E., D.A. Dierig, and R. Kleiman. 1994b. Variation in *Vernonia galamensis* flowering characteristics, seed oil, and vernolic acid contents. Indust. Crops Prod. 2:175–183.

Thompson, A.E., D.A. Dierig, and R. Kleiman. 1994c. Characterization of *Vernonia galamensis* germplasm for seed oil content, fatty acid composition, seed weight, and chromosome number. Indust. Crops Prod. 2:299–305.

In Vitro Characterization of Apomictic Reproduction in Guayule

Roy N. Keys, Dennis T. Ray, and David A. Dierig

Guayule (*Parthenium argentatum* Gray, Asteraceae) is a latex-producing perennial desert shrub native to southwestern Texas and north central Mexico. It is potentially an economically viable new crop for the desert Southwest, with the advantages of having low water requirements and producing non-allergenic latex (Thompson and Ray 1988). Diploid guayule plants reproduce predominantly sexually, and possess a sporophytic self-incompatibility system (Powers and Rollins 1945; Gerstel and Riner 1950; Gerstel 1950). Polyploid plants are self-compatible, and can also reproduce through facultative or obligate pseudogamous apomixis, predominantly through mitotic diplospory (Powers and Rollins 1945; Esau 1946). The frequency of apomictic reproduction can vary from plant to plant, and even from flower to flower on a single plant (Esau 1946). The factors that control the mode of reproduction are not known.

Breeding programs for *P. argentatum* would be facilitated if a relatively rapid and easy technique were available to characterize plants as to their mode of reproduction. Such a technique has been developed and tested on several members of the Poaceae that reproduce both apomictically and sexually (Matzk 1991; Mazzucato et al. 1996). Their test consists of the application of an auxin to the developing flowers. The hypothesis is that auxin will stimulate development of apomictic embryos in plants that are genetically disposed toward apomixis, but not in those that are truly sexual. In the Poaceae, apomictic embryos developed to apparent maturity in the absence of endosperm after auxin treatment. Such flowers were visually distinguished from normal filled and empty flowers (Matzk 1991; Mazzucato et al. 1996).

We tried the above auxin test in field trials in order to characterize the reproductive systems of *P. argentatum* breeding lines (data not presented in this paper). Initial field trials using 2,4-dichlorophenoxyacetic acid (2,4-D) showed inhibition of embryo production, so other auxins and concentrations were tested. Results of these trials were ambiguous, mainly because of problems with pollen release in isolation bags that could have resulted in self-pollination and the uncertainty that the auxins were actually penetrating the ovule. Therefore, a technique using in vitro floral culture was developed which provided better control of environmental factors and a better means of applying growth regulators. Auxin and other growth regulators inhibited embryo production in vitro, so that the best technique appeared to be in vitro floral culture without growth regulators, and this paper presents the results of these experiments. Seven breeding lines and a sexual diploid control were characterized for reproductive mode using this technique, and the results were substantiated with RAPD analyzes of progeny arrays of selected crosses. Thus, in vitro floral culture provided good characterization of reproductive mode in breeding lines of *P. argentatum*.

METHODOLOGY

Plant Materials

Plant materials were obtained from one-year-old plants being grown at The University of Arizona, Marana Agricultural Center, 50 km north of Tucson, Arizona. Seven lines were tested (G7-11, G7-14, G7-15, N7-11, N9-5, P2-BK, and P10-13). Four of these lines have been released and registered as improved germplasm (Ray et al. 1999). These plants are all putative tetraploids. Although they have not been characterized as to ploidy level, Cho and Ray (unpublished data) found that other lines derived from the same parental stocks as the plants used in this study were predominantly tetraploid ($2n = 72$), with a small proportion being either triploid or polyhaploid. Six known diploid plants were used as the sexual control in these studies. The plants were grown in potting soil in 3- to 5-gallon pots and under greenhouse conditions. Irrigation was provided by an overhead mist system. Grolux supplemental lighting was used to provide a 14-h photoperiod during winter months to stimulate flowering. Osmocote fertilizer was applied as needed.

Culture Procedures

Inflorescences were collected on the day on which cultures were to be made. Individual flower heads were selected based on the stage of development. The best stage was considered to be when the ray flowers had begun to open and the stigma was at least partially visible, but before the stigma began to spread open into the two surfaces it presents at maturity. As an added safeguard against stray pollination, each flower head was examined under a dissecting scope for the presence of pollen on the stigmas. The entire flower head was then sterilized by placing it in 0.075% sodium hypochlorite (15% commercial bleach) and 0.1% Tween-20 for 3 min, with agitation. The heads were rinsed in sterile, deionized, distilled water and placed in the culture medium. Each flower head was cultured in 2.5 ml of liquid medium in 2.5 cm × 6.5 cm glass vials with plastic lids. The cultures were placed on a shaker at 80 rpm, and grown at room temperature and under room lighting.

In Vitro Tests

In Vitro Experiment 1. This was a preliminary test on the effect of growth regulators. Nitsch and Nitsch (1969) (NN) medium was used, either alone or supplemented with NAA, NAA with gibberellic acid (GA_3), and NAA with kinetin (all growth regulators at a concentration of 1×10^{-6} M). The NN medium consisted of (mg/L): KNO_3, 950; NH_4NO_3, 720; $MgSO_4 \cdot 7H_2O$, 185; $CaCl_2$, 166; KH_2PO_4, 68; $FeSO_4 \cdot 7H_2O$, 27.8; $Na_2 \cdot EDTA$, 37.3; $MnSO_4 \cdot 4H_2O$, 25; H_3BO_3, 10; $ZnSO_4 \cdot 7H_2O$, 10; $CuSO_4 \cdot 5H_2O$, 0.025; $Na_2MoO_4 \cdot 2H_2O$, 0.25; *myo*-inositol, 100; nicotinic acid, 5; thiamine·HCl, 0.5; pyridoxine·HCl, 0.5. Sucrose (60 g/L) was added, and pH adjusted to 5.8 prior to sterilization. This test was non-replicated, using one flower head per treatment from each of five lines.

In Vitro Experiment 2. Two media were tested: NN and Woody Plant Medium (Lloyd and McCown 1980) (WP). WP medium consisted of (mg/L): NH_4NO_3, 400; $Ca(NO_3)_2 \cdot 4H_2O$, 556; K_2SO_4, 990; $CaCl_2 \cdot H_2O$, 96; KH_2PO_4, 170, H_3BO_3, 6.2; $Na_2MoO_4 \cdot 2H_2O$, 0.25; $MgSO_4 \cdot 7H_2O$, 370; $MnSO_4 \cdot H_2O$, 22.3; $ZnSO_4 \cdot 7H_2O$, 8.6; $CuSO_4 \cdot 5H_2O$, 0.25; $FeSO_4 \cdot 7H_2O$, 27.8; $Na_2 \cdot EDTA$, 37.3; *myo*-inositol, 100; thiamine·HCl, 1.0; nicotinic acid, 0.5; and pyridoxine·HCl, 0.5. Both media contained 60 g/L sucrose and pH was adjusted to 5.8 prior to sterilization. This experiment used flower heads from two to seven plants in six of the polyploid lines and five diploid plants, for a total of 31 flower heads per treatment.

In Vitro Experiment 3. The reproductive mode of the seven breeding lines were characterized using NN medium without growth regulators. From 4 to 12 plants were sampled within lines, and the experiment was repeated 2 to 5 times per plant, depending on the availability of flowers.

All experiments used a completely randomized design. The flower heads were collected after 14 days in culture.

Analytical Technique

The ray flowers were removed from the flower heads, with note being made of the presence of pollen on the stigmas. For in vitro-grown flowers, floral development was recorded as: 0 (no obvious development); 1 (some development of either the ray or disk flowers); 2 (apparently complete development of both ray and disk flowers, even if pollen had not been released). The single ovule was excised from each flower and placed in water on a microscope slide. After covering with a cover slip, the ovule was carefully squashed and the contents examined using phase contrast microscopy. Proembryos and embryos could be detected visually, and were recorded when present.

Making Crosses

Crosses were made on plants that were considered to be predominantly apomictic, partially apomictic, and predominantly sexual, based on results of the in vitro trials. After 14 days, the crosses were harvested and the mature embryos excised, with care being taken not to include any parental tissue or endosperm.

DNA Extraction and Evaluation with RAPD Markers

DNA was extracted from newly developing leaves of all of the plants used in the study. The miniprep technique of Stewart and Via (1993) was used in the extraction. The extraction buffer consisted of 2% w/v

CTAB, 2% w/v PVP-40, 1.42 M NaCl, 100 mM Tris·HCl, 20.0 mM EDTA, 5.0 mM ascorbic acid, 4.0 mM DIECA. A single leaf was placed in 450 μl of extraction buffer and 2.5 μl of 2-mercaptoethanol in a 1.5 ml microcentrifuge tube. The tissue was ground using a Kontes pestle with a handheld drive unit. After a 15 min incubation at 70°C, 375 μl of chloroform:isoamyl alcohol (24:1) were added and the mixture shaken for 5 min. The fractions were separated by centrifugation at 1000× g for 5 min. Isopropanol (0.7 vol) was added to the water fraction and the mixture was placed at 4°C overnight to allow precipitation of the DNA. The precipitate was pelleted at 14000× g for 20 min, the supernatant was discarded, and the pellets were air-dried overnight. After being dissolved in 60 μl of sterile water, the DNA was stained with Hoescht 33258 dye and quantified using a Hoeffer TKO-100 DNA fluorometer. All samples were diluted to 25 μg/ml. DNA of embryos derived from crosses was extracted in the same way, except that the initial amount of extraction buffer was 250 μl with 1.0 μl 2-mercaptoethanol, and the volume of all other components were adjusted accordingly.

RAPD markers were generated using Amersham Life Science components, and Operon 10-mer primer set OPB (Operon Technologies, Inc.). Each 50 μl reaction mix consisted of 1X PCR reaction buffer, with a total of 25 mM $MgCl_2$, 100 mM each of dNTPs, 0.2 mM primer, 2 U *Taq* DNA polymerase and 50 ng template in a reaction. The reactions were heated to 95°C for 2 min, then subjected to 45 cycles of 1 min at 93°C, 1 min at 36°C, and 2 min at 72°C, followed by a 5 min extention cycle at 72°C using a Techne PHC-3 thermal cycler. A single master mix was used for each progeny array and its two parents. 20 μl of the reaction was mixed with 6 μl of loading buffer and run on a 1.2% agarose gel in 1X TBE at 3.5 V/cm for 4 hr, and stained with ethidium bromide for UV visualization. The particular primer used for each array was determined from prior screening of the parental templates for RAPDs. A primer was informative if it produced a band in the male parent that was absent in the female. Progeny were scored as apomictic or outcrossed, depending on the presence or absence of the male band, respectively.

Statistical Analyzes

The embryo counts were transformed using the arcsine of the square root to permit analysis of variance. Data were analyzed using the General Linear Models Procedure (SAS Institute, Inc. 1988). Mean separations among lines were accomplished using the Duncan Multiple Range Test, and orthogonal contrasts were performed for each breeding line against the bulked data for the diploid controls. Correlation coefficients were calculated between the percentage of outcrossed progeny based on RAPD analyzes and the percent of embryo production in vitro.

RESULTS

In Vitro Experiment 1

Flowers developed normally in vitro, even exhibiting the dark brown pigmentation of epidermal layers that occurs in vitro as the flowers mature. Although pollen developed in the anthers, the high humidity conditions in culture prevented pollen release. There were no significant differences among treatments, probably because of the small sample size (Table 1). However, addition of NAA to the medium reduced embryo production by half. Combination of either GA_3 or kinetin with NAA resulted in a further four-fold reduction in embryo formation.

In Vitro Experiment 2

There were no significant differences between medium means for embryo production. On WP medium, 23% (standard error = 0.06) of the flowers produced embryos, while on NN medium 22% (standard error = 0.05) produced embryos.

In Vitro Experiment 3

There were significant differences among lines for embryo production (Table 2). Orthogonal contrasts of polyploid lines against the bulked diploid controls, in combination with separations based on a Duncan Multiple Range Test and the actual mean values enabled characterization of lines as predominantly apomictic, facultatively apomictic, or predominantly sexual.

Although not statistically significant, there was important within-line variation in expression of apomixis. For example, in line N9-5, which was characterized as highly apomictic, one plant never produced embryos in vitro. In contrast, in line P10-13, characterized as being predominantly sexual, there were a few plants that produced embryos in vitro.

Of the flower heads placed in culture 37% did not develop. To determine if this phenomenon was related to reproductive mode, an χ^2 test for heterogeneity among the lines was conducted and was nonsignificant ($\chi^2 = 11.141$, p = 0.084).

RAPD Analyzes

Primer OPB-15 produced one polymorphic band that was useful for determining outcrossing rates in progeny arrays. Outcrossing rates varied from 0% to 100%. The correlation of outcrossing rate with embryo production in vitro was –0.73 (p = 0.16).

DISCUSSION

The inhibitory effects of auxin application on embryo production in *P. argentatum* flowers grown in vitro were evident with the inclusion of NAA in the culture medium. Inclusion of NAA alone reduced this level by half, and in combination with other growth regulators resulted in a further four-fold decline in embryo production. Apparently, endogenous levels of growth regulators in the floral tissues are sufficient for embryo production. External application of growth regulators may cause either an inhibitory threshold to be exceeded, or a disruption of the balance of endogenous growth regulators so that the apomictic pathway is disturbed.

The lack of differences in embryo production when flowers were grown on the two media that were tested suggests that the particular nutrient combinations and concentrations are not crucial factors. NN and WP media differ considerably in their formulations. This lack of a medium effect was also reported for ovule and embryo culture of *Helianthus* hybrids in tests that also used NN medium as one of the media (Espinasse et al. 1991).

In the in vitro trials with NAA and other growth regulators, the embryos never developed to the extent that microscopic examination would not have been necessary. The embryos were allowed ample time for development in

Table 1. Effects of various growth regulators on embryo production in flowers of *Parthenium argentatum* grown in vitro.

Treatment	No. flower heads	Mean % of flowers with embryos[z]	S.E.
No growth regulator	5	32.0	16.2
NAA (10^{-6}M)	5	12.0	12.0
NAA (10^{-6}M) + GA$_3$ (10^{-6}M)	5	4.0	4.0
NAA (10^{-6}M) + kinetin (10^{-6}M)	5	4.0	4.0

[z]Differences among means non significant.

Table 2. Embryo production in flowers in vitro and classification of apomictic potential of polyploid and diploid breeding lines of *Parthenium argentatum*.

Line	No. flower heads	Mean % of flowers with embryos[z]	S.E.	Orthogonal contrasts against diploid control (Pr > F)	Apomictic potential
N9-5	15	21.7a	7.4	0.027	High
N7-11	29	21.0a	4.9	0.005	High
G7-11	14	20.0ab	9.4	0.013	High
P2-BK	42	9.0abc	2.5	0.064	Moderate
G7-15	29	9.0abc	3.8	0.150	Moderate
P10-13	22	5.4bc	3.3	0.406	Low
G7-14	8	0.0c	0.0	0.900	Low
Diploid	15	0.0c	0.0		Sexual

[z]Mean separation by Duncan Multiple Range Test (a = 0.05).

vitro, considering that embryos were mature enough after 14 days to fill the seed cavity in controlled crosses. Matzk (1991) and Mazzucato et al. (1996) reported that embryos in several members of the Poaceae developed enough after auxin treatment for the visual distinction of apomictic seeds, even without the presence of endosperm. The Poaceae possess an aposporic reproductive system, which may respond differently to auxin application than the predominantly mitotic diplosporic system in *P. argentatum*. Espinasse et al. (1991) incorporated NAA into the nutrient media in embryo rescue of hybrids in the related genus *Helianthus* to stimulate embryo elongation. *Helianthus annuus* also exhibited parthenogenesis in vitro (Yan et al. 1989). In this case, the highest rate of embryo formation occurred also without growth regulators in the medium and development advanced only to the early heart-shaped stage. It may be that parthenogenesis and subsequent embryo development have differing growth regulator requirements in the Asteraceae, as the result of being controlled by different genes or by different signals to the same genes.

The results with different media, and inclusion or exclusion of growth regulators, led us to the decision that in vitro floral culture in NN medium without growth regulators would be an adequate technique for the estimation of apomictic potential in *P. argentatum*. Using this technique, the reproductive systems of the seven lines used in this study could be characterized as being highly apomictic, moderately apomictic, or predominantly sexual. We expected the diploid controls to have a sexual reproductive mode (Powers and Rollins 1945; Gerstel and Riner 1950; Gerstel 1950). This proved to be the case, because they never produced embryos in vitro. RAPD analyzes of progeny arrays of selected matings of polyploid parents further substantiated the accuracy of this technique.

To conclude, the environmental factors that control the mode of reproduction in *P. argentatum* are unknown. The in vitro technique described here provides a means for studying the effects of environmental factors on floral development and expression of apomixis under controlled conditions. The presence of apomictic plants in predominantly sexual lines, and sexual plants in predominantly apomictic lines, may provide a pool of plant material adequate for the study of the molecular control of reproductive mode in *P. argentatum*.

REFERENCES

Esau, K. 1946. Morphology of reproduction in guayule and certain other species of *Parthenium*. Hilgardia 7:61–120

Espinasse, A., J. Volin, C.D. Dybing, and C. Lay. 1991. Embryo rescue through in ovulo culture in *Helianthus*. Crop Sci. 31:102–108.

Gerstel, D.U. 1950. Self-incompatibility studies in guayule II. Inheritance. Genetics 35:482–506.

Gerstel, D.U. and M.E. Riner. 1950. Self-incompatibility studies in guayule I. Pollen tube behavior. J. Hered. 41:49–55.

Lloyd, G.B. and B.H. McCown. 1980. Commercially-feasible micropropagation of mountain laurel, *Kalmia latifolia*, by use of shoot-tip culture. Proc. Int. Plant Prop. Soc. 30:421–437.

Matzk, F. 1991. A novel approach to differentiated embryos in the absence of endosperm. Sexual Plant. Repro. 4:88–94.

Mazzucato, A., A.P.M. den Nijs, and M. Falcinelli. 1996. Estimation of parthenogenesis frequency in Kentucky bluegrass with auxin-induced parthenogenetic seeds. Crop Sci. 36:9–16.

Nitsch, J.P. and C. Nitsch. 1969. Haploid plants from pollen grains. Science 163:85–87.

Powers, L. and R.C. Rollins. 1945. Reproduction and pollination studies in *Parthenium argentatum* Gray and *P. incanum* H. B. K. J. Am. Soc. Agron. 37:184–193.

Ray, D. T., D.A. Dierig, A.E. Thompson, and T.A. Coffelt. 1999. Registration of six germplasm lines of guayule with high yielding ability. Crop Sci. 39:300.

SAS Institute, Inc. 1988. SAS/STAT User's Guide, Release 6.03 Edition. SAS Institute, Inc., Cary, NC.

Stewart, C.N., Jr. and L.E. Via. 1993. A rapid CTAB DNA isolation technique useful for RAPD fingerprinting and other PCR applications. Bio Techniques 14:748–751.

Thompson, A.E. and D.T. Ray. 1988. Breeding guayule. Plant Breed Rev. 6:93–165.

Yan, H., H-Y. Yang, and W.A. Jensen. 1989. An electron microscope study on in vitro parthenogenesis in sunflower. Sexual Plant Repro. 2:154–166.

FIBER & ENERGY CROPS

Developing Switchgrass as a Bioenergy Crop*

S. McLaughlin, J. Bouton, D. Bransby, B. Conger, W. Ocumpaugh, D. Parrish, C. Taliaferro, K. Vogel, and S. Wullschleger

The utilization of energy crops produced on American farms as a source of renewable fuels is a concept with great relevance to current ecological and economic issues at both national and global scales. Development of a significant national capacity to utilize perennial forage crops, such as switchgrass (*Panicum virgatum*, L., Poaceae) as biofuels could benefit our agricultural economy by providing an important new source of income for farmers. In addition energy production from perennial cropping systems, which are compatible with conventional farming practices, would help reduce degradation of agricultural soils, lower national dependence on foreign oil supplies, and reduce emissions of greenhouse gases and toxic pollutants to the atmosphere (McLaughlin 1998).

Interestingly, on-farm energy production is a very old concept, extending back to 19th century America when both transportation and work on the farm were powered by approximately 27 million draft animals and fueled by 34 million hectares of grasslands (Vogel 1996). Today a new form of energy production is envisioned for some of this same area. The method of energy production is exactly the same—solar energy captured in photosynthesis, but the subsequent modes of energy conversion are vastly different, leading to the production of electricity, transportation fuels, and chemicals from the renewable feedstocks.

While energy prices in the United States are among the cheapest in the world, the issues of high dependency on imported oil, the uncertainties of maintaining stable supplies of imported oil from finite reserves, and the environmental costs associated with mining, processing, and combusting fossil fuels have been important drivers in the search for cleaner burning fuels that can be produced and renewed from the landscape. At present biomass and bioenergy combine provide only about 4% of the total primary energy used in the US (Overend 1997). By contrast, imported oil accounts for approximately 44% of the foreign trade deficit in the US and about 45% of the total annual US oil consumption of 34 quads (1 quad = 10^{15} Btu, Lynd et al. 1991). The 22 quads of oil consumed by transportation represents approximately 25% of all energy use in the US and exceeds total oil imports to the US by about 50%. This oil has environmental and social costs, which go well beyond the purchase price of around $15 per barrel.

Renewable energy from biomass has the potential to reduce dependency on fossil fuels, though not to totally replace them. Realizing this potential will require the simultaneous development of high yielding biomass production systems and bioconversion technologies that efficiently convert biomass energy into the forms of energy and chemicals usable by industry. The endpoint criterion for success is economic gain for both agricultural and industrial sectors at reduced environmental cost and reduced political risk. This paper reviews progress made in a program of research aimed at evaluating and developing a perennial forage crop, switchgrass as a regional bioenergy crop. We will highlight here aspects of research progress that most closely relate to the issues that will determine when and how extensively switchgrass is used in commercial bioenergy production.

THE HERBACEOUS ENERGY CROPS RESEARCH PROGRAM

The Bioenergy Feedstock Development Program (BFDP) at Oak Ridge National Laboratory has been conducting research for the Department of Energy since 1978 to identify and develop fast growing trees and herbaceous crops as well as to evaluate the potential crop residues as sources of renewable energy the nation's future energy needs (Ferrell et al. 1995). The program is comprised of both a woody crops component, which has developed short rotation forest production techniques for selected woody species such as hybrid

*The research described in this paper was sponsored by the Biofuels Systems Division of the U.S. Department of Energy, under contract No. DE-ACO5-96OR22464 with Lockheed Martin Energy Research Corp. In addition, we would like to acknowledge the contributions of M.A. Sanderson (original principal investigator at Texas A&M, now with USDA/ARS) and two new task leaders who joined this effort in 1997: M.D. Casler of Univ. WI and D.W. Burgdorf of USDA/NRCS.

poplar, willow, and sycamore; and an herbaceous crops program that has focused primarily on switchgrass, which we will discuss here.

After screening more than 30 herbaceous crops species during the 1980s (Wright 1994), a decision was made in 1991 to focus the future BFDP herbaceous crops research on a high yielding perennial grass species, switchgrass, which combined excellent conservation attributes and good compatibility with conventional farming practices (McLaughlin 1992). Switchgrass is a sod-forming, warm season grass, which combines good forage attributes and soil conservation benefits typical of perennial grasses (Moser and Vogel 1995). Switchgrass was an important part of the native, highly productive North American Tallgrass Prairie (Weaver 1968; Risser et al. 1981). While the original tall grass prairies have been severely reduced by cultivation of prairie soils, remnant populations of switchgrass are still widely distributed geographically within North America (Stubbendick et al. 1981). Switchgrass tolerates diverse growing conditions, ranging from arid sites in the shortgrass prairie to brackish marshes and open woods (Hitchcock 1951). Its range extends from Quebec to Central America. Two major ecotypes of switchgrass occur, a thicker stemmed lowland type better adapted to warmer, more moist habitats of its southern range, and a finer stemmed upland type, more typical of mid to northern areas (Vogel et al. 1985). The ecological diversity of switchgrass can be attributed to three principal characteristics, genetic diversity associated with its open pollinated reproductive mode, a very deep, well-developed rooting system, and efficient physiological metabolism.

As an open pollinated species, switchgrass expresses tremendous genetic diversity, with wide variations in its basic chromosome number ($2n = 18$), typically ranging from tetraploid to octoploid (Moser and Vogel 1995). Morphologically switchgrass in its southern range can grow to more than 3 m in height, but what is most distinctive is the deep, vigorous root system, which may extend to depths of more than 3.5 m (Weaver 1968). It reproduces both by seeds and vegetatively and, with its perennial life form, a stand can last indefinitely once established. Standing biomass in root systems may exceed that found aboveground (Shifflet and Darby 1985), giving perennial grasses such as switchgrass, an advantage in water and nutrient aquisition even under stressful growing conditions.

Physiologically, switchgrass, like maize, is a C4 species, fixing carbon by multiple metabolic pathways with a high water use efficiency (Moss et al. 1969; Koshi et al. 1982). In general C4 plants such as grasses will produce 30% more food per unit of water than C3 species such as trees and broadleaved crops and grasses and are well adapted to the more arid production areas of the mid-western US where growth is more limited by moisture supply (Samson et al. 1993).

The challenges of the herbaceous crops research program have been to combine the near-term objectives of maximizing potential current economic yields with the longer term objectives of improving and protecting yields through breeding and biotechnology (Sanderson et al. 1996). Included among the former are evaluating performance of the best currently available cultivars and determining optimum management regimes for increased production efficiency and environmental benefits. Breeding, tissue culture research, physiology, and molecular biology are components of the longer term objectives of improving switchgrass genetically.

The current switchgrass research program was initiated in 1992 with 7 projects implemented through collaborative research agreements at 5 university and two government laboratories. These projects have been augmented with additional breeding research at two locations, and with additional field testing sites at 6 US Department of Agriculture, National Materials Testing Centers, which are implemented through the Natural Resources Conservation Service. The current network of research sites encompassing regional field trails and testing sites, breeding activities, and basic research on tissue culture and physiology/genetics is shown in Fig. 1. In the following sections, we highlight progress towards the objectives of evaluating and improving production of switchgrass for use as a bioenergy crop.

Field Trials and Management Research

Through a network of 18 field sites established in 1992, yields of a total of 9 switchgrass cultivars have been evaluated. These tests have included two basic harvest regimes, a single cut late in the growing season versus a 2-cut system with the first cut typically at the date of formation of seed heads, around July.

Comparisons of yield performance among cultivars indicate that the most promising cultivars for bioenergy production are 'Alamo', for the deep South, 'Kanlow', for mid latitudes, and 'Cave-in-Rock' for the central and northern states. 'Alamo' and 'Kanlow' are lowland ecotypes, while 'Cave-in-Rock' is an upland ecotype. Yield data from three years of field trials (Fig. 2) emphasize the regional specificity of optimum cutting practices. In field trials, highest yields have typically occurred with the 2-cut system at VPI and Auburn study sites, while in Texas, where drought has been a frequent problem, the 1-cut system has been superior. Average yields of the best cultivar at each location were approximately 16 Mg/ha (7 dry tons/acre), while maximum yields at any plot within each of the 3 testing regions were typically ≥ 20 Mg/ha. We believe that the poor performance of the 2-cut system in Texas reflects the effects of cutting on persistence of deep roots, which would represent an impediment to late season water uptake under drier growing conditions.

Maintenance of a deep rooting system appears to be a key consideration in the management of switchgrass and a potential source of ultimate superiority of the 1-cut system in variable climatic regimes and over time. Our continuing research with 1- and 2-cut systems in Alabama indicates that, proper timing of the 1-cut system can be critical to yields attained as shown in Table 1. Here 1-cut and 2-cut systems yielded essentially the same biomass over a five year cutting cycle (approximately 27 Mg/ha-year), and the influence of timing of the first cut of the 2-cut system on yield attained is readily apparent. Additionally, the later harvests of switchgrass have generally lower ash contents (Sanderson and Wolf 1995). This is apparently associated with retranslocation of mobile nutrients, such as K, P, and N, and carbohydrates and storage in crowns and root systems later in the growing season as plants approach senescence. This apparently contributes to the relatively low nutrient requirements of switchgrass. It also reduces ash content of the feedstock making it more acceptable for use for combustion endpoints where boiler slagging of high ash fuels can be a problem (Miles et al. 1993; McLaughlin et al. 1996). The reduction in ash content, which includes parallel reductions in potassium, an important contributor to slagging, can also be attributed to increasing proportions of stem relative to leaf mass later in the growing season. Changes in tissue ash content are shown in relationship to the length of growth period (expressed as degree days) in Fig. 3.

Attaining consistent establishment results with switchgrass is a prerequisite for rapid scale-up of switchgrass production and therefore an important research priority. As a light seeded species, it is sensitive to proper planting depth (approximately 0.6–1.2 cm), firm seed bed establishment, and control of weeds during the first growing season, particularly if planted before warm temperatures allow it to compete well with cool season weeds (Wolf and Fiske 1995). Weed control is typically attained by a single application of

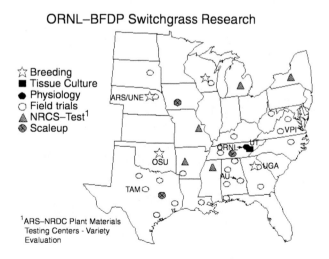

ORNL–BFDP Switchgrass Research

☆ Breeding
■ Tissue Culture
● Physiology
○ Field trials
▲ NRCS–Test[1]
⬡ Scaleup

[1]ARS–NRDC Plant Materials
Testing Centers - Variety
Evaluation

Fig. 1. Map of research activities by location in the BFDP herbaceous energy crops research task.

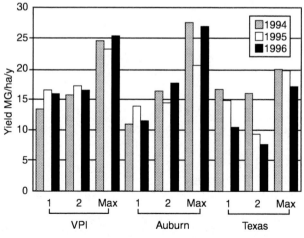

Fig. 2. Yield data for 1- and 2-cut harvest systems for 1994–1996 for the best cultivars averaged across all plots and for the best individual plots within each of three regional yield and management research centers (VPI, Auburn, and Texas A&M).

herbicide during the establishment year. In addition, high levels of seed dormancy, which can be removed by stratification or allowing adequate time for after ripening, must be considered in providing adequate germinable seed at planting (Wolfe and Fiske 1995). Because switchgrass allocates so much energy to root establishment, stands typically are not harvested during the first growing season, reach 2/3 of their capacity during the second year and full yield potential by the third year. Research at Virginia Polytechnic Institute has identified significant advantages to early establishment success by incorporating an insecticide at planting. With this as a component of prescribed planting instructions VPI was able to achieve 100% establishment success in a 20 farm field trial.

Table 1. Effects of harvest timing on yield of two switchgrass cultivars in research plots at Auburn, AL under single and two-cut harvest management schemes. Yield in megagrams per hectare (tonne/ha). Plots were established in 1992. Data from S. Sladden and D. Bransby.

No. of harvests	Timing	Cultivar	Yield (Mg/ha-year)	
			1997	1993–1997
2	May + Nov.	Alamo	8.4	56.0
		Cave-in-Rock	3.7	32.8
2	June + Nov.	Alamo	6.0	69.6
		Cave-in-Rock	0.9	43.8
2	July + Nov.	Alamo	23.8	135.8
		Cave-in-Rock	9.5	69.0
1	Aug.	Alamo	23.2	135.5
		Cave-in-Rock	7.5	58.5
1	Sept.	Alamo	40.3	135.7
		Cave-in-Rock	12.4	62.3
1	Oct.	Alamo	23.9	114.4
		Cave-in-Rock	12.4	61.9

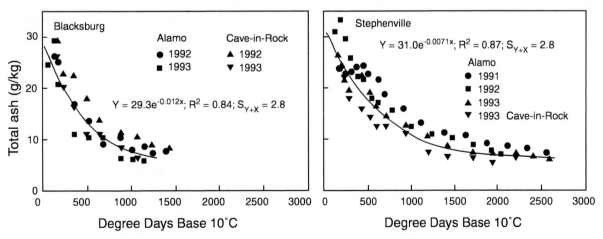

Fig. 3. Changes in ash content of harvested switchgrass (dry weight basis) with maturity level, expressed as growing degree days) for locations in Texas and Virginia. The data indicate that improved feedstock quality is realized by delaying harvest dates until nutrients can be retranslocated or leached from leaf tissue and mineral ash becomes a smaller fraction of total harvested biomass (after Sanderson and Wolf 1996).

Nitrogen management is an important component of any non-leguminous cropping system, but it is particularly important for bioenergy systems as nitrogen is an important cost energetically, economically and, potentially ecologically, as a contributor to air and stream pollution. "Standard" practices for switchgrass have called for ≤ 50 kg of N during the first year after switchgrass emergence, followed by 80–100 kg/ha thereafter (Wolf and Fiske 1995). Our research has included these and much higher rates in the search for an optimum balance between costs and yield. To date positive yield responses have been found up to and including 224 kg N/ha-year, however we suspect that long term yield stability and economics will be favored by lower annual or even longer interval applications of N, particularly where a single annual harvest is used. Data in Table 2, compare yields and nitrogen utilization rates of 1- and 2-cut systems from the Knoxville test location within the VPI system. An important feature of these data is that total N use was reduced by approximately 2/3 with the single cut system in all three cultivars examined. Interestingly, while single cut harvests removed only about half of the 110 kg/ha N supplied, two cut systems removed 50% more than was supplied. Since some weakening of the integrity of the single cut stands was beginning to appear during the 1995 season, nitrogen has been withheld during each of the last two growing seasons. Data analyses are not yet complete, however, early indications are that much lower N levels can be applied once stands are fully developed. This can be attributed in part to the level of root growth and the accumulation of soil carbon under perennial grasses as noted in studies in the Soil Conservation Program (Gebhart et al. 1994).

An additional consideration is the establishment of soil flora and fauna that are a part of nutrient cycles of perennial agricultural systems. Switchgrass is a mycorrhizal species, with dominant site-adapted mycorrhizal populations that stimulate growth, as BFDP-sponsored studies in Nebraska have indicated (Brejda 1996; Brejda et al. 1998). How long these take to establish to their full potential is not known and how important other microbial components involved in mineralization of soil organic carbon produced by switchgrass root turnover is not known. There is evidence that switchgrass can also apparently gain some nitrogen through fixation by associative soil bacteria (Tjepkema and Burris 1975). Thus establishment of an active mycorrhizal root system and the associated microbial community may take time to develop and may also be inhibited by high nitrogen application. We consider improved understanding of the nitrogen economy of switchgrass to be an important research priority in energy crops production and are pursuing ways to improve nitrogen utilization rates with lower and optimally-timed nitrogen application rates.

Breeding for Improved Yield

To date switchgrass has been bred primarily to enhance its nutritional value as a forage crop for livestock (Vogel et al. 1989). Thus, it has been managed primarily as a hay crop for which high leaf to stem ratio and high nutrient content are important. These targets are quite different from the criteria for biofuels crops for which high cellulose and low ash content are important for high energy conversion and low contamination of combustion systems, respectively. Earlier efforts to optimize both productivity and forage

Table 2. A comparison of nitrogen uptake and yield with three switchgrass cultivars reveals much lower N use in the single-cut systems. Nitrogen was applied at 110 kg/ha to all plots in the spring. Source: J. Reynolds and D. Wolf, VPI studies at Knoxville during 1995.

| | N Content (kg/Mg) | | | Totals | | | |
| | 2-cut system | | Single cut | Yields (Mg/ha) | | N-use (kg/ha) | |
Cultivar	1	2		2 cut	1 cut	2 cut	1 cut
Alamo	8.3	4.5	2.6	24.2	20.2	160	52
Kanlow	8.9	3.2	2.5	22.4	20.1	152	51
Cave-In-Rock	8.8	3.1	1.7	22.9	16.3	157	28

quality for livestock, and hence versatility of switchgrass led to the release of a new cultivar 'Shawnee' (Vogel et al. 1996). More recently, switchgrass breeding in the BFDP has included basic research on the phenology, genetics, and breeding characteristics, combined with multiple breeding approaches designed to improve switchgrass productivity as rapidly as practical. A nursery containing 110 switchgrass accessions from both existing genetic reservoirs as well as wild collections around the country has been assembled and characterized at Oklahoma State University (OSU). Genetic characterization has revealed that lowland switchgrass cultivars are predominantly tetraploid ($2n = 4x = 36$ chromosomes) while upland cultivars are predominantly octoploid ($2n = 8x = 72$ chromosomes) (Hopkins et al. 1996). In addition two cytoplasmic types occur and are differentiated between upland and lowland ecotypes (Elmore et al. 1993; Hultquist et al. 1997, 1998). Our studies have documented that self fertilization in switchgrass is very low (around 1–2% seed set), that hybridization potential between plants of different ploidy levels is very low (< 0.5%), but that plants of the same ploidy level can usually easily be intercrossed regardless of ecotype (Taliaferro and Hopkins 1996; Taliaferro et al. 1996). Switchgrass cultivars tested to date have been found to have a high tolerance of acid soil conditions with a low heritability of this trait (Hopkins and Taliaferro 1997), thus breeding for acid tolerance has not been pursued. Asexual seed production in switchgrass has not been detected to date in our collection of over 100 accessions.

Switchgrass breeding research over the past few years has provided new information to help identify the approaches most likely to provide the greatest genetic gains. In an effort to maximize and systematically evaluate the rate of genetic gains, initial breeding research at OSU focused on completing annual cycles of Recurrent Restricted Phenotypic Selection (RRPS) within four breeding populations. RRPS is a breeding system that has been highly effective in improving forage yields of other warm-season grasses (Vogel and Pederson 1993). It provides a means of increasing the representation of genes associated with phenotypic expression of desirable trait, for example, high yield. Populations selected for RRPS were comprised of the best available cultivars of upland and lowland ecotypes selected primarily from the central and southern Great Plains. Earliest efforts to speed the breeding process used detached flowering shoots to intercross 30–50 plants (5% selection index) visually judged to have the highest biomass production. The plants were selected and intercrossed near the end of the establishment growing season. Field stratification was used to minimize the effects of soil variation. Half sib seed from 25% of the visually selected plants was discarded after biomass yields of the plants that had been visually selected and crossed were subsequently verified at harvest. Three RRPS cyclic generations plus a narrow-genetic base synthetic cultivars were evaluated for performance for 2 years at Perkins, Oklahoma (Table 3). Genetic gains were indicated in all Northern Lowland (NL) RRPS cyclic populations and in the NL and southern upland (SU) synthetics.

Table 3. Comparative gains in breeding for improved yield of switchgrass with 2 cycles of Recurrent Restricted Phenotypic Selection (RRPS) and with a single cycle of narrow base synthetic selection. Four breeding populations—northern lowland (NL), southern lowland (SL), northern upland (NU), and southern upland (SU) are compared to the starting parent population in solid seeded stands in the first and second years after establishment at Perkins (OK). Source: C.C. Taliaferro.

Cycle	Yield (Mg/ha)			
	Kanlow (NL)	Alamo (SL)	Pathfinder (NU)	Caddo (SU)
Parent population	8.9 (100)[z]	10.4 (100)	6.6 (100)	7.5 (100)
RRPS Cycle 1	10.2 (115)**	10.3 (99)	5.8 (88)	7.7 (102)
RRPS Cycle 2	10.8 (121)***	9.9 (96)	6.8 (105)	7.9 (105)
Synthetic	11.7 (131)***	11.8 (113)	6.6 (101)	8.8 (118)*

[z]% of parent population
*P ≤ 0.10, **P ≤ 0.05, ***P ≤ 0.01 for comparisons with the parent population.

However deficiencies in the RRPS procedure, as a tool for yield improvement in switchgrass, were identified early in the program. A major constraint was inconsistency in attaining adequate seed set on detached flowering stems of selected plants. A second constraint was relatively low correlation between biomass yields of plants in the establishment year vs subsequent years. Though correlations were positive and often statistically significant, strong environmental influences on plant development were evident, particularly during the year of establishment. The importance and costs of developing a strong root system to growth and survival of young switchgrass plants apparently results in variations in aboveground growth during the establishment year that reduce its effectiveness as an indicator of longer term yield potential. In addition, studies indicated that the performance of half-sib progeny lines is a stronger indicator of the breeding value of individual plants than is the yield performance of the plants themselves. This realization has led the OSU program to place a greater emphasis on evaluation of genotype performance relative to the phenotype performance upon which RRPS is based.

Current breeding approaches include both genotypic and phenotypic recurrent selection in broad-genetic base populations combined with the development of narrow-genetic based synthetic cultivars. The synthetic cultivars are produced by intercrossing two or more elite parental plants. The parental plants are usually extracted from the broader base breeding populations. Genotypic recurrent selection under low- and high-yield environments is being evaluated at OSU to determine the effects of selection environment on the performance stability of cultivars developed from those breeding populations. This is an important issue in providing switchgrass cultivars that can be planted on the marginal lands where conventional crops provide only marginal economic returns. Research is underway in Oklahoma and Nebraska to determine if heterosis occurs for biomass yield in first generation single- and double-cross progeny populations. The development of the laboratory culture techniques later described, plus the strong self-incompatibility of switchgrass, make possible the development of hybrid cultivars. The economic feasibility of doing so depends on the added performance of the cultivars relative to standard cultivars. High performing hybrid cultivars capitalize on heterosis conditioned by dominance and by the control of multiple traits by a single gene (epistasis). Additional breeding approaches include a novel honeycomb selection design (Fasoulas and Fasoula 1995) being used at the University of Georgia that allows one to select superior plants in the field while considering both genetic and plant-environment interactions.

Basic Research on Propagation Techniques, Physiology, and Molecular Genetics

Research on tissue culture techniques for clonal reproduction of parent plants, physiological measurements of differences in foliar gas exchanges rates, and molecular fingerprinting constitute the tools with which we are trying to augment breeding activities, by gaining basic understanding of fundamental attributes of switchgrass biology.

Tissue Culture Technology. From a starting point of having no existing published protocols for switchgrass tissue culture regeneration, significant advances in developing such technology have been made in this program (Denchev and Conger 1993, 1994, 1995; Alexandrova et al. 1996a,b). At present hundreds of plantlets can be produced from a single parent plant and brought to field-ready status in a period of three months. The recent development of techniques for production of suspension cultures should significantly enhance that capability (Dutta et al. 1999). These techniques now make possible rapid development of isolated breeding blocks of superior plants for developing narrow genetic base synthetics as well as F_1 hybrids. This technique is currently being used at the University of Tennessee to test genetic gains from crosses involving clonal breeding blocks derived from 2, 4, or 20 elite parents of the cultivar Alamo.

In addition to its use in clonal propagation for breeding, production of tissue culture plantets and organ-specific differentiating tissues, including flowers as shown in Fig. 4 (Alexandrova et al. 1996a), provides new tools to explore genetic transformation in switchgrass (Denchev and Conger 1996). Having such tools in place will be important in incorporating into existing switchgrass cultivars new genes that can protect and improve growth or increase resistance to environmental stresses. Collaborative research has already begun to evaluate the feasibility of incorporating growth-enhancing promoters into switchgrass through transformation.

Molecular Genetics. Characterizing the molecular biology of switchgrass through the use of Randomly Amplified Polymorphic DNA (RAPD) markers is being used to provide a tool with which we can develop genetic fingerprints of existing and newly developing switchgrass lines (Gunter et al. 1996). The use of DNA markers as a tool has several uses including defining resident locations of genetic variability within the existing population of switchgrass, tracking the effects of genetic enhancement through breeding on genetic signatures, and examining the genetic stability and variability of switchgrass among commercial seed sources. A final application is the definition of the genetic stability of switchgrass stands as they develop over time in field plantings. To date, RAPD analyses have been used to develop a phylogenic diagram delineating genetic linkages among 18 existing switchgrass accessions in the OSU germplasm nurseries and to identify one low-fidelity commercial seed source for 'Alamo' switchgrass. Studies of genetic drift in field plantings and of seed source variability are continuing.

Gas Exchange Physiology. Measurements of switchgrass foliar physiology, isotopic fractionation of carbon isotopes, and nitrogen uptake have been examined in an effort to provide indicators of resource utilization efficiency that could be used to improve understanding of the capacity of switchgrass to adapt to diverse environmental conditions. Such information can also be used as a potential tool for screening accessions in breeding research. To evaluate variability in leaf physiological potential, gas exchange measurements were obtained for individuals leaves of 25 native accessions of switchgrass and two commercial cultivars planted at the OSU germplasm nursery. Significant differences were observed among populations for photosynthesis ($P = 0.003$), transpiration ($P = 0.001$), and water use efficiency ($P = 0.001$) (Wullschleger et al. 1997). Leaf level photosynthetic rates varied by almost a factor of two across accessions, from a high of 30.8 μmole m^{-2} s^{-1} to a minimum of 17.5 μmole m^{-2} s^{-1}. Water use efficiency, a potential indicator of growth potential under reduced water supply ranged from 2.08–3.77 μmole CO_2/mmole · H_2O. While higher leaf photosynthetic rates have been found with the faster-growing lowland cultivars in field studies in Texas and Virginia (Sanderson et al. 1995), Wullschleger et al. (1996) determined that differences between upland and lowland ecotypes were seasonally and environmentally dependent. During a very dry period in 1993, upland cultivars showed less reduction in photosynthetic rates than their lowland counterparts, reversing the differences among ecotypes observed under more favorable moisture supply earlier in the season.

In subsequent studies, carbon isotope discrimination d13C values were found to be rather constant (−14.6 per mil to −13.1 per mil) across accessions with no statistically significant differences among ploidy levels or ecotypes (Wullschleger et al. 1998), gas exchange physiology was not found to be statistically different between ploidy levels or ecotypes. Additionally leaf nitrogen levels, while they varied widely across accessions (1.33% to 2.25%), did not differ significantly among ecotypes or ploidy levels.

Collectively these measurements indicate that, while differences in single leaf physiological attributes are related to growth potential, the way in which they are integrated at the whole plant level in switchgrass is complex. They are most likely controlled strongly by the seasonal dynamics and plant-to-plant variations in plant and stand-level canopy architecture and allocation patterns that control the distribution of resources between shoots and roots.

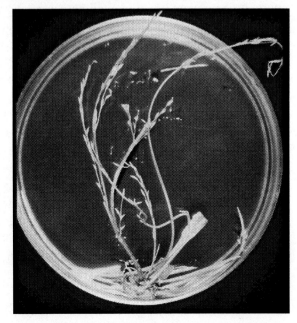

Fig. 4. Advanced regeneration techniques have been developed for switchgrass at the University of Tennessee, including production of flowers from tissue (node) culture. Clonal plantlets produced from tissue culture can be used to scale up numbers of plants from selected parents for breeding under field conditions or for genetic transformation or controlled pollination studies in the laboratory. Photos courtesy of B. Conger).

Belowground Biomass and Soil Carbon and Nitrogen Dynamics. Allocation of energy to an extensive rooting system is an extremely important aspect of the ecological adaptability, yield potential on marginal sites, and the soil conservation attributes of switchgrass. These attributes are related to nutrient and water uptake potential on degraded agricultural soils, nutrient use efficiency in capturing the benefits of applied fertilizers, and the effects of root growth and turnover on increasing soil carbon, improving soil texture and reducing soil erosion. Profiles of root distribution across 8 locations examined by VPI indicate that live root mass averaged 14.9 Mg/ha in the top 30 cm of the soil profile—approximately 2/3 of the annual harvest of aboveground biomass from these same plots. The maximum root biomass was found in shallow soils in plots in West Tennessee, and at 18.6 Mg/ha exceeded aboveground production at that site by approximately 50%.

As with most plant systems, root production in the surface soils is a predominant feature of switchgrass development and VPI studies indicate that approximately 50% and 75% of switchgrass roots in the top 90 cm of soil can be found in the top 15 cm and 30 cm of the soil profile, respectively. We have analyzed soil carbon gains in the surface horizons across a total of 13 research plots to date to document anticipated increases associated with root turnover and mineralization by switchgrass. These include measurements made after the first 3 years of cultivation in Texas, and after 5 years of cultivation in plots in Virginia and surrounding states. Preliminary analyses indicate that carbon gains will be comparable to, or greater than the 1.1 MgC/ha-year gains reported for perennial grasses, which included switchgrass, in studies in the Conservation Reserve Program (Gebhart et al. 1994). Additional studies are ongoing to document gains across deeper profiles and to standardize measurement protocols and minimize sampling variability across sites.

The issue of soil carbon gains and carbon turnover rates has become one of particular importance to energy crops for several reasons. First, soil carbon is well recognized as an extremely important determinant of soil fertility, as it controls both water and nutrient retention and lightens the texture of soils thereby promoting aeration, drainage of excess water, and root growth (Reeves 1997). This is an important issue because energy crops have the potential to improve the quality of agricultural soils depleted by decades of poor cropping management (McLaughlin et al. 1994). In this capacity they qualify as appropriate vegetative cover for fulfilling the soil conservation objectives of the Conservation Reserve Program. For this reason permission had been sought from the USDA and granted for use of existing grasses on 1600 ha of CRP lands as feedstock for power production co-firing tests in Iowa (the Chariton Valley Project). Ultimately plans call for replanting additional acreages to high yielding switchgrass cultivars that will improve the economics of bioenergy production on these lands.

The second issue tied to carbon increments and stability in the soil is that of global climate change. The approximately 30% increase in atmospheric carbon that has occurred during the last century is an important component of a global climate warming trend that is now well established (Thompson 1995; IPCC 1998). The importance of global climate change and the approximately 6 Gigatons (10^9 Mg) of anthropogenic inputs of carbon that enter the atmosphere each year has led to an international plan of reduced carbon emissions formulated in Kyoto, Japan, in 1997. Among the strategies being considered for reducing net increases in atmospheric carbon is increased reliance on energy crops, such as switchgrass, which can both displace fossil carbon inputs, with contemporary carbon removed from the atmosphere, as well as sequestering carbon in soils (McLaughlin and Walsh 1998; Romm et al. 1998).

Increases in soil carbon storage with perennial species offers the possibility of achieving added economic incentives derived from soil carbon storage credits. Isotopic and soil carbon fractionation studies being conducted at Oak Ridge National Laboratory (Garten and Wullschleger 1998) are being conducted to document and better understand the relationship of such gains in soil carbon to long term carbon storage potential. Characterization of the relative amounts of labile and non-labile carbon in the soil pools is being used to provide an indication of the expected longevity of incremental additions to soil carbon pools. In these studies, estimation of the rate of addition of root-derived carbon identified by its isotopic signature, to existing soil carbon pools has provided a preliminary estimate of 25–45 years for the turnover time for carbon derived from switchgrass roots.

EVALUATING THE COMMERCIAL POTENTIAL OF SWITCHGRASS

Bioenergy Markets and Sources

Production of transportation fuels, such as ethanol, and generation of electrical power are the two primary markets for bioenergy. The current biofuels industry in the US is based almost entirely (98%) on conversion of corn (maize) to ethanol (Petrulis et al. 1993). At present 1.3 million ha, approximately 6% of the US corn, is used in production of approximately 1 billion (B) gallons of ethanol each year. Estimates of future corn production of ethanol have been as high as 5 B gallons/year with potential net benefits to agricultural income of over $1 billion to US farmers (House et al 1993). However, recent analyses suggest that it is unlikely that corn can supply more than 2–2.5 B gallons of ethanol annually because of competing demands for corn. How much this figure will actually increase in the future depends largely on the success of agricultural, economic, and industrial research currently underway, including the development of markets for energy crops. A major consideration in the role that energy crops can play in achieving the goals of improved energy self sufficiency is their efficiency in displacing fossil fuels, a value closely tied to net energy returns. Recent calculations of the net energy gains from ethanol production from a forage crop like switchgrass indicate that both net energy savings and net carbon savings will be achieved much more rapidly than with more energy-expensive processes such as conversion of corn grain to ethanol (McLaughlin and Walsh 1998).

There are two principal sources of biomass-based renewable energy for these fuels—wastes and residues from agriculture and forestry and dedicated energy crops. Wastes such as wood and agricultural residues, municipal wastes, and poultry litter are now typically less expensive to supply to endpoint users, and will likely play an important role in early development of renewable energy supplies. However analyses of future demand for renewable energy indicate that these wastes may be capable of supplying only 14–30% of the total potential production of cellulosic ethanol and only approximately 18–60% of the production potential that could be derived from producing energy crops on currently idled or potentially available agricultural lands (Lynd et al. 1991). Thus dedicated energy crops will be required to meet the demands of a growing renewable energy market. Such crops, grown in the vicinity of the endpoint industrial user and specifically for the conversion process being used, offer important advantages of more systematic control of fuel quality, supply, and price stability than wastes derived from dispersed sources, which will be subject to alternate competitive endpoint uses and associated price fluctuations.

If these bioenergy crops are to realize their potential as a component of the national energy strategy, they must successfully compete both as crops and as fuels. Landowners will only produce those crops which provide a net economic return that is at least equivalent to conventional crops that they could produce on the same land for an equivalent level of effort. Low management intensity and positive effects on soil quality are important to landowners, but a stable source of income to supplement traditional crop returns will be a major determinant of their willingness to become involved. From the industrial perspective, both fuel cost and quality relative to alternate fossil fuels are essential considerations. An important function of the BFDP strategic plans has been to contribute to national efforts to analyze and continuously update inventories of available land, economic production costs of biofuels, and the levels and fuel characteristics of various biofuels produced within the program (Graham et al. 1995).

Potential Land Availability

At present the US has approximately 178 million (M) hectares categorized as "arable and permanent crop land" (FAO 1996). A smaller fraction of this land has been estimated to be capable of providing yields high enough to compete economically in bioenergy production. Graham (1994) established a baseline production potential of 11.2 Mg /ha-year as the criterion and used national crop production statistics to estimate that up to 131 M ha of crop land would qualify for herbaceous energy crops, such as switchgrass. Of that total, it was estimated that 91 M ha would also be suitable for fast growing, short rotation tree crops (Graham 1994).

291

The amount of land that will actually be used and the rate at which this will occur will be driven by the economics of bioenergy demand. Based on the existence of an estimated 35 M ha of idled land in 1988, the potential land area that could be incorporated into bioenergy crop production by 2012 was estimated at 60 M ha (Lynd et al. 1991). Based on a purchase price of $35/Mg, Ferrell et al. (1995) more recently projected a potential bioenergy crop area of 12 M ha by 2010. More recent estimates made using an agricultural supply and competitive pricing model (POLYSYS) are described below and indicate that energy crops would be competitive on 3.9 M ha at this price (Walsh 1998).

Bioenergy Crop Prices

Estimates of expected prices for bioenergy crops vary widely by crop, region, and estimation methods, including notably whether transportation costs are included. Walsh et al. (1998) estimated production costs to vary from $22/dry Mg to $110/Mg and transportation costs to range from $5/Mg to $8/Mg for a 25 mile transport distance. On a national scale ORNL estimates of bioenergy supply prices were $30–40/Mg at low (near term demand rates). A more detailed estimate of both bioenergy supply rates and prices has now been provided through the use of optimization models which consider the comparative economics of production of bioenergy and conventional crops (Walsh 1998). This approach has the benefit of integrating many factors that determine prices of specific bioenergy crops and their capacity to compete with conventional crops within their regions, as well as evaluating the regional differences in land availability and cost. These are factors that will be important in determining the feasibility of locating bioenergy facilities of various sizes in a particular region. These analyses have been used to provide estimates for the year 2007 of the total cropping area over which energy crops could compete successfully with conventional crops for two pricing options and three energy crops, switchgrass, hybrid poplar, and willow, as shown in Table 4. These analyses indicate that switchgrass would compete successfully with conventional crops on approximately 98% of the 3.9 M ha that would be available at a price of $38.5/Mg ($35/ton) for switchgrass (and identical prices per unit of energy content for the two other crops). As that price increased to $55/Mg, total area increases to 7 M ha and the representation of switchgrass remained at a the same relatively high percentage (97%) of the total. Total production of energy crops was estimated at 45 Mt and 79 Mt at the two respective pricing options. Longer term projections have indicated that other species such as hybrid polar may become increasingly competitive economically over time as production and harvesting technology improves (Walsh 1998).

From the perspective of impacts on current agricultural production, the crops that are most likely to be supplanted by the more competitive economics of switchgrass production on 3.9 M ha of agricultural land (pricing option 1 in Table 4) are predominantly non-alfalfa hay (2044 k ha), wheat (672 k ha), oats (328 k ha), and alfalfa hay (255 k ha). Only 85 k ha of corn would be displaced in this scenario.

Bioenergy Conversion

There are three principal technological endpoints for bioenergy crops: conversion to liquid fuels, combustion alone or in combination with fossil fuels to produce heat, steam, or electricity, and finally gasifica-

Table 4. Comparative land area projected by an econometric model to be available for energy crop production for each of three candidate energy crops at two prices levels paid to the producer (farmgate). Prices are based on a uniform cost per unit of energy and were set at $ the same per unit of energy for each of the three cropping options compared. Source: Walsh et al. 1998.

Energy crop	Price ($/Mg)	Land area (M ha)	Biofuel quantity (MMg)	Price ($/Mg)	Land area (M ha)	Biofuel quantity (MMg)
Switchgrass	38.5	3.9	45	55	7.0	79
Hybrid poplar	42.3	0.024	0.63	60.3	0.32	1.2
Willow	37.0	0.032	0.69	58.2	0.77	1.3
Total production		3.95	46.3	8.1	81.5	

tion to simpler gas products that can be used in a variety of endpoint processes. Energy crops such as switchgrass and hybrid poplar are classified as lignocellulosic crops because it is primarily the cell walls that are digested to form sugars, which can subsequently be fermented to produce liquid fuels. This is in contrast to energy recovery from corn grain, where digestion and fermentation of starch to produce sugars and ethanol is a well established technology (Wyman 1993). The rationale for developing lignocellulosic crops for energy is that less intensive production techniques and poorer quality land can be used for these crops, thereby avoiding competition with food production on better quality land. A potential limitation of some biofuels is that biochemical composition, energy content, and contamination with alkalai metals can limit their usefulness for some industrial applications (Miles et al. 1993). An analysis of the energy content and the level of alkalai and ash and combustion properties of switchgrass (Table 5) indicates that switchgrass is a versatile feedstock that is well suited to be used in combustion, gasification, and liquid fuel production (McLaughlin et al. 1996).

Fermentation to Fuels. Much of the early emphasis on biofuels has been on production of ethanol as a transportation fuel (Lynd et al. 1991). DOE has sponsored a significant research effort to produce ethanol

Table 5. Chemical and physical properties of switchgrass as a biofuel relative to selected alternate fuels (Source: McLaughlin et al. 1996).

Fuel property	Units	Switchgrass value	Alternate fuel	
			Value	Fuel type
Energy content (dry)	$Gj \cdot Mg^{-1}$	18.4	19.6	Wood
			27.4	Coal
Moisture content (harvest)	%	15	45	Poplar
Energy density (harvest)	$Gj \cdot Mg^{-1}$	15.6	10.8	Poplar
Net energy recovery	$Gj \cdot Mg^{-1}$	18	17.3	Poplar
Storage density				
(6' × 5') round bale	$kg \cdot m^{-3}$	133	150	Poplar chips
(4' × 5') round bale	(dry weight)	105		
Chopped		108		
Holocellulose	%	54–67	49–66	Poplar
Ethanol recovery	$L \cdot kg^{-1}$	280	205	Poplar
Combustion ash	%	4.5–5.8	1.6	Poplar
Ash fusion temperature	°C	1016	1350	Poplar
			1287	Coal
Sulfur content	%	0.12	0.03	Wood coal
		1.8		

Notes. Energy content of switchgrass was determined from 6 samples from Iowa. Bale density and chopped density of switchgrass are from Alabama (D. Bransby, Auburn). Poplar chip density is from studies of White et al. (1984). Poplar energy moisture content, combustion ash, and ash fusion temperatures are from NREL, as are ash fusion temperatures and sulfur contents of all fuels. Energy density is the energy per unit of wet harvest weight. Net energy recovery considers energy lost in drying fuel prior to combustion. Holocellulose content of switchgrass is from 7 cultivars in AL (Sladden et al. 1991) and from 7 hybrid poplar cultivars in P.A. (Bowersox et al. 1979). Ethanol yields are averages of SSF recovery on 3 analyses per species using a standard recovery procedure for all feedstocks. Ethanol yields can likely be improved somewhat by tailoring reaction mixtures to each specific feedstock, thus those should be considered preliminary measures of potential recovery.

from lignocellulosic crops through the SSF (Simultaneous Saccharification and Fermentation) processs (Wyman 1993), a combination of chemical and/or physical digestion followed by microbial fermentation. In the SSF process biofuels are broken down to structurally less complex organic residues that can be enzymatically converted to sugars and then fermented by microbes to produce ethanol. This is a relatively expensive technology because of the costs of acids and enzymes used in digestion. Ethanol yields are limited to 50–80% of possible levels, partly because lignin cannot be broken down by this process. On the other hand, recovered lignin has a high energy content and can be used as an energy input to the ethanol recovery process in SSF (Tyson et al. 1994). Pilot scale testing of this technology is underway with the first commercial plant targeted for the year 2005.

Gasification. Another technology for producing both ethanol and a variety of other liquid fuels and chemical products is gasification. Gasification is a process that has been available for many years as a means of converting coal, natural gas, or solid wastes into simpler synthesis gases, primarily hydrogen and carbon monoxide. Syngas can then be burned to produce heat or chemically synthesized into a wide variety of secondary products, including ethanol, diesel fuels, and chemical solvents used in industrial processes. Gasification has the benefit of converting essentially all of the carbon in biomass, including lignin, into synthetic gases.

While the conventional syngas technology, the Fischer–Tropsch system, uses high heat and temperature to synthesize secondary products, there are newer systems currently under development that use the biological capacity of microorganisms in reaction cells to produce synthetic products such as acetic acid, ethanol, and many other useful organic chemicals (Kaufman 1996). The advantage of these biological reaction cells, is that they operate at near ambient temperatures and pressures, resulting in greatly reduced costs, and with greater chemical specificity than the Fischer–Tropsch process. Talks have now begun with private industry to incorporate this new technology in to ethanol production plants and several locations in the Southeastern US are currently under consideration for an initial smaller-scale commercial facility.

Combustion. The final category of biofuel use is in combustion to produce heat or electrical power. At present there are approximately 7000 megawatts (MW) of power produced from biomass in the US (DOE 1996). This is derived largely (90%) from wood wastes at wood processing plants operated by the timber industry around the US. A much broader use of wood and other biomass energy from dedicated feedstocks is envisioned in the future. Factors that will be important to the quantities and types of feedstocks utilized are fuels quality (low ash), energy content per unit cost, and regional availability. Dedicated forage crops also offer a source of high energy feedstock for power production (Bransby 1996). For switchgrass production, the cost per unit of renewable energy produced has been estimated to be lowest in the Southeastern US [$1.78–2.03 per million BTU (MBTU)] and in the southern plains ($1.95–2.50 per MBTU, Walsh 1994).

To help achieve a concerted national program to promote the development and use of biomass power, the US DOE in 1991 formed the National Biomass Power Program. This program is strongly based on collaboration with the US Department of Agriculture (USDA) and private industry to form the government-industry partnerships necessary to achieve success. This effort promotes both direct combustion of biomass feedstocks co-fired in electric boilers with coal and other fuels, as well as gasification, an energetically more efficient process.

While wood wastes form the greatest fraction of current biomass-derived power production, there is great interest in using forage grasses for biopower as well. The DOE has recently embarked on three cooperative efforts to evaluate forage crops for power production. One involves the use of switchgrass in power generation (6 MW) with the Chariton Valley (Iowa) Resource Conservation and Development Agency. The second involves a joint effort with the Minnesota Valley Alfalfa Producers to produce electricity (from alfalfa stems) and animal feed from alfalfa leaves. A third supports tests by the Southern Research Institute in Alabama to evaluate cofiring switchgrass with coal in power production. These first commercial scale implementation efforts should provide valuable information on agricultural, sociological, and economic issues involved in a regional biomass power program.

SUMMARY AND CONCLUSIONS

Over the past 7 years research designed to evaluate and improve switchgrass as a bioenergy crop has been conducted by a team of government and university researchers in the southeast and central US. This effort is part of the DOE-sponsored Bioenergy Feedstock Development Program at Oak Ridge National Laboratory and has been focused in the areas of yield improvement through management and breeding, physiological and genetic characterization, and applications of biotechnology for regeneration and breeding research. Switchgrass, a warm season prairie grass, was chosen as the model species because of its perennial growth habit, high yield potential, compatibility with conventional farming practices, and high value in improving soil conservation and quality. Cultivar trials centered in Virginia, Alabama, and Texas have identified three excellent high-yielding switchgrass cultivars. These include 'Alamo', in the deep South, 'Kanlow' at intermediate latitudes, and 'Cave-in-Rock' for the upper Midwest. Yields of fully established stands of best adapted cultivars have averaged approximately 16 Mg/ha in research plots across 18 testing sites, and minimum costs of $1.78–2.03/MBtu have been estimated for farm-scale production in the Southeast. Management research has been directed at documenting nitrogen, row spacing, and cutting regimes to maximize sustained yields. Significant gains in soil carbon have been documented for switchgrass across a wide range of sites and associated gains in soil quality and erosion control are anticipated in connection with long term production of this species. Breeding research has focused on developing and characterizing an extensive germplasm collection, characterizing breeding behavior traits, and both narrow and broader base selection for yield improvement for both marginal and better quality soils. Tissue culture techniques have been developed to permit rapid clonal propagation of select switchgrass lines and to offer opportunities for application of advanced biotechnological tools. Energy budgets indicate that significant gains in energy return and carbon emissions reduction can be achieved with switchgrass as a biofuel.

The bioenergy industry is still in its infancy in terms of its impacts on national energy use. However the potential of biofuels to contribute to a national energy strategy is substantial. The benefits to the nation of providing cleaner burning fuels that improve both regional and global air quality while improving soil and water quality should be obvious. Combined with the improvements in farm economy, which can be expected with the production of energy on American farms and increased income for American farmers, bioenergy crops offer a "win-win" option for the planners of Americas future energy strategy. Bioenergy crops can be expected to become increasingly competitive in the future as the diversity of products possible from reformulation of biochemical constituents is developed through processes such as gasification and bioreactor technology. There are promising signs that the utility industry has recognized the value of cleaner burning renewable fuels which reduce environmental and political liabilities associated with relying totally on fossil fuels. Attainment of the potential for significantly greater participation of biofuels in national energy supply curves will require continued research on producing and improving energy crops more economically, continued improvement of the bioconversion technology to increase the diversity and value of end products, and a commitment of policy makers to improvement of environmental quality, which is measured in both long and short-term time frames.

REFERENCES

Alexandrova, K.S., P.D. Denchev, and B.V. Conger. 1996a. In vitro development of inflorescences from switchgrass nodal segments. Crop Sci. 36:175–178.

Alexandrova, K.S., P.D. Denchev, and B.V. Conger. 1996b. Micropropogation of switchgrass by node culture. Crop Sci. 36:1709–1711.

Bransby, D.I., R. Rodriguez-Kabana, and S.E. Sladden. 1993. Compatibility of switchgrass as an energy crop in farming systems of the southeastern USA. Proc. 1st Biomass Conf. of the Americas, Aug. 30–Sept. 2, Burlington, VT. p. 229–234.

Bowersox, T.W., P.R. Blankenhorn, and W.K. Murphey. 1979. Heat of combustion, ash content, nutrient content, and chemical content of populus hybrids. Wood Sci. 11:257–262.

Brejda, J.J. 1996. Evaluation of arbuscular mycorrhiza populations for enhancing switchgrass yield and nutrient uptake. Ph.D. Dissertation. Univ. Nebraska–Lincoln, Lincoln.

Brejda, J.J., L.E. Moser, and K.P. Vogel. 1998. Evaluation of switchgrass rhizosphere microflora for enhancing yield and nutrient gain. Agron. J. 90:753–758.

Denchev, P.D. and B.V. Conger. 1993. Development of in vitro regeneration systems for switchgrass. Vitro Cell. Dev. Biol. 29A No. 3, Part II. p. 63A.

Denchev, P.D. and B.V. Conger. 1995. In vitro culture of switchgrass: Influence of 2,4-D and picloram in combination with benzyl adenine on callus initiation and regeneration. Plant Cell Tissue Organ Cult. 40:43–48.

Denchev. P.D. and B.V. Conger. 1996. Gene transfer by microprojectile bombardment of switchgrass tissues. Vitro Cell Dev. Biol. 32(3) Part II. p. 98A.

Denchev, P.D. and B.V. Conger. 1994. Plant regeneration from callus cultures of switchgrass. Crop Sci. 34:1623–1627.

Department of Energy/ Environmental Information Administration. 1996. Annual energy outlook. 1996. National Energy Information, EI-231, Washington, DC.

Elmore, S.J., D. Lee, and K.P. Vogel. 1993. Chloroplast DNA variations in *Panicium virgatum* L. Proc. Am. Forage Grassland Council p. 216–219.

Food and Agriculture Organization of the United Nations (FAO). 1996. Global land use statistics. Available under AGRICULTURAL DATA at website - FAOSTAT HOMEPAGE. http://apps.fao.org/lim500/nph-wrap.pl?LandUse&Domain=LUI&servlet=1

Fasoulas, A.C. and V.A. Fasoula. 1995. Honeycomb selection designs. Plant Breed Rev. 13:87–139.

Ferrell, J.E., L.L. Wright, and G.A. Tuskan. 1995. Research to develop improved production methods for woody and herbaceous biomass crops. p. 197–206.

Fleish, T., C. McCarthey, A. Basu, C. Udovich, P. Charbonneau, W. Slodowske, S.-E. Mikkelsen, and J. McCandless. 1995. A new clean diesel technology, SAE Technical Paper Series, Int. Cong. And Exposition, Detroit, MI.

Garten, C.T., Jr. and S.D. Wullschleger. 1998. Soil carbon turnover and the potential for storage beneath long-term planting of switchgrass. Annual Meeting Soil Sci. Soc. Am., Baltimore, MD, Oct. 18–22, 1998.

Gebhart, D.L., H.B. Johnson, H.S. Mayeux, and H.W. Polley. 1994. CRP increases soil organic carbon. J. Soil Water Conserv. 49:488–492.

Graham, R.L., E. Lichtenberg, V.O. Roningen, H. Shapouri, and M. Walsh. 1995. The economics of biomass production in the United States. Proc. Second Biomass of the Americas Conference, Portland, OR., Aug. 21–24, 1995 (in press).

Graham, R.L. 1994. An analysis of the potential land base for energy crops in the conterminous United States. Biomass Bioenergy 6:175–189.

Gunter, L.E., G.A. Tuskan, and S.D. Wullschleger. 1996. Diversity among populations of switchgrass based on RAPD markers. Crop Sci. 36:1017–1022.

Hansen, J.B., B. Voss, F. Joensen, and I.D. Siguroardottir. 1995. Large scale manufacture of dimethyl ether: A new alternative diesel fuel from natural gas. Proc. SAE Int. Congr. and Exp. Detroit, MI, SAE Paper 950063.

Hohmann, N. and C.M. Rendleman. 1993. Emerging technologies in ethanol production. US Department of Agriculture, Economic Research Service, Agr. Inform. Bul. 663.

Hopkins, A.A. and C.M. Taliaferro. 1995. A comparison of selection strategies in switchgrass. Proc. 1995 American Forage and Grassland Council, Lexington, KY, March 12–14, 1995. p. 190–192.

Hopkins, A.A., C.M. Taliaferro, and C.D. Christian. 1996. Chromosome number and nuclear DNA content of several switchgrass populations. Crop Sci. 36:1192–1195.

Hopkins, A.A. and C.M. Taliaferro. 1997. Genetic variation within switchgrass populations for acid soil tolerance. Crop Sci. 37:1719–1722.

House, R., M. Peters, H. Baumes, and W.T. Disney. 1993. Ethanol and agriculture: Effect of increased production on crop and livestock sectors. US Department of Agriculture, Economic Research Service, Agricultural Economic Rep. 667.

Hultquist, S., K.P. Vogel, D.E. Lee, K. Arumuganathan, and S. Kaeppler. 1996. Chloroplast DNA and nuclear DNA content variations among cultivars of switchgrass, *Panicum virgatum* L. Crop Sci. 36:049–1052.

Hultquist, S., K.P. Vogel, D.E. Lee, K. Arumuganathan, and S. Kaeppler. 1997. DNA content and chloroplast DNA polymorphisms among accessions of switchgrass from remnant Midwestern prairies. Crop Sci. 37:595–598.

Intergovernmental Panel on Climate Change (IPCC). 1998. The regional impacts of climate change. An assessment of vulnerability. p. 267–279. In: R.T. Watson, M.C. Zinyowooa, R.H. Moss, and D.J. Dokken (eds.), Cambridge Univ. Press.

Kaufman, E.N. 1996. Biological production of chemicals from fossil fuel, municipal waste, and agricultural derived synthesis gas—a coordinated programmatic approach. A White Paper prepared for the US Department of Energy. Bioprocessing Research and Development Center, Chemical Technology Division, Oak Ridge National Laboratory.

Koshi, P.T., J. Stubbendieck, H.V. Eck, and W.G. McCully. 1982. Switchgrass: Forage yield, forage quality, and water use efficiency. J. Range Mgt. 35:623–627.

Lynd, LL., J.H. Cushman, R.J. Nichols, and C.F. Wyman. 1991. Fuel ethanol from cellulosic biomass. Science 231:1318–1323.

McLaughlin, S.B. 1992. New switchgrass biofuels research program for the Southeast. Proc. Ann. Auto. Tech. Dev. Contract. Mtng., Dearborn, MI, Nov. 2–5, 1992. p. 111–115.

McLaughlin, S.B. 1994. CRP Lands for biofuels production. Energy Crops Forum. US DOE, Biofuels Feedstock Development Program, Oak Ridge National Laboratory, Summer, 1994.

McLaughlin, S.B., D.I. Bransby, and D. Parrish. 1994. Perennial grass production for biofuels: soil conservation considerations. Proc. Bioenergy 94, Reno, NV, Oct. 1, 1994. p. 359–370.

McLaughlin, S.B., R. Samson, D. Bransby, and A. Weislogel. 1996. Evaluating physical, chemical, and energetic properties of perennial grasses as biofuels. Proc. Bioenergy 96, Nashville, TN, Sept. 1996. p. 1–8.

McLaughlin, S.B. and M.E. Walsh. 1998. Evaluating environmental consequences of producing herbaceous crops for bioenergy. Biomass Bioenergy 14:317–324.

McLaughlin, S.B. 1998. Forage crops as bienergy fuels: Evaluating the status and potential. Proc. VXIII International Grassland Congress, June 8–10, 1997, Winnipeg Canada (in press).

Miles, T.R et al. 1993. Alkalai slagging problems in biomass fuels. First Biomass Conf. Amer., Burlington, VT. p. 406–421.

Moser, L.E. and K.P. Vogel. 1995. Switchgrass, big bluestem, and indiangrass. p. 409–420. In: R.F. Barnes, D.A. Miller, and C.J. Nelson (eds.), Forages. Vol 1, An introduction to grassland agriculture. Iowa State Univ. Press, Ames.

Moss, D.N., E.G. Krenzer, and W.A. Brun. 1969. Carbon dioxide compensation points in related plant species. Science 164:187–188.

Overend, R. 1997. USA—biomass and bioenergy 1996. p. 36–39. In: The world directory of energy supplies.

Paine, L.K., T.L. Pederson, D.J. Undersander, K.C. Rineer, G.A. Bartelt, S.A. Temple, D.W. Sample, and R.M. Klemme. 1996. Some ecological and socioeconomic considerations for biomass energy crop production. Biomass Bioenergy 10:231–242.

Parrish, D.J., D.D. Wolf, and W.L. Daniels. 1997. Switchgrass as a biofuels crop for the upper Southeast: Variety trials and cultural improvements. Final report from VPI for 1992–97. Oak Ridge National Laboratory, Oak Ridge, TN, March, 1997.

Petrulis, M., J. Sommer, and F. Hines. 1993. Ethanol production and employment. USDA Agricultural Research Service. Agr. Inform. Bul. 678.

Redfern, D.D, K.J. Moore, K.P. Vogel, S.S. Waller, and R.B. Mitchell. 1997. Canopy architectural and morphological development traits of switchgrass and relationships to forage yield. Agron. J. 89:262–269.

Reeves, D.W. 1997. The role of soil organic matter in maintaining soil quality in continuous cropping systems. Soil Tillage Res. 43:131–167.

Risser, P.G., E.C. Birney, H.D. Blocker, S.W. May, W.J. Parton, and J.A. Wiens. 1981. The true prairie ecosystem. US/IBP Synthesis Series 16. Hutchinson Ross Pub.

Robertson, T. and H. Shapouri. 1993. Biomass: An overview in the United States of America. Proc. First Biomass Conf. of Americas, Burlington, VT, Aug. 30–Sept. 2, 1993. p. 1–17.

Romm, J., M. Levine, M. Brown, and E. Petersen. 1998. A road map for US carbon reductions. Science 279:669–670.

Sanderson, M.A. and D.D. Wolf. 1996. Morphological development of switchgrass in diverse environments. Agron. J. 88:908–915.

Samson, F. and F. Knopf. 1994. Prairie conservation in North America. Bioscience 44:418–421.

Sanderson, M.A., R.L. Reed, S.B. McLaughlin, S.D. Wullschleger, B.V. Conger, D.J. Parrish, D.D. Wolf, C. Taliaferro, A.A. Hopkins, W.R. Ocumpaugh, M.A. Hussey, J.C. Read, and C.A. Tischler. 1996. Switchgrass as sustainable energy crop. Bioresource Technol. 56:87–93.

Shapouri, H., J. Duffield, and M.S. Graboski. 1995. Estimating the energy value of corn-ethanol. Second Biomass Conference of the Americas: Energy, Environment, Agricultural, and Industry. National Renewable Energy Laboratory, Golden, CO.

Shiflet, T.N. and G.M. Darby. 1985. Forages and soil conservation. p. 21–32. In: M.E. Heath, R.F. Barnes, and D.S. Metcalf (eds.), Forages: The science of grassland culture. Iowa State Univ. Press, Ames.

Sladden, S.E., D.L. Bransby, and G.E. Aiken. 1991. Biomass yield, composition, and production costs for eight switchgrass varieties in Alabama. Biomass Bioenergy 1:119–122.

Stubbendieck, J., S.L. Hatch, and C.H. Butterfield. 1981. North American range plants. Univ. Nebraska Press, Lincoln.

Taliaferro, C.M., and A.A. Hopkins. 1996. Breeding characteristics and improvement potential of switchgrass. Proc. Third Liquid Fuel Conference: Liquid Fuels and Industrial Products from Renewable Resources. Nashville, TN, Sept. 15–17, 1996. p. 2–9.

Taliaferro, C.M., A.A. Hopkins, M.P. Anderson, and J.A. Anderson. 1996. Breeding and genetic studies in bermudagrass and switchgrass. Proc. 52nd Southern Pasture & Forage Crop Improvement Conference, Oklahoma City, OK, March 30–April 2, 1996. p. 41–52.

Taliaferro, C.M. and M. Das. 1998. Breeding and selection of new switchgrass varieties for increased biomass production. 1997 Annual report from Oklahoma State Univ. to Oak Ridge National Laboratory , March 1998, Oak Ridge, TN.

Thomson, D.J. 1995. The seasons, global temperature, and precession. Science 268:59–68.

Tjepkema, J. 1995. Nitrogenase activity in the rhizosphere of *Panicum virgatum*. Soil Biol. Biochem. 7:179–180.

Tyson, K.S., C.J. Riley, and K.K. Humphreys. 1994. Fuel cycle evaluations of biomass: Ethanol and reformulated gasoline. Vol. I. National Renewable Energy Laboratory, Golden, CO. NREL-TP-463-4950.

Vogel, K.P., H.J. Gorz, and F.A. Haskins. 1989. Breeding grasses for the future. Crop Sci. Soc. Am. Contrib. from Breeding Forage and Turf Grasses, CSSA Special Publ. 15.

Vogel, K.P., C.I. Dewald, H.J., H.J. Gorz, and F.A. Haskins. 1985. Development of switchgrass, Indiangrass, and eastern gamagrass: Current status and future. Range improvement in Western North America. Proc. Mtg. Soc. Range Mgt., Salt Lake City, Utah, Feb.14, 1985. p. 51–62.

Vogel, K.P. and J.F. Pederson. 1993. Breeding systems for cross-pollinated perennial grasses. Plant Breed. Rev. 2:252–257.

Vogel, K.P., A.A. Hopkins, K.J. Moore, K.D. Johnson, and I.T. Carlson. 1996. Registration of 'Shawnee' switchgrass. Crop Sci. 36:1713

Vogel, K.P. 1996. Energy production from forages or American agriculture: Back to the future. J. Soil Water Conserv. 51:137–139.

Walsh, M.E. 1994. The cost of producing switchgrass as a dedicated energy crop. Oak Ridge National Laboratory Working Paper, June 1994.

Walsh, M.E., D.D.T. Ugarte, S. Slinsky, R.L. Graham, H. Shapouri, and D. Ray. 1998. Economic analysis of energy crop production in the U.S.: Location, quantities, price, and impacts on traditional agricultural crops. Proc. Bioenergy 98, Oct 9. Madison WI, .

Walsh, M.E. 1998. US bioenergy crop economic analyses: status and needs. Biomass Bioenergy 14:341–350.

Weaver, J.E. 1968. Prairie plants and their environment. Univ. Nebraska Press, Lincoln.

White, M.S., M.C. Vodak, and D.C. Cupp. 1984. Effect of surface compaction on the moisture content of piled green hardwood chips. Forest. Prod. J. 34:59–60.

Wolf, D.D. and D.A. Fiske. 1995. Planting and managing switchgrass for forage, wildlife, and conservation. Virginia Cooperative Extension Publ. 418–013, Virginia Polytechnic Institute and State Univ., Blacksburg.

Wright, L.L. 1994. Production status of woody and herbaceous crops. Biomass Bioenergy 6:191–209.

Wullschleger, S.D., L.E. Gunter, and G.A. Tuskan. 1997. Research on improving the productivity of switchgrass (*Panicum virgatum*) as a biofuels crop: Assessing genotype diversity with physiological and molecular indicators. Annual report for 1996, Feb. 1997, Oak Ridge National Laboratory, Oak Ridge, TN.

Wullschleger, S.D., M.A. Sanderson, S.B. McLaughlin, D.P. Biradar, and A.L. Rayburn. 1996. Photosynthetic rates and ploidy levels among populations of switchgrass. Crop Sci. 36:306–312.

Wullschleger, S.D., L.E. Gunter, and C.T. Garten. 1998. Genetic diversity, carbon/nitrogen cycling, and long-term sustainability of yield in the bioenergy crop switchgrass (*Panicum virgatum*). Annual report for 1996, March 1998, Oak Ridge National Laboratory, Oak Ridge, TN.

Agronomic Research on Hemp in Manitoba*

Jack Moes, Allen Sturko, and Roman Przybylski

Hemp (*Cannabis sativa* L.) has been the subject of great interest in Canada (and in many other countries) in recent years, focused on the potential of hemp as a "new" crop. Ironically, hemp is of course a very old crop, and, as in many other western countries, it has a long history in Canada.

In the late 19th and early 20th centuries, hemp was a familiar sight at the premises of many eastern European immigrants—the plants often served as windbreaks and the seed was used for food. In the 1920s, the Dominion Experimental Farm (predecessor of today's Agriculture and AgriFood Canada) was involved in hemp research, as part of their fiber crops program which focused mainly on fiber flax. Efforts focused on cultivar evaluation, production agronomy, harvest, and retting procedures. There was some commercial area of hemp in Manitoba at that time, rising to a peak of over 400 ha in the Portage la Prairie area in 1928, used to manufacture cordage. By the mid-1930s area had declined to zero due to economic factors.

Hemp cultivation was banned in 1938 under new federal narcotics regulations. However, in response to a resurgence of interest in the cultivation, processing, and marketing of hemp products, and after a period of cultivation and evaluation under research licenses, the Canadian government was successfully lobbied for change. Industrial Hemp Regulations were introduced in March 1998, enabling licensed commercial cultivation of appropriate cultivars. Approximately 2000 ha were sown in Canada in the 1998 growing season, about 600 ha of which was in the province of Manitoba.

Industrial Hemp Regulations are administered by Health Canada. These regulations require that individuals or companies be licensed to import and export, grow, process, or sell hemp seed or products. For example, a license allows a grower to buy seed from a licensed importer or seed grower, grow the crop at a given location, harvest fiber or grain, and sell the grain to a licensed processor. Anyone more than 18 years of age with no drug-related convictions in the previous 10 years may apply for a license.

Health Canada will not license cultivation of less than 4 ha, except in special circumstances. Growers must give the G.P.S. (ie. Global Positioning System) coordinates of the location where they plan to grow hemp. The location must be at least 1 km from school grounds or other place frequented by persons less than 18 years of age.

Growers must use pedigreed seed of a cultivar approved by Health Canada. The OECD (Organization for Economic Cooperation and Development) List of Cultivars Eligible for Certification is the main basis for approving cultivars. Health Canada's main concern is that cultivars should produce less than 0.3% THC in leaves and flower parts. Growers are required to have a sample tested by an approved lab to determine the THC content under their conditions. Also, growers must keep detailed records of hemp purchases, sales, and other movements.

MANITOBA HEMP RESEARCH 1995–1998

Hemp research trials were conducted in Manitoba in 1995–1998, with the following objectives: (1) evaluate cultivars of hemp for their agronomic suitability for Manitoba; (2) evaluate seed and stalk yield in small-plot and large-scale (> 0.5 ha) trials; (3) gain experience in sowing and harvesting the crop; (4) evaluate quality of seed and fiber for various potential uses; (5) observe for potential disease, insect or other agronomic problems; (6) estimate economic feasibility of hemp production in Manitoba, at farm-gate level. Trials took place at various locations representing several different agroclimatic regions in Manitoba.

Agronomic Research—Small-plot Trials

Small-plot, randomized complete block trials were utilized to examine cultivar performance and suitability in 1995–1997. Unreplicated applications of herbicide and seed-placed fertilizer were used on occasion to provide an initial indication of suitability. In 1998, cultivar by seeding rate, and herbicide tolerance trials were initiated.

* This research was supported by the Manitoba Sustainable Development Innovations Fund.

Table 1. Agronomic characteristics and Δ^9-THC content for hemp cultivars grown at fiber density (400 seeds/m^2) at two Manitoba locations in 1996.

Cultivar	Wawanesa MB 1996			Morden MB 1996		
	Height (m)	Stalk yield (kg/ha)	Δ^9-THC (%)	Height (m)	Stalk yield (kg/ha)	Δ^9-THC (%)
Zolotonosha 11	2.11 c[z]	7860 bc	<0.05	2.18 b	5160 b	<0.05
Zolotonosha 13	2.12 bc	7920 bc	<0.05	2.30 bc	4540 b	<0.05
Polish 1[y]	2.07 c	8150 bc	0.13	2.27 bc	5200 b	0.06
Polish 2[y]	2.07 c	7260 c	0.17	2.23 c	4400 b	<0.05
Fedora 19	2.08 c	8440 b	0.10	2.20 c	4890 b	0.08
Felina 34	2.19 b	10710 a	0.06	2.37 b	7290 a	0.07
Uniko B	2.35 a	10520 a	0.30			
Kompolti	2.33 a	10450 a	0.21	2.52 a	7200 a	0.18
Futura 77	2.31a	11160 a	0.15	2.48 a	7330 a	0.06
Average	2.18	9160		2.32	5750	
CV (%)	2.6	7.2		3.4	13.4	

[z]Mean separation by protected LSD (5%).
[y]Lines believed to be of Polish origin but of unverified pedigree.

Table 2. Agronomic characteristics and Δ^9-THC content for hemp cultivars grown at seed density (100 seeds/m^2) at two Manitoba locations in 1996.

Cultivar	Wawanesa MB 1996		Morden MB 1996	
	Seed yield (kg/ha)	Δ^9-THC (%)	Seed yield (kg/ha)	Δ^9-THC (%)
Zolotonosha 11	930 bc[z]	<0.05	1056	<0.05
Zolotonosha 13	885 bc	<0.05	1074	<0.05
Polish 1[y]	820 c	0.05	1528	0.07
Polish 2[y]	691 c	0.11	1561	0.22
Fedora 19	1567 a	0.12	1783	0.11
Felina 34	1228 ab	0.08	1963	0.09
Average	1020		1494	
CV (%)	20.4		26.9	

[z] Mean separation by protected LSD (5%).
[y] Lines believed to be of Polish origin but of unverified pedigree.

Cultivar trial data from two locations in 1996 are presented (Tables 1 and 2). Separate cultivar trials were conducted for stalk and seed yield evaluations, with seeding rates of 400 and 100 seeds/m^2, respectively. The first five cultivars listed (Table 1) were similar in flowering date and were cut for stalk yield evaluation in the early flowering stage on August 7 (57 days after sowing on June 10). The next four cultivars were later to flower and were all cut on August 15 (65 days after sowing). In general, the later flowering cultivars were taller and yielded considerably higher, since they had an extra eight days of growth. If high quality fiber production is the objective, clearly the late cultivars are preferred from a productivity perspective. However, the issue of relative fiber quality among these cultivars has yet to be explored. Further trials are necessary to determine optimum seeding rate for fiber production in Manitoba. All cultivars were sampled at early flowering to determine Δ^9 - THC content; in this particular trial, all tested below the maximum allowable level of 0.30%, except 'Uniko B' which tested at 0.30%.

It was recognized early on that growing hemp as an oilseed crop would have significant potential for Manitoba. Therefore, cultivar trials were conducted with the specific aim of evaluating several earlier maturing cultivars for their seed production potential (Table 2). To promote branching, flowering and seed set on these fiber type cultivars, the plots were sown at a much lower density–100 seeds/m². The stands were acceptable but variable, and the average stand at harvest time was only about 30% of the seeds sown (not shown). Considering the late sowing, seed yields were excellent, with the 'Fedora 19' and 'Felina 34' exhibiting the highest yields.

With most commercial production interest in 1998 focussing on hemp for grain or dual-purpose (grain and fiber), trials (factorial design with four replications) were established to determine optimum seeding rates for four cultivars of interest. Limited data is available at the time of this writing, but yield data for two locations is presented (Table 3). The cultivar by seeding rate interaction was not significant, and within the range tested, seeding rates did not differ in yield. At one location (Carman), cultivars differentiated themselves with respect to yield, with the slightly later-maturing and taller 'Fedora 19' yielding the most grain.

Additional Field Observations

Weed pressure. A good stand emerging ahead of weeds is crucial to the ability of hemp to compete with weeds, especially with the low seeding rate (100 seeds/m²) used for grain hemp production. In areas where stand was poor (sown too shallow for moisture conditions or excessive moisture in low areas of field), competition from weeds was severe (no quantitative estimate was made).

Disease incidence. Particular weeds may cause more than just competitive loss. Wild mustard (*Sinapis arvensis*) infected with *Sclerotinia sclerotiorum* sometimes provided the source of inoculum for stem rot lesions on adjacent hemp plants. These lesions caused wilting and death of the upper portion of the plant and rendered the plant more susceptible to lodging due to stem breakage; well-developed sclerotia approximately 3 mm in diameter became evident in the hollow of affected stalks. Hemp has shown itself to be quite susceptible to *Sclerotinia*, which will certainly have rotational implications for Manitoba where canola and other susceptible crops are major considerations. *Botrytis cinerea* has also been observed causing a grey moldy infection in the inflorescence, but the incidence has been negligible.

Insect incidence. The 1995 growing season found areas of the province being severely affected by Bertha armyworm (*Mamestra configurata*), a cyclical pest of canola and other crops. One of the hemp research plot locations was in an affected area, and Bertha armyworm caused severe defoliation of hemp plants. This pest is currently at low population levels, but in six to eight years the population in Manitoba will rise again, and hemp is certain to be affected. Other problem insects observed include *Lygus* spp. plant bugs, grasshoppers, and European corn borer (*Ostrinia nubilalis*). Feeding and damage was observed, but not at what would be considered economic levels. Whether or not any of these will be of economic consequence is yet to be determined.

Pollinators. Honey bees were observed foraging for pollen in hemp plots, even though there appeared to be no honey bee colonies within the immediate area (i.e. 2 km²). Honey bees were by far the most abundant, pol-

Table 3. Seed yield of hemp in cultivar × seeding rate trials at Manitoba locations in 1998.

Location Cultivar	Yield (kg/ha)[z]			
	Seeding rate (seeds/m²)			
	75	100	125	Mean
Carman MB				
USO 14	1260	1160	1130	1180
USO 31	1330	990	1180	1170
Fasamo	1190	1360	1350	1300
Fedora 19	1770	1720	1850	1780
Mean	1390	1310	1380	1360
Laurier MB				
USO 14	900	850	930	890
USO 31	660	850	840	780
Fasamo	900	660	790	780
Fedora 19	870	720	890	820
Mean	830	770	860	820

[z] Cultivar main effect significant at 1% at Carman MB, seeding rate or cultivar × seeding rate not significant in either trial.

len foragers in the plots during our field observation. Hemp pollen was collected from the hemp flowers and compared to the pollen collected from the honey bees verifying that the pollen collected by the honey bees was indeed hemp. Bumble bees were observed in the vicinity of the hemp plots, but no active foraging behavior was recorded during our observations. Although no bumble bee foraging was observed in the sites, bumble bees were observed gathering pollen in a hemp field elsewhere in the province. A sample of pollen was collected from the bumble bees and found to contain hemp pollen. Although some of the insects observed were in direct contact with hemp pollen, the question that remains to be answered, is whether these same insects will also visit the female parts (or flowers) of the hemp plant.

Seed Oil Quality Evaluations

Manitoba hemp research initiated in 1995 was based on the premise that hemp represented a potential new fiber crop opportunity. However, it became clear that another opportunity was perhaps more readily accessible—that of growing hemp as an oil seed crop, with the stalk/fiber as a secondary harvest. Such a dual purpose scenario is familiar on the Canadian prairies, particularly with oil seed flax, and more recently wheat (strawboard). Therefore, the main focus of subsequent research shifted to examining seed oil quality, on the premise that hemp could become more quickly established in Manitoba has oil seed crop, and from that position the longer-term development of fiber processing and marketing could take place.

Hempseed samples were analyzed for fatty acids, antioxidants, and sterols (Table 4). The data confirm that hempseed oil is of merit nutritionally because of its relatively high level of polyunstaurates, the approximate 3:1 ratio of omega-6 and omega-3 fatty acids, and the presence of significant gamma-linolenic acid (GLA) and antioxidant levels. Without more data, we cannot be conclusive with respect to varietal or environmental differences. However, an examination of data to date suggests that varietal differences are present for some components and may be more significant than environmental differences (Fig. 1). Several cultivars sourced in the Ukraine or Poland had levels of GLA in the range of 2.5–3.0% or more, for the original source seed and for seed grown from it at two locations in Manitoba. Three cultivars sourced in Romania ('Secueni 1', 'Lovrin 110', and 'Irene') had lower GLA (1.2–1.5%).

Cost of Production Analysis

Cost of production and breakeven analyses are provided in Table 5 for the situation we believe to be most readily accessible for Manitoba: growing hemp for grain, with secondary harvest of the low-grade stalks remaining after the grain has been harvested.

Table 4. Fatty acid, anti-oxidant, and sterol profiles, and composite seed analysis for Manitoba hempseed samples taken in 1996 and 1997; sample size is 36 except for composite seed analysis where sample size is 12.

Oil/Composite component	Range	Mean	SD
Fatty acids (% of oil)			
16:0	5.9–6.6	6.2	0.20
16:1	0.1–0.2	0.2	0.04
18:0	2.4–3.4	2.7	0.27
18:1	10.5–16.3	12.9	1.34
18:2ω6	54–57.7	55.6	0.85
α-18:3ω3	15.1–17.9	16.7	0.74
γ-18:3	1.2–3.8	2.6	0.62
20:0	0.7–1	0.8	0.06
20:1	0.3–0.4	0.4	0.02
22:0	0.3–0.4	0.3	0.05
24:0	0.1–0.2	0.2	0.03
Saturates	9.6–11.1	10.3	0.43
Monounsaturates	11–16.9	13.4	1.36
Polyunsaturates	71.1–78	75.0	1.58
ω6 to ω3 ratio	3.1–3.7	3.3	0.14
Antioxidants (ppm)			
α-tocopherol	4.3–25.6	11.3	4.70
β-tocopherol	6.8–25.8	11.4	4.25
γ-tocopherol	678.4–1101.3	829.6	102.72
δ-tocopherol	19.8–68.3	41.3	11.65
α-tocotrienol	15.4–51.6	26.9	8.14
Total antioxidants	748.1–1231.8	920.5	122.38
Sterols (ppm)			
β-sitosterol	2384.2–4203.6	3113.5	445.49
Stigmasterol	101.2–242.3	183.6	30.81
Campestrol	726.3–1401.3	1062.2	140.76
Brassicasterol	31.8–122.3	65.9	22.95
Composite seed analysis (%DM)			
Protein	22.8–26.1	24.1	1.09
Fat	20.8–25.1	22.5	1.46
Fiber	23.8–28.6	26.3	1.48
Soluble fiber	4.9–5.9	5.3	0.27

Production cost for hemp in this scenario is relatively high, even compared with canola, which among the annual field crops in Manitoba is among the more expensive to grow. Several factors contribute to the higher production costs. Seed, which currently must be imported annually as pedigreed seed, is very expensive—when pedigreed seed production can take place locally, we would hope that seed costs would go down. Machinery operating costs are estimated as higher, based on the toll that the tough fibrous stalks will take on bearings and chains and knives. Industrial hemp regulations require that samples be taken an analyzed for THC content, and while there was no licensing cost in the first season, it is expected that Health Canada will introduce a licensing fee for subsequent seasons.

Even though the production cost is high, this does not render hemp unworthy of consideration from a grower's perspective. The break even analysis suggests that growers can expect to make a sufficient return. Even though this season's hemp grain price of $Cdn1.10–1.30/kg is certain to the reduced for subsequent seasons, grain yields in 1998 often exceeded 1100 kg/ha. Therefore, it appears that it will not be difficult to sustain growers' interest in hemp despite the significant "hassle" factors associated with growing hemp vs. growing other annual field crops.

FIRST COMMERCIAL PRODUCTION EXPERIENCE 1998

Approximately 70 growers sowed a total of about 600 ha of hemp in Manitoba in 1998, primarily under contract to three companies. On this, about 100 ha was seed multiplication of a German cultivar, 'Fasamo'. The balance was grown as dual purpose crop. The area was dominated by French and Ukrainian cultivars. In general, this first experience was encouraging—but it was definitely a learning experience.

Unusually heavy rainfall in June resulted in conditions of excess moisture stress in some heavier textured fields with poor internal drainage. Hemp plants in affected areas became chlorotic and stunted, and in some cases died altogether. While it had been recommended to growers to avoid sowing in such fields, the experience reinforces the sensitivity of hemp to excess moisture stress especially in the early stages of growth. An additional consequence is that some of the moderately affected areas where the stand was reduced and the growth stunted also suffered from additional weed growth and competition.

Otherwise crop growth was excellent, and the real challenge came at harvesting time. Numerous approaches were taken, many successfully given a crop averaging 2.2 to 3 m in height. Most approaches were much less successful when the crop exceeded 3 m in height. The basic approach was to direct combine using a conventional combine with the header raised to maximum height. For threshing, this approach was the most consistently successful, although in some circumstances, rotary combines were also used successfully. Generally speaking, threshing was easier when grain moisture was still relatively high (> 25%), although grain quality (fewer immature seeds evident in the dried sample) improved with lower moistures at harvest. As the stalk material dried, it became more prone to wrapping on chains and bearings in the combine, and growers quickly learned to watch for such problems developing—wrapped material is prone to burning, putting the entire combine at risk. For crops of less than 2 m in height, swathing and combining with a pick up header about a week or ten days later was also successful. After combining, remaining stalk was cut with a mower or a swather, and baled with round balers. Wrapping of fiber was also a problem at this stage, necessitating the installation of guards to protect bearings, belts, and chains.

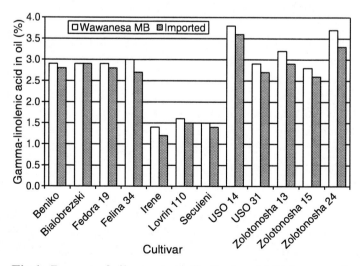

Fig 1. Percent of oil as gamma-linolenic acid for hempseed; samples drawn from seed imported for planting, and from resulting seed grown at Wawanesa MB in 1997.

Grain yields exceeded expectations, typically ranging from 800 to 1200 kg/ha on a clean, dry (10% moisture) basis. Yields as high as 1900 kg/ha were reported.

HEMP RESEARCH NEEDS

Although we have arrived at the stage of commercial cultivation, much research is still required in order for hemp to become successfully established as a specialty crop in Manitoba. It is envisioned that industry, growers, provincial and federal governments, and university researchers will all have a role in this research. This applies both to hemp as a grain/dual purpose crop and as a fiber crop.

Several cultivars are available and perform reasonably well. However there is a need for further cultivar evaluation and the development of cultivars specifically adapted to western Canada and to the production practices and market requirements for Prairie production. Current agronomic recommendations are based more on rules of thumb than on solid research, so seeding rate, seeding date, and fertility research is required. Weed management is another area of required research, primarily with respect to strategies to maximize the competitive ability of the crop, but also in the evaluation of herbicides as a production tool.

Experience in Manitoba and elsewhere have shown that the crop is not disease- and insect-free as has been sometimes claimed. Therefore, disease and insect management strategies also require examination. Harvest timing and techniques require both research and the ingenuity of growers. Management effects on quality of grain and its derived products, and stalk and its derived products must also be examined.

Finally it is important to recognize that much of the talk of the marketability of hemp derived products is in fact talk of *potential* market. Successful establishment of the crop will also require a great deal of investment of energy in product and market development.

Table 5. Estimated cost of production and break-even analysis for hemp grown as a dual purpose crop in Manitoba; typical canola values supplied for comparison. Assumes field of 8 ha in size for per field costs (licensing, sampling, and analytical fees).

Operating proforma	Hempseed ($Cdn/ha)	Residual stalk ($Cdn/ha)	Canola ($Cdn/ha)
Operating costs			
Seed	197.60		59.28
Fertilizer	64.96		79.16
Chemicals			72.87
Fuel	27.17	9.88	27.17
Machinery operating costs	49.40	14.82	24.70
Crop/hail insurance	14.82		30.63
Other costs	18.53		18.53
Land taxes	13.59		13.59
Licensing fee	30.88		
Sampling and analytical fees	30.88		
Drying costs	5.29		
Cleaning costs	34.58		
Interest on operating	16.31	0.36	11.73
Total operating costs	503.99	25.59	337.65
Fixed costs			
Land investment costs	43.97		43.97
Machinery depreciation	43.23	12.35	43.23
Machinery investment costs	17.29	18.53	17.29
Storage cost	5.29	5.29	5.29
Total fixed costs	109.77	36.16	109.77
Total operating and fixed	613.76	61.75	447.42
Labor	37.05	37.05	37.05
Total costs	650.81	98.80	484.47
Breakeven analysis—price			
Expected yield (kg/ha)	800	5000	1600
Breakeven price (/kg)			
Operating	0.63	0.01	0.21
Operating and fixed	0.77	0.01	0.28
Total	0.81	0.02	0.30
Breakeven analysis—yield			
Expected price (/kg)	0.66	0.03	0.31
Breakeven yield (kg/ha)			
Operating	762	853	1094
Operating and fixed	928	2058	1449
Total	984	3293	1569

Low-THC Hemp Research in the Black and Brown Soil Zones of Alberta, Canada

S.F. Blade, R.G. Gaudiel, and N. Kerr

In the past three years, there has been a renewed interest in the production of fiber hemp (*Cannabis sativa* L., Cannabinaceae) in Alberta. This effort has sparked a great deal of interest and speculation regarding the crop's potential. Since 1995 fiber hemp research plots have been grown in various parts of Alberta to evaluate the potential of this crop for both seed and fiber production (Blade 1998). Research licenses were granted by the Health Protection Branch of Health Canada, the branch of the federal government charged with enforcing legislation concerning this species. In March, 1998, Health Canada announced that commercial production of low-THC hemp would be allowed through a licensing process.

Hemp is an annual herbaceous plant which flourishes in temperate regions (Bosca and Karus 1998). All cultivars tested in Alberta have been low-THC (delta-9-tetrahydrocannabinol) genotypes. Canada has adopted the 0.3% THC standard established by the European Union and the OECD as the concentration which separates non-psychoactive strains suitable for legal fiber production from those which are illegally grown for their properties of intoxication. The 0.3% THC designation is very conservative. Most narcotic strains range from 8–10% THC, with cleaned, high potency material reaching as high as 30% THC. It is postulated that THC was useful to the plant by providing protection from UV-B exposure. The cannabinoid complex (which includes THC) of compounds is secreted by epidermal resin glands which are most numerous on and around the reproductive structures. This makes sense, since the reproductive structures require the highest level of protection. Low-THC cultivars secrete resin, but it is composed of non-intoxicating substances.

Plant growth is very vigorous (Fig. 1, 2). Fiber hemp can reach heights of up to 9 m, but the usual average under commercial production is 2–4 m. The crop has been subject to intensive breeding programs in Poland, Rumania, Hungary, France, the Ukraine, and several other European countries. Breeders have attempted to increase bast fiber yield and quality. One method has been to select monoecious strains (male and female reproductive organs located on the same plant) which eliminates the problem of different maturities between male and female plants. Due to the fact that flowering is dictated by day length, most land races have been selected to mature in early autumn, to take full advantage of the temperate growing season. Breeders also prefer to select for long internodes and a hollow stem, which increase the quality of the fiber.

The plant consists of a single main stalk, with an external sheath of bast fiber and an interior core of white, fibrous hurd. The plant has been used for a wide cultivar of purposes including rope-making, textiles, paper production and construction materials; the seed has been used as a source of high quality

Fig. 1. Morley Blanch and Prof. John Toogood inspect the spectacular growth of low-THC hemp in the research plots at the Blanch farm.

Fig. 2. The distinctive leaf morphology of *Cannabis sativa* at three weeks after emergence.

oil (both for industrial and edible uses) and protein (22%).

HISTORY

It is believed that hemp originated somewhere in Central Asia. There is evidence of hemp use as a food grain as well as for fish nets, clothing and rope 6500 years ago in China. The crop gradually moved into Europe, and the first reports of hemp growing in the New World began in Chile in 1545. Hemp was so important to the navies of England and the American colonies that farmers were legislated to dedicate a portion of their land to the production of the crop. George Washington and Thomas Jefferson were hemp growers; shortly after independence, US taxes could be paid with hemp if farmers chose to do so. Area of hemp in the former Soviet Union peaked at almost a million hectares in the early part of this century, but has declined to approximately 60,000 ha in the Ukraine and western parts of Russia. In 1992, the world's largest producers of hemp fiber were India (45,000 t) and China (24,000 t).

Hemp was a popular crop in Eastern and Central Canada throughout the 18th and 19th centuries. During the period 1923–1942 there was an extensive research effort by the Canada Department of Agriculture to test agronomic management, processing, and some crop improvement at approximately 30 locations across Canada. A concise report on the trials including: seeding methods, date of planting, plant breeding efforts, harvesting methods, and processing procedures is available. The hemp acreage in Canada decreased during this time period due to the high cost of production and because of strong competition by other fiber crops produced in the tropics.

In 1938 the Canadian government made the cultivation of *Cannabis sativa* illegal (through the Opium and Narcotics Act), although a small amount of production was allowed during the war years. In 1994, the first license in decades was granted for a low-THC hemp research plot in southern Ontario. This prompted the current research effort taking place in Alberta since 1995.

METHODOLOGY

Cultivar screening tests were done (1996–97) in Brooks (dryland and irrigated) and Gwynne (dryland). Precipitation at Brooks was 207 mm during the growing season; at Gwynne during the same period the rainfall was 658 mm. The trials were seeded as soon as the required research licenses were received from the Health Protection Branch of Health Canada. The Brooks trials were planted on June 12 and the Gwynne trials were seeded on May 16. Cultivars tested included (by country of origin): Poland (Beniko, Bialobrezeski), Ukraine (Zolotonosha lines, USO lines), Hungary (Kompolti, Unico), Romania (Irene, Lovrin 110, Secuieni) and France (Fedora 19, Felina 34, Futura 77) (Fig. 3).

Harvesting at each location was done by hand using sharp knives at physiological maturity. Weights were taken and samples of stalk were dried in a forced air drier for moisture determination. Seed was hand-threshed, or run through a stationary small-plot combine. Selected small seed samples were selected for oil extraction and analysis. Analysis of variance was done using SAS, and LSD at the 95% probability level was used for mean separation.

The evaluation of the textile properties were done by determining the linear density of fibers. Common units of linear density are tex (weight in grams of 1,000 m of fiber) or denier (weight in grams of 9,000 m of fiber). The tensile strength of single fibers 70 mm long is measured with an Instron Universal tester using a gauge length of 40 mm and test speed of 200 mm/min. The force (Newtons) needed to break each fiber is determined, as will as the extension (%) at break. The average tenacity (N tex^{-1}) of 25 fibers from each hemp sample was determined.

Fig. 3. Dr. Stan Blade surveys a French low-THC hemp cultivar in a varietal evaluation trial.

RESULTS

THC levels

The Health Canada regulations require that all hemp have a THC level of less than 0.3%. In 1996 THC levels were < 0.05–0.30. In 1997, THC levels were < 0.01–0.65%. The only cultivar above the required minimum was re-evaluated in 1998, since there were concerns about seed quality and purity of the cultivar in question.

Cultivar Evaluation

The Gwynne site (Table 1) had much higher productivity than Brooks. This was due to higher levels of precipitation and an extended period of warm weather. The total biomass ranged between 8.92–17.32 t/ha. The low yield of Zolotonosha 24 was caused by poor germination which resulted in a reduced plant stand. Following 72 h of drying at 100°C the moisture content was calculated to be approximately 57%. The significant difference in biomass yield between cultivars indicated that further screening may identify specific cultivars which are well-suited to Alberta conditions.

The seed yields at Gwynne ranged between 563–1341 kg/ha. This result was very interesting, because one goal of the trial was to investigate whether seed production was possible in Alberta. The results indicated that seed produc-

Table 1. Total biomass, seed weight, individual plant weight and height of ten low-THC hemp cultivars grown at Gwynne, AB in 1997.

Cultivar	Total biomass (t/ha)	Seed weight (kg/ha)	Plant weight (g/plant)	Height (cm)
Beniko	13.74	1016	70.6	229
Irene	16.17	1169	50.1	220
Lovrin 110	11.82	563	52.2	227
Secuieni	17.33	1341	72.1	233
USO 31	11.8	1024	59.7	203
USO 14	11.91	925	50.1	189
Zolotonosha 15	15.97	1094	90.6	228
Zolotonosha 24	8.92	685	109.4	202
Zolotonosha 13	15.53	1070	60.2	201
Bialobrezeski	11.33	973	42.5	196
LSD (0.05)	3.62	511	26.9	32

Table 2. Total biomass and stalk yield of hemp cultivars grown at Brooks in 1996.

Cultivar	Total biomass (t/ha dry matter)				Stalk yield	
	Sept. 12		Oct. 28		(Sept. 12, Irrig.)	
	Irrigated	Dryland	Irrigated	Dryland	% of total biomass	Dry matter (kg/ha)
Zolotonosha 11 (LR)[z]	7.3	7.4	9.0	8.4	46.9	3450
Zolotonosha 11	8.6	7.5	9.1	7.8	51.8	4469
Zolotonosha 13 (LR)	7.2	6.8	7.3	6.8	46.9	3368
Zolotonosha 13	8.4	7.9	10.4	7.7	50.9	4271
Kompolti (LR)	11.5	8.8	11.9	10.8	66.7	7648
Kompolti	11.2	9.3	14.1	12.1	68.3	7616
Beniko	9.4	8.4	7.1	8.4	51.1	4787
Fedora 19	9.0	8.7	8.8	9.2	53.1	4767
Felina 34	9.8	8.4	7.8	8.0	51.9	5077
Futura 77	10.1	8.9	10.5	9.4	60.6	6145
Unico	10.2	-	7.8		63.4	6448
Mean	9.3	8.2	9.4	8.8	55.6	
LSD .05	2.2	1.5	4.2	2.6	3.7	
C.V. (%)	13.9	10.4	26.5	17.0	3.9	

[z]LR = Low rate of seeding

308

tion was possible. The seed was harvested from only the upper portion of the stem (80 cm) to make harvesting and threshing easier. It is estimated that, averaged across all cultivars, the sampling method collected approximately 95% of total seed production.

At the Brooks site the cultivar Kompolti produced the highest total above ground biomass yield both under irrigated and dryland conditions (Table 2). Plants seeded at higher rates and under irrigation, produced higher biomass yield than those plants seeded at the low seeding rate and under dryland, respectively. The total biomass yields ranged from 7–14 t/ha of dry matter. The stalk yields from this total biomass ranged from 3400–7600 kg/ha. These yields are somewhat less than the higher biomass and stalk yields reported from Europe. This is, however, quite comparable to some European yields in spite of the hail damage suffered by the plants in mid-July. The difference between irrigated and dryland biomass yields were less than expected, possibly due to residual water available in the summerfallow for the dryland crop.

Most of the cultivars matured early enough to produce viable seeds. The irrigated plants yielded better than their dryland counterparts (Table 3). Potential seed yields ranged from 216–1322 kg/ha with 'Fedora 19', 'Zolotonosha 11' and 'Zolotonosha 13' giving the highest yields. 'Kompolti' which gave the highest biomass and stalk yield was the lowest seed yielder. Plants seeded at the lower rates appeared to yield better than those seeded at the higher seeding rates. Plants in the outside guard rows yielded much higher than those plants in the inside rows indicating that the plants potential can be increased given more favorable growing conditions.

Oil Analysis

The hemp seed contains approximately 25–35% oil. Analysis of seed oil from the different cultivars together with a typical canola oil is shown in Table 4. The predominant fatty acid was linoleic 54.6–56.1%, followed by linolenic 17.8–19.2%, and oleic 11.8–12.8%. It is interesting to note that the oil also contains 1.5–2.2% of gamma linolenic acid. Only slight differences in oil composition was noted among the cultivars. The high percentage (73.7–74.6%) of the polyunsaturated fatty acids (linoleic + linolenic) indicates a good, nutritious but unstable oil. This compares with canola oil that has high (55.6%) monounsaturated fatty acid in the oil. The presence of the gamma linolenic acid makes hemp oil even more nutritionally desirable. This fatty acid is the important component found in evening primrose and borage seed oils. The analysis also indicated high levels of various anti-oxidants, including tocopherols and sterols.

Fiber Analysis

The hemp samples tested from Alberta grown hemp (Table 5) were evaluated for some of the primary characteristics associated with fiber quality. The tensile strength was measured to be 1.6 g d-tex^{-1} at 33 days after planting, and reached an optimum (3.0 g d-tex^{-1}) at the August 7 measurement. This coincided

Table 3. Seed yield of different hemp cultivars grown at Brooks in 1996.

| | Seed yield (kg/ha) | | | |
| | Samples harvested from inside rows with borders | | Samples harvested from outside guard rows | |
Cultivar	Irrigated	Dryland	Irrigated	Dryland
Zolotonosha 11 (LR)^y	1326	750	1339	1417
Zolotonosha 11	1186	694	1470	1301
Zolotonosha 13 (LR)	1322	801	1593	1741
Zolotonosha 13	988	834	1591	1556
Kompolti (LR)	216	377	475	441
Kompolti	347	329	647	750
Beniko	787	685	2170	1670
Fedora 19	1491	1145	2900	2676
Felina 34	890	1066	2123	2762
Futura 77	863	848	2235	2155
Unico	596	--	1130	--
Mean	910	753	1607	1647
LSD .05	467	369	634	1030
C.V. (%)	30.1	28.6	23.1	36.5

^{y}LR = Low rate of seeding

Table 4. Fatty acid composition of the seed oil of different hemp cultivars grown at Brooks in 1996. Canola data obtained from Dr. Terry Rachuk, Canadian Agra Foods Inc., Nisku, Alberta.

Fatty acid	Fatty acid composition (%)						
	Felina 34	Beniko	Fedora 19	Futura 77	Zolotonosha 11(LR)	Zolotonosha 13(LR)	Typical canola
Palmitic (16:0)	5.4	5.5	5.5	5.6	5.4	5.6	3.2
Palmitoleic (16:1)	0.2	0.2	0.2	0.2	0.2	0.2	0.2
Stearic (18:0)	2.8	2.8	2.8	3.1	2.9	2.9	1.2
Oleic (18:1)	11.8	12.2	12.0	12.8	12.1	12.2	55.6
Linoleic (18:2)	56.1	56.4	56.1	54.6	55.6	55.6	21.7
Gamma linolenic (γ 18:3)	2.1	2.1	2.0	1.5	2.2	2.2	--
Linolenic (α18:3)	18.5	17.8	18.1	19.2	18.4	18.1	12.9
Arachidic (20:0)	0.8	0.8	0.8	0.8	0.4	0.8	0.6
Eicosenic (20:1)	0.4	0.4	0.4	0.4	0.4	0.4	2.2
Behenic (22:0)	0.3	0.3	0.3	0.6	0.3	0.3	0.3
Lignoceric (24:0)	0.2	0.2	0.2	0.2	0.2	0.2	0.4
Others	1.4	1.3	1.6	1.3	1.5	1.5	--
Erucic	--	--	--	--	--	--	1.5

with the time of flowering. As described in the literature, the tensile strength of the Alberta-grown fiber decreased following flowering (Batra 1985). The parameters for hemp fiber were well within the range generally observed for hemp. The results indicated similar characteristics to flax fiber. The extension at break was lower than reported figures for cotton.

Can hemp compete with flax or cotton for fine textiles? There are many factors to consider, not the least of which is the cost and difficulty of processing the fiber into fine yarns. Considering only fiber properties, hemp is remarkably similar to flax, a fiber that is prized for many of its characteristics—its beautiful luster, light resistance, absorbency and dyeability.

Table 5. Fiber yield, tex, tenacity, and breaking extension of 'Kompolti' hemp.

Retting time (d)	Fiber yield (%)	Tex (g/km)	Tenacity (cN tx^{-1})	Extension at break (%)
4	30	4	31 ± 5	2.7
4	31	4	41 ± 9	2.4
5	27	3	40 ±9	0.8
5	17	3	40 ± 6	2.2

SUMMARY

The preliminary research reported here indicated that low-THC hemp can be produced in Alberta. Current European cultivars are available that have acceptable levels of THC which are low enough to be acceptable for industrial purposes. Several cultivars were identified as excellent seed or fiber lines. Lower seeding densities resulted in higher seed yields; higher planting rates resulted in greater fiber yields. The composition of the seed oil indicated a high level of polyunsaturated fatty acids, as well as high levels of anti-oxidants. Fiber testing indicated that the time of harvest will determine the quality of the fiber, and that hemp grown in Alberta has interesting fiber characteristics.

REFERENCES

Batra, S. 1985. Other long vegetable fibers: abaca, banana, sisal, henequen, flax, ramie, hemp, sunn and coir. p. 727–807. In: M. Lewin and E.M. Pearce (eds.), Fiber chemistry. Marcel Dekker, New York.

Blade, S.F. 1998. Industrial hemp in Alberta. p. 2–20. In: S.F. Blade (ed.), Alberta Hemp Symposia Proc., Edmonton: Alberta Agriculture, Food and Rural Development. Edmonton, Alberta.

Bosca, I. and M. Karus. 1998. The cultivation of hemp. HempTech Publishers, Sebastopol, CA.

Feasibility of Adopting Kenaf on the Eastern Shore of Virginia*

Altin Kalo, Susan B. Sterrett, Paul H. Hoepner, Fred Diem, and Daniel B. Taylor

Since the 1940s, kenaf (*Hibiscus cannabinus* L., Malvaceae) has been viewed as a potentially attractive source of fiber, mainly for newsprint and high quality paper. Kenaf research has drawn increased interest due to the establishment of the Kenaf Demonstration Project in 1986 and the involvement of numerous researchers of the Agricultural Research Service Stations in Mississippi, Oklahoma, and Texas.

Much of the kenaf research, however, has focused on the establishment of breeding, genetics, and agronomy programs (Kugler 1996) and less attention has been given to the economic and environmental issues related to the adoption of this crop. This paper seeks to address the void in the current literature by discussing the economic and environmental impact of growing kenaf on the Eastern Shore of Virginia. This research is part of a more comprehensive study that focuses on the adoption of alternative crops in order to provide local producers with more sustainable farming strategies. To prove successful, potential alternatives need to have an economic or environmental record that is equal or superior to that of traditional crops in the area. Only after these new crops pass this test, can we then decide to address the numerous barriers that accompany their commercial production.

Kenaf is no exception. Initially, it seemed that kenaf could be a feasible addition to the current crop mix on the Eastern Shore. Closer scrutiny, however, revealed that our assumption was not correct. Currently, kenaf fails to be an economically viable alternative and it offers only marginal environmental benefits. The value of this computer model approach is two-fold. This project was able to effectively investigate many options for incorporating kenaf into existing whole farm plans on the Eastern Shore of Virginia without the growers actually risking their economic investment of inputs, land, and labor. The model is also available for investigating other crops, which may be of interest for consideration by local growers.

METHODOLOGY

Project Development

An integrated systems approach was developed at the request of growers on the Eastern Shore of Virginia to evaluate potential new or non-traditional crops. This approach examines the production, economic, and marketing feasibility and the potential environmental impact of both traditional crops and proposed alternatives. Agriculture is a major source of income and employment opportunities for the population of this predominately rural area (Center for Public Service 1995). This environmentally sensitive growing area is located on the southern end of the Delmarva Peninsula, between the Atlantic Ocean and the Chesapeake Bay.

Economic Analysis

A linear programming model was designed to perform the economic analysis (Kalo et al. 1997). The model evaluated the economic performance of two representative farms with distinct cropping systems, grain or vegetable production, typical of the region. Both farms were assumed to be commercial operations with 255 ha of arable land available and limited machinery and labor resources. The grain farm had no irrigation and it could grow a combination of full season soybeans or winter wheat, double-cropped with soybeans. The vegetable farm had limited water resources and irrigation equipment but could grow various crops. The economic performance of kenaf was evaluated under each cropping system. Rotational constraints were established in order to prevent planting the same crop in a field in two consecutive years.

The model identified the optimum grain and vegetable farm plans by maximizing net returns to land,

*This project was financially supported by the Southern Region Sustainable Agricultural Research and Education Project (SARE LS96-80).

management, and fixed capital. Local farm management agents and extension specialists developed detailed calendars of resource utilization and budget estimates for production costs. Table 1 summarizes the kenaf cost of production budget estimates for the Eastern Shore.

The lack of an established market for kenaf makes it impossible to collect historical price data for kenaf. Instead, we used prices specified in the proposed contracts for kenaf in the mid-Atlantic region. Table 2 summarizes the profit ha for kenaf given various yield and price estimates. Kenaf was profitable only if price exceeded $75/t and/or yield was more than 12 t/ha.

Environmental Analysis

The computer program PLANETOR (Center for Farm Financial Management 1995) was used to measure environmental parameters regarding soil erosion, pesticide leaching and runoff, pesticide toxicity, and nitrogen leaching. This program was customized for the study area in order to evaluate the environmental performance of the optimal farm plans generated by the economic model.

ECONOMIC ANALYSIS

The results of the initial economic model showed that kenaf could not compete with the alternatives available in either cropping system. Simply stated, kenaf did not provide the same returns to investment given the limited resources of each farm plan. To understand what it would take to make kenaf a viable alternative, we evaluated three important variables: (1) an increase in price; (2) an increase in yield; and (3) a decrease in transportation costs (marketing).

The results of the sensitivity analysis indicate that kenaf becomes economically feasible if we could institute the following changes:

1. Kenaf enters the optimal solution for the representative grain farm when its price increases to $120/t. This represents a 42% increase in the price of kenaf from the $75/t assumption used in the base farm models. Net returns for the grain farm, however, increase only 13.5%, from $72,910 to $102,110. For the representative vegetable farm, the price of kenaf has to increase to $100 in order to grow 81 ha of the crop. However, at $100/t of kenaf, net returns increase by only 1%.

2. The yield of the kenaf crop needs to increase from around 12 t/ha to 19 t/ha in the grain farm model and to 17 t/ha in the vegetable farm model. This change increases net returns by 24% for the grain farm and 9% for the vegetable farm.

3. Kenaf enters the optimal solution if the transportation distance is cut from 241 to 80 km, in both models.

Table 1. Kenaf cost of production budget*.

Costs	Quantity	Price	Cost/ha
Variable			
Seeding costs:	15.71 kg	$6.61	$103.82
Fertilizer:	109.92 kg	$0.70	$76.94
Fertilizer application costs:	1.00 ha	$14.83	$14.83
Lime	0.82 t	$38.00	$30.99
Spray materials:			
Herbicides	0.35 L	$23.62	$8.17
Insecticides	0.00 L	0.00	$0.00
Fungicides	0.00 L	0.00	$0.00
Irrigation:			$0.00
Machinery			
Production, repairs			$30.36
Production, fuel			$19.44
Harvest, repairs			$0.00
Harvest, fuel			$0.00
Miscellaneous:			$70.00
Interest:	$354.53	8.00%	$28.36
Custom rate harvest			$86.45
Marketing:	241.40 km	$3.07	$741.00
Labor charges:	7.41 hr	$6.00	$44.46
Fixed			
Machinery cost:			
Production			$58.02
Harvest			$0.00
Total			$1,312.83
Cost per sale unit	12.35 t		$106.30

*No land cost included, return is calculated as net returns to land and management.

Given that the cost of transporting a trailer load of kenaf is calculated at $3 per loaded km, transportation costs fall from $741/ha (destination being 241 km away) to $245.6/ha (destination only 80 km away). This change has a more pronounced effect than the previous changes, as it results in a 33% increase of net returns for the grain farm and a 12% increase of returns for the vegetable farm.

Challenges and Reality

The price, yield, or transportation cost adjustments required to include kenaf in the model's profit maximizing solutions pose a number of problems. The most difficult to overcome is the increase of average yield from 12 to 17 or 19 t/ha. Agronomic research (Hallmark et al. 1994; Hovermale 1995, 1994a,b; Kurtz 1996) demonstrated that yields of 17 or 19 t/ha are difficult to achieve. Such yields become even harder to attain with the inexperience of kenaf producers. Finally, most of the data on kenaf yields is provided via experiments conducted at experiment stations in the Deep South. Realizing that farmers are unlikely to replicate the results achieved by controlled experiments, it is prudent to assume even lower yields. Hence, the attainment of economic feasibility based on higher yield expectations would be difficult at best.

Price and transportation requirements are also problematic. Economic feasibility studies (Zhang and Dicks 1992) suggest that the best way to sell kenaf would be through processor forward contract agreements. Each contract would have built-in specific conditions relating to price and transportation agreements. Initially, some of the farmers on the Eastern Shore entered into negotiations with a kenaf processing company and developed one such contract. The contract specified that farmers would receive $75/t (dry) delivered to the processor. Furthermore, they would also receive $25/t towards their harvesting and storage costs. This increased the potential farm price to $100/t, which makes it economically feasible even when transportation costs are estimated at $3 per loaded km. Considering that the contract also offered to subsidize transportation costs by almost 50%, the agreement seemed quite attractive.

However, the contract has not yet been implemented. The processor was unable to reimburse farmers $25/t in harvest costs or the 50% transportation subsidy. This is reflected in the present kenaf production budget that now a price of $75/t and charges full transportation costs at $3 per loaded km for the 241 km that it is necessary to move the product to the processing facility (Table 1).

The relationship between processors and potential kenaf growers has deteriorated over the last year. Farmers are fearful of the dangers inherent in dealing with a single buyer and they suspect that they will be used by processors until the latter establish themselves in the market and develop more profitable partnerships with raw product supply sources closer to their plant. Farmers realize that there is nothing to prevent the processing company from changing the terms of the contract or even discontinuing their Eastern Shore activity. On the other hand, the processors are suspicious that the farmers are only using their company as a risk minimization tool to field test the economic feasibility of the crop. They fear that once the farmers gain some experience and expertise in growing kenaf they will seek to built their own processing plant, given the simplicity of the technology required.

The feasibility analysis for kenaf illustrates some of the problems in introducing a new crop alternative. In this section we only focused on three potential barriers to the production of kenaf: prices, yield, and transportation costs. However, other issues may pose difficulties. The acquisition of specific machinery for harvesting the product is a case in point. The production budgets assume a lump-sum custom harvesting cost of $86.45/ha. This assumption may be valid if an individual farmer could rent the machinery. Production costs would be much higher, however, if farmers had to purchase specialized equipment to harvest 81 ha, especially considering the fact that it cannot be used for harvesting either wheat or soybeans.

It is not economically feasible to ship the product over a long distance. Thus, the Eastern Shore is at a significant disadvantage in this regard. Because of its peninsular geographic location, growers

Table 2. Net return to land and management for various yield and price assumptions*.

| Yield | Profit ($/ha) | | |
| | Kenaf price/t | | |
(t/ha)	$65.00	$75.00	$85.00
7.4	482	556	630
12.4	803	826	1,050
17.3	1,124	1,297	1,480

*Total cost/ha = $1,312.83 (see Table 1)

313

face the risk of dealing with monopsonistic partners or building the needed infrastructure. Having a processing plant on the Eastern Shore would greatly reduce transportation costs, at or below the 80-km constraint discussed above. The investment, however, would only shift the marketing problems from the farmers to the kenaf processors. They would now have to deal with the few paper mills using kenaf and it is not clear whether the processing activity offers enough value-added to the profit margins to make the entire operation economically feasible. The region, on its part, has neither the land or water resources to accommodate a paper mill. Given these barriers, kenaf appears to have no near-term potential for the Eastern Shore.

Environmental Sustainability

Despite the lack of economic viability, one main question needs to be addressed. Does kenaf offer sufficient environmental benefits to warrant any intervention into eliminating the barriers discussed above? Our conclusion, based on the environmental analysis, is that kenaf offers only marginal environmental improvement, mainly because it uses fewer pesticides than wheat and soybean, the main crops with which it would compete. The effect of including 81 ha of kenaf in the optimal solution of the representative grain farm. Net returns under this scenario fall by almost 30%, while the weighted average of soil erosion from 255 ha increases from 1.9 to 4.2 t/ha. Consequently, kenaf fails to provide the economic or environmental incentives to justify its adoption in a wheat and soybean operation.

Including kenaf in the crop mix of the representative vegetable farm provides similar results. The introduction of kenaf increases the weighted average soil erosion from 6.3 to 7.7 t/ha on the 255 ha farm. Net returns also decrease from about $625,000 to $530,000.

When the economic model was forced to incorporate 81 ha of kenaf in its farm plan, the two crops that were eliminated were wheat and soybeans. While kenaf has higher soil erosion parameters than wheat and soybeans, it could provide some relief in terms of potential pesticide leaching and runoff compared to the grain crops. On the other hand, kenaf offers little relief in terms of potential nitrate leaching. The potential for nitrate leaching is 319 kg/ha for kenaf compared to 291 kg/ha for romaine lettuce. These results and the results for soil erosion show that kenaf is no magic bullet that can solve the potential risk of environmental pollution from agricultural production on the Eastern Shore. Moreover, these results show that there are few incentives to justify research into eliminating the barriers for the production and marketing of kenaf.

CONCLUSIONS

Our analysis indicates that presently kenaf is not a viable crop for the Eastern Shore and its future commercialization faces serious constraints. Current knowledge of the economic and environmental impacts of this crop shows that it lacks both the economic and environmental incentives to justify further research. Kenaf cannot compete with other crops available to producers because it does not provide competitive profit margins and it lacks an established marketing infrastructure for delivering the product. Unattainable yields, low expected price levels, and dealing with a monopsonistic buyer make the production of kenaf on the Eastern Shore of Virginia a very risky venture. Furthermore, its mixed environmental record removes any justification for subsidizing further efforts for its commercialization.

REFERENCES

Center for Farm Financial Management. 1995. PLANETOR: Users manual. Dept. Agr. and Applied Economics, Minnesota Ext. Service, Univ. Minnesota, St. Paul.

Center for Public Service. 1995. Economic profile of the Accomac-Northampton Counties. Univ. Virginia, Charlottesville.

Goforth, C. and E.J.M. Fuller. 1994. A summary of kenaf production and product development research 1989–1993. Mississippi Agr. Forestry Expt. Sta., Res. Rep. 19(10).

Hallmark W.B., L.P. Brown, H.P. Viator, R.J. Habetz, W.D. Caldwell, and C.G. Cook. 1994. Kenaf: A new crop for Louisiana. Louisiana Agr. 37(2).

Hovermale, C.H. 1994a. The effect of nitrogen rate on kenaf planted at three row widths. Mississippi Agr. Forestry Expt. Sta., Res. Rep. 19(13).

Hovermale, C.H. 1994b. Effect of tillage method, cover crop, and nitrogen rate on yield of kenaf. Mississippi Agr. Forestry Expt. Sta., Res. Rep. 19(11).

Hovermale, C.H. 1995. Effect of date of planting on four varieties of kenaf. Mississippi Agr. Forestry Expt. Sta., Res. Rep. 20(9).

Kalo, A. 1997 Analyzing the economic and environmental impacts of agricultural alternatives: The case of the Eastern Shore. MS Thesis. Dept. Agr. and Applied Economics, Virginia Tech., Blacksburg VA.

Kugler, D.E. 1996. Kenaf commercialization: 1986–1995. p. 129–133. In: J. Janick (ed.), Progress in new crops. ASHS Press, Alexandria, VA.

Kurtz, M.E. 1996. Pre-emergence herbicide trials in kenaf. Mississipi Agr. Forestry Expt. Sta. Bul. 1053, August.

Effect of Kenaf and Soybean Rotations on Yield Components

Charles L. Webber III

As kenaf (*Hibiscus cannabinus* L., *Malvaceae*) production in the US continues to increase it is essential to integrate this alternative fiber crop into existing cropping systems. Soybean [*Glycine max* (L.) Merr., Fabaceae] is now grown widely in the same production areas where kenaf can be successfully produced. A kenaf/soybean rotational system could have long term economic and pest control advantages, if there are no adverse effects of rotating these two crops. A three-year field study was conducted at Haskell, Oklahoma to determine the effect of six kenaf/soybean rotations on kenaf and soybean yield components. The kenaf cultivar 'Everglades 41' and soybean cultivar 'Forrest' were planted on a Taloka silt loam soil in mid May and harvested each October. The crops received no irrigation, rainfall was the only source of moisture. The individual kenaf/soybean rotations did not adversely affect the kenaf stalk yields or soybean seed yields. Kenaf stalk yields across all rotational combinations and years averaged 7.9 t/ha, while soybean seed yields averaged 866 kg/ha. Seasonal rainfall affected soybean growth and yields more than any effects due to the cropping sequence. A continuous kenaf rotation produced the greatest kenaf yields (9.4 t/ha) in the final year. It was determined that either a three-year continuous or rotational cropping system can be used for kenaf and soybean production without reducing crop yields.

Kenaf is a warm season annual fiber crop closely related to cotton and okra that can be successfully produced in a large portion of the US, particularly in the southern states. For the last 3000 years kenaf has been used as a cordage crop to produce twine, rope, and sackcloth (Wilson et al. 1965). Kenaf was first domesticated and used in north Africa. India has produced and used kenaf for the last 200 years, while Russia started producing kenaf in 1902 and introduced the crop to China in 1935. The US started kenaf research and production during World War II to supply ropes for the war effort and developed high-yielding anthracnose-resistant varieties, cultural practices, and harvesting machinery (Nieschlag et al. 1960; Wilson et al. 1965; White et al. 1970). Then in the 1950s and early 1960s USDA researchers determined that kenaf was an excellent source for cellulose fibers for a large range of paper products (newsprint, bond paper, and corrugated liner board) with less energy and chemical requirements than standard wood sources. More recent research and development work in the 1990s has demonstrated the plant's use in building materials (particle boards of various densities, thicknesses, and fire and insect resistance), adsorbents, textiles, livestock feed, and fibers in new and recycled plastics (injected molded and extruded).

As kenaf production in the US continues to increase, it is essential to integrate this alternative fiber crop into existing cropping systems. One practical means of integrating kenaf into existing cropping systems would be the introduction of kenaf into the rotational crop sequences. Crop rotation is an important management strategy which research has demonstrated can provide numerous crop production advantages, including increased pest control (Benson 1985; Edwards et al. 1988), improved soil aggregation resulting in greater wet aggregate stability (Baldock and Kay 1987; Raimbault and Vyn 1991), increased nutrient availability (Adams et al. 1970; Baldock and Musgrave 1980; Asghari and Hanson 1984; Peterson and Varvel 1989b; Varvel and Peterson 1990), increased grain quality (Asghari and Hanson 1984), and increased yields (Crookston et al. 1988; Edwards et al. 1988; Peterson and Varvel 1989a,b,c; Varvel and Peterson 1990; Crookston et al. 1991; Lund et al. 1993).

Many factors can be responsible for the positive yield response to crop rotation, but increased nitrogen availability following a legume crop is often the most recognized benefit (Adams et al. 1970; Baldock and Musgrave 1980; Asghari and Hanson 1984; Peterson and Varvel 1989b; Varvel and Peterson 1990). Crop rotations involving legumes such as alfalfa (*Medicago sativa* L.) (Adams et al. 1970; Baldock and Musgrave 1980; Asghari and Hanson 1984; Raimbault and Vyn 1991), red clover (*Trifolium pratense* L.) (Peterson and Varvel 1989b,c; Raimbault and Vyn 1991), and soybean [*Glycine max* (L.) Merr.] (Crookston et al. 1988; Edwards et al. 1988; Peterson and Varvel 1989b,c; Crookston et al. 1991; Lund et al. 1993) have often demonstrated their benefit to non-legume crops such as maize (*Zea mays* L.) (Crookston et al. 1988; Edwards et al. 1988; Peterson and Varvel 1989c; Crookston et al. 1991), and grain sorghum [*Sorghum bicolor* (L.) Merr.] (Peterson and Varvel 1989b).

Soybean, a legume crop grown for its grain, is now grown widely in the same production areas where kenaf can be successfully produced. A kenaf/soybean rotation system could have long term production and economic advantages, if there are no significantly adverse affects of rotating these two crops. At present there is insufficient research concerning the affect of including kenaf into a crop rotation with other existing crops such as soybean. The objective of this research was to determine the affect of a kenaf and soybean rotations on kenaf and soybean yield components.

METHODOLOGY

A 3-yr field study was conducted at Oklahoma State University's Eastern Research Station, Haskell, Oklahoma, to determine the effect of six kenaf/soybean rotation combinations on kenaf and soybean yield components. The 3-yr experimental design was a randomized complete block with four replications.

Plots were 7 m wide (eight 76-cm rows) and 7 m long, and oriented in an east-west direction. Each year prior to planting, fertilizer was applied and incorporated to the kenaf plots at a rate of 168, 22, and 42 kg/ha of N, P, and K, respectively and to the soybean plots at a rate of 0, 22, and 168 kg/ha of N, P, and K, respectively. The kenaf cultivar 'Everglades 41' and soybean cultivar 'Forrest' were planted on a Taloka silt loam soil (fine mixed, thermic Mollic Albaqualf, 0–1% slope) in mid May (May 11, 1989, May 25, 1990, and May 10, 1991) and harvested each October (Oct. 19, 1989, Oct. 26, 1990, and Oct. 16, 1991). The soybean seeds were inoculated with the appropriate *Rhizobium* culture and each crop was planted at a rate of 370,000 seeds/ha.

The day after planting all plots received a pre-emergence application of metolachlor [2-chloro-N- (2-ethyl-6-methylphenyl)-'(2-methoxy-1-methylethyl) acetamide] at 1.7 kg/ha to control weeds. The herbicide was applied using a tractor-mounted sprayer delivering 187 liters/ha at 275 kPa pressure. Fan-tip sprayer nozzles were 0.5 m apart on a 7-m boom. All plots were kept weed free during the growing season by handweeding. The crops received no irrigation, rainfall was the only source of moisture.

The research plots were harvested 161 days after planting (DAP) in 1989, 154 DAP in 1990, and 159 DAP in 1991. A 2.25 m^2 (1.50 by 1.50 m) quadrant was harvested from the center two rows of each plot. All plants were harvested by hand and cut at ground level. Plant counts from the harvest quadrant were used to determine plant populations. Three kenaf plants were selected randomly from the harvested material to determine plant heights and stalk diameters. Calipers were used to measure stalk diameters at 1 m above the base of the kenaf stem.

The kenaf leaves, flowers, and flower buds were removed from the stalks and weighed separately to determine the fresh weight. Plant samples were oven dried at 66°C for 48 h and weighed to determine dry weight. All kenaf weights and percentages were based on oven dry weights. The soybean seeds were weighed and yields adjusted to the standard 13% moisture.

In the final year of the rotational study soil samples (0–15 cm) were collected at planting and harvest to determine soil pH, nitrogen (N), phosphorus (P), and potassium (K) content. Soil and root samples were also collected at planting, mid-season (76 DAP), and harvest to determine nematode populations in the soil. The nematode samples were collected with a soil sampling tube with an inside diameter of 5.2 cm. Soil and nematode samples were processed at Oklahoma State University's analytical and diagnostic laboratories.

When the F-test indicated statistical significance ($P < 0.05$), the least significant difference (LSD) test was used to separate means (Snedecor and Cochran 1967). When no significant year by cultivar interactions were indicated, results are reported as averages across years.

EXPERIMENTAL RESULTS

Precipitation was measured during each growing season from May 1 to harvest and totaled 580 mm in 1989, 571 mm in 1990, and 422 mm in 1991 (Table 1). The 20-yr average precipitation (1 May–31 Oct.) for the location was 567 mm. Precipitation in 1989 and 1990 was just above the 20-yr average precipitation of

567 mm, compared to the 1991 precipitation which was 25% less than the 20-yr average. A greater percentage (77%) of seasonal precipitation was received in the first 3 months of 1989 than the first 3 months of either 1990 (48%) and 1991 (51%).

Kenaf Yield Components

First Year. The kenaf stalk yields for the first year ranged from 6.0 to 7.0 t/ha (Table 2). Even though there were no significant differences for kenaf stalk yields, plant populations, and stalk diameters among treatments during the first year, there was a significant difference for plant heights. The first kenaf treatment had significantly greater plant height compared to the third kenaf treatment. A general trend existed between decreasing plant heights and increasing plant populations and stalk yields.

Second Year. The kenaf stalk yields for the second year ranged from 8.0 to 8.3 t/ha (Table 3) with no significant differences among kenaf treatments for stalk yields, plant height, and stalk diameter. Plant population for the kenaf following soybean was significantly less than kenaf following kenaf production. It is unclear why there would be such a significant decrease in kenaf plant population when kenaf was planted the year following a soybean planting. Even with lower plant populations the kenaf treatment following soybean did not have significantly less yields. These results are consistent with earlier research by Higgins and White (1970) and Webber (1993) who reported no significant yield loss from population differences as long as the plant populations remain above a certain limit, 99,000 plants/ha. The kenaf plants often compensate for decreased plant populations by increasing plant height and/or stalk diameters. Although the plant population was the only yield parameter with a significant difference there was a slight inverse trend between decreased plant populations and increased stalk diameters.

Table 1. Precipitation for 1998, 1990, and 1991 for Haskell, Oklahoma. The 20-yr average seasonal (May 1–Oct. 31) precipitation for the research location is 567 mm.

	Precipitation (mm)		
Period	1989	1990	1991
May	214	187	63
June	143	22	145
July	89	67	7
Aug.	65	36	45
Sept.	52	213	135
Oct. 1 to harvest	17	45	28
Total[z]	580	570	423

[z]May 1 to harvest.

Table 2. Kenaf yield components for the first year (1989) of the rotational study.

Plant population (plants/ha)	Plant height (cm)	Stalk diameter (mm)	Stalk yield (t/ha)
213,000	200a[z]	11.4	6.0
282,000	195ab	11.6	6.9
309,000	181b	10.8	7.0

[z]Mean separation in columns by LSD (5% level).

Table 3. Kenaf yield components for the second year (1990) of the rotational study.

Previous crop	Plant population (plants/ha)	Plant height (cm)	Stalk diameter (mm)	Stalk yield (t/ha)
Soybean	151,000b[z]	220	14.8	8.1
Kenaf	244,000a	220	14.0	8.3
Kenaf	274,000a	240	13.8	8.0

[z]Mean separation in columns by LSD (5% level).

Table 4. Kenaf yield components for the third year (1991) of the rotational study.

Previous crops	Plant population (plants/ha)	Plant height (cm)	Stalk diameter (mm)	Stalk yield (t/ha)
Soybean, soybean	381000	184	11.2	8.5
Soybean, kenaf	410000	185	11.0	8.7
Kenaf, kenaf	420000	193	11.9	9.4

Third Year. The kenaf stalk yields for the third and final year of the rotational study ranged from 8.5 to 9.4 t/ha with no significant differences among kenaf treatments for stalk yields, plant population, plant height, and stalk diameter (Table 4). Plant population for the kenaf crops which followed soybean were slightly less, but not significantly less than the continuous kenaf for three years. There was a general trend of increasing stalk yields with increasing plant populations, plant height, and stalk diameter.

Soybean Yield Components

Soybean seed yields among treatments within years were not significantly different during any year of the 3-yr rotational study and averaged 1.2, 0.73, and 0.67 t/ha, for the first, second, and third year, respectively (Table 5). Soybean yields decreased after the first year independently of the cropping sequence. The soybean yields reflect the greater water available in 1989 during the soybeans initial growth through flowering and podfill, compared to less available water in the first half of the 1990 and 1991 seasons (Table 1).

Soil Analysis

Initial soil analysis of the research plots prior to the first fertilizer application and planting in 1989 had an analysis of 9.0 N, 64 P, and 170 K kg/ha. In the final year of the research, 1991, the soil K content at planting and harvest was significantly different among cropping sequences, while the pH and N were only significantly different for samples collected at harvest (Table 6). When comparing K values between planting and harvest, the cropping sequences with kenaf had the greatest drop in soil K values planting to harvest. Not only did the kenaf crops deplete a greater amount of K during the growing season, but the slight

Table 5. Soybean and kenaf yields by cropping sequence for 1989, 1990, and 1991.

Cropping sequence			Yields (t/ha)					
			1989		1990		1991	
1989	1990	1991	Soybean	Kenaf	Soybean	Kenaf	Soybean	Kenaf
Soybean	Soybean	Soybean	1.2	--	0.8	--	0.6	--
Soybean	Soybean	Kenaf	1.2	--	0.7	--	--	8.5
Soybean	Kenaf	Kenaf	1.2	--	--	8.1	--	8.7
Kenaf	Kenaf	Kenaf	--	6.0	--	8.3	-	9.4
Kenaf	Kenaf	Soybean	--	6.9	--	8.0	0.7	--
Kenaf	Soybean	Soybean	--	7.0	0.7	--	0.7	--

Table 6. Effect of crop sequence on soil analysis (pH, N, P, and K) at planting and harvest during the final year (1991) of rotational research at Haskell, Oklahoma.

Cropping sequence			Planting				Harvest			
1989	1990	1991	pH	N (kg/ha)	P (kg/ha)	K (kg/ha)	pH	N (kg/ha)	P (kg/ha)	K (kg/ha)
Soybean	Soybean	Soybean	5.53	40.9	73.9	170.8ab[z]	5.82ab	5.9b	44.5	159.6b
Soybean	Soybean	Kenaf	5.55	44.0	59.6	209.2a	5.62bc	6.4b	43.7	136.4d
Soybean	Kenaf	Kenaf	5.45	46.8	72.0	180.6ab	5.27d	12.6a	43.1	143.1cd
Kenaf	Kenaf	Kenaf	5.33	35.6	62.4	175.6ab	5.27d	5.6b	46.2	155.7bc
Kenaf	Kenaf	Soybean	5.35	40.3	71.4	172.2ab	5.60c	4.2b	49.3	181.2a
Kenaf	Soybean	Soybean	5.55	36.7	62.2	162.4b	5.85a	5.3b	46.8	165.8b
Main effects		Soybean[y]	5.54a	40.5	65.2	180.8	5.76a	5.1	46.9	168.8a
		Kenaf	5.38b	39.8	68.6	176.1	5.39b	8.2	44.3	145.0b

[z]Means in the column followed by the same letter are not significantly different based on a LSD, 5% level.
[y]Means averaged across the final year's crop.

net gain in K for two of the three soybean sequences indicates that the addition of 42 kg/ha of K at planting served as an adequate soil maintenance application for the soybean crops.

When the cropping sequences were grouped for analysis by their 1991 crop, soybean versus kenaf, the analysis indicated significant differences for pH at planting and harvest, and for K at harvest. The soil analysis at planting in 1991 was an indication of the effect of the proceeding crops, especially the crop grown in 1990. The pH data among cropping sequences had a general trend for lower pH values for those cropping sequences which had kenaf the previous year. The

Table 7. Stunt nematode (*Tylenchorhynchus* spp.) means at planting, mid-season (76 DAP), and harvest during the final year.

Cropping sequence			Nematode count (No/100 cm³)		
1989	1990	1991	Planting	Mid-season	Harvest
Soybean	Soybean	Soybean	44ab²	87ab	147ab
Soybean	Soybean	Kenaf	50ab	45ab	15bc
Soybean	Kenaf	Kenaf	14b	41ab	6c
Kenaf	Kenaf	Kenaf	12b	69ab	11bc
Kenaf	Kenaf	Soybean	14b	40ab	68abc
Kenaf	Soybean	Soybean	65a	149a	175a
Main effects		Soybean	41.0a	92.0a	130.0a
(final year)		Kenaf	25.3a	51.7a	10.7b

²Means in the column followed by the same letter are not significantly different based on a LSD, 5% level.

pH at planting compared to harvest is evidence that the cropping sequences influence the soil pH values. The most likely factor within the cropping sequences that would decrease the pH for the kenaf rather than the soybean was the difference in the fertilizer applications for the specific crops. The symbiotic N_2 fixation relationship of soybean and *Rhizobium japonicum* precluded the need N fertilizer applications to the soybean plots, whereas the kenaf production system involved the addition of N fertilizer. The addition of N in the form of ammonium nitrate (NH_4NO_3) to the kenaf plots prior to planting would most likely be responsible for the lower and decreasing pH values in the kenaf plots resulting from nitrification of the ammonium. Another issue that should be investigated is whether kenaf production in itself may reduce soil pH independently of fertilizer applications.

Nematodes

A small number of various nematodes (*Pratylenchus* spp., *Helicotylenchus* spp., *Xiphinema americanum*, and *Hoplolaimus galeatus*) were identified on the soybean and kenaf roots or in soil samples collected from the research plots, but stunt (*Tylenchorhynchus* spp.) nematode was the only nematode present throughout all the cropping sequences. Stunt nematode is an ectoparasitic nematode which feeds on a large range of different types of plants including soybean, maize, grain sorghum, alfalfa, wheat (*Triticum aestivum* L.), tobacco (*Nicotiana tabacum*), peanut (*Arachis hypogaea* L.), pearl millet [*Pennisetum glaucum* (L.) R. Br.], and lawn grasses. Stunt nematode populations at planting were the lowest for cropping sequences which followed kenaf (Table 7). When averaged across the previous season's crop, stunt populations where significantly less for kenaf (13.3 nematodes/100 cm³) than for the soybean (52.7 nematodes/100 cm³). Stunt nematode populations in the soybean treatments increased from planting through mid-season to harvest, compared to stunt nematode populations in the kenaf treatments which increased from planting to mid-season and then decreased at harvest to levels at or below initial stunt nematode populations. The stunt populations averaged across crops (soybean versus kenaf) followed the same pattern as the individual cropping sequences and resulted in significantly fewer stunt nematodes in kenaf treatments compared to the soybean treatments. These results are consistent with earlier rotational research (Edwards et al. 1988) which reported reduced stunt nematode populations in corn when included in a rotation compared to a continuous corn cropping system.

CONCLUSIONS

The kenaf stalk yields and soybean seed yields in the third and final year of the rotational study were not significantly different as a result of 3-yr cropping sequences. Seasonal rainfall affected soybean growth and yields more than any effects due to the cropping sequence. The pH values in cropping sequences following kenaf were lower at planting and continued to decrease though the growing season for those crop sequences planted in kenaf. The lower pH values did not adversely affect crop yields and could be increased by liming the soil. Including kenaf in a soybean rotation decreased stunt nematodes the following year and produced lower stunt nematodes during a kenaf growing season compared to soybean production. The 3-yr study determined that a kenaf/soybean rotation was successful and did not adversely affect kenaf or soybean yields, but did reduce stunt nematode populations.

REFERENCES

Adams, E., H.D. Morris, and R.N. Dawson. 1970. Effect of cropping systems and nitrogen levels on corn (*Zea mays*) yields in the southern piedmont region. Agron. J. 62:655–659.

Asghari, M. and R.G. Hanson. 1984. Nitrogen, climate, and previous crop effect on corn yield and grain N. Agron. J. 76:536–542.

Benson, G.O. 1985. Why the reduced yields when corn follows corn and possible management responses? p. 161–174. In: D. Wilkerson (ed.), Proc. 40th Annual Corn and Sorghum Research Conference. Chicago, IL., Dec. 11–12, 1985. Am. Seed Trade Assoc., Washington, DC.

Baldock, J.A. and B.D. Kay. 1987. Influence of cropping history and chemical treatments on the water-stable aggregation of a silt loam soil. Can. J. Soil Sci. 76:501–511.

Baldock, J.O. and R.B. Musgrave. 1980. Manure and mineral fertilizer effects in continuous and rotational crop sequences in Central New York. Agron. J. 72:511–518.

Crookston, R.K., J.E. Kurle, P.J. Copeland, J.H. Ford, and W.E. Lueshen. 1991. Rotational cropping sequence affects yield of corn and soybean. Agron. J. 83:108–113.

Crookston, R.K., J.E. Kurle, P.J. Copeland, and W.E. Lueshen. 1988. Relative ability of soybean, fallow, and triacontanol to alleviate yield reductions associated with growing corn continuously. Crop Sci. 28:145–147.

Edwards, J.H., D.L. Thurlow, and J.T. Eason. 1988. Influence of tillage and crop rotation in yields of corn, soybeans, and wheat. Agron. J. 80:76–80.

Higgins, J.J. and G.A. White. 1970. Effects of plant populations and harvest date on stem yield and growth components of kenaf in Maryland. Agron. J. 62:667–668.

Lund, M.G., P.R. Carter, and E.S. Oplinger. 1993. Tillage and crop rotation affect corn, soybean, and winter wheat yield. J. Prod. Agr. 6:207–213.

Nieschlag, H.J., G.H. Nelson, I.A. Wolff, and R.E. Perdue, Jr. 1960. A search for new fiber crops. TAPPI 43:193–201.

Peterson, T.A. and G.E. Varvel. 1989a. Crop yield as affected by rotation and nitrogen rate. I. Soybean. Agron. J. 81:727–731.

Peterson, T.A. and G.E. Varvel. 1989b. Crop yield as affected by rotation and nitrogen rate. II. Grain sorghum. Agron. J. 81:731–734.

Peterson, T.A. and G.E. Varvel. 1989c. Crop yield as affected by rotation and nitrogen rate. III. Corn. Agron. J. 81:735–738

Raimbault, B.A. and T.J. Vyn. 1991. Crop rotation and tillage effects on corn growth and soil structural stability. Agron. J. 83:979–985.

Snedecor, G.W. and W.G. Cochran. 1967. Statistical methods. 6th ed. Iowa State Univ. Press, Ames.

Varvel, G.E. and T.A. Peterson. 1990. Residual soil nitrogen as affected by continuous, two-year, and four-year crop rotation systems. Agron. J. 82:958–962.

Webber, C.L. III. 1993. Yield components of five kenaf cultivars. Agron. J. 85:533–35.

White, G.A., D.G. Cummins, E.L. Whiteley, W.T. Fike, J.K. Greig, J.A. Martin, G.B. Killinger, J.J. Higgins, and T.F. Clark. 1970. Cultural and harvesting methods for kenaf. USDA Prod. Res. Report 113. Washington, DC.

Wilson, F.D., T.E. Summers, J.F. Joyner, D.W. Fishler, and C.C. Seale. 1965. 'Everglades 41' and 'Everglades 71', two new varieties of kenaf (*Hibiscus cannabinus* L.) for fiber and seed. Florida Agr. Expt. Sta. Cir. S-168.

Feasibility of Cotton as a Crop for Pennsylvania

Polly S. Leonhard

Cotton (*Gossypium hirsutum* L., Malvaceae) is grown primarily in the southern and southwestern United States with an upper limit of about 37° latitude. A study was undertaken to determine if cotton might be adapted to Pennsylvania (40°–42°), specifically under the conditions of Lancaster County where farmers are seeking an alternate crop to replace or supplement tobacco which is currently under disease stress (blue mold) and price constraints. Although ordinary white cotton is a questionable crop under current pricing, higher prices can be obtained for colored, "organically grown" cotton. The rationale for considering cotton production in this area is that (1) existing tobacco nursery transplant facilities could be used to start cotton plantings; and (2) there may be a niche market for organic, colored cotton. The lack of any cotton production in the area facilitates the production of naturally colored cotton, which cannot be grown in the presence of white cotton due to cross-pollination problems.

EXPERIMENTAL

In order to evaluate the feasibility of production of cotton in Pennsylvania, several types of cotton, including naturally colored cotton ones, were evaluated for three years in Akron, Pennsylvania to determine agronomic performance and cotton quality. These included white cotton originally from Arkansas, and brown, green, red, mocha, and natural cotton originally from Texas. Fiber analyses were carried out on all cotton samples including fiber strength, elongation, upper length, short fiber content and micronaire (the measure of resistance of a plug of cotton to air flow to determine fineness and maturity). Fiber wax content was also evaluated by Soxhlet extraction with ethyl alcohol (Conrad 1994).

Few problems were faced in growing cotton in Pennsylvania, with the exception of some insect pests, including bud/bollworms and Japanese beetles, but they were controlled with insecticidal soap. Some of the brown cottons had the shortest seasons and hybrids made with brown cottons were among the faster growing types. Estimated lint yields ranged from 1,122 to 7,051 kg/ha. The average lint percent among all of the plants was 33.2 which falls within the "normal" range of 30 to 45%. The average number of bolls per plant was 14.9 and the average weight of the seed cotton per boll was 4.1 g. Based on fiber evaluations, the cotton produced in Pennsylvania appeared to be competitive in most respects. The majority of the samples had excellent strength, elongation, upper length, short fiber content, and microniare values. The white, natural, and brown samples had the best fiber characteristics and high yields.

CONCLUSION

These results indicate that it is potentially feasible to produce cotton with suitable fiber qualities in northern areas. It is uncertain if so-called organic cotton could be produced economically based on a three-year study. The use of *Bt* cottons would be helpful to reduce insect damage but it would not be clear if such genetically transformed cottons would be accepted in the "organic" market. The absence of infrastructure for cotton production (such as on-farm harvesting equipment, and ginning facilities) is an issue that would have to be addressed to make cotton an economically viable crop. The major constraints to cotton production would appear to be economic, i.e. would the prices received in light of the expected yields be sufficient to establish a niche industry for established tobacco growers.

REFERENCES

Conrad, C.M. 1994. Determination of wax in cotton fiber: A new alcohol extraction method. Indust. Engin. Chem. p. 748–749.

FRUITS

Temperate Berry Crops

Chad Finn

Wherever humans have lived, they have made berries a part of their diet. Most of these have never been developed beyond local markets but some have become economically important crops. In this paper, the berry crops have been divided into four groups based on their current international popularity and potential future value. An overview of the status of development, current production, and future potential for these crops is presented with an American perspective. The discussion is limited to temperate "berry" crops that are produced on a shrub, a perennial herbaceous plant, or a vine, which excludes many of the cherry/plum (*Prunus* sp., Rosaceae) relatives; jujube (*Ziziphus jujuba* Mill., Rhamnaceae); *Cornus* sp. Cornaceae; *Sorbus* sp., Rosaceae; and many other tree fruit.

MAJOR BERRY CROPS

The most economically important and best described berry crops worldwide include strawberry (*Fragaria ×ananassa* Duch., Rosaceae) (Galletta and Bringhurst 1990; Hancock et al. 1996); blueberry (*Vaccinium corymbosum* L., *V. angustifolium* Ait., *V. ashei* Reade, Ericaceae) (Eck et al. 1990; Pritts et al. 1992; Galletta and Ballington 1996); cranberry (*V. macrocarpon* Ait., Ericaceae) (Dana 1990; Eck 1990; Roper and Vorsa 1997); black currant (*Ribes nigrum* L., Grossulariaceae) (Harmat et al. 1990; Brennan 1996); table and wine grapes (*Vitis* spp., Vitaceae) (Ahmedullah and Himelrick 1990); raspberry (*Rubus idaeus* L., Rosaceae) (Jennings 1988; Crandall and Daubeny 1990; Pritts and Handley 1989; Daubeny 1996); and blackberry (*Rubus* sp., Rosaceae) (Pritts and Handley 1989; Hall 1990; Moore and Skirvin 1990; Crandall 1995). These need no further discussion as information is widely available on each.

Other major berry crops have large production areas worldwide but for a variety of reasons have not reached the stature and importance of the above. These include the hybrid berries such as 'Logan' and 'Boysen' (*Rubus* sp., Rosaceae); black raspberry (*R. occidentalis* L., Rosaceae); lingonberry (*Vaccinium vitis-idaea* L., Ericacea); gooseberry (*Ribes uva-crispa* L. Grossulariaceae); and red currant (*Ribes rubrum* L., Grossulariaceae). These will be discussed in turn below.

Hybridberries

'Logan' is a result of cross between red raspberry (*R. idaeus* L.) and a blackberry (*R. ursinus* Cham. & Schldl., Rosaceae derivative) (Logan 1909; Brown 1916; Logan 1955; Jennings 1980). 'Logan' fruit are similar in color and appearance to red raspberry but the torus remains with the fruit like a blackberry and they have a distinctive flavor. The fruit are excellent for processing and are dried, juiced, and canned. In the late 1800s to mid 1900s, 'Logan' was planted on thousands of hectares and accounted for millions of dollars in sales. Today approximately 40 ha remain in commercial production in Oregon. Many factors have led to 'Logan's' decline in popularity including: the difficulty of picking the fruit especially with a mechanical harvester, relatively low yields, and a decline in popularity in a younger generation of consumers.

'Boysen' was discovered on Rudolph Boysen's farm in California. This red raspberry (or 'Logan') × blackberry hybrid was the basis for the initial development of the Knott's Berry Farm fruit and entertainment empire. 'Boysen' has the growth habit of trailing blackberry and the fruit are similar in appearance, larger on average, with larger drupelets, and a purple fruit color. 'Boysen' was widely produced in California, Oregon, and New Zealand into the 1980s. Currently, a few thousand hectares of 'Boysen' are grown in Oregon and New Zealand but California production has largely disappeared. The market for 'Boysen' remains strong.

'Logan' and 'Boysen' can be grown wherever trailing blackberries such as 'Marion' can be grown. Information is available on cultural practices at the Northwest Berry and Grape Infonet (http://osu.orst.edu/dept/infonet/).

Lingonberry

Stang et al. (1990) specifically addressed lingonberries (*Vaccinium vitis-idaea* L., Ericaceae) at the First New Crops Symposium in 1988. Lingonberry continues to be largely a European crop. However, the Pacific

Northwest has seen a substantial increase in plantings the last 3 years. Lingonberry is harvested from native stands in northeast China and in some localities a substantial quantity of juice is produced. Lingonberry is found natively on acidic soils in northern temperate zones and can range to near the Arctic Circle, but in many of these northern areas they are protected by snow cover. In addition to Stang et al. (1990), St.-Pierre (1996) has published production information. Since lingonberry is largely a processed crop, either better cultivars or better machines must be developed that will make mechanical harvest viable.

Black Raspberry

Black raspberry (*Rubus occidentalis* L., Rosaceae) production is concentrated in Oregon where 400–600 ha are harvested for processing into juice and jam. Ohio growers are planning on doubling their crop area to 250 ha in the next few years (J. Scheerens, pers. commun.). The juice is valuable as a natural colorant. In other regions, particularly regions of Ohio and Pennsylvania, black raspberry is harvested fresh as a pick-your-own crop. 'Munger' is the most important cultivar worldwide. Black raspberries are relatively easy to grow, however, they are short-lived. Plantings often last only 2–3 harvest seasons due to virus and disease infestation. Poor pollination from rain during bloom can limit the crop. The plants are trellised in the eastern US for fresh harvest but in the western US the plants are "hedged" at about 1 m for mechanical harvesting and processing. The biggest challenge with large-scale production is the fluctuation in fruit price. In 1997, the fruit sold for $US 4.18/kg whereas the price in 1995 had only been $US 1.32/kg (USDA-NASS-ERS 1998). Black raspberries can be established relatively quickly and cheaply so growers are constantly getting into and out of production in response to the fruit price.

Gooseberry and Red Currant

These two members of the *Ribes* have a long history of cultivation. They are widely adapted to temperate regions and many soil types. While they are popular in Europe as a fresh market and processed product, they have had limited success in the US, in part due to white pine blister rust (*Cronartium ribicola* Fisher) restrictions. Both are grown throughout the eastern US primarily as a fresh market crop for local sales. In the Pacific Northwest, they are grown and shipped nationally on the fresh market and Washington has about 40 ha in production. Both of these crops are primarily processed into pies or preserves in the case of gooseberries, and juice or jelly in the case of red currants. Since 1966, when federal legislation was deregulated, 17 states in the US continue to restrict the production of some or all of the *Ribes* species because they are a cohost for white pine blister rust. Some of these states are considering repealing their restrictions so that *Ribes* can be grown. Some *Ribes* genotypes are resistant to this disease and some are immune (Hummer and Finn 1998b). Gooseberry production in the US is often limited by powdery mildew (*Sphaerotheca mors-uvae* [Schwein.] Berk. & Curt), which can regularly cause defoliation in plants. While the primary cultivars grown in the US have been 'Oregon Champion', 'Poorman', and 'Pixwell' due to their reliable production, other cultivars are suitable and are mildew resistant (Hummer and Finn 1998a). The main limitation to these crops appears to be consumer education and acceptance. A number of hybrids between gooseberry and black currant are larger fruited and milder flavored than black currant; 'Josta' is the best known example (Reich 1991). While not widely planted, there are some very small commercial plantings and market development seems to be the biggest drawback to further expansion.

NEGLECTED BERRIES

Neglected berries include those that are regionally important such as elderberry (*Sambucus canadensis* L., Caprifoliaceae); aronia (*Aronia melanocarpa* [Michx.] Elliott, Rosaceae); cloudberry (*R. chamaemorus* L., Rosaceae); arctic raspberry (*R. arcticus* L., *R. stellatus* Sm., and their hybrids, Rosaceae); mora (*R. glaucus* Benth., Rosaceae); alpine strawberry (*F. vesca* L., Rosaceae); muscadine grape (*Vitis rotundifolia* Mich., Vitaceae); Juneberry/saskatoon (*Amelanchier* sp., Rosaceae); hardy kiwi (*Actinidia arguta* [Siebold & Zucc.] Planch. ex Miq., Actinidiaceae); edible honeysuckle (*Lonicera caerulea* L., Caprifoliaceae); sea buckthorn (*Hippophae rhamnoides* L., Elaeagnaceae); and schisandra (*Schisandra chinensis* [Turcz.] Baill., Schisandraceae).

Elderberry

The juice and preserves of *Sambucus canadensis* L., native to eastern North America, and *S. nigra* L., native to Europe, have often been mainstays of rural pantries and Native Americans and early settlers used them as a dried and medicinal crop. Elderberry was seldom cultivated because it was so common in fence rows and along roadsides. While limited, information is available on commercial production (Way 1981; Stang 1990). Selections of superior plants from the wild have traditionally been used locally but high quality cultivars were developed from *S. canadensis* by breeding programs in New York, Pennsylvania, and Nova Scotia (Ritter and McKee 1964; Way 1964; Craig 1966; Darrow 1975). Pennsylvania and Oregon have a few fairly large plantings, Kansas has a small elderberry wine industry and the Austrians have substantial plantings. 'Haschberg', a wild selection of *S. nigra* from near Vienna, is the main European cultivar (R. Wrolstad pers. commun.). The fruit is in demand for processing in preserves, as a natural colorant, and for wine making. In Europe, a company has just released an anthocyanin/flavonoid enriched extract primarily from elderberry for colorant and nutraceutical use. While the crop would benefit from further breeding, it is generally adapted to most locations although viruses can be a problem in the northwestern US. Incorporating the desirable acylated anthocyanin pigments of *S. canadensis* into *S. nigra* and improving pigment stability in processed products (R. Wrolstad pers. commun.) are improvements desired by processors.

Baby Kiwi, Hardy Kiwi, Tara, Wild Fig, Wee-kee

Actinidia arguta (Siebold & Zucc.) Planch. ex Miq., a smooth-skinned, winter tolerant relative of the kiwifruit (*A. chinensis* Planch./ *A. deliciosa* [A. Chev.] C. F. Liang & A. R. Ferguson, Actinidiaceae), has many common names. Recently, it has been developed from a novelty into an economically important crop (Ferguson 1990; Strik and Cahn 1998). The fruit are small, about the size of a large table grape, and are packaged multiply in "clam shell" containers rather than as single fruit. Darrow, in 1937, presented this species as a potential crop but it was not until the 1990s that it has become a small scale commercial crop. As the New Zealanders brought the fuzzy Chinese gooseberry (renamed the kiwifruit) to world attention, homeowner enthusiasts spread the more winter tolerant *A. arguta* across North America. It might have remained in the realm of enthusiasts until Hurst's Berry (Sheridan, Oregon) decided it would fit well in a diverse fresh berry product line. Their interest and development of this crop demonstrate the impact a single company, with good marketing savvy, can have on a relatively obscure crop. While the consistent demand for this crop is undetermined, more than 35 ha of fruit have been planted in Oregon since 1994 and there are a few growers with substantial plantings in Pennsylvania. *Growing Kiwifruit* is an excellent guide to growing *A. arguta* commercially (Strik and Cahn 1998). The major drawbacks to this crop are the expense of establishing a planting and the length of time to the first crop. *Actinidia arguta* requires a substantial trellis system, irrigation, and takes three years until a small crop is produced. Once in production, the biggest problems are frosts that kill newly emerged shoots that would produce the flowers, and abrasion on the fruit surface due to wind. Irrigation and other forms of frost protection reduce this problem. The fruit is harvested and put into storage when it is mature but before it begins to soften. When it is ready to ship it is treated with ethylene to begin the final ripening process. In storage, the pedicels, which do not form an abscission zone with the fruit, dry out and harden which can be of concern to consumers. This problem has yet to be solved but mechanical or genetic solutions may be possible.

Alpine Strawberry

Fragaria vesca L. production seems to have reached its peak prior to the development of *F.* ×*ananassa* as the primary commercial strawberry. There seems to be a constant interest in producing these small but aromatic fruit (Reich 1991). Bakers like to use these small fruit in products such as muffins where an entire berry is desirable. Homeowners often write passionately about them. They are not likely to have a major commercial impact but could be grown and marketed successfully to niche markets if the labor costs of harvest and low yields can be justified. While the plants are relatively easy to grow and can be raised from seed, they are short-lived where virus pressure is high. In breeding programs, *F. vesca* is often used as in indicator plant for strawberry viruses.

Rubus

Cloudberry (*R. chamaemorus*) and arctic raspberry (*R. arcticus, R. stellatus* and their hybrids) in northern Europe and mora (*R. glaucus*) in Andean South America are regionally extremely important and valuable crops. As a group, their perishable nature lends them to processing as juice, preserves, and liqueurs. Rapp et al. (1993) addressed the potential of *R. chamaemorus* at the 1992 New Crops Symposium.

Arctic raspberries are native to the colder regions of the northern hemisphere and are renowned for the strong aromatic character of their fruit. Breeding programs in Finland and Sweden have developed cultivars from these species (Jennings 1988). The cultivars are largely self-sterile so more than a single cultivar must be planted. The cultivars are apparently widely and successfully grown in Scandinavia. Production at this point in time appears to be limited to Scandinavia.

Rubus glaucus is commonly sold in the Andean countries of South America. This crop is typically grown in small, up to 0.5 ha plantings for local sale. Large bottles (2 L) of mora carbonated soda were available in grocery stores suggesting larger scale commercial production is viable. A large fruit processor seriously considered commercial production of this crop in the US in the 1970s but pulled out just before the plantings were to be established. This crop may be similar to the 'Marion' blackberry, which is renowned for its flavor and aroma and excellent processing characteristics. However, as with 'Marion', *R. glaucus* fruit are too perishable for the fresh market. This crop appears to be a developed "land-race;" cultivars have not been developed but the species has commercial qualities. Plants require irrigation on their native volcanic soils and are often trellised (Gaitoni ~1970; Federación Nacional de Cafeteros de Columbia ~1984 [The exact date of these publications is unknown but they are available upon request]). Because the species is native to high elevation near the equator (little change in photoperiod and moderate temperatures year round), widely adapted types must be developed if this crop were to be more widely planted. Commercial production has been reported in Mexico and Central and South America (Gaitoni ~1970; Federación Nacional de Cafeteros de Columbia ~1984; Rincon 1987).

Aronia, Chokeberry

Aronia melanocarpa is native to the eastern US, however, this crop was popularized and is commonly planted in Eastern Europe and the former Soviet Union. Prior to World War II, aronia was primarily used as an ornamental. Seeds were later imported from Germany to the former Soviet Union where cultivars were developed for fruit production (Kask 1987). By 1971, 5400 ha were planted in the former Soviet Union, 4000 of which were in Siberia. The original species is diploid, however, most of the cultivars are tetraploid. The 4*x* cultivars are sometimes designated as *Aronia mitshurinii* Skvorsov et Majjtulina, Rosaceae. The fruit was designated as a "healing plant" in the former Soviet Union. Experimental plantings of cultivars have been established in Czechoslovakia, Scandinavia, and Germany. The fruit is valued for its juice which is very high in anthocyanins, blends well with other fruit juices and is reputed as a source of "phenols, leucoanthocyanins, catachines, flavonoles, and flavones" that are considered to be bioactive in humans. The plants have no special cultural or site requirements. In the Pacific Northwest, the extremely vigorous plants will bear a small crop one year after rooted cuttings are planted. The plants resemble *Amelanchier*, Rosaceae in many respects. Yields of up to 17 kg/bush with 10 kg/bush average are reported in Eastern Europe. The fruit is often hand harvested by cutting the fruit clusters, but they can be mechanically harvested. In Oregon, an unidentified spring rust on the fruit caused some yield loss but its effect appeared to be minimal. Currently, there is interest in establishing commercial production of *Aronia* in the US. (Much of this information is from an unpublished document of an unknown source but is available upon request.)

Muscadine Grape

Vitis rotundifolia, a southeastern US native, seems to primarily have consumer appeal in that region. While evolutionarily related to the "bunch grapes" such as *V. vinifera* L., Vitaceae and *V. labrusca* L., Vitaceae, the muscadines differ in many ways including chromosome number, vine and berry anatomy and morphology, and physical and chemical characteristics of the fruit and juice (Olien 1990). The fruit with its distinctive musky or fruity aroma is eaten fresh but is even more commonly made into juice, wines, pies, jellies, and other processed products. While the fruit has been cultivated by indigenous peoples for more than 400 years, production has been limited to the South. Most of the 1600 ha in production is concentrated in the coastal states from North Carolina to Louisiana (Olien and Hegwood 1990). Traditionally, muscadines have been grown where Pierce's disease has limited the production of American and French-hybrid grapes (Olien 1990). Lack of research on improved cultivars, cultural techniques, and processing methods has also limited commercial expansion of the industry (Olien 1990). Currently, the price for fruit has been good. In 1998, muscadines were being sold at 1.6 times the price of 'Thompson Seedless' on the fresh market (J. Clark, pers. commun.). Although more research is needed, information is available about production practices (Dearing 1947; Hegwood et al. 1983; Olien 1990; Anderson 1996) and breeding of muscadines has been reviewed (Goldy 1992)

Juneberry/Saskatoon

A number of *Amelanchier* sp. have been harvested for their fruit and included in breeding programs (Reich 1991). The Saskatchewan government has developed excellent production guides for saskatoons (St.-Pierre 1997). While this crop has widespread commercial potential, because the purple fruit appear somewhat similar to blueberries (*Vaccinium* sp., Ericaceae), success in the fresh market will probably be limited to areas where blueberry cannot be grown due to extremely cold winter temperatures or alkaline soils (Stang 1990). However, there is certainly the potential to develop processed products that are uniquely different from blueberry. Currently, the industry is concentrated in the Canadian Provinces of Alberta (500 ha), Saskatchewan (200 ha) and Manitoba (80 ha) where growers feel they do not have enough production to meet the demand (Mazza and Davidson 1993; Delidais 1998).

Edible Honeysuckle

Lonicera caerulea (Synonym of *Lonicera caerulea* var. *edulis* Turcz. ex Herder, Caprifoliaceae) is widely harvested in regions of China and northern Eurasia. Superior Russian cultivars have been developed. The cylindrical, blue fruit ripen extremely early in the season on 1–2 m tall bushes. As with *Amelanchier*, its similar appearance to blueberry will make it difficult to establish a marketing foothold. However, it is extremely winter hardy and can grow in regions where blueberry cannot. In trial plantings in the US, it has been extremely susceptible to leaf disease in the Midwest (M. Widrlechner pers. commun.) and, as with apricot, it flowers very quickly given warm temperatures and can be severely damaged by frost. However, the flowers are reported to be able to survive temperatures several degrees below freezing.

Schisandra

A native of northeastern China and the former Soviet Union, *Schisandra chinensis* (Turcz.) Baill. vines produces red fruit that are high in vitamin C. It is harvested from the wild for local consumption and has received a great deal of attention for its reputed medicinal qualities. While it is mentioned as a cure for a large number of maladies, most refereed publications seem to focus on its effects on liver function (Ahumada et al. 1989; Mizoguchi et al. 1991; Ko et al. 1995a,b). Managed plantings would be most comparable to grape production. Currently, most production is from wild harvested fruit and commercial viability is unknown. I could find no references that discussed cultural management. Improved selections from the wild are currently being evaluated in Jilin Province by the Chinese Academy of Agricultural Science (Shen Yugie pers. commun.).

Sea Buckthorn

Hippophae rhamnoides, a native of the colder regions of Eurasia, has been harvested on a large scale in eastern Europe as a vitamin C rich fruit for processing into jellies, juices, and liqueurs. Cultural management (Li and Schroeder 1986; Bernáth and Földesi 1992; Pietilä 1998) and breeding potential (Anderson and Wahlberg 1994) have recently been reviewed for sea buckthorn. This crop could be a valuable crop in North America. Plants can tolerate extremely harsh winters and poor soils. The main limitations are the development of an infrastructure for a processing market. The "Catch-22" with this crop, and others that are only suitable for processing, is that processors are not likely to get interested unless there is a market and conversely, there is not likely to be a market developed unless there are processed products in place.

POTENTIAL NEW BERRIES

Many locally harvested crops from indigenous and introduced fruit could become economically important crops. For example, several companies in the western and northwestern US are hiring pickers to harvest wild fruit for regionally, nationally, and internationally distributed fresh and processed products. In Europe and Asia, native fruit are also commonly harvested to supply a growing nutraceutical industry. Native *Vaccinium* and *Rubus* are the most common examples of these "wild" harvested crops. Potential new crops often follow a natural progression. Interest is first generated in the crop from "wild" harvested plants. When the suppliers run into difficulty with erratic supply, interest in stabilizing the supply of the crop through cultivation follows. If the crop can be adapted to cultivation economically and the market remains in place the crop can become a new crop with commercial potential. In this group I include examples of *Rubus* and *Vaccinium* that are currently harvested form the wild.

Vaccinium "Huckleberries"

"Huckleberry" is a confusing common name. Most accurately huckleberries would describe species in the genus *Gaylussacia* Kunth., Ericaceae. However, in the commerce, this name is often used regionally as a name for the local wild *Vaccinium* species. In late summer and early fall, pickers fan out over the Northwest from the Cascade Mountains as far east as the mountains of Montana to primarily pick *V. membranaceum* Douglas ex Torr. (syn. *V. globulare* Rydb.), Ericaceae as well as *V. ovalifolium* Smith and *V. deliciosum* Piper. The fruit is sold on roadsides and to wholesalers. The restaurant trade is a major consumer of fruit and a company in Montana has become well known nationally for their "chocolate-covered huckleberries." Stark and Baker (1992) present a great deal of information on the biology of the species and propose how they could be raised commercially. While their suggested production practices may be valid they are not practical for large scale production. Recently, our US Dept. of Agriculture laboratory has begun to evaluate populations of these species at a low elevation location using cultural practices that are similar to those used for highbush blueberry (*V. corymbosum*). While the planting is only three years old, it has begun to fruit, and some genotypes appear to be adapted to low elevation production. Whether this crop will prove to be commercially viable in cultivated stands has yet to be determined. In the Northwest, some fruit is also harvested in the Coastal Range from *V. ovatum* Pursh which is more commonly cut as an evergreen "green" for floral arrangements or used as an ornamental plant in the landscape. In the past two years, more than 3 ha have been planted for commercial fruit production. While *V. ovatum* has been grown successfully in the landscape for decades, the fruit are much lower quality, particularly aromatic components to fruit flavor, than the other species discussed.

Mortiño

Vaccinium floribundum Kunth., Ericaceae grows profusely in northern South America (Popenoe 1924). The evergreen nature of this species and its fruit are similar to *V. ovatum* of North America and *V. confertum* Kunth of Mexico and *V. consanguineum* Klotzch of southern Mexico and Central America. In Ecuador, baskets of fruit are commonly available in the market. Mortiño could be more widely grown as it is a popular crop and should be amenable to cultivation. However, if production is expanded it will most likely be successful in a niche market similar to the *Vaccinium* huckleberries of North America. Since, commercial highbush blueberries, which are relatives and produce a somewhat similar fruit, are in such wide production in North America, Chile, and Argentina, it would be difficult for *V. floribundum* to displace this market.

Bilberry

Bilberry (*Vaccinium myrtillus* L., Ericaceae) has a long history in European folk medicine (Morazzoni and Bombardelli 1996) and while the fruit is largely harvested from the wild, commercial production is not unknown (Dierking and Dierking 1993). Recently attention has been focused on bilberry as efforts to determine whether cultivated North American blueberries have similar nutraceutical characteristics to *V. myrtillus* (Kalt and McDonald 1996; Kalt 1997; Prior et al. 1998). These preliminary studies indicate that while *V. myrtillus* has higher levels of antioxidants than the North American commercial blueberries, the commercial blueberries do contain high levels. Although there is potential for commercial production of *V. myrtillus*,

Dierking and Dierking (1993) would not recommend it from a horticultural point of view. From a nutraceutical point of view, it would appear that the commercial blueberries, produced in such great abundance and at a relatively low cost, could satisfy this demand.

Bog Bilberry

Millions of hectares of *Vaccinium uliginosum* L., Ericaceae stretch across circumboreal regions of the Northern Hemisphere. In China, we saw these fruit harvested and sold locally and the fruit pressed for juice that was marketed around China and in the Western US. In Scandinavia, the fruit are sold from wild harvested plants and cultivars have been developed for commercial production (Hiirsalmi 1989; Hiirsalmi and Lehmushovi 1993). Unfortunately, this crop has the same niche as North America's commercial blueberry industry. While it would be hard to justify cultivation of this crop if it is to be marketed in competition with the North American blueberries there would seem to be ample justification for improved management of the huge expanses of this species for the processing market.

Trailing Blackberry

Rubus ursinus Cham & Schldl., Rosaceae is an early colonizer of disturbed sites throughout the Northwest. In this era where timber is clear-cut and agriculture has disturbed many sites, the northwest trailing blackberry is very common. As with the *Vaccinium* huckleberries, pickers harvest these fruit from native stands for the restaurant trade and specialty markets. The species is dioecious and the fruit are medium sized, soft, and have a very aromatic flavor. 'Marion' blackberry is remarkable for its flavor and this can be traced to the *R. ursinus* in its pedigree. Trailing species can be grown like the commercial trailing cultivars. However, the only justification for establishing a managed planting of the species as opposed to cultivars is if the "wild" label is critical to marketing. In general, the species is much more susceptible to foliar diseases than the cultivars. Our first generation hybrids between cultivars and this species have yielded disease tolerant, thornless, and early ripening genotypes that retain the species flavor. Whether these are commercially viable will be determined in the future.

Miscellaneous *Rubus*

As you move to different regions, there is often a *Rubus* species that is harvested from the wild for local sales. These species are very similar in their place in the market to the *R. ursinus* just described. Examples worldwide would include the southern dewberries (*Rubus trivialis* Michaux) in the US; *Rubus parvifolius* L. of China and Japan; *R. phoenicolasius* Maxim. in Japan; *R. crataegifolius* Bunge; *R. niveus* Thunb. and *R. coreanus* Miq. in China; and the many blackberry species (*Rubus*) in Europe (Anon. 1912a,b; Card 1915; Williams and Darrow 1940; Sherman and Sharpe 1971; Jennings 1988; Finn et al. 1998). Any of these crops could be developed into a major crop. However, as with mortiño, they will be competing against a well established, productive industry that produces a somewhat similar crop, i.e. red (*R. idaeus*) and black raspberry (*R. occidentalis*), and blackberry (*Rubus* sp.).

POTENTIAL CROPS WITH UNMET POTENTIAL

Finally, many crops that have previously been mentioned in these sorts of forums have not been further developed. Usually, market drives the production, either a large market does not exist or another berry is filling that market niche. A few examples of previously described potential crops in this class include: Nanking cherry (*Prunus tomentosa* Thunb., Rosaceae); cranberry bush (*Viburnum opulus* L., Caprifoliaceae); and buffalo berry *(Shepherdia argentea* Nutt. Elaeagnaceae).

Nanking Cherry/Hansen Bush Cherry

Prunus tomentosa is native throughout temperate regions of eastern Asia and is widely sold in the local markets. At least a century ago, seed lots were brought to the US and breeding programs released a few cultivars in the first half of this century (Darrow 1937; Fogle 1975; Kask 1989). Hansen in South Dakota recognized that the superior cold hardiness of this productive, but small fruited, cherry might be valuable for the northern Great Plains (Kask 1989). While the species has not developed into a commercial crop, it is commonly sold through catalogues to homeowners (Reich 1991). This crop will remain in the realm of homeowners in North America and in local markets of Asia because the fruit are inferior to the commercially available sour cherries (*P. cerasus* L.).

Cranberry Bush

Viburnum opulus (syn. *V. trilobum* Marsh.) is found in northern temperate regions. While popular as an ornamental plant in the landscape, the species has never developed into a commercial fruit crop. The fruit is unpalatable fresh and must be processed into jellies or juice (Card 1915). Cultivars with superior fruit and processing characteristics have been released (Darrow 1975). Cranberry bush may have potential for small scale, local production, or for a unique processed product (Stang 1990) but, in general, the fruit is too similar to cranberry (*Vaccinium macrocarpon*) or red currant (*Ribes rubrum*) as a processed product (Darrow 1975) to justify large scale production.

Buffalo Berry

Shepherdia argentea has been stuck in the "potential" class for nearly two centuries. In 1915, Card wrote, "The buffalo berry has enjoyed the distinction of remaining a new fruit for a very long time.... Yet we are still talking about buffalo berry as a new fruit which ought to be introduced." He cites references back to 1841 that state that buffalo berry was widely grown. This species is native to the Great Plains of the US (Darrow 1975). I think horticulturists continue to return to this crop because the plants are productive, extremely winter hardy, and drought tolerant, the flowers are frost tolerant, and the scarlet fruit are very high in Vitamin C. Despite being dioecious and spiny, it would appear that it might be well adapted for mechanical harvesting and processing as a juice. Buffalo berry is likely to retain its "potential" label until someone aggressively develops the market for the fruit.

CONCLUSION

Berries have been an important part of the diet of indigenous people. The world is now a global market which will have adverse and beneficial effects on crops that have developed from a specific region. Some crops that formerly had only local interest will develop demand worldwide. Other crops may be lost as similar crops from other regions will displace them. These crops will remain regionally important but will not develop worldwide importance.

Blueberry, strawberry, and grape production will continue to expand worldwide. Cranberry production will likely expand rapidly where the proper soil and water requirements can be found or where "wetlands regulations" are not as stringent as they are in the US. Red raspberry, blackberry/hybridberry, lingonberry, and black currant will steadily increase in production. It becomes more difficult to predict the future of "neglected" berries, as there seems to be serious problems with each crop except *Actinidia arguta*. *Actinidia arguta* shows tremendous promise and will see increased production worldwide. Elderberry, aronia, and sea buckthorn production will increase if their unique anthocyanin characteristics are desirable for the colorant and nutraceutical markets. Alpine strawberry, muscadine grape, juneberry, cloudberry, arctic raspberry, and mora will continue to play important roles in regional or niche markets but are not likely to join the lists of major crops worldwide. Similarly, we hope that the "potential new crops" will develop stable crop areas with consistent production as they solidify their standing as important regional or niche market crops. I do not see great potential in Nanking cherry, cranberry bush, or buffalo berry unless someone energetically develops markets for them. The rising interest in the nutraceutical characteristics of foods has carried over to berries. Schisandra and bilberry are two examples of crops primarily harvested for their nutraceutical potential. Whether this is a trend or a fad may impact which new crops will develop a large commercial industry.

Each era has their surprises as to which new crops develop into important crops. In 1915, Card had blueberries listed in the miscellaneous section of his book. By the second half of this century blueberries had become an important crop and today they are one of the major berry crops worldwide. Who would have thought a few years ago that a major chain store in the US would be promoting "Aronia Berry Juice Cocktail?" George Darrow in 1975, gave equal space to *Actinidia arguta* and *Viburnum trilobum* (syn. *V. opulus*) in a chapter on minor temperate fruit. Twenty-five years later, *A. arguta* is on the verge of becoming an important fruit crop while *V. trilobum* remains "only" a beautiful ornamental for the landscape. Let us hope in the future that we continue to be surprised by the neglected or unknown crops.

REFERENCES

Ahmedullah, M. and D.G. Himelrick. 1990. Grape management. p. 383–471. In: G.J. Galletta and D.G. Himelrick (eds.), Small fruit crop management. Prentice Hall, Englewood Cliffs, NJ.

Ahumada F., J. Hermosilla, R. Hola, R. Pena, F. Wittwer, E. Hegmann, J. Hancke, and G. Wikman. 1989. Studies on the effect of *Schizandra chinensis* extract on horses submitted to exercise and maximum effort. Phytotherapy Res. 3:175–179.

Anderson, M.M. and K. Wahlberg. 1994. The breeding potential of sea buckthorn (*Hippophae* L.). New Plantsman Dec. p. 207–217.

Anderson, P.C. 1996. Yield and berry quality of 23 cultivars and selections of muscadine grapes in North Florida. NFREC-Monticello Research Report 91-6 (http://128.227.103.58/txt/fairs/52408).

Anonymous. 1912a. Chinese brambles; for shrubberies, pergolas, and pleasure ground. Gardeners' Chron. March 16, 1912. 147–149, 165–167.

Anonymous. 1912b. New brambles. Am. Florist Oct. 19, 1912.

Bernáth J. and D. Földesi. 1992. Sea buckthorn (*Hippophae rhamnoides* L.): A promising new medicinal and food crop. J. Herbs Spices Med. Plants 1:27–35.

Brennan, R.M. 1996. Currants and gooseberries. p. 191–296. In: J. Janick and J.N. Moore (eds.), Fruit breeding Vol. II. Vine and small fruits. Wiley, New York.

Brown, W.S. 1916. The loganberry. Extension Bul. 165. Oregon Agr. College Ext. Serv., Corvallis.

Card, F.W. 1915. Bush fruits: A horticultural monograph of raspberries, blackberries, dewberries, currants, gooseberries, and other shrub-like fruits. Macmillan Co., London.

Craig, D.L. 1966. Elderberry culture in eastern Canada. Can. Dept. Agr. Publ. 1280.

Crandall, P.C. and H.A. Daubeny. 1990. Raspberry management. p. 157–213. In: G.J. Galletta and D.G. Himelrick (eds.), Small fruit crop management. Prentice Hall, Englewood Cliffs, NJ.

Crandall, P.C. 1995. Bramble production: The management and marketing of raspberries and blackberries. Food Products Press, New York.

Dana, M.N. 1990. Cranberry management. p. 334–362. In: G.J. Galletta and D.G. Himelrick (eds.), Small fruit crop management. Prentice Hall, Englewood Cliffs, NJ.

Darrow, G.M. 1937. Some unusual opportunities in plant breeding. p. 545–558. 1937 Yearb. Agr. U.S. Dept. Agr., Washington, D.C.

Darrow, G.M. 1975. Minor temperate fruits. p. 269–284. In: J. Janick and J.N. Moore (eds.), Advances in fruit breeding. Purdue Univ. Press, W. Lafayette, IN.

Daubeny, H.A. 1996. Brambles. p. 109–190. In: J. Janick and J.N. Moore (eds.). Fruit breeding Vol. II. Vine and small fruits. Wiley, New York.

Dearing, C. 1947. Muscadine grapes. U.S. Dept. Agr. Farmers' Bul. 1785.

Delidais, A. 1998. The fruit of the nineties: Saskatoons. Joonas Int. News Forum for Professional Berry Growers, Joensuu, Finland.

Dierking Jr., W. and S. Dierking. 1993. European *Vaccinium* species. p. 299–304. In: K.A. Clayton-Greene (ed.), Fifth Int. Symp. on Vaccinium Culture. Acta Hort. 346.

Eck, P. 1990. The American cranberry. Rutgers Univ. Press, New Brunswick, NJ.

Eck, P., R.E. Gough, I.V. Hall, and J.M. Spiers. 1990. p. 273–333. In: G.J. Galletta and D.G. Himelrick (eds.), Small fruit crop management. Prentice Hall, Englewood Cliffs, NJ.

Federación Nacional de Cafeteros de Columbia. ~1984. El cultivo de la mora de castilla.

Ferguson, A.R. 1990. Kiwifruit management. p. 472–503. In: G.J. Galletta and D.G. Himelrick (eds.), Small fruit crop management. Prentice Hall, Englewood Cliffs, NJ.

Finn, C.E., K. Wennstrom, and K. Hummer. 1998. Crossability of Eurasian *Rubus* species with red raspberry and blackberry. In: G. McGregor and H. Hall (eds.), Seventh Int. Rubus and Ribes Symp. Acta Hort. (in press).

Fogle, H.W. 1975. Cherries. p. 348–366. In: J. Janick and J.N. Moore (eds.), Advances in fruit breeding. Purdue Univ. Press, W. Lafayette, IN.

Gaitoni, L.A. ~1970. Una mora silvestre cultivada (*Rubus glaucus*). Horticultura p. 3–8.

Galletta, G.J. and J.R. Ballington. 1996. Blueberries, cranberries and lingonberries. p. 1–108. In: J. Janick and J.N. Moore (eds.), Fruit breeding Vol. II. Vine and small fruits. Wiley, New York.

Galletta, G.J. and R.S. Bringhurst. 1990. Strawberry management. p. 83–156. In: G.J. Galletta and D.G. Himelrick (eds.), Small fruit crop management. Prentice Hall, Englewood Cliffs, NJ.

Goldy, R.G. 1992. Breeding muscadine grapes. Hort. Rev. 14:357–405.

Hall, H.K. 1990. Blackberry breeding. Plant Breed. Rev. 8:249–312.

Hancock, J.F., D.H. Scott, and F.J. Lawrence. 1996. Strawberries. p. 419–470. In: J. Janick and J.N. Moore (eds.), Fruit breeding Vol. II. Vine and small fruits. Wiley, New York.

Harmat, L., A. Porpaczy, D.G. Himelrick, and G.J. Galletta. 1990. Currant and gooseberry management. p. 245–272. In: G.J. Galletta and D.G. Himelrick (eds.), Small fruit crop management. Prentice Hall, Englewood Cliffs, NJ.

Hegwood, C.P., R.H. Mellenaz, R.A. Haygood, T.S. Brook, and J.L. Peeples. 1983. Establishment and maintenance of muscadine vineyards. Mississippi Coop. Ext. Serv. Bul. 913.

Hiirsalmi, H.M. 1989. Research into *Vaccinium* cultivation in Finland. p. 175–184. In: E.J. Stang (ed.), Fourth Int. Symp. on Vaccinium Culture. Acta Hort. 241.

Hiirsalmi, H. and A. Lehmushovi. 1993. Occurrence and utilization of wild *Vaccinium* species in Finland. p. 315–321. In: K.A. Clayton-Greene (ed.), Fifth Int. Symp. on Vaccinium Culture. Acta Hort. 346.

Hummer, K.E. and C. Finn. 1998a. Recent *Rubus* and *Ribes* acquisitions at the USDA ARS National Clonal Germplasm Repository. In: G. McGregor and H. Hall (eds.), Seventh Int. Rubus and Ribes Symp. Acta Hort. (in press).

Hummer, K.E. and C. Finn. 1998b. Susceptibility of *Ribes* genotypes to white pine blister rust. In: G. McGregor and H. Hall (eds.), Seventh Int. Rubus and Ribes Symp. Acta Hort. (in press).

Jennings, D.L. 1980. A hundred years of Loganberries. Fruit Var. J. 35:34–37.

Jennings, D.L. 1988. Raspberries and blackberries: Their breeding, diseases and growth. Academic Press, London.

Kalt, W. 1997. Health functionality of blueberries. HortTechnology 7:216–221.

Kalt, W. and J.E. McDonald. 1996. Chemical composition of lowbush blueberry cultivars. J. Am. Soc. Hort. Sci. 121:142–146.

Kask, K. 1987. Large-fruited black chokeberry (*Aronia melanocarpa*). Fruit Var. J. 41:47.

Kask, K. 1989. The Tomentosa cherry, *Prunus tomentosa* Thunb. Fruit Var. J. 43:50–51.

Ko, K.M., S.P. Ip, M.K.T. Poon, C.T. Che, K.H. Ng, and Y.C. Kong. 1995a. Effect of lignin-enriched Fructus Schisandrae extract on hepatic glutathione status in rats: protection against carbon tetrachloride toxicity. Planta Med. 61:134–137.

Ko, K.M., P.K. Yick, M.K.T. Poon, C.T. Che, K.H. Ng, and Y.C. Kong. 1995b. *Schisandra chinensis*-derived antioxidant activities in 'Sheng Mai San', a compound formulation, in vivo and in vitro. Phytotherapy Res. 9:203–206.

Li, T.S.C. and W.R. Schroeder. 1986. Sea buckthorn (*Hippophae rhamnoides* L.): a multipurpose plant. HortTechnology 6:370–380.

Logan, J.H. 1909. Loganberry, Logan blackberry and Mammoth blackberry. The Pacific Rural Press and California Fruit Bul. 78:1–2.

Logan, M.E. 1955. The Loganberry. Published by Mary E. Logan, Oakland, CA.

Mazza, G. and C.G. Davidson. 1993. Saskatoon berry: A fruit crop for the prairies. p. 516–519. In: J. Janick and J.E. Simon (eds.), New crops. Wiley, New York.

Mizoguchi, Y., N. Kawada, Y. Ichikawa, and H. Tsutsui. 1991. Effect of gomisin A in the prevention of acute hepatic failure induction. Planta Med. 57:320–324.

Moore, J.N. and R. M. Skirvin. 1990. Blackberry management. p. 214–244. In: G.J. Galletta and D.G. Himelrick (eds.), Small fruit crop management. Prentice Hall, Englewood Cliffs, NJ.

Morazzoni, P. and E. Bombardelli. 1996. *Vaccinium myrtillus* L. Fitoterapia 67:3–29.

Olien, W.C. and C.P. Hegwood. 1990. Muscadine-A classic southeastern fruit. HortScience 25:726, 831.

Olien, W.C. 1990. The muscadine grape: Botany, viticulture, history and current industry. HortScience 25:732–739.

Pietilä, M. 1998. Buckthorn, the plant for tomorrow's healthy living. Joonas International news forum for professional berry growers, Joensuu, Finland.

Popenoe, W. 1924. Hunting new fruits in Ecuador. Nat. History 24:455–466.

Prior, R.L., G. Cao, A. Martin, E. Sofic, J. McEwen, C. O'Brian, N. Lischner, M. Ehlenfeldt, W. Kalt, G. Krewer, and C.M. Mainland. 1998. Antioxidant capacity as influenced by total phenolic and anthocyanin content, maturity, and variety of *Vaccinium* species. J. Agr. Food Chem. 46:2686–2693.

Pritts, M. and D. Handley. 1989. Bramble production guide. NRAES-35. Northeast Reg. Agr. Eng. Serv., Ithaca, NY.

Pritts, M.P., J.F. Hancock, and B.C. Strik. 1992. Highbush blueberry production guide. NRAES-55. Northeast Reg. Agr. Eng. Serv., Ithaca, NY.

Rapp, K., S.K. Næss, and H.J. Swartz. 1993. Commercialization of the cloudberry (*Rubus chamaemorus* L.) in Norway. p. 524–526. In: J. Janick and J.E. Simon (eds.), New crops. Wiley, New York.

Reich, L. 1991. Uncommon fruits worthy of attention: A gardener's guide. Addison Wesley Publ., New York.

Rincon, T.A.R. and M.J.A. Salas. 1987. Influence of the levels of N, P, and K on the yield of blackberry. Acta Hort. 188: 183–185.

Ritter, C.M. and G.W. McKee. 1964. The elderberry. Penn. Agr. Expt. Stat. Bul. 709.

Roper, T.R. and N. Vorsa. 1997. Cranberry: Botany and horticulture. Hort. Rev. 21: 215–249.

St.-Pierre, R. 1996. The lingonberry. A versatile wild cranberry. 3rd ed. Univ. Saskatchewan.

St.-Pierre, R.G. 1997. Growing saskatoons: A manual for orchardists. Agr. Dev. Fund. Government of Saskatchewan, Saskatoon.

Sherman, W.B. and R.H. Sharpe. 1971. Breeding *Rubus* for warm climates. HortScience 6:147–149.

Stang, E.J. 1990. Elderberry, highbush cranberry and juneberry management. p. 363–382. In: G.J. Galletta and D.G. Himelrick (eds.), Small fruit crop management. Prentice Hall, Englewood Cliffs, NJ.

Stang, E.J., G.G. Weis, and J. Klueh. 1990. Lingonberry: Potential new fruit for the northern United States. p. 321–323. In: J. Janick and J.E. Simon (eds.), Advances in new crops. Timber Press, Portland, OR.

Stark, N. and S. Baker. 1992. The ecology and culture of Montana huckleberries: A guide for growers and researchers. Montana Forest & Conservation Expt. Sta. Misc. Publ. 52. Missoula.

Strik, B. and H. Cahn. 1998. Growing kiwifruit. PNW 507. Oregon State Univ. Ext. Serv., Corvallis.

USDA-NASS-ERS. 1998. USDA Agricultural Prices—1997 Summary. Pr1-3(98). Washington, DC.

Way, R.D. 1964. Elderberry varieties and cultural practices. N.Y. State. Hort. Soc. Proc. 110:233–236.

Way, R.D. 1981. Elderberry culture in New York State. Food Life Sci. Bul. 91, N.Y. State Agr. Expt. Sta., Geneva.

Williams, C.F. and G.M. Darrow. 1940. The trailing raspberry: *Rubus parvifolius* L.: Characteristics and breeding. North Carolina Agr. Expt. Sta. Tech. Bul. 65.

Sea Buckthorn: New Crop Opportunity

Thomas S.C. Li

Sea buckthorn (*Hippophae rhamnoides* L., Elaeagnaceae) is a winter hardy, deciduous shrub with yellow or orange berries (Bailey and Bailey 1978). It develops an extensive root system rapidly and is therefore an ideal plant for preventing soil erosion and land reclamation. It can withstand temperatures from –43° to 40°C (Lu 1992). It is considered to be drought resistant (Heinze and Fiedler 1981); however, irrigation is needed in regions receiving <400 mm of rainfall per year for better growth (Li and Schroeder 1996).

BOTANY

Sea buckthorn is a dioecious multi-branched, thorny shrub, reaching 2 to 4 m in height with stout branches forming a round often symmetrical head. It has brown or black rough bark and a thick grayish-green crown (Rousi 1971). Staminate and pollinate flowers are inconspicuous appearing before the leaves. Leaves are alternate, narrow 4 to 6 cm long, and lanceolate with a silver-grey color on the upper side (Synge 1974). Flower buds are formed mostly on 2-year-old wood, differentiated during the previous growing season. Fruit is subglobose, 6 to 10 mm long and 4 to 6 mm in diameter, turning yellow to orange when mature in mid Sept. The root system is characterized by nitrogen fixing nodules (Akkermans et al. 1983).

NUTRACEUTICAL VALUES

Sea buckthorn can be used for many purposes and has momentous economic potential. It has been used for centuries in Europe and Asia. Recently, it has attracted considerable attention from researchers around the world, including North America, mainly for its nutritional and medicinal value. The fruits are rich in carbohydrates, protein, organic acids, amino acids and vitamins. The concentration of vitamin C in sea buckthorn fruit, ranged from 100–300 mg/100 g fruit, is higher than strawberry, kiwi, orange, tomato, carrot, and hawthorn (Bernath and Foldesi 1992; Lu 1992). Sea buckthorn is also high in protein, especially globulins and albumins, and fatty acids such as linoleic and linolenic acids. Vitamin E content in sea buckthorn (202.9 mg/ 100 g fruit) is higher than wheat embryo, safflower, maize, and soybean.

Medicinal uses of sea buckthorn are well documented in Asia and Europe. Clinical tests on medicinal uses were first initiated in Russia during the 1950s (Gurevick 1956). Sea buckthorn oil was formally listed in the Pharmacopoeia in 1977 and clinically tested in Russia and China (Xu 1994). The most important pharmacological functions attributed to sea buckthorn oil are: anti-inflammatory, antimicrobial, pain relief, and promoting regeneration of tissues. Sea buckthorn oil is also touted as a treatment for oral mucositis, rectum mucositis, vaginal mucositis, cervical erosion, radiation damage, burns, scalds, duodenal ulcers, gastric ulcers, chilblains, skin ulcers caused by malnutrition, and other skin damage. More than ten different drugs have been developed from sea buckthorn in Asia and Europe and are available in different forms, such as liquids, powders, plasters, films, pastes, pills, liniments, suppositories, and aerosols. Sea buckthorn oil extracted from seeds is popular in cosmetic preparations, such as facial cream (Li and Wang 1998). In Europe and Asia, there are numerous products made from sea buckthorn, such as tea from leaves, beverages and jam from fruits, fermented products from pulp, and animal feeds from leaves, pulp, and seed residues.

NEW CULTIVAR: INDIAN-SUMMER

A new cultivar in Canada, 'Indian-Summer', is being released by the Prairie Farm Rehabilititation Administration (PFRA), Shelterbelt Center at Indian Head, Saskatchewan in co-operation with the Pacific Agri-Food Research Center, Agriculture and Agri-Food Canada, at Summerland, British Columbia (Schroeder et al. 1996). It is a seed-propagated cultivar that originates from a hedge located at the PFRA Shelterbelt Center. This accession was obtained from the Morden Research Station, Morden, Manitoba in 1963 as rooted cuttings. The original plants are growing in a seed block at the Shelterbelt Center. The progeny has been tested as Accession E5098 in advanced evaluation and field plantings to determine soil and climatic adaptation.

'Indian-Summer' is well adapted to growing conditions on the Canadian prairies. In the last 20 years, it was tested at 12 sites in Manitoba and Saskatchewan and 'Indian-Summer' performed well on a variety of soils including moderately saline sites. It shows above average drought tolerance and is fully hardy (Ouellet and Sherk 1967). The most limiting factor affecting growth and survival was weed competition. After twenty growing seasons, the height of established plants ranged from 3 to 4 m (Schroeder and Walker 1994). Indian-Summer will not tolerate prolonged flooding or poorly drained soils. Fruit production ranged from 4 to 5 kg per shrub. Analysis of fruit samples showed average ascorbic acid content of 165 mg/100 g of fruit. Seed oil content averaged 11.9% and contained 31.2% protein, 88.3% to 89.1% unsaturated fatty acids, particularity linolenic acid (32.3%), linoleic acid (40.8%), and oleic acid (15%). Other constituents of the seed oil included gamma and alpha tocopherol.

Based on performance studies conducted by the PFRA Shelterbelt Center the projected area of adaptation for 'Indian-Summer' sea buckthorn was determined to be the prairie and boreal plain ecozones of Alberta, Saskatchewan and Manitoba (Ecological Stratification Working Group 1994). The average annual precipitation of the area of adaptation ranges from 300 to 500 mm. The average annual temperature ranges from 0°C to 3.5°C, average frost-free period is 85 to 130 days. The plant hardiness zones (Ouellet and Sherk 1967) include zones 1a, 1b, 2a, 2b, 3a, and 3b. The annual minimum temperatures range from –40°C to –50°C.

SEA BUCKTHORN CULTIVATION

Recently, sea buckthorn has been recommended for orchard-type cultivation in British Columbia and the prairie provinces. Sea buckthorn normally is transplanted or directly seeded in the spring. It grows best in deep, well drained, sandy loam soil with ample organic matter. In arid or semiarid areas, water must be supplied for establishment. Soil acidity and alkalinity, except at extreme levels, are not limiting factors, although it thrives best at pH 6 to 7. Sea buckthorn is sensitive to severe soil moisture deficits, especially in spring when plants are flowering and young fruits are beginning to develop (Li and McLoughlin 1997). Sea buckthorn, like other crops, requires adequate soil nutrients for a high yield of good-quality fruit. It responds well to phosphorus fertilizer. Nitrogen fertilization can adversely affect root nodulation and it delays the development of nodules after inoculation with *Frankia*.

Recommended plant spacing for sea buckthorn is 1 m within the row and 4 m between rows to allow equipment access, with rows oriented in a north-south direction to provide maximum light. The ratio of male to female plants is important for maximizing the number of fruit-bearing trees. Recommendations for male:female ratios vary with plant density and region. In British Columbia, with an orchard planting of 4000 trees/ha, a 1:6 to 1:8 male:female ratio is considered adequate. Moderate pruning of sea buckthorn will increase the yield and reduce fluctuation of fruiting from year to year. The crown should be pruned annually to remove overlapping branches, and long branches should be headed to encourage development of lateral shoots. Weed control is very important in sea buckthorn plantings, especially for promoting growth of newly planted seedlings (Li and McLoughlin 1997).

CONCLUSION

Sea buckthorn is an unique and valuable plant species currently being domesticated in various parts of the world. The species has been used to a limited extent in North America for conservation plantings, but the use of food and non-food sea buckthorn products has not been pursued. The plants are easily propagated and yields are relatively high, and production is reliable, with the potential market mainly in Europe at the moment. Most sea buckthorn research has been conducted in Asia and Europe in the past and Canada has increased its research recently. Unique plant products, especially those with proven nutritional quality, are gaining popularity in North America. Development of a North American sea buckthorn industry presents a unique opportunity for agricultural production of a value-added crop on marginal land.

REFERENCES

Akkermans, A.D.L., W. Roelofsen, J. Blom, K. Hussdanell, and R. Harkink. 1983. Utilization of carbon and nitrogen compounds of *Frankia* in synthetic media and in root nodules of *Alnus glutinosa*, *Hippophae rhamnoides* and *Datisca cannabina*. Can. J. Bot. 61:2793–2800.

Bailey, L.H. and E.Z. Bailey. 1978. Hortus third, A concise dictionary of plants cultivated in the United States and Canada. McMillan Publ. Co., New York.

Bernath, J. and D. Foldesi. 1992. Sea buckthorn (*Hippophae rhamnoides* L.): a promising new medicinal and food crop. J. Herbs Spices Med. Plants 1:27–35.

Gurevick, S.K. 1956. The application of sea buckthorn oil on ophthalmology. Vesttn. Ottamologu 2:30–33.

Heinze, M. and H.J. Fiedler. 1981. Experimental planting of potash waste dumps. I. Communication: Pot experiments with trees and shrubs under various water and nutrient conditions. Archiv Acker Pflanzen. Bodenkunde 25:315–322.

Li, T.S.C. and C. McLoughlin. 1997. Sea buckthorn production guide. Canada Sea Buckthorn Enterprises Ltd. Peachland, British Columbia.

Li, T.S.C. and W.R. Schroeder. 1996. Sea buckthorn (*Hippophae rhamnoides* L.): A multipurpose plant. HortTechnology 6:370–380.

Li, T.S.C. and L.C.H. Wang. 1998. Physiological components and health effects of ginseng, echinacea and sea buckthorn. In: G. Mazza (ed.), Functional foods, biochemical & processing aspects. Technomic Publ. Co. Inc., Lancaster, PA.

Lu, R. 1992. Sea buckthorn: A multipurpose plant species for fragile mountains. Int. Centre for Integrated Mountain Development, Katmandu, Nepal.

Ouellet, C.E. and L. Sherk. 1967. Woody ornamental plant zonation. III. Suitability map for probable winter survival of ornamental trees and shrubs. Can. J. Plant Sci. 47:351–358.

Rousi, A. 1971. The genus *Hippophae* L. A taxonomic study. Ann. Bot. Fennici 8:177–227.

Schroeder, W.R. and D.S. Walker. 1994. Performance of twenty-four tree and shrub species after twenty-five growing seasons. Report of the PFRA Shelterbelt Centre, Indian Head, SK. p. 14–15.

Schroeder, W. R., T.S.C. Li, and D.S. Walker. 1996. Indian-Summer sea buckthorn. Agriculture & Agri-Food Canada, PFRA Shelterbelt Centre. Supplementary Report.

Synge, P.M. 1974. Dictionary of gardening: A practical and scientific encyclopaedia of horticulture. 2nd ed. Clarendon Press, Oxford.

Xu, M. 1994. The medical research and exploitation of sea buckthorn. Hippophae 7:32–34.

DNA Analysis as a Tool in Sea Buckthorn Breeding

N. Jeppsson, I.V. Bartish, and H.A. Persson

Sea buckthorn (*Hippophae rhamnoides* L., Elaeagnaceae) is a dioecious windpollinated shrub with nitrogen fixing ability. The berries have bright colors, varying from yellow, orange to red. Domestication of sea buckthorn started in Siberia in the 1930s (Kalinina and Panteleyeva 1987). Local germplasm (ssp. *mongolica*) from the Altai mountains was used in the onset of the breeding. Breeding projects have, later on, been initiated also in other countries such as Germany (Albrecht 1990), Finland (Yao and Tigerstedt 1994), China (Huang 1995), and Canada (Li and Schroeder 1996). At SLU-Balsgård, Sweden, breeding of sea buckthorn started in 1986. Conventional breeding methods, including germplasm evaluation, hybridization and selection, are used (Trajkovski and Jeppsson 1999). The Swedish food industry is very interested in the berries as a raw material for various food products as marmalade, beverages and as a flavor. Sea buckthorn is not yet grown at a commercial scale in Sweden. The main objective is to develop cultivars suitable for large scale production of berries in Sweden. We work mainly with adaptation, disease resistance, growth habit that will permit machine harvesting, and improvement of fruit quality and yield. The plant material used in our breeding program has been derived from domesticated Russian forms (ssp. *mongolica*) and native Scandinavian forms (ssp. *rhamnoides*). Initially, 28 genotypes were selected. Nine superior selections are now being tested in full scale orchards both in northern and southern Sweden. Recently we started to use the PCR-based method RAPD (Random Amplified Polymorphic DNA) to study population structure in the ssp. *rhamnoides* and to find a marker linked to gender determination.

RAPD METHODOLOGY

We use a standard method with the following steps (Weising et al. 1995):
1. DNA isolation and purification, using fresh or frozen leaves.
2. Enzymatic amplification of certain segments of DNA by use of thermostable DNA polymerase, deoxynucleotides and oligonucleotide primers. The DNA fragments are multiplicated during 40 cycles of temperature shuttling between +94°C, +36°C and +72°C.
3. Electrophoretic separation of the fragments based on size followed by staining with ethidium bromide. Visual detection of bands under UV light.
4. The presence of a band is scored as 1 and the absence as 0. The resulting matrix is evaluated by statistical procedures, and parameters as e.g. genetic relatedness and heterozygosity are calculated.

GENETIC VARIABILITY

During the initial evaluation of germplasm, ssp. *mongolica* was shown to be susceptible to various diseases in Sweden. In comparison, native Scandinavian germplasm of ssp. *rhamnoides* seemed less susceptible. Progenies between these two taxa are now under evaluation at Balsgård.

The need for native germplasm from ssp. *rhamnoides* prompted the research on how the genetic variation is partitioned within and among native populations in northern Europe. RAPD variation was recently studied in ten populations (Fig. 1) from native stands (Bartish et al. 1999). Samples were collected from almost the whole distribution range, with the most southwestern sample from the Netherlands and the most northeastern sample from the arctic circle in Sweden. A population of seedlings derived from ssp. *mongolica* (population C) was used as an outgroup. Based on data from the RAPD analysis, a dendrogram showing genetic relatedness was calculated by use of cluster analysis (UPGMA). The population of ssp. *mongolica* was found to be the most remote population (Fig. 2). The ssp. *rhamnoides* populations grouped into two main clusters, one with populations Ea, Eb, Fb, Hb, Ib, and another with populations Fa, Ga, Gb, Ha. The population Ia clustered more distantly. The data was also analyzed with Principal Co-Ordinate analysis (PCO), and a plot of the first two principal co-ordinate axis was created which showed similar results, with the ssp. *mongolica* population separated from the ssp. *rhamnoides*. Using Analysis of Molecular Variance (AMOVA) to measure the partitioning of molecular variance within the ssp. *rhamnoides*, 85% of the variation was found within

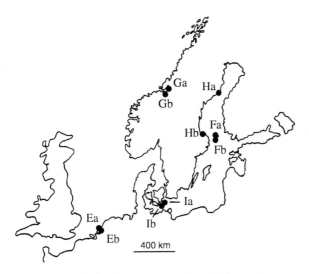

Fig. 1. Sampling sites for *Hippophae rhamnoides* ssp. *rhamnoides*.

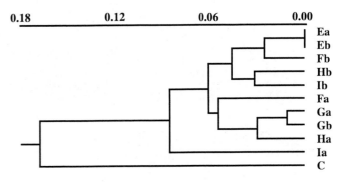

Fig. 2. UPGMA dendrogram of the genetic relationship among populations of *Hippophae rhamnoides*. (For population origins, see Fig. 1.)

populations and 15% among populations. That means that 85% of the RAPD variation can be captured by sampling within one or a few of the populations. If more than 85% of the variation is requested, sampling must be performed in more populations. However, to what extent RAPD variation is a good predictor for variation in important morphological and physiological characters is as yet unknown.

Since sea buckthorn is dioecious, wind pollinated, and obligately outcrossing, one would expect to find much variation within populations and very little variation between populations. In trembling aspen (*Populus tremuloides* Michx.), another dioecious species, 97% of the RAPD variability (Table 1) was found within the populations (Yeh et al. 1995). In Norway spruce (*Picea abies*), an outcrossing species, 95% of the isoenzyme variability was found within populations (Lagercrantz and Ryman 1990). Both trembling aspen and Norway spruce are widespread and continuously distributed. In Brazilwood (*Caesalpinia echinata* Lam.), another outbreeding species, RAPD variability within populations accounted for only 42% (Cardosos et al. 1998). This was suggested to be the result of discontinuous distribution that have lead to the isolation of small populations. Inbreeding and/or genetic drift has then increased the differentiation between populations.

Sea buckthorn appears to have differentiated more than trembling aspen and Norway spruce. The history of sea buckthorn in northern Europe actually supports this theory. Sea buckthorn of the ssp. *rhamnoides* has been extensively studied in Scandinavia. Fossil records of pollen have shown that it colonized the land shortly after the ice retreat in the late glacial period and showed a wide distribution. Later on, the distribution declined and became fragmented, resulting in the present discontinuous distribution (Sandegren 1943). During this time period differentiation due to isolation and restricted population size may have occurred.

A previous study based on isozymes showed differences in partitioning of genetic variation when ssp. *sinensis* and ssp. *rhamnoides* were compared, with the latter showing less similarity between populations (Yao and Tigerstedt 1993). Here too, the differences may be explained by differences in population sizes and/or the continuity in spatial distribution.

Table 1. Comparison of the within population variability component in different outcrossing species.

Species	Variability within populations (%)	Distribution	Population size
Trembling aspen *Populus tremuloides*	97	continuous	large
Norway spruce *Picea abies*	95	continuous	large
Sea buckthorn *Hippophae rhamnoides* ssp. *rhamnoides*	85	discontinuous	??
Brazilwood *Caesalpinia echinata*	42	discontinuous	small

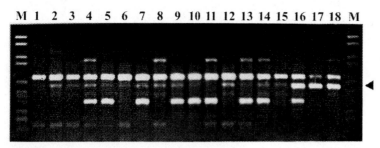

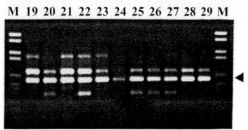

Fig. 3. Sex-linked RAPD marker amplified with primer OPD15 in individual plants from the cross 'Leikora' × 'Pollmix 1'. Lanes 1 to 15 are females, lanes 16 to 29 are males. The male specific band OPD15-600 is indicated with an arrow. M stands for molecular size marker.

GENDER DETERMINATION

Since sea buckthorn is dioecious, plant breeding projects aim at producing both female and male cultivars. However, breeding objectives for female and male cultivars differ and, generally, there are more quality criteria to be met in a female cultivar. Therefore, the selection pressure is higher on female cultivars and as a consequence, larger seedling population size is needed to obtain a certain number of female selections than to obtain the same number of male selections. Much of work and money could thus be saved if a large proportion of the males could be discarded already at an early stage in the evaluation process.

Gender is most often genetically determined in dioecious plants, either by distinguishable sex chromosomes or by alleles at one or several loci on non-distinguishable chromosomes (Irish and Nelson 1989; Durand and Durand 1990). The presence of distinguishable sex chromosomes in sea buckthorn was suggested by Shchapov (1979), although this has not been substantiated in other studies. Previously gender determination based on RAPD markers has been successful in *Silene latifolia* (Mulcahy et al. 1992), *Pistacia vera* (Hormaza et al. 1994), *Asparagus officinalis* (Jiang and Sink 1997), and *Atriplex garettii* (Ruas et al. 1998). Recently, we studied the usefulness of RAPD markers for gender determination in sea buckthorn (Persson and Nybom 1998). Two F_1 progenies were investigated (34 plants derived from the cross 'Leikora' and 'Pollmix 1' and 22 plants derived from the cross BHi 10224 and 2-24). When flowering, the gender of the plants were determined. The analyses were performed as a bulked segregant analysis (BSA). DNA was extracted from each individual and two bulks were produced in each cross, one from the males and the other from the females. Out of 78 primers tested, four seemed to yield partitioning between male and female bulks and these four primers were chosen for further amplification at individual plant level. The band OPD15-600 (Fig. 3) was present in all males in the offspring of 'Leikora' × 'Pollmix 1' as well as in the father, 'Pollmix 1' and absent in all female offspring of the same cross as well as in the mother, 'Leikora'. However, this band was present in only one of the males and in none of the females in the other cross. Unfortunately, this marker was not therefore universal and could be used for gender determination in only one of our progenies.

CONCLUSIONS

We used RAPD markers to determine genetic variation between and within native populations and to measure relatedness between populations. Such investigations can be used as guidelines for collection of germplasm in native stands but also to develop strategies for a breeding program. One marker was found to be present only in male genotypes of a progeny and in the father while it was absent in all females and the mother. This marker was useful only in one hybrid progeny out of two tested, and search for a more general marker will be undertaken, now that we have at least obtained evidence that gender is genetically determined in sea buckthorn. In a dioecious crop as sea buckthorn, molecular markers for fruit quality traits could also be a powerful tool when selecting the male parent to be used in crosses. RAPD band patterns can serve as finger-prints for genotype identification in vegetatively propagated crops, and can therefore be useful also for gene bank management.

REFERENCES

Albrecht, H.-J. 1990. Sortenentwicklung bei Sanddorn. Gartenbauwissenshaft 37:207–208.

Bartish, I.V., N. Jeppsson, and H. Nybom. 1999. Population genetic structure in the dioecious pioneer plant species *Hippophae rhamnoides* investigated by RAPD markers. Molecular Ecol. 8:791–802.

Cardosos, M.A., J. Provan, W. Powell, C.G. Ferreira, and D.E. De Oliveira. 1998. High genetic differentiation among remnant population of the endangered *Caesalpinia echinata* Lam. (Leguminose-Caesalpinioideae). Molecular Ecol. 7:601–608.

Durand, R. and B. Durand. 1990. Sexual determination and sexual differentiation. Crit. Rev. Plant Sci. 9:295–316.

Hormaza, J.I., L. Dollo, and V.S. Polito. 1994. Identification of a RAPD marker linked to sex determination in *Pistacia vera* using bulked segregant analysis. Theor. Appl. Genet. 89:9–13.

Huang, Q. 1995. A review of seabuckthorn breeding in China. Proc. Int. Workshop on Seabuckthorn, Beijing, China. p.111–117.

Irish, E.E. and T. Nelson. 1989. Sex determination in monoecious and dioecious plants. Plant Cell 1:737–744.

Jiang, C. and K.C. Sink. 1997. RAPD and SCAR markers linked to the sex expression locus *M* in asparagus. Euphytica 94: 329–333.

Lagercrantz, U. and N. Ryman. 1990. Genetic structure of Norway spruce (*Picea abies*): Concordance of morphological and allozyme variation. Evolution 44:38–53.

Li, T.S.C. and W.R. Schroeder. 1996. Sea buckthorn (*Hippophae rhamnoides* L.): A multipurpose plant. HortTechnology 6:370–380.

Kalinina, I.P. and Y.I. Panteleyeva. 1987. Breeding of sea buckthorn in the Altai. In: Advances in Agricultural Science. Moscow, Russia.

Mulcahy, D.L., N.F. Weeden, R. Kesseli, and S.B. Carroll. 1992. DNA probes for the Y-chromosome of *Silene latifolia*, a dioecious angiosperm. Sexual Plant Reprod. 5:86–88.

Persson, H.A. and H. Nybom. 1998. Genetic sex determination and RAPD marker segregation in the dioecious species sea buckthorn (*Hippophae rhamnoides* L.). Hereditas 129:45–51.

Ruas, C.F., D.J. Fairbanks, R.P. Evans, H.C. Stutz, W.R. Andersen, and P.M. Ruas. 1998. Male-specific DNA in the dioecious species *Atriplex garrettii* (Chenopodiaceae). Am. J. Bot. 85:162–167.

Sandegren, R. 1943. *Hippophae rhamnoides* L. i Sverige under senkvartär tid. Svensk Botanisk Tidskrift. 37:1–26. (in Swedish with German summary).

Shchapov, N.S. 1979. On the caryology of *Hippophae rhamnoides* L. Tsitologiya i Genetika. 13:45–47 (in Russian with English summary).

Trajkovski, V. and N. Jeppsson. 1999. Domestication of sea buckthorn. Botanica Lithuanica Suppl. 2:37–46.

Weising, K., H. Nybom, K. Wolff, and W. Meyer. 1995. DNA fingerprinting in plants and fungi. CRC Press, Boca Raton, FL.

Yao, Y. and P.M.A. Tigerstedt. 1993. Isozyme studies of genetic diversity and evolution in *Hippophae*. Genetic Resources Crop Evol. 40:153–164.

Yao, Y. and P. Tigerstedt. 1994. Genetic diversity in *Hippophae* L. and its use in plant breeding. Euphytica 77:165–169.

Yeh, F.C., D.K.X. Chong, and R.C. Yang. 1995. RAPD variation within and among natural populations of trembling aspen (*Populus tremuloides* Michx.) from Alberta. J. Hered. 86:454–460.

New Temperate Fruits: *Actinidia chinensis* and *Actinidia deliciosa*

A.R. Ferguson*

"The kiwifruit industry is unique among global fruit industries in being so totally dominated by one variety, the Hayward variety. Most consumers are not even aware that other varieties exist."

(World Kiwifruit Review 1998)

The above statement may accurately represent the current situation but will very soon be out of date. The last 30 years has seen the emergence of the kiwifruit from being a small but increasing crop in one country to being a crop of worldwide significance. The next few years will see the entry of new and quite different *Actinidia* fruit into international trade.

The kiwifruit (*Actinidia deliciosa* [A. Chev.] C.F. Liang et A.R. Ferguson, Actinidiaceae) is one of the very few temperate fruit crops to have been domesticated this century. At the turn of the century it was still just a wild plant in China, but by 1970 it had been developed into a new fruit crop in New Zealand. Today, as we approach the next century, the kiwifruit is an important commercial crop grown in different parts of the world: it has also become uniquely dependent on international trade as the three biggest producer countries export most of the kiwifruit they produce. Total world production now exceeds a million tonnes per year, more than that of well-established crops such as raspberries and currants, and is soon likely to exceed production of strawberries or of apricots (World Kiwifruit Review 1998).

Until recently, the international success of kiwifruit could really be considered as the success of one fruiting cultivar, 'Hayward', and associated male (pollenizer) cultivars. World trade has become restricted to this one cultivar and the name 'Hayward' could almost be taken as synonymous with "kiwifruit." Now, however, different cultivars of *A. deliciosa* are being grown; different males of *A. deliciosa* have been selected for the specific climatic conditions of different kiwifruit regions; and a start has been made to the commercial cultivation of different *Actinidia* species with quite distinct fruit.

THE GENUS *ACTINIDIA*

The genus *Actinidia* Lindl. contains about 60 species (Ferguson 1984, 1990). The genus is variable, as are the individual species, and much of this diversity is potentially useful (Ferguson 1990). There is thus great scope for crop improvement through breeding programs (Ferguson et al. 1996). All *Actinidia* species are perennial, climbing or scrambling plants, mostly deciduous although a few from warmer areas are evergreen. All species appear to be dioecious: the flowers on male vines produce viable pollen but lack a properly developed ovary, ovules, or styles; the flowers of female vines appear perfect but the pollen they release is shrivelled and non-viable. Botanically, the fruits of the various *Actinidia* species are all defined as berries in that they are fleshy, they have many seed embedded in the flesh, and they do not split open at maturity. Horticulturally, however, they display great diversity, often in those attributes that are commercially important. The fruit can occur singly, in small bunches of three to five fruit, or sometimes in larger bunches or infructescences containing up to 30 or more fruit (Fig. 1). They vary in size, shape, hairiness, and external color. Some change color as they ripen. The flesh can also vary in color, juiciness, texture, and composition. The fruits of some species are basically inedible or, at best, unpalatable, whereas the flavor of the fruit of other *Actinidia* species is considered by many to be much superior to that of the kiwifruit.

ORIGIN OF PRESENT KIWIFRUIT CULTIVARS

Almost all the kiwifruit cultivars grown in commercial orchards outside of China are descended from two female plants and one male plant, themselves derived from a single introduction of seed to New Zealand in 1904 (Ferguson and Bollard 1990). The provenance of this seed is unknown but it was probably collected

*I thank R. Beatson, R.G. Lowe, M.A. McNeilage, R. Meyer, A.G. Seal, and R. Testolin for useful discussions.

by E.H. Wilson from Hubei or Sichuan in China. Despite such limited sampling of the gene pool, the first kiwifruit plantings in New Zealand contained considerable variation in the fruit carried by individual plants. Grafted plants were first sold in the 1920s and this allowed the propagation of good strains with large, oval or elongated fruits having bright green flesh but lacking a hard or woody core. Such strains are only one or two generations removed from the original introduction of seed from China.

'Hayward'

'Hayward' was selected from a small number of seedlings, possibly only 40, likewise only a couple of generations from the original seed, and likewise descended from those two female and one male plants. 'Hayward' eventually became the cultivar of choice because its fruit were larger, had a better appearance, and their flavor was considered by many to be superior. These qualities meant that the fruit of 'Hayward' were preferred by consumers in both New Zealand and overseas to that of other kiwifruit cultivars then available. Initially, 'Hayward' had been planted on only a small scale and most early exports of kiwifruit from New Zealand were of other cultivars. However, "the customer is always right": the clear preference by consumers and marketers, especially in the developing markets in the US, for 'Hayward' kiwifruit meant that by the mid 1960s new kiwifruit plantings in New Zealand were almost exclusively of 'Hayward' and by 1975, only 'Hayward' fruit were accepted for export. 'Hayward' may have been less easily managed and less productive than some of the other kiwifruit cultivars but these disadvantages were convincingly outweighed by the qualities of the fruit. Perhaps most important for the New Zealand growers was the remarkably long storage life of 'Hayward' fruit, as this allowed the development of a kiwifruit industry based on exports by ship to distant markets.

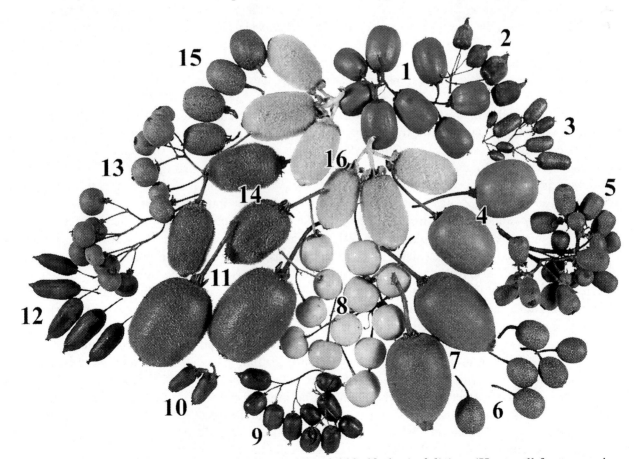

Fig. 1. Fruit diversity in *Actinidia*, with the commercial kiwifruit, *A. deliciosa* 'Hayward' for comparison.

1 *A. rufa*	5 *A. latifolia*	9 *A. arguta*	13 *A. guilinensis*
2 *A. melanandra*	6 *A. indochinensis*	10 *A. fulvicoma*	14 *A. setosa*
3 *A. glaucophylla*	7 *A. chinensis* 'Hort16A'	11 *A. deliciosa* 'Hayward'	15 *A. chrysantha*
4 *A. chinensis*	8 *A. macrosperma*	12 *A. arguta* var. *purpurea*	16 *A. eriantha*

When growers in other countries were inspired to emulate the New Zealand kiwifruit growers, it is not surprising that they too chose to grow 'Hayward'. In California, initial plantings were based on 'Chico', an early importation from New Zealand and later recognized as being indistinguishable from 'Hayward'. The commercial plantings in Europe were also of 'Hayward'. This reliance of a fruit industry on a single cultivar in most parts of the world is most unusual, possibly unique. Such reliance does have advantages: it facilitates standardization amongst suppliers from different countries. There are, however, also disadvantages in that standardization can lead to greater competition in individual markets and makes branding more difficult. Furthermore, experience with fruits such as apples has shown that different cultivars are not necessarily competitive but may allow market segmentation. Another consideration is that any monoculture is undesirable in principle.

Other Fruiting Cultivars of *A. deliciosa*

The special qualities of 'Hayward' make it likely that it will be an important cultivar for many years to come. Indeed, it has become the standard against which customers will assess any new cultivars. 'Hayward' is, however, by no means perfect (Ferguson et al. 1990), and there is room in the marketplace for other kiwifruit cultivars.

Numerous selections have been made in China from wild-growing populations of *A. deliciosa*. Few of these have yet proved themselves and 'Hayward' has been taken back to China and is now widely cultivated there. The best known of the Chinese selections is 'Qinmei', selected from the Qinling Mountains, Shaanxi. It has been enthusiastically promoted in China and has been extensively planted, perhaps more extensively than justified by its qualities. The fruit are a reasonable size but have only a mild flavor and a comparatively short shelf life. None of the new Chinese selections of *A. deliciosa* has yet been grown on a commercial scale outside of China.

'Koryoku', a seedling produced by open pollination of 'Hayward', is being grown on a limited scale in Japan, mainly because it reaches harvest maturity earlier and because the fruit, when ripe, are considered to be sweeter.

Top Star® is a bud mutation of 'Hayward' which has smooth, essentially hairless fruit (Bergamini 1991). The reduced vigor of the vine allows for easier management and less summer pruning. Plantings of Top Star® have been established in Italy but the reduction in management costs is too small and the fruit are unlikely to receive sufficient a market premium to justify the reworking of existing 'Hayward' orchards.

'Tomua' is a selection recently released in New Zealand (Muggleston et al. 1998). It is the first female kiwifruit cultivar from New Zealand not derived solely from the 1904 introduction of seed. It comes from a cross between 'Hayward' and males from an accession of *A. deliciosa* seed, originally collected in the Qinling Mountains and introduced into New Zealand in 1975. As far as we know this was the second ever introduction of *A. deliciosa* seed into New Zealand. This accession was notable for flowering early in the season and the fruit, although small, also maturing early. The aim of the cross was to combine the size and fruit quality of 'Hayward' with this early fruit maturation. 'Tomua' has large fruit of good, sweet flavor but of limited storage life: its main advantage is that, depending on location, the fruit can be harvested up to a month earlier than 'Hayward' fruit and yet, when ripened, are very acceptable to consumers. (Tomua means "early" in Maori. This fruiting cultivar should not be confused with the old pollenizer cultivar, 'Tomuri'; tomuri means "late" in Maori.) 'Tomua' is therefore complementary to 'Hayward', filling a special if restricted market niche by allowing the export of kiwifruit from New Zealand earlier in the season.

In some areas with warmer winters, e.g., southern California, 'Hayward' crops poorly because its winter chilling requirements are not satisfied. Adequate crops are instead obtained by using some of the older New Zealand selections, e.g., 'Abbott' or 'Allison', or newer selections, e.g., 'Tewi' (from the Canary Islands) or 'Vincent' (from Orange County, California), but these generally have fruit of mediocre quality (Meyer 1992; Ferguson 1997) and cannot compete commercially when 'Hayward' fruit are regularly available. 'Donné' was specifically selected for the warmer climatic conditions of South Africa.

Male (Pollenizer) Cultivars of *A. deliciosa*

Most male cultivars have been selected to coincide in flowering with 'Hayward', the predominant fruiting cultivar. Effective pollination, seed set, and full-sized fruit require coincident flowering of males and females but weather conditions during spring have a major effect on the time of flowering of males, relative to that of 'Hayward'. Different males are therefore being selected for the different growing countries. A series of males has been selected in New Zealand for 'Hayward' from seedlings derived from the 1904 seed introduction. Appropriate males, e.g., 'Autari', have also been selected in Italy for local growing conditions (Testolin et al. 1995) and these plants are probably also originally derived from the 1904 seed accession into New Zealand. The 'Californian Male' or 'Chico Male' appears to have come from a separate E.H. Wilson introduction of seed from China.

Males, 'King' and 'Ranger', have now been selected for 'Tomua' as it flowers several weeks ahead of 'Hayward'. These males are from the cross that produced 'Tomua'.

Market Diversification Through Cultural Practices

'Hayward' kiwifruit appear to appeal to special or discerning customers, particularly in Europe. In New Zealand, much more emphasis is now being placed on sustainable methods of production, especially with minimal use of insecticides and fungicides. Organically-produced kiwifruit are now being marketed separately and currently return a 70% premium to growers over conventional first-class fruit.

Actinidia chinensis Planch.

Of all the various *Actinidia* species, *A. chinensis* is most like the kiwifruit, *A. deliciosa* (Fig. 2). Indeed, until about 15 years ago, *A. chinensis* and *A. deliciosa* were classified together in the one species. There are good botanical reasons for separating the two species, but more important for horticulturists, their fruit are distinctly different. In wild plants of *A. chinensis*, the fruit are generally much smaller, more rounded, and less cylindrical than those of cultivated kiwifruit and initially it was feared that fruit size would be too small for commercial development. However, there is considerable variation in fruit size and shape and the fruit of good selections of *A. chinensis* can approach or even exceed the average size of 'Hayward' fruit. The fruit is almost hairless at maturity, and what hair remains is usually much shorter and finer than kiwifruit hair, more like the fuzz of a peach. Flesh color can vary greatly from bright green, shading through lime green to a clear, intense yellow. Possibly the most attractive fruit are those in which the inner pericarp flesh is red, the outer yellow. More important, the flavor of good selections of *A. chinensis* is thought by many to be much better than that of 'Hayward'. The fruit are sweeter, more aromatic with a flavor reminiscent of some subtropical fruit.

A. chinensis has already proved itself in China. The Chinese collect large quantities of *Actinidia* fruit from the wild, and they generally consider the fruit of *A. chinensis* to be superior to those of *A. deliciosa*. Most of their *Actinidia* selections from the wild are of *A. chinensis* and some of the better selections are now being planted extensively. The fruit produced is so far purely for local consumption

Chinese Cultivars of *A. chinensis* Being Grown Outside of China

Material of some of the better Chinese selections of *A. chinensis* have been available through Japanese nurseries from 1989 onwards. Cultivar names are often confused because Japanese importers have given the cultivars new names and yet other names have been given when material was subsequently imported into Europe. The female selections available include: 'Lushanxiang' (syn. '79-2', 'ACC 226', 'Elizabeth', 'First Emperor', 'K189', 'Yellow Joy'); 'Jiangxi 79-1' (syn. 'Koushin', 'Kosuei 79-1', 'Lushan 79-1', 'Red Princess'); 'Kuimi' (syn. 'F.Y. 79-1', 'ACC 211', 'Apple Sensation', 'Kamitsu', 'Turandot', and, possibly, 'K191'); 'Jinfeng' (syn. 'FT79-3', 'Golden Yellow', 'Kinpo').

These selections have been grown in Italy, New Zealand, and the US. The only commercial production outside of China appears to be in

Fig. 2. The old and the new: *A. deliciosa* 'Hayward' (*left*), the present kiwifruit of commerce, and *A. chinensis* 'Hort16A' (*right*), to be sold under the name ZESPRI™ GOLD kiwifruit.

California, with 'Lushanxiang', 'Jiangxi-79-1', and 'Golden Yellow' fruit having been sold in the Los Angeles market since 1995. There is obviously keen demand by consumers with growers receiving $US16 per 3.0 kg tray.

Cultivars of *A. chinensis* Selected Outside of China

Seed of *A. chinensis* was first introduced to New Zealand in 1977, and research workers there were soon convinced that this species had the greatest commercial potential of any of the *Actinidia* species other than the kiwifruit itself. The first New Zealand bred cultivar of *A. chinensis* has been formally released and full-scale marketing will commence in the 1999 harvest season. In 1987 a cross was made between plants from two accessions of *A. chinensis* with the aim of combining fruit size, good flavor, and yellow flesh. One seedling was identified in 1991 as having particularly good fruit which have a very characteristic, pointed shape, quite different in appearance to 'Hayward' fruit. The skin is covered with very soft downy hair, which is easily rubbed off. The flesh is a bright yellow when fruit are harvested at the right maturity and the flavor is much preferred by most consumers to that of 'Hayward' fruit. This selection has been registered under the PVR name of 'Hort16A' and will be marketed under the commercial name ZESPRI™ GOLD Kiwifruit. (ZESPRI™ is the name of the marketing subsidiary of Kiwifruit New Zealand, the successor to the New Zealand Kiwifruit Marketing Board.)

Although the fruit of 'Hort16A' are recognizably kiwifruits (i.e., are considered by most consumers to be related to 'Hayward' kiwifruit) they will need to be handled differently, especially as the fruit skin is more tender and more easily damaged than that of 'Hayward' kiwifruit. The ancestors of the plant come from different parts of China and 'Hort16A' will probably differ from 'Hayward' in its climatic requirements and in its response to management practices. Futhermore, the kiwifruit is hexaploid whereas 'Hort16A' is diploid and flowers almost a month earlier: diploid males of *A. chinensis* have therefore had to be selected as pollenizers. Two such plants, 'Meteor' and 'Sparkler', have so far been registered.

Many of the large-fruited Chinese selections of *A. chinensis* are tetraploid. A cultivar of *A. chinensis*, also suggested to be tetraploid, has been selected in France. ChinaBelle® has fruit averaging between 80 and 100 g, and it too has yellow flesh (Blanchet and Chartier 1998). A pollenizer, PolliChina®, has also been selected.

CULTIVATION OF OTHER *ACTINIDIA* SPECIES

Two other *Actinidia* species have fruit of obvious commercial potential: these are *A. arguta* (Sieb. et Zucc.) Planch. ex Miq. and *A. kolomikta* (Maxim. et Rupr.) Maxim. The potential of *A. arguta* has been discussed at length for more than 80 years, but it is only over the past few years that it has shown any signs of being anything more than just a novel or experimental crop. The fruit is about the size of a European gooseberry or grape, with a skin that is smooth and polished, completely hairless and generally palatable (Fig. 3). The flesh is sweet but the best feature is the unique, "sophisticated" flavor. Good selections of *A. arguta* show that this species has the potential to be much more than just a crop for those areas that are too cold for kiwifruit cultivation.

Small commercial plantings of *A. arguta* have been established in Canada, France, Germany, Italy, New Zealand and the United States, with local selections being promoted. These have been selected for size, for color (red or green skins), sweetness and flavor. The most widely planted, about 25 ha in Oregon, is what has been identified as the cultivar 'Anna' (correct name 'Anansnaya'), but this material seems not true to label (R. Meyer pers. commun.).

Fig. 3. Fruit of *A. arguta*. The fruit are small (about 10–20 g), but good selections have a fine, sweet flavor and the skins are smooth, hairless, and edible.

Fruit of *A. arguta* has had a good reception in the San Francisco and Los Angeles markets, fetching remarkably high prices, and increasing amounts of both fresh fruit and processed products are being exported.

The true 'Ananasnaya' is possibly a hybrid of *A. arguta* and *A. kolomikta* (Evreinoff 1949) selected many years ago by Michurin. *A. kolomikta* is a particularly cold-hardy *Actinidia* species with small but sweet fruit, very rich in vitamin C. It is of very limited commercial potential except for very cold areas.

THE NEW ZEALAND TRADITION OF NEW FRUITS AND NEW FRUIT CULTIVARS

The success of the kiwifruit is basically attributable to the inherent qualities of the fruit, to its appeal to the consumer and to its ease of handling. This appeal to the consumer—which is due to the beautiful green flesh, the flavor, the texture and the high content of vitamin C—has played an essential part in the kiwifruit's popularization. Sustained and imaginative marketing has undoubtedly also helped the successful marketing of 'Hayward' fruit. New kiwifruit cultivars will require similar promotion.

It is less certain why it was in New Zealand that kiwifruit should have first been developed as a commercial crop. Chance was undoubtedly important as was the possession of a particularly good cultivar. However, even an outstandingly good cultivar such as 'Hayward' does not necessarily result in the development of an industry unless growers are aware of its potential: 'Hayward' was sent to California in 1935 but its true worth was not appreciated until the first shipment of New Zealand kiwifruit to the United States quarter of a century later. The New Zealand tradition of innovation in fruit growing was also probably important.

New Zealand has only a small population and its fruit growing industries have therefore become largely reliant on exports. As a consequence, the industries are particularly aware of customers' needs and of the desire for novelty. Some of our novelty crops such as horned melons (kiwanos), tamarillo, and feijoas have remained novelties. Much of the success of the New Zealand apple industry is due to its reliance on new or novel cultivars with 'Royal Gala'/'Gala' and 'Braeburn', both of local origin, now making up more than half of all the apples exported. The 'Hayward' kiwifruit is a good example of a novel crop that has "made good" and has become established throughout the world. It is our hope that 'Hayward' will be joined by other, equally successful new *Actinidia* cultivars.

REFERENCES

Bergamini, A. 1991. "Top Star"®: il nuovo kiwi a frutti liscio. L'Informatore Agrario 47(30):43–45.

Blanchet, P. and J. Chartier. 1998. Sélection de kiwis chinois pour les zones chaudes: Chinabelle® et Pollichina®. L'Arboriculture Fruitière 513:37–40.

Evreinoff, V.A. 1949. Notes sur les variétés d'*Actinidia*. Revue Hort. 121:155–158.

Ferguson, A.R. 1984. Kiwifruit: A botanical review. Hort. Rev. 6:1–64.

Ferguson, A.R. 1990. Kiwifruit (*Actinidia*). p. 601–653. In: J.N. Moore and J.R. Ballington, Jr. (eds.), Genetic resources of temperate fruit and nut crops. (Acta Hort. 290) Int. Soc. Hort. Sci., Wageningen.

Ferguson, A.R. and E.G. Bollard. 1990. Domestication of the kiwifruit. p. 165–246 + 3 plates. In: I.J. Warrington and G.C. Weston (eds.), Kiwifruit: Science and management. Ray Richards, Publisher in association with the New Zealand Soc. Hort. Sci., Auckland.

Ferguson, A.R., A.G. Seal, and R.M. Davison. 1990. Cultivar improvement, genetics and breeding of kiwifruit. Acta Hort. 282:335–34.

Ferguson, A.R., A.G. Seal, M.A. McNeilage, L.G. Fraser, C.F. Harvey, and R.A. Beatson. 1996. p. 371–417. In: J. Janick and J.N. Moore (eds.), Fruit breeding. Vol. 2. Vine and small fruits. Wiley, New York.

Meyer, R. 1992. Searching for warm-winter kiwifruit plants: Updating a decade-long research project. Kiwifruit Enthusiasts J. 6:57, 58.

Muggleston, S., M. McNeilage, R. Lowe, and H. Marsh. 1998. Breeding new kiwifruit cultivars: the creation of Hort16A and Tomua. Orchardist of New Zealand 71(8):38–40.

Testolin, R., G. Cipriani, L. Gottardo, and G. Costa. 1995. Valutazione de selezioni maschili di actinidia come impollinatori per la cv. "Hayward." Riv. Frutticoltura 57(4):63–68.

World Kiwifruit Review 1998. 1998. Belrose, Inc., Pulham, WA.

Phalsa: A Potential New Small Fruit for Georgia

Anand K. Yadav

Phalsa (*Grewia asiatica* L., Tiliaceae) is an exotic bush plant considered horticulturally as a small fruit crop but also used as a folk medicine. The ripe phalsa fruits (Fig. 1) are consumed fresh, in desserts, or processed into refreshing fruit and soft drinks enjoyed during hot summer months in India (Salunkhe and Desai 1984). However, phalsa fruit has a short shelf life and is considered suitable only for local marketing (Anand 1960).

The phalsa plant (Fig. 2) is native to the Indian subcontinent and Southeast Asia (Hays 1953; Chundawat and Singh 1980) but is cultivated on a commercial scale mainly in the northern and the western states of India (Hays 1953; Sastri 1956). Around the beginning of the 20th century, it was introduced into the East Indies including the Philippines where it is naturalized at low elevations in dry zones of the island of Luzon. It reached western countries much later in the century. In the United States, a few specimens have been established at the Agricultural Experiment Station at Mayaguez, Puerto Rico, the Tropical Research and Education Center of the University of Florida at Homestead, Florida, and the USDA, ARS, Tropical Horticultural Research laboratory at Miami, Florida but there are no commercial plantings.

BOTANY

Grewia asiatica L.,*G. subinaequalis* DC. (syn. *G. asiatica* Mast.), and *G. hainesiana* Hole are the only members of Tiliaceae that yield edible fruits. According to Sastri (1956), phalsa is the most commonly used vernacular name for these fruits in India but there are several names in customary usage including *dhamin*, *parusha*, and *shukri* in Hindi, *dhaman* in Punjabi, *man-bijal* in Assamese, *phalsa* and *shukri* in Begali, *mirgi chara* and *pharasakoli* in Oriya, *phalsa* in Gujrati, *phalsi* in Maharashtra, *jana, nallajana, phutiki* in Telagu, *palisa, tadachi* in Tamil, *buttiyudippe* and *tadasala* in Kannada, and *falsa* in Pakistan.

Morphology

The phalsa plant is a large, shaggy shrub (Fig. 2) or a small tree reaching 4 m or more in height (Sastri 1956). The phalsa plants grow to become straggling tall shrubs with rough bark on the stem, and have numerous long, slender, drooping branches where the young branchlets are densely covered with a coating of hairs. The alternate, deciduous, widely spaced, thick, and large leaves are broadly heart-shaped or ovate, pointed at the apex, oblique at the base, measure up to 20 cm in length and 15 cm in width, and coarsely toothed, with a

Fig. 1. Freshly harvested phalsa fruits produced at the Agricultural Research Station of the Fort Valley State University, Fort Valley, Georgia

Fig. 2. A healthy phalsa bush growing in the field inside a cold protected house at the Agricultural Research Station of the Fort Valley State University, Fort Valley, Georgia

light, whitish blush on the underside. Small (10 to 19 mm across), bright orange-yellow flowers are borne in dense cymes in the leaf axils in late spring. The small fruits, almost round drupes like blueberry and purple, crimson or cherry red in color when ripe, borne on a 2- to 3-cm-long peduncle, are produced in great numbers in open, branched clusters. Individual fruits measure from 1.0 to 1.9 cm in diameter, 0.8 to 1.6 cm in vertical height, and 0.5 to 2.2 g in weight. Fruits ripen gradually on bushes during the summer months. While ripening, the fruit skin turns from light green to cherry red or purplish red finally becoming dark purple or nearly black (Fig. 3). The ripe fruit is covered with a very thin, whitish blush, and becomes soft and tender. The delicate, fibrous flesh is light greenish-white becoming colored purplish-red from seed reaching near the skin. Overripe fruit flesh becomes suffused with purple color later followed by shriveling of fruit skin due to moisture loss. The phalsa flavor is pleasantly astringent but delicious due to very appropriate sugar-acid blend. Large fruits have two hemispherical, hard, buff-colored seeds up to 5 mm in diameter while small fruits are generally single-seeded.

Adaptation

The phalsa is a warm climate fruit plant. In India, this plant grows satisfactorily and produces well up to an elevation of 1,000 m. It can tolerate light frost although at the cost of defoliation; the plant is deciduous and normally loses its leaves slowly in those areas with mild winter season. The phalsa planting flourishes well under variable climatic conditions, requires protection from the freezing cold temperatures. Adequate sunlight and warm or hot temperatures are required for fruit ripening, development of appropriate fruit color, and good eating quality.

The phalsa plant grows vigorously and produces satisfactorily under variable soil types including fine sand, clay or even limestone, when soil fertility is not very poor. Phalsa is often grown in marginal lands close to city markets to facilitate prompt marketing of fruit. Fertilizer application although casual, helps plant health and production. The plant is drought-tolerant, but occasional irrigation during the fruiting season and in dry periods, is profitable for growers.

HORTICULTURE

The tall-growing wild phalsa plants produce fruits which are of marginal quality and are not relished by most consumers. The low-growing dwarf and/or bushy type of phalsa plants which develop a good blend of sugar and acid in the fruit flesh, are preferred for cultivation (Hays 1953). There are no well-known cultivars available but there are local favorites for different growing regions (Sastri 1956; Yadav 1998).

Fig. 3. Various stages in the ripening of the phalsa fruits ranging from fully developed green to fully develped ripe (center).

Propagation

Conventional propagation of phalsa plant is by seed. The phalsa plant is readily propagated by rooting of hardwood cuttings as well as layering (Samson 1986). Seedlings produce the first crop of well evolved fruits 12 to 15 months from planting. Seeds stay viable for years and germinate in less than three weeks (Sastri 1956). Wood type and planting date influence rooting of phalsa (Singh et al.1961). Treatment with auxins (IAA, IBA, NAA) improve rooting of difficult-to-root hardwood cuttings of phalsa (Yadava and Rajput 1969), ground layers, and air layers (Mohammed and Chauhan 1970).

Culture and Management

One-year-old seedlings are usually spaced 2 to 4.5 m apart. Bushes flower progressively during the spring months. Since phalsa bears fruit on current season's growth there is a need for regular but severe annual pruning before the on-set of spring. Annual pruning to a height of about 1 m encourages new shoots and higher yield of marketable fruit than does more drastic trimming (Singh and Sharma 1961). Gibberellic acid has been reported to improve fruit set and increase fruit size (Randhawa et al. 1959). The phalsa plant shows good response to nitrogen applications. High levels of phosphorus supply increase sugar content in the fruit while higher potassium suppresses sugar and promotes acidity. Phalsa is considered stress-tolerant and is commonly grown under neglect (Hays 1953).

Gradual but steady fruit ripening during the summer of only a few fruits in a cluster necessitates frequent harvesting (Salunkhe and Desai 1984). The fruits which resemble blueberry have a very short shelf life, and must be marketed within a day or two. Average fruit yield per bush from a well managed planting varies from 7 to 10 kg during one season.

Biotic Stresses

Leaf-cutting caterpillars attack the foliage at night. A blackish caterpillar causes galls on the growing shoots. Termites often damage the roots. In some areas, leaf spot is caused by *Cercospora grewiae*. At the Fort Valley State University, however, there has been no serious biotic stresses observed on phalsa plants or fruits except that ants were found on some bushes during the fruit ripening periods.

Medicinal and Other Uses

The fruit is astringent and stomachic. Morton (1987) reported that when unripe, phalsa fruit alleviates inflammation and is administered in respiratory, cardiac, and blood disorders, as well as in fever reduction. Furthermore, an infusion of the bark is given as a demulcent, febrifuge, and treatment for diarrhea. The root bark is employed in treating rheumatism. The leaves are applied on skin eruptions and they are known to have antibiotic action.

The fresh leaves are valued as animal fodder. The bark is used as a soap substitute in Burma. A mucilaginous extract of the bark is useful in clarifying sugar. Fiber extracted from the bark is made into rope. The wood is yellowish-white, fine-grained, strong, and flexible. It is used for archers' bows, spear handles, shingles, and poles for carrying loads on the shoulders. Stems that are pruned serve as garden poles and for basket-making.

The flowers have been found to contain grewinol, a long chain keto-alcohol, tetratricontane 22-ol 13-one (Lakshmi and Chauhan 1976). The phalsa seeds produce approximately 5% yield of a bright yellow oil that contains 8% palmitic acid, 11% stearic acid, 13.5% oleic acid, and 64.5% linoleic acid with 3% unsaponifiable (Morton 1987).

EVALUATIONS IN GEORGIA

Beginning in 1994, we have been investigating feasibility of phalsa growing at the Fort Valley State University Agricultural Research Station. The phalsa seeds were obtained from India through the USDA Plant Introduction division. The greenhouse-raised seedlings of Indian phalsa, and seedlings and hardwood cuttings of MIA-12489 phalsa that we received from the USDA-ARS Tropical Horticulture Research Laboratory, Miami, Florida, were established in the field plots during spring 1995. The experimental field plots for phalsa were established inside a 25 m × 40 m cold protection wooden house (Fig. 2) which was covered with

350

Table 1. Plant characteristics of two phalsa lines as observed at Fort Valley, Georgia.

Plant type	Leaves	Flowers	Fruit Diam. (cm)	Weight (g)	Seed weight (g)
Indian Shaggy shrub or small tree, bushy appearance, numerous branches	Large thick dark green ovate leaves, pointed apex, coarse teeth, prominent leaf veins	Small flowers, prominent orange petals, light yellow sepals, 10.5 mm diam.	10.45±0.25	0.98±0.43	0.16±0.015
Miami 12489 Tall shrub or small tree with fewer and stronger branches, light green	Large thick leaf blades, broadly ovate, base oblique, coarsely toothed, white blush underside	small inconspicuous flowers, orange petals, bright yellow sepals, 18.40 mm diam.	14.26±0.38	1.15 ± 0.61	0.23±0.02

6-mil clear polyethylene and was equipped with 6 high-speed fans and 4 electric heaters. Plants were fertilized and irrigated as needed; no controls have been required for insects and disease. Bushes were heavily pruned each February to a height of 30 cm (Yadav 1998).

The green-house raised plants of both Indian and MIA-12489 phalsa lines planted during spring 1995, were established satisfactorily and increased in vigor each year. Plant survival was 70% for the Indian phalsa compared to 50% for MIA-12489 germplasm from Miami, Florida. Selected vegetative and fruiting characters of two phalsa lines are compared in Table 1.

The Indian phalsa bushes had greater flowering and fruiting intensity than the MIA-12489. The peak fruiting period for the 1998 season ranged from July 15 to August 5. The average fruit production from the Indian type was 2.7 kg fruits per bush compared to only a few dozen fruits (approx 25 g) on the MIA-12489 during 1998. Mean fruit diameter for Indian type was 10.5 mm vs. 14.3 mm for MIA-12489, fruit height for Indian phalsa measured from the blossom-end to the stem-end, was 10.1 mm (ranging from 6.8 to 12.8 mm). Mean berry weight was 1.0 g for Indian phalsa and 1.2 g for Miami phalsa. The astringent fruits of the Indian type phalsa had more pleasing flavor but the fruit of Miami phalsa (MIA-12489) was sweeter. The majority of MIA-12489 phalsa fruits had double seeds while only a few large fruits of the Indian phalsa had double seeds. The Indian phalsa had smaller seeds than MIA-12489, which is preferred. Fruits of Indian phalsa stayed fresh for 7 days or longer under refrigeration but spoiled in a few days at room temperatures.

Phalsa juice ferments so readily that sodium benzoate must be added as a preservative (Anand 1960). Analyses made long ago in the Philippines established the following values as reported by Morton (1987): 725 calories/kg edible fruit; moisture, 81.13%; protein, 1.58%; fat, 1.82%; crude fiber, 1.77%; and sugar, 10.27%. Freshly harvested phalsa fruits were frozen and analyzed for 17 nutritional items by a local commercial analytical laboratory (Table 2).

Table 2. Nutrient content of phalsa fruits produced at Fort Valley, Georgia.

Nutrients analyzed in 1998	Nutrient values/ 100 g fruit
Calories (Kcal)	90.5
Calories from fat (Kcal)	0.0
Moisture (%)	76.3
Fat (g)	<0.1
Protein (g)	1.57
Carbohydrates (g)	21.1
Dietary Fiber (g)	5.53
Ash) (g)	1.1
Calcium (mg)	136
Phosphorus (mg)	24.2
Iron (mg)	1.08
Potassium (mg)	372
Sodium (mg)	17.3
Vitamin A (µg)	16.11
Vitamin B_1, Thiamin (mg)	0.02
Vitamin B_2, Riboflavin (mg)	0.264
Vitamin B_3, Niacin (mg)	0.825
Vitamin C, Ascorbic acid (mg)	4.385

SUMMARY

Greenhouse-raised seedlings as well as rooted cuttings of phalsa plants were successfully established inside a cold protected polyethylene-covered field structure. Our observations in the Middle Georgia area indicated that growing phalsa is feasible in temperate areas if cold protection is assured. Indian phalsa was more fruitful and had tastier fruits than the MIA-12489 line. Studies to improve plant regeneration and cold hardiness are under way.

REFERENCES

Anand, J.C. 1960. Efficacy of sodium benzoate to control yeast fermentation in phalsa (*Grewia asiatica* L.) juice. Indian J. Hort. 17:138–141.

Chundawat, B.S. and R. Singh. 1980. Effect of growth regulators on phalsa (*Grewia asiatica* L.). I. Growth and fruiting. Indian J. Hort. 37:124–131.

Hays, W.B. 1953. Fruit growing in India. 2nd Revised edition. Kitabistan, Allahabad, India.

Lakshmi, V. and J.S. Chauhan. 1976. Grewinol, a keto-alcohol from the flowers of *Grewia asiatica*. Lloydia 39:372–374.

Mohammed, S. and K.S. Chauhan. 1970. Vegetative propagation of phalsa (*Grewia asiatica* L.). Indian J. Ag. Sci. 40:581–586.

Morton, J.F. 1987. Phalsa. p 276–277. In: Fruits of warm climates. Julia Morton, Miami, FL.

Randhawa, G.S., R.S. Malik, and J.P. Singh. 1959. Note on clonal propagation in the phalsa (*Grewia asiatica* L.). Indian J. Hort. 16:119–120.

Randhawa, G.S., J.P. Singh, and S.S. Khanna. 1959. Effect of gibberellic acid and some other plant growth regulators on fruit set, size, total yield and quality of phalsa (*Grewia asiatica* L.). Indian J. Hort. 16:202–205.

Salunkhe, D.K. and B.B. Desai. 1984. Phalsa. p. 129. In: Salunkhe and Desai (eds.), Postharvest biotechnology of fruits. Vol. 2. CRC Press, Boca Raton, FL.

Samson, J.A. 1986. The minor tropical fruits. p. 316. In: Tropical fruits. Longman Inc., New York.

Sastri, B.N. 1956. The wealth of India: Raw Materials #4. *Grewia* Linn. Tiliaceae. p. 260–266. In: Council of Scientific and Industrial Research, New Delhi, India.

Singh, J.P. and H.C. Sharma. 1961. Effect of time and severity of pruning on growth, yield and fruit quality of phalsa (*Grewia asiatica* L.). Indian J. Hort. 18(1):20–28.

Singh, J.P., P.S. Godara, and R.P. Singh. 1961. Effect of type of wood and planting dates on the rooting of phalsa (*Grewia asiatica* L.). Indian J. Hort. 18(1):46–50.

Yadav, A.K. 1998. Phalsa. Commodity sheet FVSU-004 of the Fort Valley State Univ., Ag. Res. Sta.

Yadava, U.L. and C.B.S. Rajput. 1969. Anatomical studies of rooting in stem cuttings of phalsa (*Grewia asiatica* L.). Hort. Sci. 1:19–22.

The Pawpaw Regional Variety Trial

Kirk W. Pomper, Desmond R. Layne, and R. Neal Peterson

The pawpaw [*Asimina triloba* (L.) Dunal, Annonaceae] tree produces the largest edible tree fruit native to the United States (Darrow 1975; Layne 1996). This fruit, known commonly as the "poor man's banana," may reach up to 1 kg in size. Pawpaws grow wild in the mesic hardwood forests of 26 states in the eastern United States, ranging from northern Florida to southern Ontario (Canada) and as far west as eastern Nebraska (Kral 1960); as shown in Fig. 1. Pawpaws are hardy to USDA growing zone 5 (–15°F, –9.4°C) and flourish in the deep, rich fertile soils of river-bottom lands where they grow as understory trees or thicket-shrubs (Sargent 1890). The unique qualities of the pawpaw fruit, ornamental value of the tree, and the natural compounds in the leaf, bark, and twig tissues that possess insecticidal and anti-cancer properties (McLaughlin 1997), suggest that pawpaw has great potential as an alternative high-value crop.

THE PAWPAW

Taxonomy

Pawpaw is the only temperate member of the tropical Custard Apple family (Bailey 1960), which includes several delicious tropical fruits such as the custard apple (*Annona reticulata* L.), cherimoya (*A. cherimola* Mill.), sweetsop or sugar apple (*A. squamosa* L.), atemoya (*A. squamosa* × *A. cherimola*), and soursop (*A. muricata* L.). There are also eight other members of the *Asimina* genus that are native to the extreme southeastern states of Florida and Georgia. These include *A. incarna* (Bartr.) Exell. (flag pawpaw); *A. longifolia* Kral; *A. obovata* (Willd.) Nash; *A. parviflora* (Michx.) Dunal (dwarf pawpaw); *A. pygmaea* (Bartr.) Dunal; *A. reticulata* Shuttlw. ex Chapman; *A. tetramera* Small (opossum pawpaw); and *A.* × *nashii* Kral (Kral 1960).

Description of the Plant

Pawpaw is a small, deciduous tree that may attain 5 to 10 m in height and tends to be found in patches due to root suckering (Layne 1996). In sunny locations, trees typically assume a pyramidal habit, with a straight trunk and lush, dark green, long, drooping leaves (Fig. 2). The blossoms occur singly on the previous year's wood, reaching up to 5 cm in diameter, emerging before leaves in mid spring (about April in Kentucky). Flowers have a globular androecium and a gynoecium composed of 3–7 carpels or 3–7 fruited clusters (Kral 1960). Flowers are strongly protogynous, self-incompatible and require cross-pollination (Wilson and Schemske 1980), although some trees may be self-compatible. Pollination may be by flies (Wilson and Schemske 1980) and beetles (Kral 1960).

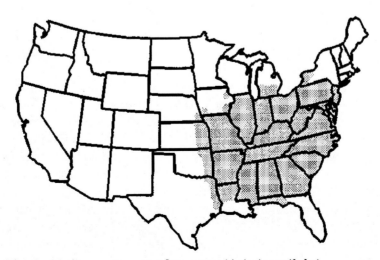

Fig. 1. Native range map of pawpaw (*Asimina triloba*).

353

Fig. 2. 25 yr. old pawpaw tree in full sun location. Photograph by R.N. Peterson.

Fig. 3. Pawpaw fruit from the cultivar Zimmerman. Photograph by S.C. Jones. Plate diam. = 18 cm.

Fig. 4. Pawpaw fruit from the PawPaw Foundation advanced selection 1-7-1. Photograph by S.C. Jones. Plate diam. = 18 cm.

Fig. 5. Pawpaw fruit cluster. Photograph by D.R. Layne.

Description of the Fruit

Fruit set in the wild is usually low and may be pollinator or resource-limited (Wilson and Schemske 1980), but under cultivation, tremendous fruit loads have been observed. The fruits are oblong-cylindrical berries that are typically 3 to 15 cm long, 3 to 10 cm wide and weigh from 200 to 400 g (Fig. 3 and 4). They may be borne singly or in clusters(Fig. 5), which resemble the "hands" of a banana plant (*Musa* spp.). When ripe, skin color ranges from green, to yellow, to brownish black, and flesh color ranges from creamy white through bright yellow to shades of orange (Fig. 3 and 4). The flavor of the pawpaw fruit resembles a combination of banana, mango, and pineapple. However, flavor varies among cultivars, with some fruit displaying more complex flavor profiles. The flavor of a pawpaw fruit can intensify as it over-ripens, as with banana, resulting in pulp that is excellent for use in cooking. The skin should not be eaten. Shelf life of a tree-ripened fruit stored at room temperature is 2–3 days. With refrigeration, fruit can be held up to 3 weeks while maintaining good eating quality.

COMMERCIAL POTENTIAL OF PAWPAW

In Kentucky and the southeastern United States, alternative and potentially high-value cash crops are being examined for their potential to help supplement the incomes of small farmers who are currently dependent upon growing and selling tobacco. Identifying alternative high-value crops would allow small farmers to diversify to these alternative crops and generate new sources of farm income. This plant is well adapted to the climate of the southeastern United States where tobacco is now commonly grown.

Pawpaw fruit have a unique tropical flavor and powerful aroma. There is commercial potential for use of pawpaw fruit in juices, ice creams, yogurts, and baby food. The fruit are very nutritious, being higher in some vitamins, minerals, and amino acids than apple, grape, and peach (Peterson, et al. 1982). Pawpaw trees are also attractive to the homeowner as an ornamental planting in edible landscapes. Natural compounds (annonaceous acetogenins) in leaf, bark, and twig tissue possess insecticidal and anti-cancer properties (Rupprecht et al. 1986; Zhao et al. 1994; McLaughlin 1997). Harvesting the leaves and twigs for the extraction of these compounds may also represent a lucrative opportunity for small farmers.

There is already a potentially profitable market for those interested in establishing pawpaw orchards. Most fruit for sale are currently collected from wild stands in the forest. Bray Orchards of Bedford, Kentucky has developed a pawpaw ice cream, in cooperation with the pawpaw project at Kentucky State University. The ice cream has been selling for $10.00 per gallon, and Bray Orchards has been paying up to $5.00 per lb. for pawpaw pulp. Pawpaws also show potential as fresh market fruit. Pawpaw fruit were sold at the Lexington Farmers Market this year (1998) for $1 each (or about $3.00 per pound) and sellers had a difficult time finding enough fruit to market.

Chefs at the Oakroom, a AAA five-diamond restaurant at Louisville's Seelbach Hilton, have developed a menu of Appalachian cuisine, which includes pawpaw. The menu includes a pawpaw and green tomato relish on a French rib pork chop, pawpaw sorbet, and a Pawpaw Foster with pawpaw ice cream (made by Bray Orchards) for dessert. The Pawpaw Foster consists of flambéed pawpaw and banana using ginger liqueur (wild ginger is indigenous to Kentucky) over pawpaw ice cream. An Associated Press article was carried in newspapers across the country in the last week of September 1998 concerning the development of this menu and the use of pawpaws. Hopefully, this media coverage will continue to increase public awareness of the pawpaw fruit and its uses.

Table 1. List of pawpaw selections in Regional Variety Trial (RVT).

Clone	Orchard source	Genetic background[z]
1-7	Wye, MD	'Overleese'
1-23	Wye, MD	'Taytwo'
2-54	Wye, MD	GAZ-VA
8-58	Wye, MD	BEF-30
9-47	Wye, MD	BEF-49
9-58	Wye, MD	BEF-50
10-35	Wye, MD	BEF-49
11-5	Wye, MD	BEF-53
11-13	Wye, MD	BEF-53
'Middletown'		From Ohio
'Mitchell'		From Illinois
'NC-1'		'Davis' × 'Overleese'
'Overleese'		From Indiana
'PA-Golden'		From Pennsylvania
'Sunflower'		From Kansas
'Taylor'		From Michigan
'Taytwo'		From Michigan
'Wells'		From Indiana
'Wilson'		From Kentucky
1-7	Keedysville, MD	BEF-30
1-68	Keedysville, MD	'Overleese'
2-10	Keedysville, MD	BEF-30
3-11	Keedysville, MD	BEF-33
3-21	Keedysville, MD	BEF-43
4-2	Keedysville, MD	BEF-53
5-5	Keedysville, MD	BEF-54
7-90	Keedysville, MD	RS-2
8-20	Keedysville, MD	'Sunflower'

[z]BEF = Blandy Experimental Farm Collection, Boyce VA.
GAZ = George A. Zimmerman Collection., Linglestown, PA.
RS = Ray Schlaanstine Collection, West Chester, PA.

Table 2. Regional Variety Trial cooperators and cooperating institutions by location.

State	Cooperator	Institution	Location
Indiana	Bruce Bordelon	Purdue University	West Lafayette, IN
Iowa	Tom Wahl	Iowa State University	Wapello, IA
Kentucky	Kirk Pomper	Kentucky State Univ.	Frankfort, KY
Kentucky	Jerry Brown	Univ. of Kentucky	Princeton, KY
Maryland	Chris Walsh	Univ. of Maryland	Keedysville, MD
Michigan	Dennis Fulbright	Michigan State Univ.	Jackson, MI
Nebraska	Bill Gustafson & Stan Matzke	Univ. of Nebraska	Lincoln, NE
New York	Ian Merwin	Cornell University	Ithaca, NY
North Carolina	Mike Parker	N.C. State University	Raleigh, NC
Ohio	Tom Wall	Ohio State University	Piketon, OH
Oregon[z]	Kim Hummer	USDA-NCGR	Corvallis, OR
	Anita Azarenko	Oregon State University	Corvallis, OR
South Carolina	Greg Reighard	Clemson University	Clemson, SC
Country			
China[z]	Xi Sheng-Ke	Chinese Academy of Forestry	Beijing, China
China[z]	Hongwen Huang	Chinese Academy of Sciences	Wuhan, Hubei, China
Chile[z]	Luis Luchsinger	University of Chile	Santiago, Chile

[z] Outside pawpaw native range.

PAWPAW RESEARCH AT KYSU

Kentucky State University (KYSU) has had a comprehensive pawpaw research program since 1990 directed toward developing pawpaw as a new commercial tree fruit crop for Kentucky and the United States. Current research efforts at KYSU have been directed at developing cultural recommendations (Layne 1996), improving propagation methods (Finneseth et al. 1998), culinary development (see the Commercial Potential of Pawpaw section above), and germplasm collection and molecular characterization of genetic diversity in existing cultivars (Huang et al. 1997, 1998). Germplasm collection and characterization is an important continuing mission for the pawpaw program, since KYSU is also the site of the USDA National Clonal Germplasm Repository for *Asimina* spp.

There is a great deal of public interest in pawpaw. KYSU usually receives hundreds of inquiries each year, from the US and around the world, from interested individuals concerning pawpaw. To help meet public interest in pawpaw, two extension bulletins have been developed and are now available from KYSU: "*Cooking With Pawpaws*" by Jones and Layne (1997), and "*Pawpaw Planting Guide: Cultivars and Nursery Sources*" by Jones et al. (1998). A web site (http://www.pawpaw.kysu.edu) has been developed to disseminate pawpaw information and it contains the above extension bulletins, updates on pawpaw research at KYSU, pawpaw photographs, a bibliography, and information on the PawPaw Foundation. This site will also serve as a medium for the delivery of new information concerning orchard culture and available genetic resources.

THE REGIONAL VARIETY TRIAL (RVT)

In 1993, the PawPaw Foundation (PPF) and Desmond Layne, former Principal Investigator of Horticulture at KYSU, embarked on a joint venture to test within pawpaw's native range many of the commercially available, named pawpaw cultivars and PPF's advanced selections. Orchards for the Regional Variety Trial (RVT) were planted in 15 different locations from Fall 1995 through Spring 1999, each consisting of 300 trees. At each RVT site, 8 replicate trees of each of the 28 grafted scion varieties will be tested in a randomized complete block design. Named cultivars that are secured for testing include: 'Middletown', 'Mitchell', 'NC-1', 'Overleese', 'PA-Golden', 'Sunflower', 'Taylor', 'Taytwo', 'Wells', and 'Wilson'. The other 18 clones to be evaluated were selected from PPF orchards at the University of Maryland Experiment Stations at Wye, MD and Keedysville, MD (Table 1). These "advanced selections" were selected based on superior horticultural traits including fruit size and taste, flesh-to-seed ratio, resistance to pests and diseases, and overall productivity on a year-to-year basis. Seedling trees from local native sources were planted around the perimeter as a buffer against edge effects and to allow comparisons with local germplasm. Identical orchards will be or have been planted at the institutions and locations presented in Table 2.

THE FUTURE OF THE RVT

Orchard performance to be examined at each RVT site will include climatic factors, culture, pests, growth, flowering, yield, and fruit characteristics. Trees will be evaluated for several years for such characteristics as yield, year-to-year consistency, and regional suitability. At the end of several fruiting seasons, regional recommendations will be made. Each site will serve as a regional demonstration for growers and nursery operators, and they will serve public education/extension purposes. Kirk Pomper will serve as the coordinator of all RVT plantings. Once trees begin fruiting, annual RVT cooperator meetings will rotate from site to site for research updates and plot evaluations/tours. Final results of the RVT and regional recommendations will be available on the KYSU web site (http://www.pawpaw.kysu.edu). At KYSU, the RVT was planted in late March of 1998. By July, only 6 trees had failed to survive. Approximately 40 trees have initiated flower buds during the summer. These trees will be starting their third growth cycle next spring, and with the presence of flower buds, there is at least the potential that some fruit may be produced next year.

REFERENCES

Bailey, L.H. 1960. The standard cyclopedia of horticulture, Vol. I. MacMillan Co., New York.

Darrow, G.M. 1975. Minor temperate fruits. p. 276–277. In: J. Janick and J.N. Moore (eds.), Advances in fruit breeding. Purdue Univ. Press, West Lafayette, IN.

Finneseth, C.L.H., D.R. Layne, and R.L. Geneve. 1998. Morphological development of pawpaw [*Asimina triloba* (L.) Dunal] during seed germination and seedling emergence. HortScience 33:802–805.

Huang, H., D.R. Layne, and R.N. Peterson. 1997. Using isozyme polymorphisms for identifying and assessing genetic variation in cultivated pawpaw [*Asimina triloba* (L.)Dunal]. J. Am. Soc. Hort. Sci. 122:504–511.

Huang, H., D.R. Layne, and D.E. Riemenschneide. 1998. Genetic diversity and geographic differentiation in pawpaw [*Asimina triloba* (L.) Dunal] populations from nine states as revealed by alozyme analysis. J. Am. Soc. Hort. Sci. 123:635–641.

Jones, S.C. and D.R. Layne. 1997. Cooking with pawpaws. KSU Pawpaw Ext. Bul.-001.

Jones, S.C, R.N. Peterson, T. Turner, K.W. Pomper, and D.R. Layne. 1998. Pawpaw planting guide: cultivars and nursery sources. KSU Pawpaw Ext. Bul.-002.

Kral, R. 1960. A revision of *Asimina* and *Deeringothamnus* (Annonaceae). Brittonia 12:233–278.

Layne, D.R. 1996. The pawpaw [*Asimina triloba* (L.) Dunal]: A new fruit crop for Kentucky and the United States. HortScience 31:777–784.

McLaughlin, J.L. 1997. Anticancer and pesticidal components of pawpaw (*Asimina triloba*). Annu. Rpt. N. Nut Growers Assoc. 88:97–106.

Peterson, R.N., J.P. Cherry, and J.G. Simmons. 1982. Composition of pawpaw (*Asimina triloba*) fruit. Annu. Rpt. N. Nut Growers Assoc. 73:97–106.

Rupprecht, J.K., C.-J. Chang, J.M. Cassady, and J.L. McLaughlin. 1986. Asimicin, a new cytotoxic and pesticidal acetogenin from the pawpaw, *Asimina triloba* (Annonaceae). Heterocycles 24:1197–1201.

Sargent, C.S. 1890. Silva of North America. Houghton Mifflin Co., New York.

Wilson, M.F. and D.W. Schemske. 1980. Pollinator limitation, fruit production, and floral display in pawpaw (*Asimina triloba*). Bul. Torrey Bot. Club. 107:401–408.

Zhao, G.X., L.R. Miesbauer, D.L. Smith, and J.L. McLaughlin. 1994. Asimin, asiminacin, and asiminecin: Novel highly cytotoxic asimicin isomers from *Asimina triloba*. J. Med. Chem. 37:1971–1976.

Climbing and Columnar Cacti: New Arid Land Fruit Crops

Yosef Mizrahi and Avinoam Nerd*

In Israel, scarcity of water, high input prices, and market competition limit the number of orchard crops that can be grown profitably. Our approach to the further development of the horticultural industry in the dry regions of Israel—the Negev and Judean deserts—is thus to establish new crops that will demand high prices in the export markets (Mizrahi and Nerd 1996). To this end, about 40 species of rare or wild fruit trees were introduced by us into these dry regions in a number of locations that differed in terms of soil, water, and climate (Nerd et al. 1990; Mizrahi and Nerd 1996). Emphasis was placed on candidates of the Cactaceae because of their high water-use efficiency (5–10 times higher than that of most conventional crops), resulting in low water requirement (Nobel 1988, 1994). The high water-use efficiency of cacti is provided by their unique photosynthetic pathway—crassulacean acid metabolism (CAM). In CAM plants, the stomata open and CO_2 uptake takes place during the night when evaporation is low. Among the Cactaceae, there are about 35 species that have a potential for cultivation as fruit, vegetable, or forage crop species (Nobel 1994; Mizrahi et al. 1997).

Starting in 1984, we have introduced, for investigation as potential crop species, 17 members of the subfamily Cactoidae (Nerd et al. 1990; Mizrahi and Nerd 1996). Among these, four climbing (epiphytic) species and one columnar species have already been planted as commercial crops, and their fruits are being exported successfully to European markets as exotic fruits from Israel. The main reasons that these crops have made their way onto the market within so short a time after introduction are their precocious early yielding (three to four years after seeding or one to three years after propagation from cuttings) and their acceptability in the markets.

At present, our studies are aimed at examining the environmental adaptations of the species and their reproductive biology mode and at developing appropriate agrotechnological practices. In addition, a breeding program accompanied by cytological and molecular studies is being carried out in order to develop improved clones for cultivation.

This review is divided into two parts. The first part deals with the climbing cacti of the genera *Selenicereus* and *Hylocereus* and the other with the columnar cactus *Cereus peruvianus*.

CLIMBING (EPIPHYTIC) CACTI

Taxonomy

We collected wild or cultivated types of climbing cacti from a variety of sources—amateur cactus lovers, growers, botanical gardens, and backyards. We very soon realized that there is tremendous confusion about the taxonomic identity of these cacti: accessions with the same name were found to be of different species. We are currently applying cytological and molecular techniques to determine the proper taxonomic identities of the species that we have introduced (Lichtenzveig 1997). These species belong to at least to two different genera, *Selenicereus* and *Hylocereus*.

From the genus *Selenicereus* we will elaborate here only on one species *S. megalanthus,* currently grown in Israel and in Colombia, where it is known as yellow pitaya (Hunt 1989; Barthlott and Hunt 1993), Accessions of *S. megalanthus* were introduced by us as *H. triangularis* or *H. undatus* and were later classified as *S. megalanthus* (Weiss et al. 1995; Mizrahi et al. 1997). We have 37 selected clones from this species.

From the genus *Hylocereus*, we have introduced the following species, some with a number of clones (Table 1): *H. undatus, H. polyrhizus, H. purpusii, H. ocamponis,* and *H. costaricensis* (Britton and Rose 1963; Barthlott and Hunt 1993). In addition, we have introduced some promising unidentified clones of *Hylocereus* (*Hylocereus* sp.), the best of which was designated as 10487. Of these species, only the ones that are currently being grown in Israel for export are described in this paper, as follows: *H. undatus, H. polyrhizus,* and *Hylocereus*

*The authors thank the Fleischer Foundation and Harry-Stern & Hellen-Zoref Fund for Applied Research at BGU, for supporting this program. Special thanks to Mrs. Inez Mureinik for editing the manuscript.

sp. The later two species are not cultivated any where else in the world to the best of our knowledge. Some of these and other species are grown elsewhere: *H. costaricensis*, (several commercial clones) as grown in Nicaragua (known as red pitaya); *H. undatus*, in Mexico (known as pitahaya), in other Latin American countries (known as pitaya), in Vietnam (known as dragon pearl fruit or *thang loy*) (Mizrahi et al. 1997), and according to colleagues there, in Guatemala.

Horticulture

At the beginning of our program, there was very little information available in the scientific literature on cultivation and biological background of these cacti. This information was mainly in Spanish in the form of hard-to-get dissertations and professional brochures (Mizrahi et al. 1997). We thus set out to investigate both horticultural and physiological aspects of climbing cacti and the results of our studies have been published in the professional literature, as follows: reproductive biology (Weiss et al. 1991, 1994a,b; Nerd and Mizrahi 1997), shading requirements (Raveh et al. 1993, 1996, 1998), and fruit development, ripening, and post-harvest handling (Nerd and Mizrahi 1998, 1999). Here, we will summarize some of the results and give details of new unpublished data to provide an up-to-date picture of the state-of-the-art know-how and marketing.

Light tolerance. The climbing cacti originate in shady habitats of subtropical and tropical America. In Israel, the canopy suffers from bleaching and die back when these species are grown outdoors as a result of the intensive irradiation (noon photosynthetic photon flux densities can reach as much as 2200 mmol photons $m^{-2} s^{-1}$). Our studies showed that for optimal development they have to be planted in nethouses and the required shade level (ranging between 30–60%) depends on the particular species as well as on the location (Fig. 1) (Raveh et al. 1996, 1998). *H. polyrhizus* and *H. costaricensis* are the most light tolerant, probably because of their unique skin characteristics (a wax cover and a thick skin). The radiation stress is exacerbated by high temperatures, as discussed below.

Temperature tolerance. Sub-freezing temperatures damage the climbing cacti, and for most species 0°C is the minimal threshold for cultivation. Among the investigated species, *Hylocereus* sp. (10487) was the most sensitive to low temperatures and suffered cold injury when the temperature fell below 4°C. In the areas of the Negev with low night temperatures, the climbing cacti have to be cultivated in plastic- or glass-houses. Symptoms of cold injury are round lesions that expand along the stems. Plants recover easily when temperature increases.

Our long-term observations showed that in the hottest parts of the Negev (Arava and Jordan valleys), where extreme summer temperatures (Fig. 3, 4) may rise up to 45°C, (average 39°C), annual flower produc-

Fig. 1. Cactus grown in a nethouse.

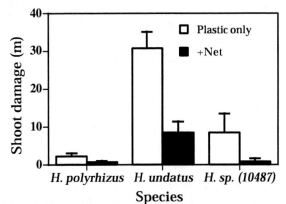

Fig. 2. Heat and radiation damage to three *Hylocereus* species growing in a greenhouse in Beer-Sheva in the summer of 1998. The damage was estimated as the length (m) of stem that was liquefied along the trellis system. The net provided over 60% shade. *Selenicereus megalanthus* was not damaged by the high temperature. The numbers are averages per plant ± SE.

tion was very low, being about 15–20% of that obtained in areas with more moderate temperatures (where the average summer temperatures are lower by approximately 7°C). The timing of flowering was also affected by temperature. In areas with more moderate temperatures, flushes of flowers appeared in *Hylocereus* species from May to November and in *S. megalanthus* from September to December. In the hotter areas, flowering of both genera was restricted mainly to the cool seasons, May and Oct./Nov. for *Hylocereus* species and Nov./Dec. for *S. megalanthus*. In physical terms, *H. undatus* showed the greatest sensitivity to the extremely high temperatures of the hot valleys: segments of stems at the surface of the shrubs turned brown and became liquefied. The spell of unusually high temperatures during the past summer in Beer-Sheva (4–5°C above the multiannual average) (Fig. 3, 4) resulted in extensive damage to *H. undatus*, but very small to the other species and nil to *Selenicereus megalanthus* (Fig 2). The damage becomes more intensive when combined with high light radiation (Fig 2). Raveh et al. (1995) also reported physiological damage to *Hylocereus undatus* when grown under 35/45°C night/day temperature regime. The results of these studies indicate that these climbing cacti should not be planted in extremely hot areas. *H. undatus* should be avoided, others may be manipulated with different shading regimes and/or other agrotechniques, the feasibility of which should be tested.

Reproductive biology. Studies on the reproductive biology of these cacti, including the work of Weiss et al. (1994a,b) on flowering and pollination, have previously been reviewed by us (Nerd and Mizrahi 1997). The results may be summarized as follows: Flowers are nocturnal and open only once. All species, with the exception of *S. megalanthus*, are self-incompatible and thus require cross pollination. Due to a lack of local pollinators in Israel, hand pollination is necessary to obtain fruits; this factor results in a tremendous increase in labor costs for the producers. All tested species were able to pollinate each other, and some pollinators produced bigger fruits than others. The fruits develop from both the ovary (pulp) and the receptacle that sur-

Table 1. Species of the crawling cacti *Hylocereus* and *Selenicereus* introduced by Ben-Gurion University of the Negev

	No. of clones	
Species	Introduced	Grown commercially
H. costaricensis (Weber) Britton & Rose	1	
H. ocamponis (Salm-Dyck) Britton & Rose	1	
H. polyrhizus Weber	7	2
H. purpusii Weingart	1	
Hylocereus sp.	8	2
H. undatus (Haworth) Briton & Rose	27	3
S. megalanthus (Schum.) Britton & Rose	37	6
Total	82	13

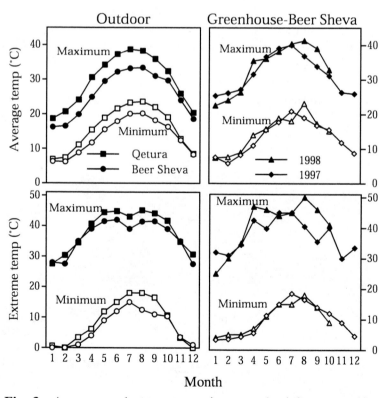

Fig. 3. Average and extreme maximum and minimum outdoor temperatures in Beer-Sheva and Qetura 1994–1997 and greenhouse temperatures in Beer-Sheva for 1997 and 1998.

rounds the ovary (peel). The weight of the fruit correlates with the number of seeds (Weiss et al. 1994; Nerd and Mizrahi 1997), and with proper pollination the weight of *Hylocereus* fruits can reach about 800 g and that of *S. megalanthus* fruits, about 350 g. The *Hylocereus* species flower in waves, each wave lasting about one week, and hence ripening also occurs in waves. The number of waves varies among the genera species and clones—from one to eight per season. This characteristic creates marketing problems: the fruits are available on the markets in short waves, whereas the buyers want them spread out evenly over longer periods. Some species produce flowers continuously, and hence fruits are available throughout the season from June to December (Fig. 5–7), which is highly desirable for the fresh-fruit market. We have already produced hybrids between this continuously flowering species (*Hylocereus* sp. clone 10487) and others with two- three waves of flowering (e.g., *H. undatus* clone 88-027), but these hybrids have not yet started to flower. Studies are underway in an attempt to solve this problem. All clones of *S. megalanthus* are tetraploids; they flower mainly in the autumn, and they are self-compatible (Weiss et al. 1994; Lichtenzveig 1997). The time elapsing between flowering and ripening is about 30 days for the *Hylocereus* species, and that for and *S. megalanthus* is 90 and 180 days for the early (late Sept.) and late flowers (late Nov.), respectively (Nerd and Mizrahi 1998). This means that ripe fruits of the *Hylocereus* species may be ready for marketing from late May to early Jan. while those of *S. megalanthus* are available from Jan. to mid-May.

Fruit ripening and post-harvest behavior. Some ripening characteristics have already been determined for a number of species, but studies on ripening and post-harvest behavior are still under way (Nerd and Mizrahi 1998, 1999). In general, the fruits are non-climacteric and are sensitive to chilling injury. They may be stored for 10 days at room temperature if the proper maturation stage had been reached before harvest.

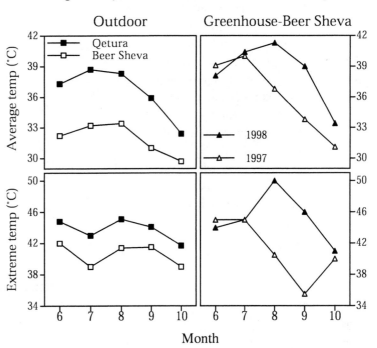

Fig. 4. Average and extreme maximum outdoor temperatures during the hot summer months in Beer-Sheva and Qetura 1994–1997 and greenhouse temperatures in Beer-Sheva for 1997 and 1998.

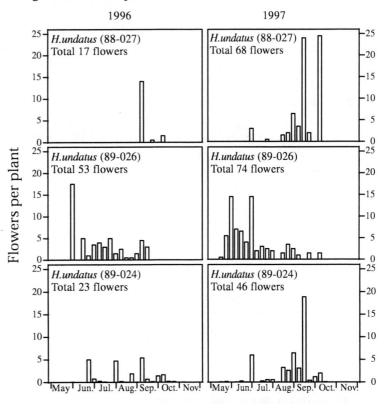

Fig. 5. Flowering waves of three clones of *Hylocereus undatus* during 1996 and 1997. The numbers are flowers per week.

361

Irrigation and fertilization. To date, no systematic research has been performed on irrigation and fertilization requirements. In the meantime, we recommend that the climbing cacti be irrigated with 150 mm water/year and fertilized with 35 ppm N from 23N-7P-23K fertilizer. Some farmers use their own formulas and may irrigate with as much as 250 mm/year. Some preliminary experiments have demonstrated large differences among species in response to water regimes. In a research project for undergraduate students in the Department of Life Sciences of Ben-Gurion University, Mr. A'ssa'el Ram found that *H. polyrhizus* exhibited the greatest tolerance to lack of irrigation (drought treatments) and *S. megalanthus* the least, with *H. undatus* falling between the two extremes. These findings were paralleled by the tolerance of the three species to high photon flux densities (Raveh et al. 1996, 1998), the tolerance being related to the xeromorphic traits of the species as follows: *H. polyrhizus* has wax layer over the "skin," the stomata are sunk into the epidermis, and the stem tissue contains a considerable volume of parenchyma; *H. undatus* has similar characteristics but lacks the wax layer; and *S. megalanthus* has no parenchyma, no sunken stomata and no wax layer and is thus most sensitive to water deprivation. This "drought experiment" was performed over a short period (three months in winter), and the findings should be confirmed in a long-term experiment, since these three species of cactus are perennials. An understanding of the effect of water regimes on fruiting and fruit quality is obviously of the utmost importance.

Pests and diseases. To date, no significant problems of pests or diseases have arisen. In areas with high relative humidity during the day (around 65%), some black knot may develop on fruits of *H. polyrhizus*, which excrete sugars (glucose, fructose, and sucrose) from the scales at the fruit tip. Ants are occasionally found on the fruits, fruits buds, and stems (which also excrete sugars), but no major damage has been found.

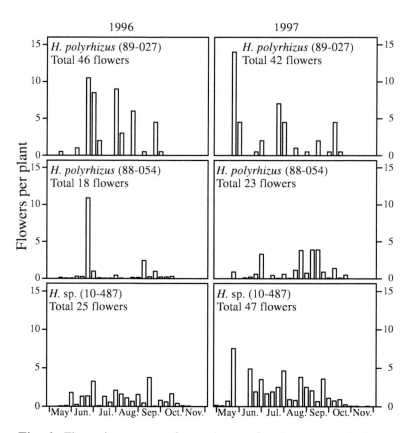

Fig. 6. Flowering waves of two clones of *Hylocereus polyrhizus* and clone 10487 of *Hylocereus* sp. during 1996 and 1997. The numbers are flowers per week.

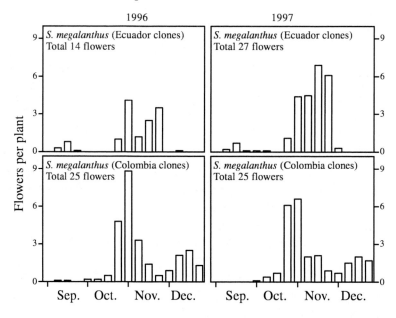

Fig. 7. Number of flowers per week of *Selenicereus megalanthus* during 1996 and 1997.

362

Commercialization and Marketing

From the moment we started our R&D project of introducing rare or wild fruit species to the Negev, we were told by the agricultural establishment that we were "playing with botany" in a project that had no real agricultural value. By 1993, however, we felt that we had accumulated sufficient know-how to start commercial evaluation of both input and output and to test the market in both Israel and Europe. Most of the farmers, government R&D agencies and AGREXCO were not interested in our new fruit crops (AGREXCO is an agency owned by the government and the farmers unions, which handles the export, under the brand label "Carmel," of more than 90% of Israel's agricultural produce). However, we did find three farmers who were fortunately prepared to risk their money and invest time and effort to try these unique hitherto unknown crops. The first three orchards (around 0.4 ha each) were planted in three different ecozones in the autumn of 1993. Among these growers was small private Israeli firm, Tropigarden, which also was the first to sell these fruits in Europe. Sellers of the fruits claim that it is the beauty of the fruits that sell them. In 1995, a few hundred fruits were sent to the local market and 400 kg were sold in Europe, both for prime prices. In 1996, 10 tonnes were sold mainly in Europe and some in the local market, again for prime prices. In 1997, about 25 tonnes were sold, mainly in Europe. In that year, AGREXCO eventually realized the potential of these fruits, and offered the farmers a few more shekels (at that time 3.2 shekels = 1$US) per box (5 kg) than Tropigarden and sold a few tonnes in Europe. This year (1998), the "giant" AGREXCO is competing with the small Tropigarden to market the fruit for export.

A number of problems have yet to be solved before the fruit can be commercially merchandised. One of the more important problems is that of nomenclature. At present, AGREXCO and Tropigarden are selling the fruits under two different names. Tropigarden sells the fruit under the name "Eden," with pitaya in small letters, while AGREXCO markets it under the name pitahaya, the Mexican term. As already mentioned, this fruit is also supplied to the world markets (Asia+Europe) from Vietnam under the name dragon fruit. We suggest that a different name be used for each species to prevent confusion on the market. In Israel, we market three species of *Hylocereus* under the same name—whether Eden or pitahaya—but specify red or white flesh on the box. Since these three species differ from each other in many characteristics, including taste, each should have its own name. In particular, we should avoid using the term pitaya, which is a common name for many genera and species of the Cactaceae (Pimienta–Barrios and Nobel 1994; Mizrahi et al. 1997) differing from each other as do, for example, the rosaceae fruits such as peach, cherry, apple, nectarine, and plum. We should keep the name "yellow pitaya" for *S. megalanthus,* which has already been available for some years in Europe, and dragon fruit for the white-fleshed *H. undatus.* For all other species, we should find new names. Even *H. polyrhizus* and *Hylocereus* sp. 10487, which both have red flesh, are not identical in taste or pigmentation: the pigment in 10487 is betalain (like that in beet-root), while the more purple pigment of *H. polyrhizus* has not yet been identified.

Other problems of introducing the new fruits to the market are "educating" the buying public and ensuring that the produce reaches the shops at the correct stage of ripening. For example, we have seen overripe fruits in the local markets without leaflets to explain the nature of the fruit and its nutritional value, or how to eat it and where to store it. The fact that such problems can hamper the introduction of a new product to the market has been confirmed by Ms. Frieda Caplan, who was instrumental in introducing kiwi fruit and many other exotics to the American market.

There are also horticultural problems that are still to be solved. Ways have to be found to manipulate the cacti to flower throughout the season, rather than in two or three waves, leaving most of the season without fruits—a major hurdle to successful marketing. At present, the crop is pollinated by hand, both the extraction of pollen and the cross pollination process itself being laborious and expensive. The development of pollen storage techniques and the subsequent creation of a pollen bank would solve the problem of the lack of proper pollen during the flowering period. In some species, taste should be improved. The long-term effects of environmental conditions, including irrigation and fertilization, have not yet been studied. Nothing is known about pruning and shaping of the plants, and the trellis system requires optimization, since such systems are quite expensive, the currently used one costing $12,800/acre. Breeding is also an important issue and is probably an easily achievable aim, since all existing clones are simply wild types awaiting genetic manipulation.

363

THE COLUMNAR CACTUS, *CEREUS PERUVIANUS*

Origin and Taxonomy

We introduced seven species of columnar cacti of three genera, *Pachycereus*, *Stenocereus*, and *Cereus*. Among these, *C. peruvianus* (Britton and Rose, 1963) appeared to be most promising in terms of its rapid growth and precocious early yielding (Nerd et al., 1993; Weiss et al., 1993). *C. peruvianus* is not known in the wild and is always found as a planted ornamental (Nerd and Mizrahi, 1997). Dr. Leiah Scheinvar, a cactus taxonomist from the National University of Mexico (UNAM), drew our attention to the similarity between this species and *C. jamacaru*, which is native to the north-east of Brazil. Dr. Scheinvar thought it likely that the two were identical, and our recent study in Brazil supported this idea. However, seedlings raised by us from seeds of *C. jamacaru* obtained from Brazil exhibited some morphological traits different from those of *C. peruvianus* (Fig. 8). Studies are now being performed by us to determine the relationship between the two species, since *C. jamacaru* may be an important candidate for domestication and/or for breeding *C. peruvianus*.

The first accession of *C. peruvianus* seeds was sent to us from Camarillo (southern California) by Mr. Ron Kadish, who collected the seeds from private gardens. *C. peruvianus* has already attracted attention in the US as a potential fruit-crop. The species, known as apple cactus, is mentioned in the excellent book of the late J. Morton, entitled *Fruits of Warm Climates* (Morton 1987). However, to date, the only research and development on this species is that performed in Israel by our group (Mizrahi et al. 1997), as described below.

Horticulture

We do not intend to reiterate our findings published in the literature, but rather to summarize them together with unpublished scientific observations and information contained in student dissertations (Nerd et al. 1993; Weiss et al. 1993, 1994a,b; Wang 1997). The plants are being grown outdoors in all the ecozones tested in our introduction program (Nerd et al. 1993; Weiss et al. 1993).

C. peruvianus is a precocious yielder from an early age—three to five years from seeds and two to three years from cuttings. The flower is nocturnal, and since it is self-incompatible, requires cross pollination. Clones should be mixed together in the orchard to guarantee pollination and fruit set. Pollination is performed by the honey bee *Aphis mellifera*, which is active during the day-time in the early and late hours of the flower opening. Low temperatures of –6 to –7°C resulted in significant damage to the plants, and these temperatures can thus be considered as minimum low for cultivation. We did not notice any damage from temperatures as high as 45°C. Water use is low, being 150 mm/year, as expected from cacti. The species is sensitive to salinity particularly when Na and Cl are the main salinity ions. Higher than normal concentrations of Ca, Mg and sulfates cause long-term damage, which precludes cultivation of *C. peruvianus* under salinity of 4 dS/m. The effect of low salinity (2.5 dS/m) on this crop is currently being tested.

Approximately 30 days elapse from anthesis to ripening. Fruits have to be harvested for marketing at the stage at which the peel becomes smooth and fully colored. Fruits harvested prior to this stage (a very common mistake) have an inferior taste and do not ripen properly in storage. The fruit is nonclimacteric and can be stored for 14 days provided that it is harvested at the proper stage of ripening (Wang 1997).

Breeding is now in progress from seeds of known parents. We have several thousand seedlings in our orchards from which to select the future cultivars. Orchards for this species are much cheaper to establish and maintain than those for the crawling cacti described above. Needless to say, the findings described above are only the beginning of the R&D required to bring this species to the stage at which it will be a fully exploited crop.

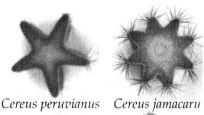

Cereus peruvianus   *Cereus jamacaru*

Fig. 8. *Cereus peruvianus* and *Cereus jamacaru*.

Commercialization and Marketing

From 350 seedlings, we have selected eight clones that seemed to us suitable for cultivation. They were planted in 1993 in two ecozones in the Negev, i.e., Qetura, which has a harsh hot climate with saline water of 4 dS/m, and in the Besor, with fresh water of 1 dS/m and moderate temperatures (Nerd et al. 1993). To date, the orchard at Qetura has not yielded significant quantities (probably due to salinity), although export has already started from the Besor region. In 1995, the first fruits were obtained in small quantities. In 1996, a significant yield was obtained, but the farmers did not harvest the fruit, because "nobody knows what is this fruit all about" and were about to uproot the "useless orchard." Not until mid-Sept. (half way through the season), did the farmers realize the potential of the crop, after the first boxes had been sold by Mrs. Dovrat Schwab to local restaurants, expensive hotels and exotic shops. She sold the fruit for 32 shekels/kg and gave the farmers half the proceeds (at that time 3.0 shekels = 1$US). This convinced the farmers that the crop might indeed be worthwhile. In 1997, AGREXCO started to sell this fruit under the name Koubo (to avoid the name pitaya) in Europe, where it was accepted very well. In 1997, 4 t were sold, both in Europe and on the local market, with the demand far exceeding the available supply. In Europe, the fruit is marketed with an excellent leaflet, giving information about the fruit and its uses. In the local market, some growers provide the leaflet but others do not. We found fruits in the supermarkets that had been harvested well before they reached the proper maturity stage and without accompanying leaflets. This can hamper the promotion of this new fruit. We hope that the Fruit Growers Association will take over introduction of the fruit to the local market proper way in a way that will ensure market penetration. We are sure that the export market for which AGREXCO is promoting the fruit for the first time as real new crop, will give the proper return to farmers and encourage further planting.

SUMMARY AND CONCLUSIONS

The Negev Desert is the only place in Israel in which further development of the agricultural industry is still possible. Due to market competition, very high input prices and scarcity of water, conventional crops cannot ensure the future of agricultural development in Israel. Against this background, a variety of wild and rare fruit tree species were introduced by our group into the Negev Desert in a number of locations that differ in their environmental conditions. Among these species, two cacti—one climbing and the other columnar—have already made their way on to the European market. Two genera of climbing cacti from tropical and subtropical shady habitats—*Selenicereus* and *Hylocereus*—were introduced and are being grown either in greenhouses to prevent exposure to subfreezing temperatures or in shade-houses to prevent damage by high photon flux densities. These cacti are *S. megalanthus* (known in Colombia as yellow pitaya) and *H. undatus, H. polyrhizus,* and an unidentified species *Hylocereus*, all known as red pitayas. The fruit of the latter three species was sold, under the name of Eden fruit, in local Israeli markets and exported to Europe for the first time in 1996. Total yields exported were 10 and 25 t in 1996 and 1997, respectively, with the fruit commanding the highest prices ever obtained from fruits exported from Israel. To enable efficient production, studies of all aspects of horticulture, including agromanagement and breeding, being carried out at Ben-Gurion University. *Cereus peruvianus,* a columnar cactus grown outdoors, went through a similar process of domestication. This fruit was sold for the first time in Europe in 1997 under the name Koubo. The farm-gate-price of 4 US$/kg was far beyond that commanded by any common fruit crop exported from Israel. The names Eden and Koubo were given the new fruits to avoid the use of the name pitaya, which covers many species and genera. The success of these new fruits supports the hypothesis that new crops can serve as a remedy to the troubled Israeli export market and that a viable agricultural industry is indeed feasible under the harsh conditions of the Negev Desert.

REFERENCES

Barthlott, W. and D.R. Hunt. 1993. Cactaceae. In: K. Kubitzki, J.G. Rohwer, and V. Bittrich. (eds), The families and genera of vascular plants Vol II Flowering plants. Dicotyledons. Springer–Verlag, Berlin.

Britton N.L. and J.N. Rose. 1963. The Cactaceae: Description and illustrations of plants of the Cactus family, Vols. 1 and 2. Dover, New York.

Hunt, D. 1989. Notes on *Selenicereus* (A. Berger) Britton & Rose and *Aporocactus* Lemaire (Cactaceae-Hylocereinae). Bradleya 7:89-96.

Lichtenzveig, J. 1997. Occurrence of self-incompatibility and semi-sterility in climbing cacti of the genera *Hylocereus* and *Selenicereus*. M.Sc. Thesis. Department of Life Sciences, Ben–Gurion University of the Negev, Beer-Sheva. Israel

Mizrahi, Y. and A. Nerd. 1996. New crops as a possible solution to the troubled Israeli export market. p. 56–64. In: J. Janick (ed.), Progress in new crops. ASHS Press, Alexandria, VA.

Mizrahi, Y., A. Nerd, and P.S. Nobel. 1997. Cacti as crops. Hort. Rev. 18:291–320.

Morton, J.F. 1987. Cactaceae. p. 347–348. In: Fruits of warm climates. J.F. Morton, Miami, FL.

Nerd, A., J.A. Aronson, and Y. Mizrahi. 1990. Introduction and domestication of rare fruits and nut trees for desert areas. p. 355–363. In: J. Janick and J.E. Simon (eds.), Advances in new crops. Timber Press, Portland, OR.

Nerd, A., E. Raveh, and Y. Mizrahi. 1993. Adaptation of five columnar cactus species to various conditions in the Negev Desert of Israel. Econ. Bot. 43:31–41

Nerd, A. and Y. Mizrahi. 1997. Reproductive biology of cactus fruit crops. Hort. Rev. 18:321–346.

Nerd, A. and Y. Mizrahi. 1998. Fruit development and ripening in yellow pitaya. J. Am. Soc. Hort. Sci. 123.:560–562.

Nerd, A. and Y. Mizrahi. 1999. The effect of the ripening stage on fruit quality after storage of yellow pitaya. Postharvest Biol. Tech. 15(2):99–105.

Nobel, P.S. 1988. Environmental biology of agavi and cacti. Cambridge Univ., Cambridge, MA.

Nobel, P.S. 1994. Remarkable agaves and cacti. Oxford Univ. Press, New York.

Pimienta-Barrios, E. and P.S. Nobel. 1994. Pitaya (*Stenocereus* spp., Cactaceae): An ancient and modern fruit crop of Mexico. Econ. Bot. 48:76–83.

Raveh, E., J. Weiss, A. Nerd, and Y. Mizrahi. 1993. Pitayas (genus *Hylocereus*) new fruit crop for the Negev Desert of Israel. p. 491–495. In: J. Janick and J.E. Simon, (eds.), New crops, Wiley, New York.

Raveh, E., M. Gersani, and P.S. Nobel. 1995. CO_2 uptake and fluorescence response for a shade-tolerant cactus *Hylocereus undatus* under current and double CO_2 concentration. Physiol. Plant. 93(3):505–511.

Raveh, E., A. Nerd, and Y. Mizrahi. 1996. Responses of climbing cacti to different levels of shade and to carbon dioxide enrichment. Acta Hort. 434: 271–278.

Raveh, E., A. Nerd, and Y. Mizrahi. 1998. Responses of two hemiepiphytic fruit-crop cacti to different degrees of shade. Scientia Hort. 73(2,3):151–164.

Wang, X. 1997. Biology of ripening of apple cactus [*Cereus peruvianus* (L.) Miller]. Ph.D. Thesis, Ben-Gurion Univ. of the Negev, Beer-Sheva, Israel.

Weiss, J., A. Nerd, and Y. Mizrahi. 1993. Development of the apple cactus (*Cereus peruvianus*) as a new crop to the Negev Desert of Israel. P. 486–491. In: J. Janick and J.E. Simon (eds.), New crops. Wiley, New York.

Weiss, J., A. Nerd, and Y. Mizrahi. 1994a. Flowering and pollination requirements in *Cereus peruvianus* cultivated in Israel. Israel J. Plant Sci. 42:149–158.

Weiss, J., A. Nerd, and Y. Mizrahi. 1994b. Flowering behavior and pollination requirements in climbing cacti with fruit crop potential. HortScience 29:1487–1492.

Weiss, J., L. Scheinvar, and Y. Mizrahi. 1995. *Selenicereus megalanthus* (the yellow pitaya), a climbing cactus from Colombia. Cactus Succulent J. 67:280–283.

Low Input Agricultural Systems Based on Cactus Pear for Subtropical Semiarid Environments

Candelario Mondragon Jacobo

The biological productivity of semiarid lands is typically restricted by the amount of rain. Agricultural systems relying in C4 and C3 plants are present throughout the world. In Central Mexico maize and dry beans have been the cornerstones for centuries. Some CAM plants can provide an opportunity to improve the productivity in these regions, due to their higher water use efficiency. According to Nobel (1988), the water use efficiency, averaged over a season, in mmol CO_2 per mol H_2O typically is 1.0 to 1.5 for C3 plants, 2 to 3 for C4 plants, and 4 to 10 for CAM plants. Moreover, certain irrigated CAM plants can have an annual productivity that exceeds that of nearly all cultivated C3 and C4 species. In particular, *Opuntia amychlaea* Tenore and *O. ficus-indica* Mill. can produce annual aboveground yields of at least 45 t/ha. The inclusion of cactus pear (*Opuntia* spp. Cactaceae), a perennial crop plant for fruit, vegetable and/or forage production can enhance sustainability in the long term, however its inception as a sole crop represents an important financial drain to the typical farmer of this regions.

With the idea of improving overall productivity several combinations of cactus pear and annual crops have been attempted. Annual crops represent competition for the cactus pear both below and aboveground. Crops which have a low canopy planted in rows and are short lived could provide less competition for the perennial cactus pear. On the contrary dense tall crops can cripple cactus pear growth and yield, because it is very sensitive to shading in the first two years. However growing cactus pear alone produces no income before the first commercial harvest is obtained.

Mixed cropping has been reported as an efficient practice to improve overall yield and reduce risk. The combination of plants which have different morphology, depending on different soil depths and exploiting several aboveground strata allow for a more effective tapping of the resources (Cannell et al. 1996).

Cactus pear can be grown in dense planting layouts of up to 66,000 plants/ha under rainfed conditions. It also tolerates heavy and repeated pruning when managed as a perennial shrub, producing tender pads which can be consumed as a vegetable. Under restrictive moisture environments the last flush of cladodes can be allowed to mature, collected, and used as a survival forage for livestock during the dry winter season.

THE REGION

The North of Guanajuato Research Station (Instituto Nacional de Investigaciones Forestales y Agropecuarias, Mexico) is located in Central Mexico at the Southern tip of the Chihuahua–Sonoran desert 21°N, characterized for its limitative rainfall pattern, an average of 548±112 mm a year, shallow sandy-clay soils (less than 40 cm), the traditional cultivation of corn and beans, and the intensive use of overgrazed natural pastureland.

The presence of a few heavy rains followed by numerous events of short duration and low intensity is typical of the Northern semiarid areas of the Mexico. Fig. 1 presents the rainfall recorded during 1990 and 1991 years in which we conducted the experiments on multiple cropping.

Water Harvesting

Intelligent use of rainfall is a prerequisite for successful farming in semiarid areas. Rainwater should be handled in such a way that at least the water that falls upon the plot of interest is fully kept in the soil profile.

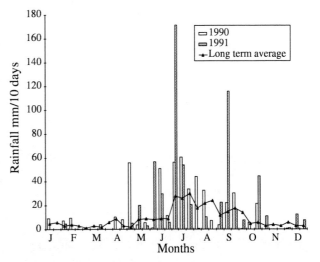

Fig. 1. Precipitation pattern on the experimental site.

More sophisticated systems, but also more expensive, can be designed preparing runoff collection and storage areas. Fig. 2 shows a water harvesting system chosen for its simplicity for our experiments and suitable for farmers with limited financial resources. The land is plowed at the end of winter and "microcatchments" are constructed by means of a simple mechanical device attached to the small tractor (60–80 HP) common in the area. Water is impounded allowing its penetration into the soil in order to be later used by the annual crop. The actual cost of this practice is about $30/ha, it ensures an effective intercrop germination, a better control of weeds and higher response to fertilizer, therefore higher yields of both the perennial and the annual crop.

Multiple Cropping of Annuals and Cactus Pear

The water needs for cactus pear cultivation are satisfied by 500–600 mm of annual rainfall (Pimienta 1990). The distribution of the rainfall will determine the possibility of mixed cropping.

A fully mature cactus pear plant is expected to take up a circular area measuring anywhere from 2 to 4 m in diameter depending upon the cultivar and crop management. A typical planting layout will have rows 4 to 7 m apart and 2 to 4 m in between plants. Plants will need from 5 to 6 years to reach full size and yield. In the first year the cactus plant will produce a single set of pads (in wet years it can produce up to two) which does not interfere with the growth or the operations associated with the field crops. Beyond the second year the plant will cover a larger area, reducing the number of rows that can be interplanted and interfering with operations, spiny cultivars such as 'Cristalina' which was used in the experiment pose special difficulties. As all orchards are started from small vegetative pieces (either single or double pads) there will be some temporary empty space in between rows. In mechanized orchards pruning is considered a regular practice to get a clear space in which machinery can transit. A secondary advantage of the mixed crop systems is that better weed control is accomplished. Controlling weeds on the annual crop, either mechanically or with herbicides, automatically reduces competition for the cactus pear, as well. This is important considering that in practice the commercial orchards in Mexico are left almost abandoned during the juvenile stage of the crop.

In 1990 a new cactus pear orchard was planted in a 4 × 3 m layout using the native greenish pulp cultivar 'Cristalina' (Mondragon and Perez 1996). Maize, dry beans, and forage sorghum were planted in between rows on top of the furrows in late May. Planting densities were 40,000, 100,000 and 60,000 plants/ha for these row crops. Four rows 85 cm apart were placed in between the cactus plants. Small cereals and canola (Fig. 3) were broadcasted on 3.5 m of the empty space using early cultivars of all the crops (100–120 days): 'Cerro Prieto' barley, 'BN-00' canola (*Brassica napus*), 'VS-201' maize, and 'FMBajio' and 'Canario' dry beans. For cereals and canola 120 and 6 kg of seed/ha respectively, were planted and 25N–20P–0K fertilizer was applied simultaneously, with an additional 25 units of N fertilizer applied 45 days later. Weeds were controlled either mechanically for row crops or by herbicides in cereals. The additional cost for having an intercrop varied according to the crop. Small cereals represented an investment of US $185/ha whereas for row crops $207/ha were needed.

Agronomic Performance of Annual Crops

In 1990 the site received 497 mm of total rain, slightly below the long term average. In 1991 there was 625 mm, 14% above average. Both years were characterized by the presence of a single heavy rainfall (55 and 56 mm in less than an hour) early in the season which allowed the collection and storage of water in the soil profile used for germination by the annual crops.

Yields (Table 1) were comparable to those observed in the region for these crops after adjusting for the area taken up by the cactus pear. In the second year all the crops had reduced yields which ranged from 30% for wheat to 43% for dry beans,

Fig. 2. Water harvesting system. Two months old cactus pear orchard with "microcatchments" after a heavy rain (45 mm) early in the rainy season in Northern Guanajuato, Mexico.

Fig. 3. Canola and rye intercropped in a two year old cactus pear orchard under rainfed conditions.

Table 1. Performance of annual crops planted in between rows of a cactus pear orchard.

Annual crop	1990 yield (kg/ha)	1991 yield (kg/ha)	1st year net income (US$/ha)
Wheat	1354	947	-50
Canola (BN-00)	480	392	-147
Barley (Cerro Prieto)	872	--	-115
Barley + canola	564 + 294	376 +110	-99
Foxtail millet	12364	--	-2
Maize (VS-201)	936	--	-136
Dry beans (Canario)^z	439	250	+10
Dry beans (FMBajio)^y	1337	426	+99
Sorghum (forage)	17,608	--	+54

^{z}110–130 days to maturity.
^{y}90–110 days to maturity.

presumably attributed to the competition effects of the cactus pear. Foxtail millet and forage sorghum are unknown in the area but the yields observed were encouraging considering the climatic limitations, additional difficulties for these potential crops are their acceptability by ranchers which in turn will define their adoption in the market. The dry bean 'FMBajio' performed better than 'Canario' due to its longer flowering season with more than one flush of flowers.

An economic analysis indicated that none of the cereals produced enough to cover expenses, a fairly common situation in the area for all the traditional rainfed crops, due to the low prices usually obtained from these grains in a government controlled market. The best option is dry beans which have a steady market as a result of their status of staple grain for the Mexican population. Both cultivars were associated with positive net profits, but 'FMBajio' is a favorite and commands a higher price than 'Canario', US $0.90/kg vs $0.60/kg.

Growth of the cactus pear plant reduced the plowable area about 30% in the second year and planted only two rows of dry beans and half of the original area with some cereals. The cost of cultivation and yields were also diminished by approximately 60%. According to our data, only high value crops such as dry beans are a sound alternative for two consecutive years. Overall they represented less competition for the perennial crop which is associated with nonsignificant effects on fruit yield.

Effects of Intercropping on Yield of the First Commercial Harvest of Cactus Pear

Intercropping in general reduced the yield of cactus pear from 19% to 176% (Table 2). The effect can be attributed to competition for light. Shaded plants tend to grow slower and had a lower number of new buds. Small cereals which are planted in dense stands, grow faster and closer to the cactus pads, had the most significant effect. The lowest loss, 19.1% was observed in dry beans probably due to its small size and short growing period.

Intensive Production of Mature Cactus Pear Pads for Forage

Spineless cactus pear were planted on raised broadbeds 150 cm wide (Fig. 4). Year old cuttings were obtained from mature plants. Before planting they are placed in a shady and dry spot for 2–4 weeks to pro-

369

Table 2. Effect of intercropping on first commercial harvest (1995) of cactus pear.

Intercrop	Fruit yield (kg/ha)	Gross income ($/ha)	Income difference[z] (%)
Barley	1684	69	-177
Canola	2000	82	-149
Barley + canola	2438	122	-0.6
Foxtail hay	2221	91	-34
Dry beans (FMBajio)	2517	102	-19
Maize	1771	68	-80
Cactus pear (microcatchments, no irrigation)	3503	175	+44
Cactus pear (limited irrigation, no microcatchments)	5835	285	+233
Cactus pear (no microcatchments, no irrigation)	2982	122	--

[z]Compared to rainfed cactus pear as a sole crop.

Fig. 4. Intensive production of cactus pear pads (cv. Seleccion Pabellon) for vegetable or forage on raised broadbeds.

mote wound suberization and partial dehydration. Cladodes are planted on a 40 × 30 cm layout (up to 66,000 pads/ha can be used depending upon the space between broadbeds), burying the cladode halfway into the ground. On low sloped land microcatchments can be created in the ditches between beds at regular distances to promote even distribution of rain. Response to manure and fertilizers have been reported, for our location a maximum 6 tonnes of manure and 60 kg/ha N and 16 kg/ha P (on annual basis) can be used for rainfed conditions. A residual effect of manure, up to three years has also been observed.

Supplementary irrigation greatly increase yields, and therefore higher dosages of manure and fertilizers can be used. Up to 90 kg N and 16 kg P/ha fertilizer plus 9 t/ha of dry manure, assuming that additional 30 cm of water are applied throughout the year, as needed, can be applied. The plants are maintained as bushy perennials, keeping only two to three pads per plant in the second story. The plant responds to pruning during the wet season, quickly producing new shoots. Depending on the length of the rainy season up to two harvests of tender pads suitable for human consumption as a vegetable are possible, leaving the third flush to ripen for forage. Mature pads can be harvested at the end of the season (October through December), and stored in a dry and shady location for later use. The pads can be stacked on piles 30–40 cm high, or placed in rows, they can keep their appearance and palatability for up to six months, provided that they are flipped over one or two times to avoid rooting. The best results are obtained by selecting a spot near to the farm household, which reduces the damage by wild rodents. In our trials yields of fresh pads ranged from 43 t/ha obtained with the control to 53 t/ha associated to the fertilized treatment.

This planting layout has been adopted in Central Mexico to produce vegetable cactus pear. It is labor intensive because the tender shoots are collected at least once a week. It is practiced in small plots (0.5 ha). However, if adapted for forage production the need for labor is greatly reduced because the mature pads are to be collected in a single harvest. It has been estimated that 1 ha of this system could cost up to $4600, 95% attributed to the cost of the planting stock.

The initial cost of planting can be greatly reduced if the planting stock is produced on site. This requires the acquisition of a fraction of the total needed to establish a nursery which in turn will provide for the needs of the commercial plantation. The size of the nursery will depend on the area and the timeframe of the operation for forage. Calculation is done considering that each plant will produce at least 3 pads per season, or more if it is irrigated and fertilized.

In the second year of the commercial planting the costs are minimal, fertilization and labor for harvest. The perennial nature of the plant and their tolerance for continuous pruning allows the system to be considered a long term investment. There are commercial as well as experimental plots in Mexico which have been under continuous cultivation for more than ten years. Considering a minimum price of US $30/t of fresh pads, up to 28% of the cost of planting can be recovered in the first year for the cheapest option, that is using no fertilizer. Some of advantages of the system include low maintenance, continuous supply of fresh animal forage during the dry season, and easy storage.

Grassland and Cactus Pear Association

Combining two perennials in semiarid conditions in the same piece of land does not seem to be a sound alternative. However, under natural conditions the association of xerophyte shrub vegetation, grasses, and cacti form successful communities. Both are perennial plants which are unaffected by long dry spells and can grow quickly after a heavy rain, which make them a good alternative for areas of scant rainfall pattern.

We designed a system in which the empty space among cactus pear rows was occupied by buffel grass (*Cenchrus ciliaris*), a native perennial grass (Fig. 5). In the first three years the grass was planted from seed and allowed to grow without any disturbance. After the prairie was fully established the orchard was laid out. Minimal perturbation of the grass canopy was accomplished by a subsoiler which opened a ditch just wide enough to place the cactus pear pads. Loosening the soil with this tool also allowed better establishment of the cactus pear crop. The pastureland has been managed according to the following criteria: in "good" years (with rains early in the season) the grass cover can be expected to produce a single harvest of fresh grass. The grass is clipped as soon as it reaches 10–20% of flowering and hay bales can be prepared to be used at will.

Fig. 5. One year old cactus pear orchard on a induced pastureland. The empty space among cactus pear rows is occupied by buffel grass (*Cenchrus ciliaris*) a perennial grass.

The time left at the end of the season allows the plant to recover and accumulate reserves for the next dry season. In "bad" years no harvest of the grass is obtained. The pastureland is left undisturbed, the grass flowers and produces seed. Natural dehiscence of the grass spikes insures reseeding of the pastureland maintaining the vegetative cover.

Additional advantages of this system are that dicot weeds are reduced due to the cyclic clipping which normally is done before they reach maturity, thus preventing seed formation. If machinery to harvest and preserve grass as a hay is not available direct rotational grazing can be an alternative. The yield of grass does not have economic value, due to the availability of other feedstuff but it has possibilities for a complex system with limited sheep grazing. Additionally, its ecological value is high because the area has been subjected to overgrazing and is prone to eolic erosion.

CONCLUSIONS

Intercropping decisions are dependent on factors such as economic value and potential on-farm utilization of crop intercrop. Small cereals are not favored because of low prices, but maize and dry beans can be readily utilized and marketed because they are the staple food of the Mexican population. Dry beans can be also considered a cash crop due to the demand and steady high price (up to $1/kg depending on the cultivar). They represent the top choice to include in this system. Cultivars with short plant type and extended flowering such as 'FMBajio' produced better results on these regions with scant rainfall.

The production of mature cactus pear pads for forage is best suited for an integrated system in which the forage will be utilized in a dairy or beef production operation in those locations in which traditional forages are not readily available. Intensive utilization of wild populations of cactus pear is routine in Northern Mexico (Dela Cruz 1994); they are over exploited and left to recover without human intervention leading to ecological devastation. The total cost of the cactus pear crop can be reduced if a small nursery is prepared in advance. The inception of on farm production systems of cactus pear could alleviate the pressure on natural populations. The combination of pasture and fruit production system is an example of complementation of two perennial plants exploiting different strata. Both species endure drought and are flexible in their water needs. Integrating small ruminants to use the grass on site may improve the benefits.

REFERENCES

Cannell R.G.M., M. Van Noordwijk, and C.K. Ong. 1996. The central agroforestry hypothesis: the trees must acquire resources that the crop would not otherwise acquire. Agroforestry Systems 34:27–31.

Dela Cruz, C.A. 1994. Prickly pear cactus for forage in Mexico. In: Proc. 5th Annual Texas Prickly Pear Council. Kingsville, TX.

Mizrahi, Y., A. Nerd, and P.S. Nobel. 1997. Cacti as crops. Hort. Rev. 18:291–320.

Mondragon-Jacobo, C. and S. Perez-Gonzalez. 1996. Native cultivars of cactus pear in México. p. 446–450. In: J. Janick (ed.), Progress in new crops. ASHS Press, Arlington, VA.

Pimienta B.E. 1990. El nopal tunero. Coleccion Tiempos de Ciencia. Universidad de Guadalajara, Mexico.

VEGETABLES

Advances in New Alliums

Michael J. Havey

The genus *Allium* L., Amaryllidaceae includes numerous economically important vegetables used primarily for their unique flavors. Most people recognize the main vegetable alliums, such as bulb onion and shallot (*A. cepa*), garlic (*A. sativum*), leek (*A. ampeloprasum* var. *porrum*), and chive (*A. schoenoprasum*). However, there are numerous cultivated vegetable alliums of more regional importance, including kurrat (*A. ampeloprasum* var. *kurrat*) which is eaten as a pickled leaf primarily in Egypt, elephant garlic (*A. ampeloprasum* var. holmense), rakkyo (*A. chinense*), and Chinese chives (*A. tuberosum*) (Jones and Mann 1963). Other minor use *Allium* species were thoroughly listed by Fenwick and Hanley (1985).

Alliums are also cultivated as ornamentals. Well known to gardeners are *A. gigantheum, A. aflatunense*, and *A. caesium*, which are prized for their large umbels and many brilliantly colored flowers. Wild and less well-known alliums offer potential as perennial ornamentals, including *A. unifolium, A. moly*, or *A. schubertii*. Japanese researchers at the Hokkaido Agricultural Experiment Station have collected these and other unique alliums towards the goal of developing new and beautiful ornamentals.

FERTILITY IN GARLIC

Garlic is an obligate apomictic plant and is primarily propagated asexually by cloves (Jones and Mann 1963). Flowering garlics can also be propagated by small bulbils that form in the umbel. One of the most exciting *Allium* research developments over the last 10 years is the identification and selection of fertile garlic plants. Etoh (1986) described male-fertile garlic plants from Soviet Central Asia. This stimulated research on the environmental conditions conducive to flower induction and the production of true seed in garlic (Pooler and Simon 1994). Of course, plants grown from this first generation of seed were weak and commercially unacceptable. Private companies producing dehydrated onion and garlic products have used these results to produce copious amounts of true garlic seed. The economic potential of this seed is enormous. Plants grown from true seed are either virus-free or have significantly reduced virus titers, eliminating the need for meristem culture to cleanse the plant of viruses (Walkey et al. 1987). The phenotypic diversity among garlic strains is well known (Engeland 1991). It is not known how much of this diversity is due to mutations or rearrangements in the garlic genome versus independent domestication of garlic strains from progenitor populations. Imagine the phenotypic diversity that could be generated from crossing among garlic strains and selection for unique appearance or flavor, higher solids for dehydration, etc. Over the long term, efficient seed production could reduce the time required to produce seed cloves from individual garlic plants for commercial level production. On average the increase of garlic per generation is approximately seven, meaning that one garlic clove will produce seven cloves of sufficient size to produce the next generation. If hybrid garlic could be sexually produced from true seed, the number of phenotypically uniform plants could be increased to reduce the time to commercial production. Over my lifetime, I predict that sexually produced garlic will be the most exciting new commercially used allium.

FLAVOR STUDIES

Another major new product will be alliums with defined flavor and health-enhancing attributes. The rest of this manuscript will concentrate on the genetic and environmental factors that influence the flavor of onion, generally with selection or environmental manipulations towards low pungency, and the antithrombotic and anticarcinogenic potential of onion and garlic.

The alliums produce unique flavors savored by almost all of the world's cultures. These flavor compounds are formed when the enzyme alliinase acts upon the flavor precursors, alk(en)yl-L-cysteine-sulfoxides (ACSOs), to produce thiosulfinates (Block 1992). Alliinase is located in the vacuoles and is released when the alliums are cut, damaged, or bruised (Lancaster and Collin 1981). The enzyme acts quickly upon the cytoplasmic ASCOs to produce a suite of thiosulfinates. Different ASCOs exist among *Allium* species and different amounts of ASCOs may exist among onion cultivars (Lancaster and Kelley 1983; Lancaster and Boland 1990; Randle et al. 1995). The total flavor experience is the result of the action of alliinase on the different amounts of individual flavor precursors (Lancaster and Boland 1990). However, little is known about which specific ASCOs, or combinations there of, give specific flavors.

Onion Pungency

The relative pungency of onion has both genetic and environmental components. The onion is very efficient to uptake and sequester sulfur from the soil. The acquisition of sulfur and its distribution to various metabolic pathways may have been an important survival mechanism for the alliums; grazing animals maybe learned to avoid the alliums because of their strong taste. Production of onion in a low sulfur environment reduces the pungency of onion (Platenius and Knott 1941; Freeman and Mossadeghi 1970). A case in point is the Vidalia onion, produced on extremely low sulfur soils in 13 counties and parts of 7 others in the state of Georgia, US. The federal marketing order mandates that the Vidalia onion must be a Granex-type onion, meaning that it must have the phenotype of Granex, a hybrid released in 1952 by the USDA and the Texas Agricultural Experiment Station. The distinctive flat shape of the Vidalia onion comes from the Yellow Bermuda 986A female used to produce the hybrid Granex seed. Granex-type cultivars are generally low in pungency, but reach extremely low levels grown under the low sulfur soils of the Vidalia valley.

A genetic component for low (or high) pungency also exists (Lin et al. 1995; Simon 1995; Wall et al. 1996). Onion breeders routinely measure enzymatically derived pyruvate as an indicator of onion pungency. When the enzyme alliinase converts the ACSOs to the thiosulfinates, a byproduct is pyruvate (Schwimmer and Weston 1961). The amount of pyruvate produced is directly related to the onion pungency, as determined by taste panels (Schwimmer and Guadagni 1962; Wall and Corgan 1992). Higher pyruvate production indicates higher pungency, either due to greater concentrations of the substrate ACSOs, greater activity or concentration of alliinase, or both. Selection for less enzymatically derived pyruvate reduces the pungency of the onion (Wall et al. 1996). Another potential trait for selection would be the sugar content of onion; however, little work has been directed towards the genetic bases of sugars in the onion bulb.

A very interesting and potentially economically important molecular manipulation of onion is being attempted by researchers at Crop and Food Research, Lincoln, New Zealand. Drs. Jane Lancaster and Colin Eady, among others, are attempting to use antisense technology to manipulate onion pungency. The gene encoding alliinase has been cloned (Damme et al. 1992). The New Zealand group is trying to introduce a backward (antisense) copy of this gene into the onion genome. Antisense genes are known to reduce the expression of the wild-type (sense) copy of the gene (Eguchi et al. 1991). In onion, antisense of the alliinase gene should reduce the amount of functional alliinase present in the plant, but hopefully not eliminate all alliinase and allow the production of some onion flavor. The result could be low pungency onions that possess important production characteristics such as firmness and long dormancy. However if the same thiosulfinates were responsible for flavor and antiplatelet activity (see below), antisense alliinase could reduce both.

Antiplatelet Activity of Onion

It is well known that the alliums possess unique thiosulfinates that condition antithrombotic benefits, including antioxidant activity (Yang et al. 1993; Yin and Cheng 1998), reduced serum cholesterol (Bakhsh and Khan 1990), and enhance in vitro antiplatelet activity (Ariga et al. 1981; Block et al. 1984; Lawson et al. 1992; Goldman et al. 1995; Bordia et al. 1996). This latter effect is important for cardiovascular heath by reducing the probability that platelets aggregate in the blood, a major cause of heart attacks and strokes. The antiplatelet activity of onion has been primarily measured in vitro using platelet aggregometry. Correlations between measurements in vitro and in vivo have not been reported.

There is a significant environmental effect on the antiplatelet activity of onion. Goldman et al. (1996) demonstrated that growing onions in the presence of high sulfur increased both the pungency and antiplatelet activity of the bulb. This means that high sulfur soils or medium should produce highly pungent bulbs that tend to show greater antiplatelet activity. Because most people will not eat highly pungent onions fresh, it becomes an important research goal to determine which compounds condition each trait and whether the high pungency can be separated from greater antiplatelet activity.

Garlic-Onion Hybrids

An interesting combination of flavors and health-enhancing compounds may be available from a sexual hybrid between garlic and onion. Ohsumi et al. (1993) generated this interspecific hybrid using garlic as the female and pollinating with bulb onion. The ovules were extracted and cultured to rescue the embryo. The interspecific hybrid was intermediate in phenotype between garlic and onion. This plant possesses unique combinations of the flavor compounds from onion and garlic. Interestingly, combinations of thiosulfinates from various alliums may show more health-enhancing effects than those from individual alliums (Morimitsu et al. 1992). I expect that this garlic-onion hybrid will be asexually propagated because of extreme sterility. Nevertheless, the garlic-onion hybrid could represent a new commercially produced allium with potentially unique flavors and antiplatelet activities.

Flavonoids

Flavonoids are a second class of health-enhancing compound produced by the alliums. These compounds are complex carbohydrates that exist in many different forms (Crozier et al. 1997). The flavonoids tend to accumulate in the outer cell layers exposed to sunlight and protect the photosynthetic compounds from auto-oxidation. Flavonoids in the alliums have been shown to have significant anticarcinogenic and antithrombotic activities (Leighton et al. 1992). In onion, the primary flavonoid is quercitin (Price and Rhodes 1997). Dr. Leonard Pike and colleagues at Texas A&M University has studied the relative amounts of quercitin in onion bulbs. As expected, the highest amount of quercitin is present in the outer rings of the bulbs and decreases significantly towards the interior of the bulb. Red onions have greater amounts of quercitin than yellow; yellow onions possess greater amounts than white bulbs (Bilyk et al. 1984; Trammell and Peterson 1976; Patil and Pike 1995). Even though we discard the outer dried or partially dried scales of onion bulbs, red onions still possess potentially significant amounts of quercitin. Patil et al. (1995) demonstrated that the environment has an effect on the amount of quercitin in the onion bulb, indicating that the production environment is an important consideration for quercitin-enhanced onions. There is no information on the genetic control of the amount or distribution of quercitin in onion bulbs.

Selenium

Selenium is a micronutrient required in small amounts in the human diet (Young 1981). Selenium has been shown to reduce tumor growth in rats and may possess anticarcinogenic properties in humans (Ip et al. 1992). Selenium is closely related to sulfur, is efficiently taken up by the alliums (Morris 1970), and may be substituted for sulfur in metabolic pathways. When onion or garlic is grown under higher selenium concentrations, the amount of this element relative to sulfur is increased and selenium-enhanced onion and garlic can be artificially produced.

CONCLUSION

It is an exciting time for the alliums. The research efforts of many scientists around the world will bring new alliums to the marketplace. New ornamental alliums are available and may become more common in our gardens. New onion or garlic cultivars with unique flavor and defined health-enhancing attributes are on the horizon and should increase consumption. Sexually produced garlics may reduce the cost of production and create new garlics to open up new and unique markets. Because onion, garlic, and other cultivated alliums are common part of our meals, I expect that these new alliums will increase consumption and not significantly reduce the production of more classical types presently used by us all.

REFERENCES

Ariga, T., S. Oshiba, and T. Tamada. 1981. Platelet aggregation inhibitor in garlic. Lancet 2:150.

Bakhsh, R. and S. Khan. 1990. Influence of onion and chaunga on serum cholesterol, triglycerides, and total lipids in human subject. Sarhad J. Agr. 6:425–428.

Bilyk, A., P.L. Cooper, and G.M. Sapers. 1984. Varietal differences in distribution of quercetin and kaempferolin onion tissues. J. Agr. Food Chem. 32:274–276.

Block, E. 1992. The organosulfur chemistry of the genus *Allium*—implications for the organic chemistry of sulfur. Angew. Chem. Int. Ed. Engl. 31:1135–1178.

Block, E., S. Ahmad, M.K. Jain, R.W. Crecely, and R. Apitz-Castro. 1984. (E,Z)-Ajoene, a potent antithrombotic agent from garlic. J. Am. Chem. Soc. 106:8295–8296.

Bordia, T., N. Mohammed, M. Thomson, and M. Ali. 1996. An evaluation of garlic and onion as antithrombotic agents. Prostaglandins Leukotrienes Essential Fatty Acids 54:183–186.

Crozier, A., M.E.J. Lean, M.S. McDonald, and C. Black. 1997. Quantitative analysis of the flavonoid content of commercial tomatoes, onions, lettuce, and celery. J. Agr. Food Chem. 45:590–595.

Damme, E.J.M. van, K. Smeets, S. Torrekens, F. van Leuven, and W.J. Peumans. 1992. Isolation and characterization of alliinase cDNA clones from garlic (*Allium sativum* L.) and related species. Eur. J. Biochem. 209:751–757.

Eguchi, Y., T. Itoh, and J. Tomizawa. 1991. Antisense RNA. Ann. Rev. Biochem. 60:631–652.

Engeland, R.L. 1991. Growing great garlic. Filaree Prod., Okanogan, WA.

Etoh, T. 1986. Fertility of garlic clones collected in Soviet Central Asia. J. Jpn. Soc. Hort. Sci. 55:312.

Fenwick, G. and A. Hanley. 1985. The genus *Allium*. Crit. Rev. Food Sci. Nutr. 22:199–271.

Freeman, G.G. and N. Mossadeghi. 1970. Effect of sulphate nutrition on flavour components of onion (*Allium cepa* L.). J. Sci. Food Agr. 21:610–615.

Goldman, I.L, B.S. Schwartz, and M. Kopelberg. 1995. Variability in blood platelet inhibitory activity of *Allium* (Alliaceae) species accessions. Am. J. Bot. 82:827–832.

Goldman, I.L, M. Kopelberg, J.P. Debaene, and B.S. Schwartz. 1996. Antiplatelet activity in onion (*Allium cepa* L.) is sulfur dependent. Thrombosis Homeostasis 76:450–452.

Ip, C., D.J. Lisk, and G.S. Stoewsand. 1992. Mammary cancer prevention by regular garlic and selenium enriched garlic. Nutr. Cancer 17:279–286.

Jones, H. and L. Mann. 1963. Onions and their allies. Interscience Publishers, New York.

Lancaster, J.E. and M.J. Boland. 1990. Flavor biochemistry. p. 33–72. In: H.D. Rabinowitch and J.L. Brewster (eds.), Onion and allied crops, Vol III. CRC Press, Boca Raton, FL.

Lancaster, J. and H. Collin. 1981. Presence of alliinase in isolated vacuoles and of alkyl cysteine sulphoxides in the cytoplasm of bulbs of onion (*Allium cepa*). Plant Sci. Lett. 22:169–176.

Lancaster, J. and K. Kelly. 1983. Quantitative analysis of the S-alk(en)yl-L-cysteine sulphoxides in onion (*Allium cepa* L.). J. Sci. Food Agr. 34:1229–1235.

Lawson, L.D., D.K. Ransom, and B.G. Hughes. 1992. Inhibition of whole blood platelet-aggregation by compounds in garlic clove extracts and commercial garlic products. Thrombosis Research 65:141–156.

Leighton, T., C. Ginther, L. Fluss, W.K. Harter, J. Cansado, and V. Notario. 1992. Molecular characterization of quercetin and quercetin glycosides in *Allium* vegetables. p. 220–238. In: Phenolic compounds and their effects on human health II. Am. Chem. Society.

Lin, M.W., J.F. Watson, and J.R. Baggett. 1995. Inheritance of soluble solids and pyruvic acid content of bulb onions. J. Am. Soc. Hort. Sci. 120:199–120.

Morimitsu, Y., Y. Morioka, and S. Kawakishi. 1992. Inhibitors of platelet aggregation generated by mixtures of Allium species and/or S-alk(en)nyl-L-cysteine sulfoxides. J. Agr. Food Chem. 40:368–372.

Morris, V.C. 1970. Selenium content of foods. J. Nutr. 100:1385–1386.

Ohsumi, C., A. Kojima, K. Hinata, T. Etoh, and T. Hayashi. 1993. Interspecific hybrid between *Allium cepa* and *Allium sativum*. Theor. Appl. Genet. 85:969–975.

Patil, B.S. and L.M. Pike. 1995. Distribution of quercitin content in different rings of various coloured onion cultivars. J. Hort. Sci. 70:643–650.

Patil, B.S., L.M. Pike, and B.K. Hamilton. 1995. Changes in quercetin concentration in onion owing to location, growth stage, and soil type. New Phytol. 130:349–355.

Platenius, H. and J.E. Knott. 1941. Factors affecting onion pungency. J. Agr. Res. 62:371–379.

Pooler, M.R. and P.W. Simon. 1994. True seed production in garlic. Sex. Plant Reprod. 7:282–286.

Price, K.R. and M.J.C. Rhodes. 1997. Analysis of the major flavonol glycosides present in four varieties of onion and changes in composition resulting from autolysis. J. Sci. Food Agr. 74:331–339.

Randle, W.M., J.E. Lancaster, M.L. Shaw, K.H. Sutton, R.L. Hay, and M.L. Bussard. 1995. Quantifying onion flavor components responding to sulfur fertility-sulfur increases levels of and biosynthetic intermediates. J. Am. Soc. Hort. Sci. 120:1075–1081.

Schwimmer, S. and D.G. Guadagni. 1962. Relation between olfactory threshold concentration and pyruvic acid content of onion juice. J. Food Sci. 27:94–97.

Schwimmer, S. and W.J. Weston. 1961. Enzymatic development of pyruvic acid in onion as a measure of pungency. J. Agr. Food Chem. 9:301–304.

Simon, P.W. 1995. Genetic analysis of pungency and soluble solids in long-storage onions. Euphytica 82:1–8.

Trammell, K.W. and C.E. Peterson. 1976. Quantitative differences in the flavonol content of yellow onion. J. Am. Soc. Hort. Sci. 101:205–207.

Walkey, D.G.A., M.J.W. Webb, C.J. Bolland, and A. Miller. 1987. Production of virus-free garlic (*Allium sativum* L.) and shallot (*A. ascalonium* L.) by meristem-tip culture. J. Hort. Sci. 62:211–220.

Wall, M.M., A. Mohammad, and J.N. Corgan. 1996. Heritability estimates and response to selection for the pungency and single center traits in onion. Euphytica 87:133–139.

Wall, M.M. and J.N. Corgan. 1992. Relationship between pyruvate analysis and flavor perception for onion pungency determination. HortScience 27:1029–1030.

Yang, G.C., P.M. Yasaei, and S.W. Page. 1993. Garlic as anti-oxidants and free radical scavengers. J. Food Drug Anal. 1:357–364.

Yin, M.C. and W.S. Cheng. 1998. Antioxidant activity of several *Allium* members. J. Agr. Food Chem. 46:4097–4101.

Young, V.R. 1981. Selenium: A case for its essentiality in man. New Engl. J. Med. 304:1228–1230.

"New" Solanums

Charles Heiser and Gregory Anderson*

The nightshade family (Solanaceae) has provided many plants to humankind. The genus *Solanum* has been particularly significant for it has furnished us with one of our basic food plants, the Irish potato (*S. tuberosum*), as well as the widely used eggplant (*S. melongena*). The genus has recently become even more important with the addition of two other food plants—both because of recent changes in their names. The tomato, which has long been known as *Lycopersicon esculentum*, is now recognized as *S. lycopersicum* and the tree tomato, previously known as *Cyphomandra betacea*, has become *S. betaceum*. Three other Solanums— the pepino (*S. muricatum*) the cocona (*S. sessiliflorum*) and the naranjilla (*S. quitoense*), long valued for their fruits in South America, deserve greater attention.

TOMATO

The tomato was originally named *Solanum lycopersicum* by Linnaeus in 1753. Philip Miller in *The Gardeners Dictionary* in 1768 used *Lycopersicon esculentum* and this became the accepted name until very recently. Various lines of evidence had suggested that the tomato belonged to the genus *Solanum* and molecular studies provided the compelling evidence that has led to the readoption of *S. lycopersicum* (Spooner et al. 1993). The other species of *Lycopersicon* have also been assigned or re-assigned to *Solanum*. More recent studies (Bohs and Olmstead 1997; Olmstead and Palmer 1997) have provided additional evidence for the inclusion of the tomato in *Solanum*. The tomatoes are placed in *Solanum* in groups very close to the tuberous and non-tuberous potatoes. This has not led to major changes in tomatoes as yet, but it does make very clear the relationship of tomatoes and potatoes, and should promote even greater efforts at interbreeding and genetic transfer among these two important groups.

TREE TOMATO

The tree tomato, originally named *Solanum betaceum* by Cavanilles in 1799, was transferred by Sendtner in 1845 to the genus *Cyphomandra* where it remained until recently when Bohs (1995) returned it to *Solanum*. As in the tomato, molecular studies prompted the recent change in the name (Olmstead and Palmer 1992; Spooner et al. 1993; Bohs and Olmstead 1998). More recent molecular studies have furnished additional support for its inclusion in *Solanum* (Bohs and Olmstead 1997; Olmstead and Palmer 1997).

The tree tomato is hardly a new crop, for it has long been popular in the Andes, where it is native, and in many other parts of the tropics. It has been grown successfully in New Zealand and has been exported to the north temperate zone for around two decades. In New Zealand, the plant was renamed tamarillo, a made-up name. The new name does have some advantages, for although many people have claimed a resemblance of the tree tomato to the tomato it is mostly superficial. Both the aroma and taste are quite different from that of the tomato. The fruit is eaten raw as well as for the juice, preserves, jellies, and as a vegetable, either cooked or raw in salads. Some improvement of the plant has taken place in New Zealand in recent years, and on a visit to Ecuador in 1983 we were happy to learn that New Zealand had sent improved forms to South America.

Although the tree tomato now reaches the United States from New Zealand and South America, it probably first became familiar to many people here some 30 years ago through advertisements of a nursery company offering plants for sale. A few years later we saw another advertisement stating "Does 60 pounds of tomatoes from one yield sound unreasonable—not if you own the tree tomato." Other advertisements with extravagant claims have appeared from time to time, but we hadn't seen any for several years until this past spring when two companies offered it again claiming yields of 60 pounds. This time however, we saw a release from the Associated Press in a local newspaper headlined, "Gardeners should beware of catalog exaggerations" and the tree tomato was one of the plants featured.

---

*We would like to thank Jorge Soria, Mario Lobo, Lynn Bohs, Jan Salick, and Donald Burton for their help and acknowledge the assistance of the National Science Foundation which has supported much of our research on the Solanaceae over the years.

Several wild species of the tree tomato in South America also have edible fruits. In addition, various parts of the plants, particularly the leaves, of the tree tomato and several wild species have been used in folk medicine in Latin America. More information on the ethnobotany of the tree tomato may be found in Bohs (1989); see also National Research Council (1989). A wild ancestor of the tree tomato has not yet been identified unequivocally but recent studies indicate that it most likely had its origin in the southern Andes, perhaps Bolivia, and that the cultivated tree tomato was derived from a wild form of the same species (Bohs 1998).

PEPINO

The pepino (*cachum* in Quechua), *Solanum muricatum*, shows some similarities to the tree tomato in that it is also native to the Andes where it is an important fruit; it has also recently been grown in New Zealand (Hewett 1993) whence it is exported to the United States, Europe and Japan. Like the tree tomato, its future as a crop outside of the Andes is uncertain. There are many differences—the pepino is eaten raw and it was probably domesticated much earlier than the tree tomato; in addition, its origin is more complex and the two plants are very different in appearance. Pepino is the Spanish word for cucumber, and the plant was so named by the Spanish because of a slight resemblance to some cucumbers. The typical fruit of the pepino is usually broadly ovate, white or ivory colored, often with a few purple stripes, when mature. However, there are many fruit color and shape varieties among the 'indigenous' cultivars in South America e.g., see the cover of the 1993 *American Journal of Botany* 80(6). It is generally eaten out-of-hand, frequently after peeling, for its sweet pleasant taste. The plant itself somewhat resembles the Irish potato; in fact, both species are placed in the subgenus *Potatoe*. There is great variation in the pepino not only in its fruits but in vegetative features as well; the leaves may be simple or compound.

Some improvement of the plant has taken place in New Zealand, and at present it is being tried in a number of countries. The history of the plant, both ancient and modern, has been treated in some detail by Prohens et al. (1996).

Contrary to the situation in the tree tomato, and similar to the naranjilla (see following) there is no clearly defined wild ancestral form of the pepino. This highly variable cultigen is know only from cultivation. Extensive field studies have failed to indicate a wild form. However, many studies and approaches have identified three wild species that are most closely related to the pepino, and that have been treated as possible progenitors (e.g., Anderson 1979; Anderson et al. 1996; Anderson and Jansen 1998; Jansen et al. 1998): the widespread *S. caripense*, which ranges from Costa Rica to Peru, *S. tabanoense*, which is found in a few localities in Colombia, Ecuador, and Peru, and *S. basendopogon*, which has a very limited distribution in Peru. Recent molecular studies (Anderson et al. 1996; Anderson and Jansen 1998; Jansen et al. 1998) suggest *S. basendopogon* as a less likely element in the origin of the pepino. The chloroplast DNA evidence is quite strong in support of *S. tabanoense* as central to its origin; over 85% of the pepino collections from South America are linked in a cladistic study with this species (Anderson et al. 1996). The remaining pepino collections are associated with *S. caripense*, implying either a second origin, or possibly post-origin hybridization. Although multiple origins are known for a few other domesticated plants, in each of these, only one wild species is involved; none is known involving more than one species. The molecular results, however, do not rule out the possibility that the pepino evolved from a single species, most likely *S. tabanoense*, followed by hybridization with *S. caripense*, and possibly other closely related wild species (perhaps including *S. basendopogon*). Hybridization among the wild species, and with the pepino, is not obviously rampant in the wild but, is relatively easily performed in plants grown in the greenhouse. Either the unprecedented multiple origin hypothesis, or post-origin hybridization, would help explain the great variability in the fruits, flowers, and leaves of this polymorphic species. Work now in progress should help explain more about the origin of the pepino, including its place of origin. Southern Colombia and northern Ecuador, at present, would seem to be the most likely places.

COCONA

Although not yet grown commercially outside of Latin America, two other South American *Solanums* deserve wider cultivation. The cocona or tupiro, as it is known in Spanish speaking countries, or cubiu in Brazil (*S. sessiliflorum*) is a shrubby perennial, generally a meter or more tall with extremely large leaves (to

9.5 cm long). It bears maroon, orange-red or yellow fruits up to 10 cm in diameter with yellow flesh. The fruits are covered with small, soft readily deciduous hairs. The plant is widely cultivated, usually in very small plots, in the upper Amazon basin from sea-level to 700 m or higher. The berries have a pleasant acidulous flavor, somewhat like citrus, and are used as fruits, chiefly for the juice, and also cooked as a vegetable (Whalen et al. 1981). Parts of the plant have also been used in medicine. The species is quite variable, particularly in size, shape, and flavor of the berries, as might be expected in a species domesticated for its fruits. A wild type (*S. sessiliflorum* var. *georgicum*) with smaller (3–4 cm), globose, orange berries and prickles on the stem and leaves is the likely ancestor of the domesticate (Heiser 1972). The nutritional value, cultivation and production of the cocona have been treated by Salick (1989, 1992) who believes that its wider cultivation could significantly improve human subsistence in tropical lowlands.

NARANJILLA

The naranjilla or lulo (*S. quitoense*) belongs to the same section of *Solanum* (*Lasiocarpa*) as the cocona. It is a taller plant (ca. 2 m), less branched and with slightly smaller leaves (Whalen et al. 1981). The leaves, particularly younger ones, the veins and petioles are often purplish, making the plant sufficiently attractive to be grown occasionally as an indoor ornamental in the United States. The globose berries are around 5 cm in diameter, orange and covered with short stiff hairs which have usually rubbed off by the time the fruits have reached the markets. The pulp is green and gives a green juice, the form in which it is usually used. The taste is unique, but has been described by some as like that of a mixture of pineapples and strawberries. The plant was found by the Spanish in Ecuador and Colombia (Patiño 1963) where most of it is still grown. It was introduced to Panama and Costa Rica in the middle of this century where it is also grown today. Introductions to other places, including Florida, have been largely unsuccessful (Heiser 1985). Recently Bernal et al. (1998a) have postulated that the form similar to the cultivated plant that grows spontaneously as a weed in Colombia is the original wild type. We feel, however, that these plants may be nothing more than escapes from cultivation, no different from escapes in Costa Rica where the plant was introduced a half century ago. Moreover, it is difficult to imagine no change occurring in a plant that has been cultivated for several hundred years. On the other hand, no other wild ancestor has been reported and it is most unlikely that *S. quitoense* descends directly from any other species in section *Lasiocarpa*. For a domesticated plant it shows very little variability. In Colombia and Central America the plants usually have prickles on the stems and leaves whereas in Ecuador plants are unarmed. Several cultivars are recognized in Ecuador, based on slight differences in fruits.

Over 20 years ago Heiser (1969) wrote that "the fruit [of the naranjilla] yields one of the most delicious beverages known." Why then hasn't it become more widely appreciated? Many early visitors to Colombia and Ecuador spoke highly of the juice (Patiño 1963) so it is difficult to believe that the Spanish didn't carry seeds to other regions, but it was not successfully introduced elsewhere. The naranjilla grows best at between 1200 and 2300 m and it is neither a tropical plant (pollen aborts at high temperatures) nor a temperate one (it is killed by freezing and fruits need four months to mature). The juice has been canned but its unique flavor is lost in doing so. Freeze-drying techniques have shown promise in Ecuador but no attempts have been made to use them to make juice for export, probably because no certain markets are known. Fresh fruits, of course, could also be exported if produced in sufficient quantities. Root knot nematodes and various insect pests and fungal diseases limit its production in all of the countries where it is presently grown (National Research Council 1989). There are two promising recent developments toward its improvement: the use of interspecific hybrids with the cocona and the introduction of a nematode resistant variety.

In Ecuador a local farmer, Raul Viteri, crossed *S. quitoense* with the wild variety of *S. sessiliflorum* (Torre and Camacho 1981) and the resulting hybrids were vigorous and highly productive. The hybrid was propagated vegetatively. The chief disadvantage of the hybrid was that the fruits were much smaller than those of the naranjilla. This was overcome by spraying plants with dilute solutions of 2,4-D after it was learned by accident that such spraying would cause enlargement of the fruits. From the late 1980s to the present most of the naranjilla in Ecuador was from this hybrid which came to be know as the "Puyo hybrid." In the markets they could be distinguished from the pure naranjillas by the lighter green color of the pulp and the absence of filled seeds.

After learning of Viteri's hybrid, Heiser who had previously made many attempts to secure this hybrid, again attempted to do so. No fruit was ever secured with *S. quitoense* as the female parent[1]. Several fruits without seeds were secured with *S. sessiliflorum* as the maternal parent, and finally in 1989, using a large fruited cocona, a fruit was secured with two partially filled seeds. These were germinated on nutrient agar and gave rise to vigorous hybrids, differing primarily from the Puyo hybrid in the much larger berry (7 cm in diameter) having orange colored flesh (Heiser 1993)[2]. Dr. Jorge Soria carried cuttings of the hybrids to Ecuador where they were planted at the experiment station of the Instituto Nacional de Investigaciones Agropecuarias at Palora. After multiplication they were distributed to a few farmers for trial and became known as the "Palora hybrids."

The most recent reports from Ecuador (J. Soria pers. commun.) state that the cultivation of the Palora hybrid is increasing rapidly in Ecuador and has spread into southern Colombia. In addition to the larger size of the fruit the Palora hybrid has other advantages over the Puyo hybrid. It survives for three years in cultivation in contrast to one and a half years for Puyo, produces about twice as much annually and does not require spraying with 2,4-D. Moreover, the juice of Palora takes nearly 24 h to start oxidizing, whereas the juice of Puyo and that of pure naranjilla both begin oxidizing in less than an hour. Some buyers, however, are paying more for Puyo than Palora because of the color of the juice (greenish vs. orange respectively) and the thicker rinds of the latter. The thick rind, however, would be an advantage in shipping. Perhaps the cultivation of Palora will increase even more after it is formally released and its advantages become more widely known. The Palora hybrid may fare well in other countries in competition with the cocona, a fruit that is not very important in Ecuador.

One of the most serious pests—if not the most serious—of the naranjilla is root knot nematode, various species of *Meloidogyne* (National Research Council 1989). Although the naranjilla is a perennial and ought to yield for several years, because of these parasites it is often grown as an annual in many places. Resistance to nematodes was discovered in plants of *S. hirtum* grown in the greenhouse at Indiana University (Heiser 1971). Hybrid seed of *S. hirtum* × *quitoense* (for B_1 and F_2) were send to Soria at CATIE in Costa Rica who found that many of these plants were resistant to nematodes there. Subsequently seeds of hybrids were sent to plant breeders in several countries and a resistant cultivar has now been produced in Colombia from F_2 seeds sent to Dr. Mario Lobo in 1984 (Bernal et al. 1998a,b). The plant was backcrosssed to *S. quitoense* twice, and material from F_2 plants of the BC_2 was multiplied by tissue culture. Tests in various parts of the country showed good performance. In addition the new plant, called "lulo la selva" grows better in full sun, has fruit of better quality, outyields the traditional lulo and is also spineless. Moreover, the juice has a longer life because of delayed oxidation (M. Lobo pers. commun.). The resistant cultivar is propagated vegetatively. It was formally released in June of 1998. There are still problems to overcome but if work on the improvement of the naranjilla continues, the juice eventually may become more widely available.

LOOKING BACKWARDS

A book on the nightshade family (Heiser 1969) appeared 30 years ago. To us, at least, it is interesting to see some of the changes that have taken place with various numbers of the family since then. Neither the tree tomato nor the pepino were available in stores in the United States at that time. In the book it was written that "several attempts have been made to make it (the tomatillo, *Physalis ixocarpa*) more popular in the United States, but it has never really caught on." It has now definitely caught on and has become a standard item in grocery stores and many gardens, not only in the Southwest but in most of the rest of the country as well. *Capsicum* peppers, of course, were widely available in 1969 but mostly the sweet kinds of *C. annuum*. The pungent sorts were seldom seen in those days and what was available belonged to a single species, *C. annuum*. Now pungent varieties are common. Furthermore, today more than one species is represented; the extremely

---

[1]Viteri claimed that *S. quitoense* was the female parent of his hybrid. Heiser gave material of Viteri's hybrid and the parental species to Richard Olmstead for analysis. Examination of the cpDNA indicated that *S. sessiliflorum* was the female parent of his hybrid (J.P. Whitney and R.G. Olmstead, unpubl.).

[2]Another hybrid from different parents but again with the cocona as the female was produced the following year but it had smaller fruits than either parent.

pungent *C. chinense* is now well accepted by many people. Along with this a huge increase in the number of pepper sauces has become available. Tomatoes, of course, were as popular in 1969 as they are today and then, as now, they were in demand in the winter as well as the summer. Although not mentioned in the book, we remember that some of us referred to the tomatoes then available in the winter as "little pink rocks." For several years now we have been fortunate in having more edible tomatoes in the winter. Tobacco, of course, received a lengthy chapter in the book where it was stated that "probably more has been written about tobacco than any other plant." That is still true. In fact, we can think of no other plant that has been featured in headlines recently more than tobacco. In spite of much adverse publicity over the centuries the plant still shows no signs of disappearing.

REFERENCES

Anderson, G. 1979. Systematic and evolutionary consideration of species of *Solanum*, section *Basarthrum*. p. 549–562. In J. Hawkes, R. Lester, and A. Shelding (eds.), The Biology and taxonomy of the Solanaceae. Linnean Society Symposium Series 7.

Anderson, G. and R. Jansen. 1998. Biosystematic and molecular systematic studies of *Solanum* section *Basarthrum* and the origin and relationship of the pepino dulce (*S. muricatum*). Ann. Missouri Bot. Gard. In Press.

Anderson, G., R. Jansen, and Y. Kim. 1996. The origin and relationships of the Pepino (Solanaceae): DNA restriction fragment evidence. Econ. Bot. 50:369–380.

Bernal E.J., M. Londoño B., G. Franco, and M. Lobo A. 1998a. Lulo la selva. Corpoica, Rionegro, Colombia.

Bernal E.J., M. Lobo A., and M. Londoño B. 1998b. Documento presentación del material "Lulo la selva." Corpoica, Rionegro, Colombia.

Bohs, L. 1989. Ethnobotany of the genus *Cyphomandra* (Solanaceae). Econ. Bot. 4:143–163.

Bohs, L. 1995. Transfer of *Cyphomandra* (Solanaceae) and its species to *Solanum*. Taxon 44:583–587.

Bohs, L. 1998. Evolutionary relationships of the tree tomato, *Solanum betaceum*, (Solanaceae) and its wild relatives. Am. J. Bot. 85:116. (Abstr.).

Bohs, L. and R. Olmstead. 1997. Phylogenetic relationships in *Solanum* (Solanaceae) based on ndhF sequences. Sys. Bot. 22:5–17.

Bohs, L. and R. Olmstead. 1998. *Solanum* phylogeny inferred from chloroplast DNA sequence data. In: Proc. Fourth International Solanaceae Conference. Adelaide, Australia (in press).

Heiser, C. 1969. Nightshades, the paradoxical plants. W.H. Freeman and Co., San Francisco.

Heiser, C. 1971. Notes on some species of *Solanum (Leptostemonum)* in Latin America. Baileya 18:59–65.

Heiser, C. 1972. The relationship of the naranjilla, *Solanum quitoense*. Biotropica 4:77–84.

Heiser, C. 1985. Of plants and people. Univ. Oklahoma Press, Norman.

Heiser, C. 1993. The naranjilla (*Solanum quitoense*), the cocona (*Solanum sessiliflorum*) and their hybrid. p. 29–34. In: J. Gustafson et al. (eds.), Gene conservation and exploitation. Plenum Press, New York.

Hewett, E. 1993. New horticultural crops in New Zealand. p. 57–64. In: J. Janick and J. Simon (eds.), New crops. Wiley, New York.

Jansen, R., G. Anderson, and S. Chen. 1998. Origin and relationships of "Pepino" (*Solanum muricatum*) based on ITS sequence data. Am. J. Bot. 85:137. (Abstr.).

National Research Council. 1989. Lost crops of the Incas: Little-known plants of the Andes with promise of the worldwide cultivation. National Academy Press, Washington, DC.

Olmstead, R. and J. Palmer. 1992. A chloroplast DNA phylogeny of the Solanaceae: subfamilial relationships and character evolution. Ann. Missouri Bot. Gard. 79:346–60.

Olmstead, R. and J. Palmer. 1997. Implications for the phylogeny, classification, and biogeography of *Solanum* from cpDNA restriction site variation. Sys. Bot. 22:19–29.

Patiño, V. 1963. Plantas cultivadas y animales domesticas en America equinoccial. Tomo I. Frutales. Imprenta Departmental. Cali, Colombia.

Prohens, J., J. Ruiz, and F. Nuez. 1996. The Pepino (*Solanum muricatum*, Solanaceae): a "new" crop with a history. Econ. Bot. 50:355–368.

Salick, J. 1989. Cocona (*Solanum sessiliflorum*): an overview of production and breeding potentials of the peach-tomato. In: G. Wickens and P. Day (eds.), New crops for food and industry. Chapman and Hall, London.

Salick, J. 1992. Crop domestication and the evolutionary ecology of cocona (*Solanum sessiliflorum* Dunal). Evol. Bio. 26:247–285.

Spooner, D., G. Anderson, and R. Jansen. 1993. Chloroplast DNA evidence for the interrelationships of tomatoes, potatoes, and pepino (Solanaceae). Am. J. Bot. 80:676–698.

Torre, R. and S. Camacho. 1981. Campesino fitomejorador de naranjilla. Carta de Frutales no. 14, Instituto Nacional de Investigaciones Agropecuarias, Quito, Ecuador.

Whalen, M., D. Costich, and C. Heiser. 1981. Taxonomy of *Solanum* section *Lasiocarpa*. Gentes Herb. 12:41–129.

Edamame: A Vegetable Soybean for Colorado

Duane Johnson, Shaoke Wang, and Akio Suzuki

Edamame, or vegetable soybean (Fig. 1), has a long history in many Asian cultures as a side dish or snack (Lumpkin et al. 1993). Japan has been consuming edamame for over 400 years. National consumption in Japan has averaged 110,000 tonnes (t) annually (Nakano 1991). These vegetable soybeans are generally sold in the pod as fresh or frozen beans (Fig. 2). Beans are harvested when bean pods are green and Brix readings (soluble solids) are generally between 8.5 and 12.0. For consumption, edamame is boiled for 5 to 7 minutes in highly salted water, drained and are served either hot or cold. Other vegetable soybean products are a shelled version of edamame called *mukimame* and a green bean paste, *zunda-mochi* (Masuda 1991).

Edamame quality is measured in Japan with three primary concerns: flavor, sweetness, and texture. To accomplish these concerns, breeders have based cultivar selection on five criteria: appearance, taste, texture, flavor, and nutritional value. Taste is determined by sucrose, glutamic acid, and alanine. Flavor is most desirable when it is "flower-like" and "beany" (Masuda 1991). The boiled beans are a good source of vitamin C (ascorbic acid), vitamin E (tocopherol), and dietary fiber. Trypsin inhibitors and other antinutrional factors do exist in edamame. To market, the pods should be bright green, have a light (white to grey) colored pubescence, be free of defects and contain a minimum of two beans per pod. Texture studies at Colorado State University indicate a preference for a "buttery" texture attainable by delaying harvest (J.A. Maga 1996, pers. commun.). The changes in texture preferred by US consumers will decrease the concentration of cis-jasmone and hexenyl-acetate responsible for the flowery flavor Masuda (1991). Boiling induces production of furans and ketones. In the lipid fraction, as the soybean mature, the fraction of linoleic fatty acids increases. Linolenic and palmitic fractions decreased. Monounsaturates tend to dominate in these immature soybeans. Nutritionally, edamame is very sound (Table 1).

Fig 1. Edamame soybean.

Fig 2. Edamame soybean plants ready for market.

Table 1. Proximate analysis of vegetable soybean in Japan (Masuda 1991) and Colorado.

Composition	Value/100 g	
	Japan	Colorado
Energy (Kcal)	582	573
Water (g)	71.1	71.1
Protein (g)	11.4	12.4
Lipids (g)	6.6	7.1
Carbohydrates g)	7.4	8.3
Fiber (g)	1.9	3.2
Dietary fiber (g)	15.6	13.8
Ash (g)	1.6	1.6
Calcium (mg)	70.0	72.0
Phosphorus (mg)	140	148
Iron (mg)	1.7	1.2
Sodium (mg)	1.0	1.5
Potassium (mg)	140	145
Carotene (mg)	100	89
Vitamin B1 (mg)	0.27	0.27
Vitamin B2 (mg)	0.14	0.14
Niacin (mg)	1.0	1.0
Ascorbic Acid (g)	27.0	17.0

THE COLORADO EDAMAME PROJECT

Colorado, due to elevation and latitude, is generally characterized as a Group 1-2 in the north and a Group 4 in the southern half of the state. Its extremes of climate, limited rainfall, and distance to market have restricted its introduction to the state. The dry climate and isolation, however, have also provided levels of yield and quality that stimulated interest in specialty markets for soybeans. Developing specialty soybeans required identifying the market. The developing immigrant and tourism industries for Asians on the West and East Coasts of the United States showed interest in purchases exceeding 2,200 t annually.

Edamame cultivars were obtained from Red Hen Corp. (Colorado Springs, Colorado) and Washington State University (Pullman, Washington) in 1992 and from Seedex, Inc. (Longmont, Colorado) in 1994. Initial results positive for yields and quality and additional yield trials were conducted from 1995 through 1998 utilizing Japanese cultivars of edamame provided by Seedex, Inc. Cultivar entries varied from year to year but five remained constant from 1994 through 1998. Cultivars were numbered as SE 1 through SE 5 and maintained under those designations throughout this timeframe. In 1995, a Colorado derived selection from SE 4 was relabeled as SE 7. SE 7 data were not significantly different from SE 4 and were not included in this analysis.

All entries were planted annually at Rocky Ford, Colorado from May 15 to June 1 and in Ft. Collins, Colorado from May 20 to June 10. All trials were planted in a randomized complete block design in four-row plots 7 m long and 1.6 m wide. Four replications were used at Rocky Ford and three replications at Fort Collins. Yields were divided into green bean and dry seed harvests using a split plot design. Green bean yields were based on harvests of two center rows, 2 m long and included total yield, salable yield (two or more beans/pod with no defects). Green harvest was based on physiological maturity of the beans (85–90% pod fill) and a Brix reading of parboiled beans in excess of 8.5. All plots were hand harvested. All green bean yields were adjusted to salable beans and analyzed using ANOVA protocols for Brix, texture, and salable beans. Texture was measured by sensory panel analyses in the Department of Food Sciences, Colorado State University. Dry bean harvests were conducted at the initiation of pod shatter (>10%). Dry bean plots harvested were two meters in length and were the two center rows in each plot. Dry beans were cut and tarped until dry and threshed using a modified Hege 125B combine. All dry bean yields were adjusted to 10% moisture and analyzed using ANOVA protocols.

Table 2. Yield results for edible beans at Rocky Ford and in Ft. Collins, Colorado, 1994–1998.

Cultivar	Whole yield (t/ha)	Salable yield (t/ha)	Brix (%)	Texture score[y] (1–10)
		Rocky Fork		
SE 4	10.2a[z]	6.8a	8.4b	8a
SE 2	8.4b	4.6b	8.4b	7b
SE 5	7.8b	5.2b	9.3a	9a
SE 1	5.9c	4.0c	9.1a	6b
SE 3	4.1c	2.1d	8.1c	4c
		Ft. Collins		
SE 4	8.1a	4.8a	8.3b	6a
SE 2	6.1b	3.9b	10.1a	7b
SE 5	6.9b	3.4b	10.5a	8a
SE 1	4.1c	2.8b	9.1a	8b
SE 3	2.2c	1.0c	7.1c	4c

[z]Mean separation by Duncans Multiple Range test, 5% level.
[y]1 = poor, 10 = execellent

Table 3. Yield of dry soybeans at Rocky Ford and Ft. Collins, Colorado, 1994–1997.

Cultivar	Whole yield (t/ha)	Salable yield (t/ha)	Germination (%)
		Rocky Fork	
SE 4	2.2a[z]	1.7a	82b
SE 2	1.5b	0.9a	85a
SE 5	1.3b	0.7a	85a
SE 1	0.9b	0.4b	76b
SE 3	0.4c	0.1c	78b
		Ft. Collins	
SE 4	1.6a	1.1a	78a
SE 2	0.9b	0.6a	75a
SE 5	0.6b	0.5a	65b
SE 1	0.3c	0.2b	62b
SE 3	0.1d	0.06c	72b

[z]Mean separation by Duncans Multiple Range test, 5% level.

Table 4. Brix quality of fresh green vegetable soybeans from Ft. Collins, Colorado, 1997.

	Brix (%)										
	Day										
Cultivar	0	2	4	6	8	10	12	14	16	18	20
SE 4	10a[z]	10a	10a	10a	10a	9a	9a	8a	6a	4a	3a
SE 2	11a	10a	10a	9a	9a	8a	8a	8a	6a	4a	3a
SE 5	11a	10a	10a	9a	8a	7b	7b	7a	6a	4a	3a
SE 1	9a	9a	9a	8b	8b	7b	7b	6b	6a	4a	3a
SE 3	6b	6b	6b	6b	5b	5c	5c	4c	4b	3a	3a

[z]Mean separation by Duncans Multiple Range test, 5% level.

Green Bean Yields

Green bean yields were significant for location effect and yield quality as measured by Brix. Results are summarized in Table 2. No location, year or cultivar interactions were noted. Data are summarized over years.

Yields of dry beans reflected similar variation with variance for cultivars, years, and locations being significant. Data are summarized in Table 3 for locations. Dry bean production was used to supplement potential green bean production and as an alternative seed market.

Processing Edamame

Edamame is marketed fresh or frozen. Fresh beans were harvested and selected for salable quality. Fresh beans of SE 4 from Ft. Collins were packaged in resealable plastic bags in lots of 2 kg/package. Plastic packs were flooded with air, 20:80 CO_2:N_2 gas and 40:60 CO_2: N_2. Replicated packs were refrigerated at 3°–5°C and sampled at 2 day intervals with subsamples of 0.10 kg/sample removed and prepared for consumption. Brix (%) were used as quality measures. Analysis of replicated trials are shown in Table 4. High sugar cultivars tended to have a shorter shelf life than the higher yielding lines. It appears for commercial edamame production, most cultivars are capable of 10 to 14 day storage without significant loss in quality.

Frozen soybeans were capable of long term storage and were commercially processed using IQF (Instant Quick Frozen) technology in Rocky Ford, Colorado. Frozen samples were distributed to commercial outlets for evaluation in 2 kg bags. Production and Instant Quick Freeze (IQF) processing technology utilized are illustrated in Fig. 1 and 2.

SUMMARY

Edible vegetable soybeans are feasible for production in Colorado. The dry climate and high altitude provide a low pathogen, high quality environment. Markets for this new crop exist within the United States and could conceivably require 13,000 ha to fill that market niche.

The crop can be processed either as fresh (chilled) product or as frozen product. New consumers in the United states appear to prefer a more mature bean which has a more "buttery" flavor and texture as opposed to traditional Japanese consumes who prefer a sweeter, more flowery flavor, and crisper texture.

REFERENCES

Lumpkin, T.A., J.C. Konovsky, K.J. Larson, and D.C. McClary. 1993. Potential new specialty crops from Asia: Azuki bean, edamame soybean, and astragalus. p. 45–51. In: J. Janick and J.E. Simon (eds.), New crops. Wiley, New York.

Masuda, R. 1991. Effect of holding time before freezing on the constituents and flavor of frozen green beans (edamame). In: R. MacIntyre and K. Lopez (eds.), Vegetable soybean: Research needs for production and quality improvement. Asian vegetable Research and Development Center. Taipei, Taiwan.

Nakano, H. 1991. Vegetable soybean area, production, demand, supply, domestic and foreign trade in Japan. In: R. MacIntyre and K. Lopez (eds.), Vegetable soybean: Research needs for production and quality improvement. Asian vegetable Research and Development Center. Taipei. Taiwan.

Evaluation of Tropical Leaf Vegetables in the Virgin Islands

Manuel C. Palada and Stafford M.A. Crossman

Tropical leaf vegetables are grown in the tropics and are rich sources of nutrients, particularly minerals, and vitamins (Oomen and Grubben 1978). A number of species and cultivars have been introduced and grown in the continental US on a limited-scale, particularly in the southern region (Lamberts 1993). The US is a major market for tropical and specialty greens and most of the shipments come from the Caribbean and Latin America. For example, in 1998, total US imports for dasheen leaves *(Colocasia esculenta* L. Schott, Araceae) was over 90 t. From this total, 70% came from Jamaica and 30% from the Dominican Republic (Pearrow 1991). In the same year, the US imported amaranthus (*Amaranthus* spp. L., Amaranthaceae) at 27 t from the same countries. In 1988, shipments of Oriental, Mexican, tropical, and exotic produce, including specialty leafy greens, accounted for about 5% of fresh vegetable shipments, whereas in previous years the volumes have been too low to track (Cook 1990; Lamberts 1990, 1993).

There are several reasons for the increasing demand of tropical and specialty leafy greens in the US. Growth in ethnic populations contributes to demand for product diversity within the produce section (Cook 1990) and food, previously considered ethnic or regional in nature is increasingly being consumed by a broader portion of the population. This trend will likely continue as the ethnic population continues to grow and more Americans become familiar with and develop the taste for the new crops.

This research study is undertaken with the following objectives: (1) collect and describe growth characteristics of minor tropical leaf vegetables; and (2) evaluate yield performance and commercial potentials in the Virgin Islands and the Caribbean.

AMARANTHACEAE

Amaranth

Amaranthus spp. are common, short-lived annuals, the leaves of which are used as potherb. Some species are cultivated in home gardens and for marketing. Several species exist depending on the region in the tropics. For example, *A. tricolor* L. is mostly found in East Asia, while *A. cruentus* (L.) Sauer is common in Africa and *A. dubius* Mart. ex Thell. in the Caribbean. Amaranths are probably the most important leaf vegetables of the lowland tropics of Africa and Asia, but scarcely known in South America. The nutritional value is high where vitamins A and C, and calcium and iron are found in good quantity. However, the high oxalic acid content may decrease the availability of calcium (Oomen and Grubben 1978; Martin and Ruberte 1979). Boiling produces a very acceptable spinach. Some Indian cultivars are markedly short-day plants, so market growers plant them in the beginning of the summer and harvest over several months by repeated pruning until plants flower at the end of the season.

Amaranths are upright and branch sparsely. The leaves are relatively small (5–10 cm long) but quite variable among cultivars. The flowers are small, and are borne in abundance in terminal or axillary spikes. The seeds are born in large numbers, small and edible. The flowers are not edible. The leaves, petioles, and young tips are used in salads and as potherb.

Amaranth is a suitable plant in crop rotation. It is not affected by common soil diseases such as nematodes, fungal, and bacterial wilt. Serious pests and diseases are damping off, wet rot, caterpillars, and stemborers. Early flowering may occur as a consequence of a short daylength or as a result of a short period of water stress.

Considerable differences exist between the three main species. The African cultivars of *A. cruentus* are originally grain-amaranths. They have a long stem and a high dry matter content in the leaves and bear large inflorescence. *A. dubius* and *A. tricolor* cultivars have a much lower seed production and their habit is similar to spinach, with a short stem and succulent leaves. Commercial cultivars exist in India, Taiwan, the Caribbean, and the US.

Eight cultivars were evaluated at the experiment station during the summer-fall season of 1997. The cultivars consisted mainly of *A. tricolor* compared against the local amaranth (*A. dubius*). Most of them performed well in terms of plant establishment, but differed in seedling or plant vigor. Most of the *A. tricolor* cultivars had poor plant vigor and were susceptible to damage by cutworms and leafrollers (Lepidoptera: *Pyralidae*). They also produced seed head (bolting) early before producing considerable leaf area. The number of days from planting to first harvest ranged from 40 to 47 days (Table 1). Edible leaf fresh yield was highest (1158 g/m^2) for amaranth cv. Callaloo and lowest (240 g/m) for amaranth cv. Greenleaf. It appears that 'Callaloo' is suitable for production in the Virgin Islands.

Celosia

Celosia (*Celosia argentia* L.) is present in Africa and Asia both as a weed and as a cultivated leaf vegetable resembling amaranth. Some species and cultivars with a wide variation in leaf color are grown as ornamental plants. The vegetable type celosia is the most important leaf vegetable of Southern Nigeria and is popular in Benin, Zaire, and Indonesia. It is grown in home gardens and small farms both for home consumption and marketing. The plants are vigorous annuals that grow rapidly from seed. They are upright with alternate leaves and few branches until near flowering time. The flowers are borne in dense heads that yield large numbers of edible seeds. The flowers are often brilliantly colored, and the green foliage may contain large amounts of anthocyanin pigments. The leaves, young stems, and young flowers are eaten as a pot herb. Much of the pigment is lost on cooking, but the leaves retain a pleasant green color.

Three cultivars were evaluated during the 1997 spring season. *Celosia argentia* cv. USA produced the highest yield and productivity (Table 1). *C. argentia* cv. India and *C. argentia* cv. Quailgrass have similar yield of edible leaves. All cultivars were resistant to pest and diseases. Celosia appears to be a good alternative leaf vegetable to local amaranth which is very susceptible to many insect pests.

Table 1. Yield and productivity of warm season tropical leaf vegetables in the Virgin Islands, Summer-Fall, 1997.

Common name	Botanical name	Days to first harvest	Edible leaf fresh wt. ±SE (g/m^2)	Daily productivity ±SE (g/m^2)
Amaranth				
Local	*Amaranthus dubius*	42	365±2.5	8.70±0.2
Tigerleaf	*Amaranthus tricolor*	41	455±2	11.1±0.1
Callallo	*Amaranthus cruentus*	40	1158±15	29.0±0.4
Greenleaf	*Amaranthus tricolor*	47	240±7	5.1±0.2
Gangeticus	*Amaranthus tricolor*	41	295±12	7.2±0.3
Merah	*Amaranthus tricolor*	41	432±8.5	10.5±0.2
Pinang	*Amaranthus tricolor*	42	430±3	10.2±0.1
Puteh	*Amaranthus tricolor*	42	367±14	8.7±0.3
Celosia				
USA	*Celosia argentia*	41	1604±584	39.1±14.2
India	*Celosia argentia*	41	650±17	15.9±0.4
Quailgrass	*Celosia argentia*	42	615±8	14.6±0.2
Bush Okra	*Corchorus olitorius*	43	735±18	17.1±0.4
Malabar				
Green	*Basella alba*	57	344±4.5	6.04±0.1
Red	*Basella rubra*	57	385±81.5	6.75±1.43
Sweet potato	*Ipomoea batatas*	42	821±98.0	14.7±1.8
Water spinach	*Ipomoea reptans*	57	412±68.0	7.23±1.2

BRASSICACEAE

Arugula

Arugula (*Eruca sativa* Mill.) is a low growing, annual leaf vegetable with dull green, deeply cut, compound leaves. The edible leaves are characterized by a distinctive spicy, pungent flavor resembling horseradish. The leaves are used in a young tender stage for salads and sometimes cooked as a potherb. The plant was considered by early writers as a good salad herb, but not to be eaten alone. Ancient Egyptians and Romans both have considered the leaves in salads to be an aphrodisiac. Arugula is a very minor crop in the US. In Florida, it is grown to a limited extent commercially and in home gardens where it seems to do quite well (Stephens 1988).

In the Virgin Islands, arugula grows best during fall planting, where it takes 61 days from planting to first harvest. Average edible fresh yield is 840 g/m^2 (Table 2). When grown during the hot summer months, arugula tends to produce flower heads (bolting) and susceptible to insect pest damage. It is a suitable leaf vegetable for the Virgin Islands where there is demand from local food stores, restaurants, and hotels.

Chinese Mustard

Chinese mustard [*Brassica juncea* (L.) Czern.], is a popular leaf vegetable in the Far East. In contrast to Chinese cabbage the petioles of mustard have no wings and are not swollen, instead the dented leaf blades are thin and crispy, and the taste is sharp. Some cultivars have a strong pungent taste. Leaves of Chinese mustard are deeply notched, narrow, and feathery. A single plant may have as many as 20–50 leaves clustered together in a compact bunch. Local mustard cultivars are used as leaf vegetables in tropical Asia. The leaves may be eaten raw, as in a salad. As a potherb it is prepared in many ways: as a steamed or boiled well-seasoned green, stir-fried, in soups, or mixed with other vegetables. Like other mustard, Chinese mustard is rich in vitamins and minerals.

Chinese mustard grows well when planted in the fall season in Virgin Islands. However, it is not as productive as the common mustard greens. Average edible leaf yield is only 5–10% of common mustard greens cvs. Florida Broadleaf and Savanna (Table 2). Nevertheless, Chinese mustard is a promising specialty crop in the Virgin Islands.

Pak Choi

Pak choi (*Brassica rapa* L. var. *chinensis*) is a very popular tropical leaf vegetable. It is a non-heading Chinese cabbage with prominent white, fleshy petioles and upstanding glabrous leaves forming a loose rosette as in swiss chard. The large leaves are glossy and dark green. Pak choi flowers and sets seed very easily at high temperatures and long days are favorable for flower development (bolting). It is a quick maturing plant which can be harvested 30 to 45 days after planting. Individual leaves or entire heads are harvested, used raw or cooked. The popularity of pak choi as a summer vegetable in temperate zones and as an all-year leaf vegetable in the humid tropics is increasing. In the Virgin Islands, it is one of the most productive leaf vegetables grown during the fall season with average edible leaf yield of 3577 g/m^2 (Table 2). It is being grown by many home gardeners in St. Croix and St. Thomas. It is seen in local markets and on farmers' market.

Table 2. Yield and productivity of cool season tropical leaf vegetables in the Virgin Islands, fall, 1997.

Common name	Botanical name	Days to first harvest	Edible leaf fresh wt. ±SE (g/m^2)	Daily productivity ±SE (g/m^2)
Arugula	*Eruca sativa*	61	840±14	9.2±0.2
Chinese mustard	*Brassica juncea*	38	451±42	12.9±1.2
Mustard				
Florida	*Brassica juncea*	33	2118±141	176.5±11.5
Savanna	*Brassica juncea*	34	4169±1612	720.5±105
Pak choi	*Brassica rapa* var. *chinensis*	38	3577±400	32.3±4.3
Komatsuna	*Brassica rapa* var. *perviridis*	51	3843±131	19.5±0.3

Komatsuna

Otherwise known as Japanese mustard, komatsuna (*Brassica rapa* L. var. *perviridis*) is an annual cool season leaf vegetable. The plant appears similar to common mustard, but grows faster and bigger than mustard. Leaves are broad and oval in shape with dark green color. It has the combined flavor of mustard and spinach and remains tender in dry and hot weather. It can be grown year-round and tolerates cold weather. It is the most productive leaf vegetable in the evaluation trial at the experiment station. It matures in 51 days after planting with average edible leaf yield of 3843 g/m^2 (Table 2).

BASELLACEAE

Malabar Spinach

Malabar spinach (*Basella* spp. L.) is also known as Ceylon spinach, vine spinach or Malabar nightshade. It is a climbing perennial plant, mostly cultivated as an annual vegetable against a support in home gardens but in some areas as a vine like market vegetable without staking. There are two common species of Malabar spinach, the red stem and leaves (*Basella rubra* L.) and the green leaves and white stem (*Basella alba* L.). Malabar spinach is not a true spinach (*Spinacia oleracea* L., chenopodiaceae), but its leaves, which form on a vine, resemble spinach, and are used in the same way. The plant is a native of the East Indies, and found its way to the New World from China. It has spread throughout the tropical world and it is one of the best tropical spinach widely adapted to a variety of soils and climates. It is particularly abundant in India, Malaysia, and the Philippines, but it is also seen throughout tropical Africa, the Caribbean, and tropical South America.

Malabar spinach has thick tender stems and the leaves are almost circular to ovate, alternate, and short petioled. They are thick, rugose, succulent, and colored from green to purple. The flowers, borne on axillary spikes or branching peduncles are bisexual and inconspicuous. The fruits are fleshy and purplish black and the juice is sometimes used as a dye.

The succulent young and mature leaves, and the stems are eaten. The most common method of cooking is as a pot herb, mixed with stew or other vegetables. On cooking, the green stem/leaf species retains its fresh green color. The red species loses much pigment to the water and is less attractive. The leaves have mild flavor or are almost tasteless. The stems may be somewhat bitter, and become gelatinous or mucilaginous especially when overcooked. Malabar spinach is a good source of vitamins A and C, calcium, and iron.

Malabar spinach is a perennial that tends to extend itself over time. Seeds can be sown directly or vines may be established directly from stem cuttings. These need a little shade on transplanting, but root readily. Malabar spinach can thrive under conditions of moderate soil fertility, but is quite responsive to nitrogen fertilizer. Evaluation trial at the experiment station indicated that plants can be harvested at 57 days after planting. The red species is slightly more productive than the green species (Table 1). Edible leaf yield was 385 g/m^2 for the red species compared to 344 g/m^2 for the green species. Malabar spinach is one of the rapidly growing tropical leaf vegetables in the Virgin Islands, responds well to pruning and nitrogen fertilizer. In addition, it is tolerant to insect pests and diseases. It is definitely one of the minor tropical leaf vegetables with market potential in the Virgin Islands.

CONVOLVULACEAE

Water Spinach

Water spinach, kangkong, swamp cabbage, or water convolvulus (*Ipomoea aquatica* Forsk., or *Ipomoea reptans* Poir) is an important green leaf vegetable in most of Southeast Asia. It is a trailing tropical lowland plant, related to sweet potato. Two main cultivar groups can be distinguished: var. *aquatica* and var. *reptans*. The first is an aquatic plant or paddy vegetable in the Southern part of India and Southeast Asia, propagated by cuttings and growing in the wild or cultivated in fish ponds and water courses. The second is an upland vegetable, cultivated on dry or marshy land and propagated by cutting or seeds. Both types are an important market vegetable in Malaysia, Indonesia, and other Southeast Asian countries. Several cultivars are known, but the most important distinction is between upland (dry) forms and paddy (swamp) forms.

Water spinach develops a trailing vine that spreads rapidly by rooting at the nodes. Vertical branches arise from the leaf axils. It is quite glabrous, with sagitate, alternate leaves. The leaves are somewhat succulent, particularly in the wet land form, and has a pleasant light green color. A white flower is produced which matures into a 4-seeded pod.

Almost all parts of the young plant are eaten. Older stems, especially from plants cultivated on dry land, contain considerable fiber. Therefore, cultural methods emphasize the production of young succulent tips. These can be eaten fresh in salads. Often they are cooked as spinach. The flavor is bland and some spicy ingredients or salt are added to enhance flavor. The leaves maintain much of their green color, but the stems are yellowish when cooked (Martin and Ruberte 1979).

Water spinach is planted either from seed or from cuttings. Seeds do not germinate well under water, but can be direct seeded. Plants are normally grown in nursery beds for later transplanting in the field. In evaluation trial conducted at the experiment station, the upland type of water spinach was harvested 57 days after planting. The average edible leaf yield was 412 g/m^2 (Table 1). Productivity was about similar with Malabar spinach. Under Virgin Islands climatic conditions, water spinach grows well during summer-fall season and is a suitable leafy green vegetable with market potential.

Sweet Potato

The leaves of sweet potato (*Ipomoea batatas* L.) are used as a potherb in Southeast Asia, the Pacific, and locally in Latin America. It is an important foodstuff for the highland population of New Guinea. Sweet potato leaves are considered as a cheap and coarse vegetable. Stems and leaves are used as forage. Often considered a poor man's food, sweet potato leaf has a rich protein content that helps fill the nutritional gap left by eating principally the protein-poor tubers. Sweet potato leaves are particularly important, and cultivars have been developed that are used only for the leaves. These cultivars are rich in calcium. However, cultivars differ in general appearance, flavor, and bitterness. Many cultivars have a resinous flavor that is acceptable unless quite strong.

Sweet potato merits a place in tropical gardens because it is easy to culture and yields edible tubers as well as leaves. Leaves and tubers can be produced year round and plants are adapted to a wide range in climatic conditions. Most soils are suitable for sweet potato, but soils rich in organic matter promote lush growth of leaves. Sweet potato is adapted to calcareous soils of the Virgin Islands. Leaves and young shoots can be harvested in 42 days after planting (Table 1). It is more productive than amaranth, Malabar spinach, and water spinach. Frequent harvest stimulates development of side shoots and vines. Although it is a perennial, its succulent nature restricts its cultivation to relatively short growing seasons of 3 to 5 months. It is definitely a suitable leaf vegetable for the Virgin Islands and a good alternative to local spinach.

TILIACEAE

Bush Okra

Bush okra, jew's mallow, or jute mallow (*Corchorus olitorius* L.) is primarily known as a fiber crop, however, special types with shorter and more branched stems are frequently cultivated as a mucilaginous tropical leaf vegetable. Bush okra is one of the popular tropical leaf vegetables in Africa, Asia, and some parts of the Middle East. The plant belongs to the Tiliaceae and is characterized as an annual upright, branching, glabrous, slightly woody herb. Leaves are narrow and serrate, about 5–13 cm in length. Flowers are small, yellow-petioled, and borne in small clusters in the leaf axils. The cylindrical capsules of 2–5 cm are produced in large numbers, especially during the short days (Martin and Ruberte 1979). Seeds are dark bluish-green, angular, and about 2 mm long.

Bush okra is one of the leading leaf vegetables in West Africa and is often stored dry. It is also commonly used in Malaysia, the Philippines, and parts of Latin America. It is the most important leaf vegetable in Egypt, where it is cultivated from March to Nov. (Oomen and Grubben 1978). The nutritional value of bush okra compares very well with other common tropical leaf vegetables. It is high in protein, fiber, calcium, iron, and carotene. The edible shoot tips and leaves are always eaten and cooked as a potherb. Their edible qualities are widely appreciated in West Africa where the shoots and leaves are combined in stews to be eaten as a starchy paste. In India the shoots are cooked with rice.

Although bush okra is a popular leaf vegetable in many countries of the tropics, little research and development work have been done to improve its culture and production. According to Oomen and Grubben (1978) seed yields of bush okra are low, and germination is often very poor due to dormancy which can be overcome by soaking in hot water. Leaf production is also low compared to other tropical leaf vegetables, but dry matter content is high. Trials at the experiment showed that bush okra can be harvested in 33 days after planting, however, edible leaf yield is very low (Table 1). Yield and productivity can be increased by increasing planting density. Studies by Palada and Crossman (1998) indicated that a planting density of 98,522 plants/ha or a plant spacing of 50 × 20 cm was optimum for maximum yield of bush okra. Bush okra is resistant to damage by pests and diseases. It is one of the most suitable leaf vegetables for growing in the Virgin Islands.

SUMMARY

The germplasm evaluation trials indicate that under Virgin Islands conditions, most of the cool season *Brassica* spp., including the Oriental greens show potentials for adaptability and higher productivity. The warm season species such as the Malabar spinach, celosia, and sweet potato performed better than amaranth, bush okra, and water spinach. Planting density study with bush okra indicated that yield and productivity can be increased with closer spacing. Crop management trials involving plant spacing and fertilizer application are on-going to improve the yield of the common species including amaranth, Malabar spinach, celosia, and water spinach. When outstanding species and cultivars are identified and improved cultural management practices are developed, local growers will be able to adopt these recommendations to enhance production of tropical leaf vegetables. Future efforts will be focused on product development and marketing of these specialty vegetables.

REFERENCES

Cook, R. 1990. Catering to the American consumer. Fresh trends '91. Packer 54:12,14,16,18,20,22,24,26.

Lamberts, M. 1990. Latin American vegetables. p. 378–387. In: J. Janick and J.E. Simon (eds.), Advances in new crops. Timber Press, Portland, OR.

Lamberts, M. 1993. New horticultural crops for the Southeastern United States. p. 82–92. In: J. Janick and J.E. Simon (eds.), New crops. Wiley, New York.

Martin, F.W. and R.M. Ruberte. 1979. Edible leaves of the tropics. Antillan College Press, Mayaguez, Puerto Rico.

Oomen, H.A.P.C. and G.J.H. Grubben. 1978. Tropical leaf vegetables in human nutrition. Communication 69, Dept. of Agr. Research, Royal Tropical Institute, Amsterdam, Netherlands. Orphan Publishing Co., Willemstad, Curacao.

Palada, M.C. and S.M.A. Crossman. 1998. Planting density affects growth and yield of bush okra. Proc. Caribbean Food Crops Soc. 34:(in press).

Pearrow, J. 1991. U.S. imports of fruits and vegetables under plant quarantine regulations, Fiscal year 1988. U.S. Dept. of Agriculture. Economic Research Service. Washington, DC.

Stephens, J.M. 1988. Manual of minor vegetables. Bul. SP-40. Co-op. Extension Service, Univ. of Florida, Gainesville, FL.

Evaluation of Macabo Cocoyam Germplasm in Cameroon

O.U. Onokpise, J.G. Wutoh, X. Ndzana, J.T. Tambong, M.M. Meboka, A.E. Sama, L. Nyochembeng, A. Aguegia, S. Nzietchueng, J.G. Wilson, and M. Burns*

Macabo cocoyam [*Xanthosoma sagittigolium* (L.) Schott], Araceae is an important food crop for more than 400 million people worldwide, especially in the tropics and subtropics. Ethnic groups in Cameroon prepare, process, and consume cocoyams in many forms (Tandehnije 1990). These include: (1) cormels peeled, boiled, and eaten with a vegetable sauce; (2) *ekwan*, a delicacy obtained by tying grated peeled corms (underground stem) or cormels with younger leaves plus some palm oil, fish and crayfish, salt and pepper; (3) *belbach*, a special thick sauce prepared from young tender unopened leaves and tender petioles; (4) *nyeh* bell soup in which the young tender leaves are used as vegetables; (5) *kohki-beans*, cowpea cake prepared by mixing ground cowpea, palm oil, salt and pepper, and young leaves, tying with plantain leaves, and eaten with boiled cormels or plantains; (6) *kohki-corn*, corn cake prepared by mixing ground corn, palm oil, young tender cocoyam leaves, salt and pepper, and tying with plantain leaves; (7) the corms or cormels peeled, boiled, and pounded into futu; (8) cormels transformed into a porridge; and (9) *akwacoco*, a delicacy obtained by grating peeled cormels, mixing with pieces of crayfish, palm oil, salt and pepper, and boiling. In addition, in some locations the large petioles and leaves and some of the roots find widespread use in the local medicinal industry.

For many years, production of macabo cocoyams declined significantly in Cameroon and other cocoyam producing countries, due largely to a root rot disease principally caused by *Pythium myriotylum*. The United States Agency for International Development (USAID) and the Government of the Republic of Cameroon (GRC) funded a Root and Tubers Research Project (ROTREP) in Cameroon with the main objective of developing tolerant/resistant cultivars with acceptable agronomic and sociological characteristics. Assemblage of cocoyam germplasm from different agroecological zones of Cameroon was a major part of the breeding program. Additional accessions were collected from other places like Gabon, Ghana, and Puerto Rico. The root rot disease of macabo cocoyam caused by *Pythium myriotylum* (Steiner 1981; Nzietchueng 1985; Pacumbaba et al. 1992) and dashen mosaic virus (DMV) (Anon. 1991) have been largely responsible for the significant decline of macabo or tannia cocoyam. The crop is widely consumed throughout the tropical regions of the world. In fact, in some places such as the Cameroon, this crop ranks second only to cassava (*Manihot utilisima*) as an important source of energy in daily diets of the people (Lyonga 1980; Onokpise et al. 1992a).

The lack of adequate supply of planting materials associated with the root rot disease is also responsible for the overall low production of cocoyams. Studies on seed production have been undertaken (Alamu 1978; Wilson 1984; Onokpise et al. 1992b), and in vitro plant regeneration of cocoyams have now been demonstrated (Nyochembeng and Garton 1998). Yet the original problems that led to the funding of ROTREP by USAID and GRC, have yet to be resolved. This paper presents preliminary results from several studies whose ultimate objective is the development of cocoyam cultivars with acceptable agronomic and sociological characteristics.

THE GERMPLASM COLLECTION

Between 1986 and 1991, over 300 accessions of cocoyam were assembled from three agroecological zones in Cameroon, from Equatorial Guinea, Gabon, Ghana, Puerto Rico, and Togo, by teams assembled at the Ekona Research Center (Onokpise et al. 1993). Much of this cocoyam population in Cameroon and the rest of the West African Region, are land races of previous introductions especially by the Portuguese. Collections from Puerto Rico were obtained through formal requests sent to scientists working on *X. sagittifolium* in that region. Data was collected on source of planting materials, disease incidence using the Cameroonian National Scoring System(CNSS) (Nzietchueng 1985; Anon. 1991).

---

*Funds provided by USAID and GRC under project No. 631-0051 for these studies are gratefully acknowledged. We thank the executive committee of ROTREP for the logistical support given to these studies. Special thanks are extended to all technical staff and drivers at ROTREP/IRAD for their technical assistance.

All accessions brought back from the expeditions were initially planted in the project's greenhouse prior to being transplanted to the field. Field evaluation included petiole length, yield, and stand, and disease in incidence evaluated by a scale of 0 (no symptom) to 4 (76–100% yellowing) using Nzietchueng (1985) and Pacumbaba et al. (1992).

Detailed observations have been reported for collections made in Ndian Division of the Southwest Province, in agro-ecological zone I of Cameroon (Onokpise et al. 1993). The high prevalence of root rot disease remains a serious limitation to cocoyam production. Contrary to previously undocumented reports "yellow" cocoyams produced significant amounts of cormels, suggesting resistance to the root rot disease. Cocoyam farmers carried out several cultural practices such as multiple cropping, hill planting, early planting, and early harvesting in order to reduce the disease incidence and its devastating effects on yield. A fungicide, "Ridomil Plus 72," was very effective in controlling the disease, but its cost may be prohibitive to the small scale farmer and the supply is unsteady. A combination of cultural management and fungicidal application could enhance local production until tolerant/resistant cultivars are developed and released by cocoyam breeders. Conventional plant breeding and biotechnology procedures are underway to develop cultivars tolerant or resistant to the disease.

PROTEIN EVALUATION

Tuber protein was determined by the micro-Kjedahl process, and leaf protein involved a block digester method. The LKB horizontal electrophoretic unit and ampholine PAG plates (LKB, Producer, Sweden) were used for electrophoretic characterization. Protein extracts were obtained from fresh cormels by grinding and centrifuging in distilled water and using the supernatant as source of protein extract for electrophoretic identification of accessions. Standard proteins were included in the electrophoretic runs of various accessions. Each electrophoretic run was repeated three to four times. Using the anodic ends of the ampholine PAG plates, isoelectric points were determined and accessions were classified into acidic and alkaline protein types.

Mean tuber proteins in accessions ranged from 2.5% to 9.4%. The mean tuber protein contents were 5.1% (white), 5.2% (yellow), and 5.4% (red) cocoyams. Leaf proteins were significantly higher than tuber proteins and ranged from 11.5% to 25.6% crude protein. Younger leaves had higher protein content than the older ones. Mean protein contents of some of the accessions is shown in Table 1. Crude protein of cocoyam cormels compares favorably with other root and tuber crops such as cassava (1–2%), yams (1.1–2.8%), and sweetpotato (0.95–2.4%), but is lower than taro corms (up to 7%) (Onwueme 1978).

Table 1. Mean protein content (%) of some cocoyam germplasm sources in Cameroon.

Accession	Type	Geographic location of collection	Crude peeled tuber protein (%)
048	Red	Douala, Km 21	7.78a[z]
005	White	Njombe, Km 7	7.20a
XAN 141	Yellow	Njombe	5.01b
037	White	Tombel	4.78b
004	White	Sanje	4.29b
010	White	Bonakanda	4.23b
022	White	Bafia	4.20b
045	Red	Bonaberi	4.14bc
051	Red	Mloboyek	3.85bc
046	Red	Duala Km 10	3.58c
009	White	Upper Bokova	3.55c
041	Red	Yato-17 km to Douala	3.50c
023	White	Banga Bakundu	3.24c
001	White	Bobende	3.18c

[z]Mean separation by Duncan's Multiple range test, 5% level.

The alkaline proteins with isoelectric points of 8.53 and 9.41 corresponded to the standard protein L-lactic-dehydrogenase whose isoelectric points range from 8.30 to 8.55. The acidic proteins with isoelectric points ranging from 4.24 to 4.53, are related to the standard soybean trypsin inhibitor (SBTI), because the banding patterns were found in the region of 1.0 to 1.5 cm, the distance from the anode where the SBTI is usually found during electrofocusing.

FLOWER INDUCTION AND FRUIT FORMATION

Flower induction was initiated by spraying gibberellic acid (GA$_3$) at 750 ppm, to the cocoyam leaves until the petiole cup was full. More than 700 crosses were made over a five-year period.

Plants flowered 50 to 70 days after GA$_3$ was applied, and fruits matured in 40 to 60 days. The fruit is dome shape and is made up of a dense cluster of berries. The mean number of berries per fruiting head was 244 while the weight of each head was 16 g. An average of 15 seeds were found in each berry, with 100 seeds weighing 26.6 mg. Thus, a single cocoyam plant producing an average of five inflorescences from which five successful fruits are harvested, can give rise to an average of 20,000 seeds.

Hybridization resulted in the production of more than 10,000 seeds from "white" × "white and "white" × "red" crosses. Virtually no viable seeds were produced from the "white" × "yellow" or "red" × "yellow" crosses perhaps due to ploidy differences.

REFERENCES

Alamu, S. and C.R. McDavid. 1978. Production of flowering in edible aroids by gibberellic acid. Trop. Agr. 55:81–86.

Anon. 1991. Tropical Roots and Tubers Research Project. Fifth Annual Report. Institute of Agronomic Research, Cameroon.

Lyonga, S.N. 1980. Cocoyam production in Cameroon. International Foundation for Science, Stockholm, Provisional Rep. 5.

Nyochembeng, L.M. and S. Garton. 1998. Plant regeneration from cocoyam callus derived from shoot tips and petioles. Plant Tissue Cell Culture 53: 127–134.

Nzietchueng, S. 1985. Genre *Xanthosoma* (macabo) et contrainte de production: cas particulier de la pourriture racinaire cause par *Pythium myriotylum* au Cameroon. These Doctorat es Sciences Naturelles, Universite de Yaunde.

Onokpise, O.U., J.T. Tambong, L. Nyochembeng, and J.T. Wutoh. 1992a. Acclimatization and flower induction of tissue culture derived cocoyam (*Xanthosoma sagittifolium*, Schott) plants. Agronomie 12:193–199.

Onokpise, O.U., M. Boya-Meboka, and J.T. Wutoh. 1992b. Hybridization and fruit formation in macabo cocoyam (*Xanthosoma sagittifolium* (L) Schott). Ann. Appl. Biol. 120:527–535.

Onokpise, O.U., M.M. Meboka, and A.S. Eyango. 1993. Germplasm collection of macabo cocoyams in Cameroon. African Tech. Forum 6:28–31.

Onwueme, L.C. 1978. The tropical tuber crops, yams, cassava, sweet potato, and cocoyam. Wiley, Chichester, UK.

Pacumbaba, R.P., J.G. Wutoh, A.E. Sama, J.T. Tambong, and L.M. Nyochembeng. 1992. Isolation and pathenogenicity of rhyzophere fungi of cocoyam in relation to the cocoyam root rot disease. J. Phytopath. 135:265–273.

Steiner, K.G. 1981. A root rot of macabo (*Xanthosoma* sp.) in Cameroon associated with *Pythium myriotylum*. Plant Distr. Protection 88:608–613.

Tandehnjie, J. 1990. Comparison of the crude protein and moisture contents of leaves and petioles among cocoyam (*Xanthosoma sagittifolium*) accessions. Research Report submitted to ITA, University Center, Dschang, June, 1990.

Wilson, J.E. 1984. Cocoyam. Chapter 18. In: P.R. Goldsworthy and N.M. Fisher (eds.), Physiology of tropical field crops. Wiley, New York.

Pointed Gourd: Potential for Temperate Climates

Bharat P. Singh and Wayne F. Whitehead

Pointed gourd (*Trichosanthes dioica* Roxb., Cucurbitaceae) is a tropical vegetable crop with origin in the Indian subcontinent. It is known by the name of *parwal*, *palwal*, or *parmal* in different parts of India and Bangladesh and is one of the important vegetables of this region. The fruit is the edible part of the plant which is cooked in various ways either alone or in combination with other vegetables or meats. Pointed gourd is rich in vitamin and contains 9.0 mg Mg, 2.6 mg Na, 83.0 mg K, 1.1 mg Cu, and 17.0 mg S per 100 g edible part (Singh 1989). It is purported that pointed gourd possesses the medicinal property of lowering total cholesterol and blood sugar. These claims are supported by preliminary clinical trials with rats (Chandra-Sekar et al. 1988) and rabbits (Sharma and Pant 1988; Sharma et al. 1988).

BOTANY

The plant is a perennial, dioecious, and grows as a vine (Fig. 1). Roots are tuberous with long taproot system. Vines are pencil thick in size with dark green cordate simple leaves. Flowers are tubular white with 16–19 days initiation to anthesis time for pistillate flowers and 10–14 days for staminate flowers. Stigma remains viable for approximately 14 hours and 40–70% of flowers set fruit (Singh et al. 1989). Based on shape, size and striation, fruits can be grouped into 4 categories: (1) long, dark green with white stripes, 10–13 cm long, (2) thick, dark green with very pale green stripes, 10–16 cm long, (3) roundish, dark green with white stripe, 5–8 cm long, and (4) tapering, green and striped, 5–8 cm long (Singh 1989).

CULTIVATION

The pointed gourd is usually propagated through vine cuttings and root suckers. Seeds are not used in planting because of poor germination and inability to determine the sex of plants before flowering. As a result, crop established from seed may contain 50% nonfruiting male plants. To propagate from root suckers, tuberous roots of pointed gourd are dug in the early spring, subdivided, and replanted. Both pre-rooted and fresh vine cuttings are used for propagation. Vine cuttings made in the fall of previous year and rooted during winter are planted when danger from frost is over in the spring in order to obtain a crop in the same year. Current year vine cuttings are also planted to establish the crop during the summer, but optimum plant yield is only obtained during the next year. Fresh vines used for field planting should have 8–10 nodes per cutting and should be partially or fully defoliated to check transpiration. The distance between plants is kept between 1.5–2.0 m × 1.5–2.0 m depending on the method of training of vines (Singh 1989; Yadav 1989). A female:male ratio of 9:1 is optimum for ensuring maximum fruit set (Maurya et al. 1985).

Pointed gourd prefers a well-drained sandy loam soil with good fertility. Das et al. (1987) reported maximum early as well as total yield at N:P rates of 90:60 kg/ha, while Kumar et al. (1990) obtained maximum number of fruits/plant when both N and P were applied at the rate of 60 kg/ha.

Vines require training on some form of aerial support system to achieve maximum fruit production (Prasad and Singh 1987; Yadav et al. 1989). Singh (1989) reported 14% higher yield on vines trained on bower system compared to those growing on the ground. In tropics, pointed gourd produces maximum yields for 3–4 years, after which yielding potential gradually declines (Samalo and Parida 1983).

To determine whether pointed gourd can be grown successfully in temperate climates, a study was initiated in 1994 at the Fort Valley State University in Georgia. Cuttings of male and female vines were ob-

Fig. 1. Pointed gourd trained on a fence wire trellis

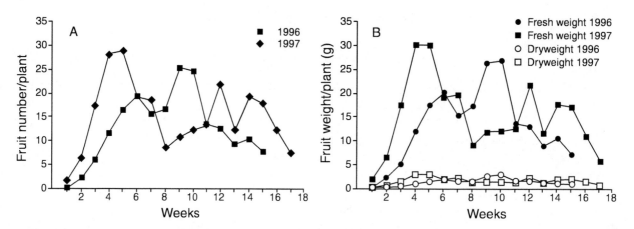

Fig. 2. (A) Per plant fruit number, and (B) fresh and dry weight of pointed gourd on a weekly basis in Georgia during 1996 and 1997.

Table 1. Total fruit number, fruit fresh and dry yields of pointed gourd in Georgia during 1996 and 1997.

Year	Total fruit number		Fresh fruit yield		Dry fruit yield	
	Per plant	Per ha	Per plant (kg)	Per ha (t)	Per plant (kg)	Per ha (t)
1996	190	614,840	5.0	16.2	0.5	1.6
1997	254	821,944	6.5	21.0	0.7	2.3

tained from the Department of Vegetable Crops, Narendra Deva University of Agriculture and Technology, Faizabad, India with the assistance of National Bureau of Plant Genetic Resources, New Delhi, India, and National Germplasm Resources Laboratory, Beltsville, Maryland. Cuttings were first multiplied in the greenhouse under mist. Rooted cuttings were planted in the field in six 12 m long rows during April 1995. Planting distance was 1.52 m in the row and 1.83 m between rows. Thus, planting density amounted to 3,595 plants/ha, of which 3,236 were female and 359 male (female:male ratio 9:1). Vines were trained on trellises made of 1.52 m high fence wire. Plants were cut back to the ground level before frost and roots covered with straw to safeguard from cold during winter. Vines sprouted from over-wintered roots during subsequent years in spring when the average soil temperature reached above 12.5°C.

HARVESTING AND YIELD

Pointed gourd vines produced limited number of fruits during 1995, however, full scale fruiting only began in 1996. Fruits were produced for harvest from the beginning of July and continued to the middle of October. There was a continuous increase in the number of fruits produced during the first 4 weeks, thereafter, variation among weekly fruit numbers was dependent on the environmental conditions (Fig. 2). Plants produced fruits for harvesting for 15 weeks in 1996 and 17 weeks in 1997. Harvesting was carried out twice a week to obtain fruits at proper maturity for cooking. Over matured fruits developed hard seeds, rendering them less desirable. It took approximately 15 days for fruits to reach the marketable size from fruit-set. Seasonal pattern for the fresh and dry fruit yield /plant was similar to the fruit number/plant. Total fruit number, fresh and dry yields on per plant basis were higher in 1997 than 1996 (Table 1). Fresh fruit yields/ ha for 1996 and 1997 were 16.2 and 21.0 t, respectively. These yield levels compare favorably to those reported from the Indian subcontinent (Singh 1989).

PROSPECT FOR THE CROP IN THE UNITED STATES

Two main factors will determine the prospect of pointed gourd in the United States: (1) the ability of the plant to adapt to the temperate climate, and (2) market demand for the crop. Research conducted in Georgia clearly demonstrate that pointed gourd can overwinter successfully and produce fruit for approximately 16 weeks during summer with yield comparable to that reported from India. The demand for this vegetable from ethnic minorities from the Indian subcontinent is also high and occasional imports are sold for US $9–10/kg in International Farmers Markets and ethnic grocery stores. However, since fruit harvest is spread over a long period, pointed gourd appears only suited for production on small areas. Price of this commodity produced locally would probably be similar or higher than the imports. The labor employed in manually harvesting these small sized fruits will have to be paid at a higher rate in the US than the exporting countries offsetting any transportation cost advantage from producing the crop locally. Therefore, pointed gourd provides most opportunity to small farmers living close to metropolitan cities where it can be grown as U-Pick or for supply to International Farmers Markets and ethnic grocery stores. The crop also has potential for production in home gardens where it can provide a nutritious vegetable for an extended period.

REFERENCES

Chandrasekar, B., B. Mukherjee, and S.K. Mukherjee. 1988. Blood sugar lowering effect of *Trichosanthes dioica* Roxb. in experimental rat models. Int. J. Crude Drug Res. 26:102–106.

Das, M.K., T.K. Maity, and M.G. Som. 1987. Growth and yield of pointed gourd (*Trichosanthes dioica* Roxb.) as influenced by nitrogen and phosphorus fertilizer. Vegetable Sci. 14:18–26.

Kumar, R., R.K. Singh, M.M. Pujari, and R. Kumar. 1990. Effect of nitrogen and phosphorus on pointed gourd (*Trichosanthes dioica* Roxb.): A note. Haryana J. Hort. Sci. 19:368–370.

Maurya, K.R., S. Barooah, R.K. Bhattacharya, and R.K. Goswami. 1985. Standardization of male plant population in pointed gourd. Ann. Agr. Sci. 30:1405–1411.

Prasad, V.S.R.K. and D.P. Singh. 1987. Effect of training on pointed gourd (*Trichosanthes diocia* Roxb.) for growth and yield. Progressive Hort. 19:47–49.

Samalo, A.P. and P.B. Parida. 1983. Cultivation of pointed gourd in Orissa. Indian Farming 33:29–31.

Singh, K. 1989. Pointed gourd (*Trichosanthes dioica Roxb.*). Indian Hort. 33:35–38.

Singh, A.K., R.D. Singh, and J.P. Singh. 1989. Studies on floral biology of pointed gourd (*Trichosanthes dioica* Roxb.). Vegetable Sci. 16:185–190.

Sharma, G., and M.C. Pant. 1988. Effects of feeding *Trichosanthes dioica* (parval) on blood glucose, serum triglyceride, phospholipid, cholesterol, and high density lipoprotein-cholesterol levels in the normal albino rabbit. Current Sci. 57:1085–1087.

Sharma, G., M.C. Pant, and G. Sharma. 1988. Preliminary observations on serum biochemical parameters of albino rabbits fed on *Trichosanthes dioica* (Roxb.). Indian J. Medical Res. 87:398–400.

Yadav, J.P., K. Singh, R.C. Jaiswal, and K. Singh. 1989. Influence of various spacings and methods of training on growth and yield of pointed gourd (*Trichosanthus dioica* Roxb.). Vegetable Sci. 16:113–18.

Cucurbit Resources in Namibia*

Vassilios Sarafis

Namibia has several cucurbits with potential for development into commercial crops either through selection or through the introduction of genes into known crops. *Acanthosicyos horrida* Welw. ex J.D. Hook., wild *Citrullus ecirrhosus* Cogn., and *C. lanatus* (Thunb.) Matsum. & Nakai in the Cucurbitaceae are examples of gene sources. The areas from which these plants come are arid and the plants derive their water needs from dew precipitation in the mornings, very occasional rains every few years, and deep ground water (Seely 1987; Lovegrove 1993).

ACANTHOSICYOS HORRIDA

Acanthosicyos horrida forms clumps of vegetation in the dunes of the Sossuvlei region near Walvis Bay (Fig. 1) (Craven and Marais 1986; Lovegrove 1993; Klopatek and Stock 1994). *Acanthosicyos horrida* is a dioecious perennial cucurbit attaining a height of about 1.5 m (Fig. 2). It forms plants of one sex in single clumps which may touch plants of the same or other sex nearby (Fig. 1). It bears deep water table seeking roots (G. Wardell–Johnson, pers. commun. 1998). The plants are totally leafless (Fig. 2) and have a fruiting habit of oblong spherical fruits reaching up to 25 cm average diameter. The plants are able to build up sand deposits around themselves and continuously grow to be above these sand deposits. New plants establish only when rain falls and quickly form deeply growing roots that seek the water table (G. Wardell–Johnson, pers. commun. 1998).

The fruit may not be spaced apart and may occur in clusters of several touching each other. The fruits are spiny (Fig. 3). Maturation of the fruits occurs between February and April. The fruits do not change color and remain green on the outside but the flesh surrounding the seeds dissociates from the skin, turns orange in color (Fig. 4), extremely sweet in taste and strongly aromatic. Maturational changes are easily detected by the bushmen living in the area without breaking the fruit in any way. The fruits are used by the bushmen for two main purposes. The first is for the extraction of the seed which are consumed as pips by splitting in the mouth and the second is for pulp processing where the flesh is boiled and poured to form a fruit leather. This fruit leather is eaten throughout the year and is considerably less flavorful than the pulp. The plant thus forms an important food resource because of the easy storage of both the seeds and the dried pulp (leather). The fruits are eaten also when immature by animals including jackals and rodents who do not seem to be bothered by the bitter taste of the fruits caused by cucurbitacins (Hylands and Magd 1986).

The mature pulp has a flavor which is aromatic and maybe due in part to sulphur components as in some types of *Cucumis melo* L. No trace can be tasted of cucurbitacins in the mature pulp. The pulp could be commercialized and used to make ice-cream, and could be freeze dried and chocolate coated. The seeds which are already sold to an European population in Walvis Bay can have their market expanded by selling the seeds either whole or dehusked in packaging developed for nuts. Their rarity should provide a premium price and help the economic existence of the bushmen in this area. Ice-cream manufacture and freeze drying facilities are only within 30 km of the bushmen. Partnerships with firms interested in commericalizing the unique, aromatic pulp of *Acanthosicyos horrida* could be fostered to further improve the economic existence of the native people in the area.

CITRULLUS ECIRRHOSUS

Citrullus ecirrhosus is a desert perennial (Fig. 5, 6) which is monoecious. Fruits mature (Fig. 7, 8) February to March. The leaves form an annual stems which die back each year. The leaves have a special feature where the lamina is curved over the mid-rib and the lateral veins so that when viewed from above the top surface is only visible in the vein regions and the leaves have a greenish white appearance due to the lower epidermis being reflected up as the upper surface of the leaf. This lower epidermis is covered with warts and hairs which account for the whitening effect. Both lower and upper epidermis contain similar amounts of

*This project was supported by the Centre for Microscopy and Microanalysis, The University of Queensland and the Centre for Horticulture and Plant Sciences, University of Western Sydney, Hawkesbury.

Fig. 1. View of *Acanthosicyos horrida* in sand dunes at the Sossuvlei region near Walvis Bay Namibia.

Fig. 2. A close up of *Acanthosicyos horrida* plants. Note the leaflessness.

Fig. 3. Back of a mature fruit of *Acanthosicyos horrida* showing the large spines on the surface of the fruit. The distances separating the spines are small in young fruits.

Fig. 4. Cross section through three fruits of *Acanthosicyos horrida*. The one on the extreme right is a bitter immature fruit of full size. The one on the top an almost mature fruit with only a little bitterness. The bottom left hand fruit a fully mature fruit with a flesh having an orange color, no bitterness and very aromatic in flavor.

Fig. 5. *Citrullus ecirrhosus* perennial plant growing approximately 20 km inland from Walvis Bay, showing a mature fruit on current years growth and brown dead stems from last years growth.

Fig. 6. *Citrullus ecirrhosus* perennial plant showing young developing fruit in the foreground and the bending of the leaves over the mid-rib and lateral veins.

Fig. 9. *Citrullus lanatus* mature fruit from a plant growing on a dry river bed approximately 20 km inland from Walvis Bay, cut to show chlorophyll in the flesh and browny-black seeds. The more deeply colored regions of the flesh are green. The flesh is more juicy than in *Citrullus ecirrhosus*.

Fig. 7. Mature *Citrullus ecirrhosus* showing folded nature of the leaves of the mid-rib and lateral veins.

Fig. 8. Fruit of *Citrullus ecirrhosus* cut showing white creamy flesh which is non juicy and brown seeds.

stomata. The water relations of this plant are reliant on a deep water layer in the ground which the roots reach and possibly some water availability from morning fogs and the very occasional rainfall. The fruit and seeds contain cucurbitacins but the seeds are harvested in times of need and processed by crushing and decantation to remove the bitter substances. *Citrulls ecirrhosus* plants may be a source of drought tolerance genes for *Citrulls lanatus*. Successful crossability of *Citrulls ecirrhosus* and *C. lanatus* is discussed in Navot and Zamir (1986) and Navot et al. (1990). They have shown the way for breeding *Citrullus lanatus* containing genes from *C. ecirrhosus*.

CITRULLUS LANATUS

Citrullus lanatus wild plants seen near Walvis Bay have green fleshed fruit unknown from domesticated watermelons (Fig. 9). The genetics of fruit color in the watermelon, *Citrullus colocynthis* and *ecirrhosus* are discussed by Navot et al. (1990). White, yellow, orange, pink, red, and crimson flesh types are known. The green flesh color of this wild *Citrullus lanatus* (Fig. 9) is a unique feature which can be transferred to domestic watermelon due to the crossability of wild and domestic watermelons. This would offer a new fruit type for consumers to enjoy. A red flesh cultivated watermelon from the north of Namibia has some green zone within the fruit suggesting that the green flesh character can be easily introduced. However, the wild watermelon has cucurbitacins which would render them unfit for human consumption. Drought tolerance and green flesh color from *C. ecirrhosus* and wild *Citrullus lanatus,* could be vaiable traits for watermelon improvement.

REFERENCES

Craven, P. and C. Marais. 1986. Namib Flora Swakopmund to the Giant Welwitschia via Goanikontes. Gamsberg MacMillan Publishers: Windhoek. p. 80–83.

Hylands, P.J and M.S. Magd. 1986. Cucurbitacins from *Acanthosicyos horridus*. Phytochemistry 25:1681–1684.

Klopatek J.M. and W.D. Stock. 1994. Partitioning of nutrients in *Acanthosicyos horrida*, a keystone endemic species in the Namib Desert. J. Arid Environments 26:233–240.Lovegrove, B. 1993. The living deserts of Southern Africa. Fernwood Press, Vlaeberg, South Africa. p. 30, 47, 71, 158, 190.

Navot, N. and D. Zamir. 1987. Isozyme and seed protein phylogeny of the genus *Citrullus* (Cucurbitaceae). Plant Syst. Evol. 156:61–68.

Navot, N., M. Sarfatti, and D. Zamir. 1990. Linkage relationships of genes affecting bitterness and flesh colour in watermelon. J. Hered. 81:162–165.

Seely, M. 1986. The Namib. Shell Namibia: Namibia. 2nd ed. 19, 43–45, 50, 84, 90.

FLORAL & LANDSCAPE CROPS

Ornamentals: Where Diversity is King—the Israeli Experience

Abraham H. Halevy

In the increasingly competitive international cut flower and pot-plant market, novelty of crops plays an important role in maintaining and expanding market share. The ornamental industry is unique among the agricultural industries in that novelty is an important attribute. Customers always seek "something new." Although the standard major ornamental crops will continue to constitute an important part of the market, a distinct trend towards increasing the share of "new crops" is clearly evident in recent years. These new products normally fetch higher prices than the traditional crops for a certain period, but quite often the prices drop when the market is saturated, and the attraction novelty lessens. By that time new products should be ready to enter the market. Research on introduction of new ornamental crops is therefore an endless project.

The floriculture industry in Israel is relatively new. Until about 30 years ago cut flowers and pot-plants were only produced on low-scale for the limited domestic market. In recent years ornamental plants became a major agricultural exportable product of over 250 million US$ per annum. Israel is now second only to Holland in flower export in Europe.

Initially Israel produced and exported mainly the major traditional cut flowers, such as carnations, roses, and gladiolus. Gradually the share of these crops declined and those of new minor crops increased, so that the "new crops" now constitute over 60% of the exportable cut flowers (Fig. 1). None of these "new crops" has become a major crop as roses or carnations, but together they are and will certainly continue to be the major part of our exportable ornamental products.

Introduction of new crops includes many research stages that begins with the initial search and screening and is concluded when the product is introduced commercially.

The introduction and adaptation of new exportable crops normally includes the following stages:

1. Searching for optional crops.
2. Selection and improvement.
3. Developing propagation methods.
4. Studying the growth and flowering physiology and developing practical means for their control.
5. Evaluation of horticultural practices.
6. Studying postharvest physiology and developing practical methods for postharvest handling, transport, and storage.
7. Semi-commercial export shipments to markets abroad.

Some important cut flowers (in European markets), which we introduced and developed in Israel, were "new crops" about 25 years ago such as Gypsophila and Geraldton wax flower. The development of these and other crops are described in the following examples of successful introduction projects.

GYPSOPHILA (BABY'S BREATH)

Gypsophila (*Gypsophila paniculata* L., Caryophyllaceae) is really not a new crop. It has been cultivated for many years as a minor field crop for harvesting in the natural summer flowering season. Today, Gypsophila is a major cut

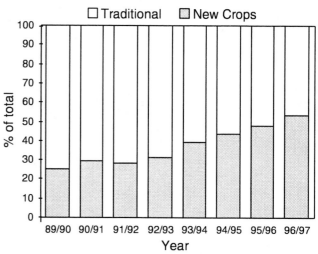

Fig. 1. Relative percentage of "traditional" and "new" cut flowers exported from Israel from 1989/90 to 1966/7.

404

flower in Israel, grown in over 200 ha of greenhouses for the autumn to spring export season. It is now the main flowering shoot used as a "filler" in flower arrangements, in both Europe and the US.

The introduction and development of this crop involved many aspects (Shillo and Halevy 1982; Shillo 1985b; Shillo et al. 1985).

1. Selection of superior clones for controlled cultivation.
2. Development of in vitro meristem culture method to obtain disease-free mother plants.
3. Propagation of clean, uniform commercial cuttings in controlled insect-free greenhouses.
4. Study the physiology of flowering and development of practical methods to control flowering. It was found that Gypsophila is an absolute long day plant, with quantitative response to vernalization. The methods developed to control flowering include cold storage of rooted cuttings prior to planting and supplementary night illumination. Spraying with gibberellic acid is also used to promote flower and shoot breaking and elongation.
5. Development of specific cultivation methods, such as pruning, watering, and feeding.
6. Study the postharvest physiology of the flower and development of methods for postharvest treatment of the flowers prior to shipment, to ensure their longevity and quality. When 15% to 30% of the florets are open, flowering shoots are harvested, treated with silver thiosulfate (STS) to protect them from internal and external ethylene, followed by pulsing the flowers in sugar (7% to 10%) and germicides until about 2/3 of the florets are open.

The flowers of commercial *G. paniculata* plants are sterile and do not produce seeds. This prevented real breeding of this plant. Recently this obstacle was overcome and real new varieties were introduced by Dan Nursery in Israel. A major success is the cultivar Million Stars that was introduced last year, and is already grown this year on over 40 ha in Israel and over 100 ha worldwide.

GERALDTON WAX-FLOWER

Geraldton wax-flower (*Chamaelaucian uncinatum* Schauer, Myrtaceae) is a native shrub in Western Australia. It was introduced to southern California and grown outdoors as a minor cut flower. Most of the initial physiological and horticultural research was carried out in Israel, a fact that facilitated the rapid development of the plant as an important commercial crop (Shillo 1985a; Shillo et al. 1985; Halevy 1994). More recently important research is also conducted in its native country, Australia. In Israel this plant is currently a major commercial ornamental crop, grown on ca. 300 ha. It is used mainly for cut flower production, but also for cut shoots with flower buds, cut foliage, and flowering pot plants. Israel became the main exporter of wax flowers to Europe in the winter.

Plant material of native plants in Australia, as well as breeding, enabled the establishment of a wide assortment of various plant colors (pink, purple, white, lilac, and bicolors), which bloom from November to May. The selected plants are propagated by semi-woody vegetative cuttings in order to form uniform varieties. Recently, virus-free mother plants have been produced by meristem in vitro culture.

Studies on the physiology of flowering revealed that the wax flower is an absolute short-day plant under conditions of mild temperatures. At very high and very low temperatures no flowers are produced. At medium-low temperature some flowers are formed regardless of photoperiod. To advance flowering in the autumn, plants of several cultivars are covered in the field at the end of the summer to create artificial short days. An interesting physiological phenomenon, revealed for the first time in this plant, was that the young flower buds produce a factor that inhibits the formation of new flowers, even under inductive conditions (Shillo et al. 1984).

For production of flowering pot plants, plants are heavily pruned to promote branching and then treated with growth retardants (CCC or paclobutrazol). Controlled photoperiod is employed to extend the flowering period.

Abscission of individual flowers during shipment and handling is a problem of the cut flowering shoots. This problem can be ameliorated by dipping the cut flowering shoots in auxin (NAA) solution and hydrating them in cold water.

PEONY

Peony (*Paeonia lactiflora* Pall., Paeoniaceae) has been in cultivation in China for thousands of years, and have been grown as garden and outdoor cut flower plants in Europe and the US for many years. Cut flowers were, however, available only for a few weeks a year during the natural flowering season in late spring. Until recently, however, very little was known on the flowering physiology of the plant. We have found that flower bud initiation starts after the old leaves senescence in the summer and continues until late autumn when they become dormant. Release from dormancy requires a period of low temperatures, and can be accelerated by GA treatment. After the release from dormancy the plants may start growing and blooming under mild-warm temperatures (Wilkins and Halevy 1985; Byrne and Halevy 1986). This basic information enabled the development of a practical method for extending the flowering season and obtaining cut flower production in the winter, 2–3 months before the natural flowering season (Halevy et al. 1995). Plants are grown under ambient natural cold temperatures of the early winter. After sufficient cold units are accumulated, the structures are covered with polyethylene at mid-winter and the plants are drenched with GA solution. Sprouting and flowering soon follow.

The introduction and improvement of this "new crop" is actually developing new horticultural techniques for flowering control of a very old ornamental plant. One of the obstacles of rapid development of peony as a commercial crop is the slow rate of natural propagation by division of crowns. We are now developing a tissue culture propagation method that should solve this problem.

REFERENCES

Byrne, T.G. and A.H. Halevy. 1986. Forcing herbaceous peonies. J. Am. Soc. Hort. Sci. 111:379–383.

Halevy, A.H. 1994. Introduction and development of Geraldton Wax flower as a commercial cut flower in Israel. Israel Agresearch 7:45–54.

Halevy, A.H., D. Weiss, V. Naor, M. Cohen, M. Levi, and D. Skuler. 1995. Introduction of herbaceous peony as new cut flower in Israel (in Hebrew). Dapei Meida 5:58–62.

Shillo, R. 1985a. *Chamelaucium uncinatum.* p. 185–189. In: A.H. Halevy (ed.), Handbook of flowering, Vol. II. CRC Press, Boca Raton, FL.

Shillo, R. 1985b. *Gypsophila paniculata.* p. 83–87. In: A.H. Halevy (ed.), Handbook of flowering, Vol. II. CRC Press, Boca Raton, FL.

Shillo, R. and A.H. Halevy. 1982. Interaction of photoperiod and temperature in flowering-control of *Gypsophila paniculata.* Scientia Hort. 16:385–393.

Shillo, R., A. Weiner, and A.H. Halevy. 1984. Inhibition imposed by developing flowers on further flower bud initiation in *Chamelaucium uncinatum* Schauer. Planta 160:508–513.

Shillo, R., A. Weiner, and A.H. Halevy. 1985. Environmental and chemical control of growth and flowering in *Chamelaucium uncinatum* Schauer. Scient. Hort. 25:287–297.

Shlomo, E., R. Shillo, and A.H. Halevy. 1985. Gibberellin substitution for the high night temperature required for the long-day promotion of flowering in *Gypsophila paniculata* L. Scientia. Hort. 26:69–76.

Wilkins, H.F. and A.H. Halevy. 1985. *Paeonia.* p. 2–4. In: A.H. Halevy (ed.), Handbook of flowering, Vol. IV. CRC Press, Boca Raton, FL.

New Flower Crops

Abraham H. Halevy

Most edible crops have been introduced into cultivation thousands of years ago. There are only a few new edible plants in the contemporary western horticulture, such as pecan, blueberry, and kiwifruit, but even these plants have been cultivated since ancient days by local farmers in their native region. This is not the case with ornamental crops. Many of the commercial cut flowers and pot-plants grown today have not been cultivated commercially until several years ago.

The ornamental plant industry is characterized by its great diversity. There are more ornamental species cultivated today than all other agricultural and horticultural crops combined. In some ways the introduction of new ornamental crops is easier that of edible crops. Neither their nutritional value nor their general toxicity to human has to be considered, as evident in plants such as *Aconitum, Diffenbachia, Oleander*, and many others. Our main considerations in the introduction of new ornamental crops are the esthetic value, production costs, postproduction longevity, quality, and marketability.

The introduction of new crops includes many research stages, that start with the initial search and screening and concludes when the product is introduced commercially, as detailed in my other presentation in this proceedings (Halevy 1999).

Ten years ago the traditional major crops constituted over 60% of the cut flowers grown in and exported from Israel. This year over 60% of the exportable flowers are "new crops," most of them have not been grown commercially 10 years ago as shown in Table 1. Many of these new commercial flower crops are not even mentioned in a recently published textbook on floriculture (Dole and Wilkins 1999). There are several sources that serve for the introduction of new plant material as potential plant crops.

MINOR OUTDOOR GROWN CROPS

Many of the new greenhouse floral crops, grown and exported during the winter, are the so called "Summer Flowers." They are field grown plants that were used in Europe during their natural flowering season in the summer. Their introduction as a year round crop requires developing physiological and horticultural techniques for out of season production. *Gypsophila* and peony described above (Halevy 1999) are typical examples of such crops. Other examples are listed below.

Aconitum napellus L., Ranunculaceae (Monk's hood)

This is a tuberous plant native to Europe. For winter flowering, tubers are cold stored during the summer and pretreated with gibberellic acid before planting.

Asclepias tuberosa L., Asclepiadaceae (Butterfly weed) and *A. incarnata* L. (Swamp milkweed)

Both plants are native to the US and considered as weeds there. They are absolute long day (LD) plants that require warm temperature during their growth and flowering. For winter production they are grown in heated greenhouses and provided with supplementary light at night.

Achillea filipendulina Lam., Asteraceae (Yarrow)

Native to East Asia it is used mainly for summer harvest as dry flowers. Year round production is obtained by digging the crowns and cold storing them for a few weeks before replanting.

Table 1. quantities of various exportable cut flowers from Israel in the 1996/7 export season.

Flowers	Exportable flowers (millions of stems)
Roses	453
Carnation	144
Gypsophila	116
Solidago	105
Ruscus	78
Wax flower	74
Hypericum	48
Gerbera	45
Limonium	41
Aster	35
Helianthus	32
Asclepias	27
Anemone	27
Safari sunset	20
Anigozanthos	17
Phlox	9
Others	210

Liatris spicata Willd., Asteraceae

Native to Eastern US, for winter production, tubers are cold-stored during the summer and plants are lighted in the field.

Phlox paniculata L., Polemoniaceae

Native to Eastern US, this herbaceous summer perennial is now grown for year round production in greenhouses. It is a LD plant, requiring supplementary night lighting.

Solidago sp. L., Asteraceae (Goldenrod)

Species of goldenrod native to North America are considered as weeds there. New interspecific hybrids turned this plant into an important cut flowers. For winter production plants first receive LD to extend their stems and then are exposed to the natural winter short days (SD) for flower initiation and development.

Trachelium caeruleum L., Campanulaceae

Native to South Europe, it is an absolute LD plant and grown in the warmer parts of Israel for winter production.

NEW CULTIVARS OF ORNAMENTAL FIELD PLANTS

Plants of this group have been grown as minor cut flowers, but recent introduction of new cultivars, with modified and improved horticultural traits, turned them into important floral crops. Examples are:

Anigozanthos hyb., Haemodoraceae (Kangaroo Paw)

This Australian plant was grown mainly outdoors until a few years ago. Recently introduced highly yielding interspecific hybrids are now grown indoors for year round production. These new hybrids are propagated by in vitro tissue culture.

Aster hyb., Asteraceae

New interspecific hybrids of A. novi-belgii and other species native to Eastern North America turned these herbaceous perennial, late summer garden plant, into an important greenhouse crop. This is a LD-SD plant, requiring at first LD, until the stems reach a certain desired height and then it is exposed to natural winter SD.

Campanula medium L., Campanulaceae

This plant, native to south Europe, was used only as garden and pot-plant until recently. The original species required a long cold period followed by LD for flowering (Wellensiek 1985). However, new varieties, introduced recently, have long flowering stems and require only LD for flower induction. This enables growing the plant as a commercial cut flower crop.

Clarkia amoena Nels. & Macbr., Onagraceae (Godetia, Satin flower)

This Western North American plant was mainly a garden plant until the recent introduction of improved cultivars for use as cut flowers. The plant requires mild temperature and moderate watering and feeding. It is a facultative LD plant.

Eustoma grandiflorum Shinn (Syn. Lisianthus russellienus), Gentianaceae

Native to Southern US, it was used sparsely as garden and cut flower plant. Newly introduced F_1 hybrids turned the plant into an important greenhouse cut flower crop for year round production. Seed propagated, it requires mild-low temperatures in the first growing stage, followed by warmer temperatures.

Leucadendron Hyb., Proteaceae

This South African shrub became an important outdoor crop for cut flowering shoots with the introduction of new hybrid cultivars. The 'Safari Sunset' cultivar is now grown on over 200 hectares in Israel.

Limonium hyb., Plumfaginaceae

Interspecific hybrid cultivars of several perennial limoniums became important greenhouse cut flower crop, used as "filler."

GARDEN AND LANDSCAPING PLANTS

These are mainly woody or herbaceous perennials, used for many years in gardens and introduced recently into the floral trade. Examples are:

Cotinus coggygria Scop., Anacardiaceae (Smoke Tree)

A deciduous shrub, native to South Europe, used for many years as a garden plant. The cultivar 'Royal Purple' is now grown for cut foliage. LD is applied to prevent plants from entering dormancy.

Hypericum sp. (Hypericaceae)

Several species and hybrids of these shrubby plants, native to the Mediterranean and the Canary Islands, have recently became important floral crop grown both outdoors and in greenhouses for cut shoots with fruits of various colors. This is an absolute LD plant that requires night lighting for winter production.

Ruscus hypoglossum L. (Liliaceae)

This herbaceous perennial has been grown in Israel as a garden plants for many years. It is now the main cut foliage crop in Israel, grown exclusively in shaded houses.

ORNAMENTAL CULTIVARS OF FIELD CROPS

In some plants, grown mainly as field crops, new ornamental cultivars have been introduced and used as cut flowers. Examples are: sunflower (*Helianthus annuus* L., Asteraceae), cotton (*Gossypium hirsutum* L., Malvaceae), and safflower (*Carthamus tinctorius* L., Asteraceae).

PLANTS GROWN IN BOTANICAL GARDENS

Botanical gardens and specialized plant collections are rich sources for plant material, some of which can be used for introduction as potential floral crops. Some examples are the bulbous plants of the Liliaceae: *Eremurus* sp. of Central Asia, the South African *Bulbinella kookerri* of yellow, orange, and white flowers, and *Ornithogalum dubium* of yellow and orange flowers, and the South Asian *Curcuma alismatifolia* (Zingiberaceae).

WILD PLANTS IN THEIR NATIVE HABITAT

The introduction and development of Geraldton wax-flower described above (Halevy 1999) is an example of such introduction. Some such plants are currently under intensive developmental stages. They include plants originated from remote areas, but also plants native to Israel and California.

REFERENCES

Dole, J.M. and H.F. Wilkins. 1999. Floriculture principles and species. Prentice Hall, Upper Saddle River, New Jersey.

Halevy, A.H. 1999. Ornamentals: where diversity is king—the Israeli experience. p. 398–400. In: J. Janick (ed.), Perspectives in new crops and new uses. ASHS Press, Alexandria, VA.

Wellensiek, S.J. 1985. *Campanula medium*. p. 123–126. In: A.H. Halevy (ed.), Handbook of flowering, Vol. II, CRC Press, Boca Raton, FL.

Proteaceae Floral Crops: Cultivar Development and Underexploited Uses*

Kenneth W. Leonhardt and Richard A. Criley

The Proteaceae apparently originated on the southern supercontinent Gondwana long before it divided and began drifting apart during the Mesozoic era, accounting for the presence of the Proteaceae on all of the southern continents (Brits 1984a). The Protea family comprises about 1400 species in over 60 genera, of which over 800 species in 45 genera are from Australia. Africa claims about 400 species, including 330 species in 14 genera from the western Cape. About 90 species occur in Central and South America, 80 on islands east of New Guinea, and 45 in New Caledonia. Madagascar, New Guinea, New Zealand, and Southeast Asia host small numbers of species (Rebelo 1995).

Proteas are neither herbaceous nor annual, and they are always woody. Their structural habit is variable from groundcover forms with creeping stems, and those with underground stems, to vertical to spreading shrubs, to tree forms. The leaves are generally large, lignified, hard, and leathery. A mature leaf will generally snap rather than fold when bent. The leaf anatomy is specially adapted for water conservation and drought resistance. These characteristics and the high leaf carbon to nitrogen ratio render the leaves indigestible to most insect pests (Rebelo 1995), accounting for the relatively pest-free status of most commercial protea plantings.

The distribution of the family is linked to the occurrence of soils that are extremely deficient in plant nutrients (Brits 1984a). An accommodating characteristic of the family is the presence of proteoid roots. These are dense clusters of hairy rootlets that form a 2–5 cm thick mat at the soil surface which enhances nutrient uptake from the nutrient-sparse soils on which the Proteaceae evolved. All species examined so far, in at least ten genera, possess proteoid roots (Lamont 1986).

The capitulum, in which the flowers are borne on a flat or pointed receptacle, is the most common type of flowerhead. Involucral bracts surround the receptacle and may be prominent, as in *Protea*, or inconspicuous, as in *Leucospermum*. Involucral bracts in *Leucadendron* are also inconspicuous, but the floral bracts of female plants are large and develop into woody cones. An important taxonomic feature of the family is that its flowers do not have separate sepals and petals. The perianth is made up of a single set of four segments called tepals. As a bud opens, the perianth segments curl back to expose the style which extends from the superior ovary to the stigma. In some species, small floral nectaries at the base of the ovary secrete nectar to attract pollinators (Rebelo 1995). Another adaptation of evolution that has become important in the utilization of *Leucospermum* and *Protea* as cutflowers is bird pollination. The large, usually solitary, terminal flowers of these genera, and their predominant colors of creamy white, and blends of yellow, orange, and red, colors birds are attracted to, are probably adaptations to pollination by the Cape Sugarbird, *Promerops cafer*, and other nectar eating native birds. This is thought to be the prime reason why the South African species, and perhaps particularly *Leucospermum* and *Protea*, are the most attractive of the family and have excellent potential as cutflowers (Brits 1984a).

The floral biology of proteas is protandrous, with anthesis occurring prior to the stigma becoming receptive; a mechanism to help insure cross pollination. Most *Protea* seem incapable of self-pollination, although certain *Leucospermum* and *Serruria* species will produce seed when self pollinated. There are four types of pollinators, or pollen delivery; rodents, birds, insects, and wind. The flowerheads of certain *Leucospermum* and *Protea* species are visited by several species of gerbils, mice, rats, and shrews. Many *Leucospermum*, *Mimetes*, and *Protea* species are pollinated by birds. Since birds do not rely on smell, bird-pollinated species have little if any scent. The small species of *Leucospermum* and *Protea* are pollinated by bees and wasps and a few other insects; *Leucadendron* are pollinated by several beetle species, and most small-flowered genera are visited by a number of beetle, fly, and wasp species. Ten *Leucadendron* species are the only wind-pollinated proteas in southern Africa (Rebelo 1995).

*College of Tropical Agriculture and Human Resources Journal Series No. 4431

The hermaphrodite species of protea are extremely low seed producers, with only one to 30% of flowers resulting in seed. It is surmised that a large percentage of hermaphrodite flowers function only as males. Another reason for low seed production may be the plant's need to produce nutrient rich seeds to reduce seedling mortality in a nutrient-poor environment. However, dioecious species generally have a high seed set, possibly because all flowers on a female plant are reproductively functional (Rebelo 1995).

HISTORICAL

In 1771, the great Swedish botanist Carl Linnaeus wrote a colleague in Amsterdam, "*Inexhaustum credo Cap. B. spei esse plantarum speciebus; certe nulla Flora ditior erit.*" Freely translated this reads, "I believe the Cape of Good Hope is by no means exhausted of plant species; surely no other flora could be richer..." The Australian Proteaceae contain more genera, but it is the South African ones that have attracted the interest of the world, commencing with the attempts by Joseph Knight to cultivate them under artificial conditions in the late 1700s. They proved to be difficult subjects, but the first flowers produced outside of South Africa were shown at the Royal Botanical Gardens at Kew in 1774 and 21 species had bloomed there by 1810. In the 18th and 19th centuries they became a "patrician indulgence" as collections were found in the royal conservatories from St. Petersburg to Paris (Parvin 1984).

Although there has been much exploration of the Cape Floral Kingdom over the past 400 years, new species are still being found, many of them in the Proteaceae. The interest in this family began with the 1605 description by botanist Carolus Clusius of the flowerhead of *Protea neriifolia* as thistle-like, graceful, and unique. The Proteaceae are often the most prominent elements of the Cape fynbos with large, stiffly erect flowerheads. A cut flower industry developed around flowers harvested from the fynbos (Brits 1984b). A far-sighted Stellenbosch farmer, Mr. Frank Batchelor pioneered the development of the commercial protea industry in South Africa as he retired from deciduous fruit production. From wild-collected materials, he moved into selection, hybridization, and vegetative propagation, recognizing that superior quality was required for the marketplace (Soutter 1984).

Contemporaneously, Marie Vogts took up the challenge to learn more about the habits and production of proteas. Wild plants in their natural habitat were her reference books as she established cultivated plantings which could be studied and compared. By the late 1950s, she had acquired enough knowledge to conclude that proteas could be cultivated, and her book "Proteas: Know them and grow them" (1958) was published. The book focused attention on the economic possibilities of proteas. Continuing her work, she sought out the natural variants that occurred in the mountains and brought them into cultivation. Seeing horticultural potential as well as marketability in these variants, she recognized the variations in flowering times at different sites as important to extending the marketing season (Vogts 1984). She was instrumental in founding the Protea Research Unit of the South African Department of Agriculture and focusing its attention on the genetic variability of wild populations and interspecific hybrids, and initiating a research program to improve their adaptability to cultivation (Brits 1984b). Her four decades of research and investigation laid a foundation for an industry that benefits not only South Africa's landed establishment, but also their rural peoples, and that has reached beyond the Cape Floral Kingdom to Australia, New Zealand, Israel, California and Hawaii, Zimbabwe, El Salvador, Chile, the Canary Islands, France, and other distant lands. Thus, this "alternative crop/new crop" has a long and honorable history predating the 1990s interest.

The University of Hawaii became involved in 1964 when a visiting professor of Horticulture, Dr. Sam McFadden, imported seed of various Proteaceae for trial in Honolulu. The next year additional seed were planted at the Maui Agricultural Research Center, 1000 m up the slope of Haleakala crater, where they flourished. Evaluations were made in 1969 separating 34 species into those suitable for cutflowers, for landscaping, and those unsuitable for either purpose. In 1970, 63 species of nine genera were imported, and a research program with emphasis on propagation and nutrition began to take form (Parvin et al. 1973). It was soon recognized that *Leucospermum* flowered on Maui as much as two months earlier than in California, suggesting the potential for a competitive advantage for a Hawaii export industry (Parvin 1984) if early flowering cultivars of horticultural merit could be obtained or created. The first pollinations in a new breeding program were made in 1972. The current research program emphasizes breeding for improved cultivars for cutflower production, the physiology of flowering, disease management, and postharvest handling and storage for *Leucospermum* and *Protea*.

The results from protea research programs in South Africa and Hawaii, and to an extent, Israel, have helped to stimulate protea cultivation in those and several other countries around the world, primarily for the international cutflower trade. Europe has been the traditional market for protea, but the United States and Japan have significantly expanded floral consumption and increased purchases of protea in recent years. Israeli market research recently reported that the world-wide cutflower markets can still absorb large quantities of proteas without lowering prices, but that the market is in need of new cultivars to refresh existing selections (Danziger 1997).

The emphasis on cultivar development is put into perspective by Soutter (1984); "It is generally said that a horticultural industry is only as good as its cultivars, and certainly in the case of floriculture, one can add the rate at which new cultivars are placed on the market." Numerous cultivars have been introduced by scientists and commercial breeders, hobbyists, and plant and flower collectors gathering from the native fynbos. *The International Protea Register*, Stellenbosch, South Africa, now in its fourth edition, keeps track of named cultivars along with their origin and brief descriptions to the extent it can obtain the information. This valuable resource allows researchers and producers to exchange plant materials and communications about known cultivars and their adaptability and performance in cultivated situations around the world. Table 1 summarizes the registered cultivars, and cultivars recognized but not yet registered, of four protea genera.

The register also lists cultivars of unknown origin, including 14 *Leucadendron*, 14 *Leucospermum*, and 24 *Protea*. It lists only one intergeneric hybrid, (*Mimetes* × *Leucospermum*) 'Splendidus'. The 86 registered cultivars are authored by 20 registrants from five countries; Australia, New Zealand, South Africa, United States, and Zimbabwe (Sadie 1997).

The International Protea Association (IPA) supports the promotion and commercial production of protea as well as scientific research and conservation of native germplasm (Mathews 1984). The IPA held its 9th biennial conference in Cape Town, South Africa, in August 1998 with representatives from 12 nations participating. Regional production and marketing reports were made by industry leaders, while academics and graduate students gave oral and poster presentations on research progress in areas of conservation, pest and disease management, cultivar development, propagation, pruning, irrigation, nutrition, and postharvest physiology.

Economically, the most important protea species is *Macadamia integrifolia*, the only native food plant in Australia to achieve international status as a commercial nut crop. Macadamia breeding and selection work was initiated in 1934 by the University of Hawaii, and over the next 50 years 13 cultivars were introduced from 120,000 seedlings evaluated. Commercial development of this crop began in Hawaii in the 1940s (Hamilton and Ito 1984). Hawaii was the world's leading producer until Australia recently claimed that spot. Today, this gourmet dessert nut is cultivated commercially in Australia, Brazil, Costa Rica, Guatemala, Hawaii, Kenya, Malawi, South Africa, and Zimbabwe. Recently, China established commercial plantings on Hainan island.

Selected Proteaceae floral crops were analyzed for their potential profitability, on a hypothetical 4 ha farm, and determined to be profitable, given adequate farm management and marketing. A computerized spreadsheet model of protea production in Hawaii enables one to estimate profitability over a wide range of conditions (Fleming et al. 1991, 1994). The computerized program is available free by contacting fleming@hawaii.edu.

Table 1. Summary of cultivars in The International Protea Register (the number of interspecific hybrids is in parentheses).

| Genus | Named cultivars | | |
	Registered	Recognized but not registered	Total
Leucadendron	12 (5)	101 (18)	113 (23)
Leucospermum	30 (19)	58 (15)	88 (34)
Protea	42 (24)	139 (29)	181 (53)
Serruria	2 (2)	4 (1)	6 (3)
Total (4 genera)	86 (50)	302 (63)	388 (113)

A variety of protea species have uses within or near their habitat. The seeds of *Brabeium stellatifolium* of South Africa are roasted and eaten or used as a coffee substitute. The seeds of *Finschia chloraxantha* from New Guinea and *Gevuina avellana* from South America are eaten by natives. The timbers of the Australian silky oaks *Cardwellia sublimis*, *Orites excelsa*, and *Grevillea robusta* are used in furniture and panelling. The Australian *Oreocallis wickhami* and *Banksia serrata* are used for yokes and boat knees, and the wood of *B. verticillata* for railway carriages and furniture. *Hakea leucopteris* and *H. vittata* wood is used for smoking pipes. The barks of *Faurea saligna* and *Leucospermum conocarpum* are used for tanning leather in South Africa. Species reported to have medicinal uses are *Faurea speciosa* for ear drops (root and leaf extract), and *F. saligna* to treat dysentery and diarrhea (root extract). Most protea flowers are of value to apiarists for their abundant nectar production (Rao 1971).

PROTEA

Protea is a large genus with 136 species of which 70 are distributed in the southern hemisphere temperate zones and the balance distributed in southern hemisphere sub-tropical to tropical zones, with 3 extending above the equator into the northern tropics (Rao 1971). Of the 117 species native to the African continent, 82 are from South Africa (Vogts 1982). A recent account of *Protea* species in Southern Africa lists 90 species and numerous subspecies (Robelo 1995). Linnaeus named the genus *Protea* in 1735 after the Greek god Proteus, who, according to legend, was able to transform his shape and appearance into numerous animate and inanimate forms at will (Robelo 1995). It was from the name *Protea* that the family name Proteaceae was assigned by the French botanist Jussieu (Rousseau 1970).

The natural habitat of *Protea* ranges in elevation from sea level to over 2000 m. In South Africa a rich diversity of species inhabit the well-drained, moderately-acid, low fertility, granite soils from Cape Town to the Table Mountain areas up to 1300 m (Parvin et al. 1973).

The genus is characterized by large bracts, often brightly colored, surrounding a composite type flower. The bracts are smooth or pubescent, with many species having bracts fringed with a dark "fur" lending a tactile as well as visual appeal. The range of colors includes red, pink, yellow, white, and occasionally green (Watson and Parvin 1973). The most widely recognized species in the genus is *Protea cynaroides*, the King Protea, the national flower of South Africa. It has flower heads up to 30 cm across, with widely spaced bracts arranged around a peak of flowers that vary in color from near white to soft silvery-pink to deep rose pink to crimson, in a few selected cultivars.

Many natural variants of *P. cynaroides* can be placed into three South African ecotypes. Those from the eastern cape and southern coastal plain have long leaves on long stems that terminate with relatively small but wide-open flower heads. They are very attractive but difficult to pack. The plants are vigorous and bear 10 to 20 heads per plant. Variants from the Outeniqua mountains region bear large bowl-shaped rose colored flower heads on thick stems. The heads are more easily packed because the bracts do not flare out. Average flower head yield is five to eight per plant. The variants from the Western Cape region are slow growing and average only about four heads per plant, but are described by Vogts (1980) as beautifully goblet-shaped. A miniature form of *P. cynaroides*, with flower heads the size of a typical pincushion protea, offers much promise for expanded florist use of this species. *Protea cynaroides* generally show good resistance to *Phytophthora* root rot (von Broembsen and Brits 1986).

Several clonal selections of *P. neriifolia* are grown commercially for their prolific production of fall and winter blooms. The plant becomes a large shrub with foliage resembling that of oleander. Flower heads range in color from light pink to rose to dark red, with some white selections known. Some selections have silvery hairs subtending tufts of black hairs at the bract tips. Plants of 20 years age have been reported to bear commercial quality flowers (Vogts 1980). This species also shows good resistance to *Phytophthora* root rot (von Broembsen and Brits 1986).

The grey-leaf sugarbush, *P. laurifolia* (formerly *P. marginata*), is similar to *P. neriifolia* for many characteristics, although their natural distributions do not overlap and natural hybrids between the two do not occur. To the non-taxonomist, foliage characteristics may be the most distinguishing feature between the two species. The leaves of *P. laurifolia* are grey-green to blue-green, elliptic and broader than the bright green leaves of *P. neriifolia* (Vogts 1982). If left unmanaged, *P. neriifolia* will grow to become an erect shrub 3 m tall while *P. laurifolia* will become an 8 m tall tree (Robelo 1995).

Protea magnifica, the Queen protea, is somewhat susceptible to *Phytophthora* (von Broembsen and Brits 1986) but is still grown for its large 15 to 20 cm flower heads of white to rose pink to salmon colors. Many cultivars have black-and-blond tufts of hair on the bract margins (Vogts 1980) and are sometimes referred to as woolly-beard protea.

The rose-spoon protea, *P. eximia*, gets its common name from the long spathulate inner bracts that are widely splayed and easily distinguish this large flowered species from others (Vogts 1982). These bracts range in color from pink to orange-brown. Awns extending from the perianth have purple-black velvety hairs. Plants generally range in height from 2 to 5 m, are sparsely branched, and flower from early winter through late spring (Robelo 1995). A tall tree-like variant reaches peak flowering in summer (Vogts 1982).

Long, narrow leaves, and flower heads with pointed bracts characterize *P. longifolia*. Bract color is variable from white to pink and green. Plant stature and growth habit are also variable in its native stands, where many natural hybrids with other *Protea* species overlapping its range have been found.

Protea grandiceps, also somewhat susceptible to *Phytophthora* (von Broembsen and Brits 1986) comes from high elevation mountainous regions that are snow-covered in winter. It is a slow growing long-lived plant that bears up to 40 salmon-colored flower heads of 10 to 15 cm. Plants can be cultivated for more than 20 years (Vogts 1980).

The sugar bush, *P. repens*, is widely grown commercially for its white to pink to deep red colors and long flowering season (Vogts 1980). It is another species with good resistance to *Phytophthora* root rot and can be used to replant *Phytophthora* infested fields (von Broembsen and Brits 1986).

Protea compacta has lanky flower stems on a stiffly upright, sparsely branched shrub that grows to 3.5 m tall. The rich pink bracts, with their light-reflecting fine-hair-fringed margins are longer than the cup-shaped flower heads. The prominent flower heads, unobscured by foliage, make fine winter cutflowers (Vogts 1982).

Numerous selections of these and other species, and of naturally occurring hybrids that have been identified from the South African fynbos, are cultivated by commercial growers. Cuttings of 22 named selections of *Protea* of South African origin were imported by the University of Hawaii in 1988, propagated, field planted, and evaluated for adaptability and plant growth characteristics at its Kula Agricultural Research Center on Maui. Yield, seasonality of bloom, and keeping quality were recorded from 1989 to 1993. The yield and seasonality over a 12 month period (August 1992 through July 1993) of the seven cultivars that produced 30 flowers or more are reported in Table 2. By careful evaluation of seasonality in localized climates, it is possible to select cultivars to cover a large portion of the year, although some months may not be well represented (Criley et al. 1996).

Protea can be propagated from seed, with the resulting variation expected of cross-pollinated heterozygous materials. Given the availability of clonal selections, the method of choice among progressive com-

Table 2. Yields and seasonality on Maui for selected South African *Protea* accessions.

Cultivar	Yield (12 mo)	Aug.	Sept.	Oct.	Nov.	Dec.	Jan.	Feb.	Mar.	Apr.	May	Jun.	Jul.
Annette	37					✓	✓	✓	✓	✓	✓	✓	
Brenda	210			✓	✓	✓	✓						
Cardinal	31	✓					✓	✓	✓	✓	✓	✓	✓
Guerna	86	✓	✓	✓	✓	✓	✓	✓	✓	✓	✓	✓	✓
Heibrech	45	✓	✓	✓	✓	✓	✓	✓	✓		✓	✓	
Red Baron	86	✓	✓	✓	✓	✓	✓	✓	✓		✓		
Sylvia	66	✓	✓	✓	✓	✓	✓	✓	✓	✓	✓	✓	✓

mercial growers is to propagate from cuttings. Terminal and sub-terminal cuttings are made from the current season's mature growth. Robust cuttings of 20 to 25 cm in length root readily, for most cultivars. Longer cuttings may be taken if the grower desires to have lower branches well above ground level. Removing half of each leaf of long leaf cultivars is a common practice. An IBA auxin treatment of 4000 to 8000 ppm is beneficial. The rooting medium should be very well aerated but not allowed to dry. Mixtures of 25% to 50% peat with the balance being polystyrene or perlite has given good results. Rooting is generally done under standard mistbed conditions. An approved fungicide sprayed over the cuttings following planting can prevent infections. Rooting time is variable among species, with *P. cynaroides* rooting quickly and *P. neriifolia* often taking many weeks (Mathews 1981).

When selecting a production site, good soil drainage is the most important requirement for protea production. Deep soils that allow expanded root development and can store a good supply of water and nutrients are preferred, but shallow soils can be suitable if drainage is rapid and frequent irrigation can be provided (Claassens 1981).

Relatively low concentrations of nutrients are required for normal growth of proteas. Most species react favorably to nitrogen, particularly in the ammonium form, while most are intolerant of amounts of phosphorus that would be considered moderate for non-proteaceous plants. Protea plants seem to have a very effective mechanism to scavenge phosphorous from soils with low phosphorus status (Claassens 1981, 1986).

Cresswell (1991) produced a tissue analysis standard for assessing the appropriate phosphorus status for two *Protea* cultivars. Only the desirable ranges are reported in Table 3. Values lower or higher than those reported here were considered low to deficient or high to toxic, respectively.

The recommended developmental stage for harvesting most *Protea*, to insure market quality and acceptable postharvest life, is the so-called soft-tip stage when bracts have lost their firmness and begin to loosen but still cohere (Meynhardt 1976). At this stage, few insects are present because anthesis has not yet occurred, so there is little to attract them. However, flowers picked too early will not open (Coetzee and Wright 1991). A serious problem with marketing several species of *Protea* is the undesirable discoloration of leaves soon after harvesting. The problem is most pronounced in *P. eximia* and *P. neriifolia*, and their hybrids, and to a lesser extent with *P. compacta* and its hybrids (Ferreira 1986; Paull 1988).

Leaf blackening, or browning as it is sometimes called, in *Protea* is caused by carbohydrate depletion due primarily to the sugar demand by the inflorescence for nectar production (Dai and Paull 1995). Low availability of mobile carbohydrates in the leaves, combined with the high respiratory demand of the inflorescence, resulted in a 70% decline in mobile leaf carbohydrate levels in *P. neriifolia* within 24 hours of harvest (Jones et al. 1995). Warm temperatures and low light in postharvest storage have been correlated with increased rates of leaf blackening (Ferreira 1986). Refrigeration, especially during postharvest storage, packaging, and shipping periods, has a most significant effect on delaying the onset of leaf blackening (Paull 1988). Refrigeration will slow respiration, reduce water stress, and slow nectar production by the inflorescence, thereby conserving carbohydrate reserves in the stem and foliage. A storage and transport temperature of 2° to 5°C will help ensure bloom quality and postharvest life (Coetzee and Wright 1991). Pulsing cut *Protea* stems in a 1% sucrose solution, or a floral preservative solution, before packing and especially post-unpacking is an effective treatment to delay the onset of leaf blackening (Brink and de Swardt 1986; Paull 1988). Another form of leaf browning results from fumigation with methyl bromide (Coetzee and Wright 1991), which imported shipments are often subjected to if insects are present. Growers should practice good field sanitation and appropriate postharvest disinfestation practices prior to packing, so that agriculture inspectors at the receiving end will not fumigate the shipment.

Table 3. Desirable ranges of phosphorus in tissues of two *Protea* cultivars.

| Cultivar | Desirable phosphorus (ppm) | | |
	Stems	Recently matured leaves	Old leaves
Satin Pink	0.19–0.35	0.19–0.29	0.21–0.44
Pink Ice	0.06–0.29	0.06–0.27	0.16–0.46

Natural hybrids within the genus *Protea* are not an uncommon occurrence where the geographic ranges of two or more species overlap. Scientists at the Fynbos Research Unit, Elsenburg, South Africa, have made a collection of such natural hybrids and in most cases have been able to determine their parentage (Table 4). Some of these meet standards of commercial horticulturists and have been released to South Africa's protea industry. Dr. Littlejohn, the protea breeder at Elsenburg, recently initiated a program to produce controlled hybrids in the genus *Protea* to further benefit the protea industry.

LEUCOSPERMUM

Leucospermum species are evergreen woody perennials with growth habits that range from small trees to spreading shrubs to prostrate ground covers. The most widely grown species are floriferous, spreading shrubs on which relatively short-stemmed inflorescences are borne in the spring. Horticulturists have had to develop management practices to improve stem length and straightness for their use as cut flowers.

Rourke (1972) and Jacobs (1985) describe the inflorescence as a capitulum that develops from an axillary rather than a terminal bud, but that appears to arise distally. Inflorescences may be solitary, as in *L. cordifolium*, *L. lineare*, and *L. vestitum*, or in clusters (conflorescences), as in *L. oleifolium*, *L. tottum*, and *L. mundii*. The individual florets consist of a perianth formed by four fused perianth segments, one of which separates from the other three as the flower opens. The perianth curls back to display a prominent style; the striking appearance of the whole inflorescence of open flowers resembles a pincushion—thus one of the common names is pincushion protea. The styles, perianth, and involucral bracts may be white, yellow, pink, orange, or red and the combinations are responsible for the popularity of the pincushion proteas as cutflowers.

Although most of the *Leucospermums* are indigenous to nutrient-poor, coarse, acidic, sandstone-derived soils, they seem adaptable to a variety of soil types within a narrow range of pH and fertility levels. This is evidenced by their culture in several regions of southern Africa, southern California, Israel, Australia, and in the volcanic soils of Hawaii and the Canary Islands (Criley 1998).

Propagation of the commercial cultivars of *Leucospermum* is by cuttings, of which most root readily. While cuttings can be rooted at almost any physiological stage of development, a preferred cutting is the recently matured new growth, known as a semi-hardwood cutting (Malan 1992). This type of material is gathered in autumn after shoot growth terminates.

A tissue culture protocol for *Leucospermum* was developed using axillary bud explants induced to proliferate on a basal medium of half-strength Murashige and Skoog inorganic salts supplemented with sucrose and benzyl adenine (Kunisaki 1989, 1990). *Leucospermum* Hawaii Gold, propagated from tissue cultures, is flowering at the Kula Agriculture Research Center and appear identical to the type cultivar from which the explants were taken.

Grafting is often viewed as a solution to problems of root system adaptation to low or high pH soils, or soil-borne diseases. The selection of rootstock plays a significant role in improving adaptability and yield of *Leucospermum* (Van der Merwe 1985). Grafting onto lime-tolerant rootstocks, such as *L. patersonii*, has been recommended as an approach to problems of protea production on soils of neutral to slightly basic pH (Brits 1984b). The standard grafting technique is wedge-grafting of leafy semi-hardwood scions onto selected rootstocks (Rousseau 1966; Vogts et al. 1976). Cutting grafts, where the graft union develops while the cutting roots, is also recommended (Brits 1990b). Brits (1990c), in screening 19 species and several hybrids for their potential as rootstocks, determined that *L.* 'Spider' (a primary hybrid of *L. formosum* × *L. tottum*) has a degree of tolerance to *Phytophthora cinnamomi*. *Leucospermum* 'Spider' is pres-

Table 4. Recent *Protea* hybrids of South African origin.

Sheila	(*P. magnifica* × *P. burchelli*)
Venetia	(*P. magnifica* × *P. neriifolia*)
Pink Duke	(*P. compacta* × *P. susannae*)
Candida	(*P. magnifica* × *P. obtusifolia*)
Valentine	(*P. cynaroides* × *P. compacta*)
King Grand	(*P. cynaroides* × *P. grandiceps*)
Venus	(*P. repens* × *P. aristata*)
Liebencherry	(*P. repens* × *P. longifoli*)
unnamed	(*P. cynaroides* × *P. nitida*)
unnamed	(*P. cynaroides* × *P. repens*)

ently being used as a rootstock by several commercial producers in South Africa. *Leucospermum saxosum* has also been determined to have low susceptibility to *Phytophthora* (Moffat and Turnbull 1994), and a selection of *L. patersonii*, designated 'Nemastrong', has tolerance to nematodes (Ackerman et al. 1995). Such rootstocks may have the potential to expand and increase yields of plantings where *Phytophthora* and nematodes are a problem.

The production period for *Leucospermum* is late winter to late spring. Parvin (1974) reported that 65% to 75% of the total crop of *L. cordifolium* 'Hawaiian Sunburst' was harvested from Dec. through Feb. in Hawaii. They are also high yielding. During a three-year study, beginning with 6-year-old plants, the per plant yields averaged 600 to 650 flowers.

Research on postharvest handling practices has shown that the pincushion protea will tolerate cool, dry, long-term storage and still provide a useful vaselife. *L. cordifolium* flowers that were cooled and hydrated at 1°C in water, wrapped in newsprint and bagged in plastic film withstood periods of three and four weeks at 1°C storage, and after rehydration, possessed an average vaselife of 8 days (Jones and Faragher 1990). Downs and Reihana (1986) found significant varietal differences in vaselife following a period of simulated transport, with the New Zealand cultivar 'Harry Chittick' at 35.5 days, a Hawaii hybrid of *L. lineare* × *L. cordifolium* at 29.7 days, and 'Veldfire', a South African hybrid at 16.9 days.

Parvin (1978) improved vaselife with 2% to 4% sucrose plus 200 to 600 ppm hydroxyquinoline citrate solutions. Silver nitrate at 1000 ppm did not benefit cultivars of *L. cordifolium* but improved vaselife for the hybrid *L.* 'Hawaii Gold' (Parvin and Leonhardt 1982). Criley investigated revival of wilted flowers with extruded styles, in order to increase packing densities for export shipments. Flowers pulsed with a preservative prior to partial dehydration (20% loss of FW) and storage (24 h at 13°C) could be revived, although vaselife was not as long as with fresh cut flowers (Criley et al. 1978a,b). Flowers cut in bud (7 cm diam.) offered better promise, with full development and less loss of vaselife than flowers cut at a younger stage (Criley et al. 1978a; Parvin and Leonhardt 1982).

While a number of Proteaceae may be grown as potted plants, the *Leucospermums*, with their relative ease of rooting and attractive floral display, have the greatest potential (Sacks and Resendiz 1996). Criley (1998) reported that budded cuttings flowered soon after rooting, adding confirmation to their potential as potted plants, and proposed that stock plants be manipulated to achieve stronger branches for this use.

Research on photoperiod responsiveness of *Leucospermum* (Wallerstein 1989; Malan and Jacobs 1990) indicates that daylength manipulation may have implications for potted flowering plant production. High light intensity was shown to be necessary for flowering (Jacobs and Minnaar 1980; Napier and Jacobs 1989; Ackerman et al. 1995) and to promote rapid rooting of cuttings.

Leucospermum species suitable for potted plants are of two types: those having a single large inflorescence, such as *L. cordifolium*, *L. lineare*, and *L. tottum*; and those with small multiple inflorescences (conflorescences) such as *L. oleifolium*, *L. muirii*, and *L. mundii* (Brits et al. 1992; Ackerman et al. 1995; Brits 1995a). It is important to select material that will root rapidly and support flower initiation and development on a young root system (Ackermen and Brits 1991; Brits et al. 1992).

Although the genus *Leucospermum* consists of 48 species (Rourke 1972), little genetic improvement through hybridization has taken place until relatively recently. Jacobs (1985) reported that only a few species were utilized as cut flowers (*L. cordifolium*, *L. patersonii*, *L. lineare*, *L. conocarpodendron*, *L. vestitum*), but that natural and man-made interspecific hybrids exist as clonal selections. Collection and introduction of natural interspecific hybrids has occurred (Brits and van den Berg 1991), and controlled crosses were made between species in efforts to produce later flowering, improve color and shape, and to introduce tolerance to *Phytophthora cinniamomi* (Brits 1992a). Today, active breeding programs are being conducted at the Fynbos Research Station, Elsenburg, South Africa (Brits 1992a,b; Littlejohn et al. 1995) and at the Maui Agricultural Research Station of the University of Hawaii (Ito et al. 1978, 1979, 1990; Leonhardt et al. 1995), and in Israel (Shchori et al. 1995).

As of the fourth edition of the *International Protea Register* (*International Registration Authority: Proteas 1997*), 30 cultivar names have been registered and another 58 have been noted but not registered for selections and interspecific hybrids (Criley 1998). Among the hybrids registered, only three are advanced hybrids (having more than two species in their genealogy). These hybrids, developed and registered by the University of Hawaii, are:

L. 'Rachel', with parentage (*L. lineare* × *L. vestitum*) × *L. glabrum*

L. 'Hawaii Moon', with parentage (*L. lineare* × *L. cordifolium*) × *L. conocarpodendron*

L. 'Kathryn', with parentage (*L. lineare* × *L. cordifolium*) × *L. conocarpodendron*

The criteria for developing new *Leucospermum* cultivars must consider the needs of growers, handlers, retailers, and consumers. The criteria developed for the Hawaii breeding program includes disease resistance, earliness to flower, an extended flowering season, long slender and straight stems, slender leaves, reduced leaf pubescence, ease of propagation, high yields, good postharvest characteristics, new and improved colors, and market acceptance (Leonhardt et al. 1995). Leaf and stem characteristics, and disease resistance are given emphasis.

Many commercial *Leucospermum* cultivars are bulky, heavy, and cumbersome to pack due to large stem diameters and large heavy-textured leaves. These are undesirable characteristics, particularly to exporters, because freight charges are based on a formula that considers cubic dimensions and weight of the box. A densely packed, light-weight box reduces the per-bloom freight charge and allows exporters to compete more favorably in overseas markets. The species *L. lineare*, and particularly the selection *L. lineare* 'Starlight', has slender, light-weight yet strong stems with narrow, nearly needle-like foliage. Breeding has demonstrated that these characteristics are heritable and that *L. lineare* hybrids have improved leaf and stem characteristics. *Leucospermum lineare* is also free of foliar pubescence. Foliar pubescence attracts and retains moisture, which provides an environment for fungal spore germination and infection (Leonhardt et al. 1995).

The most important diseases occurring on protea in Hawaii are root and collar rots caused by *Phytophthora cinnamomi*, *P. nicotianae*, and *Cylindrocladium* sp., stem and leaf scab caused by *Sphaceloma* (Elsinoe) sp., leaf spots and blights caused by *Drechslera biseptata* and *D. dematioidea*, leaf spec caused by *Alternaria alternata*, and root knot galls caused by *Meloidogyne incognita*, the root-knot nematode (Nagata and Ferreira 1993). Root-knot nematodes can severely limit growth and productivity of *Leucospermum*. Heavily infected plants show stunting and chlorosis, followed by death of the plant (Cho et al. 1976; Cho and Apt 1977). Wu reported that this nematode can reduce cut flower yields by at least 25% in infected fields compared to fumigated fields (Wu et al. 1978).

The Hawaii breeding program has utilized ten species and numerous F_1, F_2, and F_3 hybrids to produce seedling populations that are evaluated for disease resistance and other horticultural characteristics (Leonhardt et al. 1995). Some of the parental materials are used for very specific purposes. A selection of *L. saxosum* for example, was determined to be immune to *Sphaceloma* (Elsinoe scab disease) (Nagata et al. 1995) and has been used to impart resistance into commercial hybrids. Among hybrids, L. 'Rachel' has demonstrated a good level of resistance to *Sphaceloma*, and has also shown a good level of resistance to two isolates causing *Botrytis* blight and moderate resistance to two isolates causing *Drechslera* blight. The hybrid L. 'Ka Hoku Hawaii' (Hawaii Star), *L. cordifolium* × (*L. lineare* × *L. vestitum*), and the unnamed hybrids No. 36, *L. lineare* × [*L. conocarpodendron* × (*L. lineare* × *L. cordifolium*)], and No. 49, [*L. conocarpodendron* × (*L. lineare* × *L. cordifolium*)] × *L. cordifolium* 'Sweet Lemon', have shown a good level of resistance to both *Drechslera* isolates. The hybrids L. 'Pohaka La Hawaii' (Hawaii Sunbeam), (*L. lineare* × *L. glabrum*), and No. 36 have shown a good level of resistance to *Botrytis* isolates (Nagata et al. 1995; Leonhardt et al. 1995).

Commercial producers in Hawaii compete in North American markets with California producers. The flowering season for *Leucospermum* begins several weeks earlier in Hawaii than in California. Producers in Hawaii could enjoy this advantage for a longer period if earlier flowering cultivars could be developed. The species *L. patersonii* and *L. pluridens* are among the earliest-flowering, and are being used in breeding for that quality. Two accessions of *L. pluridens* × (*L. lineare* × *L. cordifolium*) flower earlier at the Maui Agriculture Research Station than the hybrid parent (Leonhardt et al. 1995).

LEUCADENDRON

The South African genus *Leucadendron* contains about 60 species, collectively referred to as the conebushes. They are easily identified since they are dioecious, having plants of separate male and female sexes. Both sexes have terminal flowerheads. Female plants produce woody cones containing fruits and seeds while male plants do not produce cones. The cones on female plants consist of spirally arranged floral bracts which partially cover the cone. Male plants are often larger and more heavily branched and may have smaller leaves than female plants (Robelo 1995).

As with most Proteaceae, the *Leucadendrons* grow best in areas with light, well-drained soils with low concentrations of dissolved salts, an adequate supply of fresh water, temperatures in the range of 7°C to 27°C, and frequent if not regular light winds. *Leucadendrons* require an acid soil with a pH not exceeding 5.0. Sandy soils with some humus provide the best growing medium (Vogts 1980).

Several species are cultivated commercially for their decorative foliage, including *L. argenteum, L. discolor, L. galpinii, L. laureolum, L. salicifolium, L. salignum, L. tinctum,* and *L. uliginosum. Leucadendron argenteum,* the 'Silver Tree', can grow to a 8 m tall tree if left unmanaged. Its leaves are grey-green with abundant fine satiny silver hairs that glisten in sunlight. It is grown for its long-lasting cut foliage, and also makes an attractive landscape plant. Its natural habitat is arid, and in cultivation it will succumb to overwatering, soil fungi, and nematodes. *Leucadendron discolor,* harvested in winter and spring, ranges in color from light to dark green to yellow to red, and is a spreading bush up to 1.5 m high. *Leucadendron laureolum* is chartreuse to bright yellow when flowering in winter and spring while *L. salignum* (formerly *L. adscendens*) is bright red, becoming more intensely colored as temperatures decrease. A particularly outstanding cultivar is the female selection *L. salignum* 'Safari Sunset'. *Leucadendron uliginosum* has elegant, slender shoots covered with numerous shiny, silvery leaves (Vogts 1980, Kepler 1988).

The genetic variation in *Leucadendron* is vast and largely untapped for breeding purposes (Littlejohn et al. 1995), although a few hybrids have been introduced to the commercial trade, mostly from South Africa. Hybrids cultivated by commercial growers include *L.* 'Silvan Red' and *L.* 'Inca Gold', both (*L. laureolum* × *L. salignum*), *L.* 'Kam-ee-lion' (*L. salignum* × *L. eucalyptifolium*), and the recent South African introduction *L.* 'Rosette' (*L. laureolum* × *L. elimense* ssp. *salterii*), which can be harvested as a green, yellow, or red-brown product, depending on the season (Littlejohn et al. 1998).

In addition to their highly colorful, easily packaged, long and long-lasting cut stems, a characteristic of many *Leucadendrons* that makes them commercially important is their potential for very high yields. Pruned and managed *L.* 'Silvan Red', in a 3 year study at 3 locations averaged 265 marketable stems per plant per year (Barth et al. 1996). This cultivar can be harvested in the fall as a red-foliaged stem, and in the winter as a tricolor stem with yellow, red, and green foliar bracts. *Leucadendron* 'Safari Sunset', a selection of New Zealand origin, is probably the most widely grown commercial cultivar, with extensive plantings in Australia, New Zealand, South Africa, and Israel. The erect bushy plant is vigorous and fast growing. Its deep wine-colored bracts have excellent keeping quality, lasting up to 60 days (Tija 1986). Dr. Ben-Jaacov, in his presentation at the International Protea Association Conference in Cape Town in 1998 reported that 'Safari Sunset', under intensive management in Israel, has given yields in excess of 600,000 marketable stems per hectare per year. Recent research in South Africa compared *L.* 'Rosette' with *L.* 'Safari Sunset' for yield and stem length. In the third harvest year *L.* 'Rosette' yielded 44.5 stems per plant with 20 stems 80 cm or longer while *L.* 'Safari Sunset' yielded 37.0 stems per plant with 10 stems 80 cm or longer (Littlejohn et al. 1998). Both cultivars are exceptional commercial materials.

Although more widely known as commercial cut foliages and landscape plants, *Leucadendrons* can be grown as colorful potted "flowering" plants. The male *L. discolor* 'Sunset' naturally flowers profusely in early spring with colorful flower-heads. Israeli research has demonstrated that flowering potted plants of 'Sunset' can be produced in 3–5 months by rooting large branched cuttings with initiated flowers. The basal stems of branched 15 cm long cuttings were dipped in a 4,000 ppm IBA solution prior to sticking in a styrofoam/peat medium under intermittent mist and 25% reduced natural light. Rooting began in 4 weeks. The stage of development of the flower-head at rooting was critical for the cutting's further development into a flowering potted plant. If not fully initiated as floral buds, the meristem aborted or reverted to the vegetative state. However, when cuttings were taken at the right stage of floral initiation, colorful flowering potted plants were produced in 3–5 months. Conventional technology for producing potted flowering plants of *L. discolor* by rooting small unbranched vegetative cuttings, growing them to the appropriate size, retarding them chemically, and bringing them to flower, would take 2 years or longer (Ben-Jaacov et al., 1986). The potential for using this technology to produce attractive *Leucadendron* flowering potted plants for the commercial nursery trade is significant.

BANKSIA

The fourth largest export wildflower crop of Australia (Sedgley 1996), the genus *Banksia* is named for the famous botanist, Sir Joseph Banks. Seventy-six taxa have been described under 2 sub-genera, 3 sections, and 13 series (Sedgley 1998). *Banksia* are evergreen, woody perennials with growth habits that range from prostrate ground-huggers to trees. Most of the species are found in the south-west with the remainder along the southern and eastern coasts and tablelands. Nearly all have ornamental features that confer horticultural potential, whether as fresh or dried cut flowers, cut foliages, or in the landscape (Elliott and Jones 1982; Joyce 1998; Parvin et al. 1973; Sedgley 1998; Wrigley and Fagg 1996).

Species widely grown for cut flowers and foliages are shown in Table 5. The most popular cut flower types bear their cylindrical flower spikes terminally, but a few terminal-flowering selections have been made of axillary bearers (Sedgely 1998). Some species produce attractive flowerheads upright on horizontal branches and would need considerable management to be suitable for the commercial markets. Although many commercial plantings are produced from seed and show considerable variability, progress has been made in cultivar development (Fuss and Sedgley 1991; Sedgley et al. 1991; Sedgley 1991, 1995a,b,c,d).

Concurrent with these developments is a need to improve the vegetative propagation systems, as cutting propagation often results in development of a large knob of callus (Hocking 1976). Cutting propagation yields variable results, but the use of intermittent mist (allowing some drying between cycles) and auxin stimulates better root development on semi-hardwood terminal cuttings (Bennell and Barth 1986; Sedgley 1995c). Grafting onto various disease resistant species such as *B. robur* and *B. spinulosa* offers some promise, but additional research is needed to establish successful techniques and timing and to determine compatibility relationships and tolerance to stresses. It is necessary to avoid rootstocks that form lignotubers as these may sucker and compete with the scion (Elliott and Jones 1982). Cutting grafts have been successful with a few species (Elliot and Jones 1982).

Seed germination is reliable, but not for the hybrids, and seed supplies are limited. *Banksia* seed is produced in a hard follicle that often requires heat or heat followed by immersion in water to cause it to open. Seedlings are susceptible to damping off and should be germinated in a sterile well-drained medium. Germination requires 21 to 90 days at 20°–25°C (Elliot and Jones 1982). The optimum medium temperature can range from a constant 10° to 25°C or fluctuate by 10° to 15°C (Bennell and Barth 1986). Transplanting is done as soon as the seedling is large enough to handle.

All species grow best in light, sandy soils of acid pH. They are adapted to soils of low fertility, but benefit from a supply of calcium and application of nitrogen, potassium and iron (Sedgley 1996, 1998). Like other Proteaceae, banksias tend to be intolerant to high levels of phosphorus which interfere with iron uptake (Handreck 1991). In cultivation, pruning is necessary to remove shoots that will not flower and to encourage development of

Table 5. Some Banksia species suitable for cut flower or cut foliage production. Sources: Parvin et al. 1973; Elliot and Jones 1982; Salinger 1985; Sedgley 1998.

Banksia species	Cut flower	Cut foliage
ashbyi	✓	
baxteri	✓	✓
burdettii	✓	
coccinea	✓	
ericifolia		✓
grandis	✓	✓
hookeriana	✓	
integrifolia		✓
menziesii	✓	
occidentalis		✓
prionotes	✓	
speciosa	✓	✓
victoriae	✓	

shoots with sufficient diameter to initiate the inflorescence (Fuss et al. 1992; Sedgley and Fuss 1992; Rohl et al. 1994; Sedgely 1996).

Sedgley (1996, 1998) notes that there has been little published research on postharvest care of cut *Banksia*. Sucrose pulses did not improve the 15 days vase life of *B. coccinea* (Delaporte et al. 1997), and anti-bacterials such as 0.01% chlorine, acidifiers such as 0.01% citric acid, and 0.02% aluminum sulfate have been recommended as a matter of course, but without verified results (Sedgely 1996).

GREVILLEA

A large genus (more than 340 species) of shrubs and trees from dry sclerophyll forests and heaths in Australia (5 species are found in New Caledonia, Sulewesi, and Papua New Guinea), grevilleas have many ornamental uses, especially in landscapes. Growth habits of the most popular species range from prostrate ground covers to mounded shrubs. Some have unusual, asymmetric, or layered habits. A tree form, *Grevillea robusta*, flourishes in sub-tropic climates and has potential as an invasive species because of its abundant seed production.

Grevillea have been in cultivation outside of Australia for over 200 years, with the earliest record of introduction of 3 species to England in 1791 and another 15 species in the 1820s (Elliot and Jones 1990). Nurseries in New Zealand and California also grow a wide range of species for landscape uses. As these locations and their native habitats suggest, many grevillea are frost and cold tolerant to –4°C (Elliot and Jones 1990). Drought tolerance is another quality to recommend many species in areas with dry summers.

Grevillea inflorescences are mostly toothbrush-like clusters, about 5 to 12 cm in length, and running the color gamut from white and greenish through yellow, orange, purplish, and red, and include some multi-colored forms. Other inflorescence categories include upright spider-like, pendant, or terminal cylindrical clusters. Some have very strong aromas while others are pleasant and sweet. Flowering tends to be seasonal, depending upon moisture, temperature, and daylength. Flowering is strongest in sunny locations and diminished in shade (Wrigley and Fagg 1996). There are also reports of skin irritation and rash from handling some prickly as well as non-prickly species.

While a number of hybrid grevilleas and more tropical species have been selected for large colorful inflorescences (Tully 1977), they are not widely marketed as cut flowers because of short vaselife and a tendency for floret abscission. Although the best can achieve a vase life of 7 to 10 days following cutting, production is said to be low (Olde and Marriott 1995). They are also alleged to be difficult to pack. The potential for their use is good if the problems can be overcome through the use of postharvest treatments, improved packaging, and breeding and selection (Joyce et al. 1996). A sugar, 2% citric acid, and bleach mixture has been recommended for home use, together with maintaining turgor by placing the cut stems in water.

Grevillea foliage displays a wide range of textures, colors, and shapes. The textures and shapes range from deeply divided, fern-like and fishbone-like leaves to entire or broadly-lobed shapes and pinnately-divided and regularly toothed and holly-like foliage. Colors range from silver-grey to dark glossy green. Many species have attractive undersurface of silver or bronze indumentum. These elements contribute to their value as landscape plantings, but also to their use as cut foliages. Cut as a growth flush matures, the foliage may last 30 days in water or commercial floral preservatives (Parvin 1991; Criley and Parvin 1993). It is this use that has potential in the floriculture trade. Management practices for cut foliage production need to be developed.

Potted grevillea plants have wide acceptance for patio and garden use both because of their foliage and flowers. They are easily rooted and managed, both by pruning and with growth regulators (Ben-Jaacov et al. 1989). The development of tissue culture techniques enables greater availability of attractive, but difficult-to-root cultivars such as 'Robyn Gordon' (Gorst et al. 1978, Watad et al. 1992). Several potted grevilleas have been introduced into the trade by Israel producers (Ben-Jaacov et al. 1989).

Grevillea may be propagated by seed, cuttings, layers, grafting, and tissue culture. Propagation by cuttings is said to be easy with mid to late summer matured growth (Wrigley and Fagg 1996). A Hawaii study indicated that quick dips in liquid auxin formulations (2000 to 4000 ppm) applied to terminal or immediately sub-terminal growth gave satisfactory results in 5 to 6 weeks (Groesbeck and Rauch 1985). Commercial liquid and powder auxin formulations provided good rooting for a number of species. Bottom heat of 29°C with no auxin stimulated 90% take on older wood of *G.* 'Robyn Gordon' (Dupee and Clemens 1981). Interesting landscape forms have been produced by approach-grafting weeping or prostrate forms onto rootstocks of *G. robusta* (Crossen 1990) or *G. banksii* or *G.* 'Poorinda Royal Mantle'(Wrigley and Fagg 1996). The cleft graft was reported successful as well (Dupee and Clemens 1981). However, it is recommended that a healthy top bud be left near the cut to prevent dieback below the graft (Elliot and Jones 1990). Air layering is reported as regularly successful (Tully 1977). Seed germination is enhanced by a presoak in 0.2% potassium nitrate for 12–24 hours, sowing in a sandy medium, and subjecting the seed to alternating warm (25°–33°C) temperatures (Heslehurst 1977). Germination required 4 to 5 weeks. Scarification or seedcoat removal also improves germination (Dupee and Clemens 1981).

More than one-half of the species have been tried in horticulture because of their wide adaptability to a range of soil conditions (Molyneux 1978). Many of the Western Australia species are found on infertile, non-calcareous soils, sands, and leached lateritic soils. Many of the eastern species can be found in clay or clay-loam soils. A few inhabit deserts or rainforests, and some tolerate slightly saline or alkaline soils (Elliot and Jones 1990; Olde and Marriott 1995). Many of the Western Australian species are not demanding of substrate as long as it is well-drained, although there are a few that will even tolerate poor drainage (Olde and Marriott 1995). While they respond to good fertility, high nutrient levels, especially phosphorus, are not required. Controlled release fertilization with careful attention to the form provided is recommended (Bowden 1987).

Longevity under well-fertilized conditions appears to be a problem, especially where the plants are also well-watered (Specht 1978). However healthy 10 to 15 year old plants can be found. These have generally benefited from regular pruning (Elliot and Jones 1990). *Phytophthora cinnamomi* is very devastating to *grevilleas* (Molyneux 1978).

ISOPOGON

Isopogon is native to temperate Australian regions, with the main distribution in southwestern Australia. Many are coastal or near-coastal in habitat and grow in well-drained, highly-leached sandy or lateritic soils and gravels or clay loams. They range from sea level to moderate altitudes and cope with a wide temperature range down to –7°C, where damage occurs. Full sun is the preferred light environment, but some tolerate semi-shade (Elliot and Jones 1990).

They offer some interesting, hardy plant materials for the landscape, and possibly for the cutflower trade. Most of the 35 described species are temperate zone shrubs of 1 to 2.5 m tall, but a few can grow into small trees (Foreman 1997). Most species are small to medium-sized shrubs while others are dwarf, spreading undershrubs. A number of species are adapted to container culture.

Cone- or drumstick-shaped flower clusters, of white to yellow to pink to mauve, are borne terminally or in the upper leaf axils. Flowering is chiefly in the spring months. A few species have good vase life and are grown commercially in Australia. Among species with cutflower potential are the winter-flowering *I. cuneatus*, and spring-flowering *I. latifolius* and *I. formosus* (Salinger 1985; Elliot and Jones 1990; Foreman 1997). While their cones are decorative, the scales are often shed with the seed.

Seed is not plentiful because it is often lost when the scales dehisce from the cone. Fresh seed, sown shallowly in a moist medium, germinates in 20 to 90 days. One pregermination recommendation is to lightly singe the seed with a flame (Elliot and Jones 1990). Cutting propagation is usually successful when aided by hormone rooting powders.

In culture, controlled release fertilizers with a low phosphorus content are recommended. Established plants are fairly drought tolerant, and over-watering contributes to loss of plants because the wet conditions favor *Phytophthora cinnamomi* infection. Tip pruning following flowering stimulates bushy growth. Species with lignotubers tolerate severe pruning.

DRYANDRA

These Australian natives (120 species) have a variable growth habit, ranging from prostrate shrubs to small trees. Foliage characteristics range from soft and needlelike to tough and prickly. The flowerhead resembles a shaving brush surrounded by basal bracts. Flower colors are mainly in the yellow to orange to bronze shades. Their potential as cutflowers needs further evaluation, but some species, such as *D. formosa*, dry nicely and could be added to this niche market (Joyce 1998). Among the recommended cut flower species are *D. formosa*, *D. praemorsa*, and *D. quercifolia* (Elliot and Jones 1984). Many species have foliage so spiny that they are not suitable for floral purposes (Salinger 1985).

Propagation is generally by sowing fresh or stored seed into a well-drained, loose medium; however, seed-feeding insects often render viable seed scarce (Elliot and Jones 1984). Pregermination treatments do not seem necessary (Cavanaugh 1994). Germination times range from 3 weeks to 3 months with an average of 5 to 8 weeks. Transplanting can be done fairly early, when seedlings have attained 50–75 mm in height. Cutting and grafting propagation successes have been reported for some species (Cavanaugh 1994), but neither practice is widely used. Softwood cuttings taken during winter and treated with rooting hormones have yielded some success (Elliot and Jones 1984).

Their native habitats include lateritic gravel, sandy, or granitic soils, always well-drained (Elliot and Jones 1984). Once established, they are said to be more drought tolerant than *Banksia* species (Elliot and Jones, 1984). Many species are fairly cold tolerant, tolerating light frosts or short-lived snowfalls. Cultural conditions for success include full sun, good drainage, and good air circulation. As with other Proteaceae, low fertility is adequate and high phosphorus levels are to be avoided. Chlorosis is a problem on some soils and may be countered with weak iron chelate drenches. A few species are recommended as container plants: *D. ferruginea*, *D. polycephala*, and *D. speciosa* (Elliot and Jones 1984).

TELOPEA

Five species of *Telopea* have been described, and a number of hybrids have been released (Dennis 1991; Nixon and Payne 1996; Wrigley and Fagg 1996). They are native to acid, infertile, well-drained soils in New South Wales, Victoria, and Tasmania. As small trees or managed shrubs, they have both landscape value and commercial cutflower use. All species produce terminal, brilliant to rose red (occasional pink, white or yellow) inflorescences up to 15 cm in diameter on stems of up to 1 m length. The florets are arranged spirally on elongated cones subtended by an involucre of similarly colored bracts. *Telopea speciosissima* and *T. oreades* are the principal species for commercial flower production. Unlike other proteas, there is little by-pass by lower shoots. Plants tend to be upright and vigorous.

The most important species, *T. speciosissima* is known as the waratah and is the floral emblem of New South Wales, Australia. Although blooms were originally wild-collected, commercial production has increased in Australia as well as in New Zealand, US (Hawaii), Israel, and South Africa (Offord 1996). Australian production has been reported at 20,000 to 50,000 stems/ha five years after establishment in high density plantings (Worrall 1994), with annual production estimated at 0.6 to 1.7 million stems (Worrall, cited in Offord 1996). The plants are long-lived and capable of production for many years with good management.

While telopeas occur in woodland situations, they flower best in full sun or light shade. Flowering occurs in spring to early summer, but the bloom period is only 4 to 5 weeks duration. Choice of location can influence flowering time as can selection of hybrids (Matthews 1993; Offord 1996). Floral display life is about 7 to 13 days (Dennis 1991) and browning of the bracts can be a problem. The inflorescence is usually harvested before all florets have matured, and is discarded when one-third have turned blue-red (Faragher 1986). Vase life could be extended 3 to 5 days by harvesting when only the first cycle of flowers has matured, by the use of 5% sucrose and a germicidal compound in the water, and by refrigerating the cut flowers at 2°C after hydration (Lill and Dennis 1986). Ethylene does not appear to be a critical factor in senescence (Faragher 1986). Selection for lack of bract browning and low nectar production is a consideration in developing commercial types (Salinger 1981).

Telopea are readily propagated from fresh seed with germination occurring in 2.5–4 weeks at 25°C (Worrall 1994; Wrigley and Fagg 1996). Seed can be stored at room temperature for 6 months and for at least 2 years at 5°C (Worrall 1994; Offord 1996).

Semi-hardwood terminal cuttings (20 cm length with 5–6 leaves) of *T. speciosissima*, treated with 2000 to 4000 ppm IBA as a quick basal dip rooted with success rates of 50 to 75% after 8 weeks (Worrall 1976). A talc dust of 0.3% IBA is also satisfactory. Bottom heat of 24°C enhances rooting as does intermittent mist (Worrall 1994). Response varied with season, and cuttings from actively growing mother plants responded better to the low levels of IBA than to cuttings taken in winter from dormant plants. Leaf bud cuttings have been used to increase selected plants when propagative material is limited (Ellyard and Butler 1985). Tissue culture has been successful as well (Seelye et al. 1986; Offord and Campbell 1992; Offord et al. 1992).

Telopea culture requires well-drained soils, full sun, and freedom from frost. While their native soils are deep sands, they also thrive on well-drained basaltic clays (Offord 1996). Water requirements are high during summer flower bud initiation. Established plants tolerate a temperature range from 3° to 24°C. Plants are spaced at 1.5 to 3 m in rows with 3 m between rows (Dennis 1991). Pruning at or soon after harvest is practiced to encourage new stems of suitable length for cutflowers in the next season. Rejuvenation pruning is practiced periodically to reduce plant height and encourage production of longer stems (Worral 1994). Pot culture is also possible, but the lignotuber produced by the plants requires a fairly large container (Offord 1996). Potting media need to be well-drained with a pH of 5.5 and low phosphorus content.

SERRURIA

Like many other South African proteas, the genus *Serruria* (50 species) occur in well-drained nutrient-poor soils of the winter-rainfall area (1000 mm) of the Cape Floral Kingdom of South Africa. Their distribution is limited to small, specific localities within this region (Rebelo 1995), and many are endangered because of loss of habitat (Worth and van Wilgen 1988).

The serrurias are small shrubs (prostrate habit to 2 m tall) with fine, feathery foliage and prominent, white to pink bracts subtending the individual flowers borne multiply on one to 11 capitula. Commonly called spiderheads in their native South Africa, serruria inflorescences may be solitary or consist of clusters of small flowerheads. The principal species in commercial culture are *S. florida* (Blushing Bride) and *S. rosea* and their hybrids.

The cutflower serrurias tend to be upright growers. The globose flowerheads range from 3 to 5.5 cm in diameter but appear larger because of the bracts. Flowering occurs in the late winter to early spring, and is known to be stimulated by the long days of the preceding summer and fall (Malan and Brits 1990). Initiation and early development required about 6 weeks and another 10 weeks was required to reach anthesis. Little work has been reported on improving vase life, which is about 7 to 10 days following cutting. The flowers also dry well (Matthews and Carter 1993).

Serruria potted plants have good floral display qualities and can be produced in less than one year (Malan and Brits 1990). Cuttings should be taken during the high light, long days of early spring and summer as induced cuttings taken in the fall had low rooting percentages (Ackerman et al. 1995). The flowering period ranges from 30 to 55 days under outdoor conditions. Short durations of darkness as in shipping are not damaging to the post-harvest life of potted plants. Growth retardants such as paclobutrazol inhibit shoot elongation, while ethephon increases branching and branch angle (Brits 1995).

Propagation of serrurias to establish desirable clones is by mainly by cuttings. Ten weeks is required for acceptable rooting, but up to 20 weeks may be required if cuttings are taken during late fall or winter. Cutting bases are dipped for 10 seconds into a potassium salt formulation of IBA at the rate of 4000 mg/L (Ackerman et al. 1995). Techniques to establish and proliferate *Serruria* in vitro have been reported, but the rooting of plantlets from such cultures was not described (Ben-Jaacov and Jacobs 1995). Seed is reportedly long-lived and germinates in response to soil temperature fluctuations following clearing of the understory by fire (Brits 1986a; Worth and Wilgen 1988), but soaking in 1% hydrogen peroxide has been shown to stimulate germination in the laboratory (Brits 1986b).

MIMETES

Known as the Pagoda flowers in their native South Africa, *Mimetes* species bear large terminal flowerheads containing smaller headlets (capitula) bearing few to many flowers. Leaves and bracts subtending these headlets are often brightly colored and may curl around to clasp the flowers. Some species in the Silver Pagoda group bear silvery hairs, making them attractive for this character rather than for colored bracts.

Most of the 13 species of *Mimetes* are rare and found in isolated habitats of the south and southwestern Cape (Rourke 1984), frequently at high elevations in low to moderate rainfall areas. Most are found on sandstone-derived soils, but a few are found in moist peaty soils along marshes and swamps (Rebelo 1995). Coastal species such as *M. cucullatus* also withstand salt winds. Repeated burning maintains the shrub in a rounded form with numerous upright unbranched stems arising from a woody, persistent lignotuber (Rourke 1984). Flowering is most profuse on the vigorous young growth, suggesting that commercial flower production will be dependent upon efficient pruning.

Mimetes cucullatus is one of the more widely distributed species. It flowers year around but most heavily during the fall and winter months and offers potential as a cutflower as it produces 30 cm stems tipped with scarlet red and yellow bracts. It is said to have good vase life as a cutflower (Matthews and Carter 1993). It is a long-lived shrub once established and tolerates heavy pruning. Other *Mimetes* species with attractive flowerheads are reportedly short-lived although the seed remains viable for many years, ready to germinate when the natural habitat is cleared by fire (Rebelo 1995).

Cultivation of *Mimetes* requires well-drained, acid soils with some organic matter. Studies of plant management are still needed. Propagation is by seed or semi-hardwood cuttings taken in the fall (Matthews and Carter 1993).

REFERENCES

Ackerman, A., J. Ben-Jaacov, G.J. Brits, D.G. Malan, J.H. Coetzee, and E. Tal. 1995. The development of *Leucospermum* and *Serruria* as flowering pot plants. Acta Hort. 387:33–46.

Ackerman, A. and G.J. Brits. 1991. Research and development of protea pot plants for export under South African and Israeli conditions. Protea News 11:9–12.

Ackerman, A., Y. Shchori, S. Gilad, B. Mitchnik, K. Pinta, and J. Ben-Jaacov. 1995. Development of *Leucospermum* tolerant to calcareous soils for the protea industry. J. Int. Protea Assoc. 29:24–29.

Barth, G.E., N.A. Maier, J.S. Cecil, W.L. Chyvl, and M.N. Bartetzko. 1996. Yield and seasonal growth flushing of *Protea* 'Pink Ice' and *Leucadendron* 'Silvan Red' in South Australia. Austral. J. Expt. Agr. 36:869–875.

Ben-Jaacov, J., A. Ackerman, and E. Tal. 1989. Development of new woody flowering pot plants: A comprehensive approach. Acta Hort. 252:51–58.

Ben-Jaacov, J. and G. Jacobs. 1986. Establishing *Protea, Leucospermum* and *Serruria* in vitro. Acta Hort. 185:39–52.

Ben-Jaacov, J., R. Shillo, and M. Avishai. 1986. Technology for rapid production of flowering pot-plants of *Leucadendron discolor* E. Phillips, S. Hutch. Elsevier Sci. Publ. B.V.

Bennell M. and G. Barth. 1986. Propagation of *Banksia coccinea* by cuttings and seed. Proc. Int. Plant Prop. Soc. 36:148–152.

Blomerus, L.M., G.J. Brits, and G.M. Littlejohn. 1998. 'High Gold', a yellow, vigorous *Leucospermum*. HortScience 33:1094–1095.

Bowden, A. 1987. Application of phosphorus to proteaceous plants. Proc. Int. Plant Prop. Soc. 37:138–141.

Brink, J.A. and G.H. de Swardt. 1986. The effect of sucrose in a vase solution on leaf browning of *Protea*. Acta Hort. 185:111–119.

Brits, G.J. 1984a. Historical review of the South African protea cut-flower industry. Roodeplaat Bul. (5):18–19.

Brits, G.J. 1984b. Production research on South African proteas: Current trends. Veld & Flora 70(4):103–105.

Brits, G.J. 1984c. Protea production in the United States of America. Protea News 1:10–11.

Brits, G.J. 1986a. Influence of fluctuating temperatures and H_2O_2 treatment on germination of *Leucospermum cordifolium* and *Serruria florida* (Proteaceae) seeds. S. Afr. J. Bot. 52:286–290.

Brits, G.J. 1986b. The effect of hydrogen peroxide treatment on germination in Proteacae species with serotinus and nut-like achenes. S. Afr. J. Bot. 52:291–293.

Brits, G.J. 1990b. Rootstock production research in *Leucospermum* and *Protea*: I. Techniques. Acta Hort. 264:9–25

Brits, G.J. 1990c. Rootstock production research in *Leucospermum* and *Protea*: II. Gene sources. Acta Hort. 264:27–40.

Brits, G.J. 1992a. Breeding programmes for Proteaceae cultivar development. Acta Hort. 316:9–18.

Brits, G.J. 1992b. The VOPI diversifies its protea cultivar releases. J. Int. Protea Assoc. 24:19–25.

Brits, G. J., E. Tal, J. Ben-Jaacov, and A. Ackerman. 1992. Cooperative production of protea flowering pot plants selected for rapid production. Acta Hort. 316:107–118.

Brits, G.J. and G.C. van den Berg. 1991. Interspecific hybridization in *Protea, Leucospermum*, and *Leucadendron* (Proteaceae). Protea News 10:12–13.

Brits, G.J. 1995. Selection criteria for protea flowering pot plants. Acta Hort. 387:47–54.

Cavanaugh, T. 1994. *Dryandra*, a growing guide for beginners. Austral. Plants 18:11–17.

Cho, J.J. and W.J. Apt. 1977. Susceptibility of proteas to *Meloidogyne incognita*. Plant Dis. Rptr. 61:489–492.

Cho, J.J., W.J. Apt, and O.V. Holtzmann. 1976. The occurrence of *Meloidogyne incognita* on members of the Proteaceae family in Hawaii. Plant Dis. Rptr. 60:814–817.

Claassens, A.S. 1986. Some aspects of the nutrition of *Proteas*. Acta Hort. 185:171–179.

Claassens, A.S. 1981. Soil preparation and fertilisation of proteas. Farming in South Africa Ser.: Flowers and Ornamental Shrubs, B14. Dept. Soil Sci. and Plant Nutrition, Univ. of Pretoria, S. Afr.

Coetzee, J.H. and M.G. Wright. 1991. Post-harvest treatment and disinsectation of protea cut flowers. Proc. Int. Protea Assn. 6th Biennial Conf. Perth, Western Australia.

Cresswell, G.C. 1991. Assessing the phosphorus status of proteas using plant analysis. Proc. Int. Protea Assoc. Sixth Biennial Conf., Perth, West. Australia. p. 303–310.

Criley, R.A., P.E. Parvin, and P. Shingaki. 1996. Evaluation of *Protea* accessions from South Africa for time of flowering on Maui. Hort. Digest 110:1–2. Col. of Trop. Agr. and Human Resources, Univ. of Hawaii.

Criley, R.A. 1998. *Leucospermum*: Botany and horticulture. Hort. Rev. 22:27–90.

Criley, R.A. and P.E. Parvin. 1993. New cut foliages from Australia, New Zealand, and South Africa. Acta Hort. 337:95–98.

Criley, R. A., P.E. Parvin, and C. Williamson. 1978. Keeping quality of *Leucospermum cordifolium*. First Ann. Orn. Sem. Proc. Univ. Hawaii, CES Misc. Publ. 172:34–36.

Criley, R.A., C. Williamson, and R. Masutani. 1978. Vaselife determination for *Leucospermum cordifolium*. Flor. Rev. 163(4228):37–38, 80–81.

Crossen, T. 1990. Approach grafting grevilleas. Proc. Int. Plant Prop. Soc. 40:69–72.

Dai, J. and R.E. Paull. 1995. Source-sink relationship and *Protea* postharvest leaf blackening. J. Am. Soc. Hort. Sci. 120(3):475–480.

Danziger, D. 1997. Producing Israel's unique protea. Grower Talks 61(10):51.

Delaporte, K.L., A. Klieber, and M. Sedgely. 1997. To pulse or not to pulse? Improving the vase life of *Banksia coccinea* by postharvest treatments. Austral. Hort. 95(8):79–84.

Downs, C. and M. Reihana. 1986. The quality of *Leucospermum* cultivars after simulated transport. J. Int. Protea Assoc. 10:5–8.

Dupee, S.A. and J. Clemens. 1981. Propagation of ornamental grevillea. Proc. Int. Plant Prop. Soc. 31:198–208.

Elliot, W.R. and D.L. Jones. 1982. p. 281–286. In: Encyclopedia of Australian Plants Suitable for Cultivation. Vol. 2. Lothian Publ. Co. Pty. Ltd., Sydney, Australia.

Elliot, W.R. and D.L. Jones. 1984. p. 349–353. In: Encyclopedia of Australian Plants Suitable for Cultivation. Vol. 3. Lothian Publ. Co. Pty, Ltd., Melbourne, Australia.

Elliot, W.R. and D. L. Jones. 1990. P. 7–16. In: Encyclopedia of Australian Plants Suitable for Cultivation. Vol 5. Lothian Publ. Co. Pty. Ltd., Melbourne, Australia.

Ellyard R.K. and G. Butler. 1985. Breakthrough in waratah propagation. Austral. Hort. 83(3):27–28.

Faragher, J.D. 1986. Post-harvest physiology of waratah inflorescences (*Telopea speciosissima,* Proteaceae). Scientia Hort. 28:271–279.

Ferreira, D.I. 1986. The influence of temperature on the respiration rate and browning of *Protea neriifolia* R Br inflorescences. Acta Hort. 185: 121–125.

Fleming, K.D., K.W. Leonhardt, and J.M. Halloran. 1991. The economics of protea production in Hawaii. Proc. Sixth Biennial Conf., Int. Protea Assoc. Perth, Aust.

Fleming, K.D., K.W. Leonhardt, and J.M. Halloran. 1994. An economic model of protea production in Hawaii. Proc. Hawaii Trop. Cutflower and Ornamental Plant Industry Conf. Kailua-Kona, Hawaii.

Foreman, D.B. 1997. *Isopogon,* drumsticks: the genus. Austral. Plants 19(150):63–78, 80, 96.

Fuss, A.M. and M. Sedgley. 1991. The development of hybridisation techniques for *Banksia menziesii* for cut flower production. J. Hort. Sci. 66:357–365.

Fuss, A.M., S.J. Pattison, D. Aspinall, and M. Sedgley. 1992. Shoot growth in relation to cut flower production of *Banksia coccinea* and *Banksia menziesii* (Proteaceae). Scientia Hort. 49:323–334.

Gorst, J.R., R.A. Bourne, S.E. Hardaker, A.E. Richards, S. Dircks, and R. A. de Fossard. 1978. Tissue culture propagation of two *Grevillea* hybrids. Proc. Int. Plant Prop. Soc. 28:435–446.

Groesbeck, S.J. and F.D. Rauch. 1985. *Grevilleas*: propagation and potential use in Hawaii. p. 77–90. In: Proc. 2nd Fert. and Orn. Short Course, March 1984. Univ. Hawaii; CTAHR, HITAHR 01.04.85.

Hamilton, R.A. and P.J. Ito. 1984. Macadamia nut cultivars for Hawaii. Info. Text Series 023. College Trop. Agr. Human Resources, Univ. of Hawaii, Manoa.

Handreck, K.A. 1991. Interactions between iron and phosphorus in the nutrition of *Banksia integrifolia* L.f. var. *ericifolia* (Proteaceae) in soil-less potting media. Austral. J. Bot. 39:373–384.

Heslehurst, M.R. 1977. Germination of *Grevillea banksii* (R.Br.). Austral. Plants 9:206–208.

Hocking, F.D. 1976. *Banksias*: propagation and culture. Austral. Plants 9:95–103.

Ito, P.J., K.W. Leonhardt, P.E. Parvin, T. Murakami, and D.W. Oka. 1991. New hybrid *Leucospermum* (Proteaceae) introductions. The Hawaii Tropical Cut Flower Industry Conference March 29–31, 1990. Coll. Trop. Agr. & Human Resources, Univ. Hawaii, Res.-Ext. Ser. 124:164–165.

Ito, P.J., T. Murakami, D. Oka, and P.E. Parvin. 1990. New protea hybrids developed by breeding. p. 91–92. Proc. Third Fert. and Orn. Short Course. Coop. Ext. Serv. Coll. Trop. Agr. Human Resources, Univ. Hawaii. HITAHR 01.01.90.

Ito, P.J., T. Murakami, and P.E. Parvin. 1978. Protea improvement by breeding. First Ann. Orn. Sem. Proc. Coop. Ext. Serv. Coll. Trop. Agr. Human Resources, Univ. Hawaii. Misc. Pub. 172:1–4.

Ito, P.J., T. Murakami, and P.E. Parvin. 1979. New *Leucospermum* hybrids. Seventh Ann. Protea Workshop Proc. Coop. Ext. Serv. Coll. Trop. Agr. Human Resources, Univ. Hawaii. Misc. Pub. 176:1–4.

Jacobs, G. 1985. *Leucospermum.* p. 283–286. In: A.H. Halevy (ed.), Handbook of flowering. CRC Press, Boca Raton, FL.

Jacobs, G. and H.R. Minnaar. 1980. Light intensity and flower development of *Leucospermum cordifolium.* HortScience 15:644–645.

Jones, R. and J. Faragher. 1990. The viability of transporting selected cutflower species by seafreight. J. Int. Protea Assoc. 19:43–44.

Jones, B.J., R. McConchie, W.G. van Doorn, and M.S. Reid 1995. Leaf blackening in cut *Protea* flowers. Hort. Rev. 17: 173–201.

Joyce, D.C. 1998. Dried and preserved ornamental plant material: not new but often overlooked and underrated. Acta Hort. 454:133–145.

Joyce, D.C., P. Beal, and A.J. Shorter. 1996. Vase life characteristics of selected *Grevillea*. Austral. J. Exp. Agr. 36:379–382.

Kepler, A.K. 1988. Proteas in Hawaii. Mutual Publishing. Honolulu, Hawaii.

Kunisaki, J.T. 1989. *In vitro* propagation of *Leucospermum* hybrid, 'Hawaii Gold'. HortScience 24:686–687.

Kunisaki, J.T. 1990. Micropropagation of *Leucospermum*. Acta Hort. 264: 45–48.

Lamont, B.B. 1986. The significance of proteoid roots in proteas. Acta Hort. 185:163–170.

Leonhardt, K., S. Ferreira, N. Nagata, D. Oka, J. Kunisaki, P. Ito, and P. Shingaki. 1995. Hybridizing *Leucospermum* (Proteaceae) for disease resistance and improved horticultural characteristics. p. 76–77. Proc. Third Multicommodity Cutflower Industry Conf. College Trop. Agr. Human Resources, HITAHR 03.04.95

Lill, R.E. and D.J. Dennis. 1986. Post-harvest studies with NSW waratah. Acta Hort. 185:267–271.

Littlejohn, G.M., G.C. van den Berg, and G.J. Brits. 1998. A yellow, single-stem *Leucadendron*, 'Rosette'. Hort. Sci. 33(6):1092–1093.

Littlejohn, G.M., I.D. van der Walt, G.C. van den Berg, W.G. de Waal, and G. Brits. 1995. 'Marketable product' approach to breeding Proteaceae in South Africa. Acta Hort. 387:171–175.

Malan, D.G. 1992. Propagation of Proteaceae. Acta Hort. 316:27–34.

Malan, D.G. and G.J. Brits. 1990. Flower structure and the influence of daylength on flower initiation of *Serruria* Florida Knight (Proteaceae). Acta Hort. 185:87–92.

Malan, D.G. and G. Jacobs. 1990. Effect of photoperiod and shoot decapitation on flowering of *Leucospermum* 'Red Sunset'. J. Am. Soc. Hort. Sci. 115:131–135.

Matthews, L. 1993. The protea growers handbook. Tradewinds Press (Pty) Ltd., Durban North, S. Africa.

Matthews, L.J. and Z. Carter. 1993 Proteas of the world. Timber Press, Portland, OR.

Mathews, P. 1984. The history of IPA. Veld & Flora 70(4):115–116.

Mathews, P. 1981. The growing and marketing of proteas. Report of the First Int. Conf. of Protea growers. Proteaflora Enterprises Pty. Ltd., Melbourne, Australia.

Meynhardt, J.T. 1976. Proteas, picking and handling. Farming in South Africa Ser.: Flowers, Ornamental Shrubs and Trees, B5. Hort. Res. Inst., Pretoria, South Africa.

Moffat, J. and L. Turnbull. 1994. Production of *Phytophthora* tolerant rootstocks: II. Grafting compatibility studies and field testing of grafted proteas. J. Int. Protea Assoc. 27:14–22.

Molyneux, W. 1978. *Grevillea* in horticulture. Austral. Plants 9:287–298.

Nagata, N. and S. Ferreira. 1993. The protea disease letter, 2. College of Trop. Agr. and Human Resources, Univ. of Hawaii, Manoa.

Nagata, N., S. Ferreira, and K. Leonhardt. 1995. The protea disease letter, 3. College of Trop. Agr. and Human Resources, Univ. of Hawaii, Manoa.

Napier, D.R. and G. Jacobs. 1989. Growth regulators and shading reduce flowering of *Leucospermum* cv. Red Sunset. HortScience 24:966–968.

Nixon, P. and W.H. Payne. 1996. Waratahs: The new breed to the year 2000. Austral. Plants. 18:303–311.

Offord, C.A. 1996. *Telopea* (Waratah), Family Proteaceae. p. 68–81. In: K.A. Johnson and M. Burchett (eds.), Native Australian plants: Horticulture and uses. Univ. S. Wales Press, Sydney, Australia.

Offord, C.A. and L.C. Campbell. 1992. Micropropagation of *Telopea speciosissima* R.Br. (Proteaceae). 2: Rhizogenesis and acclimatisation in ex vitro conditions. Plant Cell Tissue Organ Cult. 29:223–230.

Offord, C.A. and L.C. Campbell. 1992. Micropropagation of *Telopea speciosissima* R.Br. (Proteaceae). 1: Explant establishment and proliferation. Plant Cell Tissue Organ Cult. 29:215–221.

Olde, P. and N. Marriott. 1995. The grevillea book. Timber Press, Portland, OR.

Parvin, P.E. 1974. Project: Proteas—Hawaii. Proc. Int. Plant Prop. Soc. 24:41–42.

Parvin, P.E. 1978. The influence of preservative solutions on bud opening and vase life of selected Proteaceae. Proc. 1st Ann. Orn. Sem. UHCES Misc. Publ. 173: 29–33.

Parvin, P.E. 1984. Proteas—from 'curiosity' to 'commodity'. Veld Flora 70(4):109–111.

Parvin, P.E. 1991. Potential cut flower, cut foliage production of Australian and South African flora in Florida. Proc. Fla. State Hort. Soc. 104:296–298.

Parvin, P.E., R.A. Criley, and R.M. Bullock. 1973. Proteas: developmental research for a new cut flower crop. HortScience 8:299–303.

Parvin, P.E. and K.W. Leonhardt. 1982. Care and handling of cut protea flowers. Proc. 8th and 9th Ann. Protea Workshop. Univ. Hawaii, HITAHR Res.-Ext. Ser 018:20–23.

Paull, R.E. 1988. Protea postharvest black leaf. Hort. Digest. 88:3–4. Col. of Trop. Agr. and Human Resources, Univ. of Hawaii, Manoa.

Rao, C.V. 1971. Proteaceae, botanical monograph No. 6. Sree Saraswaty Press Lt., Calcutta, India.

Rebelo, T. 1995. SASOL Proteas: A field guide to the proteas of Southern Africa. Fernwood Press, Vlaeberg, S. Africa.

Rohl, L.J., A.M. Fuss, J.A. Dhalwal, M.B. Webb, and B.B. Lamont. 1994. Investigation of flowering in *Banksia baxteri* and *B. hookeriana* Meissner for improving pruning practices. Austral. J. Expt. Agr. 34:1209–1216.

Rourke, J.P. 1972. Taxonomic studies on *Leucospermum* R. Br. S. Afr. J. Bot. (Suppl.) 8:1–194.

Rourke, J.P. 1984. A revision of the genus *Mimetes* Salisb. (Proteaceae). J. S. Afr. Bot. 50:171–236.

Rousseau, F. 1970. The Proteaceae of South Africa. Purnell and Sons, Cape Town, S. Africa.

Rousseau, G.G. 1966. Proteas can be grafted. Farming S. Afr. 42(6):53–55.

Sacks, P.V. and I. Resendiz. 1996. Protea pot plants: production, distribution and sales in Southern California. J. Int. Protea Assoc. 31:34.

Sadie, J. 1997. The international protea register. Stellenbosch, S. Africa.

Salinger, J.P. 1985. Commercial flower growing. Butterworths of New Zealand, Wellington, NZ.

Sedgley, M. 1991. *Banksia* (*Banksia hookeriana* hybrid). Plant Var. J. 4(2):9–11.

Sedgley, M. 1995a. *Banksia coccinea* 'Waite Crimson'. Plant Var. J. 8(2):8–9, 18.

Sedgley, M. 1995b. *Banksia coccinea* 'Waite Flame'. Plant Var. J. 8(2):9, 18–19.

Sedgley, M. 1995c. Cultivar development of ornamental members of the Proteaceae. Acta Hort. 387:163–169.

Sedgley, M. 1995d. Breeding biology of *Banksia* species for floriculture. Acta Hort. 397:155–162.

Sedgely, M. 1996. *Banksia*, family Proteaceae. p. 18–35. In: K. Johnson and M. Burchett (eds.), Native Australian plants horticulture and uses. Univ. South Wales Press, Sydney, Australia.

Sedgley, M. 1998. *Banksia*: New proteaceous cut flower crop. Hort. Rev. 22:1–25.

Sedgley, M. and A.M. Fuss. 1992. Correct pruning lifts banksia yields. Austral. Hort. 90:50–53.

Sedgley, M., M. Wirthensohn, and A.M. Fuss. 1991. Selection and breeding of banksias. p. 193–200. In: Proc. Sixth Biennial Conf. Int. Protea Assoc., Monbulk, Australia.

Seelye, J.F., S.M. Butcher, and D.J. Dennis. 1986. Micropropagation of *Telopea speciosissima*. Acta Hort. 185:281–285.

Shchori, Y., J. Ben-Jaacov, A. Ackerman, S. Gilad, and B. Metchnik. 1995. Horticultural characters of intraspecific hybrids of *Leucospermum patersonii* × *L. conocarpodendron*. J. Int. Protea Assoc. 30:10–12.

Soutter, R. 1984. Protea heights—cultivation for conservation. Veld & Flora 70(4):99–101.

Specht, R.L. 1978. Conditions for cultivation of Australian plants. Austral. Plants 9:312–313.

Tija, B.O. 1986. New Zealand, a new source of unusual cut flowers. Florist's Rev. Design Portfolio 177(6):12–14.

Tully, P. 1977. The Poorinda grevilleas. Austral. Plants 9:213–215.

Van der Merwe, P. 1985. The genetic relationship between the South African Proteacea [sic]. Protea News 3:3–5.

Vogts, M. 1982. South Africa's Proteaceae, know them and grow them. Struik Publishers, Cape Town, South Africa.

Vogts, M.M. 1980. Species and variants of protea. Farming in South Africa Ser.: Flowers and Ornamental Shrubs B.1. Dept. Agr. Water Supply.

Vogts, M.M., G.G. Rousseau, and K.L.J. Blommart. 1976. Propagation of proteas. Farming in South Africa Ser.: Flowers, Ornamental Shrubs, and Trees, B.2. Dept. Agr. Water Supply.

Vogts, M. 1958. Proteas—know them and grow them. Afrikaamse pers-Boeklandel BPK, Johannesburg.

Vogts, M. 1984. Research into the South African Proteaceae. Veld Flora 70(4):101–102.

von Broembsen, S.L. and G.J. Brits. 1986. Control of *Phytophthora* root rot of proteas in South Africa. Acta Hort. 185: 201–207.

Wallerstein, I. 1989. Sequential photoperiodic requirement for flower initiation and development of *Leucospermum patersonii* (Proteaceae). Israel J. Bot. 38:24–34.

Watad, A.A., J. Ben-Jaacov, E. Tal, and H. Solomon. 1992. In vitro propagation of *Grevillea* species. Acta Hort. 316:51–53.

Watson, D.P. and P.E. Parvin. 1973. Ornamental proteas, new cut flower crop. Hort. Sci. 8(4):290.

Worrall, R.J. 1976. Effects of time of collection, growing conditions of mother plants and growth regulators on rooting of cuttings of *Telopea speciosissima* (Proteaceae). Scientia Hort 5:153–160.

Worrall, R.J. 1994. Growing waratahs. Austral. Plants 18:17–21.

Worth, S.W. and B.W. van Wilgen. 1988. The Blushing Bride: Status of an endangered species. Veld Flora 74(4):122–123.

Wrigley, J.W. and M. Fagg. 1996. Australian native plants: propagation, cultivation and use in landscaping. Reed Books, Australia.

Wu, I.-P., J.J. Cho, and P.E. Parvin. 1978. Response of Sunburst protea to irrigation inputs and root knot nematode infections. First Ann. Orn. Sem. Proc. Coop. Ext. Serv. Coll. Trop. Agr. Human Resources, Univ. Hawaii. Misc. Pub. 172:16–20.

Spiranthes: Terrestrial Orchid with Ornamental Potential

Helena Leszczyñska-Borys, Michal W. Borys, and J.L. Galván S.

Terrestrial orchids are rarely cultivated (Bailey 1914; Oszkinis 1991). The genus *Spiranthes* L.C. Rich. is represented by 150–200 species, mostly in the subtropical and tropical regions of the American Continent (Orpet 1914; Peña 1990). Some species are hardy and are found in the United States and Poland (Bailey 1914; Orpet 1914; Szafer et al. 1953; Peña 1990). None of the species has found horticultural application in Mexico, although the wealth of terrestrial orchids, their presence in a range of contrasting environmental conditions, and their esthetic values suggest various ornamental uses.

BOTANY

The genus *Spiranthes* (Orpet 1914), is characterized as "hardy native, easily transplated, flowering biennially, late summer and early autumn." Bailey (1914) concludes that "few species have any value; some of the hardy species are advertised by dealers of the native plants and by collectors." The name *Spiranthes* derives from Greek and refers to the twisted spikes (Bailey 1914), hence the common name "Ladies' Tresses." Martínez (1979) mentions a common name *ya áxchi* (Mayan) for *S. acaulis* (Sm.) Cog. and Rzedowski and Equihua (1987) report two names *coyol cimarrón* and *cutzis* for *S. aurantiaca* (Llave & Lex.) Hemsl from the Valley of Mexico. Peña (1990) describes 17 species of *Spiranthes* from the Valley of Mexico. Among them is *Spiranthes cinnabarina* (Llave & Lex.) Hemsl. [syn. *Dichromanthus cinnabarinus* (Lex.) Garay and *Stenorrhynchos cinnabarinus* (Lex.) Lindl.]. *Spiranthes cinnabarina* (Llave & Lex.) Hemsl. occurs in the Valley of Mexico, in sod and in dry brush land, at 2250–2600 m above sea level (Peña 1990). The scape and inflorescence are attractive and suggest uses as a bedding plant, pot plant, or cut flower (Fig. 1).

Wild plants of *S. cinnabarina* were examined that were growing inside the property of the Universidad Popular Autonoma del Estado de Puebla, in Cholula, Puebla, Mexico (19°N 2100 m above sea level) where the Centro de Investigación Universitaria and the Germplasm Bank of Flowering Bulb Species are located. The climate is temperate, subhumid, with rainfall in the summer. Precipitation in the driest month is below 40 mm. The winter precipitation is less than 5% of the annual total (798 mm). The temperatures of the coldest month range from –3° to 18°C. The dry season has spells of precipitation, sometimes extending for three days. The wet season presents frequent spells of hot weather, lasting from two to three weeks. Rainfall is usually brief and intense. Evaporation rate exceeds precipitation almost three times, resulting in rapid development of soil water deficit due to sandy volcanic soils.

The first group of wild *S. cinnabarina* was found in partial shade accompanied by pasture plants, under *Eucalyptus* trees. The dry remains of grasses were burned yearly, leaving ashes and unburned remains, an old system of agriculture in Mexico, still commonly used by farmers and gardeners. The second group of wild plants, in the pasture in the open field at the same location, were growing in a layer of 5–10 cm of accumulated organic matter which contained residues from a steel factory. The third group, forming part of the collection in the Germplasm Bank, was situated in the same area, in beds, in the open field, in sandy, volcanic soil containing some added organic matter.

Spiranthes cinnabarina is an herbaceous, perennial plant that is adapted to a range of soil physical conditions. This species, when found in localities with frequent rainfall or where water accumulates, gave higher shoots, larger inflorescences and more brillant orange coloration. It is adapted to xerophytic conditions and extended spells of drought during the vegetative season, suggesting landscaping uses especially in urban areas. This species develops fleshy, fasciculated roots at a depth of 20 to 30 cm. The fleshy root system may explain the species tolerance to transplanting, characteristic of the genus (Orpet 1914).

Fig. 1. Flowering stem of *Spiranthes cinnabarina*.

CULTURE AND USE

The growing season is determined by the start of the rainy season, and the dormancy of the bud located upon a short, underground shoot. The dormancy period usually begins in the second half of October and lasts until May or June.

Trimmed bare root plants in the flower bud stage or with 30% of flowers opened can be air dried for 3 days, and then successfully transplanted. At the beginning of the growing season when plants were transplanted at the rosette stage (5–6 leaves) only one out of 108 plants was lost; when transplanted at the flowering stage bare root all 50 plants survived.

Plants grown in pots, flowered yearly for three years. The plant diameter at flowering stage ranges from 20–30 cm. The scape is covered with several bracts, being longer at their base. The inflorescence is a spike whose dimensions vary with the soil, water conditions, and solar radiation.

The plant can be grown in mass in parks, gardens, and urban areas in beds or pots. The medium height flowering scapes have orange flowers that turn to coffee-black starting at the base of the spike. Although this is a disadvantage for cut flowers, in group plantings the black flowers provide a contrast to the upper orange flowers which is attractive. At the end of the growing season the dead, above ground parts, should be trimmed.

The stems are resistant to winds. Both the "bud," and the open flower are attractive in color and form. The inflorescence, slowly opens from the base to the apex but its esthetic value is lost when more than 70–75% of the spike is opened. The remains of the flowers do not abscise but remain attached to the spike. Low scape height for group planting and, for use as a pot plant would be desired. A low scape genotype has been found and is currently under observation.

The stem can be used as a cut flower and for arrangements. The stem should be cut above the third leaf. The individual flowers opened continually. The stems held in water, without preservative, in a room of northern exposition (19°C day, 16°C night), lasted 10–14 days. Inflorescences wilted, after the stem was cut at noon, in full sun, but regained normal rigidity easily when the stem was placed in water within an hour. However, the inflorescence bent if the time from cutting to placing in a vase exceeded 1 h.

REFERENCES

Bailey, L.H. 1914. *Spiranthes*. p. 3215–3216. In: The standard cyclopedia of horticulture. McMillan, NY.

Martínez, M. 1979. Catálogo de nombres vulgares y científicos de plantas mexicanas. Fondo de Cultura Económica, México, D.F.

Orpet, E.O. 1914. Orchids. p. 2403. In: Bailey L.H., The standard cyclopedia of horticulture. McMillan, NY.

Oszkinis, K. 1991. Storczyki (Orchids). Pañstwowe Wydawnictwo Rolnicze y Lesne, Warszawa, Poland.

Peña, M. 1990. *Orchidaceae*. p. 369–379. Revisado por G.C. Calderón de Rzedowski y J. Rzedowski. En: Flora Fanerogámica del Valle de México. III. Monocotyledoneae. J. Rzedowski y G.C. de Rzedowski (eds.). Instituto de Ecología, Centro Regional de Bajío, Pátzcuaro, Mich.

Rzedowski, J. and M. Equihua. 1987. Atlas cultural de México. Flora. Secretaría de Educación Pública, Instituto Nacional de Antropología e Historia, Grupo Editorial Planeta. México, D.F.

Szafer, W.L., S. Kulczyñski, and B. Pawlowski. 1953. Rosliny Polskie (Polish Plants). Pañstwowe Wydawnictwo Naukowe, Warszawa, Poland. p. 951.

Evaluation of Safflower Germplasm for Ornamental Use

Vicki L. Bradley, Robert L. Guenthner, Richard C. Johnson, and Richard M. Hannan*

Safflower (*Carthamus tinctorius* L., Asteraceae), is an annual or winter annual herbaceous plant that is adapted for growth in hot and dry environments. Plants often have many spines on leaves and bracts, although plants with few spines exist. Young plants form rosettes that tolerate temperatures as low as –7°C, but are unable to compete well with weeds. Stem elongation in spring-planted safflower begins approximately one month after emergence. After stem elongation the plants are very susceptible to frost injury (Helm et al. 1991). Florets form a capitulum at the end of each stem, usually displaying hues of yellow, orange, red, and rarely white, which darken when dry (Li and Mündel 1996). Safflower is mostly self-pollinating in the absence of pollinators (Knowles 1969).

The Western Regional Plant Introduction Station (WRPIS) is a part of the United States National Plant Germplasm System (NPGS) and maintains the US safflower collection which includes more than 2,300 accessions collected from over 50 countries. Descriptors for the crop include branching pattern, flower color, spininess, head size, plant height, date of flowering, growth habit, iodine number, lysine content, and oil content. Many of the accessions have been evaluated for one or more descriptors and the data have been entered in the Germplasm Resources Information Network (GRIN) along with "passport" data such as taxonomy, improvement status, country of origin, and other collection information. Accessions from the collection are available at no charge to scientists worldwide.

Historically, safflower has been used for many purposes such as, a dye, food coloring, a medicinal, a vegetable side dish, hay and forage, birdseed, and most commonly for edible food oil (Li and Mündel 1996). Presently, safflower is grown primarily for oil in the United States. In Europe, however, safflower is used for both fresh-cut and dried flowers (Uher 1997). Although dried safflower may be found in craft stores in the United States, it is often imported from Holland. It is possible that a market for safflower as an ornamental plant could be developed in the US and would depend on the availability of safflower of ornamental quality. The WRPIS safflower collection most likely includes material that could be used for ornamental purposes with little or no selection.

METHODOLOGY

Safflower accessions for observation were chosen from the WRPIS collection based on information listed in GRIN. All safflower accession data were reviewed for presence or absence of narrative descriptions. Of the almost 800 accessions that had such information, 80 were selected because the narrative description indicated they were used for flowers, were relatively spineless, or displayed a characteristic of interest.

The evaluation nursery was planted on March 30, 1998 at the WRPIS Central Ferry Research Farm, Washington, located in the Snake River Canyon at 46.7°N latitude, 200 m above sea level. Prior to planting approximately 67.2 kg/ha of nitrogen and 10 kg/ha of sulfur were applied and incorporated into the soil as was a pre-emergence herbicide. Seeds were treated with a fungicide prior to being planted in 3 m long single rows, on 1.5 m centers. Rosette-stage plants were thinned to 10 cm apart within a row, and irrigated at a rate of approximately 20 mm per week.

Field evaluation data were recorded and samples of 13 of the most promising were collected for fresh-flower arrangements and for drying (Table 1). Safflower stems were cut and placed in water and kept cool while transported to the WRPIS main facilities at Pullman, Washington. They were submerged up to the heads in water overnight. The following day, the stems were trimmed, lower foliage was removed, and stems were placed in a preservative solution commonly used in the floral industry. These stems were then

---

*The authors thank Cougars Flowers of Pullman, Washington, and Ms. Andrea Lawrence for assistance with fresh floral arrangements.

Table 1. Data collected on 13 safflower accessions selected for potential ornamental use.

Accession number	Country	Improvement status	Days to 50% bloom	Flower color[z] wet	dry	Head size (cm)	Plant height (cm)	Variation[y]	Spines[y]
PI 304439	Iran	wild	107	O	R	3.0	100	1	2
PI 304441	Iran	wild	108	O	R	3.0	100	2	1
PI 304442	Iran	wild	107	O	R	3.5	100	1	2
PI 314650	Kazakhstan	wild	100	DO	R	3.0	77	1	1
PI 406012	Iran	cultivated	99	Y	O	3.0	84	2	2
PI 406014	Iran	cultivated	107	Y	O	3.0	84	3	2 or 4
PI 406016	Iran	cultivated	107	Y	O	3.0	84	1	2
PI 406018	Iran	cultivated	100	Y	O	2.5	80	1	2
PI 537618	US	breeding material	98	O	R	3.0	100	1	1
PI 537619	US	breeding material	109	O	R	3.0	90	2	2
PI 544058	China	cultivated	107	O	R	2.5	86	1	1
PI 544059	China	cultivated	101	O	O or G	2.5	80	3	1 or 5
PI 560197	US	breeding material	102	Y or O	O or R	3.5	80	2	2

[z]O = orange, DO = dark orange, Y = yellow, G = gold, R = red.
[y]Rated from 1 (low) to 5 (high).

placed in several fresh arrangements, and photographed. Other stems were hung upside down until dry and then sprayed with a sealant/protectant to prevent breakage and preserve color.

RESULTS

The most important plant characteristics indicating ornamental value in safflower are a combination of appealing flower color and few spines and these descriptors were used as the major evaluation criteria. Thirteen of the 80 accessions observed showed promise for ornamental use as fresh-cut, and/or dried flowers (Table 1).

The majority of selected accessions in Table 1 had few spines except PI 544059 and PI 406014, both of which had equal numbers of plants with many and with few spines. Plants with desired characteristics could easily be selected from these two accessions.

Many flower colors and shades were observed in the eighty accessions evaluated, but accessions with few spines had either orange, red, or yellow fresh flowers. One of the 13 selected accessions, PI 560197, had plants with either yellow or orange fresh flowers, while another, PI 544059, had yellow flowers that dried to either orange or gold. None of the selected accessions had white or pink flowers.

Overall variation within each accession row was evaluated and accessions with high variation scores (4 or 5) were not considered to have ornamental value. Although several of these accessions had individual plants of interest, there are other accessions in the WRPIS safflower collection that could be grown for floral use with little selection.

CONCLUSIONS

A small percentage of the safflower collection at the WRPIS was screened for ornamental value and all 13 of the selected accessions had flowers in shades of orange, red, or yellow. Screening of the collection will continue with the major objective of finding accessions or individual plants with white, pink, or other unique flower colors and few spines. Images of each accession evaluated will be downloaded into GRIN. Other plans include working with horticulturists to evaluate potential use of safflower in home gardens and landscapes, growing safflower in the greenhouse for cut flowers, and consulting with floral industry personnel regarding the market for fresh-cut safflower in the US.

REFERENCES

Helm, J.L., A.A. Schneiter, N. Riveland, and J. Bergman. 1991. Safflower production. North Dakota State Univ., Fargo, A-870(revised), 14 AGR-6.

Knowles, P.F. 1969. Centers of plant diversity and conservation of crop germplasm: Safflower. Econ. Bot. 23:324–329.

Li, D. and H.-H. Mündel. 1996. Safflower *Carthamus tinctorius* L. Promoting the conservation and use of underutilized and neglected crops. 7. Institute of Plant Genetics and Crop Plant Research, Gatersleben/International Plant Genetic Resources Institute, Rome, Italy.

Uher, J. 1997. Safflower in European horticulture. In: A. Corleto and H.-H. Mündel (eds.), Safflower: A multipurpose species with unexploited potential and world adaptability. Proc. Fourth Int. Safflower Conference, Bari, Italy, 2–7 June, 1997. Adriatica Editrice, Bari, Italy.

New Arid Land Ornamentals: Recent Introductions for Desert Landscapes

Janet H. Rademacher

Over the past decade, water conservation has become an increasingly important issue across the southwestern United States. This concern has led local horticulturists and landscape architects to explore the use of water-thrifty ornamentals from dry climates throughout the world. The Chihuahuan and Sonoran deserts in particular have yielded a vast array of successful landscape plants. Universities, growers, and plant enthusiasts have all participated in the collection, propagation, evaluation, and promotion of new plant introductions. A group of recent proven introductions, including trees, shrubs, ground covers, and perennials are included below with information on their origins, growth habits, cultural requirements, and potential uses in the landscape.

Acacia redolens Maslin, Desert Carpet™

Native to inland areas of Western Australia, *Acacia redolens* Maslin has been used extensively in southern California and Arizona to cover large areas inexpensively. Seedlings of *Acacia redolens* vary widely in their growth habits, often reaching heights in excess of 1.8 m (6 feet). The Desert Carpet™ clone was selected from the first Phoenix freeway plantings for its prostrate growth habit, and was released by Mountain States Wholesale Nursery in 1984. Since that time, this groundcover has performed consistently on many projects, and years after installation has maintained a height of only 0.6 m (24 inches). One plant can spread to a width of 3.6 m (12 feet), although we have observed that the cutting-grown Desert Carpet™ plants are slower to establish and reach their mature size than seedlings. The slower growth rate and prostrate nature of this clone should reduce maintenance costs, since pruning is not necessary to control vertical growth. Instead of true leaves, *Acacia redolens* has thick, leathery, gray-green phyllodes. This plant blooms in the spring with small yellow flowers. Freeway acacia will tolerate low temperatures of –11.1° to –9.4°C (12°–15°F), alkaline and slightly saline soils, and does not seem to be choosy about soil types. In coastal areas it requires little or no supplemental irrigation, but does require regular irrigation in hot desert regions. Desert Carpet™ seems to be disease and pest free.

Baccharis hybrid 'Starn' (P.P.A.F.) Thompson™

When Dr. Tommy Thompson and Dr. Chi Won Lee of the University of Arizona released *Baccharis* hybrid 'Centennial', it filled a great void in our plant palette. Their research has been carried on, and now the improved Thompson™ clone is available. Since *Baccharis* 'Centennial' is a female plant, it has two undesirable characteristics. First, it produces pappus, or white "fluff," which litters the landscape and reduces the esthetic appearance of the plants for a short period of time. Also, since 'Centennial' is a female plant, it can be pollinated by nearby male *Baccharis sarothroides* Gray (Desert broom), and seedlings often result. This is why you sometimes see stands of 'Centennial' with taller *Baccharis* plants growing up through them. The Thompson™ clone is a male plant, eliminating these two negative characteristics. Also, Thompson™ was selected from the next generation after 'Centennial', and has 25% more *Baccharis sarothroides* for heat and disease resistance. The growth habits and uses of these two clones are essentially the same: both grow to about 0.9 m (3 feet) tall by 1.2–1.5 m (4-5) feet wide, are evergreen with bright green foliage and inconspicuous flowers, and provide a low-maintenance, long-lived alternative for difficult locations.

Cercidium species 'Desert Museum'

This hybrid palo verde is a three-way cross between *Parkinsonia aculeata* L., *Cercidium microphyllum* (Torr.) Rose & I.M. Johnst., and *Cercidium floridum* Benth. ex Gray, and seems to combine the best qualities of all three plants. 'Desert Museum' grows very rapidly to 6.1 m (20 feet) tall and wide in 3 to 5 years, after which it needs little or no irrigation. It is completely thornless, and produces very little litter, with few seed pods. It has a sturdy, upright growth habit which requires very little pruning or staking. It blooms over a long period of time, with the heaviest bloom from about mid-March to May 1. It also tends to bloom again in June and August. The yellow flowers are larger than any of its three "parents." It does not reseed like the messy *Parkinsonia aculeata*!

Chilopsis linearis (Cav.) Sweet, Lucretia Hamilton™

Desert willow trees occur along washes throughout the southwestern US and northern Mexico. This small deciduous tree has narrow, light green leaves that give it a weeping appearance. In the summer, the tree is covered with fragrant, trumpet-shaped flowers. In the wild, the flower colors range from white to purple, although a pale pink to lavender flower color is the most common. The Lucretia Hamilton™ clone was selected for its intense, deep pink to purple flower color, as well as its small stature. While many desert willow trees can grow to 7.6 m (25 feet) tall and wide, this clone seems to stay below 5.4–6.1 m (18-20 feet) tall and wide. After flowering, long narrow seed pods are produced. Plant *Chilopsis linearis* in full sun and well-drained soil, and in regions where temperatures do not drop below –17.8°C (0°F).

Chrysactinia mexicana Gray (Damianita)

This small, compact shrub grows to 0.6 m (2 feet) tall and wide, and bears a very strong resemblance to turpentine bush, with needle-like green leaves and yellow daisy-like flowers. However, damianita blooms from March to September, while turpentine bush blooms from September to November. Combining the two plants would be a great way to prolong the color display! Damianita has wonderful-smelling foliage, and would be a great selection for sensory gardens. Damianita is a very tough, durable plant, tolerating extreme heat and cold, down to –17.8°C (0°F). Plant in full sun, and almost any soil. If this plant starts to look woody, prune it back severely in the early spring. Damianita ranges from New Mexico to west Texas and northeastern Mexico, at elevations of 609–2134 m (2000–7000 feet).

Dalea capitata Sierra Gold™

This well-behaved ground cover grows to about 20 cm (8 inches) tall by 0.9 m (3 feet) wide. Because of its compact size, Sierra Gold™ is a good selection for tight planting areas, such as small planters or medians. Its fine-textured, light green foliage has a fresh, lemony scent. Rabbits seem to avoid it! Yellow flowers carpet Sierra Gold™ in the spring and the fall. This plant is hardy to at least –15°C (5°F), but it will be deciduous at –3.9°C (25°F). The one drawback to this plant is that the whiteflies seem to like it, so some insecticide applications will be necessary in heavily infested areas around Phoenix. Plant in full sun for best results. No soil amendments should be necessary. In hot desert regions this plant requires some supplemental irrigation from spring to fall. Although most dales native to Arizona and Mexico tend to rot out if overwatered, we have observed this plant thriving right next to turfgrass, where it receives heavy irrigation. More testing is needed to determine if it will tolerate coastal areas, or regions with high rainfall.

Dasylirion longissimum

This user-friendly accent plant is a great selection for high-traffic areas such as walkways and near entries. This grasslike plant does well in containers, and its symmetrical form provides a striking focal point. Its thin, stiff green leaves are completely unarmed, and have smooth edges. Eventually, its single trunk can grow to 1.8 m (6 feet), topped by a 1.5 m- (5-foot-) wide rounded head of leaves. The older, bottom leaves can be trimmed off to expose the trunk. *Dasylirion longissimum* is native to Mexico, and is hardy to about –8.3°C (17°F).

Euphorbia biglandulosa Desf. (Gopher Plant)

This evergreen perennial or subshrub has a very unusual form and appearance. Its arching stems angle out and up, and can reach a length of 0.6 m (2 feet). The plant grows to 0.9 m (3 feet) tall by 1.2 m (4 feet) across; with narrow, fleshy grey-green leaves. Broad clusters of chartreuse flowers occur at the tips of the arching stems, usually in the late winter and early spring. Flowers are followed by small brown seed pods that explode upon ripening. The stems usually die back after fruiting, leaving a small clump of grey-green foliage near the ground. Plant *Euphorbia biglandulosa* in full sun or light shade, in a well-draining soil. It is cold hardy to –15°C (5°F).

Hesperaloe parviflora (Torr.) J. Coult., 'Yellow' (Yellow yucca)

A clumping perennial with long, gray-green leaves, *Hesperaloe parviflora* grows slowly to form a grasslike clump 1.0–1.2 m (3–4 feet) tall and wide. From spring through fall, it produces 1.5 m- (5-foot-) tall flower spikes. Red-flowering plants have been a staple in southwestern landscapes for many years. This is simply a yellow-flowering selection. Use this tough accent plant in full sun. Since it also tolerates reflected heat, yellow yucca is a reliable plant to use along sidewalks, in parking lots, etc. Tolerant of temperature extremes, yellow yucca is cold-hardy to at least –17.8°C (0°F). Once established, it requires little or no irrigation. All in all, yellow yucca is one of the toughest and most maintenance-free plants.

Hymenoxys acaulis (Pursh) K. Parker (Angelita Daisy)

This perennial is native to the southwestern US, where it occurs most often at elevations from 1219–2134 m (4000–7000 feet), on dry rocky slopes and mesas. Angelita daisy bears a strong resemblance to *Baileya multiradiata* Harv. & A. Gray ex Torr. (desert marigold). However, the foliage is green rather than gray, and the flower is a deeper gold color. Forming rounded clumps to fifteen inches tall and wide, *Hymenoxys acaulis* is a wonderful plant to use as a border in front of larger shrubs. I f water is available, it will naturalize in the landscape. In Phoenix, this plant blooms all year, with especially heavy bloom in the spring and fall months. This prolonged bloom period results in many dried flower stalks, which can make the plants look scruffy. We recommend cutting off the old flower spikes occasionally to rejuvenate the plant and initiate new flower production. Angelita daisy seems to prefer well-drained soils and full sun. It is very cold hardy, heat tolerant, and drought tolerant.

Leucophyllum candidum I.M. Johnst. Thunder Cloud™

As with all of the other *Leucophyllum* species, this clone blooms when the humidity is high. The silver, pubescent foliage is a perfect foil for the masses of indigo flowers that appear in the summer and fall months. Thunder Cloud™ was selected and trademarked by Benny Simpson of Texas A&M University. His clone is highly valued because of its small, dense growth habit. Unlike most of the larger *Leucophyllum* species, Thunder Cloud™ remains reliably small, to three feet tall and wide. This plant is cold hardy to at least –12.2°C (10°F). Plant all of the *Leucophyllum* species in full sun and well-drained soil. Avoid overwatering.

Leucophyllum langmaniae **Rio Bravo™**

Trademarked by Mountain States Nursery, this clone has a nice, compact growth habit very similar to *L. frutescens* 'Compacta'. Rio Bravo™ has become very popular because of its bright green foliage and rounded, dense form. It has lavender flowers and will eventually grow to 1.5 m (5 feet) tall and wide. Like the *L. candidum* species, it requires well-drained soils and full sun. It is hardy to –12.2°C (10°F).

Muhlenbergia capillaris **(Lam.) Trin. Regal Mist™**

We feel that this ornamental grass shows great promise for many different regions of the country. Native to humid southeastern Texas, this grass has adapted extremely well to the hot, dry conditions of deserts in Arizona and Nevada. In fact, it has performed incredibly well in Las Vegas, which is cursed with poor soils, high winds, high summer temperatures, and cold winters. Regal Mist™ is also happy in heavy soils, with ample irrigation. In short, it has worked everywhere it has been tried, so far! It is hardy to at least –17.8°C (0°F). Regal

Mist™ has narrow, dark green, glossy leaves. It grows quickly to form a rounded clump to 0.9 m (3 feet) tall and wide. The flower spikes on this grass have attracted a lot of attention... they form misty masses of pink to purple flowers in October and November. We recommend cutting this plant back in early spring to cut off the dead flower spikes and any dormant foliage.

Penstemon **species**

There are so many wonderful *Penstemon* species to try in the garden, that is difficult to select just a few. Most of the penstemons are perennials with a basal rosette of foliage, which send up spikes of tubular flowers in the spring and early summer. They add incredible color to the landscape, and attract hummingbirds as well! Penstemons come in a wide range of colors, including blue, purple, pink, and red. After they finish blooming, allow the flower spikes to dry on the plant. Then cut off the spikes and sprinkle the seed in the garden to increase next year's mass of color. There are two new species to try: *Penstemon triflorus* Heller, which has short, 46 cm (18 inch) spikes of dark pink-purple flowers which occur along the stem in clusters of three; and *Penstemon clevelandii* Gray, native to southern and Baja California, with spikes of clear, bright pink flowers to 0.6–0.8 m (2–2.5 feet) tall.

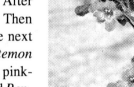

MEDICINAL, AROMATIC, SPICE, & BIOACTIVE CROPS

Herbs Affecting the Central Nervous System

Varro E. Tyler

LAWS AND REGULATIONS GOVERNING HERBAL MEDICINES IN THE US

As a result of the existing laws and regulations, most herbal remedies (botanicals, phytomedicines) are currently marketed in the United States as dietary supplements, a food category, not as the drugs they are. This situation stems directly from the passage, in 1962, of the Kefauver–Harris Amendments to the Federal Food, Drug, and Cosmetic Act of 1938. These amendments required drugs to be proven effective prior to marketing, in addition to the purity and safety requirements already in effect.

Although the law specifically exempted from this requirement all drugs marketed prior to 1938, which included almost all of the herbal remedies, the Food and Drug Administration (FDA), by a clever application of administrative law, i.e., regulations, circumvented the will of Congress. The agency declared that although sale could continue, any unproven claim of efficacy would cause the herb to be considered as misbranded which could result in confiscation.

To determine efficacy of over-the-counter products, in 1972 the FDA established 17 panels to evaluate, on the basis of submitted information only, the data supporting such claims. The panels were not permitted to utilize anecdotal information, product popularity, nor did they conduct detailed literature surveys. Because proof of efficacy studies are extremely expensive—current estimates for new chemical entities range up to one-half billion dollars—a limited amount of data was submitted to the panels for the old botanical remedies for which patent protection and market exclusivity with consequent profitability were doubtful.

In consequence, very few herbal remedies were approved as drugs. The few that were found acceptable included capsicum (*Capsicum* spp.) as a topical analgesic, slippery elm (*Ulmus rubra* Muhl.) as a demulcent, psyllium seed and husk (*Plantago* spp.) as a laxative, and distilled extract of witch hazel (*Hamamelis virginiana* L.) as an astringent (due to the 14% alcohol added to the final preparation). On the basis of inadequate data submission, the FDA declared peppermint (*Mentha × piperita* L.) to be an unsafe and ineffective digestive aid and placed other useful remedies such as the antitussive eucalyptus (*Eucalyptus globulus* Labill.) and the laxative prune (*Prunus domestica* L.) concentrate in the same category.

Sales of such herbal products declined for a time but began to increase once again in the 1980s as consumers became more health conscious and displayed a significant interest in natural products. Finally in 1993, Commissioner David Kessler of the FDA threatened additional controls on such products. The public rebelled, and submitting to the will of the people, Congress passed the Dietary Supplement Health and Education Act of 1994. That Act permits any herb sold in the United States prior to October 1994 to continue to be marketed as a dietary supplement (food) without FDA approval.

Labeling of such products is, however, restricted. Therapeutic claims are not permitted. This prompts the potential user to turn to the herbal literature, much of which is hyperbolic and advocacy in nature—especially on the internet—designed to promote sale of the product. Claims regarding the effect of the remedy on the structure or function of the human body are permitted on the label, but such a statement must be followed by another indicating the claim has not been approved by the FDA. A third statement must then follow indicating the product is not intended to diagnose, treat, cure, or prevent any disease. That, of course, is why the potential user wanted to purchase the product in the first place, so the entire situation becomes very confusing, even to the sophisticated consumer.

Another serious problem with the herbal product status quo is the total lack of quality standards for such preparations. Products and the companies selling them range from very bad to very good. Consumers have few resources to determine which is which, and published analyses of particular brands have been quite limited, probably due to the fear of litigation.

Other advanced nations have solved the herbal problem in one of two ways. The first is simply to allow botanicals to be sold as "traditional" drugs whose efficacy is based on long usage and anecdotal information, not on scientific or clinical studies. This is not a satisfactory solution because many of such claims of utility remain unverified.

The best solution, without question, is to modify existing regulations to allow herbal remedies to be sold as drugs, with appropriate quality standards, following absolute proof of safety and "reasonable" proof of efficacy. The reasonable qualification would allow the marketing of a product with a research investment of perhaps a few million dollars as opposed to the hundreds of millions now required. Such an expenditure is just a small fraction of the marketing budget of many herbal product manufacturers.

This type of drug approval is the one that exists today in Germany. It has functioned well there for decades, and no serious problems have been encountered. Safety and efficacy are determined by an independent panel of experts, designated Commission E, appointed by the German equivalent of our FDA. The Presidential Commission on Dietary Supplement Labels recommended the appointment of a similar panel in the United States to evaluate herbal safety and efficacy. So far, the FDA has not acted upon that recommendation nor has the agency given any indication that it will do so. Likewise, the FDA has not acted upon a long-standing petition that would allow the safety and utility of herbs to be evaluated as OTC remedies on the basis of the voluminous scientific and clinical studies already carried out in European countries, especially Germany.

Adherence to standards for drug approval designed for new synthetic chemical entities is obviously not realistic for long-used herbal remedies. In view of the fact that one-third of the adult population in this country now employs herbal remedies and the retail market approximates $4 billion annually, serious consideration needs to be given to modifying outdated regulations. A change is urgently required to assure that consumers will be supplied with quality products providing appropriate information regarding their therapeutic utility directly on the label.

Among the numerous botanicals marketed in the United States today are several that produce significant direct or indirect effects on the central nervous system. Some of the more prominent ones have been extensively studied with respect to their chemical composition and physiological and therapeutic activities. Others continue to be employed largely on the basis of their folkloric reputations as useful remedies.

CONSIDERATION OF SPECIFIC HERBS

Ephedra or Ma Huang (*Ephedra* spp.)

Ephedra, commonly referred to by its Chinese name *ma huang*, has received much publicity recently. The herb consists of the dried green stems of several species of *Ephedra* native to Central Asia. These include *E. distachya* L., *E. equisetina* Bunge., and *E. sinica* Stapf. Ma huang has been used in China for the treatment of bronchial asthma and related conditions for more than 5000 years.

The therapeutic use of ephedra is due to its content of several closely related alkaloids of which ephedrine is both the most active and the one present in largest amount. A typical ephedra plant contains about 1.5% total alkaloids of which some 80% is ephedrine. American species of *Ephedra*, one of which is *E. nevadensis* S. Wats., often referred to as Mormon tea, contain no active alkaloids. Ephedrine was carefully researched here in the United States during the 1920s and was a standard over-the-counter (OTC) medication for many years. Like all other effective medications, it may also produce undesirable side effects.

The alkaloid's vasoconstricting effect makes it a useful nasal decongestant, but it also raises blood pressure and increases heart rate. It is an effective bronchodilator, but it also stimulates the central nervous system (CNS) with side effects ranging from nervousness to insomnia. This stimulation is greatly increased by consumption of caffeine or caffeine beverages such as coffee, tea, or cola. Consequently, ephedrine has been replaced to a large extent in over-the-counter cold and cough products by related chemicals such as pseudoephedrine or phenylpropanolamine. These have a similar action but much reduced CNS effects.

In recent years, ephedra and caffeine combination products have been promoted as appetite depressants and metabolic stimulants for weight loss, as athletic performance enhancers and, in very large doses, legal euphoriants or intoxicants. There is a considerable literature about the effects of ephedra on weight loss with some modestly favorable results. As one of my pharmacologist friends quipped, "You're bound to lose weight if you take ephedra and caffeine because you'll be so hyper the soup will slop out of the spoon before it reaches your mouth." That is, of course, a slight exaggeration.

Detailed studies of the herb's effect on athletic performance or its euphoriant activities do not exist. But that is of little moment because ephedra should not be taken chronically for any purpose unless the consumer is under the direct care of a competent physician. Many of the herbal products do not list the concentration of ephedrine present. Some manufacturers almost certainly "spike" their dosage forms with additional quantities of synthetic ephedrine.

As a result, the Food and Drug Administration (FDA) convened a special advisory group on ephedra in October 1995. That committee of experts made a number of recommendations regarding the sale of ephedra products, including strict dosage limitations, appropriate warning labels preventing chronic use, prohibition of sale to persons under age 18, and warnings to individuals with specific health risks.

These recommendations were never implemented, although in April 1996 the FDA did issue a warning cautioning consumers not to buy any high-dose ephedra products marketed as "legal highs." By August 1996, the FDA had compiled a list of some 800 adverse reactions, including 22 deaths and a number of serious cardiovascular and nervous system effects, including heart attack, stroke, and seizures, which they attributed to the consumption of ephedra, often in combination with caffeine-containing herbs. (The numbers in both categories have increased significantly since that time.) Consequently, another special advisory group meeting was held, but after lengthy discussion, no consensus was reached, and no concrete recommendations were made.

Food and Drug Commissioner David A. Kessler, who attended the meeting, said the agency would consider the matter and act quickly on ephedra's marketability, almost certainly before the end of 1996. He subsequently resigned, and no action was taken prior to the self-imposed deadline.

Finally, in June 1997, the FDA requested comment on a series of proposed restrictions on the sale of ephedra as a dietary supplement. These included a limitation of 8 mg per dose of ephedra alkaloids, labels warning consumers not to take more than 24 mg of ephedra alkaloids per day, labels advising at-risk persons not to consume ephedra, and limitation of consumption to a maximum of 7 days. As of June 1999, these proposed regulations have never been implemented.

Several states have acted, or are considering action, to control ephedra products. Most authorities continue to believe that small doses of ephedra, equivalent to not more than 40 mg of ephedrine per day, consumed on an occasional, not a chronic, basis for the relief of bronchial asthma, are safe for otherwise normal persons.

Another concern often expressed about ephedra is the possibility that the contained ephedrine may be illegally converted to methamphetamine or "speed," a common drug of abuse, by basement chemists. Although this is possible, it is not likely because of the difficulty in separating the product from the accompanying plant material. A much more likely starting product is pseudoephedrine which is readily available in pure form. Indeed, large-scale sales of that alkaloid are now monitored by the Drug Enforcement Administration (DEA).

Ephedra remains a prime example of an age-old dilemma in medicine. Must a useful therapeutic agent be banned because of its abuse potential? In many cases, this has proven to be necessary. Whether ephedra herb falls in this category remains to be seen.

Ginkgo (*Ginkgo biloba* L.)

An important breakthrough in herbal medicine took place in October 1997 when the *Journal of the American Medical Association* published the results of the first clinical trial conducted in the United States of the effect of *Ginkgo biloba* extract (GBE) on patients suffering from mild to severe dementia caused by Alzheimer's disease or multiple clots in the blood vessels. The generally positive results showing modest improvement in the cognitive performance and social functioning of the demented patients basically confirmed results previously obtained in European studies. Some 36 clinical trials with GBE had been conducted there between 1975 and 1996.

The ginkgo tree is well-known, particularly to residents of our large cities where its resistance to pollution and its overall toughness has caused it to be much planted as an ornamental along the very busy streets. This resistance to stress has enabled the species to survive for at least 70 million years; it is often referred to as a living fossil.

The seeds of the tree, after removal of the smelly, outer fleshy layer, have been used in China, both as a medicine and a food, for thousands of years, but it is the dried green leaves, not the seeds, that yield the herbal medicine that has become so popular today. Research beginning in the 1930s identified a number of constituents in the leaves, and in 1965 a product containing 24% flavone glycosides and 6% terpenes (ginkgolides A,B,C, and bilobalide) was first marketed for the treatment of conditions resulting from cerebral and peripheral circulatory disturbances. The product subsequently became very popular as an approved drug in Germany where sales in 1996 exceeded $163 million.

GBE is used primarily to treat cognitive deficiency, a condition caused by inadequate blood flow and nerve degeneration in the brain and expressed by symptoms such as vertigo, tinnitus, headache, short-term memory loss, reduced vigilance and concentration, and confusion. It is also useful in peripheral vascular disorders, having been shown to increase pain-free walking distance in patients with intermittent claudication. Evidence is also accumulating to support the value of GBE in alleviating sexual problems in males, particularly in those whose erectile dysfunction has been induced by the consumption of various antidepressant drugs.

Just how does GBE function to produce all of these beneficial effects? As is often the case with herbs, no single compound and no one mechanism is responsible. The ginkgo flavonoids reduce harmful brain effects by preventing the activity of enzymes that produce damaging free radicals. They also act as antioxidants scavenging any free oxygen radicals that are formed. In addition, they affect calcium transport which has a powerful influence on brain metabolism.

The sesquiterpene bilobalide reduces increased water and electrolyte levels in damaged brain tissue. The diterpene ginkgolides are potent platelet activating factor (PAF) inhibitors. PAF can contribute to brain damage not only by inducing thrombosis but also by increasing the permeability of blood vessels, allowing liquid to seep through them into brain tissue. This is likely to result in nerve damage.

In contrast to many synthetic drugs (halperidol, fluoxetine, and tacrine) for the treatment of dementia, GBE exerts its beneficial action with a much lower incidence of side effects. Only 1.67% of 10,632 patients treated with ginkgo experienced mild stomach upsets, headache, or allergic reactions in comparison to 5.42% of 2,325 patients treated with synthetic drugs who suffered these or more serious reactions.

The customary daily dose of GBE is 120–240 mg taken in 2 or 3 separate doses. A minimal 8-week course of treatment is recommended for cognitive deficiency. Contraindications to ginkgo therapy are not known, but because it does inhibit PAF, the herb should be administered cautiously in patients undergoing anticoagulant therapy utilizing either other herbs, e.g., garlic (*Allium sativum* L.), ginger (*Zingiber officinale* Roscoe), feverfew [*Tanacetum parthenium* (L.) Schultz Bip.], or synthetic drugs (warfarin).

Consumption of unprocessed ginkgo leaves in any form, including teas, should be avoided because they contain several potent allergens known as ginkgolic acids. These compounds, removed during processing of GBE, are chemically related to urushiol, the active principle in poison ivy [*Toxicodendron radicans* (L.) Ktze.]. Current regulations in Germany limit the concentration of ginkgolic acids in ginkgo preparations to a maximum of 5 parts per million. No standard has been established in the United States, but quality products do not exceed that level.

The GBE preparations long marketed worldwide are concentrated about 50-fold and contain 24% flavone glycosides and 6% terpenes, often designated 24/6. Recently, one company has made available a 27/7 product prepared by a method that increases the ginkgolide B concentration. Studies show that it produces a higher concentration of ginkgolides in the blood for a longer period of time, thus enhancing the effects of the herb.

In contrast to such improved products, it is believed that there are many substandard GBE preparations on the market today. Because of the extensive processing required, ginkgo is relatively expensive to produce, and the standard 50–60 mg tablets usually retail in the $10–$15 range for 60 capsules, depending on the brand. So-called bargain herbs, sometimes selling for a dollar or two for the same quantity, are very likely not bargains at all.

Sometimes one sees ginkgo advertisements touting it as a "smart pill" that improves the cognitive function of persons in the absence of any pathological condition. There is practically no evidence to show that the herb produces significant beneficial effects in the normal human brain.

St. John's Wort (*Hypericum perforatum* L.)

Depression, at least in its milder forms, is a condition that seems to afflict many Americans, occasionally if not chronically. In this country, the disease is the fourth most likely reason for one to consult a family physician and costs our economy more than chronic respiratory illness, diabetes, arthritis, or hypertension. The treatment and rehabilitation expenses in the United States exceed $12 billion annually; but the true cost, including loss of earnings, absenteeism, and loss of productivity, totals nearly $44 billion annually.

Depression is characterized by a number of subjective complaints ranging from despondency and loss of interest to irritability and sleep and digestive disorders. One authority has noted that the depressed patient suffers from melancholia about the present, guilt about the past, and anxiety about the future. Mild to moderate depressive symptoms occur in 13–20% of the population; major depression, often characterized by suicidal tendencies, plagues 1.5–5%.

More than a dozen prescription drugs are routinely used to treat America's depression. All of them are synthetic, and they all produce more or less unpleasant side effects ranging from skin rashes to overtly violent behavior. Meanwhile, in Germany the most popular prescription drug of any type, natural or synthetic, for the treatment of depression is a concentrated extract of the flowers and leaves of St. John's wort, often simply called hypericum. More than 200,000 prescriptions per month are filled for a single brand (Jarsin) there compared to about 30,000 per month for fluoxetine (Prozac). This figure does not include sales of other hypericum products, whether prescribed or self-selected. Actually, 80–90% of the sales in Germany are prescriptions, which allows their cost to be reimbursed by the health insurance system.

During the past two years, St. John's wort has become extremely popular in the United States. In 1996, it was not even listed among the best-selling herbs. Now it is one of the five most popular botanicals. Favorable publicity on television, and subsequently in the popular press, had much to do with this phenomenal increase in use.

Many clinical trials show St. John's wort to be especially useful in treating mild to moderate depressive states. Studies in 3,250 patients found improvement or total freedom from symptoms in about 80% of the cases treated, with only 15% not responding. These results are typical of all drug therapies for depression. However, the side effects were far fewer than those observed with conventional antidepressants. The incidence was less than 2.5% and consisted mainly of gastrointestinal disturbances and allergic reactions.

The herb's multiple constituents apparently function in several different ways. Initially, St. John's wort was thought to act as a monoamine oxidase (MAO) inhibitor. This effect has now been shown to be insignificant. Some evidence supports its effect as a selective serotonin reuptake inhibitor (SSRI).

It may also inhibit COMT (catechol-*O*-methyltransferase), an enzyme capable of destroying biological amines. Still another mechanism seems to suppress interleukin-6 release, affecting mood through neurohormonal pathways. The advantage of this combined action is fewer side effects for the patient because the total response is not due to a single type of activity.

Some authorities have warned against exposure to sunlight while consuming hypericum because it may induce photosensitivity with its dermatitis and associated inflammation. However, while light-skinned animals grazing on great quantities of the herb have had such reactions, photosensitivity is very uncommon in people taking it in normal amounts. It has occurred in patients injected intravenously with very large amounts of hypericin—50 to 70 times the normal oral dose.

Although St. John's wort is marketed as a drug in Germany and has been approved there by the German equivalent of our Food and Drug Administration for the treatment of depression, anxiety, and nervous unrest, it is sold in the United States only as a dietary supplement. The most effective preparations are capsules containing an extract of the herb standardized on the basis of 0.3% hypericin. Dosage is 300–900 mg daily. Improvement of mild to moderate depression should result after 2 to 6 weeks of treatment.

Kava (*Piper methysticum* G. Forst.)

The first Europeans to observe the kava plant and its ritualistic consumption by natives of Oceania were Dutch explorers Jacob Le Maire and William Schouten. In 1616, they encountered the plant in the Hoorn Islands, now a part of the French territory Wallis and Fatuna. Later travelers in the Pacific region provided a wealth of detail regarding this highly valued and widely used pepper plant.

Long cultivated and known by a number of common names, including *kava-kava*, *ava*, *yagona*, and *yangona*, the plant is now classified by botanists as *Piper methysticum*, meaning "intoxicating pepper." In religious and social rituals that naturally vary somewhat from island to island, the rhizome of the plant is grated (originally chewed by young people with sound teeth), mixed with water in a bowl, strained, and the resulting beverage drunk to produce a feeling of well-being.

Observers and even scientists long disagreed on the effects of kava. Captain James Cook, who observed its use during his world voyage of 1768–1771, thought the symptoms resembled those of opium. Lewis Lewin, a pioneer pharmacologist in the field of mind-altering drugs, referred to it in the 1880s as a narcotic and sedative, but noted these effects followed a period of quiet euphoria. Modern authorities call kava a psychoactive agent; it reduces anxiety much like the potent, synthetic benzodiazapines (e.g., Valium) and is a potent muscle relaxant. Kava does promote relaxation and sociability, but its effects are very different from those produced by either alcohol or synthetic tranquilizers. It does not produce a hangover, and, even more significant, it does not cause dependency or addiction.

Naturally, people were interested in finding out how kava produced these interesting effects. It was once thought that chewing the root converted its starch into sugar which then fermented to produce alcohol. Although this sounds far-fetched, *chicha*, a corn-based beer brewed by natives in Peru and Bolivia, is prepared in just this fashion. Lewin said a resinous material was the active component. Finally in the 1950s and 60s, two teams of German scientists headed by H.J. Meyer in Freiburg and R. Hänsel in Berlin found that the various activities of the kava plant were due to some 15 different chemical compounds known as pyrones. Collectively named kavapyrones or kavalactones, the compounds were found to increase the sedative action of barbiturates, to have both analgesic and local anesthetic effects, to cause muscles to relax, and to have antifungal properties.

Shortly after these findings, preparations of kava extract began to appear on the European market, usually standardized to provide a daily dosage in the range of 60–120 mg of kavapyrones. German Commission E, the group responsible for evaluating the safety and efficacy of botanical medicines, reviewed the data on kava and, in 1990, approved its use for conditions such as nervous anxiety, stress, and restlessness. It is frequently marketed as an anxiolytic. Use of kava is contraindicated during pregnancy, nursing, and in cases of depression caused by internal factors.

With an herb as potent as this one, there is naturally concern about side effects. Observations on 4,049 patients consuming 105 mg of kavapyrones daily for 7 weeks noted 61 cases (1.5%) of undesired effects. These were mostly mild and reversible gastrointestinal disturbances or allergic skin reactions. A 4-week study of 3,029 patients taking 240 mg of kavapyrones daily produced a slightly greater incidence (2.3%) of similar side effects. This would be expected because of the larger dosages used.

Long-term consumption of very large quantities of kava may result in a yellow coloration of the skin, nails, and hair, allergic skin reactions, visual and oculomotor equilibrium disturbances. For this reason, Commission E recommends that kava not be consumed for longer than 3 months without medical advice. Driving and operating machinery during consumption should be avoided.

The results of 5 controlled, double-blind clinical trials carried out with a total of 410 subjects over periods ranging from 28 to 84 days, using daily doses of kavapyrones between 30 and 210 mg, were all positive. For example, in 1995, 100 patients suffering anxiety and stress symptoms were given 210 mg of kavapyrones daily. After 8 weeks, the treated subjects were clearly improved in comparison to those receiving a placebo. As for side effects, 15 persons receiving the placebo reported them in comparison to only 5 taking kava extract.

Kava products have been steady but unspectacular sellers in Europe for several decades. Until recently, no one in the United States seemed much interested in them. Ironically, when the Food and Drug Administration began to express concern over the safety of ephedra, a stimulant herbal product, herb marketers became enthusiastic about kava, a depressant. Both herbs have psychoactive properties, but the effects are almost exactly opposite.

Kava and its contained pyrones are, without question, effective medications. They are also subject to abuse. The kava scenario in this country is just beginning. It is too early to predict whether it will continue to be marketed freely or will eventually be subjected to rigid controls.

Valerian (*Valeriana officinalis* L.)

The dried rhizome and roots of this tall perennial herb have enjoyed a considerable reputation as a minor tranquilizer and sleep aid for more than 1000 years. They contain from 0.3% to 0.7% of an unpleasant-smelling volatile oil containing bornyl acetate and the sesquiterpene derivatives valerenic acid and acetoxyvalerenic acid. Also present is 0.5% to 2% of a mixture of lipophilic iridoid principles known as valepotriates. These bicyclic monoterpenes are quite unstable and occur only in the fresh plant material or in that dried at temperatures under 40°C. In addition, various sugars, amino acids, free fatty acids, and aromatic acids have been isolated from the drug.

Identity of the active principles of valerian has been a subject of controversy for many years. Initially, the calmative effect was attributed to the volatile oil; indeed, this kind of activity was long associated with most herbs containing oils with disagreeable odors. Then, beginning in 1966 with the isolation of the valepotriates, the property was attributed to them for a 20-year period. This was done in spite of the fact that they were highly unstable and were contained in most valerian preparations only in small amounts. Finally, in 1988, it was shown that although valerian did produce CNS depression, neither the tested valepotriates, nor the sesquiterpenes valerenic acid or valeranone, nor the volatile oil itself displayed any such effect in rats. Although the active principles of valerian remain unidentified, it seems possible that a combination of volatile oil, valepotriates, and possibly certain water-soluble constituents may be involved.

Because the valepotriates possess an epoxide structure, they demonstrate alkylating activity in cell cultures. This caused concern that the herb might possess potential toxicity. However, those valepotriates decompose rapidly in the stored drug and also are not readily absorbed. For these reasons, no toxicity has ever been demonstrated in intact animals or human beings, so there is no cause for concern.

Eight clinical studies conducted between 1977 and 1996 have shown that valerian is not a suitable herb for the acute treatment of insomnia. Rather, its principal utility lies in its ability to promote natural sleep after several weeks of use, without risk of dependence or adverse health effects. Users should realize the prolonged length of time required for valerian to exert its effectiveness. When properly employed, it offers a gentle alternative to synthetic hypnotics and benzodiazapines in patients with chronic sleep disorders. Recommended dosage is 2–3 g one or more times daily.

Miscellaneous CNS-Depressant Herbs

There are several botanicals with a folkloric reputation for utility in the treatment of restlessness and sleep disturbances whose efficacy is largely unproven by scientific methods. Some of them are frequently used in combination with other CNS depressants in proprietary herbal products.

Hops, the dried strobiles of *Humulus lupulus* L., is one of these. Although very small amounts of methylbutanol, a compound with sedative effects, have been detected in hops, clinical studies have not verified any such activity of the herb in human subjects.

Balm or lemon balm, the leaves of *Melissa officinalis* L., has long been purported to have sedative effects. In the only experimental study to date, its volatile oil did show some activity, but the results were not dose dependent. This suggests the effects were nonspecific.

Passion flower, the dried above-ground parts of *Passiflora incarnata* L., has an ancient reputation as a sleep aid. However, no controlled clinical trials have ever been conducted on single-herb preparations, so preliminary positive results in the few animal studies have not been verified.

Lavender, the dried flowers of *Lavandula angustifolia* Mill., yield a volatile oil, the calming and relaxing effects of which are better documented by both empirical medicine and experimental studies than the three previous herbs. Apparently, its actions are mediated by olfactory receptors, but it may possibly act directly on the CNS following systemic administration. Suitable research in human subjects is required to verify preliminary observations.

The German Commission E approved hops as a calming and sleep promoting drug, lemon balm as a sedative, passion flower for nervous restlessness, and lavender flower as a sedative. There is little or nothing in the published scientific or clinical literature to substantiate these recommendations.

This concludes a brief survey of the herbs that have either been demonstrated to have significant effects on the central nervous system or are postulated to have such effects. Because this broad overview does not present original research findings, specific references to previously established facts are not listed. They are simply too numerous to include. Instead, a general list of recommended reading in which all appropriate references may be found is presented here.

RECOMMENDED READING

Bisset, N.G., ed. 1994. Herbal drugs and phytopharmaceuticals. CRC Press, Boca Raton, FL.

Blumenthal, M. (ed.). 1998. The complete German Commission E monographs. American Botanical Council, Austin, TX and Integrative Medicine Communications, Boston, MA.

Robbers, J.E., M.K. Speedie, and V.E.Tyler. 1996. Pharmacognosy and pharmacobiotechnology. Williams & Wilkins, Baltimore, MD.

Schulz, V., R. Hänsel, and V.E. Tyler. 1998. Rational phytotherapy. Springer Verlag, Berlin.

Foster, S. and V.E. Tyler. 1999. Tyler's honest herbal. 4th ed. The Haworth Herbal Press, New York.

Robbers, J.E. and V.E. Tyler. 1999. Tyler's herbs of choice. 2nd ed. The Haworth Herbal Press, New York.

Immune Stimulants and Antiviral Botanicals:
Echinacea and Ginseng

Dennis V.C. Awang

Preparations of echinacea and ginseng likely comprise the most popular and commercially successful combination of medicinal plant products in North America. Interestingly, the development of *Echinacea* species for medicinal purposes stems from use by indigenous American people, but research has been conducted predominantly in Germany (Bauer and Wagner 1991). On the other hand, the reputation of ginseng is firmly planted in China and has spread west in modern times, research in China, Japan, and Korea, being bolstered by European investigations of mainly extracts of the Asian species, *Panax ginseng*. Following its discovery outside present-day Montreal, Quebec, in 1716 (Foster 1992), the trade in North American ginseng, *Panax quinquefolius,* grew steadily and now constitutes a considerable market in the Orient, where roughly 90% of Canada- and US-grown root is sold. Only one of the additionally recognized *Panax* species, namely *Panax notoginseng* (Sanchi or Tienchi ginseng), is of commercial significance, primarily in Asia, where it finds application as a hemostatic, cardiotonic, and anti-infective agent (Gao et al. 1996).

The traditional use of echinacea by native Americans as a topical anti-infective and snake-bite remedy, and to assuage coughs associated with colds, was presumably the impetus for investigation of its anti-microbial potential. Anti-bacterial, anti-fungal and anti-viral activities were subsequently assessed for a variety of preparations, the identity of the particular *Echinacea* species being still somewhat equivocal (Bauer and Wagner 1991). Latterly, attention has concentrated on the plant's application to the treatment of symptoms of cold and other influenza-like infections, the basis for its economic success; any effectiveness in such conditions is widely attributed to effect on the immune system—potentiation or "stimulation" being generally regarded more particularly as "immunomodulation" (Bauer and Wagner 1991). Any benefit deriving from treatment with echinacea preparations is thought to be due to such stimulation of the immune system, rather than to any direct anti-viral effect.

Ginseng (*P. ginseng*) has a traditional reputation as a tonic and drug of longevity in Asia, where it has also long been used to treat a variety of diseases, including cancer and diabetes, as well as various infections; in modern times, it has been characterized as an 'adaptogen' or anti-stress agent, exerting a non-specific normalizing effect on bodily functions. Ginsenosides, also termed panaxosides, are triterpene saponins regarded as the main active constituents of *Panax* species; hydrophilic bioactive polysaccharides are held responsible for anti-complementary and anti-tumor activities (Gao et al. 1991). However, while the attention of the herbal industry has focused almost exclusively on ginsenosides, almost all investigations of the immunomodulatory effects of ginseng have involved crude extracts of *P. ginseng* (see later) and polysaccharide fractions of *Panax* species (Yun et al. 1993; Gao et al. 1996). The results of one human study with a proprietary *P. ginseng* extract, support the claim of enhancement of immune response against influenza and colds, *after vaccination* (Scaglione et al. 1996).

ECHINACEA

The German Commission E (Blumenthal et al. 1998) approves the use of *Echinacea purpurea* herb as follows: (1) "Internally, as supportive therapy for colds and chronic infections of the respiratory tract and lower urinary tract," and (2) "Externally, on poorly healing wounds and chronic ulcerations." *E. pallida* is recognized as "supportive therapy for influenza-like infections."

Investigation towards identification of the priniciple(s) responsible for the claimed immunomodulatory effects of echinacea preparations, and related effort to establish effective standardized therapeutic formulations have focused on five classes of echinacea constituents, namely, caffeic acid derivatives, alkylamides, 'polyacetylenes' (ketoalkenes/ketoalkynes), glycoproteins, and polysaccharides (Bauer 1988).

450

Constituents/Species Variation/Activity

Echinacoside, a caffeoylic popular for "standardization" but absent in appreciable quantity from roots of *Echinacea purpurea*, "has low anti-bacterial and anti-viral activity, and does *not* show immunostimulatory effects" (Beuscher et al. 1989; Bodinet and Beuscher 1991). Echinacoside is present in the roots of *E. angustifolia* and *E. pallida* at levels between 0.3% and 1.7%, but the roots of the two species can be differentiated by the occurrence of 1,3-0-(cynarin) and 1,5-0-dicaffeoylquinic acids present only in *angustifolia*. Cichoric acid, which possesses phagocytosis stimulatory activity in vitro and in vivo, is present in *E. purpurea*, but in only trace amounts in *E. angustifolia* roots; it is also apparently subject to ready enzymatic degradation in the plant matrix and its content in commercial preparations has been observed to be extremely variable (Bauer 1998).

In echinacea, alkamides accumulate primarily in the roots and inflorescences. The highest alkamide content is found in *E. angustifolia* root, while *E. pallida* roots contain only trace amounts. Small quantities of alkamides also occur in the expressed juice of *E. purpurea*. Alkylamide content also has been reported to vary widely in commercial products (Bauer 1998). However, it is important to note that the alkylamides of *E. purpurea* and *E. angustifolia* differ in general chemical structure. All but one from *E. purpurea* have *two* double bonds in conjugation with the amide carbonyl group, whereas eight *E. angustifolia* alkamides have only *one* double bond in conjugation with the amide group; *E. angustifolia* and *E. purpurea* share five alkamides with the diene structure conjugated with the amide grouping, including the prominent isomeric tetraene pair, readily observable at 260 nm, whereas the three most prominent monene alkamides of *angustifolia* are most conveniently detected at 210 nm (Bauer and Wagner 1991).

The main *lipophilic* constituents of *E. pallida* roots are ketoalkenes and ketoalkynes, all with a carbonyl group at the C-2 position of the carbon chains. It is misleading to designate these prominent lipophilic constituents of *E. pallida* root as polyacetylenes, in differentiation from the corresponding alkamide constituents of *E. angustifolia* (15 identified) and *E. purpurea* (11 identified), since many of these latter compounds have conjugated diacetylenic (diyn-) functionality (8 so far identified). Only two ketodiynes have been identified in *E. pallida* roots (Bauer 1998).

The hydrophilic, highly polar constituents, notably polysaccharides and glycoproteins, are present most prominently in expressed juice of the plant, and in aqueous and hydroalcoholic extracts of the aerial parts. Two polysaccharides have been isolated from an aqueous extract of the aerial parts of *E. purpurea* and shown to stimulate phagocytosis in vitro and in vivo and to enhance production of oxygen radicals by macrophages in a dose-dependent way (Stimpel et al. 1984; Lohmann–Matthes and Wagner 1989). Other activities, such as anti-infective and antitumor, have also been noted (Bauer 1998). In a phase-I clinical trial, a polysaccharide fraction, isolated from *E. purpurea* tissue culture, caused an increased production of leukocytes, segmented granulocytes, and tumor necrosis factor *alpha* (TNFα). In addition, high-molecular weight proteins have been isolated from roots of *E. angustifolia* and *E. purpurea*; purified extracts of a glycoprotein-polysaccharide complex were found to exhibit B-cell stimulating activity and to induce release of interleukin I, TNFα and IFNα, β in macrophages. *E. angustifolia* and *E. purpurea* roots are reported to contain similar levels of glycoproteins, *E. pallida* less (Beuscher et al. 1995). It should be noted, however, that these activities of polysaccharide and glycoprotein-polysaccharide have been observed with *isolated fractions* of polar echinacea extracts: traditional ethanolic extracts likely contain insignificant amounts of polysaccharides, and Bauer and Wagner (1991) have stated that "...only low concentrations of polysaccharides are present in the expressed juice and they do not have the same composition as those in the extracts of aerial parts." In addition, polysaccharides are expected, for reasons of stability, to be ineffective by oral administration.

While levels and profiles of caffeoylics (Beuscher et al. 1995) and alkylamides from *E. angustifolia* and *E. purpurea*, as well as ketoalkenes and ketoalkynes from *E. pallida*, may be obtained by HPLC procedures with excellent resolution and a high level of sensitivity (Maffei Facino et al. 1995), glycoproteins are most conveniently assessed by an ELISA method (Egert and Beuscher 1992) and polysaccharides by enzymatic hydrolysis and determination of total fructose (L.D. Lawson pers. commun.).

The stimulation of macrophage activity earlier noted for the polar glycoprotein-polysaccharide complement of echinacea preparations, is paralleled by apparently similar activity of their lipophilic component. In fact, the lipophilic fractions of alcoholic extracts from *E. angustifolia* and *E. purpurea* aerial parts and roots showed the greatest activity (Bauer et al. 1989; Jacobson 1967); such extracts of aerial parts exhibited an "immunomodulating effect on the phagocytoxic, metabolic, and bactericidal activities of peritoneal macrophages in mice" (Bauer 1998).

Attempts to assess the therapeutic potential of echinacea preparations have mainly focused on measurement of phagocytoxic capacity by the carbon-clearance-test; enhancement by a factor of 1.5 to 1.7 was found in purified alkamide fractions from *E. angustifolia* and *E. purpurea* roots (Bauer et al. 1989). However, overwhelmingly the major constituent of the alkamide fraction of *E. purpurea* root, namely the dodecatetraenoic acid isobutylamide 10 E/Z isomeric pair, exhibited only weak activity, and, in the language of Bauer (1998), "the most effective constituent remains to be found." In consideration of all of the above, Bauer properly infers that immunostimulant activity of echinacea extracts resides in various classes of echinacea constituents. In acknowledgment of that reality, what remains to be appreciated is which of these constituents are responsible for the medicinal value observed in the limited number of clinical studies available to this point. Is measurement of phagocytoxic activity a sensible, reliable indicator of immunostimulant potential in terms of effectiveness in the treatment of symptoms of cold and flu, and abatement of respiratory and urinary infections?

At an even further level of reduction, is there any real benefit—beyond identity and quality control considerations—in monitoring levels of *individual* echinacea constituents? And which ones would one select? As far as total tare of the five designated classes of compounds: caffeoylics, alkamides, ketoalkenes/ketoalkynes, glycoproteins, and polysaccharides, their relative importance regarding therapeutic effect yet remains to be properly assessed.

Clinical Testing

Cold. Treatment with the dose equivalent of 900 mg root per day of an alcoholic extract of *E. purpurea* stimulated a faster reduction of cold symptoms in patients compared to the placebo group, and to a group treated with the equivalent of 450 mg per day of root (Braunig et al. 1992). An *E. purpurea* root extract, combined with vitamin C, was also found effective in treating the common cold (Scaglione and Lund 1995). A double-blind, placebo-controlled study was conducted with freshly-expressed juice of *E. purpurea* 108 volunteers with chronic upper respiratory tract infections (Schoneberger 1992). Compared to the placebo, the echinacea preparation reduced the number of infections, lengthened the time between infections, shortened the duration of the illness and lessened the severity of symptoms; patients with diminished immune response seemed to benefit most from the treatment. (Curiously, the German Commission E does not recognize the usefulness of *E. purpurea* root in treating colds, while approving *E. purpurea* "fresh above-ground parts, harvested at flowering time" as a supportive treatment for colds and chronic infections of the respiratory tract and lower urinary tract).

Infections. *E. purpurea* expressed juice promoted faster healing in bronchitis, a viral infection, than an antibiotic (Baetgen 1988) and, as adjuvant, reduced the frequency of occurrence of *Candida* infection better than econazol nitrate (Coeugniet and Kuhnast 1986).

A hydroalcoholic tincture of *E. pallida* roots produced significantly faster recovery from upper respiratory tract infections than placebo (Braunig and Knick 1993). A placebo controlled, double-blind trial conducted on 160 adults (Dorn et al. 1997) indicated that a daily dose of 900 mg of the extract of *E. pallida* root was effective in shortening the duration of upper respiratory tract infections in infected adults, whether of bacterial or viral origin; however, the precise character of the preparation cannot be ascertained from the published information.

Clearly, there is a need for more and better studies to determine the most effective and condition-appropriate echinacea preparations, with respect to species, plant part, character of preparations, and recommended dosage. Given the present state of knowledge of the pharmacology of echinacea constituents, attempts at *therapeutic* standardization based on individual chemical constituents are unrealistic—pending identification of the constituent(s) responsible for beneficial clinical effect. While such efforts continue, any meaningful standardization must rest with careful and extensive chemical profiling of proven efficacious medications, and/or *relevant* bioactivity correlation.

GINSENG

Undoubtedly because of the difficulty of confirming such diffuse and protracted actions as "tonic," rejuvenating, and "adaptogenic" effects, most modern research on *Panax* has been conducted on more specific influences, with more easily determined end-points. Such experiments have invariably utilized either gross extracts of one sort or another, fractions of extracts, or individual ginseng constituents isolated from them.

Various Extracts, Fractions and Isolated Principles

In an attempt to resolve conflicting reports of ginseng's effect on the immune system, a 70% ethanolic extract of *P. ginseng* was subjected to a comprehensive testing battery capable of detecting subtle immune changes in mice, whose immune function was suppressed by cyclophosphamide, a common chemotherapeutic agent (Kim et al. 1990). The observation that basal natural killer (NK) cell activity was stimulated, supports an immunomodulatory property for ginseng. However, while subchronic exposure helped stimulate recovery of NK function in cyclophosphamide—immunosuppressed mice, NK activity was not further stimulated in polyinosinic-polycytidilic treated mice—and other immunological parameters examined, such as T and B cell responses, were not affected.

Using a sensitive two-site enzyme immunoassay specific for human basic fibroblast growth factor (bFGF), "bFGF-like immunoreactivity" was detected in a *P. ginseng* extract prepared by methanol extraction, followed by aqueous buffer treatment which precipitated a water-insoluble fraction, centrifugation, and filtration. A bFGF-like molecule was identified (Takei et al. 1996).

The ethanol-insoluble fraction of an aqueous extract of *P. ginseng*, tested for immunomodulatory activity, was found to induce proliferation of splenocytes and to generate activated killer cells in vitro (Yun et al. 1993); the activated killer cells killed both NK cell sensitive and insensitive tumor target cells without MHC-restriction. Activation of splenocytes was mediated through endogenously produced interleukin-2. Treatment of young male mice with this ethanol-insoluble ginseng fraction significantly reduced lung tumor incidence, compared with a cohort treated only with tumor-inducing benzo [*a*] pyrene. Prior to treatment with 95% ethanol, the original aqueous extract was solubilized in methanol and the methanol-soluble fraction, which exhibited cytotoxicity to splenocytes in vitro, was excluded.

Stimulated by the observation that several polysaccharides from *P. ginseng* exhibit the immunological effects of increasing serum IgG level and activating the reticuloendothelial system (RES), Japanese researchers have isolated two acidic polysaccharides, named ginsenan PA and ginsenan PB, whose detailed structures were determined (Tomoda et al. 1993). Both polysaccharides displayed similar RES, anti-complementary, and alkaline-phosphatase-inducing activity. In this study, *P. ginseng* root was extracted, in the traditional Chinese manner, with hot water, the polysaccharides precipitated with cetyltrimethylammonium bromide, extracted with dilute sodium chloride solution and the free polysaccharides precipitated by the addition of ethanol; column chromatography, followed by dialysis and then gel filtration, yielded ginsenans PA and PB.

Immunostimulating activity of *Panax* has been observed also in polysaccharidic material extracted from the leaves of *P. ginseng* (Sum et al. 1994). In order to estimate the potential of ginseng leaves which can be harvested annually, as compared to ginseng roots which usually require at least four years of maturation, these researchers compared the immune complex binding enhancing activity of water and alkaline-soluble polysaccharide fractions of root and leaves of *P. ginseng*. Binding of glucose oxidase-anti-glucose oxidase complexes (GAG) to macrophages was enhanced by crude polysaccharide fractions GL-2 and GR-3, from leaves and root respectively, GL-2 being the most active of all four fractions; the alkaline-soluble leaf extract GLA-2 showed no activity. A pure active polysaccharide GL-4IIb2 containing 33% uronic acid, was isolated. GAG binding to macrophages by GLIIb2 increased in a dose-dependent fashion. The authors judge that their observations "... suggest that the carbohydrate moiety of GL-4IIb2 activated the mononuclear phagocytic system in vivo through an increase of Fc receptor expression mediated by de novo synthesis of the receptor protein and resulted in enhanced immune complexes clearance."

Immunostimulating polysaccharides have also been isolated from *P. notoginseng*; (Sanchi or Tienchi ginseng). A RES-active polysaccharide, sanchinan A (Ohtani et al. 1987), and four distinct homogeneous active polysaccharide fractions (Gao et al. 1991), have been isolated from the roots of *P. notoginseng*; in the latter study, one fraction showed strong anti-complementary activity, two significantly induced the production of interferon-γ in the presence of concanavalin-A, and all induced the production of TNFα, using human serum and antibody-sensitized sheep red blood cells.

One study with an isolated saponin from *P. ginseng*, ginsenoside Rg1, assessed its stimulatory effect on immune function of lymphocytes in the elderly (Liu et al. 1995). A reduced responsiveness of lymphocytes has been associated with aging as well as a lower percentage of IL-2 receptor positive cells, compared with the younger population. Rg1 was found to stimulate proliferation of lymphocytes and to increase the fluidity of lymphocyte membrane in the aged.

G115 (Ginsana®)

In the past two decades, a considerable number of studies have been conducted on the proprietary *P. ginseng* extract, G115, from Pharmaton SA, Lugano-Bioggio, Switzerland. The immunomodulatory activity of G115 was assessed in mice (Singh et al. 1984). Pretreatment with the ginseng extract led to an increase in the number of antibody forming cells, as well as of the titers of circulating antibodies. Pretreatment with ginseng extract also modulated the cell mediated immune response in test animals. Ginseng extract significantly elevated NK cell activity in test animals, as compared to control animals, and enhanced production of interferon of both the pH stable and labile types, induced by 6-MFA, an interferon inducer of fungal origin.

Subsequently, three controlled human studies, with G115 conducted in Italy, examined the immunomodulatory effects of the extract, in one case was comparing G115 with an aqueous extract of *P. ginseng*. In a controlled single-blind study (Scaglione et al. 1994) placebo was compared to 100 mg of G115 administered every 12 hours for 8 weeks to patients suffering from chronic bronchitis. The functioning of alveolar macrophages, collected by bronchoalveolar lavage, was determined by measuring the phagocytic activity and killing power towards *Candida albicans*, which were "significantly ... *increased* at the eighth week of treatment." The observed improvement in macrophage immune response was seen to indicate an important role for G115 in "the *prevention and therapy* of respiratory disorders (emphasis added).

A comparative double-blind study with healthy volunteers was conducted with G115 and an aqueous extract of *P. ginseng* root (PKD 167/79), using the same dosing regime followed in the previous study (Scaglione et al. 1990). No details are provided for preparation of the aqueous extract, as are the mode of preparation and the constituent profile of G115 not disclosed, except for a roughly 4% combined content in the seven major ginsenosides usually assayed for in ginseng materials. While both extracts demonstrated immunostimulatory activity, the two extracts differed in some parameters. A noticeable increase in phagocytosis was observed in the fourth week in the group treated with G115, whereas elevation in the PKC 167/79 group "appeared to be delayed to the end of the eighth week." A similar difference was observed regarding increase in T helper lymphocytes (T4). These researchers interpreted "the earlier effective action of G115"—also evident in an enhancement of T4/T8 ratio—as due an earlier onset of immune response, as compared with PKC 167/79. The final sentence of the publication: "It can be suggested that on the basis of its different actions on the immune system, the G115 extract may have a different composition in comparison with the PKC 167/79 extract *in terms of ginsenoside content*" (emphasis added).

The third human trial with G115 involved 227 volunteers at 3 private practices (Scaglione et al. 1996). Subjects were given either placebo or 100 mg of G115 daily, for 12 weeks, an anti-influenza polyvalent vaccination being administered after the 4th week. The frequency of influenza or common cold between 4 and 12 weeks was 42 cases in the placebo group, and only 15 in the G115 group, the difference being highly significant ($p < 0.001$). NK cell activity was also significantly higher in the ginseng group after 8 weeks, as were specific antibody titers. The researchers concluded that G115 can improve the immune response in humans and can protect against influenza and the common cold.

As with echinacea, the results of studies assessing the immunostimulatory effect of ginseng preparations cannot be confidently interpreted in terms of particular plant constituents. The relative contribution of ginsenoside, polysaccharide, or other ginseng components cannot be estimated and, therefore, inferences regarding relative effectiveness of different extracts of ginseng remain entirely speculative. Estimation of ginsenoside effect, for example, may depend not only on the total content of the most prominent 6–8 neutral ginsenosides, but also on their relative ratios and perhaps contributions from their more-than-20 other closely related triterpene saponins. As Scaglione et al. (1990) have posited, contradictory results from studies on the effect of ginseng extracts on the immune system "could arise from the difference in *dose levels, composition and duration of therapy*" (emphasis added).

REFERENCES

Baetgen, D. 1988. Behandlung der akuten bronchitis im kindesalter. Praxisstudie mit einem immunstimulans aus *Echinacea purpurea*. Therapiewoche Padiatrie 1:66–70.

Bauer, R. 1988. Echinacea: Biological effects and active principles. p. 140–157. In: L.D. Lawson and R. Bauer (eds.), Phytomedicines of Europe. Chemistry and Biological Activity. American Chemical Society, Washington, DC:

Bauer, R. 1998. Standardization of *Echinacea purpurea* expressed juice with reference to cichoric acid and alkamides. J. Herbs Spices Med. Plants loc.cit.

Bauer, R., P. Remiger, K. Jurcic, and H. Wagner. 1989. Beeinflussung der phagosytase-aktivitat durch Echinacea-extrakte. Z. Phytother. 10:43–48.

Bauer, R. and H. Wagner. 1991. *Echinacea* species as potential immunostimulatory drugs. p. 253–321. In: H. Wagner and N.R. Farnsworth (eds.), Economic and medicinal plant research. Vol. 5. Academic Press, New York.

Beuscher, N., H.U. Beuscher, and C. Bodinet. 1989. Enhanced release of interleukin-1 from mouse macrophages by glycoproteins and polysaccharides from *Baptisia tinctoria* and *Echinacea* species. Planta Med. 55:660.

Beuscher, N., C. Bodinet, I. Willigmann, and D. Egert. 1995. Immonomudulierende eigenschaften von wurzelextrakten verschiedenen echinacea-arten. Z. Phytother. 16:157–166.

Blumenthal, M., W.R. Busse, A. Goldberg, J. Gruenwald, T. Hall, C.W. Riggins, and R.S. Rister (eds.). 1998. The complete German Commission E Monographs. Therapeutic guide to herbal medicines, Austin, TX. American Botanical Council; Boston: Integrative Medicine Communications.

Bodinet, C. and N. Beuscher. 1991. Antiviral and immunological activity of glycoproteins from *Echinacea purpurea radix*. Planta Med. 57, suppl.2:A33–A34.

Braunig, B., M. Dorn, and E. Knick. 1992. *Echinacea purpurea radix*: Zur starkung der korpereingenen abwehr bei grippalen infekten. Z. Phytother. 13:7–13.

Braunig, B. and E. Knick. 1993. Therapeutische erfahrungen mit echinaceae pallidae radix bei grippalen infekten. Naturheilpraxis 1:72–75.

Coeugniet, E.G. and R. Kuhnast. 1986. Rezividierende Candidiasis. Adjuvante immuntherapie mit verschiedenen echinacin darreichengsformen. Therapiewoche 36:3352–3358.

Dorn, M, E. Knick, and G. Lewith. 1997. Placebo-controlled, double-blind study of *Echinacea pallidae radix* in upper respiratory tract infections. Complementary Therapies Medicine 5:40–42.

Egert, D. and N. Beuscher. 1992. Studies on autiglen specificity of immunoreactive arabinogalactan proteins extracted from *Baptisia tinctoria* and *Echinacea purpurea*. Planta Med. 58: 163–165.

Foster, S. 1992. Herbal emissaries. Academic Press, Rochester, VT. p. 104.

Gao, H., F. Wang, E.J. Lien, and M.D. Trousdale. 1996. Immunostimulating polysaccharides from *Panax notoginseng*. Pharm. Res. 13(8):1196–1200.

Gao, Q.P., H. Kiyohara, J.C. Cyong, and H. Yamada. 1991. Chemical properties and anti-complementary activities of heteroglycans from the leaves of *Panax ginseng*. Planta Med. 57:132–136.

Jacobson, M. 1967. The structure of echinacein, the insecticidal component of American coneflower roots. J. Org. Chem. 32:1646–1647.

Kim, J.Y., D.R. Germolec, and M.I. Luster. 1990. *Panax ginseng* as a potential immunomodulator: studies in mice. Immunopharm. Immunotox. 12(2):257–276.

Liu, J., S. Wang, H. Liu, L. Yang, and G. Nan. 1995. Stimulatory effect of saponin from *Panax ginseng* on immune function of lymphocytes in the elderly. Mechanisms Ageing Development 83:43–53.

Lohmann-Matthes, M-L. and H. Wagner. 1989. Aktivierund von makrophagen durch polysaccharide aus gewebedwituren von *Echinacea purpurea*. Z. Phytother. 10:52–59.

Maffei Facino, R., M. Carini, C. Aldini, L. Saibene, P. Pietta, and P. Mauri. 1995. Echinacoside and caffeoyl conjugates protect collagen from free-radical induced degradation. A potential use of echinacea extracts in the prevention of skin photodamage. Planta Med. 61:510–514.

Ohtani, K., K. Mizutani, S. Hatono, R. Kasai, R. Sumino, T. Shiota, M. Ushijima, J. Zhou, T. Fuwa, and O. Tanaka. 1987. Planta Med. 53:166.

Scaglione, F., G. Cattaneo, M. Alessandria, and R. Cogo. 1996. Efficacy and safety of the standardized ginseng extract G115 for potentiating vaccination against common cold and/or influenza syndrome. Drugs Exptl. Clin. Res. XXII(2):65–72.

Scaglione, F., R. Cogo, C. Cocuzza, M. Arcidiacono, and A. Beretta. 1994. Immunomodulatory effects on *Panax ginseng* C.A. Meyer (G115) on alveolar macrophages from patients suffering with chronic bronchitis. Int. J. Immunotherapy X(1):21–24.

Scaglione, F., F. Ferrara, S. Dugnani, M. Falchi, G. Santoro, and F. Fraschini. 1990. Immunomodulatory effects of two extracts of *Panax ginseng* C.A. Meyer. Drugs Expt. Clin. Res. XVI(10):537–542.

Scaglione, F. and B. Lund. 1995. Efficacy in the treatment of the common cold of a preparation containing an echinacea extract. Int. J. Immunother. 11:163–166.

Schoneberger, D. 1992. The influence of immune-stimulating effects of pressed juice from *Echinacea purpurea* on the course and severity of colds. Forum Immunologie 2:18–22.

Singh, V.K., S.S. Agarwal, and B.M. Gupta. 1984. Immunomodulatory activity of *Panax ginseng* extract. Planta Med. 48:462–465.

Stimpel, M., A. Proksch, H. Wagner, and M-L. Lohmann–Matthes. 1984. Macrophage activation and induction of macrophage cytotoxicity by purified polysaccharide fractions from the plant *Echinacea pupurea*. Infect. Immun. 46:845–849.

Sum, X.B., T. Matsumoto, and H. Yamada. 1994. Purification of immune complexes clearance enhancing polysaccharide from the leaves of *Panax ginseng*, and its biological activities. Phytomedicine 1:225–231.

Takei, Y., T. Tamamoto, H. Hagashira, and K. Hayashi. 1996. Identification of basic fibroblast growth-factor-like immunoreactivity in *Panax ginseng* extract: Investigation of its molecular properties. Biosci. Biotech. Biochem. 60(4):584–588.

Tomoda, M., K. Takeda, N. Shimizu, R. Gonda, and N. Ohara. 1993. Characterization of two acidic polysaccharides having immunological activities from the root of *Panax ginseng*. Biol. Pharm. Bul. 16(1):22–25

Yun, Y.S., Y.S. Lee, S.K. Jo, and I.S. Jung. 1993. Inhibition of autochthonous tumor by ethanol insoluble. Fraction from *Panax ginseng* as an immunomodulator. Planta Med. 59:521–524.

New Antimicrobials of Plant Origin

Maurice M. Iwu, Angela R. Duncan, and Chris O. Okunji

Infectious diseases account for approximately one-half of all deaths in tropical countries. In industrialized nations, despite the progress made in the understanding of microbiology and their control, incidents of epidemics due to drug resistant microorganisms and the emergence of hitherto unknown disease-causing microbes, pose enormous public health concerns. Historically, plants have provided a good source of antiinfective agents; emetine, quinine, and berberine remain highly effective instruments in the fight against microbial infections. Phytomedicines derived from plants have shown great promise in the treatment of intractable infectious diseases including opportunistic AIDS infections. Plants containing protoberberines and related alkaloids, picralima-type indole alkaloids and garcinia biflavonones used in traditional African system of medicine, have been found to be active against a wide variety of micro-organisms. The profile of known drugs like *Hydrastis canadensis* (goldenseal), *Garcinia kola* (bitter kola), *Polygonum* sp., *Aframomum melegueta* (grains of paradise) will be used to illustrate the enormous potential of antiinfective agents from higher plants. Newer drugs such as *Xylopia aethiopica*, *Araliopsis tabouensis*, *Cryptolepis sanguinolenta*, *Chasmanthera dependens* and *Nauclea* species will be reviewed.

INFECTIOUS DISEASE

World wide, infectious disease is the number one cause of death accounting for approximately one-half of all deaths in tropical countries. Perhaps it is not surprising to see these statistics in developing nations, but what may be remarkable is that infectious disease mortality rates are actually increasing in developed countries, such as the United States. Death from infectious disease, ranked 5th in 1981, has become the 3rd leading cause of death in 1992, an increase of 58% (Pinner et al. 1996). It is estimated that infectious disease is the underlying cause of death in 8% of the deaths occurring in the US (Pinner et al. 1996). This is alarming given that it was once believed that we would eliminate infectious disease by the end of the millenium. The increases are attributed to increases in respiratory tract infections and HIV/AIDS. Other contributing factors are an increase in antibiotic resistance in nosicomial and community acquired infections. Furthermore, the most dramatic increases are occurring in the 25–44 year old age group (Pinner et al. 1996).

These negative health trends call for a renewed interest in infectious disease in the medical and public health communities and renewed strategies on treatment and prevention. Proposed solutions are outlined by the CDC as a multi-pronged approach that includes: prevention, (such as vaccination); improved monitoring; and the development of new treatments. It is this last solution that would encompass the development of new antimicrobials (Fauci 1998).

Historic Use of Plants as Antimicrobials

Historically, plants have provided a source of inspiration for novel drug compounds, as plant derived medicines have made large contributions to human health and well-being. Their role is two fold in the development of new drugs: (1) they may become the base for the development of a medicine, a natural blueprint for the development of new drugs, or; (2) a phytomedicine to be used for the treatment of disease. There are numerous illustrations of plant derived drugs. Some selected examples, including those classified as antiinfective, are presented below.

The isoquinoline alkaloid emetine obtained from the underground part of *Cephaelis ipecacuanha*, and related species, has been used for many years as and amoebicidal drug as well as for the treatment of abscesses due to the spread of *Escherichia histolytica* infections. Another important drug of plant origin with a long history of use, is quinine. This alkaloid occurs naturally in the bark of *Cinchona* tree. Apart from its continued usefulness in the treatment of malaria, it can be also used to relieve nocturnal leg cramps. Currently, the widely prescribed drugs are analogs of quinine such as chloroquine. Some strains of malarial parasites have become resistant to the quinines, therefore antimalarial drugs with novel mode of action are required.

Similarly, higher plants have made important contributions in the areas beyond antiinfectives, such as cancer therapies. Early examples include the antileukaemic alkaloids, vinblatine and vincristine, which were both obtained from the Madagascan periwinkle (*Catharanthus roseus* syn. *Vinca roseus*) (Nelson 1982). Other cancer therapeutic agents include taxol, homoharringtonine and several derivatives of camptothein. For example, a well-known benzylisoquinoline alkaliod, papaverine, has been shown to have a potent inhibitory effect on the replication of several viruses including cytomegalovirus, measles and HIV (Turano et al. 1989). Most recently, three new atropisomeric naphthylisoquinoline alkaloid dimers, michellamines A, B, and C were isolated from a newly described species tropical liana *Ancistrocladus korupensis* from the rainforest of Cameroon. The three compounds showed potential anti-HIV with michellamine B being the most potent and abundant member of the series. These compounds were capable of complete inhibition of the cytopathic effects of HIV-1 and HIV-2 on human lymphoblastoid target cell in vitro (Boyd et al. 1994).

The Development of Phytomedicines and the Ethnomedicinal Approach

The first generation of plant drugs were usually simple botanicals employed in more or less their crude form. Several effective medicines used in their natural state such as cinchona, opium, belladonna and aloe were selected as therapeutics agents based on empirical evidence of their clinical application by traditional societies from different parts of the world. Following the industrial revolution, a second generation of plant based drugs emerged based on scientific processing of the plant extracts to isolate "their active constituents." The second-generation phytopharmaceutical agents were pure molecules and some of the compounds were even more pharmacologically active than their synthetic counterparts. Notable examples were quinine from *Cinchona*, reserpine from *Rauvolfia*, and more recently taxol from *Taxus* species. These compounds differed from the synthetic therapeutic agents only in their origin. They followed the same method of development and evaluation as other pharmaceutical agents.

The sequence for development of pharmaceuticals usually begins with the identification of active lead molecules, detailed biological assays, and formulation of dosage forms in that order, and followed by several phases of clinical studies designed to established safety, efficacy and pharmacokinetic profile of the new drug. Possible interaction with food and other medications may be discerned from the clinical trials.

In the development of "Third Generation" phytotherapeutic agents a top-bottom approach is usually adopted. This consists of first conducting a clinical evaluation of the treatment modalities and therapy as administered by traditional doctors or as used by the community as folk medicine. This evaluation is then followed by acute and chronic toxicity studies in animals. Studies should, when applicable, include cytotoxicity studies. It is only if the substance has an acceptable safety index would it be necessary to conduct detailed pharmacological/ biochemical studies.

Formulation and trial production of the dosage forms are structured to mimic the traditional use of the herb. The stability of the finished product is given careful attention during the formulation of the final dosage form. This is a unique blend of the empiricism of the earlier first *generation* botanicals with the experimental research used to prove the efficacy and safety of second *generation* isolated pure compounds. Several pharmaceuticals companies are engaged in the development of natural product drugs through the isolation of the so-called active molecules from plant extracts.

PRESENT USE OF PLANTS AS ANTIMICROBIALS

It is estimated that today, plant materials are present in, or have provided the models for 50% Western drugs (Robbers 1996). Many commercially proven drugs used in modern medicine were initially used in crude form in traditional or folk healing practices, or for other purposes that suggested potentially useful biological activity. The primary benefits of using plant derived medicines are that they are relatively safer than synthetic alternatives, offering profound therapeutic benefits and more affordable treatment.

Therapeutic Benefit

Much of the exploration and utilization of natural products as antimicrobials arise from microbial sources. It was the discovery of penicillin that led to later discoveries of antibiotics such as streptomycin, aureomycin and chloromycetin. (Trease 1972). Though most of the clinically used antibiotics are produced by soil micro-organisms or fungi, higher plants have also been a source of antibiotics (Trease 1972). Examples of these are the bacteriostatic and antifugicidal properties of *Lichens,* the antibiotic action of allinine in *Allium sativum* (garlic), or the antimicrobial action berberines in goldenseal (*Hydrastis canadensis*) (Trease 1972). Plant based antimicrobials represent a vast untapped source for medicines. Continued and further exploration of plant antimicrobials needs to occur. Plants based antimicrobials have enormous therapeutic potential. They are effective in the treatment of infectious diseases while simultaneously mitigating many of the side effects that are often associated with synthetic antimicrobials. They are effective, yet gentle. Many plants have tropisms to specific organs or systems in the body. Phytomedicines usually have multiple effects on the body. Their actions often act beyond the symptomatic treatment of disease. An example of this is *Hydrastis canadensis. Hydrastis* not only has antimicrobial activity, but also increases blood supply to the spleen promoting optimal activity of the spleen to release mediating compounds (Murray 1995).

Economic Benefit

World wide, there has been a renewed interest in natural products. This interest is a result of factors such as: consumer's belief that natural products are superior; consumer's dissatisfaction with conventional medicines; changes in laws allowing structure-function claims which results in more liberal advertising; aging baby boomers; national concerns for health care cost.

Sales of products in this market have increased dramatically in the last decade. Sales of botanical products in the United States have reached $3.1 billion of the $10.4 billion dollar dietary supplement industry 1996 (NBJ June 1998). The industry anticipates growth on the order of 15–20% into the new millenium (Herbalgram 1996). This growth rate will be maintained in an industry that is still considered to be in its infancy. Many plants that were previously wildcrafted will need to be grown domestically to meet the demands of the consumer. This represents many opportunities for the cultivation of crops for this industry.

A market based illustration of the need for plant based antimicrobials is demonstrated by the dissection of the herbal products market. In reviewing the top botanicals used as antiinfectives, the primary botanical used as an antimicrobial is *Hydrastis* with sales of 4.7% in 1995 (Gruenwald 1997). While antiinfectives agents make up 24 % of the pharmaceutical market (1992 Census of Manufactures 1994).

A similar, analysis of *Hypericum* (St. John's wort), demonstrates the value of such an evaluation. Though *Hypericum* is an antiviral, it is primarily used for its antidepressant activity. In 1995 it was not among the top selling herbs (Gruenwald 1997). However, by 1997, it had become an overnight success, with sales increasing over 20,000% in the mass market sector (Aarts 1998). The meteoric increase in the sales of *Hypericum* is multifactorial, but one factor in its in popularity was the existence of an unexploited market opportunity. In 1994 21% of pharmaceuticals sold were for the conditions affecting the central nervous system (1992 Census of Manufactures). Most of the drugs sold in this category are for depression. During this period of time, none of the top selling herbs sold had a primary indication for depression. This market hole, coupled with the media exposure produced a market success.

Many market holes exist. When using the same strategy to look at antimicrobial agents there is a similar gap. If the market dissection for antiinfectives is viewed in the same light as the *Hypericum analogy,* then perhaps this market is prime for receiving new plant based antimicrobials.

The potential for developing antimicrobials into medicines appears rewarding, from both the perspective of drug development and the perspective of phytomedicines. The immediate source of financial benefit from plants based antimicrobials is from the herbal products market. This market offers many opportunities for those cultivating new crops, as many of the plants that are wildcrafted today must be cultivated to match the demands of this market. Again *Hydrastis,* one of the top selling antimicrobials in the US herbal market, represents an example of a herb that has undergone domestication. Originally this plant, native to eastern North America, was wild crafted. *Hydrastis,* has been used by Native Americans for many conditions, including as an antimicrobial for infections. Efforts to cultivate this plant were undertaken in order to supply the demands of the herbal products market and to battle it's threatened extinction.

It is vital to be in the position to capitalize on the phytomedicine market, providing environmentally responsible solutions to public health concerns presented by new trends in infectious disease. In order to be prepared, the industry must be able to sustainably harvest and supply the herbal market. That means we must be able to anticipate the market needs and develop products to satisfy this market.

PLANTS WITH PROMISING ANTIINFECTIVE ACTIVITY

In our organizations, our major emphasis has been on drug discovery from ethnomedicinal information using the "Third Generation Approach." This method differs in that the clinical evaluation in humans takes place before the precise active constituents are known but the chemical composition and safety of the extracts are determined before formulation into dosage forms.

Plants containing protoberberines and related biflavones used in traditional African system of medicine have been found to be active against a wide variety of micro-organisms. Many medicinal plants of Africa have been investigated for their chemical components and some of the isolated compounds have been shown to posses interesting biological activity. Some of these plants are discussed below.

Garcinia kola, bitter kola (Guttiferae)

Garcinia kola, is found in moist forest and grows as a medium size tree, up to 12 m high. It is cultivated and distributed throughout west and central Africa. Medicinal uses include, purgative, antiparasitic, antimicrobial. The seeds are used in the treatment of bronchitis and throat infections. They are also used to prevent and relieve colic, cure head or chest colds and relieve cough. Also the plant is used for the treatment of liver disorders and as a chewing stick (Iwu 1993).

The constituents include—biflavonoids, xanthones and benzophenones. The antimicrobial properties of this plant are attributed to the benzophenone, flavanones. This plant has shown both anti-inflammatory, antimicrobial and antiviral properties. Studies show very good antimicrobial and antiviral properties. In addition, the plant possesses antidiabetic, and antihepatotoxic activities (Iwu 1993).

Aframomum melegueta (Zingiberaceae) Grains of Paradise

This is a spicy edible fruit that is cultivated and occurs throughout the tropics. It is a perennial herb. The medicinal uses of *Aframomum* include aphrodisiac, measles, and leprosy, taken for excessive lactation and post partem hemorrhage, purgative, galactogogue and anthelmintic, and hemostatic agent (Iwu 1993). The constituents are essential oils—such as gingerol, shagaol, paradol. Studies show antimicrobial and antifungal activity and effective against schistosomes (Iwu1993).

Xylopia aethiopica, Ethiopian Pepper (Abbibacceae)

An evergreen, aromatic tree growing up to 20 m high with peppery fruit. It is native to the lowland rainforest and moist fringe forest in the savanna zones of in Africa. Largely located in West, Central and Southern Africa. Medicinal uses of the plant are, as a carminative, as a cough remedy, and as a post partum tonic and lactation aid. Other uses are stomachache, bronchitis, biliousness and dysentery. It is also used externally as a poultice for headache and neuralgia. It is used with lemon grass for female hygiene. It is high in copper, manganese, and zinc (Smith 1996).

Key constituents are diterpenic and xylopic acid. In studies, the fruit as an extracts has been shown to be active as an antimicrobial against gram positive and negative bacteria. Though it has not been shown to be effective against *E. coli* (Iwu 1993). Xylopic acid has also demonstrated activity against *Candida albicans* (Boakye-Yiadom 1977).

Cryptolepis sanguinolenta Lindl. Schltr. (Periplocaceae)

A shrub that grows in the rainforest and the deciduous belt forest, found in the west coast of Africa. Related species appear in the east and southern regions of the continent. Its main medicinal use is for the treatment of fevers. It is used for urinary tract infections, especially *Candida*. Other uses are inflammatory conditions, malaria, hypertension, microbial infections and inflammatory conditions, stomach aches colic (Iwu 1993).

Active principals identified are indo quinoline alkaloids. Studies show inhibition against gram negative bacteria and yeast (Silva 1996). Additionally studies have shown this plant to have bactericidal activity. Clinical studies have shown extracts of the plant were effective in parasitemia. Recent in vitro study shows activity against bacteria specifically, enteric pathogens, most notably *E. coli* (but also staphylococcus, *C. coli*, *C. jejuni*, pseudomonous, salmonella, shigella, streptococcus, and vibrio) and some activity against *candida* (Sawer 1995). It has shown histamine antagonism, hypotensive, and vasodilatory activities (Iwu 1993). In addition it has demonstrative antihyperglcyemic properties (Brierer 1998).

Chasmanthera dependens Hoschst (Menispermaceae)

A woody climber that grows wild in forest margins and savanna. The plant is cultivated. It is used medicinally for venereal disease, topically on sprained joints and bruises and as a general tonic for physical and nervous debilities. The constituents include berberine type alkaloids, palmatine, colombamine, and jateorhizine. Studies show that the berberine sulfate in the plant inhibits lieshmania.

Nauclea latifolia Smith (Rubiaceae)

It is a shrub or small spreading tree that is a widely distributed savanna plant. It is found in the forest and fringe tropical forest. Medicinal uses are as a tonic and fever medicine, chewing stick, toothaches, dental caries, septic mouth and malaria., diarrhea and dysentery (Lamidi 1995).

Key constituents are indole-quinolizidine alkaloids and glycoalkaloids and sapponins. There are studies showing the root has antibacterial activity against gram positive and negative bacteria and antifungal activity (Iwu 1993). It is most effective against Corynebacterium *diphtheriae*, *Streptobacillis* sp., *Streptococcus* sp., *Neisseria* sp., *Pseudomonas aeruginosa*, *Salmonella* sp. (Deeni 1991).

Araliopsis tabouensis (Rutaceae)

It is a large evergreen tree found throughout west tropical Africa. Its medicinal use is for the treatment of sexually transmitted diseases. The bark infusion is drunk for gonorrhea in the Ivory Coast (Irvine 1961). Its major constituents are alkaloids. Seven alkaloids have been isolated from the root and stem bark (Fish 1976).

CONCLUSION

Thomas Jefferson wrote that "The greatest service which can be rendered any country is to add a useful plant to it's culture." Plants have forever been a catalyst for our healing. In order to halt the trend of increased emerging and resistant infectious disease, it will require a multi-pronged approach that includes the development of new drugs. Using plants as the inspiration for new drugs provides an infusion of novel compounds or substances for healing disease. Evaluating plants from the traditional African system of medicine, provides us with clues as to how these plants can be used in the treatment of disease. Many of the plants presented here show very promising activity in the area of antimicrobial agents, warranting further investigation.

REFERENCES

1992 Census of Manufactures. 1994. U.S. Department of Commerce, Economics and Statistics Administration, Bureau of Census.

Aarts, T. 1998. The dietary supplements industry: A market analysis. Dietary Supplements Conference, Nutritional Business International.

Bierer, D., D. Fort, C. Mendez, J. Luo, P. Imbach, L. Dubenko, S. Jolad, R. Gerber, J. Litvak, Q. Lu, P. Zhang, M. Reed, N. Waldeck, R. Bruening, B. Noamesi,, R. Hector,, T. Carlson, and S. King. 1998. Ethnobotanical-directed discovery of the antihyperglycemic properties of cryptolepine: Its isolation from *Cryptolepis sanguinolenta*, synthesis, and in vitro and and in vivo activities. J. Med. Chem. 41:894–901

Boakye-Yiadom, K., N. Fiagbe, S. Ayim. 1977. Antimicrobial properties of some West African medicinal plants IV. Antimicrobial activity of xylopic acid and other constituents of the fruits of *Xylopia aethiopica* (Annonaceae). Lloydia 40:6:543–545.

Boyd, M., Y. Hallock, J. Cardellina II, K. Manfredi, J. Blunt, J. McMahon, R. Buckheit, G. Bringmann, M. Schaffer, G. Cragg, D. Thomas, and J. Jato. 1994. Anti-HIV michellamines from *Ancistrocladus korupensis*. Med. Chem. 37:1740–1745.

Deeni, Y. and H. Hussain. 1991. Screening for antimicrobial activity and for alkaloids of *Nauclea latifolia*. J. Ethnopharmacol. 35:91–96.

Evans, W. 1996. Trease and evans pharmacognosy. W.B Saunders Company Ltd., London.

Fauci, A. 1998. New and reemerging diseases: The importance of biomedical research. Emerging Infectious Diseases. (www.cdc.gov/ncidod/EID/vol4no3/fauci). 4:3.

Fish, F., I. Meshal, and P. Waterman. 1976. Minor alkaloids of *Araliopsis tabouensis*. Planta Med. 29:310–317.

Gruenwald, J. 1997. The herbal remedies market in the US, market development, consumer, legislation and organizations. Phytopharm Consulting, Communiqué.

Irvine, F. 1961. Woody plants of Ghana. Oxford Univ. Press., London.

Iwu, M. 1993. Handbook of African medicinal plants. CRC Press, Boca Raton, FL.

Johnston, B. 1997. One-third of nation's adults use herbal remedies. HerbalGram 40:49.

Lamidi, M., E. Ollivier, R. Faure, L. Debrauwer, L. Nze-Ekekang, and G. Balansard. 1995. Quinovic acid glycosides from *Nauclea diderichii*. Planta Med. 61:280–281.

Murray, M. 1995. The healing power of herbs. Prima Publishing. Rocklin, CA. p. 162–171.

Nelson, R. 1982. The comparative clinical pharmacology and pharmacokinetics of vindesine, vincristine and vinblastine in human patients with cancer. Med. Pediatr. Oncol. 10:115–127.

Nutrition Business Journal (NBJ). 1998. Industry overview. Sept. 1998.

Pinner, R., S. Teutsch, L. Simonsen, L. Klug, J. Graber, M. Clarke, and R. Berkelman. 1996. Trends in infectious diseases mortality in the United States. J. Am. Med. Assoc. 275:189–193.

Robbers, J., M. Speedie, and V. Tyler. 1996. Pharmacognosy and pharmacobiotechnology. Williams and Wilkins, Baltimore. p. 1–14.

Sawer, I., M. Berry, M. Brown, and J. Ford. 1995. The effect of Cryptolepine on the morphology and survival of *Eschericia coli*, *Candida albicans* and *Saccharomyces cerevisiae*. J. Appl. Bacteriol. 79:314–321.

Silva, O., A. Duarte, J. Cabrita, M. Pimentel, A. Diniz, and E. Gomes. 1996. Antimicrobial activity of Guinea-Bissau traditional remedies. J. Ethnopharmacol. 50:55–59.

Smith, G., M. Clegg, C. Keen, and L. Grivetti. 1996. Mineral values of selected plant foods common to southern Burkina Faso and to Niamey, Niger, West Africa. International J. Food Sci. Nutr. 47:41–53.

Trease, G. and Evans, W. 1972. Pharmacognosy, Univ. Press, Aberdeen, Great Britain. p. 161-163

Turano, A., G. Scura, A. Caruso, C. Bonfanti, R. Luzzati, D. Basetti, and N. Manca. 1989. Inhibitory effect of papaverine on HIV replication in vitro. AIDS Res. Hum. Retrovir. 5:183–191.

Antimicrobial and Cytotoxic Activity of the Extracts of Khat Callus Cultures

Hamid Elhag, Jaber S. Mossa, and Mahmoud M. El-Olemy*

Khat, (*Catha edulis* Forssk., Celastraceae) is an evergreen tree indigenous to East Africa and Yemen. The fresh young leaves are commonly chewed, known as the Khat habit, to alleviate hunger and to produce stimulating effects (CNS). Such effects were shown to be due to phenylalkylamine alkaloids, primarily cathinone (Kalix 1990; Crombie et al. 1990). Habitual use of khat is often associated with social and medical problems (Shadan and Shellard 1972).

Previous work on khat cultures, in our laboratories, has dealt with in vitro micropropagation (Elhag 1991) and the production of secondary metabolites by micropropagated plantlets and callus cultures (El-Domiaty et al. 1994). In the course of our work with khat tissue cultures, the production of dark colored pigments was observed as a typical characteristic of the callus culture (Elhag and Mossa 1996). The present investigation deals with the isolation, identification, and biological effects of such pigments.

METHODOLOGY

Callus Culture

Leaves from young twigs of micropropagated greenhouse grown plants (Elhag 1991) were used as explants for callus induction and as fresh material for extraction and chemical analysis. As described previously by El-Domiaty et al. (1994) callus induction from leaf sections was best achieved on MSB5 basal medium (Murashige and Skoog inorganic salts with Gamborg-B5 vitamins) supplemented with 3.0 mg/L of either IBA (indolebutyric acid) or NAA (naphtalene acetic acid). PVP (polyvinylpyrrolidone) at a concentration of 0.1% in the medium was found beneficial in enhancing callus growth. However, PVP was excluded from the culture medium after the establishment of proliferating cultures. Callus tissues collected form several subcultures on IBA or NAA-containing media were used for extraction and analysis of pigments.

Extraction and Isolation

The freeze-dried powdered callus of *C. edulis* (25 g) was extracted with MeOH at room temperature. The MeOH extract was evaporated and the residue (3.9 g) was partitioned between $CHCl_3$ and H_2O. The dark brown residue (0.89 g) left after evaporation of $CHCl_3$ was chromatographed on silica gel column (2 × 20 cm) and eluted with $CHCl_3$ (300 ml) and $CHCl_3$ containing a trace of acetic acid (10:0.01, 200 ml). Two fractions were collected, fraction A (230 mg) and fraction B (110 mg). Fraction A (containing the pigments) was further chromatographed on reversed phase silica gel (C-18, 25–40 μ) using medium pressure column (2 × 18 cm, 75% MeOH in 5% aqueous acetic acid) with a flow rate of 3 ml/min. Upon evaporation, fractions eluted between 25–60 ml gave 32 mg of compound 1 (0.13%), while fractions 120–175 ml gave 120 mg of compound 2 (0.48%).

Antimicrobial Assays

Evaluation of the antimicrobial activity of callus extracts (petroleum ether, $CHCl_3$, MeOH, and aqueous successive extracts) and the isolated compounds 1 and 2 was conducted according to the disk diffusion and agar dilution methods (Mitscher et al. 1972; Jayasuria 1988). Chloramphenicol and streptomycin were used as positive controls. The solvent DMSO was used as the negative control for all the experiments. The bacteria used were obtained from the National Collection of Type Culture (NCTC), Central Public Health Laboratory, London; and the Center for Disease Control, Atlanta, Georgia, US. The minimum inhibitory concentration (MIC) of the two compounds (22β-hydroxytingenone and tingenone) was determined by the two-fold serial dilution assay (Hufford et al. 1975). The MIC was taken as the lowest concentration that inhibited growth

*The authors thank King Abdelaziz City for Science and Technology (KACST), Riyadh, Saudi Arabia, for financial support (Project No. AT-15-39.

after 48h of incubation at 37°C for *B. subtilis*, *S. aureus*, and *S. durans* and after 72 hr for *Mycobacterium* strains.

Cytotoxicity and Anti-HIV

The assays were performed according to the standard procedures of the in vitro primary screen of NCI (Weislow et al. 1989; Grever et al. 1992).

EXPERIMENTAL RESULTS

Extract Identification

Khat callus produced on MSB5 medium containing IBA or NAA continued to proliferate as compact hard clumps with snow-white top surfaces and dark-pigmented lower parts. The dark pigments partly diffused into the medium at the point of contact with the agar medium. This dark pigmentation was thought to result from the high content of polyphenols (tannins) that is known for the intact plant (El-Sissi and Abdalla 1966). However, the initial TLC (Si gel, EtOAc-HCOOH-H_2O, 90:6:5) screening of the EtOH extract of the callus detected only a trace of the polyphenolic precursors of tannins gallocatechin and epigallocatechin. On the other hand, intact khat plants did not show the same pigments detected in the callus (TLC : Si gel, EtOAc-HCOOH-toluene, 9:1:10).

The MeOH extract of the freeze-dried callus was fractionated between H_2O and $CHCl_3$. The colored chloroformic fraction yielded two orange pigments using combined normal and reversed phase chromatography. Compound 1 gave, $[\alpha]_D$ -317.4°; UV spectrum [λ_{max} MeOH; 420, 286. and 246 nm], suggested the presence of a chromophore. IR spectrum showed bands at ν_{max} 3400 (OH), 1710 (carbonyl), and 1580 cm^{-1} (conjugated C = C). The UV spectrum and IR bands, in addition to the positive Liebermann-Burchard test suggested a quinone methide triterpene structure (Gonzalez et al. 1983; Fernando et al. 1988; Likhitwitayawuid et al. 1993). This suggestion was substantiated by the presence of proton signals at δ 6.55 (d, *J* = 2, Hz, H-1), 7.07 (dd, *J* = 7 & 2 Hz, H-6), and 6.36 (d, *J* = 7 Hz, H-7) in the ^{1}H-NMR spectrum ($CDCl_3$) (Likhitwitayawuid et al. 1993). The ^{1}H-^{1}H COSY showed a proton signal at δ 4.56 (d, *J* = 5 Hz, H-22) correlated to the signal at δ 3.66 (which is not directly correlated to any carbon signal in ^{1}H-^{13}C HETCOR), while the proton at δ 4.56 correlated to the signal at δ 76.4 in the ^{13}C NMR spectrum, (C-22). Thus the proton signal at δ 4.56 should be assigned to H-22 and the signal at δ 3.66 to OH proton. Compound 1 was thus identified as 22β-hydroxytingenone; this was confirmed by direct comparison with the reported spectral data (Kutney et al. 1981; Bavovada et al. 1990; Likhitwitayawuid et al. 1993).

Compound 2 gave $[\alpha]_D$ -307.4°; its UV spectrum [λ_{max}) MeOH : 420, 286 and 252 nm], was quite similar to that of compound 1, suggesting the presence of a similar chromophore, IR spectrum of compound 2 showed the same characterstic bands of 1. ^{1}H- and ^{13}C-NMR data of compound 2 were similar to that of 1 except that 2 showed the absence of hydroxyl group at C-22, as it appeared at δ 52.6 in ^{13}C-NMR and one of the H-22 protons appeared at δ 2.92 (H, d, J= 14 Hz) in ^{1}H-NMR. Compound 2 was thus identified as tingenone; this was confirmed by comparison with reported data (Gonzalez et al. 1975; Kutney et al. 1981; Ngassapa et al. 1994).

Antimicrobial Activity

Initial antimicrobial screening of the crude callus extracts was conducted using the disk diffusion method and was confirmed with bioautography (Jayasuria 1988). The highest growth inhibition was found in the petroleum ether and chloroformic successive extracts, the latter being more active. Further fractionation and purification of the chloroformic-soluble fraction of the methanolic extract (as described in Materials and Methods), resulted in the isolation of two active compounds (1) and (2). The isolated compounds were identified as 22β-hydroxytingenone (1) and tingenone (2) by various spectral methods. Both compounds exhibited significant activities against *B. subtilis*, *S. aureus*, and *S. durans* (MIC 0.6 μg/ml), being more potent than the positive control chloramphenicol (Table 1). Both compounds were also found to be more potent against *Mycobacterium* species (MIC was 5.0 μg/ml for both compounds) than the positive controls, streptomycin and

Table 1. MIC values for 22β-hydroxytingenone (compound 1) and tingenone (compound 2) isolated from khat callus cultures.

| Microorganisms | MIC values (μg/ml) | | Streptomycin | Chloramphenicol | Isonicotinic acid hydrazide |
	(1)	(2)			
B. subtilis[z]	0.6	0.6	NT[x]	4.0	NT
S. aureus[z]	0.6	0.6	NT	8.0	NT
St. durans[z]	0.6	0.6	NT	4.0	NT
M. chelonei[y]	5.0	5.0	10.0	NT	10.0
M. smegmatis[y]	5.0	5.0	10.0	NT	10.0
M. intracellulare[y]	5.0	5.0	10.0	NT	10.0
M. xenopi[y]	5.0	5.0	10.0	NT	10.0
E. coli	inactive	inactive	NT	NT	NT
C. albicans	inactive	inactive	NT	NT	NT

[z]MIC values after 48 hr of incubation at 37° C

[y]Incubated for 72 hr at 37° C

[x]NT = not tested.

isonicotinic acid hydrazide (Table 1). However, both compounds; were found to be inactive against the gram-negative bacteria *E Coli* and the fungus *C. albicans* (Table 1).

Cytotoxic Activity

Compounds 1 and 2 have been reported to have cytotoxic activity (Kutney et al. 1981, Bavovada et al. 1990; Ngassapa et al. 1994); however, compound 1 was tested only against a few cancer cell line systems (Bavovada et al. 1990). In the present study, compound 1 was therefore tested using NCI (USA) in vitro primary anticancer and anti-HIV screening. Table 2 shows that 22β-hydroxytingenone (1) exhibited significant cytotoxic activities against leukemia (ED_{50} 0.54 μg/ml) and prostate cancer (ED_{50} 0.85 mg/ml), while the least activity was observed against non-small cell lung cancer (ED_{50} 4.4 μg/ml). While compound 1 showed non-selective broad cytotoxicity against all tested panels (Boyd and Paul 1995), it was inactive when tested against HIV virus. Tingenone 2, was also shown to exhibit strong non-selective broad cytotoxicity against several cancer cell-line systems (Ngassapa et al. 1994).

Table 2. Results of antitumor evaluation of 22β-hydroxy-tingenone in the NCI in vitro primary screen.

Panel	ED_{50} (μg/ml)
Leukemia	0.54
Non-small cell lung cancer	4.4
Colon cancer	1.31
CNS cancer	3.3
Melonama	1.71
Ovarian cancer	2.35
Renal cancer	1.61
Prostate cancer	0.85
Breast cancer	1.11

CONCLUSIONS

The cultural conditions for khat callus induction and growth were established. Best callus induction and growth occurred on MSB5 medium supplemented with 3.0 mg/L of either NAA or IBA. The production of dark pigments was observed at the start of callus induction and continued with subcultures as a typical characteristic of khat callus.

The isolation of 22β-hydroxytingenone and tingenone from khat callus cultures is reported for the first time. They could not be detected in the mother plant grown in the greenhouse. The crude callus extracts and the isolated compounds, 22β-hydroxytingenone (compound 1) and tingenone (compound 2), showed high antibacterial activity against gram positive and myocbacteria and broad cytotoxic activity against several cell-line systems.

Large scale production of khat cultures for the commercial production of such biologically active components is a promising system worthy of further investigation. It would be ironic if khat, which is considered a plant of abuse, turned out to be a miracle plant with efficacious medical properties.

REFERENCES

Bavovada, R., G. Blasko, H.L. Sheih, J.M. Pezzuto, and G.A. Cordell. 1990. Spectral assignment and cytotoxicity of 22-hydroxytingenone from *Glyptopetalum sclerocarpum*. Planta Med. 56:380–382.

Boyd, M.R. and K.D. Paul. 1995. Some practical considerations and application of the national cancer institute in vitro anticancer drug discovery screen. Drug Develop. Res. 34:91–109.

Crombie, L., W.M.L. Crombie, and D.A. Whiting. 1990. Alkaloids of khat (*Catha edulis*). Alkaloids 39:139–164.

El-Domiaty, M.M., H.M. Elhag, F.S. El-Feraly, I.A. al-Meshal, and M.M. El-Olemy. 1994. Studies on (-)-cathinone formation in micropropagated plants and tissue cultures of *Catha edulis* (Khat). Int. J. Pharmacog. 32:135–141.

Elhag, H.M. 1991. In vitro propagation of *Catha edulis*. HortScience 26:212.

Elhag, H.M. and J.S. Mossa. 1996. *Catha edulis*: In vitro culture and the production of cathinone and other secondary metabolites. p. 76–86. In: Y.P.S. Bajaj (ed.), Biotechnology in agriculture and forestry. Springer Verlag, Berlin.

El-Sissi, H.I. and M.F. Abdalla. 1966. Polyphenolics of the leaves of *Catha edulis*. Planta Med. 14:76–79.

Fernando, H.C., A.A. Leise Gunatilaka, V. Kumar, and G. Weeratunga. 1988. Two new quinone-methide from *Cassine balae*: Revised structure of balaenolol. Tetrahedron Lett. 29:387–390.

Gonzalez, A.G., C.G. Francisco, R. Friere, R. Hernandez, J.A. Salazar, and E. Surez. 1975. Inguesterin, a new quinonoid triterpene from *Catha cassinoides*. Phytochemistry 14:1067–1077.

Gonzalez, A.G., B.M. Fraga, C.M. Gonzalez, A.G. Ravelo, and E. Ferro. 1983. X-ray analysis of netzhaulcoyone, a triterpene quinone methide from *Orthosphenia mexicana*. Tetrahendron Lett. 24:3033–3036.

Grever, M.R., A.A. Sheprtz, and B.A. Chabner. 1992. The National Cancer Institute cancer drug discovery and development program. Seminars Oncol. 19:622.

Hufford, C.D., M.J. Funderburk, J.M. Morgan, and L.W. Robertson. 1975. Two antimicrobial alkaloids from heartwood of *L. tulipifera*. J. Pharm. Sci. 4:789–792.

Jayasuria, H. 1988. Biological and chemical investigations of native Mississippi plants. *H. drummondi* (grev. and T and G). Ph.D. Thesis, Univ. of Missippi, Oxford.

Kalix, P. 1990. Pharmacological properties of the stimulant khat. Pharmacol. Ther. 48:397–416.

Kutney, J.P., M.H. Beale, P.J. Salisbury, K.L. Stuart, B.R. Worth, P.M. Townsley, W.T. Chalmers, K. Nilsson, and G.G. Jacoli. 1981. Isolation and characterization of natural products from plant tissue cultures of *Maytenus bachananii*. Phytochemistry 20:653–657.

Likhitwitayawuid, K., R. Bavovada, L-Z. Lin, and G.A. Cordell. 1993. Revised structure of 20-hydroxytingenone and ^{13}C-NMR assignments of 22 B-hydroxytingenone. Phytochemistry 34:759–763.

Mitscher, L.A., R.P. Leu, M.S. Bathala, W.N. Wu, and J.L. Beal. 1972. Antimicrobial agents from higher plants. 1. Introduction, rationale and methodology. Lloydia 35:157–166.

Ngassapa, O., D.D. Soejarto, J.M. Pezzuto, and N.R. Farnsworth. 1994. Quinone-methide triterpenes and salaspermic acid from *Kokoona ochracea*. J. Nat. Prod. 57:1–8.

Shadon, P. and E.J. Shellard. 1972. An anatomical study of Ethiopian khat (leaf of *Catha edulis* Forssk.) J. Pharm. Pharmacol. 14:110–118.

Weislow, O.W., R. Kiser, D. Fine, J. Bader, R.H. Shoemaker, and M.R. Boyd. 1989. New soluble-formazan assay for HIV-1 cytopathic effects: Application to high-flux screening of synthetic and natural products for AIDS-antiviral activity. J. Nat. Cancer Inst. 81:577–586.

Biochemical and Eco-physiological Studies on *Hypericum* spp.

T.B. Kireeva, U.L. Sharanov, and W. Letchamo

St. John's wort (*Hypericum perforatum* L., Hypericaceae) is among the first medicinal plants that had been scientifically investigated for its antibacterial, spasmolytic, antiseptic, and personal care products in the former Soviet Union. In traditional Russian medicine St. John's wort is considered to be the panacea for many health problems. Preparations are taken internally for winter depression, added in the beverage of the traditional Russian soft drink, "Baikal," and externally for wound healing and other skin care formulations. In Russia, most of the commercial supply of *Hypericum* is known to originate from the wild collections, but a small quantity of *H. perforatum* is also cultivated. However, much of the wild collected St. John's wort for commercial or home use usually constitutes both *H. perforatum* and *H. maculatum*. The objective of this study was to determine the chemical composition (hypericins, tannins, and total extractive matter) of *H. perforatum* and *H. maculatum* populations at different developmental stages.

METHODOLOGY

The plant material was collected from 1994–1997 according to the methods indicated for the collection of wild medicinal plants (Anon 1962; Hammerman et al. 1983). Sampling was carried out at four different developmental stages: stage 1 = intensive vegetative growth (before flowering), stage 2 = massive flower bud formation, stage 3 = massive flowering, and stage 4 = massive seed-capsule formation. The samples were collected from seven different populations of each species grown in Udmurtia.

Total hydro-alcohol extractable matter was determined by the method described in Chemical Characterization of Medicinal Plants (1983). For tannins we used Zaprometov (1974), and for hypericins we used the method of Bilinova et al. (1986). All the chemical analysis was replicated five times. Statistical evaluation of the obtained results were determined using dispersion analysis, based on Dospechov (1985).

RESULTS

Hypericins

The highest hypericin content (1.20%) in *H. perforatum* and (1.06%) in *H. maculatum* was measured at the seed formation stage (Table 1). During massive flowering period, hypericin content in both species varied

Table 1. The content of hypericin, tannin, and total extractable matter in *H. perforatum* and *H. maculatum* in various developmental stages of the plants in the Udmurtian region of Russia (1994–1996 growing seasons).

Stage	Hypericin content (%)	Tannin content (%)	Total extractable matter (%)
Hypericum perforatum			
Vegetative	0.57	16.20	29.90
Massive flower buds	1.20	10.30	22.50
Massive flowering	0.72	10.33	27.06
Seed capsule formation	0.29	5.60	16.50
LSD 5%	0.17	1.21	2.16
Hypericum macultum			
Vegetative	0.55	9.47	30.25
Massive flower buds	1.06	10.89	29.90
Massive flowering	0.71	11.06	27.40
Seed capsule formation	0.33	12.72	17.65
LSD 5%	0.11	1.02	0.00

from 0.70% to 0.74% (Table 1). The lowest content (0.29% for *H. perforatum*, and 0.33% for *H. maculatum*) was measured at the fruiting or 4th plant developmental stage (Table 1).

Tannins

Maximum content of tannins (16.20%) in *H. perforatum* was measured during intensive vegetative growth (Table 1). With the onset of flowering, the content of tannin in *H. perforatum* declined. The lowest tannin content (5.60%) for *H. perforatum* was measured during seed capsules formation (Table 1). Unlike *H. perforatum*, the tannin content in *H. maculatum* increased with further development of the plants from the vegetative phase (9.47%) to flowering (12.72%) (Table 1).

Total Extractable Matter

The highest content of the total extractable matter (29.90% for *H. perforatum*, and 30.25% for *H. maculatum*) was found at the 1st stage (intensive vegetative development) with the lowest value of 16.50% for *H. perforatum* and 17.65% for *H. maculatum* at seed capsule formation (Table 1). In general, the mean content of the total extractable matter in *H. maculatum* was slightly higher (28.28%) than *H. perforatum* (24.92%).

Though both *H. perforatum*, and *H. maculatum* start their growth during spring at similar time, *H. maculatum* started flower bud formation, flowering, and seed ripening 7–10 days earlier than *H. perforatum* (data not shown). From our investigation, it appears that both *H. perforatum* and *H. maculatum* grown in the Udmurtian regions of Russia have similar biochemical traits. Therefore, we suggest that *H. maculatum* should be included in the State Pharmacopoeia Books, as a source of an herbal medicine.

REFERENCES
Anon. 1962. Atlas of medicinal plants of the USSR. (in Russian) State publication of the medicinal sciences. Moscow.

Blinova K.F., P.N. Shatokina, and I.I. Priakina. 1986. Methods of quantitative determination of total flavonoids in medicinal plants. Leningrad, Medicinal Institute p. 27–30.

Dospechov B.A. 1979. Methods of field experiments. (in Russian) Higher Education. Moscow.

Hammermann A.F., G.N. Kadaev, and A.A. Yasenko-Kimeleviski. 1983. Medicinal plants. (in Russian) Higher Education. Moscow.

Zaprometov M.N. 1974. Basics in the biochemistry of phenolic compounds. (in Russian) Higher Education. 230. Moscow.

Ageratum conyzoides: A Tropical Source of Medicinal and Agricultural Products

Lin Chau Ming

Ageratum conyzoides L., Asteraceae, is an annual herbaceous plant with a long history of traditional medicinal uses in several countries of the world and also has bioactivity with insecticidal and nematocidal acitivity. This tropical species appears to be a valuable agricultural resource.

BOTANY

Ageratum is derived from the Greek "*a geras*," meaning non-aging, referring to the longevity of the flowers or the whole plant. The specific epithet "*conyzoides*" is derived from "*kónyz*," the Greek name of *Inula helenium,* which it resembles (Kissmann and Groth 1993).

The synonyms of *A. conyzoides* include *A. album* Stend; *A. caeruleum* Hort. ex. Poir.; *A. coeruleum* Desf.; *A. cordifolium* Roxb.; *A. hirsutum* Lam.; *A. humile* Salisb.; *A. latifolium* Car.; *A. maritimum* H.B.K.; *A. mexicanum* Sims.; *A. obtusifolium* Lam.; *A. odoratum* Vilm. and *Cacalia mentrasto* Vell. (Jaccoud 1961). In Brazil, *A. conyzoides* has the following vernacular names: *catinga de bode, catinga de barrão, erva de são joão, maria preta, mentrasto, erva de são josé, picão roxo, erva de santa-lúcia, camará-opela, agerato, camará apeba, camará iapó, camará japê, erva de santa maria, macela de são joão, macela francesa, matruço* (Jaccoud 1961; Oliveira et al. 1993).

Ageratum ranges from Southeastern North America to Central America, but the center of origin is in Central America and the Caribbean. Most taxa are found in Mexico, Central America, the Caribbean, and Florida. *Ageratum conyzoides* now is found in several countries in tropical and sub-tropical regions, including Brazil (Baker 1965; Lorenzi 1982; Correa 1984; Cruz 1985).

Johnson (1971), classifies two subspecies, *latifolium* and *conyzoides*. Subspecies *latifolium* is found in all the Americas and subsp. *conyzoides* has a pantropical distribution. The basic chromosome number is $2n = 20$ but natural tetraploids are found. *A. conyzoides* subsp. *latifolium* is diploid and *A. conyzoides* subsp. *conyzoides* is tetraploid.

Ageratum conyzoides is an erect, herbaceous annual, 30 to 80 cm tall; stems are covered with fine white hairs, leaves are opposite, pubescent with long petioles and include glandular trichomes. The inflorescence contain 30 to 50 pink flowers arranged as a corymb and are self-incompatible (Jhansi and Ramanujam 1987; Kaul and Neelangini 1989; Ramanujam and Kalpana 1992; Kleinschimidt 1993). The fruit is an achene with an aristate pappus and is easily dispersed by wind. In some countries the species is considered a weed, and control is often difficult (Lorenzi 1982; Scheffer 1990; Kalia and Singh 1993; Lam et al. 1993, Paradkar et al. 1993; Waterhouse 1993; Kshatriya et al. 1994). Seeds are positively photoblastic, and viability is often lost within 12 months (Marlks and Nwachuku 1986; Ladeira et al. 1987). The optimum germination temperature ranges from 20 to 25°C (Sauerborn and Koch 1988). The species has great morphological variation, and appears highly adaptable to different ecological conditions.

PHYTOCHEMICAL CHARACTERISTICS

There is high variability in the secondary metabolites of *A. conzyoide* which include flavonoids, alkaloids, cumarins, essential oils, and tannins. Many of these are biologically active. Essential oil yield varies from 0.02% to 0.16% (Jaccoud 1961). Vyas and Mulchandani (1984) identified conyzorigum, a cromene. Borthakur and Baruah (1986) identified precocene I and precocene II, in a plant collected in India. These compounds have been shown to affect insect development, as antijuvenile hormones, resulting in sterile adults (Borthakur and Baruah 1987). Ekundayo et al. (1988) identified 51 terpenoid compounds, including precocene I and precocene II. Gonzales et al. (1991) found 11 cromenes in essential oils, including a new cromene, 6-angeloyloky-7-methoxy-2,2-dimethylcromen. Vera (1993), in Reunion, found ageratocromene, other cromenes, and beta cariophylene in its essential oil. Mensah et al. (1993) and Menut et al. (1993) reported similar yields of precocene I in the essential oil of plants collected in Ghana.

Vyas and Mulchandani (1986), in India, identified flavones, including some considered new such as *ageconyfavones* A, B, and C. Horrie et al. (1993) reported hexametoxyflavone. Ladeira et al. (1987) in Brazil, reported three cumarinic compounds, including 1-2 benzopirone. The species contains alkaloids, mainly the pirrolizidinic group, which suggest that it may be a good candidate for pharmacological studies. Trigo et al. (1988) found several alkaloids, including 1,2- desifropirrolizidinic and licopsamine which can have hepatotoxic activity. Alkaloids also were found by Weindenfeld and Roder (1991) in a hexane extract of *A. conyzoides* in Africa.

FOLK MEDICINAL USES AND PHARMACOLOGICAL STUDIES

A. conyzoides is widely utilized in traditional medicine by various cultures worldwide, although applications vary by region. In Central Africa it is used to treat pneumonia, but the most common use is to cure wounds and burns (Durodola 1977). Traditional communities in India use this species as a bacteriocide, antidysenteric, and antilithic (Borthakur and Baruah 1987), and in Asia, South America, and Africa, aqueous extract of this plant is used as a bacteriocide (Almagboul 1985; Ekundayo et al. 1988). In Cameroon and Congo, traditional use is to treat fever, rheumatism, headache, and colic (Menut et al. 1993; Bioka et al. 1993). In Reunion, the whole plant is used as an antidysenteric (Vera 1993). The use of this species in traditional medicine is extensive in Brazil. Aqueous extracts of leaves or whole plants have been used to treat colic, colds and fevers, diarrhea, rheumatism, spasms, or as a tonic (Penna 1921; Jaccoud 1961; Correa 1984; Cruz 1985; Marques et al. 1988; Negrelle et al. 1988; Oliveira et al. 1993). *A. conyzoides* has quick and effective action in burn wounds and is recommended by Brazilian Drugs Central as an antirheumatic (Brasil 1989).

Several pharmacological investigations have been conducted to determine efficacy. Duradola (1977) verified inhibitory activities of ether and chloroform extracts against in vitro development of *Staphylococus aureus*. Almagboul et al. (1985), using methanolic extract of the whole plant, verified inhibitory action in the development of *Staphylococus aureus*, *Bacillus subtilis*, *Eschericichia coli,* and *Pseudomonas aeruginosa*. Bioka et al. (1993) reported effective analgesic action in rats using aqueous extract of *A. conyzoides* leaves (100 to 400 mg/kg). Assays realized in Kenia, with aqueous extract of the whole plant, demonstrated muscle relaxing activities, confirming its popular use as an antispasmotic (Achola et al. 1994).

In Brazil, assays conducted by State University of Campinas and Paraiba Federal University) showed promising results. Marques Neto et al. (1988) in clinic trials with patients with arthrosis, administered aqueous extract of the whole plant, and reported analgesic effect in 66% of patients and improvement in articulation mobility in 24%, without side effect. Mattos (1988), using aqueous extract of the whole plant, verified effective clinical control of arthrosis, reporting a decrease in pain and inflammation or improvement in articulation mobility, after a week of treatment.

BIOACTIVITY

Ageratum conyzoides has bioactive activity that may have agricultural use, as shown by several research investigations in different countries. Pereira in 1929, cited by Jaccoud (1961), reported use of the leaves as an insect (moth) repellent. The insecticide activity may be the most important biological activity of this species. The terpenic compounds, mainly precocenes, with their antijuvenile hormonal activity are probably responsible for the insecticide effects.

Assays conducted in Colombia by Gonzalez et al. (1991) showed activity of this species against *Musca domestica* larvae, using whole plant hexane extract. Vyas and Mulchandani (1980) reported the action of cromenes (precocenes I and II), isolated from *Ageratum* plants, which accelerate larval metamorphosis, resulted in juvenile forms or weak and small adults.

Ekundayo et al. (1987) also demonstrated the juvenilizing hormonal action of precocene I and II in insects, the most common effect being precocious metamorphosis, producing sterile or dying adults. Raja et al (1987), using *A. conyzoides* methanolic extract from fresh leaves (250 and 500 ppm) in the fourth instar of *Chilo partellus* (Lepidoptera, Pyralidae), a sorghum pest, observed the presence of a dark stain in the insects' cuticle and immature pupae formation, both symptoms of deficiency of juvenile hormone.

A. conyzoides also induces morphogenetic abnormalities in the formation of mosquitoes larvae (*Culex quinquefasciatus, Aedes aegypt,* and *Anopheles stephensi*). This has been verified using petroleum ether extracts (5 and 10 mg/L) of the whole plants. The larvae showed intermediary stages between larvae–pupae, discolored and longer pupae, as well as incompletely developed adults (Sujatha et al. 1988). Extracts of the flowers of this species showed activity against mosquitoes (*Anopheles stephensi*), in the last instar, showing DL 50 with 138 ppm (Kamal and Mehra 1991).

Cetonic extracts of the species produced significant effects against the mosquito, *Culex quinquefasciatus*, in India, when applied to fourth instar larvae and adult females. In larvae, the extracts produced altered individuals, intermediate between larvae and pupae, unmelaninized and with inhibition of development, as well as adults with deformed wings muscles. In female adults, there was loss of fecundity, lower eggs production, and production of defective eggs (Saxena et al. 1992). Similar results were observed in larvae *of Anopheles stephensi* and *Culex quinquefasciatus* in others essays, confirming the antijuvenile potential of *A. conyzoides* (Saxena and Saxena 1992; Saxena et al. 1994).

The species also has potential use in controlling other pests. Shabana et al. (1990), using aqueous extract of the whole plant, verified reduction of larvae emergence of *Meloidogyne incognita*. Pu et al. (1990) and Liang et al. (1994), verified that plants of *A. conyzoides* in *Citrus* orchards sheltered predators of the spider *Panonychus citri*, suggesting that its development in orchards is beneficial. Other *Citrus* spiders populations, *Phyllocoptruta oleivora* and *Brevipalpus phoenicis* were decreased with maintenance of *A. conyzoides* in the orchards and a reduction of leprosy virus was noted (Gravena et al. 1993)

The presence of *A. conyzoides* can also be used as an seed inhibitor, decreasing development of several herbaceous plants. Jha and Dhakal (1990) in Nepal, reported that an aqueous extract of the aerial part or roots of this species (15 g of aerial part or 3 g of roots in 100 ml of water, during 24 h) inhibited germination of wheat and rice seeds while Prasad and Srivastava (1991) in India, reported a lower germination index in peanut seeds with aqueous extract.

CULTURAL STUDIES

Magalhaes et al. (1989) in Brazil evaluated fertilizer studies and plant density on biomass production of *A. conyzoides*. The higher the N level, the higher the biomass production (dry weight basis). Optimum spacing was 70 cm between rows and 50 cm between plants. Biomass yields was 1.3 t (dry weight)/ha.

Correa Jr. et al. (1991) obtained biomass yields of 3.3 to 5.3 t (fresh wt)/ha. Essential oil content was 0.02% (fresh wt) and 0.16 % (dry wt) in the preflowering state. Preliminary data of Ming (1998) indicated that essential oils, higher in leaves than in flowers, peaked during early-flowering.

FUTURE POTENTIAL

There are some small pharmaceutical companies in Brazil using *A. conyzoides* as a raw material for phytochemicals. The demand is increasing year by year and this situation warrants further scientific research to develop both agricultural and medical uses. Research on medicinal plants should be focused primarily on species whose pharmaceutical activities have already been demonstrated. Positive preliminary clinical assays of *A. conyzoides* clearly demonstrate that this species may be an important economic resource in several tropical countries. The use of this species as a natural biocide or agent for pest management particularly requires further investigation.

REFERENCES

Achola, K.J., R.W. Munenge, and A.M. Mwaura. 1994. Pharmacological properties of root and aerial parts extracts of *Ageratum conyzoides* on isolated ileum and heart. Fitoterapia 65:322–325.

Almagboul, A.Z., A.A. Farroq, and B.R. Tyagi. 1985. Antimicrobial activity of certain sudanese plants used in folkloric medicine: Screening for antibacterial activity, part II. Fitoterapia 56:103–109.

Baker, H.G. 1965. Characteristics and modes of origin of weeds. Academic Press, New York.

Bioka, D., F.F. Banyikwa, and M.A. Choudhuri. 1993. Analgesic effects of a crude extract of *Ageratum conyzoides* in the rat. Acta Hort. 332:171–176.

Borthakur, N. and A.K.S. Baruah. 1987. Search for precocenes in *Ageratum conyzoides* Linn. of North-East India. J. Indian Chem. Soc. 64:580–581.

Brasil, Ministério da Saúde, Central de Medicamentos. 1989. *Ageratum conyzoides*. In: Programa de pesquisas de plantas medicinais: Primeiros resultados. Brasília.

Correa Jr, C., L.C. Ming, and M.C. Scheffer. 1991. Cultivo de plantas medicinais, condimentares e aromáticas. Emater-PR, Curitiba.

Correa, M.P. 1984. Dicionario das plantas úteis do Brasil e das exóticas cultivadas. Ministério da Agricultura Rio de Janeiro, IBDF 2:139.

Cruz, G.L. 1985. Dicionário das plantas úteis do Brasil, 3 ed. Civilização Brasileira, Rio de Janeiro.

Durodola, J.J. 1977. Antibacterial property of crude extracts from herbal wound healing remedy—*Ageratum conyzoides*. Planta Med. 32:388–390.

Ekundayo, O., S. Sharma, and E.V. Rao. 1988. Essential oil of *Ageratum conyzoides*. Planta Med. 54:55–57.

Gonzales, A.G., G. Thomas, and P. Ram. 1991. Chromenes form *Ageratum conyzoides*. Phytochemistry 30:1137–1139.

Gravena, S., A. Coletti, and P.T. Yamamoto. 1993. Influence of green cover with *Ageratum conyzoides* and *Eupatorium pauciflorum* on predatory and phytophagous mites in citrus. Bul. OILB-SROP. 16(7):104–114.

Horie, T., H. Tominaga, and Y. Kawamura. 1993. Revised structure of a natural flavone from *Ageratum conyzoides*. Phytochemistry 32:1076–1077.

Jaccoud, R.J.S. 1961. Contribuição para o estudo formacognóstico do *Ageratum conyzoides* L. Rev. Bras. Farm. 42(11/12):177–97.

Jha, S. and M. Dhakal. 1990. Allelopathic effects of various extracts of some herbs on rice and wheat. J. Inst. Agr. Anim. Sci. 11:121–123.

Jhansi, P. and C.G.K. Ramanujam. 1987. Pollen analysis of extracted and squeezed honey of Hyderabad, India. Geophytology 17:237–240.

Johnson, M.F. 1971. A monograph of the genus *Ageratum* L. (Compositae, Eupatorieae). Ann. Missouri Bot. Gard. 58:6–88.

Kalia, B.D. and C.M. Singh. 1993. Studies on weed management in maize. p. 89–90. Vol. 3. In: Proc. Indian Society Weed Science Int. Symp, Hisar.

Kamal, R. and P. Mehra. 1991. Efficacy of pyrethrins extracted from *Dysodia tennifolius* and *Ageratum conyzoides* against larvae of *Anopheles stephensi*. Pyrethrum Post. 18(2):70–73.

Kaul, M.L.H. and S. Neelangini. 1989. Male sterility in diploid *Ageratum conyzoides* L. Cytologia 54: 445–448.

Kissmann, G. and D. Groth. 1993. Plantas infestantes e nocivas. Basf Brasileira, São Paulo.

Kleinschmidt, G. 1993. Colony nutrition on the Atherton tableland. Australasian Beekeeper 94:453–464.

Kshattryya, S., G.D. Sharma, and R.R. Mishra. 1994. Fungal succession and microbes on leaf litters in two degraded tropical forests of worth-east India. Pedobiologia 38(2):125–137.

Ladeira, A.M, L.B.P. Zaidan, and R.C.L. Figueiredo-Ribeiro. 1987. *Ageratum conyzoides* L. (Compositae): Germinação, floração e ocorrência de derivados fenólicos em diferentes estádios de desenvolvimento. Hoehnea 15:53–62.

Lam, C.H., J.K. Lim, and B. Jantan. 1993. Comparative studies of a paraquat mixture and glyphosate and/or its mixtures on weed succession in plantation crops. Planter 69:525–535.

Liang, W.G., W. Hui, and W.K. Lee. 1994. Influence of citrus orchard ground cover plants on arthropod communities in China: A review. Agr. Ecosyst. Environm. 50(1):29–37.

Lorenzi, H. 1982. Plantas daninhas do Brasil. H. Lorenzi, Nova Odessa.

Magalhaes, P.M., I. Montanari, and G.M. Ferreira. 1989. Large scale cultivation of *Ageratum conyzoides* L. Unicamp-Cpqba, Campinas.

Marks, M.K. and A.C. Nwachuku. 1986. Seed-bank characteristics in a group of tropical weed. Weed Res. 26(3):151–157.

Marques-Neto, J.F., A. Lapa, and M. Kubota. 1988. Efeitos do *Ageratum conyzoides* Lineé no tratamento da artrose. Rev. Bras. Reumat. 28(4):34–37.

Matos, F.J.A. 1988. Plantas medicinais: Boldo, colônia e mentrasto. O povo, Univ. Aberta, Fortaleza, 27. Jan. 1988, p. 2–3.

Mensah, M., E.V. Rao, and S.P. Singh. 1993. The essential oil of *Ageratum conyzoides* L. from Ghana. J. Essent. Oil Res. 5(1):113–115.

Menut, C., S. Sharma, and C. Luthra. 1993. Aromatic plants of tropical central Africa, Part X—Chemical composition of essential oils of *Ageratum houstonianum* Mill. and *Ageratum conyzoides* L. from Cameroon. Flavour Fragrance J. 8(1):1–4.

Ming, L.C. 1996. Produção de biomassa e teor de óleo essencial em função de fases de desenvolvimento, calagem e adubações mineral e orgânica em *Ageratum conyzoides* L. PhD Thesis, UNESP, Jaboticabal.

Negrelle, R.R.B., D. Sbalchiero, and A.C. Cervi. 1988. Espécies vegetais utilizadas na terapêutica popular no município de Curitiba, Paraná, Brasil. Univ. Federal do Paraná.

Oliveira, F., M.K. Akisue, and L.O. Garcia. 1993. Caracterização farmacognóstica da droga e do extrato fluído de mentrasto, *Ageratum conyzoides* L. Lecta 11(1):63–100.

Paradkar, V.K., R.K. Saraf, and J.P. Tiwari. 1993. Weed flora of winter vegetable of Satpura Plateau, region of Madhya Pradesh. Proc. Indian Soc. Weed Science Int. Symp., Hisa, Vol. 2.

Penna, A. 1921. Notas sobre plantas brasileiras. Araújo Penna Filhos, Rio de Janeiro.

Prasad, K. and V.C. Srivastava. 1991. Teletoxic effect of some weeds on germination and initial growth of groundnut (*Arachis hypogea*). Ind. J. Agr. Sci. 61:493–494.

Pu, T.S., K.Y. Liao, and T. Chang. 1990. Investigations on predations mite resources in Citrus orchards in Guang Xi and their utilization. Acta Phytophyarica Sin. 17:355–358.

Raja, S.S., A. Singh, and S. Rao. 1987. Effect of *Ageratum conyzoides* on *Chilo partelus* Swinhoe (Lepidoptera: Pyralidae). J. Anim. Morphol. Physiol. 34(1–2):35–37.

Ramanujam, C.G.K. and T.P. Kalpana. 1992. *Tamarindus indica* L. an important forage plant for *Apis florea* F. in south central India. Apidologie 23:403–413.

Sauerborn, J. and W. Kock. 1988. Untersuchungen zur Keimungsbiologie von sechs tropischen Segetalaten. Weed Res. 28(1):47–52.

Saxena, A. and R.C. Saxena. 1992. Effects of *Ageratum conyzoides* extracts on the developmental stages of malaria vector, *Anopheles stephensi* (Diptera: Culicidae). J. Environm. Biol. 13:207–209.

Saxena, R.C., A. Saxena, and C. Singh. 1994. Evaluation of growth disrupting activity of *Ageratum conyzoides* crude extract on *Culex quinquefasciatus* (Diptera: Culicidae). J. Environm. Biol. 15(1):67–74.

Saxena, R.C., O.P. Dixit, and P. Sukumaran. 1992. Laboratory assessment of indigenous plant extracts for anti-juvenile hormone activity in *Culex quinquefasciatu*. Indian J. Med. Res. 95:204–206.

Scheffer, M.C. 1990. Recomendações técnicas para o cultivo das plantas medicinais selecioandas pelo Projeto de Fitoterapia do SUDS-Paraná. 2nd ed. Curitiba, SESA, FCMR-GPC-PFS.

Shabana, N., S.I. Husain, and S. Nisar. 1990. Allelopathic effects of some plants on the larval emergence of *Meloidogyne incognita*. J. Indian Appl. Pure Biol. 5:129–130.

Sujatha, C.H., S. Nisar, and C. Jadhi. 1988. Evaluation of plant extracts for biological activity against mosquitoes. Int. Pest. Control. 7:122–124.

Trigo, J.R., S. Campos, and A.M. Pereira. 1988. Presença de alcalóides pirrolizidinicos em *Ageratum conyzoides* L. p. 13. In: Simposio de Plantas Medicinais do Brasil, Sao Paulo. (Resumos).

Vera, R. 1993. Chemical composition of the essential oil of *Ageratum conyzoides* L. (Asteraceae) from Reunion. Flavour Fragrance J. 8:256–260.

Vyas, A.V. and N.B. Mulchadani. 1986. Polyoxigenated flavones from *Ageratum conyzoides*. Phytochemistry 25:2625–2627.

Vyas, A.V. and N.Br. Mulchandani. 1984. Structure reinvestigation of conyzorigun, a new chromone from *Ageratum conyzoides*. J. Chem. Soc. Perkin. Trans. 1:2945–2947.

Waterhouse, D.F. 1993. Prospects for biological control of paddy weeds in southeast Asia and some recent success in the biological control of aquatic weed. Camberra. Ext. Bul. 366, Australia.

Wiedenfeld, H. and E. Roder. 1991. Pyrrozidine alkaloids form *Ageratum conyzoides*. Planta Med. 57:578–579.

Identification of the Key Aroma Compounds in Dried Fruits of *Xylopia aethiopica*

A.O. Tairu, T. Hofmann, and P. Schieberle*

West African "Peppertree" [*Xylopia aethiopica* (Dunal) A. Rich, Annonaceae] is a slim, tall tree of about 60–70 cm in diameter and up to 15–30 m high with straight stem and a slightly stripped or smooth bark. It is widely distributed in the humid forest zones of West Africa especially along rivers in the dry country sides (Irvine 1961). *Xylopia aethiopica* has a wide variety of application, the very odorous roots of the plant are employed in West Africa in tinctures, administered orally to expel worms and other parasitic animals from the intestines, or in teeth-rinsing and mouth-wash extracts against toothaches. The fruits are also used in various forms and exhibit revulsive properties, especially when mashed with grains. These properties are used advantageously in the external treatment of rheumatism. Crushed powdered fruits can also be mixed with shea butter fat and coconut oil and used as creams, cosmetic products, and perfumes (Burkill 1985), and the dried fruits are also used as spices in the preparation of two special local soups named "obe ata" and "isi-ewu" taken widely in the southwest and southeastern parts of Nigeria.

Ekong and Ogan (1968), were the first to report on the chemical composition of *Xylopia aethiopica*, and several publications have appeared subsequently on this subject. A number of diterpenes from the bark, fruits, and pericarp of the plant have been reported, Faulkner et al. (1985); Rabunmi and Pieeru (1992); Harrigan et al. (1994). Ekundayo (1989) published a review of the volatiles in a number of Annonaceae species among which includes *Xylopia aethiopica* and, reported that they consist mainly of mono and sesquiterpenoids with typical constituents being α- and β-pinene, myrcene, p-cymene, limonene, linalool, and 1,8-cineole. Recently, two new sesquiterpenes, elemol and guaiol (among other terpenes) were found in the essential oil of the fruit from the Republic of Benin (Ayedoun et al. 1996) while Jirovetz et al. (1997) gave a semblance of the aroma note from the essential oil in the fruit of *Xylopia aethiopica* from Cameroon.

No attempt had been undertaken to rank the volatiles in their flavor contribution. In order to detect these compounds, volatiles isolated from the crushed dried fruits of *Xylopia aethiopica* from Nigeria have been analyzed by High-Resolution Gas Chromatography (HRGC) and eluate sniffing.

EXPERIMENTAL PROCEDURES

Chemicals

The reference compounds of the odorants listed in the tables were obtained from the various suppliers given in parentheses: no. 2, 6, 8, 11, 16, 17, 22, and 24, (Aldrich, Steinheim, Germany); no. 7, 10, 19, 20, and 23 (Merck, Darmstadt, Germany); no. 9, 26, and 27 (Lancaster, Mühlheim, Germany); no. 4, 5, 18, and 25 (Fulka, Neu-Ulm, Germany); no. 1, 14, and 15 (Alfa Products, Karlsruhe, Germany). α-Farnesene was a gift.

Isolation of the Volatile Oil

The smoked, dried fruits of *Xylopia aethiopica* (6 g) were immediately frozen in liquid nitrogen and finely powdered by means of a commercial blendor (Janke & Kunkel, Stanfen). The powder was extracted with solvent mixture of methanol, water, and dichloromethane. The extract (organic phase) was dried over sodium sulfate and concentrated on a Vigreux column (50 cm × 1 cm internal diameter) and the volatiles isolated by sublimation in vacuo using the equipment described by Guth and Grosch (1989).

Separation of Volatiles into Acidic and Neutral/Basic Fractions

By treatment of the distillate with aqueous sodium bicarbonate (Hofmann and Schieberle 1995), a fraction of the acidic volatiles (fraction AV) and of the neutral/basic volatiles (fraction N/B) were obtained. After

---

* This work was made possible through the previlege given to one of the authors (A.O. Tairu) by the Institute For Food Chemistry, Technical University of Munich; the University of Agriculture, Abeokuta, Ogun State, Nigeria, and the generous support of the Alexander von Humboldt Foundation, Bonn, Germany.

drying over anhydrous sodium sulfate, both fractions were concentrated to 200 ml (Schieberle 1991) and the odor-active compounds evaluated by aroma extract dilution analysis (AEDA).

Capillary Gas Chromatography (HRGC)-Mass Spectrometry (MS)

HRGC was performed with a Carlo Erba gas chromatograph, (Type 5160, MEGA SERIES) using fused silica thin-film capillaries: capillary FFAP, capillary SE54 (DB-5) (each 30 m × 0.32 mm; J and W. Scientific, Fisons, Mainz, Germany; film thickness, 0.25 µm). Samples were applied by the "cold on-column" injection technique at 35°C. After 2 min, the temperature of the oven was raised by 40°C/min to 60°C, held 2 min isothermally, then raised by 6°C to 180°C and, finally, by 10°C/min to 230°C. The retention indices were calculated by using n-alkanes as the reference (Schieberle 1991). Mass spectrometry was performed on an MD800 (Fisons Instruments, Mainz, Germany) in tandem with the capillaries described above. Mass spectra in the electron impact mode (MS/EI) were generated at 70 eV and in the chemical ionization mode (MS/CI) at 110 eV with isobutane as the reagent gas.

Aroma Extract Dilution Analysis (AEDA)

The flavor dilution (FD) factors of the odorants in the fractions AV and N/B of *Xylopia aethiopica* were determined by AEDA (Schieberle 1995). An aliquot of the respective distillate (0.5 µL of 200 µL) was separated on capillary FFAP, the efluent ratio was split to an FID and a sniffing port (1 + 1 by vol.), and the odor-active regions and the odor qualities were assigned by three assessors (GC/O). The extract was stepwise diluted with diethyl ether (1 + 1 by vol.), and aliquots of the dilutions were evaluated by two assessors. The FD chromatogram (FD versus retention indices) was plotted.

RESULTS

From the FD chromatogram obtained by applying the AEDA on an extract containing the neutral/basic volatiles of *Xylopia aethiopica*, 24 odor-active compounds were sensorily detected in the flavor range of 4-8196. Among these compounds, 6 odorants showed very high FD factors of greater than or equal to 512 (nos. 2, 6, 12, 16, 17, and 21; Fig. 1). All of the key odorants could be identified on the basis of the criteria given in footnote b of Table 1.

In the fraction of the acidic volatiles, four odorants were sensorially detected (Table 2) with all of them fully identified. Among them, 3 odorants 3-ethylphenol, 4-ethy-2-methoxyphenol and hexanoic acid have not been reported as aroma compounds found in *Xylopia aethiopica*. Vanilline and 3-ethylphenol gave a reasonably high FD-factors of 128 and are considered as being important odorant in the fruit.

All the odor-active compounds detected were terpenes eliciting the characteristic flowery and terpeny notes except 3-ethylphenol, (phenollic); vanilline, (vanilla-like); hexanoic acid, (acidic); and 4-ethyl-2-methoxyphenol, (smoky). Linalool (no. 17; Table 1), with pepperish-flowery odor showed the highest FD factor while trans-β-ocimene (flowery note) and α-farnesene (sweet-flowery note) had the same FD factor

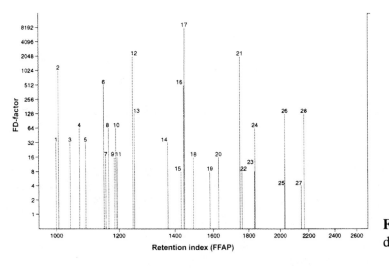

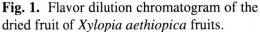

Fig. 1. Flavor dilution chromatogram of the dried fruit of *Xylopia aethiopica* fruits.

Table 1. Most odor-active compounds in the neutral/basic volatiles (FD ³ 4) in *Xylopia aethiopica*.

No.[z]	Aroma quality[y]	Retention index[x]		FD factor	Aroma compound[w]
		FFAP	SE-54		
1	Sweet, terpeny	996	919	32	α-Thujene
2	Terpeny	1007	931	1024	α-Pinene
3	Fruity	1045	855	32	Ethyl-2-methylbutanoate
4	Spicy, fruity	1075	949	64	Camphene
5	Terpeny	1095	975	32	Sabinene
6	Terpeny	1152	981	512	β-Pinene
7	Metallic	1158	988	16	Myrcene
8	Light minty	1167	1005	64	α-Phellandrene
9	Terpeny	1185	1025	16	α-Terpinene
10	Light, lemon-like	1190	1029	64	Limonene
11	Light, pepperment-like	1194	1031	16	1,8-Cineole
12	Flowery	1250	1172	2048	trans-β-Ocimene
13	Terpeny	1256	1033	128	β-Phellandrene
14	Fruity, terpeny	1375	1112	32	p-Mentha-3,8-triene
15	Terpeny	1432	1220	8	3-Carene
16	Flowery	1446	1206	512	Myrtenol
17	Flowery, pepperish	1450	1109	8192	Linalool
18	Flowery, fatty	1497	1207	16	Decanal
19	Flowery, lilac-like	1582	1195	8	α-Terpeniol
20	Light mint, terpeny	1630	1184	16	Terpinen-4-ol
21	Sweet, flowery	1750	--	2048	α-Farnesene
22	Flowery	1762	1231	8	b-Citronellol
24	Flowery, rose-like	1840	1256	64	Geraniol
27	Sweet, camphoracious	2148	--	4	Fenchone

[z]Numbers refer to Fig. 1.
[y]Odor quality perceived at the sniffing port.
[x]Retention index.
[w]Odorants were identified by comparing them with reference compounds on the basis of the following criteria: retention index (RI) on two stationary phases detailed in the table, mass spectra MS(CI), odor quality, and odor threshold (ratio of FID signal to FD factor) at the sniffing port.

(nos. 12 and 21; Table 1). α-Pinene having a typical terpeny odor note had an FD factor twice as much as β-pinene, another compound with terpeny odor note, myrtenol with flowery note which slightly resembles that of linalool (but lacks the pepperish note of linalool) had the same FD factor with β-pinene (no. 6; Table 1). Both β-phellandrene (terpeny note) and 3-ethylphenol (phenollic note) had the same FD factor of 128, the phenollic odor of 3-ethylphenol and the vanilla-like of vanilline are very easily noticed during the sniffing experiments even though these compounds appeared with very low chromatogram peaks. Camphene with spicy note had the same FD factor with limonene (bearing light citrus-like odor note) and α-phellandrene (terpeny note). Compounds with relatively low FD factors include α-thujene (terpeny note), sabinene (terpeny note), myrcene (metallic note), α-terpenene (terpeny note), 1,8-cineole (light pepperment-like note), p-mentha-3,8-triene (fruity, terpeny note), terpinen-4-ol (light minty, terpeny note). The results of the identification experiments are as shown in Table 1.

Table 2. Odor-active compounds in acidic fraction of dried *Xylopia aethiopica* fruits (FD ³ 4).

No.	Aroma quality[z]	Retention index[y]		FD factor	Aroma compound[x]
		FFAP	SE-54		
23	Acidic	1837	1019	8	Hexanoic acid
25	Smoky	2031	1285	4	4-Ethyl-2-methoxyphenol
26	Vanilla-like	2033	1402	128	Vanilline
28	Phenollic	2168	1167	128	3-Ethylphenol

[z]Odor quality perceived at the sniffing port.
[y]Retention index
[x]Cf footnote w; Table 1.

DISCUSSION

The simple chromatogram of the volatile oil of *Xylopia aethiopica* gave an indication of the presence of numerous terpenoids compounds in it, however upon application of aroma extract dilution analysis on the sample it became clear that several of these compounds have little or no contribution to the overall odor quality of the fruit volatiles. We thus paid our attention on those odorants with FD factor of 4 and above, as this range was considered to be significant enough for our studies.

Results obtained showed that linalool, β-trans-ocimene, α-farnesene, α-pinene, β-pinene, myrtenol, β-phellandrene, and 3-ethylphenol were the most important odorants present in the volatile oil of the fruit with linalool being the most intense giving the pepperish note, characteristic of the ground, dried, smoked fruits of *Xylopia aethiopica*. Linalool has also been found to be an important odor-active compound present in black pepper (*Piper nigrum* L.) by Jagella (pers. commun.). All the terpenoids detected in the volatile oil of *Xylopia aethiopica* have been reported earlier by previous workers (Ayedoun et al. 1996; Jirovetz et al. 1997). However all aroma compounds found in the acidic fraction are being reported for the first time as being present in the fruits of *Xylopia aethiopica*. Ayedoun et al. (1996) reported the presence of p-mentha-3,8-diene but, its odor note was not detected during the AEDA experiments, however, p-mentha-3,8-triene with a fruity, terpene-like odor was detected instead.

The commercially exploitable features of volatile oils of *Xylopia aethiopica* have not been investigated to any extent before, however, a close examination of the compositional features of the *Xylopia aethiopica* essential oils should reveal a promising commercial potential which could be effectively exploited. As organoleptic properties of essential oils are usually due to the oxygenated components, volatile oils of *Xylopia aethiopica* in which linalool, myrtenol, and terpinen-4-ol are very important, may be ideal resource materials as odor fixtures in toiletries and perfumes and this might explain why the locals use the dried fruits in both expectorants and disinfectant applications.

Results reported here suggest that linalool is an important odor-active compound in the fruit of *Xylopia aethiopica* and also that AEDA in combination with GC-sniffing is an excellent approach for the determination of key compounds causing the aroma in the sample as well as other similar samples.

REFERENCES

Ayedoun, A.M., B.S. Adeoti, and P.V. Sossou. 1996. Influence of fruit conservation methods on the essential oil composition of *Xylopia aethiopica* (Dunal) A. Richard from Benin. Flav. Fragr. J. 11:245.

Burkill, I. 1985. Useful plants of tropical West Africa. p. 231–237. Pergamon Press, London.

Ekong, D.E.U. and A.U. Ogan. 1968. Constituents of *Xylopia aethiopica*. Structure of xylopic acid, a diterpene acid. J. Chem. Soc. C, 1968, 311.

Ekundayo, O. 1989. A review of the volatiles of the Annonaceae. J. Essent. Oil Res. 1:223.

Faulkner, D.F., D. Lebby, and P.G. Waterman. 1985. Chemical studies in the Annonaceae. Part 19. Further diterpenes from the stem bark of *Xylopia aethiopica*. Planta Med. 51:354.

Guth, H. and W. Grosch. 1989. 3-Methylnonane-2,4-dione—an intense odor compound formed during flavor reversion of soya-bean oil. Fat. Sci. Technol. 91:225.

Harrigan, G.G, V.S. Bolzani, A.A.L. Gunatilaka, and D.I.G. Kingston. 1994. Kaurane and trachylobane diterpenes from *Xylopia aethiopica*. Phytochemistry 36:109.

Hofmann, T. and P. Schieberle. 1995. Evaluation of the key odorants in a thermally treated solution of ribose and cysteine by aroma extract dilution analysis. J. Agr. Food Chem. 43:2187.

Irvine, F.R. 1961. Woody Plants of Ghana. Oxford University Press, Oxford p. 67–76.

Jirovetz, L., G. Buchbauer, and M. Ngassoum. 1997. Investigation of the essential oils from the dried fruits of *Xylopia aethiopica* (West African "Peppertree") and *Xylopia parviflora* from Cameroun. Ernährung 21:1.

Rabunmi, R. and E. Pieeru. 1992. Japanese patent JP 92-173277. Chem. Abstr., 120:270896.

Schieberle, P. 1991. Primary odorants of popcorn. J. Agr. Food Chem. 39:1141.

Schieberle, P. 1995. Recent developments in methods for analysis of volatile flavor compounds and their precursors. p. 403–431. In: A. Goankar (ed.), Characterization of foods: emerging methods. Elsevier, Amsterdam.

Piper hispidinervum: A Sustainable Source of Safrole

Sérgio F.R. Rocha and Lin Chau Ming

Piper hispidinervum (C. DC.), Piperaceae, is a promising source of sassafras oil, the source of safrol, currently derived from endangered plants of the Lauraceae such as *Ocotea pretios* Ness (Mez.), *Cinamomum petrophilum*, *C. mollissimum*, and *Sassafras albidum* Nutt. The essential oil of *P. hispidinervum* contains high levels (83–93%) of safrole in leaves which can be easily extracted by hydrodistillation.

SAFROLE

Safrol a phenolic ether (Tyler et al. 1982), is a colorless or slightly yellow liquid, $C_{10}H_{10}O_2$, molecular wt 162.18, with density of 1.096 at 20°C, and melting point about 11°C. The oil is insoluble in water, very soluble in alcohol, miscible with ether and chloroform. The DL_{50} (oral) is 1.950 mg/kg in rats and 2.350 mg/kg in mice (Budavaris 1989).

Sassafras oil is important to several products. It was used as an ingredient in popular beverages, such as "pinga com sassafras" in Brazil, and was once used as an ingredient for "root beer" in the United States. It has been used as a topical antiseptic and pediculicide, but it may be carcinogenic.

Safrole is an important raw material for the chemical industry because of two derivatives: heliotropin, which is widely used as a fragrance and flavoring agent, and piperonyl butoxide (PBO), a vital ingredient of pyrethroid insecticides. Natural pyrethrum in particular would not be an economical insecticide without the addiction of PBO as a synergist and the future of the natural pyrethrum industry is linked to the continued availability of PBO. Safrole has many fragrance applications in household products such as floor waxes, polishes, soaps, detergents, and cleaning agents.

Markets and Demand

Japan, Italy, and the United States are the most important markets for sassafras oil, and the total annual demand is estimated to be around 2,000 tonnes. Brazil has manufacturing capacity for both heliotropin and PBO (equivalent to about 500 t of sassafras oil), although a shortage of domestically produced oil has led to importations of oil from China. The worldwide sassafras oil price is US$4 to 6/kg.

The demand for sassafras oil is determined by the market for heliotropin and PBO. Heliotropin consumption is increasing, particularly in Eastern Europe, Asia, and some developing countries, and sassafras oil is the favorite raw material for its manufacture. If price rises markedly, synthetic heliotropin would become more attractive.

Sources

In Brasil, sassafras oil has been extracted commercially from *Ocotea pretiosa*, a perennial tree native to coastal tropical rainforests from Bahia to Santa Catarina, in Mata Atlantica. This species is also found in Colombia and Paraguay (Rizzini and Mors 1995). This was based on a discovery in 1939 that wood distillation from a large tree in the state of Santa Catarina yielded a rich source of sassafras oil, containing 84% safrole. These trees were indiscriminately harvested, placing this species on the endangered list. Until the 1960s, Brazil was the major exporter of sassafras oil in the world, but production has declined with the depletion of this natural resource. Governmental restrictions in the late 1980s have resulted in a further decline in production and in the falling level of Brazilian exports. No significant replanting has ever occurred.

Vietnam has been exported sassafras oil since 1990 from wild trees of *Cinnamomum camphora*, but supplies from this source may be relatively short-lived. Current exports are estimated to be several hundred tonnes/year.

PIPER HISPIDINERVUM AS A SAFROLE SOURCE

In the early 1990s, certain forest shrubs of the Piperaceae, indigenous to the humid forests of Central America and Greater Amazonia, were found to contain high levels of safrole in their leaves. The Brazilian Amazon contains a wide variety of *Piper* species but attention had focused on *P. hispidinervum* and *P. callo-*

sum, two species with high safrole content. Subsequently, *P. callosum* has been dropped in the research work in favor of the more promising *P. hispidinervum*. This effort was carried out by Museu Paraense Emilio Goeldi in Belém in collaboration with the Center for Agroforestry Research (CPAF-EMBRAPA) in Acre.

Piper hispidinervum known as "pimenta-longa" in Brazil, has been described by Yunker (1972). It is a nodose, branching shrub with rather slender upper internodes, somewhat angular, mostly 1–3 cm long, glabrous or very sparsely pubescent, somewhat glandular dotted. Leaves are oblong-lanceolate or elliptic-oblong, with attenuately acuminate apex and inequilateral base. This species resembles *Piper aduncum* L. to some extent but differs in its scarcely scabrous leaves, glabrous stem, and short peduncle (Yunker 1972). Coppen (1995) suggests that this species is distributed throughout South America, and is especially prominent in the state of Acre in Brazil and may extend into Amazonas.

The species is most frequently found on degraded forest, bordering primary forest or farm land where it occurs as a colonizing "weed," either as a pure stand or along with other *Piper* species. On natural sites, plants develop initially into bushes and at an early stage they appear to inhibit growth of competing vegetation. As the plants age they become more tree-like, up to 10 m tall.

Pilot-scale distillations have been conduced to determine oil quality and yields and permit estimation annual productivity on a per hectare basis. The safrole content of the oil in unselected stock is about 85%, but a improvement to more than 90% is possible through selection. Leaves of *P. hispidinervum* in experimental plots of CPAF-EMBRAPA (Acre) contained 3% essential oils of which 93% is safrole.

Management and Culture

Mixed planting of *P. hispidinervum* with cash tree crops is a practical possibility and would be economically attractive as leaf harvesting would permit an early cash return. Another production possibility is sustainable management in natural vegetation since it occurs in high populations in several open areas bordering primary rainforest. Reforestation projects, in natural or deforested gaps, could take advantage of the vegetative and productive potential of *P. hispidinervum* since it is a pioneer species.

Piper hispidinervum offers excellent conditions for cultivation in areas with facilities for harvest, transport, and industrialization (Chaves 1994). Trials have been established at several sites in Brazil using both rooted cuttings and seedlings designed to provide information in both growth characteristics and biomasss yields (leaf + stem) under various planting and management regimes. Studies undertaken in laboratories of the Department of Horticulture, FCA–UNESP, Botucatu, demonstrated that this plant is positively photoblastic with 50% germination under sunlight but none in the dark.

Crop density studies in Acre state by Sousa et al. (1997) indicate that the highest biomass production was achieved at spacing of 70 × 70 cm. Essential oil yield was 0.3% in branches and 4.0% in leaves; safrole content was not evaluated.

Disease Susceptibility

In the early 1960s, Japanese immigrants to Brazil colonized the state of Pará, and initiated the culture of black-pepper (*Piper nigrum* L.) but it was an agronomic and economical disaster, due to fusarium disease caused by *Fusarium solani* f. sp. *piperis*. Poltronieri et al. (1997) evaluated the resistance of seedlings of *P. hispidinervum* to different isolates of this pathogen, and demonstrated resistance, indicating the possibility of commercial cropping in areas where black pepper had been decimated by the fusarium disease.

Cercospora is a potential pathogen of *P. hispidinervum*. In Xapuri (Acre state), plants infected by *Cercospora* had a reduction of 21% in essential oil produced, but the safrole content was unaffected (Siviero and Pimentel 1997).

FUTURE PROSPECTS

Forest preservation in developing countries is a controversial issue that involves many different interests. The worldwide demand for raw materials, such as safrole, offers an opportunity to these countries that have a source of products in their natural forests.

Forest preservation is best linked to local populations maintenance. In this context, *P. hispidinervum is* more than a new safrole source. The economical exploration of its productive potential could be an important step to assist in the maintenance of the Amazon rainforest, providing a new livelihood option. This might include culture as a crop but also sustainable management in natural vegetation, since this species occurs in high populations in several open areas such as in natural gaps in the forest and in areas bordering the primary rainforest. An integrated project based on this species among research institutes, universities, governmental, and non-governmental institutions could produce a sustainable alternative crop for the tropical rainforest.

REFERENCES

Budavaris S. (ed.). 1989. The Merck index: An encyclopedia of chemicals, drugs and biologicals. 11th ed. p. 1322. Merck & Co. Inc., Rahway, NJ.

Chaves J.L. 1994. Pimenta longa reativa o safrol. Química e Derivados. (São Paulo, SP.) 2:40–46.

Coppen, J.J.W. 1995. Sassafras oil. p. 19–25. In: Non-wood forest products 1: Flavors and fragrances of plant origin. Food and Agriculture Organization of the United Nations, Rome, Italy.

Poltronieri, L.S., F.C. de Albuquerque, and O.G.R. Neto. 1997. Reaction of *Piper hispidinervum* to isolates of *Fusarium solani* f. sp. *piperis*. Brazilian Phytopath. 22(1):112.

Rizzini, C.T. and W.B. Mors. 1995. Botância econômica brasileira. 2nd ed. Ambito Cultural Edições Ltda, Rio de Janeiro, RJ.

Siviero, A. and F.A. Pimentel. 1997. Evaluation of attack of the *Cercospora* sp. on yield and safrole grade in *Piper hispidinervum*. Brazilian Phytopath. 22:312.

Sousa, M. de M.M, F.A. Pimentel, and O.G.R. Neto. 1997. Influência da densidade de plantio sobre o crescimento de plantas de pimenta longa (*Piper hispidinervum*). Supplement of VI Brazilian Plant Physiology Congr.

Tyler, V.E., L.R. Brady, and J.E. Robbers. 1988. Pharmacognosy. 9th ed. Lea & Febiger, Philadelphia.

Yunker, T.G. 1972. Separata de hoenea: The Piperaceae of Brazil. Instituto de Botânica Press, São Paulo, SP. 2:137–139.

A Review of the Taxonomy of the Genus *Echinacea*

Kathleen A. McKeown*

Echinacea (Asteraceae), a North American genus of 11 recognized taxa (McGregor 1968), is of great economic and scientific interest. Three species, *E. angustifolia* DC. var. *angustifolia*, *E. pallida* (Nutt.) and *E. purpurea* (L.) Moench, show potential pharmacological activity (Bauer et al. 1988a; Bauer and Wagner 1991). Hydroalcoholic extracts of the roots of these three taxa stimulate phagocytosis in vitro, but there are no conclusive human clinical trials to date that fully substantiate the immune stimulating effects of such extracts upon oral administration. Although these taxa have been subjected to extensive chemical characterization (Bauer et al. 1988b; Bauer and Remiger 1989; Bauer et al. 1988, 1989), the exact chemical identity of the active constituents is unknown (Awang 1999). There is historical documentation of the medicinal uses of *Echinacea* by the North American Plains Indians (Kindscher 1989). This ethnobotanical validation has certainly been a factor in the popular herbal use of *Echinacea,* which has skyrocketed over the last few decades. The market demand and the medicinal properties are responsible for the recent interest in this genus as a new specialty crop.

All species of *Echinacea* are herbaceous, perennial flowering plants of the composite family (Asteraceae). Generally, a basal rosette of petiolate leaves and one to several annual stems arise from an underground, perennial rootstock. A single taproot is characteristic of all species except for *E. purpurea*, which has a fibrous root system. The stems terminate in a long-lasting and mildly fragrant inflorescence. The disk flowers are attached to a conical, hemispherical or occasionally flattened receptacle. The disk itself may also be flattened, conical, or hemispheric. Surrounding the capitulum is an involucre of 3 to 4 series of imbricating bracts (phyllaries). Each disk floret is subtended abaxially by a modified bract, the palea, which protrudes beyond the 5-lobed corolla. The disk florets are protandrous and development and anthesis follow the typical centripetal pattern of development of the composite family. The whorls of sharply tipped palea give a distinctive look to the capitulum; indeed, the genus name is derived from the Greek word *echinos*, for hedgehog. Pubescence on the stems, leaves and bracts ranges from hispid, hirsute and strigose to glabrous and varies among the species. The sterile ray flowers have strap-shaped ligules with colors that range from white, pink, magenta and purple to yellow.

The majority of taxa are diploid ($n = 11$). *E. pallida* and some populations of *E. angustifolia* var. *strigosa* McGregor are tetraploid ($n = 22$) (McGregor 1968). A complete understanding of the reproductive biology and breeding system is lacking; the presumed sporophytic self-incompatibility system is not perfected in this genus. Every species of *Echinacea* self pollinates to some degree, *E. purpurea* more so than the others (McGregor pers. commun. 1997). A mixed breeding system has been recently demonstrated for *E. angustifolia* var. *angustifolia* (Leuszler et al. 1996).

In addition to its possible medicinal uses, *Echinacea* has enormous ornamental potential. *E. purpurea*, the only species for which ornamental cultivars have been bred, is both productive and profitable as a field grown specialty cut flower (Starman et al. 1995). In fact, *E. purpurea* is the only species of the genus which has been domesticated thus far. It is interesting to note that the cultivars of *E. purpurea* that are now in production as source materials for herbal extracts were actually developed for ornamental purposes. Com-

---

*I would like to express my sincerest thanks to Lowell Urbatsch of Louisiana State University, who contributed helpful comments and clarified the text in many portions of this manuscript. I am honored to have worked directly with Ronald McGregor, the original monographer of the genus *Echinacea*, and I am grateful for his personal contribution to my knowledge of the taxonomy. I would like to thank the many botanists, ecologists, site managers and herbaria curators with whom I worked during the germplasm collection of 1997 and 1998. I would also like to acknowledge the following agencies that granted me access to wild *Echinacea* populations: the U.S. Forest Service, the U.S. Fish & Wildlife Service, The Nature Conservancy, the National Park Service, the Missouri Department of Natural Resources, the Nebraska Game and Parks Commission, the U.S. Army, the U.S. Army Corps of Engineers, and the Natural Heritage Programs in Arkansas, Georgia, Kansas, Louisiana, Missouri, North Carolina, Oklahoma, South Carolina, South Dakota, Tennessee, and Virginia. This work was supported by a grant from the USDA Agricultural Research Service, Island Images of Nantucket, MA and the University of Massachusetts, Amherst.

mercial field plantings of the other species in the genus have been sown from generally unimproved, wild seed. Plant breeders have an important task and a number of challenges before them.

Between 1997 and 1998, there was a six-fold increase in the number of accessions from wild populations of *Echinacea* maintained by the USDA National Plant Germplasm System. Now at 134 total accessions, this is the only available collection in the world of the primary genepool for the genus. This preliminary collection represents a valuable starting source of genetic diversity for breeders. Harlan (1984) makes a point about the evaluation and utilization of wild germplasm that is relevant to the *Echinacea* situation, namely that "plant breeders are rarely familiar with wild relatives....and are often misled by inept taxonomies." Despite the widely cited and detailed monograph on the genus *Echinacea* by McGregor (1968), there are a number of misleading generalizations regarding the taxonomy that are being repeated in the scientific literature. There is also a tendency to omit designation of taxonomic varieties. The purpose of this article is to address the problems associated with these generalizations, to review the major macroscopic characters by which the taxa can be identified, to highlight some of the morphological characters that may be of interest to breeders and in which species they are found.

COMMON MISREPRESENTATIONS

The most common misrepresentations about *Echinacea* deal with the natural range of the genus, which has implications regarding the suitability and adaptability for cultivation in different climates, and the usefulness of ligule color as a diagnostic taxonomic characteristic. These misrepresentations are addressed below.

Echinacea is not entirely a genus of the American Midwest, although the area of greatest species richness clearly lies in a band running from the Ozark Mountains of Missouri south through the east central grasslands of Oklahoma (Fig. 1). The genus as a whole is not exclusive to the Midwest or to the Great Plains, a common generic description that may be derived from the assumption that the range of the popular *E. angustifolia* var. *angustifolia* is descriptive for all species. The natural range of *Echinacea* is most accurately defined in broader terms as the Atlantic drainage region of the United States, extending into south central Canada (McGregor 1968). The southern United States is an important native area for this genus. Two of the species are endemics of the southeast, those being *E. tennesseensis* (Beadle) Small and *E. laevigata* (Boynton & Beadle) Blake, both of which are federally endangered. Half of the taxa have natural ranges that also extend well into the southern United States.

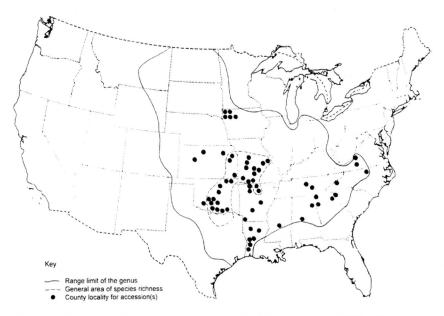

Fig. 1. The natural range of the genus *Echinacea* in the US. The outer solid line marks the limits for native populations; the inner dashed line marks the area of species richness. Accessions collected in 1997 and 1998 are mapped. Each dot may represent more than one accession.

The center of origin for *Echinacea* may well be the Ozark Mountains of Missouri and Arkansas. However, southeastern Oklahoma and the southern Appalachian Mountains as areas of diversity and speciation should not be overlooked. No native *Echinacea* populations have been discovered north or east of Pennsylvania and Virginia, although literature references to populations in Massachusetts and Maine, for example, are occasionally made without distinguishing these as probable introductions. A distinction between native and introduced populations is obviously important in any discussion of the evolution of the genus. The presence of naturalized populations in the northeastern US is a reflection of the adaptability of the genus rather than its potential or natural range. In addition to modern introductions, the existing range of the genus was impacted by the Native Americans who used and planted it; see the description of *E. angustifolia* var. *angustifolia*.

Another set of generalizations concerns ligule color. Misuse or misunderstanding of this diagnostic character is a likely factor in species misidentification and may result in the digging of rare endemics. The ligule is the strap-shaped part of the ray corolla of the composite inflorescence. Ligule color is generally a reliable character for distinguishing the yellow coneflower, *E. paradoxa* (Norton) Britton var. *paradoxa*, the only species with yellow ligules. For the other taxa of *Echinacea* ligule color is generally unreliable for a number of reasons. First, the color of the ligule is often developmentally dependent, being paler in the early stage of inflorescence development, darkening with maturation and anthesis, and fading with senescence. Therefore, one must clearly define the stage at which color is to be determined, notwithstanding the need to use a standard color system in order to avoid the subjective in the human evaluation of a physical character. The common names that refer to "pale" and "purple" coneflower are particularly misleading and ought to be avoided in the scientific literature.

A second reason to avoid using ligule color as a diagnostic character is that a geographical color cline exists for a number of the species. The ligules of *E. pallida* during mid-anthesis are nearly white in the most southern portion of its range, and take on a much deeper pink color in the north (McGregor 1968). *E. laevigata* is another species, the populations of which have very pale pink to white ligules at their most southerly boundaries and which are distinctly darker at northern latitudes, reaching a deep magenta color in some populations. The author has observed whole populations of *E. angustifolia* var. *angustifolia*, *E. paradoxa* (Norton) Britton var. *neglecta* McGregor and *E. sanguinea* Nutt. that are very pale pink to white in ligule color during anthesis. *E. pallida* is so designated because its pollen is white, not because its ligules are pale, yet it continues to be described as being *the* taxon having pale-colored ligules. In *E. sanguinea*, the disk corollas and the palea tips are blood red in color while the ligules can be quite pale.

There are two much more taxonomically useful characteristics of the ligule: the ratio of its length to width, and the ratio of its length relative to the width of the disk. This is reviewed below for each species. Obviously, with a genus as variable as this one, there will be exceptions to these features as well.

The taxonomic misconceptions are further compounded by the fact that many of the taxa are similar phenotypically and can be difficult to distinguish without familiarity with the morphological variation among them. The existence of hybrid swarms complicates this situation, especially in south central Missouri and southeastern Oklahoma. There has been no complete molecular systematic study of *Echinacea*, although restriction site investigations of the chloroplast genome suggests a close relationship among taxa (Urbatsch and Jansen 1995). Despite the very fine work of Bauer and colleagues as cited above on the identification of chemical constituents, no complete systematic chemotaxonomic study of this economic genus has been conducted. No doubt this is due, in part, to the previous lack of available material for study.

SUMMARY OF THE GENUS BY SPECIES

Ligule color, when given, is for reference to ornamental potential and is described for the period when approximately half of the florets have passed through anthesis. The reader should refer to McGregor (1968) for a complete metrical treatment of the macroscopic characters. These observations are of wild populations; variations may occur in cultivation.

E. angustifolia DC. var. *angustifolia*

This is a prairie forb that apparently thrives in the mixed and tallgrass habitats and dry, calcareous soils in which it grows. It is one of several species in the genus that has a compact growth habit, attaining 0.5 m in maximum height. Ligules are reflexed and broad relative to their length, that length being shorter than or equal to the width of the disk. Pollen color is bright yellow. These three simple macroscopic characters taken together, i.e., the short plant height, the short and broad, reflexed ligule, and the yellow pollen color, can be used to easily distinguish *E. angustifolia* var. *angustifolia* from *E. pallida*, despite reports to the contrary. The latter species is generally taller and has narrow and long ligules and white pollen. However, identification can be difficult among wild populations on the eastern edge of the range of *E. angustifolia* var. *angustifolia*, where it overlaps the range of *E. pallida* and where intermediate colonies exist. This is discussed in detail by McGregor (1968). Flowering of wild populations of *E. angustifolia* var. *angustifolia* generally occurs in June and July. This species has a hispid or hirsute pubescence, usually unbranched stems, and linear to lanceolate leaves.

The native range of *E. angustifolia* var. *angustifolia* is from south central Texas through the Great Plains north to Saskatchewan, Canada. It is the only species of the genus that is native to Canada. There is ample ethnobotanical evidence for the widespread use of this species as a favored medicinal plant of the Plains Indians (Kindscher 1989). An important point to add is that the Comanche women traditionally collected seed of this species and distributed the seed throughout their buffalo hunting grounds so as to have this useful medicinal plant at their disposal (T. Whitewolf pers. commun. 1997). The Comanche people originally ranged from Wyoming to Texas. Although more evidence from oral histories and tradition is needed, the broad distribution of this species throughout the plains region of the United States may be the result of its ethnic usage and human dispersal rather than that of natural modes of seed distribution, which for this genus include water and possibly birds and mammals.

Chemically, this species is well studied. Roots contain alkylamides and caffeic acid derivatives (Bauer et al. 1988, 1989; Bauer and Remiger 1989; Bauer and Wagner 1991).

E. angustifolia DC. var. *angustifolia* is an obvious candidate for the selection of cold hardiness because it has adapted to northern climates. McGregor (1968) reported that *E. angustifolia* var. *angustifolia* is the only species in the genus with sclereid cells in the pith, and that all hybrids with this species maintain this character, which is potentially useful in breeding for stem strength.

E. angustifolia DC. var *strigosa* McGregor

This is a taxonomic variety of *E. angustifolia* which has not been recognized in the literature or studied beyond its taxonomic identification. *E. angustifolia* DC. var. *strigosa* is an endemic that runs in a narrow band from south central Kansas through central Oklahoma and northeastern Texas. It is characterized and distinguished from its closest relative, *E. a.* var. *angustifolia*, by the presence of a strigose, rather than hirsute pubescence, frequency of branching, a wider disk and darker (reddish) ligules. These morphological characters, hybridization studies, and the species' distribution, which follows the eastern edge of *E. a.* var. *angustifolia*, and the western edge of *E. atrorubens*, led McGregor (1968) to the suggestion that this taxon is derived from hybridization between the latter two species. McGregor (1968) also found tetraploid colonies of this species, more frequently in its southern segment.

Southeastern Oklahoma is a geographical area where hybridization has been occurring and where the taxonomy of these taxa is incompletely understood (R. McGregor pers. commun. 1997). Hybridization between *E. paradoxa* var. *neglecta*, both varieties of *E. angustifolia* and *E. atrorubens* may account for some of the variation that is typical of this region.

E. atrorubens Nutt.

E. atrorubens is an endemic found in a narrow region running north to south along the eastern grasslands of Texas, Oklahoma (west of the Ouachita Mountains), and Kansas. This species, except for its taxonomic features, is essentially unstudied. In habit it has a tall, sturdy, unbranched flowering stem that grows to about 0.9 m in height, normally without lodging. The inflorescence is characterized by very broad ligules that are shorter or equal to the length of the head, which are very strongly reflexed, and which are a deep magenta or reddish purple color. *E. atrorubens* is relatively early in flowering late May and June.

A potentially useful character of *E. atrorubens* is its large seed head. Wild seed (technically, the achene) of a number of *Echinacea* species has entered the commercial market not only as a propagule but as a source of medicinally active constituents, despite the fact that there are no clinical studies to directly support this latter usage. Potentially active compounds such as cichoric acid have been extracted from the "flowers" of *E. purpurea* (Bauer and Wagner 1991), which may include the achenes in varying stages of development. Also, it has been documented that the Lakota ate the green fruit of *E. angustifolia* var. *angustifolia* (Kindscher 1989). Presumably, either the ethnobotanical usage, documented for this one species and extended to the others, or unpublished chemical profiling of the achene has been justification for its inclusion in commercial products. If medicinal value for the achene is eventually demonstrated, then breeding and selection for large achenes and seed heads that can be mechanically harvested will be desirable.

E. laevigata (Boynton & Beadle) Blake

E. laevigata is one of the two federally endangered species of *Echinacea*. It is closely related to *E. purpurea* (McGregor 1968) and is found at the eastern edge of the geographic range of the genus, in the Appalachians and piedmont running from Virginia through the Carolinas to the northeast corner of Georgia. Its distinctive characteristics are a glabrous, ovate leaf, long narrow rays that droop and a tall, usually unbranched flowering stem that can reach a meter in height. It is this latter characteristic coupled with a wide range of ligule color, which in some populations varies from white to deep magenta, that may make it a good candidate for ornamental breeding as a long stemmed cut flower. The ligule is much like that of *E. pallida* in shape, being long and thin relative to the width of the disk, and drooping. *E. laevigata* flowers in June. The presence of alkamides has been demonstrated in this species (Bauer and Wagner 1991).

E. pallida (Nutt.) Nutt.

E. pallida is the only tetraploid in the genus ($n = 22$) with the exception of occasional tetraploid colonies of *E. angustifolia* var. *strigosa* (McGregor 1968). This species is generally characterized by long, rarely branching flowering stems up to 0.9 m in height, with drooping or reflexed and slender ray flowers significantly longer than the width of the disk. Typically, this species has white pollen. However, the distinction between *E. pallida* and the closely related *E. sanguinea* at the southern boundary of its range (discussed under *E. sanguinea*) and from *E. simulata* to the east, is a matter of taxonomic controversy. Regarding the range overlap with *E. simulata*, there is an area of intergradation that is roughly one county wide which occurs in the Ozarks of north central Arkansas (P. Hyatt pers. commun. 1998) and south central Missouri. Some wild populations in this zone are of a "mixed pollen color" character, i.e., they are composed of individuals with pure white, bright yellow, and a clearly intermediate, soft yellow colored pollen. McGregor (1968) discovered sterile triploids in this area which he characterized as having an ochre-like, pale yellow pollen color (McGregor pers. commun. 1997). This interesting area of hybridization is unfortunately also an area frequented by root diggers and wild seed collectors. The flowering period of *E. pallida* follows a cline running south to north, May to July (McGregor 1968).

E. pallida is one of the three species under study as a medicinal plant. Its roots contain a number of polyacetylenes that with a few exceptions have not been found in any of the other species of the genus to date (Bauer et al. 1988b; Bauer and Wagner 1991).

E. paradoxa (Norton) Britton var. *neglecta* McGregor

This species is a narrow endemic found in the Arbuckle Mountain region of south central Oklahoma. It is characterized by strigose pubescence, a glossy, lanceolate to elliptical leaf, and drooping ray flowers that are, much like *E. pallida*, longer than they are wide and longer than the width of the disk. They are white to pink or light purple in color. This botanically unstudied species is of unknown medicinal value and could be easily confused with *E. pallida* by a wild harvester with no taxonomic background; digging of this taxon has been a problem in south central Oklahoma. Strigose pubescence, bright green stems and yellow pollen are characteristic of *E. paradoxa* var. *neglecta* and serve to distinguish it from *E. pallida*. Its closest putative relatives are *E. paradoxa* var. *paradoxa* and *E. atrorubens*.

E. paradoxa (Norton) Britton var. *paradoxa*

This is the only yellow coneflower of the genus, and is endemic to the Ozark Mountains of Missouri and Arkansas. It has the same strigose pubescence, bright green stems and yellow pollen that characterize *E.*

paradoxa var. *neglecta*. Polyacetylenes identical to those found in *E. pallida* have been identified in its roots, as well as echinacoside, a caffeic acid derivative found in *E. angustifolia* var. *angustifolia* (Bauer and Foster 1991). In the wild, *E. paradoxa* var. *paradoxa* tends to grow in the same localities as *E. pallida* or *E. simulata* and hybridization in these mixed wild populations apparently occurs (author, pers. obs.).

This species and its hybrids have ornamental value. It generally flowers in June.

E. purpurea (L.) Moench

Although the natural range of *E. purpurea* is quite broad, in the wild it is truly a plant of the ecotone, preferring the shaded edges of savannas and glades and open woodlands with partial sun exposure. Characterized by a classic discontinuous distribution, as is the genus as a whole, it is found in scattered prairie remnants of the south from Louisiana to North Carolina, and from Oklahoma and Kansas through the Midwest to Ohio, Kentucky, and Tennessee. The wild populations of *E. purpurea* in Louisiana and Mississippi are particularly striking in terms of pigmentation; the ligules are deep lavender and contrast dramatically with bright orange tipped paleae. Both soil pH and genetics may factor into ligule color, and any genetic component underlying the coloration of these populations is worthy of selection for ornamental value. *E. purpurea* flowers from late June through September.

E. purpurea is probably the most widely used and is certainly the most widely cultivated medicinal species of the genus. All parts of the plant are harvested, including the flowering heads for polysaccharides and cichoric acid (a caffeic acid derivative) and the roots for cichoric acid and various alkylamides. The polysaccharides, caffeic acid derivatives and the alkylamides are three chemical classes of constituents that are implicated in the immunostimulatory effect of this species and others (Bauer and Wagner 1991); see Bone (1997) for a good general review of these studies.

E. purpurea is a host of the aster yellows phytoplasma (Stanosz and Heimann 1997) which is a notable problem in cultivation (J. Simon pers. commun. 1998). Selection of thick pubescence to discourage leafhoppers may be achieved by working with accessions of wild populations from the western segment of the species range, as a general tendency toward the glabrous occurs to the east. Also, the transfer of an insect-deterrent pubescence from other taxa, particularly *E. tennesseensis*, which has a dense, hirsute hair type and phenological compatibility is a possibility. Such transfers have been successful for other crop plants (Harlan 1984).

E. sanguinea Nutt.

The historical range of *E. sanguinea* is from eastern Texas and western Louisiana north into Arkansas and southeastern Oklahoma. *E. sanguinea* is not cold hardy (McGregor 1968). Although it generally flowers in May, wild populations in southern Louisiana will bloom well into August.

This species can be difficult to distinguish from *E. pallida*, especially where its northern range overlaps the southern boundary of the latter, because there are close phenotypic similarities between these two. Heights of both can approach 0.9 m and the ligule of *E. sanguinea*, like that of *E. pallida*, is long and narrow and can be very pale in color. Qualitatively, *E. sanguinea* is the more delicate of the species, having more narrow rays and more slender stems (McGregor 1968). One character not noted in the McGregor monograph is that the disk corolla of *E. sanguinea* is a blood red color, whereas that of *E. pallida* is very pale (L. Urbatsch pers. commun. 1997). However, this character, as for ligule color, may be developmentally dependent and may not hold true for *E. pallida* in some parts of its range, especially where its populations intergrade with those of *E. simulata*. Some taxonomists make a geographical distinction, placing *E. sanguinea* south of the Red River in Louisiana, and *E. pallida* to the north, but there are clear exceptions to this rule. What is agreed is that *E. sanguinea* is always found in more acidic, sandy soils and open pine woodlands.

E. simulata McGregor

The range of this species runs from south central Missouri east through Tennessee and northern Georgia. It is phenotypically very similar to *E. pallida* from which it is primarily distinguished based on ploidy level (*n* = 11), pollen size and pollen color. The strictly white pollen color of *E. pallida*, as described by McGregor, may not be a reliable character in the area where the two species overlap, i.e. south central Missouri. *E. simulata* has ornamental potential in that the ligule color varies dramatically from pale pink to deep magenta in many of its populations, and it is remarkably fragrant. Ornamental, purple colored stems are also found in this species, as well as in *E. purpurea*. There is an undescribed, glabrous race of this taxon found in a few isolated prairie barrens of northeastern Alabama and northwestern Georgia.

This species is coming under the digging pressure of its close relative, *E. pallida*, probably due to misidentification. The medicinal value is unknown; however, alkamides similar to those of *E. angustifolia* var. *angustifolia* and ketoalkenynes similar to those of *E. pallida* have been identified in the roots of *E. simulata* (Bauer and Foster, 1991).

E. tennesseensis (Beadle) Small

E. tennesseensis was the second plant to be officially listed by the U.S. Fish and Wildlife Service as an endangered species (U.S. Department of the Interior 1979). Although McGregor had considered it to be possibly extinct at the time of the publication of his monograph (McGregor 1968), five populations in a 14 mile radius in central Tennessee are extant and have been the subject of a careful recovery plan (Currie and Somers 1989). In fact, *E. tennesseensis* may soon be ready for downlisting to threatened status (A. Shea pers. commun. 1997).

This species is unique from all others in the genus in that its rays neither droop nor reflex backward, but spread up and outward to form an inverted cup-shaped corolla. Ligule color varies from white to rose and purple. Its flowering occurs from June through September and is similar to that of *E. purpurea*. *E. tennesseensis* has ornamental value because of its compact habit, the multiplicity of flowering stems arising from a single rootstock, and its range of ray flower pigmentation. The medicinal value of this species is unknown. Alkamides identical to those found in the roots of *E. angustifolia* var. *angustifolia* have been identified in ethanolic extracts of the roots (Bauer et al. 1990).

CONCLUDING REMARKS

This paper was intended to provide a brief overview of the taxonomy of the genus *Echinacea*, and to highlight the macroscopic characters most useful for species identification. Another objective was to clarify the misrepresentations that have appeared in the literature, and to provide some preliminary guidelines for breeders interested in this economic genus. This review will hopefully also serve to educate a broader audience and help to prevent the digging of rare endemics, by providing a simplified version of the taxonomic descriptions. Wild harvesting is a potentially serious problem; the in situ conservation of the genus *Echinacea* is a growing concern among American scientists engaged in the monitoring of wild populations. Much research is needed in breeding and horticulture, and breeding for secondary metabolites will remain a challenge until the active constituents are unequivocally identified. Agricultural production is essential, however, and supports conservation by directly addressing the issue of supply.

REFERENCES

Awang, D.V.C. 1999. Immune stimulants and antiviral botanicals: echinacea and ginseng. p. 450–456. J. Janick (ed.), Perspectives on new crops and new uses. ASHS Press, Alexandria, VA.

Bauer, R. and S. Foster. 1991. Analysis of alkamides and caffeic acid derivatives from *Echinacea simulata* and *E. paradoxa* roots. Planta Med. 57:447–449.

Bauer, R., K. Jurcic, J. Puhlmann, and H. Wagner. 1988a. Immunological in vivo and in vitro examinations of *Echinacea* extracts. Arzneim-Forsch. 38:276–281.

Bauer, R., I.A. Khan, and H. Wagner. 1988b. TLC and HPLC analysis of *Echinacea pallida* and *E. angustifolia* roots. Planta Med. 54:426–430.

Bauer, R. and P. Remiger. 1989. TLC and HPLC analysis of alkamides in *Echinacea* drugs. Planta Med. 55:367–371.

Bauer, R., P. Remiger, and E. Alstat. 1990. Alkamides and caffeic acid derivatives from the roots of *Echinacea tennesseensis*. Planta Med. 56:533–534.

Bauer, R., P. Remiger, and H. Wagner. 1988. Alkamides from the roots of *Echinacea purpurea*. Phytochemistry 27(7):2339–2342.

Bauer, R., P. Remiger, and H. Wagner. 1989. Alkamides from the roots of *Echinacea angustifolia*. Phytochemistry 28(2):505–508.

Bauer, R. and H. Wagner. 1991. *Echinacea* species as potential immunostimulatory drugs. p. 253–321. In: H. Wagner and N. R. Farnsworth (eds.), Economic and medicinal plant research, vol. 5. Academic Press, New York.

Bone, Kerry. 1997. Echinacea: what makes it work? Alt. Med. Rev. 2(2):87–93.

Currie, R.R. and P. Somers. 1989. Tennessee coneflower recovery plan. U.S. Fish and Wildlife Service, Atlanta, GA.

Harlan, J.R. 1984. Evaluation of wild relatives of crop plants. p. 212–222. In: J.H.W. Holden and J.T. Williams (eds.), Crop genetic resources: Conservation and evaluation. George Allen & Unwin, London.

Jacobson, M. 1954. Occurrence of a pungent insecticidal principle in American coneflower roots. Science 120:1028–1029.

Kindscher, Kelly. 1989. Ethnobotany of purple coneflower (*Echinacea angustifolia*, Asteraceae) and other *Echinacea* species. Econ. Bot. 43:498–507.

Leuszler, H.K., V.J. Tepedino, and D.G. Alston. 1996. Reproductive biology of purple coneflower in southwestern North Dakota. The Prairie Naturalist 28(2):91–102.

McGregor, R.L. 1968. The taxonomy of the genus *Echinacea* (Compositae). The Univ. Kansas Sci. Bul. 48(4):113–142.

Stanosz, G.R. and M.F. Heimann. 1997. Purple coneflower is a host of the aster yellows phytoplasma. Plant Disease 81(4):424.

Starman, T., T. Cerny, and A. MacKenzie. 1995. Productivity and profitability of some field-grown specialty cut flowers. HortScience 30(6):1217–1220.

Urbatsch, L.E. and R.K. Jansen. 1995. Phylogenetic affinities among and within the coneflower genera (Asteraceae, Heliantheae), a chloroplast DNA analysis. Syst. Bot. 20(1):28–39.

U.S. Department of the Interior, Fish and Wildlife Service. 1979. Determination that *Echinacea tennesseensis* is an endangered species. Fed. Reg. 44(110):32604–32605.

Echinacea angustifolia: An Emerging Medicinal*

Ali O. Sari, Mario R. Morales and James E. Simon

The genus *Echinacea* (Asteraceae) is native to North America and consists of nine species (McGregor 1968). *E. angustifolia* DC, a narrow leafed species, is distributed on barrens and dry prairies from Minnesota to Texas, Oklahoma, Kansas, Nebraska, Iowa, the Dakotas, Colorado, Wyoming, Montana, Saskatchewan, and Manitoba (McGregor 1968; Kindscher 1989; Foster 1991). *E. angustifolia* grows lower and exhibits a smaller leaf area and biomass than *E. pallida* (Nutt.) Nutt. and *E. purpurea* (L.) Moench. *Echinacea angustifolia* grows up to 60 cm and usually branches simple not sparsely. The ligules of *E. angustifolia* are also short (to 5 cm) and the pollen is yellow (Hobbs 1989). *E. purpurea* grows taller (to 150 cm), branches more, and its pollen is also yellow. *E. purpurea* has wider leaves than *E. angustifolia* and *E. pallida*. *Echinacea pallida* is also simple branched like *E. angustifolia,* but grows taller (to 90 cm) than *E. angustifolia,* and has white pollen (Hobbs 1989). As a result of the slower growth, *E. angustifolia* is a weaker competitor to weeds than *E. pallida* (Snyder et al. 1994). The ray flowers surrounding each *Echinacea* flower head are sterile (Hobbs 1989; Foster 1991; Feghahati 1994).

MEDICINAL VALUE

Three *Echinacea* species (*E. purpurea, E. angustifolia*, and *E. pallida*) have medical properties (Schulthess et al. 1991), and are commercially traded as medicinal plants. Native Americans used echinacea extensively for the treatment of snake and other venomous bites, rabies, toothaches, coughs, soremouth, throat, dyspepsia, colds, colic, headache, and stomach cramps (Hobbs 1989; Kindscher 1989; Foster 1991). European settlers first used echinacea around the beginning of 19th century. By 1921, echinacea preparations were among the most widely sold medicines extracted from an American plant (Flannery 1998). Echinacea appears to have strong anti-inflammatory activity, wound-healing action, stimulates the immune system, and may be effective against some viral and bacterial infections (Bauer et al. 1990; Schulthess et al. 1991; Leung and Foster 1996; Burger et al. 1997; Cox 1998). Scaglione and Lund (1995) observed that an anti-cold remedy including *E. purpurea* was effective for the treatment of common cold and it shortened the healing time. Echinacin, a product made from the juice of fresh plant tops, has been produced for 50 years and subjected to more than 150 clinical studies in Germany (Hobbs 1989). Echinacea products are medically prescribed in Germany (Blumenthal 1998; Hobbs 1989; Leslie 1995, 1996). In the US, many health remedies or dietary supplements include echinacea in the form of dry root, capsules, tablets, crude extracts, tinctures, and are mainly from *E. purpurea* and *E. angustifolia* (Leung and Foster 1996; Li 1998). Injectable preparations using echinacea extracts are available only in Germany (Li 1998). Echinacea products can also be found combined with other medicinal herbs yet in the US, but the product label is not allowed to have specific therapeutic claims (Li 1998). McCaleb (1998) reported that the squeezed sap of *Echinacea purpurea* L. (Echinacinâ, Madaus AG, Cologne, Germany) is well tolerated and appropriate for long term-oral use, with no adverse reaction observed other than aversion to the taste during oral use for up to 12 weeks. Even injections were found to be safe for both adults and infants. Some rare side effects were observed such as shivering, vomiting, headache, and fever during injection treatment. Mengs et al. (1991) reported no evidence of any toxicity using *E. purpurea* in rats or mice during a 4-week application of single oral or intravenous doses, amounting to many times the human therapeutic dose. Tyler (1993) and Scaglione and Lund (1995) conclude that the use of echinacea is safe.

*J. Paper no. 15,871 of the Purdue Agricultural Research Programs, Purdue University, West Lafayette, IN. This research was supported by the Turkish Ministry of Agriculture and Rural Affairs and a grant from Quality Botanical Ingredients, Inc. We thank Larry W. Ness and Kyle W. Arvin and staff of the Office of the Indiana State Chemist and Seed Commissioner for their help and use of facilities; Jim Quinn formerly of QBI for his suggestions during the study, and to Jules Janick for his critical review.

ORNAMENTAL VALUE

Echinacea species are also grown for their ornamental value. Samfield et al. (1991) reported that Texas wild flowers, including *Echinacea* species, were desirable for their low maintenance requirements, erosion control, and esthetic qualities. *E. purpurea* is used for wildflower establishment, perennial gardens, and as a cut flower (Wartidiningish and Geneve 1994). *E. purpurea* is reported to be a good landscape plant since its flower color ranges from white to rose-pink to red-violet so that it can associate with everything from pale yellows to blues or reds (Cox 1998). *Echinacea purpurea,* with average 18 stems and 47 cm stem length per plant was reported as one of the most profitable perennial cut flowers without major pest or post harvest problems while *E. angustifolia* with lower stem yield (four) and stem length (24 cm) was unprofitable (Starman et al. 1995).

FIELD PRODUCTION

The largest areas of commercial cultivation of *Echinacea* species in North America is estimated to be in the Pacific Northwest, though actual production figures are difficult to obtain. Central and western Canada is emerging as a new production area. Commercial cultivation has increased rapidly due to increased marked demand (Li 1998), but information on cultural methods and their effect on growth, yield, and chemical composition remains limited.

One of the basic obstacles facing field production is the poor and erratic seed germination. Low seed germination percentage of *Echinacea* species was noted by Smith–Jochum and Albercht (1987, 1988). *E. pallida* and *E. angustifolia* seem to have stricter dormancy than *E. purpurea.* Shalaby et al. (1997) reported 70% germination from untreated *E. purpurea* seeds while stratification at 2°C for 40 days increased germination to 82–85% and cold dry storage to 83%. However, germination of *E. pallida* and *E. angustifolia* were 0–1% even after stratification. Baskin et al. (1992) found that fresh mature seed of *E. angustifolia* germinated only from 0% to 6% when they were placed under light and from 0% to 2% when placed in darkness, but that a 12-week cold stratification at 5°C broke dormancy. In another study, no germination was observed when *E. angustifolia* and *E. pallida* were directly sown in the spring (Smith–Jochum and Albercht 1988). Li (1998) recommended stratification of echinacea seeds in moist sand at 1°C to 4°C and for 4 to 6 weeks. Germination of *E. angustifolia* using both stratified and nonstratified seeds reached 94% after treatment with 1.0 mM ethephon and 86% after treatment with 2500 mg/L GA_3 at 25°C germination temperature (Sari 1999).

A second major field problem facing the echinacea production is the susceptibility of the plant to aster yellows. *E. purpurea* appears to be the most susceptible of all *Echinacea* species. The disease, caused by a phytoplasma transmitted by thrips, can been devastating. Growers have reported near complete losses in areas of heavy infestation. Resistant cultivars are not available, and control of the diseases can only be achieved by controlling the vector. Under field conditions thrip control has been difficult, and even more problematic under organic production systems. Research on genetic resistance to thrips, and sustainable control methods need to be pursued.

IMPROVING SEED GERMINATION WITH STRATIFICATION AND PLANT GROWTH REGULATORS

Fresh seed of *E. angustifolia* were obtained from the Prairie Moon Nursery Winoma, Minnesota. Seeds were treated as described below and stratified at 10°C for 4, 8, or 12 weeks in dishes but only the results from the 8 week stratification period are reported here. When the stratification period was complete, a subsample of all treated seeds were transferred to petri dishes containing two 90 mm Whatman filter papers (25 seeds per petri dish; with 4 petri dishes per treatment), which were placed into germination incubators at a constant temperature of 25°C. Germination from constant temperature was equal to or greater than alternating temperatures of 25° (12 h) and 15°C (12 h).

The 5 seed treatments were as follows:

1. *Stratification with water*: Seeds placed on blotter paper in the petri dish, were saturated with deionized water.

2. *Stratification after soaking*: Seeds were first placed in a cloth bag with running tap-water for 24 h. Seeds were then blotted dry on paper towels at 25°h for 24h and then transferred to petri dishes with

blotter paper saturated with deionized water.

3. *Stratification with ethephon after soaking*: Seeds were first soaked as described above and then transferred to a petri dish with blotter paper saturated with 1mM (2-chloroethyl) phosphonic acid (ethephon).

4. *Stratification with GA₃ after soaking*: Seed were first soaked as described above and then transferred to a petri dish with blotter paper saturated with 2500 mg/L GA₃.

5. *Dry stratification*: Seeds were exposed to stratification temperatures without any pre-stratification seed treatment.

Our results (Fig.1) demonstrate that germination rates of greater than 95% for *E. angustifolia* can be achieved when seeds are cold stratified and treated with either 1mM ethephon or 2500 mg/L GA₃. High germination rates were also achieved simply by soaking the seeds in water prior to or at the time of stratification. Presoaking the seeds for 24 h improved germination compared to moistening the seeds only at the time of stratification, presumably allowing increased water imbibition. Germination after dry stratification was significantly lower than all other treatments (< 50% after 20 days). Once stratified seeds were placed into incubators, germination began by day 4. The germination percentage did not significantly increase over the 20 day counting period, except for the dry stratification seeds which reached a maximum at day 16. The results of the dry stratification appear to fall within the low range that seed companies and growers report for *E. angustifolia*.

Our results clearly illustrate several options to overcome seed dormancy in this species that could be of use by growers, seed companies, and nurseries. The use of ethephon or GA₃ plus stratification increased seed germination more than when stratification alone was used, or when either growth regulator was used in the absence of stratification (Sari 1998).

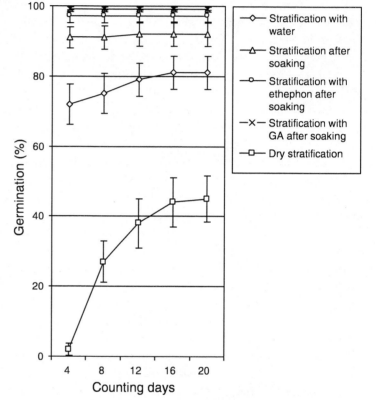

Fig. 1. Germination of *Echinacea angustifolia* following seed treatments with 1mM ethephon or 2500 mg/L GA₃ and 8 weeks of stratification at 10°C. Seed germination conditions were at a constant 25°C.

REFERENCES

Baskin, C.C., J.M. Baskin, and G.R. Hoffman. 1992. Seed dormancy in the prairie forb *Echinacea angustifolia* var. *angustifolia* (Asteraceae): Afterripening pattern during cold stratification. Int. J. Plant Sci. 153:239–243.

Baskin, C.C. and J.M. Baskin. 1988. Germination ecophysiology of herbaceous plant species in a temperate region. Am. J. Bot. 75:286–305.

Bauer, R., P. Reminger, and E. Alstat. 1990. Alkamides and caffeic acid derivatives from the roots of *Echinacea tennesseensis*. Planta Med. 56:533–534.

Blumenthal, M.. 1998. The complete German commission E monographs: Therapeutic guide to herbal medicines. American Botanical Council, Austin, Texas. Integrative Medicine Communications, Boston, MA.

Bratcher, C.B., J.M. Dole, and J.C. Cole. 1993. Stratification improves seed germination of five native wildflower species. HortScience 28:899–901.

Burger, R.A., A.R. Torres, and R.P. Warren. 1997. *Echinacea purpurea* induced cytokine production in human peripheral blood adherent mononuclear cells (PBAC). J. Allergy Clin. Immunol. 99:(1) Part 2 Suppl. 1153–1153.

Cox, J. 1998. Purple coneflower (*Echinacea purpurea*). Organic Gardening 45(5):52–54.

Feghahati, S.M.J. and R.N. Reese. 1994. Ethylene-, light-, and prechill-enhanced germination of *Echinacea angustifolia* seeds. J. Am. Soc. Hort. Sci. 119:853–858.

Flannery, M.A. 1998. John Uri Uoyd. Southern Univ. Press, Carbondale, IL.

Foster, S. 1991. Echinacea: Nature's immune enhancer. Healing Arts Press, Rochester, VT.

Hobbs, C. 1989. The echinacea handbook. Eclectic Medical Publications, Portland, OR. p. 1–7.

Kindscher, K. 1989. Ethnobotany of purple coneflower (*Echinacea angustifolia*, Asteraceae) and other *Echinacea* species. Econ. Bot. 43:498–507.

Leslie, J. 1995/96. Wildflower research blossoms into 100-acre pharmaceutical truck garden. South Dakota Farm & Home Research. 46(4):14–15.

Leung, A.Y. and S. Foster. 1996. Encyclopedia of common natural ingredients. 2nd ed. Wiley-Interscience, New York.

Li, T.S.C. 1998. Echinacea: Cultivation and medicinal value. HortTechnology 8:122–129.

McCaleb, R. 1998. Echinacea safety confirmed. Herbalgram. 42:15.

McGregor, R.L. 1968. The taxonomy of the genus *Echinacea* (Compositae). Univ. Kansas Sci. Bul. 48:113–142.

Mengs, U., C.B. Clare, and J.A. Poiley. 1991. Toxicity of *Echinacea purpurea*. Arzneimittel-Forsch./Drug Res. 41–42:1076–1081.

Parmenter, G.A., L.C. Burton, and R.P. Littlejohn. 1996. Chilling requirement of commercial echinacea seed. New Zeal. J. Crop Hort. Sci. 24:109–114.

Samfield, D.M., J.M. Zajicek, and B.G. Cobb. 1991. Rate and uniformity of herbaceous perennial seed germination and emergence as affected by priming. J. Am. Soc. Hort. Sci. 116:10–13.

Sari, A.O. 1998. Seed germination and dormancy of *Echinacea angustifolia* and *Echinacea pallida*. MS Thesis, Purdue Univ., West Lafayette, IN.

Scaglione, F. and B. Lund. 1995. Efficacy in the treatment of the common cold of a preparation containing an echinacea extract. Int. J. Immunotherapy XI(4):163–166.

Schulthess, B.H., E. Giger, and T.W. Baumann. 1991. *Echinacea*: Anatomy, phytochemical pattern, and germination of the achene. Planta Med. 57:384–388.

Seiler, G.J. 1998. Seed maturity, storage time and temperature, and media treatment effects on germination of two wild sunflowers. Agron. J. 90:221–226.

Shalaby, A.S., E.A. Agina, S.E. El-Gengaihi, A.S. El-Khayat, and S.F. Hindawy. 1997. Response of *Echinacea* to some agricultural practices. J. Herbs Spices Med. Plants 4(4):59–67.

Smith–Jochum, C. and M.L. Albercht. 1987. Field establishment of three *Echinacea* species for commercial production. Acta Hort. 208:115–120.

Smith–Jochum, C. and M.L. Albercht. 1988. Transplanting or seeding in raised beds aids field establishment of some *Echinacea* species. HortScience 23:1004–1005.

Snyder, K.M., J.M. Baskin, and C.C. Baskin. 1994. Comparative ecology of the narrow endemic *Echinacea tennesseensis* and two geographically widespread congeners: Relative competitive ability and growth characteristics. Int. J. Plant Sci. 155:57–65.

Starman, T.W., T.A. Cerny, and A.J. MacKenzie. 1995. Productivity and profitability of some field-grown specialty cut flowers. HortScience 30:1217–1220.

Tyler, V.E. 1993. Phytomedicines in western europe-potential impact on herbal medicine in the United States. ACS Symposium Series 534:25–37.

Wartidiningsih, N. and R.L. Geneve. 1994. Seed source and quality influence germination in purple coneflower (*Echinacea purpurea* (L.) Moench.). HortScience 29:1443–1444.

Cichoric Acid and Isobutylamide Content in *Echinacea purpurea* as Influenced by Flower Developmental Stages

W. Letchamo, J. Livesey, T.J. Arnason, C. Bergeron, and V.S. Krutilina

Echinacea purpurea (L.) Moench, Asteraceae, a native of North America is among the most frequently utilized *Echinacea* spp. for medicinal, cosmetic, and beverage preparations. Most of the echinacea preparations in the industry originate from wild harvesting and few from non-standardized (unregulated) cultivation and harvesting. According to the recent reports (Brevoort 1996), echinacea preparations are among the best selling herbal immunostimulant products in health food stores in the US. Currently more than 800 echinacea-containing drugs, including homeopathic preparations, are available on the German market. Most of the products are derived either from the aerial or underground parts of *Echinacea*.

Within several herbs, *Echinacea* species are among the main investigation targets, because of their value as immunostimulants (Bauer and Wagner 1991). The therapeutic effect of *Echinacea* has been assigned to the presence of caffeic acid derivatives such as cichoric acid, echinacoside, chlorogenic acid (Fig. 1), and lipophilic polyacetylene-derived compounds, such as alkylamides, constituting isobutylamides (Fig. 2), and various other components found in the hydroalcoholic extracts. The low molecular weight constituents of hydroalcoholic tinctures can be determined by HPLC or TLC analysis.

Cichoric acid has been shown to possess phagocytosis stimulatory activity in vitro and in vivo, while echinacoside has antibacterial and antiviral activity (Bauer and Wagner 1991). Cichoric acid was also recently been shown to inhibit hyaluronidase and to protect collagen type III from free radical induced degradation, while lipophilic alkamide (isobutylamides), polyacetylenes, and glycoproteins/polysaccharides have been shown to possess immunomodulatory activity (Bauer and Wagner 1991). When testing the alcoholic extracts obtained from the aerial parts and from the roots for phagocytosis stimulating activity, the lipophilic fraction alkamide (isobutylamides) showed the highest activity (Bauer and Wagner 1991). Most of the alkamides are reported to be responsible potent inhibitors of cyclooxygenase, while the requirement for inhibition of 5-lipoxigenase depends more on the particular structure of the single compound. The inhibitory properties of the alkamides on arachidonic acid metabolism are in accordance with the traditional use of the herbal drugs in the therapy of inflammatory diseases. Alkamide (isobutylamides) is also responsible for local tongue anaesthesis (tingling effect) when ingested and used as an informal test for the quality of *Echinacea* (Bauer and Wagner 1991). Purified alkamide fractions (isobutylamides) from *E. purpurea* and *E. angustifolia* were recently shown to enhance phagocytosis in the Carbon-Clearance-Test by a factor of 1.5 to 1.7 thereby contributing to the immunstimulatory activity of echinacea tinctures (Bauer and Wagner 1991). It is possible, therefore, to consider that cichoric acid, echinacoside, and alkamide (isobutylamides) are parts of the active principles and suitable for standardization of *E. purpurea* row material and finished products.

There is no published scientific information to date on the relationship between flower developmental stages and the accumulation of some of the above mentioned components of *E. purpurea*. Recent phytochemical investigation shows that there is a tremendous variation in the quality of echinacea products that were made available from various international sources (Letchamo et al. 1999). Furthermore, there is a dire need for genetic improvement and precise knowledge on the influence of ecological and devel-

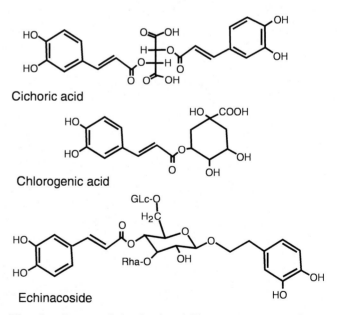

Cichoric acid

Chlorogenic acid

Echinacoside

Fig. 1. Some of the hydrophilic components of *E. purpurea* extract.

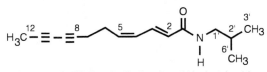

Undeca-2E, 4Z-dien-8, 10-diynoic acid isobutylamide

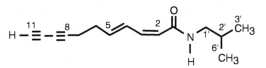

Undeca-2Z, 4E-dien-8, 10-diynoic acid isobutylamide

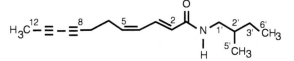

Dodeca-2E, 4Z-dien-8, 10-diynoic acid isobutylamide

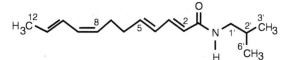

Trideca-2Z, 7Z-dien-10, 12-diynoic acid isobutylamide

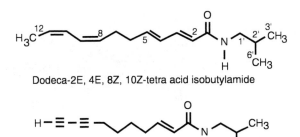

Dodeca-2E, 4Z-dien-8, 10-diynoic acid methylbutylamide

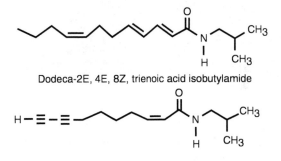

Dodeca-2E, 4E, 8Z, 10E-tetranoic acid isobutylamide

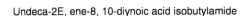

Dodeca-2E, 4E, 8Z, 10Z-tetra acid isobutylamide

Dodeca-2E, 4E, 8Z, trienoic acid isobutylamide

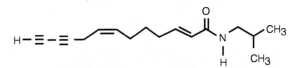

Undeca-2E, ene-8, 10-diynoic acid isobutylamide

Undeca-2E, en-8, 10-diynoic acid isobutylamide

Fig. 2. Some of the major alkamides (isobutylamides) of *E. purpurea* extracts.

opmental factors on yield and quality of *Echinacea*. The objective of this investigation was therefore to study the pattern of the accumulation of the active ingredients during different developmental stages of the flower heads of *E. purpurea* as part of the, introduction, selection, genetic improvement and agronomic programs of medicinal herbs in Trout Lake Farm since summer 1996.

METHODOLOGY

The Plant Material, Selection and Harvesting

A new selection of *E. purpurea* named 'Magical Ruth' (Fig. 3) was propagated by root division and organically grown under sandy loam soil of volcanic origin (pH = 6.5), at Trout Lake Farm, Washington. The selection procedure and identification of the promising *E. purpurea* cultivars was based on yield potentials, better morphological traits, and freedom from diseases and pest infections. Morphological traits such as suitability for mechanical harvesting (erect and non-lodging), medium plant height, number of tillers, branches and flower heads, flower size, uniformity in flowering, possession of large, smooth green leaves with minimum leaf-hairs (less potential to carry dirt and microbes) were seriously considered and recorded. In addition to this, the plants ornamental values such as possession of attractive flower heads with deep purple color, higher production of pollen grains, speed and intensity of "tongue tingling" effects of the stems based on the organoleptic tests were included (Fig. 4).

Several workers with multiple years of practical experience in the organic production of herbs volunteered to participate in the preliminary field evaluation in the summer of 1996. From a field with two million plants, about 420 plants which fulfilled most of the above mentioned criteria were selected, numbered, and flagged for further organoleptic tests and phytochemical analysis. Preliminary field organoleptic tests ("tongue tingling") were carried out on 400 plants. About 360 different lines that meet the preliminary selection criteria were selected and propagated by root divisions (Fig. 5). In autumn 1996, all the selected lines were transplanted (about 80 plants from each line with 20–30 plants planted at three different locations). Evaluation continued as above during the 1997–1998 growing seasons.

Fig. 3. A promising *E. purpurea* that had been selected from among two million plants for further propagation, Trout Lake WA, summer 1996.

Fig. 4. Organoleptic evaluation of *E. purpurea*, Trout Lake, WA, summer 1996.

Fig. 5. Left: Preparation of uprooted echinacea for root crown division. Right: Root crown divisions ready for an immediate field planting.

For phytochemical investigation, the aerial parts of 20 plants replicated three times were harvested in July–August, 1997 and 1998. Immediately after harvest, the flower heads were separated by hand from the stems and divided into four different developmental stages as described below (Fig. 6). Stage I (small or early stage): flower buds starting to develop until the beginning of the first appearance of ligulate florets without rolling out (expansion), and dehiscence of disc florets. Stage II (medium stage): the first two to three lines of the disc florets open (dehisced), while the ligulate florets just starting to roll out (expansion) and attain their typical purple color. Stage III (large or mature stage): more than three quarter of the disc florets open, with fully expanded ligulate florets and intensive release of the pollen grains. Stage IV (old or overblown stage): Ligulate florets starting to droop and slightly turning to dark-brown color, with visible seed development and sharp decline in pollen grain formation. This classification was slightly modified from earlier work for *Chamomilla recutita* (Letchamo 1996). After the separation, the samples were dried at 38°C for 72 h and sent to the University of Ottawa for phytochemical analysis.

Sample Preparation

The dried samples were milled to powder and aliquots of each (ca 0.5 g) were extracted three times in 8 mL 70% ethanol using ultrasound (5 min), followed by centrifugation and collection of the subsequent supernate. The extract volume was adjusted to 25 mL, and filtered through a 0.45 μm PTFE membrane prior to injection of 5 μL into HPLC system. The liquid extracts were centrifuged to pellet any particulate matter, and the supernate was filtered as above prior to injection. The oil sample was diluted with ethanol (ca 1/20 for the lipophilic analysis, and 1/5 for the hydrophilic analysis), and filtered as above prior to injection.

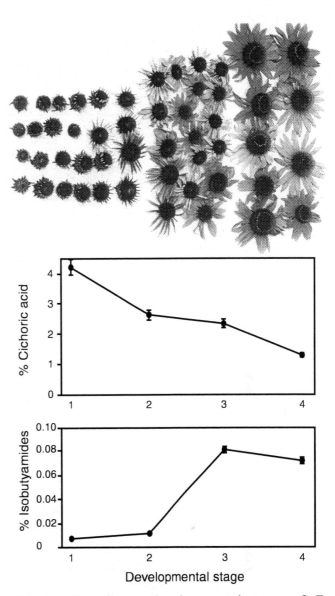

Fig. 6. Top: flower developmental stages of *E. purpurea*, cv. Magical Ruth. Bottom: variations in the content of cichoric acid and isobutylamide in flower heads at different developmental stages (1= stage I, 2= stage II, 3= stage III, 4= stage IV).

HPLC Analysis

Separation and quantitative analysis of the components was based on a modified method of Livesey et al. (1998), based on Bauer and Remiger (1989). HPLC (C-18 column, 75 × 4.6 mm; 3 μm) was used to determine the lipophilic and hydrophilic constituents.

RESULTS

Identification and Naming of the Cultivar 'Magical Ruth'

The new *E. purpurea* cultivar 'Magical Ruth' (Fig. 3) was selected because of its relatively unique agronomic traits, freedom from pests and diseases, excellent frost resistance, attractive ornamental qualities, uni-

Table 1. Variations in the content of echinacoside and chlorogenic acid in flower heads of the new *Echinacea purpurea* cultivar 'Magical Ruth'. The results are obtained from 20 plants replicated three times.

Flower stage	Hydrophilic components (%)	
	Chlorogenic acid	Echinacoside
I (early)	0.060	0.012
II (medium)	0.024	0.022
III (mature)	0.023	0.015
IV (overblown)	0.020	0.016

formity in growth, flowering, high yielding potentials, and better "tongue numbing" effect compared with other *E. purpurea* lines under investigation. There were also a dozen of other interesting lines that were simultaneously propagated in the experimental fields and meet the selection criteria.

Cichoric Acid

The highest content of cichoric acid was found at stage I (4.67%), with content declining to stage IV (1.42%) (Fig. 6).

Alkamide (Isobutylamide)

The concentration of isobutylamide was higher in stage III and IV than stage I and II with the peak at stage III, (Fig. 6).

Echinacoside and Chlorogenic Acid

Echinacoside content was maximal at stage 2 (Table 1). Maximum content of chlorogenic acid was recorded at the 1st stage, while the lowest concentration was measured at the 4th stage (Table 1).

Our results indicate that the quality of echinacea is strongly influenced by the floral developmental stage. To obtain optimum yield levels of both hydrophilic and lipophilic components, echinacea flowers should be harvested at the 3rd (mature) developmental stage. It is important that floral developmental stages be considered during screening and selection.

REFERENCES

Bauer, R. and P. Remiger. 1989. TLC and HPLC analysis of alkamides in *Echinacea* drugs. Planta Med. 55:367–371.

Bauer, R. and H. Wagner. 1991. *Echinacea* species as potential immunostimulatory drugs. Vol. 5/8. p. 253–321. In: H. Wagner and N.R. Farnsworth (eds.), Economic and medicinal plant research. Academic Press, New York.

Brevoort, P. 1996. The US botanical market. Herbalgram 36:49–57.

Letchamo, W. 1996. Developmental and seasonal variations in flavonoids of diploid and tetraploid camomile ligulate florets. J. Plant Physiol. 148:645–651.

Letchamo, W., L.V. Polydeonny, V.S. Krutilina, O.N. Gladisheva, S.A. Kodash, and T.J. Arnason. 1999. Factors affecting the yield and chemical quality of *Echinacea* and new methods of improvement for industrial processing. First AHPA International Echinacea symposium, Kansas City, MO, June 3–5.

Livesey J., C. Bergeron, J. Rana, D.V.C. Awang, W. Letchamo, and J. Arnason. 1998. Developing of a method for identity assurance and quality control in *Echinacea* spp. Paper presented at the conference of the American Society of Pharmacognosy, Orlando, Florida, July 1998.

Basil: A Source of Aroma Compounds and a Popular Culinary and Ornamental Herb*

James E. Simon, Mario R. Morales, Winthrop B. Phippen, Roberto Fontes Vieira, and Zhigang Hao

Basil, one of the most popular culinary herbs in North America is sold as a fresh-cut and dried processed product. There are over 40 cultivars available (De Baggio and Belsinger 1996), with many developed specifically for the fresh and/or ornamental markets. The popular cultivars for the fresh market and garden have dark green leaves and white flowers, with a rich spicy pungent aroma due to the presence of linalool/methylchavicol/1,8-cineole. Lesser known cultivars vary in growth habit, size, and color, and can contain a wide range of aromas including, lemon, rose, camphor, licorice, woody, and fruity. The popularity of basil has led to an infusion of many introductions into the marketplace. For example, new cultivars, both All-American selections, include the new fusarium resistant lemon basil, 'Sweet Dani', a tall (65–70 cm), up-right plant with an intense lemon aroma (Morales and Simon 1997) and 'Siam Queen', an attractive plant with purple flowers on a dense dark green foliage. This paper will address the diversity of basil in the North American market and potential new uses for the natural products of this species.

DIVERSITY IN BASIL

The genus *Ocimum*, Lamiaceae, collectively called basil, has long been acclaimed for its diversity. *Ocimum* comprises more than 30 species of herbs and shrubs from the tropical and subtropical regions of Asia, Africa, and Central and South America, but the main center of diversity appears to be Africa (Paton 1992). It is a source of essential oils and aroma compounds (Simon et al. 1984, 1990), a culinary herb, and an attractive, fragrant ornamental (Morales et al. 1993; Morales and Simon 1996). The seeds contain edible oils and a drying oil similar to linseed (Angers et al. 1996). Extracts of the plant are used in traditional medicines, and have been shown to contain biologically active constituents that are insecticidal, nematicidal, fungistatic, or antimicrobial (Simon 1990; Albuquerque 1996).

Most commercial basil cultivars available in the market belong to the species *O. basilicum*. Darrah (1980, 1984) classified the *O. basilicum* cultivars in seven types: (1) tall slender types, which include the sweet basil group; (2) large-leafed, robust types, including 'Lettuce Leaf' also called 'Italian' basil; (3) dwarf types, which are short and small leafed, such as 'Bush' basil; (4) compact types, also described, *O. basilicum* var. *thyrsiflora*, commonly called 'Thai' basil; (5) *purpurascens*, the purple-colored basil types with traditional sweet basil flavor; (6) purple types such as 'Dark Opal', a possible hybrid between *O. basilicum* and *O. forskolei*, which has lobed-leaves, with a sweet basil plus clove-like aroma; and (7) *citriodorum* types, which includes lemon-flavored basils.

In addition to the traditional types of basil, other *Ocimum* species have been introduced into the North American horticultural trade with new culinary and ornamental uses and may be potential sources of new aroma compounds. However, interspecific hybridization and polyploidy, common occurrences within this genus (Harley et al. 1992), have created taxonomic confusion making it difficult to understand the genetic relationship between many basils (Grayer et al. 1996). In addition, the taxonomy of basil is complicated by the existence of numerous botanical varieties, cultivar names, and chemotypes within the species that may not differ significantly in morphology (Simon et al. 1990). A system of standardized descriptors, which include volatile oil, has more recently been proposed by Paton and Putievsky (1996) and this should permit easy communication and identification of the different forms of *O. basilicum*. Investigations to revise the genus are underway at the Royal Botanical Garden, Kew, London (Paton 1992) and at Delaware State University.

The perfume, pharmacy, and food industries (Simon et al. 1990) use aromatic essential oils, extracted from the leaves and flowers of basil. Several aroma compounds can be found in chemotypes of basil such as citral, eugenol, linalool, methylchavicol, and methylcinnamate and are traded in the international essential oil market. These chemotypes are commonly known by names based on geographical origins such as Egyptian,

*J. Paper no. 16,045 of the Purdue Agricultural Research Programs, Purdue University, West Lafayette, IN. We thank Jules Janick for his suggestions in the preparation of this paper.

Table 1. Comparative evaluation of North American commercially available *Ocimum* basil cultivars. Plant density was 12,000 plants/ha.

				Plant characteristics				
				Color				
Cultivar	*Ocimum* species	Height (cm)	Spread (cm)	Leaf	Stem	Flower	Spike	Days to flowering
African Blue	Dark Opal × *kilimandscharicum*	44	49	green	pale-purple	pink	pale-purple	88
Anise	*basilicum*	55	44	green-purple	pale-purple	light-pink	purple	98
Bush	*minimum*	23	29	green	green	white	green	109
Camphor	*kilimandscharicum*	58	49	green	green-purple	white	gray	92
Cinnamon	*basilicum*	44	49	green	purple	pink	purple	86
Dark Opal	*basilicum*	43	36	purple	purple	pink	purple	97
Dwarf Bush Fineleaf	*minimum*	20	31	green	green	white	green	115
East Indian	*gratissimum*	52	43	green	green-purple	gray	green-purple	114
Fino Verde	*basilicum*	44	44	green	green	white	green	104
Genovese	*basilicum*	49	46	green	green	white	green	95
Green	*gratissimum*	42	25	green	green-purple	white-gray	green-purple	123
Green Bouquet	*americanium* var. *americanum*	30	32	green	green	white	green	112
Green Globe	*minimum*	23	31	green	green	white	green	115
Green Ruffles	*basilicum*	29	29	light-green	light-green	white	light-green	110
Holy	*tenuiflorum*	37	41	green	pale-purple	pale-purple	purple	98
Holy Sacred Red	*basilicum*	40	35	purple	purple	pink	purple	103
Italian Large Leaf	*basilicum*	46	42	green	green	white	green	104
Lemon	×*citriodorum*	33	53	green	green	white	green	76
Lemon Mrs. Burns	×*citriodorum*	52	55	green	green	white	green	94
Lettuce Leaf	*basilicum*	41	41	green	green	white	green	105
Licorice	*basilicum*	53	52	green-purple	green-purple	pink	purple	102
Maenglak Thai Lemon	×*citriodorum*	34	50	green	green	white	green	73
Mammoth	*basilicum*	41	38	green	green	white	green	103
Napoletano	*basilicum*	41	41	green	green	white	green	104
New Guinea	×*citriodorum*	27	35	pale-green	pale-green	pink	pale-green	78
Opal	*basilicum*	36	35	purple	purple	pink	purple	99
Osmin Purple	*basilicum*	40	32	purple	purple	pink	purple	101
Peruvian	*campechianum*	30	37	green	green	pale-purple	green	98
Purple Bush	*minimum*	24	28	purple-green	purple	pale-purple	purple	109
Purple Ruffles	*basilicum*	34	29	purple	purple	bright purple	dark-purple	134
Red Rubin Purple Leaf	*basilicum*	42	38	purple	purple	pink	purple	106
Sacred	*tenuiflorum*	32	32	green	pale-purple	pale-purple	gray	90
Spice	*americanium* var. *pilosum*	34	48	green	green	pink	green-purple	76
Spicy Globe	*americanium* × *basilicum*	21	30	green	green	white	green	115
Sweet	*basilicum*	49	42	green	green	white	green	100
Sweet Fine	*basilicum*	48	43	green	green	white	green	96
Sweet Thai	*basilicum*	35	44	green	purple	pink	purple	74
Thai (Companion Plants)	*basilicum*	41	47	green	purple	pink	purple	82
Thai (Richters)	*basilicum*	35	36	green	pale-purple	pink	purple	100
Thai (Rupp Seeds)	*basilicum*	36	52	green	green	white	green	72
Tree	*gratissimum*	48	38	green	green-purple	white-gray	green-purple	120
Tulsi or Sacred	*tenuiflorum*	45	41	green-purple	purple	pale-purple	purple-green	99

[z] 1 (highly vigorous) to 5 (lacks vigor)

[y] 1 (highly uniform) to 5 (lacks uniformity)

[x] 1 (no damage) to 5 (severe damage)

Vigor rating (1 to 5)[y]	Uniformity rating (1 to 5)[y]	Japanese beetle rating (1 to 5)[x]	Yield/plant Fresh wt. (g)	Dry wt. (g)	Essential oil yield %(vol/gdwt)	Major aroma compounds
2.0	1.0	1.0	651	121	2.34	linalool (55%), 1,8-cineole (15%), camphor (22%)
1.8	2.5	4.6	640	129	0.62	linalool (56%), methylchavicol (12%)
2.0	1.3	1.0	545	92	1.18	linalool (59%), 1,8-cineole (21%)
1.6	2.5	1.0	755	163	5.22	camphor (61%), 1,8-cineole (20%)
2.5	3.1	1.0	638	131	0.94	linalool (47%), methylcinnamate (30%)
2.5	2.7	1.0	445	74	1.08	linalool (80%), 1,8-cineole (12%)
2.0	2.0	1.3	443	75	0.95	linalool (55%), methylchavicol (4%)
3.3	2.6	1.0	409	101	0.35	eugenol (62%)
2.7	3.2	2.0	868	211	0.50	linalool (48%), methylchavicol (7%)
2.8	2.5	3.6	734	137	0.90	linalool (77%), 1,8-cineole (12%)
3.0	1.8	1.0	211	50	0.29	thymol (20%), p-cymene (33%)
2.8	3.3	1.6	533	96	0.82	linalool (36%), 1,8-cineole (31%)
1.6	2.1	1.0	573	92	0.96	linalool (54%), 1,8-cineole (25%)
3.1	2.0	3.0	478	74	0.55	linalool (33%), 1,8-cineole (18%)
2.7	2.5	3.5	236	58	0.93	β-caryophyllene (75%)
2.8	2.6	1.0	361	55	0.83	linalool (77%), 1,8-cineole (14%)
2.1	2.0	4.0	626	113	0.83	linalool (65%), methylchavicol (18%)
2.0	1.8	1.0	432	118	0.27	linalool (24%), citral (19%)
2.0	2.1	1.0	883	177	1.28	linalool (61%), citral (16%)
2.1	2.0	4.0	824	129	0.78	linalool (60%), methylchavicol (29%)
1.3	2.1	5.0	732	144	0.43	linalool (58%), methylchavicol (13%)
2.1	1.1	1.0	473	138	0.23	citral
2.3	2.1	5.0	703	111	0.77	linalool (60%), methylchavicol (32%)
2.0	1.5	4.5	635	108	0.89	linalool (66%), methylchavicol (10%)
3.3	2.0	1.0	339	75	0.69	methylchavicol (92%)
2.2	2.2	1.0	516	78	0.91	linalool (80%), 1,8-cineole (13%)
2.0	1.5	1.0	343	53	0.66	linalool (77%), 1,8-cineole (15%)
3.3	2.3	1.0	347	83	1.26	linalool (14%), 1,8-cineole (36%)
3.1	1.6	1.0	348	54	1.12	linalool (82%), 1,8-cineole (11%)
3.6	3.0	1.0	294	42	0.49	linalool (55%), 1,8-cineole (20%), methylchavicol (6%), methyleugenol (9%)
2.1	2.0	1.0	403	63	0.74	linalool (70%), 1,8-cineole (9%), methylchavicol (10%), methyleugenol (6%)
4.8	2.6	1.0	216	52	0.65	β-caryophyllene (45%)
1.3	2.6	1.0	517	132	0.22	1,8-cineole (32%), methylchavicol (16%), β-bisabolene (33%)
2.0	1.3	1.0	504	78	0.88	linalool (55%), 1,8-cineole (23%)
1.8	2.1	3.6	651	114	0.84	linalool (69%), 1,8-cineole (11%), methylchavicol (13%)
1.5	3.3	1.0	631	129	0.98	linalool (86%), 1,8-cineole (6%)
2.3	2.1	1.0	723	168	0.40	linalool (6%), methylchavicol (60%)
2.0	3.5	2.0	536	111	0.75	linalool (12%), methylchavicol (65%)
1.8	2.1	1.6	443	86	0.52	methylchavicol (90%)
1.8	1.8	1.0	463	139	0.25	linalool (15%), methylchavicol (13%)
3.0	2.3	1.0	368	89	0.51	eugenol (62%)
3.1	3.3	2.6	314	77	0.93	methyleugenol (33%), β-caryophyllene (41%)

French, European, or Reunion basil (Heath 1981; Sobti et al. 1982; Simon et al. 1990; Marotti et al. 1996). The European type, a sweet basil, is considered to have the highest quality aroma, containing linalool and methylchavicol as the major constituents (Simon et al. 1990). The Egyptian basil is very similar to the European, but contains a higher percentage of methylchavicol. The Reunion type, from the Comoro Islands, and more recently from Madagascar, Thailand, and Vietnam, is characterized by high concentrations of methylchavicol (Marotti et al. 1996). Methylcinnamate-rich basil has been commercially produced in Bulgaria (Simon et al. 1990), India, Guatemala, and Pakistan (Marotti et al. 1996). A basil from Java (Simon et al. 1990), and Russia and North Africa (Marotti et al. 1996) is rich in eugenol.

Fig. 1. Variation in basil (*Ocimum basilicum*) cultivars. (A) 'Bush' basil, a green leafed compact ornamental; (B) 'Sweet' basil, a popular, tall fresh market basil; (C) 'Dark Opal', a purple ornamental that is a rich source of anthocyanins; (D) 'Purple Bush', a compact ornamental with green-purple leaves.

CULTIVAR EVALUATION

In 1996, commercially available basils were evaluated at Lafayette, Indiana for horticultural attributes, commercial potential, essential oil yield and composition, and anthocyanin content. A total of 86 entries, including 42 different cultivars, were transplanted May 23 (Replication 1) and May 31, 1996 (Replications 2 and 3) in a randomized block design. Each plot consisted of a twin row of plants, 2 m long, 0.5 m apart, with plants spaced 0.5 m apart within the row for a density of 12,000 plants/ha. Entries were evaluated for appearance, shape, size, uniformity and vigor, biomass yield (fresh and dry weight when harvested at full bloom), extractable oil yield, and susceptibility to Japanese beetles. The harvested plant was first weighed, dried at 37°C, and oils extracted. The major volatile oil constituents were determined from GC and/or GC/MS analysis, as previously described (Charles and Simon 1990).

Basil as an Edible Ornamental Herb

The cultivars displayed a wide diversity in growth habit, flower, leaf and stem colors, and aromas (Table 1, Fig. 1). Many of the cultivars evaluated belong to the "Sweet" basil group, with 'Genovese', 'Italian large leaf', 'Mammoth', 'Napoletano', and 'Sweet' dominating the American fresh and dry culinary herb markets. Several others like 'Sweet Fine' appear similar to 'Sweet' basil though its leaves tend to be smaller. The lemon-scented cultivars ('Lemon' and 'Lemon Mrs. Burns') differed from each other in days to flower, and total oil content, but not in citral content. The 'Maenglak Thai Lemon' basil, which varied in appearance from the other lemon basils, is an attractive ornamental. Among the purple basils, 'Osmin Purple' and 'Red Rubin Purple Leaf' were the most attractive and best retained their purple leaf color. Anthocyanins in purple basils are genetically unstable leading to an undesirable random green sectoring and reversion over the growing season (Phippen and Simon 1998). Several basils with dwarf growth habit were developed as ornamental border plants including 'Bush', 'Green Globe', 'Dwarf Bush', 'Spicy Globe', and 'Purple Bush'. A group of ornamental basils were selected and named for their characteristic aroma including 'Anise' (methyl chavical), 'Cinnamon' (methylcinnamate), 'Licorice' (methylchavicol), and 'Spice' (β-bisabolene).

Many of the cultivars lacked uniformity, which suggests that further selection to improve these lines for such characters as earliness, modified appearance, and/or insect and disease resistance is feasible. Wide variation was observed on the degree of foliar damage caused by Japanese beetle.

Basil as a Source of Essential Oils and Aroma Compounds

Basil plants evaluated in this study contained a wide variety of oil compounds reflecting a diversity of available aromas and flavors. Many contained a combination of linalool and methylchavicol and/or 1,8-cineole, reflecting the traditional sweet basil aroma. Others had distinct aromas. The predominant aroma compound was eugenol (62%) in 'East Indian' and 'Tree' basil; camphor (61%) in 'Camphor' basil; thymol in 'Green' basil; β-caryophyllene in 'Holy' and 'Sacred' basil; and methylchavicol in 'New Guinea', and 'Thai - Richters' basil. Several purple cultivars ('Holy Sacred Red', 'Opal', and 'Osmin Purple') were rich in linalool (ca. 77%), as was the green cultivar 'Sweet Fine' (86%). Citral was the predominant compound in all the lemon-scented basils. However, few of these cultivars could compete as industrial sources of these compounds. Surprisingly, 'Camphor' basil, a very vigorous cultivar, was found to have relatively high oil yield (>5% dry weight).

Basil as a Source of Anthocyanins

The intensely purple pigmented basils available as culinary ornamentals prompted an examination of eight commercial cultivars as a potential new source of anthocyanins. The anthocyanins present in purple basils were analyzed utilizing high performance liquid chromatography, spectral data, and plasma-desorption mass spectrometry (Phippen and Simon 1998). Fourteen different anthocyanins were identified. Eleven of the pigments were cyanidin based with cyanidin-3-(di-p-coumarylglucoside)-5-glucoside as the major pigment. Three minor pigments based on peonidin were identified. Purple basils can be an abundant source of acylated and glycosylated anthocyanins and could provide a unique source of stable red pigments to the food industry

Table 2. Comparison of purple basil to other natural anthocyanin fruit and leaf sources for total anthocyanin yield. Modified from Phippen and Simon (1998).

Source	Plant organ sampled	Total anthocyanins (mg/100gfw ± SD)	No. of identified anthocyanins
Grape (*Vitis labrusca* 'Concord')	Fruit skin	25.2±2.9	16
Purple basil (*Ocimum basilcum* 'Purple Ruffles')	Leaf	18.8±0.7	14
Perilla (*Perilla frutescens* 'Crispa')	Leaf	18.1±0.6	6
Plum (*Prunus domestica* L. 'Stanley')	Fruit skin	15.9±2.1	2
Purple sage (*Salvia officinalis* 'Purpurea')	Leaf	7.8±1.0	5
Red cabbage (*Brassica oleracea* 'Cardinal')	Leaf	7.8±1.4	6
Red raspberry (*Rubus idaeus* 'Heritage')	Fruit	1.0±0.7	2

Table 3. Total extractable anthocyanin yields among commercial purple basil cultivars as compared to perilla. Modified from Phippen and Simon (1998).

Cultivar	Seed source	Total extractable anthocyanins[z] (mg/100 g fw ±SD)
Dark Opal	Richters	16.3±1.8
	Rupp Seeds	18.7±0.8
Holy Sacred Red	Rupp Seeds	8.8±0.6
Opal	Companion Plants	12.8±0.4
	Nichols Garden Nursery	11.7±0.6
Osmin Purple	Johnny's	18.0±0.8
Purple Bush	Richters	6.5±1.1
Purple Ruffles	Richters	18.5±0.7
	Shepard's Garden Seeds	17.9±0.8
	Johnny's	15.9±0.5
	Ball Seed	18.7±0.8
Rubin	Richters	17.4±0.8
Red Rubin	Johnny's	17.4±0.8
Red Rubin Purple Leaf	Shepard's Garden Seeds	18.3±0.5
Perilla frutescens var. *crispa*	Shepard's Garden Seeds	18.0±0.6

[z]Anthocyanins were extracted from leave tissue with acidified methanol (0.1% HCl) for 3 h at 4°C. Samples were filtered and analyzed by reverse phase HPLC.

(Table 2). The large leafed cultivars 'Purple Ruffles', 'Rubin', and 'Dark Opal' had an average extractable total anthocyanin content ranging from 16.6–18.8 mg per 100 g fresh tissue (Table 3), while the ornamental small leafed cultivar 'Purple Bush' had only 6.5 mg per 100 g fresh tissue.

CONCLUSIONS

The diversity in basil based on appearance, flavors, fragrances, industrial, edible, and drying oils, and natural pigments offers a wealth of opportunities for developing new culinary, ornamental, and industrial crops. The high variation of basil cultivars to damage by Japanese beetles suggests the presence of an active ingredient that either could be useful in commercial traps or serve as a deterrent.

A number of basils have commercial potential for the production of industrial products. 'Camphor', a cultivar from Africa, should be investigated further as a potential industrial source of camphor, while 'East Indian' and 'Tree' basils as a potential source of eugenol. Purple basils contained very high concentrations of total anthocyanins and are an abundant source of total anthocyanins and may serve as a potential new source of stable red pigments for the food industry.

REFERENCES

Albuguerque, U. 1996. Taxonomy and ethnobotany of the genus *Ocimum*. Federal Univ. Pernambuco.

Angers, P., M.R. Morales, and J.E. Simon. 1996. Fatty acid variation in seed oil among *Ocimum* species. J. Am. Oil Chem. Soc. 73:393–395.

Charles, D.J. and J.E. Simon. 1992. A new geraniol chemotype of *Ocimum gratissimum* L. J. Essential Oil Res. 4:231–234.

Charles, D.J. and J.E.Simon. 1990. Comparison of extraction methods for the rapid determination of essential oil content and composition of basil (*Ocimum* spp). J. Am. Soc. Hort. Sci. 115:458–462.

Charles, D.J., J.E Simon, and K.V. Wood. 1990. Essential oil constituents of *Ocimum micranthum* Willd. J. Agr. Food Chem. 8:120–122.

Darrah, H. 1974. Investigations of the cultivars of basils (*Ocimum*). Econ. Bot. 28:63–67.

Darrah, H.H. 1980. The cultivated basils. Buckeye Printing Co., MO.

De Baggio, T. and S. Belsinger. 1996. Basil: An herb lover's guide. Interweave Press, CO.

Heath, H.B. 1981. Source book of flavors. AVI, Westport, CT.

Grayer, R.J., G.C. Kite, F.J. Goldstone, S.E. Bryan, A. Paton, and E. Putievsky. 1996. Intraspecific taxonomy and essential oil chemotypes in sweet basil, *Ocimum basilicum*. Phytochemistry 43:1033–1039.

Harley, M.M., A. Paton, R.M. Harley, and P.G. Cade. 1992. Pollen morphological studies in the tribe *Ocimeae* (*Nepetoideae*: Labiatae): *Ocimum* L. Grana 31:161–176.

Marotti, M., R. Piccaglia, and E. Giovanelli. 1996. Differences in essential oil composition of basil (*Ocimum basilicum* L.) Italian cultivars related to morphological characteristics. J. Agr. Food Chem. 44:3926–3929.

Morales, M.R. and J.E. Simon. 1997. 'Sweet Dani': A new culinary and ornamental lemon basil. HortScience 32:148–149.

Morales, M.R. and J.E. Simon. 1996. New basil selections with compact inflorescence for the ornamental market. p. 543–546. In: J. Janick (ed.), Progress in new crops. ASHS Press, Alexandria, VA.

Morales, M.R., D.J. Charles, and J.E. Simon. 1993. New aromatic lemon basil germplasm. p. 632–635. In: J. Janick and J.E. Simon (eds.), New crops. Wiley, New York.

Paton, A. 1992. A synopsis of *Ocimum* L. (Labiatae) in Africa. Kew Bul. 47:403–435.

Paton, A. and E. Putievsky. 1996. Taxonomic problems and cytotaxonomic relationships between varieties of *Ocimum basilicum* and related species (Labiatae). Kew Bul. 5:1–16.

Phippen, W.B. and J.E. Simon. 1998. Anthocyanins in basil. J. Agr. Food Chem. 46:1734–1738.

Simon, J.E., A.F. Chadwick, and L.E. Craker. 1984. Herbs: An indexed bibliography 1971–1980. Archon Books, Hamden. p. 7–9.

Simon, J.E., J. Quinn, and R.G. Murray. 1990. Basil: a source of essential oils. p. 484–489. In: J. Janick and J.E. Simon (eds.), Advances in new crops. Timber Press, Portland, OR.

Sobti, S.N. and P. Pushpangadan. 1982. Studies in the genus *Ocimum*: Cytogenetics, breeding and production of new strains of economic importance. p. 457–472. In: C.K. Atal and B. M. Kapur (eds.), Cultivation and utilization of aromatic plants. Regional Laboratory Council of Scientific and Industrial Research, Jammu-Tawi, India.

Vieira, R.F. and J.E. Simon. 1999. Chemical characterization of basil (*Ocimum* spp.) found in the markets and used in the traditional medicine in Brazil. Econ. Bot. (in press).

Culantro: A Much Utilized, Little Understood Herb

Christopher Ramcharan

Culantro (*Eryngium foetidum* L., Apiaceae) is a biennial herb indigenous to continental Tropical America and the West Indies. Although widely used in dishes throughout the Caribbean, Latin America, and the Far East, culantro is relatively unknown in the United States and many other parts of the world and is often mistaken and misnamed for its close relative cilantro or coriander (*Coriandrum sativum* L.). Some of its common names descriptive of the plant include: spiny or serrated coriander, *shado beni* and *bhandhania* (Trinidad and Tobago), *chadron benee* (Dominica), *coulante* (Haiti), *recao* (Puerto Rico), and fit weed (Guyana).

Culantro grows naturally in shaded moist heavy soils near cultivated areas. Under cultivation, the plant thrives best under well irrigated shaded conditions. Like its close relative cilantro, culantro tends to bolt and flower profusely under hot high-light long days of summer months. Recent research at UVIAES has demonstrated that it can be kept in a vegetative mode through summer when treated with GA_3 sprays.

The plant is reportedly rich in calcium, iron, carotene, and riboflavin and its harvested leaves are widely used as a food flavoring and seasoning herb for meat and many other foods. Its medicinal value include its use as a tea for flu, diabetes, constipation, and fevers. One of its most popular use is in chutneys as an appetite stimulant. The name fitweed is derived from its supposedly anti-convulsant property. The presence of increasingly large West Indian, Latin American, and Asian immigrant communities in metropolises of the US, Canada and the UK. creates a large market for culantro and large quantities are exported from Puerto Rico and Trinidad to these areas.

Culantro is increasingly becoming a crop of international trade mainly to meet the demands of ethnic populations in the developed countries of the West. Large immigrant communities in London, New York, and Toronto represents a vast potential market for the herb. One exporter from Trinidad alone packages and air freights up to 2.4 t of fresh culantro weekly to the US. In 1988, Puerto Rico reportedly produced 165,000 kg of culantro for a value of $201,000 (Dept. of Agriculture 1988). The herb is used extensively in the Caribbean and in Asia particularly in India and Korea. It is used mainly as a seasoning in the preparation of a range of foods, including vegetable and meat dishes, chutneys, preserves, sauces, and snacks. Although used in small quantities, its pungent unique aroma gives the characteristic flavor to the dishes in which it is incorporated and this is responsible for its increasing demand among ethnic populations. Culantro is also widely used in herbal medicines and reportedly beneficial in the treatment of a number of ailments (Wong 1976).

NOMENCLATURE

The derivation of culantro and recao, by which the plant is commonly known in Central America, is unknown but many of its names outside its natural habitat compare it to the common coriander or cilantro, e.g. Hindi bhandhanya, broad dhanya, or coriander, and Thai pak chi farang "foreign coriander." The botanical genus name *Eryngium* is derived from the Greek sea holly, *Eryngium vulgare*, and its specific name comes from the Latin *foetidum* meaning stink or bad odor; its smell is sometimes equated to a crushed bedbug. Some of the common names of culantro in the Caribbean area are: *shado beni* (Trinidad), *chadron benee* (Dominica), fitweed (Guyana), *coulante* or *culantro* (Haiti), *recao* (Puerto Rico) (Seaforth et al. 1983; Morean 1988; Seaforth 1988). Names in different languages include: *langer koriander* (German); *ketumbar java* (Malay); *pak chi farang* (Thai); *ngo gai* (Vietnamese); *culantro, racao, recao* (Spanish); *bhandhanya* (Hindi), and long leaf or spiny coriander (English).

Eryngium comprises over 200 tropical and temperate species (Willis 1960). Most are spiny ornamental herbs with thick roots and fleshy waxy leaves with blue flowers in cymose heads. *Eryngium foetidum* is a tap-rooted biennial herb with long, evenly branched roots (Fig. 1). The oblanceolate leaves, arranged spirally around the short thick stem, form a basal rosette and are as much as 30 cm long and 4 cm broad. The leaf margin is serrated, each tooth of the margin containing a small yellow spine. The plant produces a well-branched cluster of flower heads in spikes forming the characteristic umbel inflorescence on a long stalk arising from the center of the leaf rosette (Morton 1981; Moran 1988). The calyx is green while the corolla is creamy white in color.

ECOLOGY AND CULTURE

Culantro is native to continental tropical America and the West Indies (Adams 1971). It grows naturally throughout many Caribbean islands including Trinidad and Tobago, where it is abundant in forests particularly in disturbed areas as in slash and burn sites. The herb is also commonly found along moist or shaded pathways and near cultivated areas where heavy soils predominate (Seaforth et al. 1983; Morean 1988). Although the plant grows well in full sun most commercial plantings occur in partially shaded moist locations. Shaded areas produce plants with larger and greener leaves that are more marketable because of their better appearance and higher pungent aroma. In a study on the effects of light intensity on growth and flowering of culantro, a significant delay in flowering and increased fresh weight of leaves were found in plants grown under 63% to 73% shade (Santiago–Santos and Cedeno–Maldonado 1991). Shaded plants also had fewer inflorescences with lower fresh weight. Although culantro grows in a wide variety of soils, it does best in moist well drained sandy loams high in organic matter particularly under full light. Precise fertilizer recommendations have not been made but high nitrogen fertilizers or manures promote leaf growth. Plants are usually started from seed which germinate in about 30 days, and for home or backyard gardens can be cultivated in containers or wooden boxes. For such cultivation, a slow release fertilizer such as Osmocote (14–14–14) can be incorporated in the soil mix at the rate of 1.8 kg/m^3.

Like many of its relatives, culantro tends to bolt and flower profusely under long day conditions resulting in reduced leaf growth and market value, and increasing costs for flower pruning. In a study to reduce bolting and increase leaf:flower size, ProGibb (PG) 4% a vegetative growth promoter was applied in increasing concentrations as a foliar spray to 1-month old culantro plants grown under 54% shade in a poly greenhouse (Ramcharan 1998). While leaf length increased with increasing levels of PG, leaf dry weight increased up to the 150 ppm PG level but was reduced at 200 ppm. Concomitantly, both fresh and dry weights of inflorescences were reduced by increasing levels of PG. Flowers produced in treated plants were less woody and spiny and leaf-like in appearance, making them easier for pruning. Pro Gibb 4% at 100 ppm concentration was therefore found to be optimum for maximizing leaf production and minimizing flower growth in culantro.

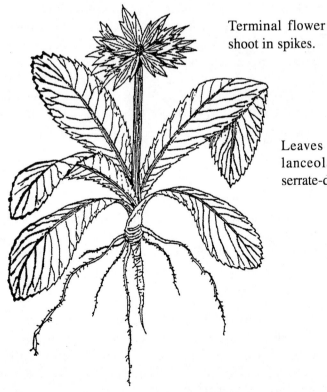

Terminal flower shoot in spikes.

Leaves in cluster; lanceolate, acute, serrate-dentate.

Fig. 1. The culantro plant, *Eryngium foetidum*.

Culantro is relatively pest- and disease-free but the author has seen root knot nematodes on plants that have been grown for 2–3 years in box containers. A leaf spot problem which appears to be bacterial black rot (*Xanthomonas* sp.) has also been observed on such long-lived plants. Anecdotal reports mention that the flower heads are attractive to ladybugs, green lacewings, and other beneficial insects. Plants around the garden have also reportedly provided excellent defense against aphids.

While there are few reports on cultivation and fertilizer requirements for culantro, there has been considerable research on postharvest techniques for the herb. In a refrigerated storage trial, Sankat and Maharaj (1991) found that unpackaged culantro became unmarketable within 4 days of storage regardless of temperature. Storage at 10°C extended shelf life up to 2 weeks and chilling injury was observed at 3°C after 8 days in storage. In another postharvest study, the combination of polyethylene packaging, gibberellic acid (GA_3) in a 200 ppm dip treatment and reduced storage temperature (20°–22°C) extended the shelf-life of culantro up to 22 days (Mohammed and Wickham 1995). Freeze drying of harvested leaves is another alternative being considered to extend postharvest life.

CULINARY USES AND NUTRITIONAL VALUE

The appearance of culantro and cilantro are different but the leaf aromas are similar, although culantro is more pungent. Because of this aroma similarity the leaves are used interchangeably in many food preparations and is the major reason for the misnaming of one herb for the other. While relatively new to American cuisine, culantro has long been used in the Far East, Latin America, and the Caribbean. In Asia, culantro is most popular in Thailand, Malaysia, and Singapore where it is commonly used with or in lieu of cilantro and topped over soups, noodle dishes, and curries. In Latin America, culantro is mostly associated with the cooking style of Puerto Rico, where recipes common to all Latin countries are enhanced with culantro. The most popular and ubiquitous example is *salsa*, a spicy sauce prepared from tomatoes, garlic, onion, lemon juice, with liberal amounts of chiles. These constituents are fried and simmered together, mixed to a smooth paste and spiced with fresh herbs including culantro. Salsa is usually consumed with tortilla chips as an appetizer. Equally popular is *sofrito* or *recaito*, the name given to the mixture of seasonings containing culantro and widely used in rice, stews, and soups (Wilson 1991). There are reportedly as many variations of the recipe as there are cooks in Puerto Rico but basically sofrito consists of garlic, onion, green pepper, small mild peppers, and both cilantro and culantro leaves. Ingredients are blended and can then be refrigerated for months. Sofrito is itself the major ingredient in a host of other recipes including eggplant pasta sauce, cilantro garlic butter, cilantro pesto, pineapple salsa, and gazpacho with herb yogurt.

Culantro is reported to be rich in calcium, iron, carotene, and riboflavin. Fresh leaves are 86–88% moisture, 3.3% protein, 0.6% fat, 6.5% carbohydrate, 1.7% ash, 0.06% phosphorus, and 0.02% iron. Leaves are an excellent source of vitamin A (10,460 I.U./100 g), B_2 (60 mg %), B_1 (0.8 mg %), and C (150–200 mg %) (Bautista et al. 1988). On a dry weight basis, leaves consist of 0.1–0.95% volatile oil, 27.7% crude fiber, 1.23% calcium, and 25 ppm boron.

MEDICINAL USES

The plant is used in traditional medicines for fevers and chills, vomiting, diarrhea, and in Jamaica for colds and convulsions in children (Honeychurch 1980). The leaves and roots are boiled and the water drunk for pneumonia, flu, diabetes, constipation, and malaria fever. The root can be eaten raw for scorpion stings and in India the root is reportedly used to alleviate stomach pains. The leaves themselves can be eaten in the form of a chutney as an appetite stimulant (Mahabir 1991).

CONCLUSION

Although used widely throughout the Caribbean, Latin America, and the Far East, culantro is still mistaken for and erroneously called cilantro. The herb is rapidly becoming an important import item into the US mainly due to the increasing ethnic immigrant populations who utilize it in their many varied dishes from around the world. It is closely related botanically to cilantro but has a distinctly different appearance and a much more potent volatile leaf oil. Recent research to prevent bolting and early flowering will increase its leaf yields and consequently its demand. Successes in prolonging its postharvest life and storage under refrigeration will undoubtedly increase its export potential and ultimately its popularity among the commonly used culinary herbs.

REFERENCES

Adams, C.D. 1971. Flowering plants of Jamaica. Univ. West Indies, Mona, Jamaica, W.I.

Bautista, O.K, S. Kosiyachinda, A.S.A. Rahman, and Soenoeadji. 1988. Traditional vegetables of Asia. Asian Food J. 4:47–58.

Honeychurch, P.N. 1980. Caribbean wild plants and their uses. Letchworth Press, Barbados, W.I.

Mahabir, K. 1991. Medicinal and edible plants used by East Indians of Trinidad and Tobago. Chackra Publ. House, El Dorado, Trinidad, W.I.

Mohammed, M. and L.D. Wickham. 1995. Postharvest retardation of senescence in shado benni (*Eryngium foetidum* L.). J. Food Quality 18:325–334.

Morean, F. 1988. Shado benni: A popular Caribbean seasoning herb and folk medicine. Trinidad Guardian, Port-of-Spain, Trinidad, W.I.

Morton, J.F. 1981. Umbelliferae. In: Atlas of medicinal plants of middle America. Springfield Illinois: C.C. Thomas.

Ramcharan, C. 1998. The effects of progibb sprays on leaf and flower growth in culantro *Eryngium foetidum* L. J. Herbs Spices Med. Plants 7:(in press).

Sankat, C.K. and V. Maharaj. 1991. Refrigerated storage of shado benni (*Eryngium foetidum* L.). Proc. Agr. Inst. of Canada. Annu. Conf. July, 1991, Fredericton, New Brunswick, Canada.

Santiago–Santos, L.R. and A. Cedeno–Maldonado. 1991. Efecto de la intensidad de la luz sobre la floracion y crecimiento del culantro, *Eryngium foetidum* L. J. Agr. Univ. P.R. 75(4):383–389.

Seaforth, C.E., C.D. Adams, and Y. Sylvester. 1983. *Eryngium foetidum*—fitweed, shado beni (Umbelliferae). In: A guide to the medicinal plants of Trinidad and Tobago. Pall Mall, London: Commonwealth Secretariat.

Seaforth, C.E. 1988. *Eryngium foetidum*—fitweed (Umbelliferae). In: Natural products in Caribbean folk Medicine. Trinidad Univ., W.I.

Willis, J.C. 1960. A dictionary of the flowering plants and ferns. Cambridge Univ. Press, Cambridge.

Wilson, P.L. 1991. Cilantro—heart of the ubiquitous sofrito. p. 12–13. Food section. San Juan Star, June 16, 1991. San Juan, Puerto Rico.

Wong, W. 1976. Some folk medicinal plants from Trinidad. Econ. Bot. 30:103–142.

Seed Germination in *Stevia rebaudiana*

Jeffrey Goettemoeller and Alejandro Ching

Stevia (*Stevia rebaudiana* Bertoni, Asteraceae) is a non-caloric natural-source alternative to artificially produced sugar substitutes. The sweet compounds pass through the digestive process without chemically breaking down, making stevia safe for those who need to control their blood sugar level (Strauss 1995). There have been no reports to date of adverse effects from the use of stevia products by humans (Brandle and Rosa 1992). Shock (1982) reported that stevia contains eight glucoside compounds, each featuring a three-carbon-ring central structure. Stevioside is the most abundant glucoside produced. An extract of one or more of these compounds may be up to 300 times sweeter than sugar (Duke 1993). Preliminary trials at Davis, California indicate that stevia could produce a sweetener equivalent to 10 t/ha of sucrose (Shock 1982).

The Guarani Indians of Paraguay, where stevia originates, have used it for centuries as a sweetener for maté tea (Brandle and Rosa 1992). Since the 1970s, stevia extracts have been widely used in many countries as a sugar substitute. In Japan, for instance, stevia extracts account for about 5.6% of the sweetener market (Strauss 1995). Stevia usage in the United States is limited at this time because the Food and Drug Administration does not allow its use as a sweetener in manufactured and processed food products. In 1991, the FDA banned stevia, claiming it was an "unsafe food additive." The FDA now allows the sale of stevia, but only as a nutritional supplement (Whitaker 1995).

HORTICULTURE

Stevia is a perennial herb with an extensive root system and brittle stems producing small, elliptic leaves. Stevia will grow well on a wide range of soils given a consistent supply of moisture and adequate drainage; plants under cultivation can reach up to 1 m or more in height (Shock 1982). Stevia is grown as a perennial in subtropical regions including parts of the United States, but must be grown as an annual in mid to high latitude regions, where longer days favor leaf yield and stevioside contents.

The tiny white florets are perfect, borne in small corymbs of 2–6 florets. Corymbs are arranged in loose panicles. Oddone (1997) considers stevia to be self-incompatible and insect pollinated. Additionally, he considers "clear" seeds to be infertile. Seeds are contained in slender achenes, about 3 mm in length. Each achene has about 20 persistent pappus bristles.

Propagation of stevia is usually by stem cuttings which root easily, but require high labor inputs. Poor seed germination is one of the factors limiting large-scale cultivation. Shock (1982), Duke (1993), and Carneiro (1997), all mention poor production of viable seeds. Propagation is a special concern for northern growers who must grow stevia as an annual.

GERMINATION STUDY

A study was undertaken to investigate the low seed germination of stevia seeds. The influence of pollination treatments as well as the effect of light and darkness during germination were evaluated. Rooted stem cuttings of a Chinese clone of stevia obtained from Dr. Ken Rohrback, University of Hawaii were transplanted into 24 cm diameter plastic pots containing silty clay as a soil medium. On Oct. 12, 1997, the plants were placed in two separate greenhouses where temperature, wind, and pollen access could be controlled. At this time, the plants were at the first stage of floral bud development. The plants were subjected to five pollination treatments: (1) cross-pollination by bumblebees in a cage; (2) cross-pollination by hand; (3) cross-pollination by wind from a fan; (4) self-pollination by hand; (5) a control group isolated from other genotypes.

Ten plants of the Chinese clone were utilized (two in each treatment group). Clone SR8 provided cross-pollination in all but the selfing by hand and control treatments (one plant in each treatment group). Pollination treatments were initiated on Oct. 30, 1997 when blossoms were beginning to open and treatments continued for the duration of anthesis (30–40 days). For the bumble bee treatment, a cage (122 cm × 91.5 cm × 183 cm) covered with wire screen was placed over a greenhouse bench. Plants were placed in the cage along with a small hive of bumble bees (*Bombus impatiens*) obtained from Koppert Biological Systems of Ann Arbor, Michigan. A large circulation fan was used for wind pollination. Plants were placed 0.9–1.7 m from the

Table 1. Seed weight and viability (by tetrazolium chloride staining) of stevia.

Seed color	Seed wt. (mg/1000 seeds ± SD)	Seed viability (% ± SD)
Black	300 ± 5	76.7 ± 23.1
Tan	178 ± 8	8.3 ± 10.4

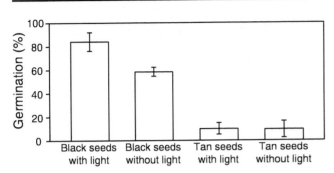

Fig. 1. Average germination rate for cross-pollinated "Chinese" stevia (± SD).

Table 2. Germination of black stevia, seeds under light derived from various pollination treatments.

Plant pollination treatment	Germination (% ± SD)
Selfing (hand)	93.3 ± 0.6
Crossing (hand)	92.0 ± 2.0
Crossing (bumblebees)	78.3 ± 14.5
Crossing (wind)	68.3 ± 22.5
Control[z]	36.3 ± 25.5[y]

[z]Control germination = 62, 36, 11
[y]Plants in isolation

intake side of the fan, providing enough breeze for gentle movement of the blossom 10–15 times daily, for 2–5 min periods. Cross-pollination and self-pollination by hand was accomplished by transferring pollen between blossoms every other day with a bumble bee thorax on the end of a toothpick. Seeds ripened during the period between Nov. 30, 1997 and Jan. 21, 1998. Seeds were judged to be ripe when they fell away with a gentle shake to the plant. Two or three harvests were done for each plant. For the final harvest, branches were cut and shaken vigorously inside a collection box. Debris was removed from the seeds by hand. Black and tan seeds were then separated.

Each germination test utilized 100 seeds placed between paper towels in a nursery flat, covered by a plastic dome. The temperature for all tests was 24°C. Fluorescent lights were placed above the domes for light treatment, 15 cm above the seeds. After 7 and 12 days, the number of seeds exhibiting normal germination were counted. For viability tests, 20 seeds were submerged in a 10% tetrazolium chloride solution at 24°C for 1 h and the stained seed counted. Each germination experiment was carried out with three replications.

There were two types of seed: black and tan. Black seed weighed more than tan seed, 0.300 vs. 0.178 mg, and viability of black seed based on tetrazolium chloride was much higher than tan seed, 76.7 vs. 8.3% (Table 1). Germination in the dark was higher for black as compared to tan seed (83.7% vs. 16.0%) while light increases the germination of black seed but not tan seed (Fig. 1). This suggests that tan seeds represent inviable seed that are produced without fertilization. There was no significant difference in black seed germination among the four pollination treatments suggesting that incompatibility is not a factor in these clones (Table 2). However, all pollination treatments increase seed germination of black seed over the control suggesting that some active manipulation of the blossoms is necessary to achieve pollination. It would appear that many of the black seeds in the control were misclassified tan seed.

REFERENCES

Brandle, J.E. and N. Rosa. 1992. Heritability for yield, leaf-stem ratio and stevioside content estimated from a landrace cultivar of *Stevia rebaudiana*. Can. J. Plant Sci. 72:1263–1266.

Carneiro, J.W.P., A.S. Muniz, and T.A. Guedes. 1997. Greenhouse bedding plant production of *Stevia rebaudiana* (Bert) bertoni. Can. J. Plant Sci. 77:473–474.

Duke, J. 1993. *Stevia rebaudiana*. p. 422–424. In: J. Duke, CRC handbook of alternative cash crops. CRC Press, Boca Raton, FL.

Oddone, B. 1997. How to grow stevia. Technical manual. Guarani Botanicals, Pawtucket, CT

Shock, C.C. 1982. Experimental cultivation of Rebaudi's stevia in California. Univ. California, Davis Agron. Progr. Rep. 122.

Strauss, S. 1995. The perfect sweetener? Technol. Rev. 98:18–20.

Whitaker, J. 1995. Sweet justice: FDA relents on stevia. Human Events 51:11.

Index of Species, Crops, and Products

Index of Authors